Citrus : Climate and Soil

A.K. Srivastava
Scientist, Sr. Scale (Soil Science)
National Research Centre for Citrus
Amaravati Road, Nagpur 440 010, Maharashtra, INDIA

Shyam Singh
Director
National Research Centre for Citrus
Amaravati Road, Nagpur 440 010, Maharashtra, INDIA

INTERNATIONAL BOOK DISTRIBUTING CO.
(Publishing Division)
Chaman Studio Building, 2nd Floor, Charbagh
Lucknow-226 004, U.P. (INDIA)

Published by
INTERNATIONAL BOOK DISTRIBUTING COMPANY
(Publishing Division)
Chaman Studio Building, 2nd Floor
Charbagh, Lucknow - 226 004 (India)
Tel. : 450004, 450007, 459058 Fax : 0522-458629
E-Mail : ibdco@sancharnet.in

First Edition, 2002

ISBN 81-85860-80-7

Composed and Designed at :
Panacea Computers
33, Nehru Road, Sadar Cantt.
Lucknow - 226 002 U.P.
Tel. : 0522-481164, 480546
E-mail : prasgupt@rediffmail.com

Printed at :
Army Printing Press
33, Nehru Road, Sadar Cantt
Lucknow - 226 002
Tel. : 0522- 481164, 480546
E-mail : armypress@sify.com

Dedicated to Citrus Workers

FOREWORD

Citrus is one of the most important fruit crops of the world with a production level of 93.75 M tons at present. Citrus is grown in many countries, of which Brazil, United States of America, China, Spain, Mexico, India, Japan, and Italy are the major producers.

The principal citrus species grown in the world are sweet orange (*Citrus sinensis* Osbeck); Sour orange (*Citrus aurautium* Linn); grapefruit (*Citrus paradisi* Macf); mandarin (*Citrus reticulata* Blanco); lemon (*Citrus limon* Burm. f); lime (*Citrus aurantifolia* Swingle). Citrus crop also provides the nutritional security to the population. The productivity of citrus varies in different countries and is showing the signs of fatigue or decline in many citrus growing environments. The reasons for citrus decline in an area can be attributed mainly to the soil and climatic limitations, besides a lesser extent to the lacunae in management practices.

Even though citrus is basically a subtropical fruit, it is however, grown under tropical and subtropical regions under semiarid to subhumid conditions. The soils under citrus in the world belong to Alfisols and Ultisols (red soils), Entisols (shallow black soils), Inceptisols (medium black soils), Vertisols (deep black soils) and Oxisols (laterites).

The soils under existing climate possess a definite set of capabilities for sustaining crop performance. The soil-site characteristics need to be matched with the crop requirements and such a parametric analysis will help in extension of citrus to similar potential areas. To have a sustainable and economically viable citriculture in the 21st century, the stresses of both biotic and edaphic origin need to be mitigated.

This book **"Citrus : climate and soil"** provides a good account of the cultivation and management aspects of citrus. The book consists of seven chapters namely, World citrus - climate and soil analysis; Soil suitability criteria for citrus; Soil fertility and production sustainability; Soil fertility management of slopy land under citrus; Citrus growing acid soils; Soil conditions and citrus decline; and Issues and strategies. Each chapter has been discussed lucidly in the light of the scientific data from different citrus growing belts of the world.

I hope that this book will be useful as a good repository of information pertaining to climatic and soil requirements of citrus and will form a reference base for the researchers in this field and citrus growers alike.

M. Velayutham.

(M. Velayutham)
President
Indian Society of Soil Science
New Delhi

September, 14, 2001

PROLOGUE

Citrus is considered to be one of the most remunerative fruit crops which has a lasting niche in international trade and world finance. The crop has gained a stature of huge industry which would be a step forward towards providing a nutritional security to the growing population. The demographic pressure has led to a dramatic reduction in per capita availability of land for citrus on one hand, and continued environmental imbalances on the other hand, compelling the citrus industry to face a twin problem. Citrus cultivation across the world is confined to 40° longitude and latitude on either side of the equator. The total citrus production, today on the global basis has registered a phenomenal improvement from 45 million tons to 96 million tons during 1971-2000.

The world citrus production is dominated by northern hemisphere with 45% contribution followed by 35% and 20% contribution from southern hemisphere and mediterranean region, respectively, with Brazil, USA, China, Mexico, Spain etc., amongst the top ranking citrus growing countries. Perennial nature of citrus allows it to be highly dependent on climatic features, through a synergism between climatological adaptability, and rootstock/variety interaction versus specific soil type. A huge climatological diversification in relation to meteorological requirement of various commercial citrus cultivars, has, unfortunately led to very poor adaptability of citrus as a whole. Over the last 50 years or so, many deviations from the normal climate have also been experienced, which have imparted a threat to the potential of citrus being under-exploited. The consequences of unprecedented changes in global climate has further infused an element of uncertainty in agricultural production, with citrus no exception. These changes have warranted to accelerate the momentum towards ecological security, besides production sustainability.

Import of citrus germplasms from one agroeco region to another, even with strict quarantine measures calls for a thorough investigation, regarding adaptability, under a given set of climate and soil. Having ventured into such an attempt, climate and soil preferences, best available at the places of their centres of origin, whether primary or secondary hold a paramount importance, to later fit into a kind of simulated growing conditions. This at the same time, prepares a blue print to adapt crop modelling and develop reproducible yield forecasting models. An accurate forecasting of yield in a specific area based on meteorological features during a given year with citrus as a test crop, is still a unreality, like other horticultural crops. The challenges which the citrus industry has to face in the 21st century, would largely depend on how well various environmental stresses are addressed in World Citriculture, to render the required flexibility, and vibrant responsiveness to the citrus industry according to consumers' changing demand, both nationally as well as internationally.

Analysis of various components of sustainability of a specific citrus cultivar, in a given citrus belt, have revealed that the climate and soil are the two most decisive elements towards success or failure of crop. Ecological adaptations of citrus, when considered worldwide, are astonishingly varied. It is not uncommon to observe an enhancement in incompatibility under specific stionic combination and period of juvenality in some growing situation. However, some of the citrus varieties such as valencia, Washington navel orange, pera belonging to sweet orange group, and satsuma mandarin have brought many revolutionary changes in the production scenario, since they possess an excellent adaptability under diverse climate and soil. Therefore, climatological delineations hold prime concern to derive the maximum production from a cultivar. Citrus trees grow best in warm climates with moist, but well drained soils having typically low water holding capacity. Drought stress is common in such environments. In cool and cold production region, specifically some of the mediterranean type areas, with cold and wet soils, in the spring especially, when coupled with a high frequency of wind, initial growth in the early years after planting is disappointingly slow. While on the other hand, in the warm production regions, the picture is quite confusing and entirely different. The tree vigour is such that tree space is occupied quite early. Such orchards have considerable proportion of declining trees. Malta type of sweet oranges, when grown under arid and semiarid areas have been witnessed showing frequent granulation, usually considered as physiological disorder, observed sometimes due to occurrence of soil fertility constraints in the form of Ca, P, B, and Zn deficiency.

Establishment of climatic norms, besides soil suitability criteria, is an utmost important exercise. Development of norms for various soil properties leads to better understanding about the best use of constraints based soil type - citrus relationship which addresses towards exploiting the productivity potential of soil. If, this is coupled with fertility management under slopy lands with citrus based land use alongwith appropriate soil conservation measures, is bound to contribute towards not only the improvised productivity, but elevated orchard productive life, and hence, the orchard efficiency in totality. The largest citrus belt existing in various countries are an exemplary piece of humanised art of citrus cultivation in soils those having inherently poor nutrient supplying capacity. To quote some prominent ones of them: Florida and California citrus industry of USA, Sao Paulo of Brazil, Valencia of Spain, Nelspruit of South Africa, Acireale of Italy, New South Wales of Australia, Bet Dagan of Israel, Concordia of Argentina, Izmir region of Turky, Sichuan of China, Cheju of Korea, Kogashima of Japan, Korat plateau of Thailand, Nile valley of Egypt, Jiroff valley of Iran, Colima of Mexico, Jaguey Grande of Cuba etc., have demonstrated an amazing success in the history of Citriculture. However, citrus is observed more comfortable in tropical and subtropical areas due to least possibility of exposure to prolonged freezing temperature. Taxonomically, a majority of citrus is grown in worlds' leading countries covering soil orders viz., Alfisols, Oxisols,

Ultisols, and Entisols. The highest quantum of production is harnessed from these soils besides quality. The book entitled **"Citrus : Climate and Soil"** has been divided into seven well demarcated chapters namely, World Citrus - Climate and Soil Analysis; Soil Suitability Criteria for Citrus; Soil Fertility and Production Sustainability; Soil Fertility Management of slopy land under citrus; Citrus Growing Acid Soils; Soil Conditions and Citrus Decline; and Issues and Strategies, are discussed comprehensively. These chapters have been discussed with best possible support of massive data base emerged from across the various citrus growing belts of world.

Authors express their optimism that such efforts in form of book, featuring climate and soil, as two elementary factors, limiting the citrus quality production would serve an exclusive guide to the students and researchers, by and large.

A.K. SRIVASTAVA

SHYAM SINGH

CONTENTS

1 World Citrus - Climate And Soil Analysis

1.1 WORLD CITRUS – A PREAMBLE

Citrus is observed native to a large area, extending from the Himalayan foot hills of northeast India to northcentral China, the Philippines in the east, and Burma, Thailand, Indonesia, and New Caledonia in southeast (Camacho-Bustos, 1981). Citrus fruits appeared on the globe after Australia separated from the continent during the Upper Cretaceous period at least 30 million years ago. The earliest form must have borne a definite inflorescence (Subgenus Archicitrus), before the entire loss of pinnae. As they adapted to cooler climate, they dropped the above characteristics by the contraction of the common peduncle, giving rise to solitary or fasicled flowered members (Subgenus Metacitrus) with more advanced variability (Tanaka, 1958). Citrus plants are claimed to have evolved from Cretaceous to Paleocene, 65-135 million years ago. Based on the continental drift theory, the ecological and geological vicissitude, citrus plants originated in south China or Cathaysian ancient continent including Sichuan, Kangdian, south of Yangtze river, and Indo-China peninsula, then dispersed into India, Africa, and Australia (Zhou, 1990).

The cultivation of citrus can be traced back to 2200 BC through a Chinese book on citrus appeared in 1178 AD. Many writings have made the reference of citrus fruits in Memoirs of Baber (1519). The only exception is the grapefruit (*Citrus paradisi* Macf.) which appeared in West Indies (Barbados) in about 1750 (Chapot, 1975). Amongst the citrus fruits known to western civilization, the citron was found cultivated in Iran some 2300 years ago by the Greek botanist. Early distribution of citrus was slow and carried out by Arab sailors and the caravans from the east. In this way, citrus spread to sea of Oman, Egypt, and later Europe. Christopher Columbus brought sweet orange, lemon, citron, and other citrus to America from Canary Islands during his second trip in 1493. The sweet orange may have reached the middle east during the time of the Crusades (1099-1291). However, many taxonomists prefer to date this arrival in Europe around 1480. Sweet orange came to South Africa in 1654. Captain George Vancouver brought orange to Hawaii in 1792. The Bahia orange or navel orange was taken to Australia from Brazil in 1824, and later to north America and other countries (Webber *et al.*, 1967). European voyagers and colonization spread citrus to south America and north America, and the Carribean Islands. Colombus is reported to have carried the sweet orange, lemon, and citron to Haiti on his second voyage in 1493. However, an earlier introduction to the south of Sahara may have been made by Arab merchants trading along the east coast of Africa. This is evident by the presence of many wild citrus trees in the Zambezi river basin. The world citrus industry (Fig. 1.1) today has a total production of 90.9 million tons during 1999-00. Productionwise, distribution of citrus on continent basis (FAO, 1998) revealed maximum production in South America (29.2 million tons) followed by Asia (28.5 million tons), North and Central America (20.9 million tons), Europe (9.2 million tons), Africa (9.0 million tons), and Australia (0.54 million ton). During the last 30 years, total citrus production has increased from 6.1 to 19.7 million tons in Brazil, 12.4 to 15.8 million tons in USA, 0.71 to 10.8 million tons in China, 1.8 to 4.7 million tons in Mexico, 2.7 to 5.6 million tons in Spain, and 2.7 to 3.2 in Italy (Table 1.1).

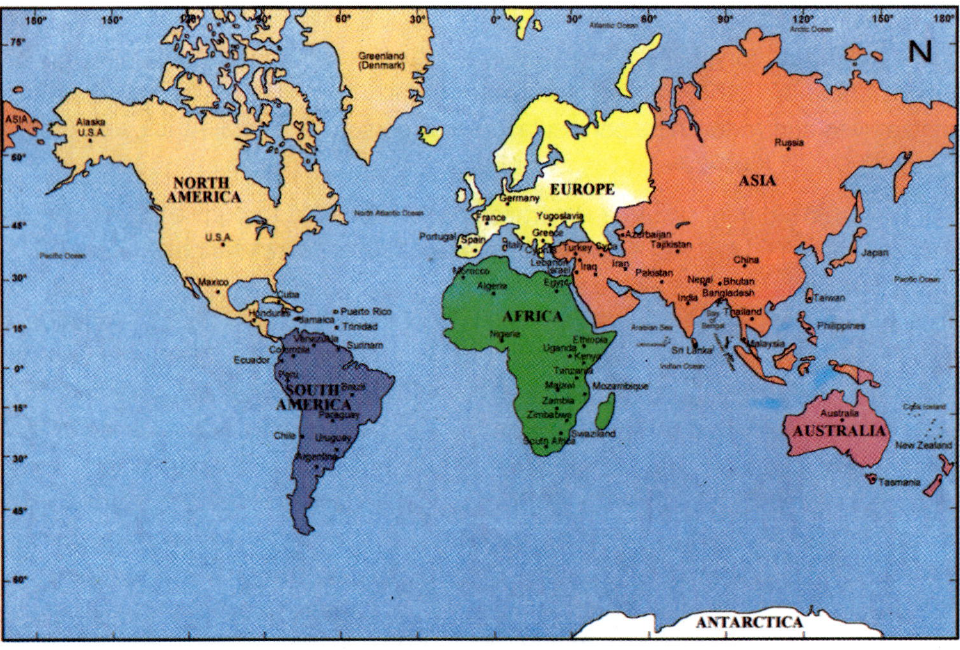

Fig. 1.1. Distribution of important citrus growing countries on world map

Table 1.1. Production, export, and processing pattern in leading citrus growing countries (thousand metric tons) over last 30 years

Country	Production		Export		Processing	
	1970-71	1999-00	1971-72	1999-00	1971-72	1999-00
World	**45385.1**	**90,886.8**	**6290.1**	**9430.5**	**14093.4**	**31381.3**
Northern Hemisphere	**34025.1**	**63891.6**	**5774.8**	**8101.4**	**11533.9**	**17071.3**
United States	11475.7	14812.1	785.1	1023.0	9048.1	11569.0
Mediterranean Region	11705.6	18291.4	4578.3	5529.8	9048.1	3579.3
Greece	668.7	1354.2	189.6	322.4	163.8	418.6
Italy	2688.6	3214.9	387.3	212.4	351.0	1422.8
Spain	2745.2	5624.5	1620.4	3221.0	276.2	924.0
Israel	1609.5	696.0	891.8	220.0	561.4	363.0
Algeria	515.5	444.9	112.6	-	-	-
Morocco	798.4	1400.0	563.3	596.8	95.8	140.4
Tunisia	119.7	269.6	28.4	27.0	-	-
Cyprus	245.0	236.8	184.5	118.5	16.6	51.0
Egypt	942.2	2497.5	164.5	226.9	3.0	102.0
Lebanon	315.4	353.5	165.6	-	-	-
Turkey	748.2	1826.2	121.9	489.5	68.7	116.5
Former USSR	84.0	93.0	-	-	-	-
Japan	3863.4	1746.0	-	-	-	159.0
Cuba	162.0	755.0	47.7	55.0	-	560.0
Mexico	1761.6	4730.0	-	254.0	-	636.0
China	719.1	10787.1	83.1	156.0	-	535.0
Southern Hemisphere	**11359.9**	**26995.2**	**515.3**	**1329.1**	**2559.5**	**14310.0**
Argentina	1434.4	2240.0	40.2	310.0	245.4	936.0
Brazil	6051.0	19700.0	53.3	102.0	1944.6	12811.0
Uruguay	73.9	-	13.4	-	-	-
Venezuela	235.2	-	-	-	-	-
United States	899.2	935.6	-	-	-	-
Australia	417.6	715.0	25.3	5.0	171.0	293.0
South Africa	604.3	1149.0	305.7	646.0	94.2	270.0

Source : FAO (2000)

Amongst leading citrus growing countries, China has recorded a phenomenal increase in production, by more than 10 folds, while, Brazil tripled its production during last 30 years. South Africa recorded two folds increase in total citrus production from 0.60 to 1.2 million tons. Mexico experienced more than double increase in its production. Other frontline citrus growing countries attained a substantial increase in total citrus production from 0.94 to 2.5 million tons in Egypt and from 0.75 to 1.8 million tons in Turkey. On the other hand, production pattern in countries like Algeria, Israel, and Japan, has decreased from 0.51 to 0.44 million ton, 1.6 to 0.70 million ton(s), and from 3.9 to 1.8 million tons, respectively. The total production in northern hemisphere is much higher than southern hemisphere. The world citrus production is expected to increase at an annual rate of 3.3 per cent during next decade to a record production. However, the total citrus during 1999-2000 was much lower compared to record production of 97.6 million tons during 1997-98. The expected increase is mainly due to a large orange crop in Brazil. Among other countries of the southern hemisphere, in South Africa, the upward trend in output experienced in recent years is likely to continue. In the northern hemisphere, large crops are expected in United States, and some countries of mediterranean region, particularly Italy and Morocco, where production of nearly all citrus varieties are expected to expand. The relatively large increase has been recorded at the rate of 10 per cent in Brazil, 2 per cent in USA, 8 per cent in Italy, 27 per cent in Morocco during last 2 years. However, noteworthy declines are anticipated for Greece, Spain, Turkey, and Mexico. The world citrus export has increased only marginally from 6.3 to 9.4 million tons during last 30 years (Table 1.1), with northern hemisphere contributing 5.8 million tons. Out of present total citrus production of 90.9 million tons during 1999-2000, only 9.4 million tons were diverted towards export and 31.4 million tons towards processing, constituting 9.6 per cent and 34.5 per cent, respectively.

1.2 AGROCLIMATIC DIVERSITY

There are three basic requirements for successful cultivation of citrus, namely, climate relatively free from frost hazards, good quality of irrigation water, and a reasonably deep and uniform fertile soil with good internal drainage (Platt, 1966; Nemec, 1986). Of these, climate is the most important component of commercial Citriculture which determines the difference in growth, yield, and quality due to differential behaviour of citrus in relation to climate-soil nexus (Cooper *et al.*, 1963; Jones and Cree, 1964; Fatta Del Bosco, 1964; Fuleihan, 1965; Kurihara, 1969; Reuther, 1973a; Yelenosky, 1985; Cherkezishvili, 1987). However, citrus is well adapted to a wide range of climate from humid tropical to desert (Levy and Syvertsen, 1981). The north island of New Zealand represents about the coolest extreme in which citrus may be grown commercially. Thailand has amongst the hottest year round climate supporting citrus. Canete, Peru, represents about the lowest extreme of rainfall, while that of Japan is amongst the highest (Reuther, 1973a). Good performance of irrigated citrus trees under low humidity explains their adaptability to arid climates, and ability to survive under severe water stress (Fereres *et al.*, 1979). Citrus is also grown in climate having worlds' largest rainfall in Cherrapunjee and Mawsynram areas of Meghalaya. It is interesting to note that Marsh seedless grapefruit trees achieved higher yields in a very dry than humid climate (Levy and Syvertsen,1981). Kapanadze (1964) claimed that citrus plants originated under semidesert conditions in monsoon areas, and as a result of their adaptation to external conditions, acquired certain specific characters. For example, citrus leaf consists of a blade joined to the petiole which is winged. During drought, plants shed the lamina, but, retain the petiole. The place of leaves is, thus, taken by the green thorny shoots and winged petioles.

Adaptability of late sweet oranges in China indicated that late orange cultivars are found in the regions between 19° and 32°N. These cultivars grow well in regions with >5400°C of annual cumulative temperature, with the annual mean temperature not less than 17.6°C and absolute minimum temperature above -3°C. Autumn and winter droughts enhance flower bud differentiation, whereas, summer drought reduces growth. Winter temperatures below -3°C usually cause heavy fruit drop (Liu, 1983). Earlier, studies by Newman (1968) showed that the annual use of water by evapotranspiration in a well watered mature citrus orchard ranged between 800 mm in a cool coastal subtropcial climate like Santa Barbera, California to 1280 mm in the hot semiarid subtropical climate of Weslaco, Texas, suggesting, thereby, the requirement of annual rainfall somewhere between 1000 to 2000 mm. Without irrigation to some tropical regions, such as Cauca valley of Columbia, with about 1000 mm of rainfall annually, having biomodal pattern of distribution, oranges tend to bloom after long and short drought periods, and the two harvests are undertaken about seven months after the dry periods (Torres-M and Rios-Castano, 1968). On the other hand, in lowland tropical areas of very high rainfall, such as in south eastern Thailand with around 3500 mm of precipitation, fairly well distributed, and high humidity, commercial citrus cultivation is marginal. A humid climate when compared to arid climate, tends to produce fruits smoother, thinner skins, and trees with more open growth habits (Reuther, 1973a).

Grigg (1969) advocated the climatic criteria as the best indicator of agricultural potentialities when world as a whole is being dealt with. But, in smaller areas, soil types and morphology, drainage conditions, and slope assume greater significance. Citrus is grown under a wide range of agroclimatic diversity, and continuous recurring efforts to exploit commercial cultivation of citrus may fail, if these exercises are not well substantiated by sound adaptability tests, and developing technology suitable for specific agroeco situation. Citrus tree growth and fruit quality both are conditioned by climate in a growing area (Cooper *et al.,* 1966; Grieson and Ting, 1978). The world is divided into three main climatic zones, viz., i. Hot climatic regions, ii. Temperate climatic regions, and iii. Cold climatic regions. The hot climatic regions are further divided into four regions, namely equatorial regions between 0° to 5° south of equator with high temperature (27°C) throughout the year and heavy rainfall of 2000-2500 mm covering West Indies, Guinea coast in Africa, parts of Sri Lanka, Indonesia, and Malaysia; summer climatic regions at both sides of equator between 5° north and 5° south of the equator to the tropics of Cancer and Capricorn having hot and moist summer, and warm and dry winters, that cover Venezuela, Kenya, south Brazil, Zimbabwe, Zambia, and Tanzania; monsoon climatic regions having high temperature throughout the year clearly marked by wet and dry seasons, and cover India, Pakistan, Thailand, parts of Sri Lanka, Philippines, and north Australia; hot desert climatic regions lie at 29-38° north and south of equator having hot days and cool nights, and cover south California, north and west Mexico, and western Australian desert.

Temperate climatic regions over mediterranean climatic regions, China type of climatic regions (temperate monsoon), steppe type climatic regions, northwest European type climatic regions, and laurentian type of climatic regions. Mediterranean climatic regions are characterised by hot and dry summers, mild and moist winters with an annual rainfall of 500-1000 mm located at 30° and 45° north and south of equator covering northern California, Portugal, Spain, Italy, Greece, Turkey, Syria, Israel, Morocco, Tunisia, Algeria, central Chile, Cape province of South Africa, and southwest Australia. China type climatic regions or temperate monsoon are found between 30°-45° on eastern side of land covering north and central China, New South Wales, eastern Victoria, northern New Zealand, and Japan. Steppe type regions are also called as temperate grasslands.

Northwest European type climatic regions and laurentian type climatic regions are confined between 45° and 60° north and south of equator and citrus is not grown in these regions. Cold climatic regions, though divided into Siverian type climatic region and Tundra type climatic region, which cover countries like Siberia, Finland, north Canada, Alaska, Greenland etc. where winters are extremely severe and irrelevant from citrus cultivation point of view. Few attempts in the past, have been made to characterise climate under which citrus is grown worldover. Based on mean annual temperature and consequently the diurnal changes, Reuther (1973a) divided climate of world citrus into three distinct thermal environments, namely, subtropical zone with low diurnal variation varying between 4.7°C and 10.8°C in cooler and hotter months, respectively, covering Ocho island in the Hiroshima representing humid climate, with well distributed rainfall; tropical zone characterised by absence of seasonal change in temperature represented by Columbia with mean maximum temperature of 32.8°C in coolest month and 34.6°C in the hottest month; and subtropical zone with moderate diurnal variation represented by San Joaquin valley of California having semidesert climate with diurnal variation up to 12°C and 22°C in coolest and hottest month, respectively.

According to Yelenosky (1977), climatically five citrus growing regions of the world are delineated :

- Moist marine areas, such as southern Japan, the black sea coast, adriatic sea coast of Yugoslavia, parts of Turkey, and New Zealand. The cold hardy cultivar, such as satsuma mandarin is usually grown in these colder locations, principally on trifoliate orange (*Poncirus trifoliata* (L.) Raf.) rootstock.
- Subtropical, mediterranean climate is typified by Spain, Israel, Italy, Turkey, parts of Australia, Lebanon, Greece, and the coastal areas of California. These areas are usually not subjected to freezing conditions and are the major areas of production for lemons as well as sweet orange cultivars. The citrus fruits produced tend to have high external and internal colour, few rind blemishes, and relatively thick peel.
- Subtropical, arid areas such as the U.S. desert regions of southern California, Arizona, and possibly the Rio Grande valley of Texas. These areas produce fruits with excellent appearance, suited primarily for fresh utilization. A sizeable amount of grapefruits

are often grown. These are occasionally subject to damaging freezes.

- Subtropical, moist areas include Florida (USA), parts of China, India, South Africa, Brazil, and Argentina. Some of the other regions are occasionally subjected to several winter freezes. The subtropical, moist areas produce fruits with generally thin peel, fair to good internal colour, but, with high soluble solids and juice content. Although, fruits produced in these areas are generally good for fresh fruit. Brazil and Florida have large processing industries, primarily for production of frozen orange juice concentrate.
- Tropical, moist areas such as Hawaii, parts of Mexico, Central America, and others include those where temperature lower than 18.3^{0}C rarely occur, and rainfall is usually heavy. Fruits rarely attain good external and internal colour. All cultivars tend to be everbearing and experience production throughout the year.

Navel orange, which has revolutionarised the world citrus industry, is known as a variety, with high quality but low yield, especially with the narrow adaptability to natural conditions, in the mediterranean region, unsuitable in humid areas due to high temperature (Chen *et al.*, 1990b). With the establishment of Fruit Tree Research Station at Okitsu, Japan in 1902, many citrus cultivars such as Washington navel, valencia, joppa, and other sweet oranges, lemons, and grapefruits were introduced from USA, and their adaptability was tested in Japan. These most worldwide citrus cultivars did not show good performance under the climatic conditions of Japan. The temperature of Japan in winter is too low to grow late maturing varieties like valencia and grapefruit, and there is too much rainfall in early summer to produce good yield of Washington navel in Japan. With these failures, satsuma mandarin was discovered and adapted well to the various climatic handicaps of Japan (Nishiura, 1981). Analysis of climatic requirement by Jackson and Looney (1999) revealed that ideal growth of citrus cultivars such as sweet oranges takes place between 13-40°C with average temperature during summer months in excess of 15°C in a latitude range from 23.5° to 40°C. While, lemons are more tolerant to cool summer temperatures than sweet orange, and do not grow well under tropical climate. Grapefruits prefer tropical and warm tropical climate. However, not many cultivar specific climatic norms are available.

The broad climatic norms developed have provided an initial means to determine whether an area is suitable for the production of citrus. The next phase of the project is to split the broad norms into ideal and marginal norms. Viability studies of new citrus production areas and suitability studies of new cultivars in existing areas require accurate climatic norms. General conditions for *Citrus sp.* have been worked out, e.g. an average temperature of less than 24°C is necessary to induce flowering with the minimum growth temperature of 13°C (Davies and Albrigo, 1994).

Viability studies of new citrus production areas require, *inter alia,* an indication of the climatic suitability of the specific area for export quality and fresh citrus production. The use of climatic norms extracted from the literature have proved unreliable by and large. Therefore, it is considered necessary to determine climatic norms to demarcate areas suitable for citrus production. In conjuction with field studies, climatic parameters are compared between known citrus production areas in order to determine the climatic norms for a specific cultivar (Barry and Veldman, 1996). Thereafter, these norms are plotted using a Geographical Information System (GIS) to relate them to citrus production areas and to compare them with those developed from field studies involving portable weather stations and measurements of growth phases. It is now possible to determine whether an area is suitable for the production of citrus, besides export marketing (Barry *et al.*, 1996).

Meteorological analysis of premier citrus growing centres from across the world such as Orlando, Florida, Waslaco, Texas, Tempe, Arizona, Campinas, Sao Paulo, Nelspruit, Valencia, Sicily, Tandono, Riverside, and Ambositra suggested a huge variation in terms of rainfall and mean temperature under which citrus is cultivated (Tables 1.2, 1.3, and 1.4). Climate plays a very important role in yield and especially in fruit size, which is the single most important factor in determining the export value of various commercial citrus cultivars (Plessis, 1982; 1983). Sys *et al.* (1993) suggested various climatic norms for lemons, oranges, grapefruits, and mandarins (Table 1.5). Comparison of various climatic parameters between cool citrus growing area (Nelspruit) and hot citrus growing area (Citrusdal) indicated a distinct climatological diversification according to adaptability of various citrus cultivar specific preferences in South Africa (Table 1.6). A multiple linear regression analysis of salinity and climate against yield of valencia and Washington navel orange orchards in citrus growing areas of Australia showed that high temperature and evaporation during flowering and fruit set (November and December) lowered the yields at Sunraysia, Berri, and Waikerie. At Mypolonga, the coolest location, high temperature has favoured the increase in yield (Cole and McCloud, 1985). Nuttonson (1965) prepared climatic analogues based on citrus – climate relationship using maximum, minimum temperature requirements, relative humidity, and winds. These analogues showed good reproducibility in other citrus growing regions of world.

The yield potential of grapefruits and oranges is as great or greater in some humid tropical areas as in humid subtropical areas such as Florida (Reuther, 1980). Climate of Okinawa resembled to that of Florida, and the soils are considered suitable when drained (Higa, 1975a). In a normal year, the trees bloom in mid March at Ishigaki and in early April at Ishikawa. But, the season may vary by 10 days depending on winter temperature and nutrient status. Yield depending upon rainfall is low when much rain occurred between bud formation and fruit drop (Higa, 1975b). Pehrson (1976) presented the annual maturity ratios (TSS:acid) of navel oranges for 22 years and the mean monthly temperatures for years in which maturity advanced or delayed, each averaged over 5

Table 1.2. Comparison of monthly mean maximum and mean minimum temperature in some important subtropical citrus growing regions of the world

Mean monthly temperature (°C)	Jan.	Feb.	Mar.	Apr.	May	Jun.	Jul.	Aug.	Sept.	Oct.	Nov.	Dec.
Orlando, Florida (1900-1959), Latitude 29°N, Elevation 30 m												
Maximum	22.8	23.6	26.4	28.9	31.9	33.2	33.6	33.6	32.2	29.4	25.6	22.8
Minimum	10.6	11.1	13.4	15.3	18.9	21.7	22.5	22.9	22.1	17.5	13.9	11.0
Weslaco, Texas (1940-1960), Latitude 26°N, Elevation 25 m												
Maximum	22.2	24.4	27.0	30.2	32.2	34.0	34.7	35.6	33.3	30.8	25.9	22.7
Minimum	10.3	12.0	14.7	18.0	20.9	22.8	23.2	23.0	21.8	18.0	13.4	10.7
Tempe, Arizona (1943-1954), Latitude 33°N, Elevation 366 m												
Maximum	18.0	20.3	23.8	28.6	33.5	37.8	39.7	38.6	36.7	30.9	23.6	19.2
Minimum	1.4	3.6	5.7	9.3	12.8	16.7	22.8	22.5	18.6	11.4	4.7	2.7
Campinas, S.P., Brazil, Latitude 23°S, Elevation 720 m												
Maximum	30.1	30.1	28.9	27.0	25.3	25.4	25.4	26.6	27.4	28.0	28.6	28.6
Minimum	18.8	18.9	18.2	15.9	13.5	11.9	13.0	12.6	14.2	16.0	16.6	17.9
Rehovot, Israel (1933-1953), Latitude 32°N, Elevation 50 m												
Maximum	18.2	19.4	21.8	25.7	29.0	30.4	31.9	32.3	30.9	29.4	24.7	20.2
Minimum	7.9	8.5	9.3	11.5	14.9	17.4	19.6	20.1	19.0	16.5	13.5	9.7
Nelspruit, Eastern Transvaal, (1926-1962), Latitude 251/2°S, Elevation 610 m												
Maximum	28.9	28.9	28.1	27.1	25.1	23.5	23.5	25.0	26.4	27.5	27.8	28.6
Minimum	18.4	18.6	17.4	15.3	10.1	6.9	6.7	8.9	12.0	15.0	16.7	17.8
Valencia, Spain (1913-1940; 1934-1947), Latitude 39°N, Elevation 0-400 m												
Maximum	14.4	15.6	17.2	19.4	22.8	25.6	28.3	28.3	26.7	22.8	18.1	10.6
Minimum	5.0	6.1	8.3	10.6	13.3	17.2	20.0	20.6	17.8	13.9	8.9	5.5
Messina, Sicily (1956-1965), Latitude 37°N, Elevation 0-200 m												
Maximum	13.9	15.0	16.7	18.9	22.8	27.2	30.0	30.0	27.8	23.3	20.0	15.6
Minimum	9.4	8.9	10.0	12.2	15.0	17.8	22.2	22.8	20.6	16.7	13.3	10.6
Tadono, Wakayama, Japan (1903-1962), Latitude 34°N, Elevation 0-400 m												
Maximum	10.8	11.7	14.4	20.0	24.1	27.0	31.1	32.4	28.9	23.5	18.6	13.6
Minimum	1.4	2.0	4.4	9.2	13.7	18.3	22.7	23.4	20.1	13.9	8.7	4.4

Source : Reuther and Rios-Castano (1969)

years. Advanced maturity is associated during the years of high spring temperature (number of degree days over 23°C).

The role played by climatic factors in year-to-year variation in yield of tree crops has been highlighted by several research workers. Sakamoto and Okuchi (1969) observed greater influence of meteorological features than fertilizers in satsuma mandarin in Japan. A detailed analysis of 12 year yield data of principal commercial sweet orange cultivars (Hamlin, Valencia, Pineapple, and Marrs) in Texas revealed that 6-7 years are required before satisfactory yields are obtained, and 8-10 years before returns becomes equal to expenditure (Melville, 1962-64; Lombard, 1965; Rouse *et al.*, 1987; Rouse and Maxwell, 1987). Similar observations have been made by Plessis and Plessis (1987) in South Africa. Another long term study at Cuba, by Frometa and Echazabal (1988) indicated that year to year variation juice quality is greater than variation between cultivars (Hamlin,

Table 1.3. Comparison of seasonal rainfall pattern in citrus growing areas of California and Columbia

Location	Indio	Riverside	Sant Paula	Lindsay	Cartagena	Medellin (Bello)	Palmira
State	Calif.	Calif.	Calif.	Calif.	Bolivar	Antioquia	Valle
Country	U.S.A.	U.S.A.	U.S.A.	U.S.A.	Columbia	Columbia	Columbia
Elevation (m).	3.4	250	80	120	3	1425	1006
Latitude, (N)	33°44′	33°57′	34°21′	36°11′	10°26′	6°26′	3°31′
Years of record	83	80	13	47	17	5	26
			Mean monthly precipitation (mm)				
January	12.5	49.2	103.0	57.5	7.7	47.1	77.2
February	10.5	57.3	53.0	52.6	0.0	59.5	67.3
March	6.3	42.7	63.2	46.4	7.0	42.0	85.5
April	2.5	24.3	42.8	30.5	18.2	115.1	132.9
May	0.3	5.5	9.3	11.5	94.6	185.0	116.6
June	0.3	1.0	0.8	3.0	93.7	84.4	67.1
July	3.0	1.5	Trace	0.3	90.6	86.1	27.9
August	8.3	4.0	0.8	0.3	140.3	103.5	34.4
September	10.8	2.3	1.0	2.7	127.3	149.5	146.1
October	5.8	13.3	6.5	10.5	257.3	149.5	146.1
November	7.3	20.8	32.0	25.0	196.6	92.8	111.8
December	17.0	48.3	53.0	50.4	85.7	60.2	87.2
Annual mean	84.5	276.0	365.0	291.0	1043.3	1146.4	1008.5

Source : Mendel (1969)

Salustiana, and Victoria). Another study by Avilan (1986) in northcentral Venezuela indicated that production efficiency of sweet orange increased from 2nd to 10th year, and then started declining. Jackson and Hamer (1980) showed that three climatic factors accounted for 64 per cent variation in yield of cox orange between seasons. Moss and Muirhead (1971) indicated a negative effect of high September temperature on yield of navel orange in Australia, but had a positive effect of high temperature during November. Models have been developed through long term field experiments in Western Georgia to separate effect of fertilizer from climate and other incidental factors (Tsanava *et al.*, 1989). Quantification analysis regarding the contribution of environment, soil, and management factors on yield of satsuma mandarin revealed that yield is more affected by physiographic environment than either climate or even fertilizer application (Egashira *et al.*, 1990). Jones and Cree (1965) reported that high temperature during the June drop period (May to June) can be very detrimental to the crop size of navels.

1.3 CLIMATE AND YIELD VARIATION

Reuther (1973a) stated that a single day with a temperature above 40°C during May-June led to a reduction in yield of navels. Occurrence of high maximum and minimum temperature from September to November during and just after flowering (September to November) imparted low yield of valencia and navel oranges in Citrusdal district (Plessis *et al.*, 1975b). The maximum daily temperature from September to November has been shown to be the main factor affecting fruit yield of Washington navel oranges in Citrusdal district of South Africa. Yields correlated negatively with the number of days

Table 1.4. Meteorological variation in some cold winter citrus countries

Month	Jan.	Feb.	Mar.	Apr.	May	Jun.	Jul.	Aug.	Sept.	Oct.	Nov.	Dec.
						Riverside – USA, Lat. 34°						
MMT	11.0	12.0	13.5	15.7	18.0	21.2	24.3	24.2	20.9	18.1	14.6	11.6
R	53.1	58.2	54.9	22.3	9.9	1.0	0.2	4.3	4.6	16.0	17.5	48.8
B	-	-	-	B	B	-	-	--	-	-	-	-
M	M	M	M	-	-	-	-	-	-	-	-	M
ET	-	-	21.7	87.0	161.2	252.0	356.5	353.4	243	164.3	54.0	-
						Mechra Bel Ksiri Maroc, Lat. 34°						
MMT	12.4	13.5	16.0	17.9	19.9	23.6	26.4	26.7	24.6	21.1	16.5	13.5
R	93.0	85.0	72.0	52.0	29.0	7.7	0.1	1.0	10.	52.0	78.0	97.0
B	-	-	B	B	-	-	-	-	-	-	-	-
M	M	M	M	-	-	-	-	-	-	-	-	M
ET	-	19.6	99.2	153	220.1	324	421.6	430.9	354	257.3	111	21.7
						Ambositra – Madagascar, Lat. 20°						
MMT	20.9	20.6	20.2	19.0	16.7	14.9	13.9	14.3	15.9	18.5	20.0	20.6
R	294.9	243.3	229.5	82.0	36.4	27.6	21.7	22.4	33.1	59.3	195.4	281.7
B	-	-	-	-	-	-	-	-	-	B	-	-
M	-	-	-	-	-	M	M	M	M	-	-	-
ET	251.1	281.4	229.4	186	120.9	63.0	34.1	46.5	93	176.7	216	241.8
						Orlando– Florida, Lat. 28°						
MMT	16.0	16.9	19.6	21.9	24.7	26.9	27.8	27.9	26.6	23.4	19.3	16.2
R	62.2	58.7	78.0	64.8	113.8	181.6	207.0	174.2	170.2	109.5	42.4	59.7
B	-	-	B	B	-	-	-	-	-	-	-	-
M	M	M	-	-	-	-	-	-	-	-	M	M
ET	99.2	114.8	210.8	273	368.9	423	465	468.1	414	328.6	195	105.4

M, Maturity ; ET, Effective temperature between midbloom and maturity; MMT, Monthly mean temperature (°C) ; R, Rainfall (mm); B, Blooming

Source : Cassin *et al.* (1969a)

with temperature above 30°C in those months based on model developed (Plessis, 1982). Hilgeman (1973) attributed annual variation in yield of Washington navel, valencia and diller oranges, and grapefruit to the variation in number of days over 35°C and maximum temperature during the 50 days at post flowering period. Under the conditions of California, Boswell and Burns (1973) observed that southwest quadrant of the tree produced appreciably higher fruits than southeast, and recommended eastwest orientation for close set rows. The best theoritical row orientation is thought to be northwest to southeast. Studying the difference in orientation of kinnow fruits in Punjab (India) within the canopy, Jawanda *et al.* (1973) observed greater fruit weight inside the canopy compared to outside and TSS:acid ratio of 12:1 and 14:1 for outer and inner fruits, respectively, as maturity standards for semiarid conditions of Punjab, India. This was confirmed by other studies of Plessis (1982) who indicated that a temperature of 41°C for one day during the flowering-fruitset period (September-November in South Africa), decreased the navel crop by approximately 6 per cent. He also showed that climatic factors accounted for 95 per cent of the year to year variation in yield of navels in the Western Cape region. Marsh grapefruits are ideally suited to hot, humid, low lying areas of the Eastern Transvaal, Swaziland, Mosambique, and northern Natal of southern Africa.

Table 1.5. Climatic norms for various citrus cultivars

Climatic characteristics	Climatic class, limitation and rating scale					
		S_1	S_2	S_3	N_1	N_2
	0	1	2	3	4	
	100	95	85	60	40	25
Annual precipitation (mm)	2300-3000	> 3000	1200-1000	1000-800	-	< 800
	2300-1500	1500-1200				
Number of dry months	2.5-3	3-4	4-5	5-6	-	>6
(P < 1/ 2PET)	2.5-2	2-0	-	-	-	-
Mean annual temp. (^{0}C)	26-30	30-33	33-36	36-39	-	>39
	26-22	22-19	19-16	16-13	-	<13
No. of months with mean temp. > 38^{0}C	0-1	1-2	2-4	4-6	-	>6
No. of months with mean temp. < 13^{0}C	0-1	1-2	2-4	4-6	-	>6
Absol. min. temp. (^{0}C)						<-9
-Lemon	> -1	-1→-2	-2→-6	-6→-9		<-9.5
-Grapefruit	> -3	-3→-4	-4→-7.5	-7.5→-9.5		<-9.5
-Oranges	> -3.5	-3.5→-4	-4→-7.5	-7.5→-9.5		<-12
-Mandarins	> -4.5	-4.5→-5.5	-5.5→-8.5	-8.5→-12		-
Mean temp. in the two	13-10	10-8	8-6	6-4	-	< 4
months after harvest (^{0}C)	13-15	15-18	18-20	20-25	-	>25
Mean temp. at flowering (^{0}C)	> 15	15-10	10-5	5 to –5	-	<-5
Relative humidity (%) in coldest month, if frost (%)	< 30	30-60	60-90	>90	-	-
Relative humidity at ripening	45-60	60-70	70-90	>90	-	-
Fruit set stage (%)	45-30	< 30		-	-	-
Relative humidity at flowering	75-70	70-50	< 50	-	-	-
(%)	75-80	80-90	> 90	-	-	-

Note: S_1 , S_2, S_3, N_1, and N_2 stand for highly suitable with no significant limitations; moderately suitable having slight limitation; marginally suitable having severe limitation; currently not suitable having correctable limitations, and permanantly not suitable having limitations beyond correction.
Source : Sys *et al.* (1993)

The hot, arid areas, situated further inland (Hoedspruit, Letsitele, and the Limpopo valley), tend to produce fruit with thicker rinds, and hence, lower juice content, but still of acceptable quality for export. This phenomenon being more prevalent in dry seasons. The so called intermediate citrus production areas are marginally suitable for export quality marsh grapefruit production. Although, the warmer microclimates in these areas may be suitable. The cold, semi-coastal areas produced fruits of high acid content with thick rind (Barry *et al.*, 1996).

Albisu (1982) reported that rain has a strong positive effect on yield of orange trees in Spain, Morocco, and Israel. Climatic factors have been shown to be mainly responsible for the year-to-year variation in yield of navel oranges (*Citrus sinensis* Osbeck)

Table 1.6. Comparison of average monthly climatic data between the cool (Nelspruit) and hot (Citrusdal) citrus growing area

Parameters	Cool climate			Hot climate		
	Aug.-Nov.	Dec.-Mar.	Apr.-Jul.	Aug.-Nov.	Dec.-Mar.	Apr.-Jul.
Max. temp. (°C)	26.7	28.8	24.7	24.2	31.6	22.0
Min. temp. (°C)	12.9	18.1	9.4	10.5	16.5	8.9
Evap. (mm day^{-1})	5.3	5.9	3.7	6.6	11.2	3.4
Wind run (km day^{-1})	83.0	70.0	68.0	154.0	191.0	106.0
Max. humidity (%)	82.0	87.0	85.0	85.0	77.0	89.0
Min. humidity (%)	37.0	42.0	32.0	37.0	32.0	43.0
Day length (hr)	12.3	13.1	10.9	12.4	13.4	10.5
Sun hours (hr day^{-1})	7.5	7.0	8.1	7.8	10.0	6.0
Heat unit (>13°C)	936.0	1400.0	616.0	661.0	1460.0	446.0
Heat unit (> 20°C)	144.0	452.0	42.0	104.0	507.0	51.0
Degree hrs. (< 13°C)	584.0	1.0	1948.0	1693.0	21.0	2798.0

Source : Plessis (1996)

in three different Navel producing areas in South Africa. Two methods have been tested for forecasting their suitability. Method one, involved use of stepwise multiple regression analysis to relate crop size over a number of seasons with climatic factors. This method can explain 59 to 88 per cent of the seasonal variation in yield, for the three areas. Crop forecasting could be made with the aid of these regression equations with 4 per cent variation of the actual yield. Method two comprises the relationship between the number of hot days during the flowering-fruit set period (3 months) and the yield. These climatic factors could explain about 80 per cent of the yearly variation in yield of oranges in the mediterranean area. Sanikidze and Mamulaishvili (1990) observed that an annual variation in fruit yield of mandarin is caused due to weather factors. These effects are more pronounced under optimum nutrition in Western Georgia. Similarly, a strong positive effect on yield of valencia and navel oranges has been observed in Spain, Morocco, and Israel.

Mean temperature and evaporation during September and November have shown as a significant predictor of the number fruits harvested from Washington navel trees (Moss and Muirhead, 1971). Gayford (1969) and Buch (1975) developed yield forecasting models for valencia orange in Florida and shamouti orange in Japan, respectively. Strong correlation has been shown between monthly rainfall and fruit yield of hamlin orange trees budded on sunki mandarin rootstock (Tubelis and Salibe, 1989a) and Rangpur lime rootstock (Tubelis and Salibe, 1989b) at the Botucatu plateau of Brazil. Tubelis and Salibe (1991) later observed that productivity of hamlin orange on *Citrus jambhiri* rootstock correlated with tree age and total rainfall during 16 months preceding the harvest season under nonirrigated condition between 7th to 17th years of orchard age at Botucatu plateau of Brazil. Base on the relationship between production of hamlin orange trees on Caipira sweet orange rootstock and monthly rainfall in Terra Roxa Estruturada soil at the plateau of Botucatu, Brazil, Dyers and Gillooly (1979) developed a mathematical model to be able to forecast the production of valencia orange one year ahead in eastern Transvaal for export, local markets, and total yields. Tubelis *et al.* (1999) observed that rainfall during sixteen months before picking season affected the quantum of fruit yield at Botucatu, Sao Paulo, Brazil. Kalma and Stanhill (1972) in Israel established a model to calculate the relevant crop and climatic parameters from standard climatic data. Okada (1972) used principal component analysis for classification of satsuma mandarin growing districts in Japan. Other studies (Brar *et al.,* 1988; Blazquez *et al.*, 1990) suggested techniques like estimating plot yield by sampling small tree and through the mathematical relationship between tree counts and citrus grove production. There are comparatively simpler methods, but, probably may lack the tree spatial reproducibility.

Human *et al.* (1994) developed a computer model to test the climatic suitability at any geographical point in the eastern Cape for growing citrus. Based on the agroecological survey of 63 stations for oranges and Mexican limes in Oaxaca state, thermal indices were prepared for both the citrus cultivars taking into consideration the average temperature, soil depth, and salinity. A total 3 categories of agroecological adaptability for oranges and 4 for Mexican limes were streamlined (Martinez Fonseca and Villegas Monter, 1994). Tubelis and Salibe (1988) observed that production is correlated with the orchard age and total rainfall received in sixteen months, before picking season. The relationship (HA/CA = -18.91 + 20.35I + 0.8190P8 – 1.3788P9 + 0.5805P10 – 0.149615, where HA/CA stands for production expressed in kg plant^{-1}; I, orchards age expressed in years; P8, P9, P10, and P15, total rainfall received, respectively, during the months of August, September, October, and March of the year before picking) showed the best determination coefficient up to 98 per cent of year to year variation in yield, having a good yield forecast ability, with substantial advance at blooming. Haggag and Maksoud (1996) later developed a mathematical model for yield forecasting in navel oranges in Egypt. Similar to above models, various other mathematical models were developed for predicting temperature requirement for Texas citrus (Chance and Rathewell, 1979) and California citrus (Pehrson, 1966).

1.4 CLIMATE AND FRUIT QUALITY

Fruit quality parameters have also been observed to be affected by the variation in climate (Turrell *et al.*, 1964a; 1964b; Hilgeman, 1966). Navel oranges grown in the year round humid, hot sunny climate in countries such as San Padro Sula, Honduras, Port of Spain, most orange varieties produce poor external quality (Reuther, 1973a). Nii *et al.* (1970) suggested that warm night temperature (25-30°C) has a stronger effect than warm day temperature in producing green fruits. While, regreening in valencia oranges is stimulated by warm temperature (Caprio, 1956; Young and Erickson, 1961; Cooper *et al.*, 1963; Cooper *et al.,* 1969b; Lee *et al.*, 1971; Surza *et al.*, 1982). Due to seasonal meteorological conditions, valencia fruits on trifoliate orange rootstock in Elimbah-Beerwah district of Queensland, Australia were larger and lower in soluble constituents than those grown in the Palmwoods-Maroochy district (Bowden, 1968). Considering the basic climate difference between inland and maritime climates, it is commonly observed that fog could be of real service in the control of fruiting as well as of low temperature problems. In inland compared to maritime areas, temperatures are higher in summer and lower in winter. Diurnal fluctuations in temperature are likewise greater. The change from low to high temperature, the winter moves into the spring more rapidly. Such a change has shown a profound influence on the yield, fruit shape, peel colour, and marketability of fruits (Finch, 1977).

Under subtropical conditions in California, most citrus varieties accumulated high sugar content during the ripening period, when vegetative growth is checked or appreciably reduced due to the onset of cooler weather in the autumn (McCarty *et al.,* 1965). While, under hot tropical conditions, such as Cauca valley, continuous vigorous vegetative growth and even flowering during ripening resulted in low total soluble solids in many varieties. Periods of dry weather and limiting soil moisture in the hot tropics induced dormancy and imparted high solids, a little before or during the fruit ripening period (Reuther and Rios-Castano, 1969). In the humid tropics, fruit size tends to be large, rinds remain thin and smooth, and juice content high compared to arid or semiarid and subtropical regions. Acid concentration in all oranges, grapefruit, pummelo, and mandarin varieties have similar in response to temperature (Reuther, 1973a).

Comparing the response of navel oranges under tropical climate of Cruz das Almas of Bahia, Brazil and subtropical climate of California, USA, Passos (1979b) observed a strong influence of climate on the behaviour of navel oranges. In California, the variation between daily and night temperature sometimes reaches 20°C and in Bahia, this variation is around 5°C. In California, under subtropical conditions, the harvest period is around six months from October to May. The fruits are well coloured, having good flavour, mainly from January to March, and have a satisfactory solid-acid ratio, fruit size, and juice content. In Bahia, under tropical conditions, the harvest priod is shorter about 3-5 months. The fruits are large, and juicy but rind colour as well as the solid-acid ratio is not so satisfactory. However, the data are compared with those from tropical areas, closer to the equator in Africa and South America, a striking difference is observed. All the lowland tropical regions from 0° to 20° do not fall into one ecological zone for citrus. Regarding the yield, it seems that navel trees showed a greater potential in Bahia where it originated.

In the hotter and more arid location, typical of the interior part of Israel, fruits developed more rapidly in weight, circumference, and volume. While, it differed little in peel thickness and N content, but had a lower naringin content from July as compared to fruit growing in the cooler and more humid location, typical of coastal region (Herzog and Monselise, 1968).

Variation in climatic parameters such as maximum and minimum monthly temperatures at various locations viz., cool coastal valley of Limoneira (25.6-27.6°C and 17.3-18.6°C) and south coast (24.2-26.9°C and 17.9-20.0°C), hot central valley of Lindcove (31.1-32.5°C and 19.3-20.7°C), and very hot interior low elevation of Thermal (33.9-35.2°C and 23.5-24.2°C) showed maximum influence on fruit quality parameters viz., fruit size, shape, rind colour, texture, thickness, flesh colour, juice percentage, and composition of red blush nucellar grapefruit, eureka, and lisbon lemons (Nauer *et al.*, 1975a , 1975b), mandarin, and valencia oranges (Nauer *et al.*, 1974), and navel oranges (Nauer *et al.*, 1972). These studies further showed a greater variance in fruit maturity and quality among climatic zones than variance among orchards within a climatic zone. Differential climate accounted for a large difference in fruit quality (total soluble solids, titrable acidity, and total soluble solids: titrable acidity) at Rungsit and Pathoomtani provinces in Thailand (Ketsa, 1988). Reuther *et al.* (1969) earlier reported a strong and consistent inverse effect of temperature level on the rate of total acid concentration in sweet oranges during maturation. Differences in fruit quality variation within a tree or an orchard have been observed in orchards of Kawano Natsuadaidai in Japan (Takahara *et al.*, 1988). German and Sardo (1988) observed a strong correlation between fruit quality and climatic parameters in sweet orange cultivars, navelina, moro, and torocco in Italy.

The physical and chemical characteristics of lemons grown across various countries representing specific agroclimate have shown a great diversity (Mc Donald and Hillebrand, 1980). Satsumas grown under plastic in New Zealand attained a standard suitable quality having 9-10 per cent soluble solids and 1-1.2 per cent acidity usually obtained in Japan (Martin, 1991). Human and Koekemoer (1991) observed low juice percentage and poor fruit colour under hot climate of Fridenheim in eastern Transvaal. The effect of temperature on growth and quality of satsuma mandarin at Kerikeri in New Zealand as described by Richardson *et al.* (1991), revealed that fruits grown outdoors, in unseasonally cool conditions did not attain an acceptable TSS:acid ratio until early June.

One of the major problems encountered in studying the effect of climate on fruit quality and maturity is often difficult to distinguish climatic effects from the impact of soil difference and differences in water quality, annual weather variation, tree age, and crop load, unless the selected orchard is maintained under healthy condition. Low winter temperature resulted in fruits with thick peel in the following year. Fruit shape (fruit diameter divided by fruit height) is affected equally by low winter temperature, summer air humidity, rate of evaporation, and continuous changes during the growing season. Summer temperature and the difference between maximum and minimum temperature in the spring and autumn has little effect on fruit shape and peel thickness (Cohen *et al.*, 1972). Fruit quality and climate in Trinidad (Weir, 1969b), especially in soluble soilds:acid ratio tested at six experimental sites, temperature above 12.8°C, and rainfall for the two months preceding the harvest correlated with quality parameters of valencia orange under humid tropical climate of Contramaestre region of eastern Cuba (Sanchez-Garcia and Fernandez, 1981). Rain during two months prior to harvesting showed an inverse effect on the total soluble solids and acidity of clementine juice in a mediterranean climate (Blondel and Cassin,1972). A much wider climatic differentiation can be achieved with marsh seedless grapefruit grown in a much wider climatic range. Relatively dried interior areas (about 25 km from the sea and shielded by mountain range) provided best fruit. A considerable range of daily temperature (May to August), a strong but not excessive evaporation during summer, medium rainfall in winter, are found in the western Yezreel valley of Israel. These conditions are linked with less advanced internal maturity (more acidity) with a thicker peel, relatively low softness, and deformation at late harvest. Other studies (Reuther *et al.,* 1969; Nauer *et al.,* 1974) showed differences in peel thickness due to variation in temperature and humidity in different climatic regions of USA.

Most striking effect of tropical temperature regions on fruit quality is on rind colour. Increase in albedo cracking and rind puffing is a common feature with increasing humidity to constant temperature and with increasing temperature at constant humidity. Rind puffing became marked towards 100 per cent relative humidity, but recorded a decimal puffing at the lowest relative humidity of 91.9-93.0 per cent (Grierson and Wardowski, 1975; Kawase, 1984; Kawase *et al.*, 1984). In subtropical climates, the cool nights during fall and winter when most of the varieties tend to gain juice and total soluble solids, caused changes in carotenoid pigments in the rind (Erickson, 1968; Cohen *et al.*, 1972; Wutscher, 1976). Sanchez *et al.* (1978) observed that heavy rainfall in the 2 months before harvest (September and October) significantly reduced the total soluble solids and acidity of Corsican clementines. In California, under subtropical conditions, the harvest period extends to about 6 months. The fruits are large, of good colour and flavour, with satisfactory solid:acid ratio and juice content. In Bahia, under tropical conditions, the fruits are large and juicy, but the rind colour and solid:acid ratio are not so satisfactory. Yield, however, is greater than in other low lands of tropical regions nearer the equator in South America and Africa (Passos, 1979a).

Difference in climate between two seasons affected the time of ripening (Damigella and Tribulato, 1973). Chilling temperatures of 15°C and low, than that, are sometimes associated with this change, which in turn is related to the ethylene, produced as a result of chilling injury effect (Cooper *et al.*, 1969a). It imparts a profound influence on shape and size of fruits (Wutscher, 1976; Fucik and Norwine, 1979). On the other hand, in the tropics, lack of chilling with a slow breakdown of chlorophyll and synthesis of carotenoids, in the rind turned the fruits pale green or at best pale yellow at maturity (Stearns *et al.*, 1942; Cooper *et al.*, 1969b). This precluded the acceptance of tropical oranges and mandarins compared to subtropical ones by the consumers. However, limes, lemons, grapefruits, and pummelos developed a quite attractive rind colour in tropical areas (Reuther, 1973a). Comparison of physical and chemical changes occurring in ripening fruit between December and February under differing climatic conditions of the Sorrento plain and the rather warmer Pellaro region (Reggio Calabria) showed higher maturation index (refractometric value: acidity) at Pellaro region of Italy (Gioffre, 1976). Response of pera sweet orange in relation to climatic variation in Brazil showed a significant difference in fruit maturity and quality between climatic zones. Total acidity of the fruit decreased with increase in temperature. Fruits produced in the coolest climate (Taquari) recorded the highest level of total acidity, by those grown in the warmer conditions registered lowest value. The juice percentage, weight, and fruit size are consistently higher in the warmer conditions of Cruz das Almas (Coelho *et al.,* 1984). Rainfall during the two months prior to harvesting has an inverse effect on the total soluble solids and acidity of clementine juice in a mediterranean climate (Blondel and Cassin, 1972; Sanchez-Garcia and Fernandez, 1981). Production of high brix and sugar-acid ratio containing sweet orange fruits has always been the major constraint in humid climate of Carribean basin countries due to absence of cool night temperature. Excessive rainfall further dilutes the brix, and cloud cover reduces the brix production (Muraro and Fairchild, 1986)

On tree life of citrus fruits is also by and large affected by variation in climate. Citrus fruits grown in hot tropical locations in Columbia deteriorated much more rapidly when retained on the tree after harvest maturity is reached than do comparable varieties produced in subtropical locations in California, primarily due to large difference in prevailing temperature after attaining harvest maturity. In California, cool winter temperatures pervail during this period. Most of the sweet oranges and mandarin varieties cannot be stored on tree more than about one month after market maturity is reached in hot tropical locations like Palmira of Columbia and Cartagena of Spain. While, in subtropical zones of California, most varieties can be retained on the tree from three to five months beyond market maturity. In cool tropical Medellin climate, most varieties can be stored for 3-4 months beyond market maturity (Reuther and Rios-Castano, 1969).

Difference in the fruit maturity time has been observed on account of prevailing temperature. In the very hot tropical, low altitude climate of Cartagena, Columbia, valencia oranges require only about 6.5 months from anthesis to each harvest maturity, but require about 10 months under high altitude, cool tropical climate of Medellin, and Columbia. At low altitudes in subtropical California, valencia matures in about 9.5 months of anthesis in the hottest, and about 14 months in the coolest climate zone (Lomas *et al.*, 1970; Monselise, 1981). The two fold spread in the time required from anthesis to market maturity shown between tropical climate of Palmira (6.5 months) and cool climate of Santa Paula (13 months) is probably the extreme limit for climates suitable for commercial Citriculture. Haro-Guzman (1979) reported no clear cut relationship between rainfall and changes in Mexican lime fruit composition.

1.5 CLIMATE AND PHYSIOLOGICAL DISORDERS

Interaction of climate and mineral nutrition bound physiological disorders are widely recognised (Li *et al.*, 1995). Trees showing N deficiency symptoms produced few flowers irrespective of temperature. Number of flowers increased with increasing P content from 40 to 160 mg kg^{-1}, provided sufficient leaf N (3.0 per cent) concentration is maintained (Inoue and Kataoka, 1992). Some of the weather related citrus fruit disorders, namely, water spot, zebra skin of tangerines, sunburn, windscar, and endoxerosis have been described by Knorr (1973) and Grierson (1981). Navel oranges have been observed to develop the problem of water spots (rupturing of epidermal cells by endosmosis) under low humidity conditions and later subjected to an anamalous cool and wet period (Riehl and Carman, 1953; Klotz, 1975; Smoot *et al.*, 1971). Zebra skin of tangerines occurs when sudden excess of soil water is consumed by tangerine trees which were earlier growing under moderate to severe drought stress. This results in excessively high turger pressure buildup in the epidermal cells, later causing blackening of peel after rupture (Grierson and Koo, 1958; Grierson *et al.*, 1965; Grierson and Brown, 1966). Mendes (1972) described Swaji spot of citrus due to Zn and Mn deficiencies. A survey of fruit sunburn in four citrus varieties in coastal and desert orchards indicated that highest sunburn in the balady mandarin and lowest in the clementine tangerine. Balady sweet orange and the sangtara mandarin are intermediate in susceptibility. Temperature of different parts of the fruit on the south side of the tree have shown signficiantly higher than those of the corresponding parts on the northern side, most of the times (Minessy *et al.*, 1970a).

Ketchie and Ballard (1968) described the environments leading to heat injury. In Japan, a physiological disorder of the rind called Kohansho due to exposure of high temperature in form of pitting on the stem end or on the periphery of the fruit. The affected fruits gradually discolour and finally become brown (Hasegawa and Iba, 1981). Fucik (1972) reported the brown, sunken, scald like spots characterizing the condition resembling oil injury due to several periods of stress after trees are subjected to conditions having ample moisture. Such a rapid drying of peel may have caused contract against the turgid pulp, strong enough to express oil from the glands and result in spotting. Holtzhausen and Plessis (1970) described skin splitting in Washington navel due to stress caused by the pressure from the endocarp cells which continue to enlarge until the fruits are ripe, especially under conditions of high rainfall and humidity.

On the other hand, citrus trees are highly prone to heat injury in form of sun burn, drying of fruit, burning and defoliation of leaves, burning and death of the bark (Ketchie, 1969). Ketchie and Furr (1968) described four types of injury to citrus fruits due to heat, namely, slight discolouration of skin, severe burn to the skin without internal injury, slight skin injury with severe internal injury, and severe skin injury with severe internal injury, may or may not be associated with granulation. Ketchie and Ballard

(1968) showed that the two environmental factors causing injury to valencia oranges are high temperature and high intensity of solar radiation. A number of researchers (Dodson, 1966; Elmer *et al.*, 1973; Freeman, 1976) are of the opinion that winds induce abrasion injury on susceptible fruits (when small) due to rubbing of leaf against fruit causing lesions that expand to large silvery or tan coloured blemishes as the fruit enlarges. In arid climate, when lemon trees are subjected to water stress develop discolouration of internal tissues in the fruit, and later disappear leaving a cavity (Smoot *et al.*, 1971; Klotz, 1978).

Leaf burn symptoms due to chloride toxicity are related to water relations of plants. Pair *et al.* (1975) pointed out that the addition of 0.4 per cent salts reduced the total available water in the soil by approximately 33 per cent. Salinity can prevent water uptake even when soil is at field capacity (Hartz, 1984). Other studies (Bielorai *et al.*, 1983; Walker *et al.*, 1983; Plessis, 1984) showed reduction in water uptake with increase in soil salinity level. Calcium has been shown to ameliorate the effect of saline conditions on the growth of citrus by: i. flocculation of dispersed soil particles by Na, ii. restricting the uptake of Na ions, and iii. maintaining the selective permeability of membranes. The beneficial effect of adding Ca depends on the anion associated with Ca salt (Koo *et al.,* 1984; Zekri and Parsons, 1990).

Climatic factors (Temperature and humidity) are involved in incidence of rumple and wrinkled condition of rind, more prevalent on larger sized fruits (Benton, 1940; Cooper, 1958; De Fossard, 1962; Sinclair, 1961; Knorr and Koo, 1969; Singh and Singh, 1980; Chanana *et al.,* 1984). Often, dryness of fruits is characterised by a lack of extractable juice in whole fruit or the affected portion of the fruit. Earlier studies have described the symptoms and occurrence of dryness due to frost damage (Bartholomew *et al.*, 1950), granulation (Bartholomew *et al.*, 1934; 1935), and endoxerosis (Bartholomew, 1937). Granulation is also considered as one of the forms of dryness. In Florida, it is called as dry end, whereas it is referred as crystallization in California (Bartholomew *et al.,* 1934), Koa Sarn in Thailand, and corkiness in West Indies (Rajput and Haribabu, 1999). The problem of granulation due to climate has also been reported from Ghana (Atubra, 1982), Japan (Matsumato, 1964; Yoshinaga *et al.,* 1986; Cai *et al.*, 1989; Takebayas *et al.*, 1993), Australia (El-Zeffawi, 1973); and China (Zhang *et al.,* 1994; Wang *et al.,* 1997; Pan *et al.,* 1998) besides India. Though, the problem of granulation under arid semiarid conditions of northwest India is more prevalent than humid subhumid tropical climate of south India (Singh, 1999c). Bartholomew *et al.* (1941) described blossom (stylar) end granulation as a kind of granulation (dryness) which can appear in the stylar and/or the central portion of the fruit. The kind of dryness which appears to have its origin in the centre of the fruit has been termed as core dryness to distinguish it from stylar end true granulation, which is located mainly at the stem end. Bruise dryness is a dryness with symptoms similar to the description of dry sac, localised in the vesicle tissue directly beneath a bruise mark on the peel. Sunburn dryness is found in fruit on the tree and after harvesting. The problem of a reduction in juice content and an excessive rag content (Bartholomew and Reed, 1943; Bitters, 1961; Chakrawar and Singh, 1977; 1978) are two distinctly different problems from the kinds of dryness in which morphological changes in the vesicle structure are visible.

Sinclair and Jolliffe (1961) demonstrated the nature of chemical changes, especially pectic substances that take place in granulated valencia orange vesicles. Evidences indicated that the development of dryness in navel fruits is related to the physiological pattern of development of fruit. This is further modified by the variation in environment (Bain, 1949; Beutel, 1964; Gerard, 1969). Pateca, another common physiological disorder in lemon, noticed in Lebanon is considered to be due to disturbance in Ca nutrition. Shallow soils and moisture stress during hot months contributed to the abnormal Ca behaviour in the developing fruits (Khalidy *et al.,* 1969a). Tal and Monselise (1965) described two types of corky (silvery) spots in shamouti oranges and marsh seedless grapefruits as spreading and nonspreading type. The spreading spots may be caused by increased pressure due to uneven growth of the endocarp including splitting of stomata and successive filling of splits by cork in a manner similar to suberization of branches. Climatic factors previously been considered as a direct cause of the problem are indirectly responsible through their action on the growth rate of the endocarp. Sand burn of citrus involves dessication of the bark and cambium of young plants at and just above soil level, is induced by high temperature and intense isolation (Vanderweyen, 1963).

1.6 CLIMATE AND FLOWERING

Citrus appears to have originated in the humid tropical regions of China, southeast Asia, the islands of Indonesia, and the Philippines (Swingle and Reece, 1967; Webber, 1967). The tropics lie between the tropics of cancer and capricorn on either side of equator. Within the tropical area, however, there are high mountains which modify the climate considerably. The term tropical areas is applied mainly to the areas where tropical climate is found without the occurrence of any frost. Citrus in the tropics is grown in areas usually within 23.5°N – 23.5° S where there is little temperature difference throughout the year. The mean monthly temperature during the coolest month is 18°C or higher, and they require substantial amount of heat for maturity (Jackson and Looney, 1999).

Citrus shoots grow in cycles, with the number of growth flushes varying from two to five in a growing season (Webber, 1948; Bain, 1949; Iwasaki and Owada, 1960; Singh and Ghosh, 1965). General knowledge of the response of citrus trees to moisture (Koo and MaCornack, 1965; Hilgeman and Reuther, 1967) suggested that in a very hot tropical climate like Cartagena and Columbia with monsoon rainfall distribution regulated the periods of moderate moisture stress and offered possibilities not only for controlling the time of flowering, but also for improving the interal fruit quality.

According to meteorological characteristics in tropical regions, no defined growth flushes occur, and the tree remains more or less in a state of continuous growth, unless interrupted by drought, which results in cessation of growth and induction of flowering. Drought plays the same role as cold does in subtropical region, leading to renewal of growth and flowering with the onset of rainy period (Bain, 1949). In subtropical climates, shoots produced during the warm summer months normally differ from those produced in the cooler spring months following the dormant period. Summer flush shoots are normally longer thicker, with large leaves and longer internodes than spring flush shoots (Smith and Reuther, 1951; Cameron *et al.,* 1951; 1952; Mendel, 1969). This is applied both on an altitudinal as well as latitudinal limit for the tropics. On the other hand, subtropics are those areas where frost occurs occasionally, and are located at latitudes towards north and south adjacent to the tropics. These areas have mean temperature in the coldest month between 13° and 18°C, distinct wet and dry season, and a seasonal change in temperature. Subtropical climates are also found in the tropics in the areas at certain altitudes.

The selection and dissemination of various commercially important citrus species are now grown in broad range of climatic and soil conditions from equator to 40^0 into the northern and southern hemisphere (Fig. 1.1) after 100 years of propagation (Burke, 1967). Two to four months of extended cool weather results in heavy spring bloom occurring in February to April in the northern hemisphere and August to October in the southern hemisphere (Sauer, 1954; Fugita and Yagi, 1955; Mendel, 1969; Reuther and Rios-Castano, 1969; Reuther, 1977; Oh *et al.*, 1981). In subtropical climates, citrus flowers in response to cold winter temperatures, but also produce periodic flushes of flowers throughout the year. Depending upon the species, up to 50 per cent of the annual production can result from this off season bloom (Manchester, 1988). A major flush of vegetative growth coincides with spring flowering flush and sporadically in the distinct flushes, throughout the warm months under subtropical conditions (Webber, 1943; Bain, 1949; Iwasaki and Owada, 1960; Singh and Ghosh, 1965; Oslund and Davenport, 1987). Reuther (1973a) pointed out the lack of genetic selection for varieties better adapted to the tropics rather than flowering habits of the trees. The flowering in *Citrus* is recurrent under tropical and subtropical conditions (Monselise, 1985), unless synchronized into a well-defined period of concentrated bloom by external conditions (Monselise and Goren, 1969; Goldschmidt and Monselise, 1972; Monselise, 1978). Soulez and Forique (1978) established a definite relationship between rainfall and flowering in Niger and Ivory Coast.

An agroclimatic model has been developed considering the effect of temperature, solar radiation on flowering time, flowering duration, number of flowers, effect of past stress evaporation, wind, rain, planting density and tree age on fruit set, maturation, colouration and, thereby, demonstrating the utility of climatic norms for yield and quality prediction of navel and valencia oranges. These criteria and functions from many parts of world are cultivar specific and independent of local climatology or other site specific effects (Ben-Mechlia and Carroll, 1989a). In later studies on verification of above model (Ben-Mechlia and Carroll, 1989b), simulation of 20 years of average yields in several climatically diverse regions compared quite well with observed yields, provided the meteorological data are representative of whole region coupled with stable and homogeneous bearing acreage.

1.6.1 Photoperiodic Response

Citrus produces flower during short days in spring and autumn, but also during long day in summer, on account of assumption that flower formation is not qualitatively affected by day length (Lenz, 1969). Light intensity is a growth retarding factor, with the growth rate decreasing with increasing light intensity (Monselise, 1951). Citrus trees react strongly to increased day length. Under conditions of illumination for 16 hours, citrus trees produced more and longer shoots than with short day (8 hours) illumination (Piringer *et al.,* 1961; Young, 1961; Warner, 1966). Dry periods which reduced the vegetative growth, favoured flower formation in Israel (Oppenheimer, 1940), California (Furr and Taylor, 1939), and Sicily (Casella, 1935).

Field experiments examining the effects of day length on flower formation in 5 year old parson brown and valencia orange trees (Furr *et al.,* 1947) and Washington navel orange cuttings (Lenz, 1964), have led to the conclusion that there is no quantitative involvement of photoperiod. Moss (1969) was also unable to discern differences in flowering response on day length. Moreover, in south Florida, only vegetative growth occurring on Tahiti limes maintained in heated glasshouses is commonly observed during the short days of winter, while flowering readily occurred in those plants exposed to low temperatures occurring either outside or in nonheated glasshouses. Increased shoot and leaf growth under long day conditions has similarly been observed (Kohl 1960; Young, 1961; Pringer *et al.*, 1961).

1.6.2 Temperature Response

It has been shown that there is no chilling requirement for sprouting of citrus (Stathakapoulos and Erickson, 1967). Sprouting takes place any time when soil temperature rises above 12°C. Bliss (1944) suggested soil temperature of 19-20°C at 10 cm depth during prebloom and bloom period in valencia orange under California conditions. In the coastal plain of Israel sprouting begins irrespective of air temperature, during the time, soil temperature reading 12-13° C (Mendel, 1969). Alteration of root and shoot growth cycles has shown to be governed by seasonal changes in soil temperature (Crider, 1927; Cameron, 1939; Cameron and Schroeder, 1945; Ducharme, 1971). A vigorous root and shoot growth takes place at soil temperature between 23.9 and 26.7°C, which ceases at a soil temperature below 10° (Wutscher, 1973). Number of other studies

(Liebig and Chapman, 1963; Cary, 1970) have suggested an optimum soil temperature of 25° C for growth of citrus roots. In citrus orchards located in inland subtropical areas, with relatively cold winters, little or no growth of root or shoot takes place in midwinter when the average temperature in the first 30 cm depth of the soil is at or just below 13°C (Ducharme, 1971; Kozyrev, 1973; Weerts and Cary, 1980). However, during this period, the roots actively absorb nutrients in anticipation of vigorous leaf and shoot growth to begin in early spring when temperatures in top 30 cm depth of soil rises to about 20°C (Weerts and Cary, 1980).

The first flush is commercially most important, since it usually includes a large number of inflorescences which flower and set fruits for the next years' harvest. During late spring and early summer, when temperature in the first 30 cm depth rises to about 25°C which is optimum for root growth, the first and most vigorous leaf and shoot growth occurs. During mid summer, when temperatures in the top 30 cm soil depth are between 25° and 32°C (Ducharme, 1971; Weerts and Cary, 1980), the second cycle of vigorous leaf and shoot growth occurs. Unless, the trees are under some stress, the flush is predominantly vegetative. From late summer to early autumn, with the temperature just below 25°C, a second type of root growth is initiated. This cycle, however, is much less vigorous than first root growth cycle occurring in late spring or early summer.

In subtropical areas, during the winter months, air and soil temperature fall below about 15°C for several months during the winter, which causes growth to cease, and the plants to become dormant for three to four months. Such a dormancy induces greater tolerance to frost damage and causes changes within the buds, which induces flowering following warmer temperatures. In a typical subtropical climate with cool winters, citrus tends to bloom profusely in the spring and set only one major crop, which matures in late fall or winter or even the following spring depending on the variety and seasonal temperature regime. Flower formation in citrus species is promoted by drought or low temperature, followed by restoration of climatic conditions favourable for growth (Monselise and Halvey, 1964; Monselise and Goren, 1969; Monselise, 1985; Southwick and Davenport, 1986; Lovatt *et al.*, 1988). The actions of these abiotic factors together with the cultural practices, that promote flowering in citrus species i.e. girdling (Monselise, 1985), graft incompatibilities resulting in weak rootstocks (Mosse, 1962), confining of root systems in small pots (Furr *et al.*, 1947), and root pruning (Monselise and Halvey, 1964; Monselise, 1985), support the idea that the root growth is an essential prerequisite to flowering in citrus as first proposed by Monselise (1947).

The stress applied in quantitative manner, provides a control system, during which changes in carbohydrates-nitrogen ratio can be monitored in relation to flower initiation. A well known example for positive effect of moderate drought stress is the improvement in the cold hardening (Yelenosky, 1979). In addition, the cessation of growth by cool winter temperature by dry periods in warm tropical areas, is important for flower induction

(Chaikiattiyos *et al.,* 1994). Bloom in citrus trees is more intense in cool winter mediterranean type of climates than in tropics. Summer drought can also enhance summer bloom in lemons after rain or irrigation (Barbera and Carmi, 1988). With the discovery of plant harmones and demonstration of their influence on flowering, carbohydrate-nitrogen status must be considered as contributory factor, but not solely regulating the flowering initiation. In subtropical climates, citrus produces flowers during spring season, following the winter rest period. Flowers are formed from the auxillary buds on shoot less than one year old (Monselise, 1985; Davenport, 1990a, 1990b). Flower formation occurs in response to chilling temperature and the number of flowers formed can increase with the duration of exposure to low temperature (Moss, 1969; Southwick and Davenport, 1986). Chilling is perceived in the buds and has two separate effects, as it releases bud dormancy and induces flowering (Garcia-Luis *et al.*, 1992).

The number of flowers formed is also inversely related to the number produced in the previous season (Moss, 1971a; Becerra and Guardiola, 1984). A reduction in crop load before winter, either through thinning of the fruits (Moss, 1971a; Goldschmidt and Golomb, 1982) or through an advancement of harvest (Garcia-Luis *et al.*, 1986), eventually increases flowering during the following season. Root temperature of 12⁰C seems to be the threshold temperature for reduction of root activity as measured by bud break (Girton, 1927; Stathakopoulos and Erickson, 1967) and root function (Elfving *et al.*, 1972). The optimum temperature for root and shoot growth of citrus is between 24⁰C and 31⁰C (Girton, 1927). Estimates of 48°C lethal temperatures for valencia orange fruits (Ketchie, 1969) and 55°C for hamlin orange leaves (Aherns and Ingram, 1988), suggested that lethal heat stress is not commonly observed on a large scale in most citrus growing areas. Possingham and Kriedemann (1969) suggested optimum temperature for citrus up to 30°C for valencia orange grown in Australia. Moss (1970) indicated optimum temperature for fruit setting in valencia orange as 21°C. While, Jones and Cree (1965) observed negative correlation between yield of Washington navel oranges and maximum temperatures recorded over a period of 38 years.

The flowering can be induced by low temperature (Lenz, 1969; Moss, 1976a; 1976b; Davenport, 1979) or water stress (Nir *et al.*, 1972; Maranto and Hake, 1985), and inhibited by applied gibberellin (Goldschmidt and Monselise, 1972; Iwamasa and Oba, 1973; Monselise and Goren, 1978; Davenport, 1983). The regulation of flowering by water stress is not common in trees and generally separated to be effective in tropical and subtropical species (Nir *et al.*, 1972). These studies, however, have been conducted under varying field conditions in the sense that the studies do not define the quantitative relationship between imposed stress and the flowering response. The increased flowering following fruit removal is inconclusive, since it involves the removal of an important carbohydrate sink, but at the same time, eliminates a major natural source of gibberellins (Moss, 1971b; Monselise, 1973). Development of fog has been observed to prevent temperatures from dropping to low or rising too high. Fog can prevent temperatures from dropping too low on cold winter nights, thereby, allowing blossom bud formation. This could lower the temperatures as warm weather comes in the spring and hence, preventing vegetative growth by avoiding excessive loss of carbohydrates and prolonging the period of blossom bud formation (Finch, 1977).

The majority of studies have clearly indicated that carbohydrate levels are not always the limiting factor for flowering of citrus. Other researchers (Webber, 1943; Mendel, 1969; Reuther and Rio-Castano, 1969; Reuther, 1973a) demonstrated the existence of link between period of chilling temperature and flowering, but such temperature invariably occurs during short days making the decisive conclusion regarding the role of both temperature and day length difficult to assess. Hield *et al.* (1966) took advantage of the observation that, prior to assuming a strictly vegetative, juvenile phase lasting from 5 to 10 years (Frost, 1943; Furr *et al.*, 1947), grapefruit seedling often form a single terminal flower at the conclusion of the first post-germination flush, if planted during the cool winter months. Temperatures of 20^0 to 26^0C day and 8^0 to 12^0C night permit the expression of up to 100 per cent of the terminal flowers whereas day temperatures of 32^0 to 39^0C prevent flowering.

Moss (1969) investigated the flowering response to varying temperatures at two photoperiods. One year old cuttings of Washington navel orange plants were kept at temperatures regimes of $27^0/22^0$, $24^0/19^0$, $18^0/13^0$, and $15^0/10^0$C (day/night, respectively). The flowering response correlated inversely with temperature with a 10 to 15 folds increase in inflorescence produced at the lowest temperature regime compared to the highest. The inflorescence type also seemed to be temperature dependent. More vegetative and mixed shoots were formed on plants treated at $24^0/19^0$ and $27^0/$ 19^0C than at $18^0/$ 13^0 and $15^0/10^0$C, which produced predominantly generative shoots. In another experiment (Moss, 1969), plants in an inductive regime of 15° day/ 80^oC night and 10 hour photoperiod for 2 to 8 weeks, before transfer to warm, growth promoting conditions showed a quantitative flowering response in relation to the length of time in the inductive regimes. Generative shoot production was 25 per cent, 20 per cent, 40 per cent, and 47 per cent of the total shoots produced after 2,4, 6, and 8, weeks, respectively, in the chilling inductive temperatures. The warm temperature control produced no flowers. However, an inductive period of about one month has been estimated sufficient for flowering. Ecological studies on *Citrus junos* growing in Utsunomiya, Japan showed that bud burst at Utsunomiya occurred 2 weeks later and crop matured later than the warm districts (Tokushima and Kohchi), since the average temperature during harvesting period was 2-5^0C lower than the warmer districts (Wakabayashi *et al.*, 1988).

Southwick and Davenport (1986) described a similar experiment conducted on one to two year old tip pruned tahiti lime plants derived from marcottage and treated for various periods of time at $18^0/$ 10^0C (day/night temperature, respectively). The response consisted of production of 45 per cent, 55 per cent, 67 per cent, and 78 per cent mixed and generative shoots following inductive periods of 2, 4, 6, and 8 weeks, respectively. The predominant shoot type was generative with balance of shoots as vegetative. Control plants maintained at 29°C day/23°C night produced only vegetative shoots. A night temperature of 22°C has been reported to be insufficiently cold to induce flowering (Moss, 1976a). Field (Webber, 1943; Cooper, 1957a) and greenhouse studies (Liebig and Chapman, 1963; Ongun and Wallace, 1958a; 1958b; Cary, 1970) have suggested that initiation of strong flush of growth in subtropical areas is mediated by warming soil temperature in the early spring. High temperature (a mean daily temperature greater than 25°C and maximum temperature of greater than 28°C) at the time of flower fall has been observed as a limiting factor associated with severe fruitlet drop (Huang *et al.*, 1993).

In the last few years, the Uruguayan citrus industry has incorporated several varieties and hybrids of mandarin for fresh fruit export. In order to characterize the phenologic reproductive behaviour of nova (Clementine x Orlando tangelo) and clementina de nules cultivars, the important differences were found in the sprouting levels, (mean values of 125 and 105 sprouts 100 nodes^{-1} for clementina de nules and nova, respectively), per cent distribution of shootkind, and flowering levels (leafless inflorescences 66 per cent of the total sprouts) with values higher than 200 flowers 100 nodes^{-1} prevailed in nova. Whereas clementina de nules' showed a more balanced sprouting pattern, with a mean of only 6 per cent of leafless inflorescences and less than 100 flowers 100 nodes^{-1} in both years. In nova, fruit set percentage related inversely to the number of flowers, but not in clementina de nules, which maintained 10 per cent fruit set in both years (Arias *et al.*, 1996). The behaviour of the ellendale tangor (*Citrus sinensis* Osbeck. x *Citrus reticulata* Blanco) was evaluated under two growing conditions of Spain and Uruguay with emphasis on flowering-fruiting relationship. There is an inverse relationship between these two processes, with profusely exceeding 230 flowers 100 nodes^{-1}, but fruit set was a very low, rarely more than 4 per cent of the flowers (Gravina *et al.*, 1996). The perspective intensity of fruit set and final harvest is further dependent upon the proportion of leafless (Fig. 1.2) to leafy

Fig. 1.2. Leafy inflorescence, more effective sink for water and nutrients

inflorescence (Fig. 1.3) produced in a bloom.

Fig. 1.3. Leafless inflorescence, less effective sink for water and nutrients

In India, citrus is grown under a variety of agroclimate, right from northern plain and central highlands having hot semiarid ecoregions with alluvium derived to central hot subhumid ecoregions with black and red soils and humid ecoregions with red and lateritic soils (Srivastava and Kohli, 1997a). The commercial cultivars, namely, kinnow mandarin in northwest India, Darjeeling mandarin, khasi mandarin in northeast India, Nagpur mandarin in central India, Coorg mandarin in Coorg area of south, sweet orange cultivars, mosambi in Marathawada region of Maharashtra and sathgudi in Andhra Pradesh, acid lime in Tamil Nadu, Kheda district of Gujarat, and Uttaranchal are commercially grown.

The flowering in citrus under Indian conditions is induced through either temperature stress or soil water deficit stress. However, none of the citrus growing regions in India experiences a freezing temperature to induce dormancy. But, at the same time, the low temperature (< 10^0C) is observed for 10-15 days in northwest India and northeast India, enough to induce low temperature stress. In central and south India, in the absence of low temperature, the soil water deficit stress is adapted to induce flowering in the cultivars such as Nagpur mandarin, mosambi, and sathgudi. There are very few commercial citrus growing belts at global level where soil water deficit is widely practised to induce flowering, which could be the reason that not much headway has been achieved to manoeuvre the quantum of stress through soil. In such areas, the problem of flowering could be addressed through a complex relationship soil-water-plant than a simple plant-climate response (Singh *et al.*, 1999).

1.7 ALTITUDE, A CLIMATE AND SOIL FERTILITY MODERATOR

Citrus culture for the commercial purposes is being practised at an altitude as high as 2400 m above sea level (Ding *et al.*, 1990), but some of the ornamental species of citrus are grown up to 2800 m above sea level in the areas close to the equator (Camacho-B, 1981). Huangpigan, a natural hybrid between *Citrus sinensis* and *Citrus reticulata* is highly adaptable to environmental conditions of Muli county of Sichuan where it is grown at an elevation of 1200-2000 m with temperature reading as low as 11°C (Yang, 1991; Xiong *et al.,* 1998). Some of the sweet orange, lemon, Meyer lemon, and limettas are grown at an altitude of 2620 m at the Nagsiche near Salcedo, Ecuador. Probably, some other citrus trees could be found at still higher altitudes in many valleys along the Andes mountains, which have varied microclimates. Changes in altitude determine the consequent fluctuations in climate, which could affect the growth of citrus. The climatic parameters such as rainfall, wind, humidity, and light intensity are directly or indirectly influenced by altitude. Regardless of the altitude, the climate at a given site close to equator in the tropical areas differs from sites located at different latitudes in two major ways. First, the photoperiod of day length, which includes the number of hours with sunlight based on the time of sunrise, sunset is affected (Francis, 1972). And second, climatic factor that changes with the altitude, is the seasonal variation in temperature. A small variation in temperature observed at Mompos, Columbia is mainly due to the influence of the rainy season and cloud cover on the heat balance. But in tropical areas represented by Orlando, Florida, and Valencia, Spain, temperatures are high in summer and low in winter months (Reuther and Rios-Castano, 1969). Imamura *et al.* (1981) reported the occurrence of yellowish fruits of satsuma mandarin on the inner and lower parts of the tree crown, particularly in the less illuminated orchards facing northeast direction in Nagasaki area of Japan. Navel orange varieties developed acceptable to good internal quality in Columbia at elevations from 1000 to 2000 m, valencia oranges from sea levels to about 1600 m, and grapefruits from sea level to about 1200 m (Reuther and Rios-Castano, 1969). Washington navels grown in Medellin, Columbia at an elevation of 1425 m had an internal quality comparable to California navels (Sniclair and Bartholomew, 1944; Tucker and Reuther, 1967).

The vegetational changes observed at various altitudes occur, because the altitudinal gradients determine environmental gradients. It is possible to find the rate of change in temperature with altitude or lapse rate for a given location taking the temperature at sea level and the altitude at which frost occurs. The average lapse rate for the atmosphere is 0.6°C per 100 m, commonly known as the environmental lapse rate. Mean air temperature decreases as altitude increases, with a lapse rate of 0.49°C per 100 m considering the lower and the higher sites, located within 0 to 9°N latitude (Table 1.7).

The temperature changes are more precisely understood when expressed in terms of degree days or heat units. These terms are synonymous to each other and calculated on the basis of base temperature of 10°C. Degree days are calculated using the formula: Degree days = (M-10) x N, where M and N stand for mean monthly temperature and number of days in the months, respectively. Heat units are also calculated on a daily basis i.e. days when temperature is above 10°C (M-10) and added to obtain an annual total, which usually remains higher than that calculated on a monthly basis (Jackson and Looney, 1999). Accordingly, heat units required during the entire growth period of valencia

Table 1.7. Variation in monthly maximum and minimum mean air temperatures (°C) at various citrus growing sites close to equator differing in altitude

Months	Jan.	Feb.	Mar	Apr.	May	Jun.	Jul	Aug.	Sept.	Oct.	Nov.	Dec.	Mean
Mompos, Columbia (23 m, 9⁰ N)													
Max.	33.7	35.3	34.5	34.7	33.1	33.6	33.9	33.8	33.1	32.6	32.7	32.6	33.6
Min.	23.8	24.2	24.6	23.8	24.4	24.2	24.8	24.8	25.0	24.0	24.1	23.7	24.3
Annual 28.5 ; Mean diurnal variation 9.3													
Armero, Brazil (300 m, 5⁰ N)													
Max.	34.6	34.6	34.2	33.5	33.4	34.0	35.3	36.1	33.4	33.0	32.9	33.7	34.0
Min.	21.6	21.5	21.5	21.6	21.3	21.3	20.7	20.9	21.1	21.2	21.5	21.5	21.4
Annual 27.6 ; Mean diurnal variation 12.6													
Palmira, Brazil (1006 m, 3⁰ 31′ N)													
Max.	30.3	30.7	30.5	29.7	29.1	29.4	30.5	30.7	30.5	29.4	29.1	28.6	29.9
Min.	17.9	18.2	18.4	18.4	18.3	18.0	17.5	17.7	17.7	17.9	18.1	18.1	18.0
Annual 23.9 ; Mean diurnal variation 11.9													
Bello, Columbia (1425 m, 6⁰26′N)													
Max.	27.4	28.4	28.5	27.7	27.2	27.5	27.9	27.4	27.4	26.2	26.3	26.4	27.4
Min.	14.7	15.2	15.7	16.1	16.2	15.8	15.5	15.1	15.3	15.5	15.2	15.4	15.5
Annual 21.4 ; Mean diurnal variation 11.9													
La Selva, Columbia (2200 m, 6⁰20′ N)													
Max.	21.9	22.1	22.1	22.0	22.1	23.0	22.1	22.3	22.2	21.9	21.9	21.5	22.1
Min.	10.7	11.1	11.3	11.7	11.9	11.7	9.5	9.8	10.6	11.3	11.5	11.0	11.0
Annual 16.5 ; Mean diurnal variation 11.1													
Tumbaco, Ecuador (2,350 m, 0⁰ 13′ S)													
Max.	25.2	24.7	24.8	24.6	25.2	25.0	25.1	25.6	25.4	25.0	24.7	25.3	25.1
Min.	9.6	9.6	9.8	9.9	9.2	8.5	7.4	7.6	8.2	9.5	9.3	9.2	9.0
Annual 17.1 ; Mean diurnal variation 16.1													

Source : Camacho-B (1981)

orange grown in South Africa under cool climate are much higher under cool climate compared to hot climate (Plessis, 1996). Similarly heat unit requirements for various citrus cultivars across tropical citrus growing region is more compared to subtropical regions (Mendel, 1969). The diurnal variation in temperature is another parameter affected by altitude (Table 1.8 and 1.9). At the higher elevations, larger variations in daily temperature occur, in contrast with the minor variation, occurring at lower altitudes due to difference in thickness of the air layer of the atmosphere at various altitudes coupled with the influence of amount and quality of solar radiant energy, reaching the earth surface. Altitudinal variation in temperature define various agricultural zones in the tropical mountains delineated as tropical or hot zone between 0-1000 m, subtropical zone at 1000-2000 m, cool zone 2000-3000 m, and cold zone at 3000 m and up. These altitudinal limits do not imply uniformly, everywhere in the tropics, since it is further affected by relief, slope orientation, wind velocity, and temperature regime. Such conditions are found in basin shaped valleys due to accumulation of cold air at night, and frost occurs at low altitude, even near the equator such as Ijen plateau in east Java, Indonesia at 900 m (Francis, 1972).

Table 1.8. Variation in temperature and annual heat units (above 12.5 °C) for various citrus growing tropical and subtropical areas

Area and location	Latitude	Elevation (m)	Annual heat units (°C)	No of months with average temperature (°C)	
				< 12.5°C	> 17.5°C
Subtropical Regions					
Spain (Valencia)	39°30′N	30	1600	3	6
California (Riverside)	34°0′ N	260	1700	3	6
California (Indio)	33°40′N	-10	3900	1	4
Israel (Degania)	32°40′ N	-200	3600	0	4
Israel (Rehovot)	31°50′ N	50	2600	1	4
Florida (Orlando)	28°40′ N	30	3700	0	2
Texas (Weslaco)	26°05′ N	40	3900	0	2
Brazil (Limeira)	22°30′ N	700	3000	0	1
Tropical Regions					
Trinidad (Piarco)	10°40′ N	10	5000	0	0
Columbia (Aracataca)	10°30′N	30	5600	0	0
Columbia (Girardot)	4°20′ N	400	5700		0
Columbia (Palmira)	3°30′ N	1000	3500	0	0
Columbia (La Florida)	4°40′ N	1800	1700	0	10
Ecuador (Ross)	3°30′ N	10	4400	0	0
Ecuador (Conoccoto)	0°15′ N	2200	1000	0	11
Kenya (Mombassa)	4°0′ N	20	5200	0	0
Kenya (Nairobi)	1°20′ N	1600	2500	0	1
Uganda (Jinja)	0°30′ N	1100	3330	0	0
Ceylon (Manner)	9°0′ N	30	5700	0	0
Ceylon (Nuwara Eliya)	7°0′ N	1900	1000	0	12

Source : Mendel (1969)

Variation in altitude has a strong influence on fruit quality. The total acidity of the fruit juice increased with the distance from the coast and increased greatly with altitude above 200 m (Matsumoti *et al.*, 1972). Response of satsuma mandarin in relation to a range of altitude (50-300 m above mean sea level) in Jeju Island of South Korea showed higher altitude imparted increased citric acid content and TSS/acid ratio, while decreased the rind colour development. The fruit quality is considered optimum up to an altitude of 200 m (Chung *et al.*, 1976). Fruit quality is better at low than high altitudes, but colour intensity and vitamin C contents are positively correlated with altitude. Altitudes above 1000 m are ideal for navel oranges of which varieties like valle, Washington, and golden nugget were found suitable for dry climate of Palmira and nativa 202 for the humid climate of Medellin. Varieties for extraction of juice suited best at lower altitudes. Valencia performed best for very dry climate of Cartagena, Indian River for Palmira and pineapple for Medellin. Low altitudes are best for tangelo, grapefruit, and acid lime. Mandarins

Table 1.9. Average monthly (°C) in various citrus growing areas during the period from first to second growth flush

Subtropical regions					Tropical regions				
Area and Location	Month				Area and Location	Month			
	March	April	May	June		March	April	May	June
Spain (Valencia)	13.4	14.5	18.0	21.4	Texas (Weslaco)	20.8	24.1	26.4	28.2
California (Riverside)	13.6	15.7	17.9	21.2	Ugnada (Jinja)	22.7	21.9	22.0	21.1
California (Indio)	18.3	22.3	26.1	31.0	Trinidad (Piarco)	24.8	24.9	25.6	26.4
Israel (Rehovot)	14.9	17.9	21.6	23.9	Euador (Ross)	27.0	27.2	26.3	24.0
Israel (Degania)	16.1	20.1	24.9	27.8	Columbia (Aracataca)	28.0	28.6	27.9	27.2
California (Indio)	18.3	22.3	26.3	31.0	Ceylon (Manner)	28.1	29.1	29.7	29.1
Florida (Orlando)	19.9	22.5	24.7	27.1	Kenya (Mombassa)	28.9	27.9	26.2	25.5
Brazil (Limeira)	20.6	21.5	22.1	22.4					

September-December (Spring in southern hemisphere)

Source : Mendel (1969)

are more adaptable than other citrus species to different altitudes (Moncada *et al.*, 1969).

Based on the suitability of citrus cultivars for Changjiang citrus belt of China, Chen (1990) recommended that satsuma mandarin is well adapted to altitude between 200 and 400 m, and sweet orange below 200m above mean sea level. In Columbia, good quality grapefruits are produced in the Cauca valley at an elevation of about 1000 m in the upper Magdalena river valley (400-500 m) in the lower Magdalena river valley and in other hot tropical areas, thereby, indicating that about 1100 to 1200 m as the upper limit of elevation in Columbia for commercial grapefruit culture. Above this altitude, the climate is likely to be too cool to reduce acidity low enough for good portability before advanced senescence is reached. Sweet oranges and mandarins grown in the Medellin valley of Columbia adapted well at an elevation between 1400 and 1500 m with a cool tropical climate (Reuther and Rios-castano, 1969). During the study visit of citrus industry of Nepal, Srivastava (2000) observed that pummelos (*Citrus grandis)* are well adapted to lower altitude of 60-100m under tropical climate, while, suntala (*Citrus reticulata* Blanco) and Junar (*Citrus sinensis* Osbeck) are confined to higher altitude of 1000-1500 m above mean sea level under subtropical climate. Altitude has also been shown to affect the time taken to attain the fruit maturity of various citrus cultivars.

Evaluation of satsuma mandarin in relation to varying altitudes at 15, 50, and 250 m above mean sea level showed trees growing at 50 m altitude cropped earlier and produced better quality fruits (Dumbadze, 1977). Qualitywise, better developed fruits are observed at lower altitudes and on southern slopes (Nakajima and Ogaki, 1972). Fruits from low hills and warmer tracts, have thicker rind, more loosely adhered to the pulp segments. The fruits from higher altitude of cooler regions are more compact, thin skinned, and the acid suger blend is better coupled with more acidic taste (Ghosh and Singh, 1993; Upadhyay and Patriam, 1994).

With some exceptions, fruits of sweet orange, grapefruit, and mandarin varieties produced in hot tropical locations in addition to maturing more rapidly and attaining larger size, have greener rinds, paler flesh, higher juice content, lower total soluble solids, and acid in juice. Lemons grown in tropical Columbia on the other hand, are as high in acid concentration in juice and yield of acid content per ton when mature as those produced in California, but are otherwise similar to other citrus fruits grown in hot tropical regions. These observations, thereby, suggested that acid citrus group responded to temperature differently than that of other major groups of cultivated citrus (Reuther and Rios-Castano, 1969). Oranges and mandarins grown in Medellin valley of Columbia at elevations between 1400 and 1500 m with a cool tropical climate are much like fruits produced Riverside, California in juice constituents when mature, but tends to be larger in size and have much more poorly coloured rinds even after the maturity is reached. Good quality oranges can be produced in the tropics at altitudes higher than 1500 m. Zhang *et al.* (1995) studied the effect of varying altitude (200-800 m) and meterological factors on the fruit quality of bearing Hongju (*Citrus reticulata*) and Jincheng (*Citrus sinensis*) at Sichuan province of China, which suggested that the quality parameters such as total sugars, acids, and sugar acid ratio increased with altitude. But, soluble soilds and ascorbic acid content showed no effect . According to Reuther and Rio-Castano (1969), navel orange varieties developed an acceptable to good internal quality in Columbia at 1600 m and grapefruit from sea level to about 1200 m altitude (Reuther, 1977).

No clear cut impact of altitude on the rainfall intensity and distribution has been observed. However, it is influenced more by the steepness of the slope, slope direction, and movement of the atmosphere (Medina, 1979). Opposite slopes of same mountain range may have different pattern of rainfall. The Cordillera de los Andes with its arid western and humid eastern slopes, is an example of mountain range affecting the large land area (Fewerda, 1979). Increased ultra violet radiation received at 2050 m altitude has been observed in comparison to sites located at 1050 m and in between (Lopez *et al.,*1972; Arcila-Pulgarin, 1974). Thus, total amount of solar energy reaching the earth surface at higher altitudes is much higher than at lower altitudes. Elfiving *et al.* (1972) and Kaufmann (1977) demonstrated that the reduction of soil temperature as low as 15°C caused water stress in the leaves during the transpiration by citrus plants. The low soil temperatures have a similar effect to that of low water potential in soil on the development of water stress in leaves. Another factor that could be limiting to citrus vegetative growth at high altitudes is higher ultraviolet radiation. Cleopatra mandarin seedlings, fully exposed to solar radiation at Tumbaco, have developed small, twisted, and deformed young leaves, with symptoms similar to those described by Lopez *et al.* (1972).

There is a definite correlation between the elevation and the particle size composition of the soils. As the elevation rose from 850 to 1500 m above sea level, the average concentration of clay fraction increased from 31 per cent to 38 per cent (Table 1.10). These differences in clay content at high altitudes and the more in northern regions of the country are due to the fairly large temperature fluctuations.

Table 1.10. Correlation of particle size composition of soils on basalts with elevation of the terrain

Province (District)	Elevation above sea level (m)	Mean annual temperature (^oC)	Precipitation (mm year^{-1})	Clay fraction < 0.001mm (%)
Bao Loc	850	21.5	2513	20.2
Di Linh	900	21.1	1841	33.0
Duc Trong	961	20.8	1626	35.0
Don Giong	1050	19.5	1580	37.7

Source : Kong-Tau (1986)

Soil fertility is also equally affected by the variation in altitude (Nakahara and Hada, 1963; Nath and Deory, 1976; Khera and Pradhan, 1976; Chakraborty, 1976; Ram *et al.*, 1987; Lahiri and Chakravarty, 1989; Jalali *et al.,* 1989; Singh *et al.*, 1990a; Walia and Chamuah, 1993; Lyngdoh and Shukla, 1993; Yashoda Avasthe and Avasthe, 1995; Lahiri and Chakravarty, 1995; Dass *et al.*, 1996; Singh *et al.*, 1997a; 1997b; Raychaudhuri and Sanyal, 1999a). While, studying the changes in soil fertility under varying altitudes in Khasi hills of Meghalaya, Singh *et al.* (1989) observed altitude variation had little influence on pH, but strongly influenced the accumulation of organic matter and available P. Organic carbon was deficient in 13 per cent soils, P in 83 per cent soils, and K in 5.5 per cent soils below 1300 m altitude. But, above 1300 m altitude, 55 per cent , 48 per cent, and 19 per cent soils registered deficient level of organic carbon, available P, and K, respectively. Relationship of changes in soil properties with altitudes showed that organic matter content decreased with fall in altitude. Low organic matter in valley soils was due to relatively higher temperature in lower reaches of the hills and regular cultivation. Better available water capacity under high elevation in strongly acidic soils of Mizoram, India has been reported by Singh and Datta (1983). The contents of SiO_2, K_2O, and P_2O_5 in these soils decreased with altitude, whereas Fe_2O_3, Al_2O_3, TiO_2, CaO, and MgO content increased with decreasing altitude (Singh *et al.,* 1991).

Singh and Datta (1989) observed decrease in pH, bulk density, soil depth, and porosity with altitude, while other soil properties like clay content in B horizon, silt content, and moisture retention increased with increasing altitude from 250 m to 1700 m in acidic hill soils of Mizoram, India. Avasthe and Avasthe (1995) showed much higher organic carbon, CEC, and sand (2.96 versus 1.38 per cent, 11.1 versus 8.64 cmol (p^+) kg^{-1}, and 79.0 versus 73.3 per cent) at an altitude of more than 2700 m compared to 300-500 m. While, other properties viz., pH, silt, EC, exchangeable Ca, and exchangeable Mg (4.9 against 6.1, 11.0 against 16.8 per cent, 0.19 against 0.63 dSm^{-1}, 2.75 against 5.04 cmol (p^+) kg^{-1}, and 1.14 against 2.14 cmol (p^+) kg^{-1}) recorded comparatively lower in soils at more than 2700 m against 300 – 500 m altitude in some acidic soils of Sikkim, India.

1.8 CITRUS GROWING BELTS FROM ACROSS THE WORLD

Citrus is grown at 40° on the either side of equator which covers most of the sub-tropical and tropical areas from across all the five continents.

1.8.1 Citrus in Asia

The oldest citrus growing region in the world comprises southeast China, south Malay penninsula, and west Burma where mandarins, pummelos, and limes are claimed to have originated (Spurling, 1969). Citrus is grown in Asian countries, namely, China, Japan, Israel, India, Nepal, Bangladesh, Bhutan, Sri Lanka, Pakistan, Malaysia, Philippines, Syria, Turkey, Lebanon, Egypt, and Thailand with equally great variation in climate, soil, and cultivar types (Table 1.11). The tropical island of Asia appeared in different periods of geological eras. Sumatra, which was ever a part of the south Cathaysian platform at the Paleozoic era and inundated by ocean at the Mesozoic era, gradually emerged from water about the late Cretaceous along with Philippines, Iran, and others. Malaysian Archipalago likely lifted in the middle of the Mesozoic era. And only have the fossils of the tertiary instead of those of the Cretaceous (Zhou, 1990).

According to Lal (1987), 31 per cent of land of south east Asia has slopes of between 8 and 30 per cent, and 20 per cent of the land has slopes greater than 30 per cent. The soils in the uplands are predominantly Ultisols, Oxisols, and Alfisols. The annual rainfall in the region is commonly higher than 1000 mm, and is usually concentrated into one or two wet season. The combination of landscape, soil, and rainfall is conducive to soil erosion, if soil surface is not protected by living plants or mulches (Saplaco and Payawan, 1985). The tropical Asia is characterised by monsoon type of climate where the rainy season alternates with dry periods. A large variation in rainfall pattern is observed. An average rainfall of 2000 mm, 375 mm, and 2500 mm has been observed in Sri Lanka, Philippines, and Indonesia, respectively, with a mean annual temperature of about 20°C. Kang (1997) focussed citrus production with respect to soil and climate in Korea. Giginejsvili (1968) highlighted in climate and soil features of citrus industry of Vietnam. In general, summers are hot and winds are mild. In the southern parts of tropical Asia, temperatures remain above 20°C during 4-11 months. While, temperatures remain between 10° and 20°C for 4-12 months in some mountanous area of India (Das and Biswas, 1998). Salinity has been focussed as one of the major citrus production problems in near and middle east countries like Greece (Crete), Turkey, Syria, Lebanon, Jordan, Saudi Arabia, Iraq, Iran, and Pakistan (Chapot, 1974).

1.8.1.1 Citrus in India

Citrus in India is cultivated under great climatic variation, right from arid and semiarid areas of northwest region to humid tropical climate of northeast India (Fig. 1.4). Area under citrus occupies 4.83 lakh ha with a production of 42.6 lakh tons (Table 1.12) and average productivity of 8.9 tons ha^{-1} (Anonymous, 1998; Kumar, 1998). The important commercial citrus cultivars in India, are the mandarin (*Citrus reticulata* Blanco), followed by sweet orange (*Citrus sinensis* Osbeck), and acid lime (*Citrus aurantifolia* Swingle). Commonly grown commercial cultivars of mandarin are Coorg,

Fig. 1.4. Distribution of various commercial citrus cultivars under diverse agroclimatic in India

Nagpur, Khasi, and Darjeeling which have ecological differences, but are akin to each other (Chadha and Singh, 1989). The states of Andhra Pradesh, Maharashtra, Karnataka, and Punjab contribute more than 10 per cent towards total production (Anonymous, 1994b; 1996a).

The area, production, and productivity trend under citrus in each citrus growing state (Table 1.12 and 1.13), provides an idea about the extent to which the cultivation is successful and sustainable. In terms of area, Maharashtra tops (1,10,839 ha) followed

Table 1.11. Major citrus growing belts across Asia

Name of the country	Citrus growing areas
India	
$8^{0}04'$- $37^{0}06$ N lat.; $68^{0}07'$- $97^{0}25'$ E long.	Gujarat, Haryana, Himachal Pradesh, Punjab, Rajasthan, Arunachal Pradesh, Assam, Manipur, Meghalaya, Mizoram, Nagaland, Sikkim, Tripura, Bihar, Orissa, Uttar Pradesh, Uttaranchal, Madhya Pradesh, Maharashtra, Andhra Pradesh, Karnataka, Tamil Nadu
Azerbaijan	
$40^{0}00'$N lat.; $47^{0}00'$ E long	
Bangladesh	
$24^{0}00$ N lat; $90^{0}00$ E long.	Sylhet, Chittagong, Sajek valley of Lushai hills, Banderban hills, Baralekha, Gazipur, Moulavibazar
Bhutan	
$27^{0}30'$ N lat.; 90 $^{0}00$ E long.	Chirang, Samchi, Gaylegphug, Samdrupjongkhar
China	
33 $^{0}00'$N lat.;105 $^{0}00$ E long.	Southern China –Gungdong (Ssettui, Riaopin, Yancun, Zhangzian), Guangxi (Qingzhou, Liuzhou, Guilin) Kwantung (Canton, Swatow), Kwangsi (Wuchow), Hainan island, Fukien (Amoy, Foochow);Western China – Hubei (Zikwei, Xiangshan, Yichang, Wuhan), Szechwan (Chengtu, Nanchung), Sekang (Tatsienlu), Kweichow; South-Western China – Yunnan (Yangbi, Weishan, Baoshan, Jungchun, Qiaojia, Yiliang, Zhengxiong, Kumming, Junning, Eshan); Central China – Chekiang (Hongchow), Kiangskii (Nanchang), Hupeh (Ichang), Hunan (Chang-sha); North-western China - (Han-chang Nan-cheng), Chamdo (Sikang), Wuhan, Jiangxi, Changjang valley, Jiangsu, Sichuan (Jiangan, Changning, Changsou, Fenjie, Kaixian), East China – Fujian (Ningde, Minho, Chantai, Xungcun).
Cyprus	
$35^{0}00'$N lat.; 33 $^{0}00'$ E long.	Lefka, Morphour, Salamis, Famagusta, Episkopi
Indonesia	
$0^{0}00'$N lat.; $118^{0}00'$ E long.	Kalimantan, Sumatra (Sumatera), Java (Jawa), Sulawasi Iran Jaya, Pontianak, (South west Kalimantan)
Iran	
$33^{0}00'$N lat.; $55^{0}00'$ E long.	Nowshar, Babol-Shahi (Ramsar), Daseroon, Jahrum, Kaseroon, Kotra, Chalus, Eisin, Minab, Shahsavar, Khuzestan
Iraq	
$32^{0}00'$N lat.; $45^{0}00'$ E long.	Bakooka, Baghdad, Diyala, Kerbela,
Israel	
$32^{0}30'$N lat; $35^{0}30'$ E long.	Mediterranean coastal plain, Yesreel, Biesan, Jordan valley, Negev, Rehovot, Kinnereth, Niryitshaq, Gilat
Japan	
$40^{0}00'$N lat.; $135^{0}00'$E long.	Kagoshima, Nagasaki, Kumamoto, Oita, Miyazaki, Wakayama, Hiroshima, Yamaguchi, Tokushima, Kagawa, Shime, Kochi, Fukuoka, Saga, Aichi, Mie, Osaka, Shizuoka, Kanagawa, Ehime
Lebanon	
$33^{0}00'$N lat.; $35^{0}00'$ E long.	Tyre, Sidon, Baoboa, Antilyas, Juniye, Tabarja, Jubeil, Tripoli, Akkar, Enfe, Damour
Indonesia	
$0^{0}00'$N lat.; $118^{0}00'$ E long.	Kalimantan, Sumatra (Sumatera), Java (Jawa), Sulawasi Iran Jaya, Pontianak, (South west Kalimantan)

Name of the country	Citrus growing areas
Iran	
33°00'N lat.; 55°00' E long.	Nowshar, Babol-Shahi (Ramsar), Daseroon, Jahrum, Kaseroon, Kotra, Chalus, Eisin, Minab, Shahsavar, Khuzestan
Iraq	
32°00'N lat.; 45°00' E long.	Bakooka, Baghdad, Diyala, Kerbela,
Israel	
32°30'N lat; 35°30' E long.	Mediterranean coastal plain, Yesreel, Biesan, Jordan valley, Negev, Rehovot, Kinnereth, Niryitshaq, Gilat
Japan	
40°00'N lat.; 135°00'E long.	Kagoshima, Nagasaki, Kumamoto, Oita, Miyazaki, Wakayama, Hiroshima, Yamaguchi, Tokushima, Kagawa, Shime, Kochi, Fukuoka, Saga, Aichi, Mie, Osaka, Shizuoka, Kanagawa, Ehime
Lebanon	
33°00'N lat.; 35°00' E long.	Tyre, Sidon, Baoboa, Antilyas, Juniye, Tabarja, Jubeil, Tripoli, Akkar, Enfe, Damour
Malaysia	
5°00'N lat.; 105°00' E long.	Kampong, Johore Bahru, Malayan peninsula, Terengganu, Kelantan
Nepal	
28 °00 N lat.; 85°00 E long.	Dhankuta, Bhojpur, Sankhuwasabha, Sindhuli, Ramechap, Kavre, Sindhupalchok, Kaski, Syangja, Baglung, Dailekh, Dadeldhura, Baitadi, Jajarkot, Gorkha, Lamjung
Pakistan	
30°00' lat.; 78°00' E long.	Rawalpindi, Lahore, Multan, Bahawalpur, Peshawar, D.I. Khan, Khairpur, Hyderabad, Punjab, Sindh, New West Frontiers, Baluchistan, Sargodha, Montogomery, Lyllapur
Philippines	
14°00'N lat.; 123°00' E long.	Southern Tagalog, Bicol, Eastern and western Oisayas, Mindanao (Davao, Davao Del, Sur), Icocos, Cagayan valley, Luzon (Sual, Pangasinan, Illagan, Isabela, Lipa, Batangas, Oriental, Laguna, Rizal, Quezon, Mindoro, Albay), Visayas (La Carlota, Negros, Occidental), Oriental, Laguna, Rizal, Quezon
Sri Lanka	
8°00'N lat.; 81°00' E long.	Anuradhapura, Budulla, Batticaloa, Bibile, Colombo, Galle, Gampaha, Hambantota, Jaffna, Kalutara, Kandy, Karunegala, Uda walawe, Matale, Monaragala, Maha Illuppallama, Nuwara Eliya, Polonnaruwa, Puttalam, Ratnapura, Vavuniya, Siyambalanduwa
Syria	
35°00'N lat.; 37°00' E long.	Latakia
Taiwan	
21°00'N lat.; 121°00' E long.	Fukien, Kwangtung, Tao Yuag, Taipei, Hsinchu, Lan, Miaoli, Taichung, Hwalien, Changhwa, Nantou, Tungtai, Yunlin, Chiayi, Tainan, Kaoh-Siungm, Taitung, Ping Tung
Thailand	
12°00'N lat.; 102°00' E long.	Chulalongkorn, Chantaburi, Nan (NHES), Phetchabun, Yala, Songkhla, Rayong, Chana, Bangkok, Chiang Mai, Trad, Pathoomtani, Kawn-Ken, Chiang-Rai, Rangsit
Turkey	
40°00'N lat.; 30°00' E long.	Ordu, Giresun, Rize, Nevsehir, Izmir, Aydin, Soke, Mugla, Antalya, Adana, Mersin, Gulf of Alexandretta, Iskenderun, Antiochm, Cukurova, Alanya, Aksu, Dortyal, Finike.

Table 1.12. Growth of citrus industry in India

Year	Area ('000 ha)	Production ('000 tonnes)	Source
1961	90.7	823.7	Directorate of Economics and Statistics, Ministry of Food and Agriculture, Govt. of India, New Delhi
1968	105.4	1211.9	India's Export Potential for Fresh and Processed Fruits and Vegetables, NHB, New Delhi
1988a	279.4	2282.2	Area and Production of Fruits and Vegetables in India – States Profile, NHB, New Delhi
1992	369.6	2820.0	Area and Production of Fruits and Vegetables in India – NHB, New Delhi
1998	482.7	4258.5	Area and Production of fruits and Vegetables in India. NHB, New Delhi

by Andhra Pradesh (67, 829 ha), Punjab (42,085 ha), Karnataka (35,406 ha), Uttar Pradesh (23,144 ha), and Bihar (20,085 ha). But, the perusal on productivity trend provides an entirely different scenario. Highest productivity is obtained in Andhra Pradesh (15.0 tons ha^{-1}), followed by Madhya Pradesh (16.0 tons ha^{-1}), Maharashtra (11.0 tons ha^{-1}), Bihar (10.0 tons ha^{-1}), Punjab (8.3 tons ha^{-1}), Assam (8.2 tons ha^{-1}), and Tamil Nadu (6.4 tons ha^{-1}). During the last 35 years, the Indian citrus industry attained a quantum expansion in area from 0.91 lakh ha in 1961 to 482.7 thousand ha in 1998 while production increased from 823.7 to 4258.5 thousand tons (Table 1.12). The indigenous species are commonly grown in homosteads and introduced species/cultivars, which have adapted in different geographical regions are the basis of commercial plantation (Chadha and Singh, 1989). The soil and climate are the two important determinants deciding the success or failure of any citrus cultivar in a given location. Batchelor and Webber (1948) expressed that judging from the habitat in which the citrus trees flourish best in nature, one can conclude that other fundamental factors, such as abundance of soil moisture, organic matter, and shade or partial protection from direct exposure to sunlight are equally important. There are distinct citrus belts having agroecological preferences. The differentiating reasons for such preferences are yet not known.

Commercial Citrus Belts

There are four major citrus growing belts, where various citrus cultivars have adapted to diverse climate and soils. A total citrus production has been recorded as 710.6 (Table 1.13), 609.4 (Table 1.14), 1126.9 (Table 1.15), and 1515.6 (Table 1.16) thousand tons in north and northwest India, east and northeast India, central India, and south India, respectively. According to benchmark soils of India, 64 soil series have been identified. If the soils under citrus have to be characterised, a total 35 soil series would cover the entire citrus belt of the country as per description of soil series outlined under Benchmark Soils of India (Table 1.17).

Approaches to Agroclimatic Delineation

Agroclimatic zone is defined as a land unit in terms of major climates suitable for

Table 1.13. Citrus belt of north and northwest India

Major citrus growing states	Cultivars	Major belts	Total area (000' ha)	Production (000' tons)
Uttaranchal	Lime, lemon and kinnow mandarin	Uttarkhashi, Chamoli, Pauri- Garhwal, Tehri-Garhwal, Pithoragarh, Almora, lower altitude of Nainital, Dehradoon valley	23.1	87.4 (3.8)*
Gujarat	Lime and lemon	Kheda, Bhavnagar, Mahesana, Khaira, Ahmadabad, Banskantha, Gandhinagar, Vadodara, Surendranagar	12.9	68.7 (5.3)
Haryana	Kinnow mandarin and sweet orange	Sirsa, Mandi, Dabarwali, Hissar, Ambala Hansi, Fatehabad, Jind, parts of Adampura	4.3	37.8 (8.8)
Himachal Pradesh	Kinnow mandarin and sweet orange	Indora, Nurpur, Kangra, Mirpur Dhaulakuan, Sirmour, Bilaspur, Mandi, Paonta valley	13.8	38.4 (2.8)
Punjab	Sweet oranges (Malta)	Hoshiarpur, Ferozepur, Faridkot, Gurdaspur, Amritsar, Jalandhar, Ludhiana, Sangrur, Patiala , Ropar, Bathinda	42.0	347.9 (8.3)
Rajasthan	Kinnow mandarin, sweet orange	Sriganganagar, Jhalawad, Bikaner, Jaipur	9.1	118.6 (13.0)
	Lime, lemon and	Bhusawar , Bharatpur		
	loose skin mandarin	Bhawani, Mandi, Jhalapatan, Jhalawad		
Jammu & Kashmir	Kinnow mandarin	Jammu	10.6	11.8 (1.1)
Total			115.8	710.6 (6.1)**

* Figures in parenthesis indicate productivity in tons ha^{-1} , ** Average productivity

certain range of crops and their cultivars. Agroecological approach is the concept developed from time to time in India for systematized agricultural research. An agroclimatic approach, which combines the effect of physiography, soil, climate, and growing period (Khanna, 1989; Ghosh, 1991; Sehgal *et al.,* 1990) has proved as a ready reckoner in characterising and evaluating the climate and soil in identifying the adaptability of different citrus cultivars.

Agroclimatic zones of the country were first divided into six climatic regions ranging from arid to pre-humid zones based on criteria of Thornthwaite (1948) system of climatic classification describing variation in rainfall from 100 to 12000 mm with 27 soil types, out of which four major soil groups namely, alluvial, black, red, and laterite soils are more prominent. Later, analysis of crop distribution and productivity was proposed in terms of climatic zones and divided the country into 40 soil climatic zones based on major soil types and moisture index. The Indian Council of Agricultural Research, New Delhi, identified 8 agroclimatic zones (Humid western himalayan region, humid Bengal, humid eastern himalayan region, subhumid Sutlez, subhumid to humid eastern and south eastern islands, western arid plains, semiarid lava plateau and central islands, and

Table 1.14. Citrus belts of east and northeast India

Major citrus growing states	Cultivars	Major belts	Total area (000' ha)	Production (000' tons)
Bihar	Pummelo, Acid lime	Palamu, Bhojpur, Samastipur	20.0	200.8 (10.0)*
Orissa	Acid lime	Gajapati (Udayagiri, Rayagada, Gomma, Nuagad.), Koraput (Laxmanpur, Koraput, Familyguda, Pottangi, Chetrahonda, Brisam, Cuttack Niamgiri), Deogarh (Deogarh, Naehol, Barkote, Kuchenda, Jammamkera.), Sundergarh (Himgiri, Lahunepada, Bonaigarh) Pyulbani (Tamudibandha, Kotagarh, Bonaigarh.), Angul (Angul, Chandi, Pada, Pallahara), Bolangir (Khaorakhole (Harisankar), Kalahandi (Theemula, Rampur, Langigarh.)	22.1	115.8 (5.2)
Uttar Pradesh (Plains)	Sweet orange, lime and lemon	Jhansi, Agra, Aligarh, Mathura, Allahabad, Fatehpur	4.0	8.4 (2.1)
West Bengal	Mandarin Acid lime	Darjeeling (Kulbong, Chegra, Soreng, Baramangmaya, Singnngtam), Kalimpong (Bhalukhop, Ecchay, Sakyong, Samsing), Pedong, Kurseong (Soureni, Pahalagaon, Ambutia)	2.4	7.8 (3.2)
Arunachal Pradesh	Mandarin	Basar, Lohit, Tirap, Upper Subansiri, Lower Subansiri East Siang West Siang, Dibang valley, East Kameng, West Kameng Itanagar.	5.5	9.0 (1.6)
Assam	Mandarin	Tinsukia (Kaka, Pathar, Hafjung divisions), North Cachar Hills Karbi anglong Nowgong, Jorhat, Barpeta, Dibrugarh, Kamrup, Lakhimpur, Sonarpur, Khetrieh	13.3	109.1 (8.2)
Manipur	Mandarin Pummelo	Tamenlong, Tipaimukh and Jiribam, Parbung, Tipaimukh, Tinsung area of Churachandpur, Thangal, Ukhnul, Senpati, Chandel	2.2	9.0 (4.1)
Meghalaya	Mandarin Pummelo	Southern slope of Khasi hills (Dawki, Pynersia, Muktapur, Nongiri, Phalgan, Iewsyiem, Pmshutia, Umwjai, Mawlong, Umsning, Mawsynram, Cherrapunjee), Northren slope of Khasi hills (Burnihat,Shella, Balat, Lumshonong, Barapani, Byrnihat, Umling, Nayabunglow), Central plateau (Mawlai and Marow), Western slope of Khasi hills (Rangalpara, Aradongda), West Garo hills (Sesatgiri, Chandigiri, Rngdengiri, Dolongiri, Dinadubi, Aguaragiri), Cental Jaintia hills (Shangpung , Jowai).	7.1	48.2 (6.8)
Mizoram	Mandarin Pummelo	Kolasib, Tawitaw, Aizawl, Thingdawal, Hachhek, Hangdung, Mawit, Pukpui, Phuldungsei (w)	9.6	34.9 (3.6)
Nagaland	Mandarin Pummelo	Longnok (Chunilinsang, Chanki, Chunthia), Mokokchung (Menkang, Tuli), Khonsa, Tuensang, Wokha, Khonsama	1.4	5.3 (3.8)
Sikkim	Mandarin	East Sikkim (Sang, Bhirkuna, Thanka, Khamdong, Nazitam, Tadong, Penlong, Rorathang, Aritar, Saramasa, Rumtek, Pangthang, and Merchak; West Sikkim (Hee Bazar, Pakyong Outok, Bermiok, Darap, and Dentam); North Sikkim (Phedong, Namli, Mangan, Dikchu, and Gyathang); South Sikkim (Tokel, Ralang, Turuk, Maniram, Lingi, Nandugaon, Mellibazar, Sodem, and Namthang)	6.3	16.6 (2.6)
Tripura	Mandarin	Dharamnagar, Kumarghat, Jampui hills (Hmunpui, Vanghmun, Kawnpui, Sabual, Bangla, Vaisan, Phuldungsei, Tlangsang, Behliangchhip, Tlaksih, and Hmawangchuan), Navincherra	12.7	44.5 (3.5)
Total			106.6	609.4 (5.7)**

Figures in parenthesis indicate productivity in tons ha^{-1}, ** Average productivity

Table 1.15. Citrus belts of central India

Major citrus growing states	Cultilvars	Major belts	Total area (000' ha)	Production (000'tons)
Madhya Pradesh	Mandrin and sweet orange	Chhindwara (Saunsar, Pandurna, Bichhua and Mohkhed), Shajapur, Mandsar, Rajgarh, Betul (Prabhat Pattan, Athner and Bhaisdehi), Hoshangabad, Ujjain	11.6	185.0 (15.9)*
	Lemon, lime	Khandwa, Indore, Dhar, Ratlam, Ujjain, Gwalior, Shivpuri		
Maharashtra	Mandarin, acid lime, and sweet orange	Wardha (Wardha, Dewli, Selu, Arvi, Karanja, Hinganghat, Samudrapur), Yavatmal (Yavatmal, Darwa, Pusad, Umerkhed, Digras, Bhindi), Nagpur (Nagpur, Hingna, Katol, Narked, Saoner, Kalameshwar, Ramtek, Parseoni, Umred.), Amravati (Amravati, Achalpur, Anjangaon, Dariapur, Tivsa, Dharni, Chandur, Chandurbazar, Dhamangaon), Aurangabad (Pachod, Karmad, Adul, Dabarwadi, Chittepimplegaon, Pimleraja, Danitabad, Paithan), Jalna (Ambad, Wadigodari, Tirthpuri, Gavrayee, Bhadrapur), Parbhani (Purna, Basmat, Waranga, Pathri), Nanded (Mudkhed, Bhokar, Himayatnagar, Limbgaon)	180.3	941.9 (5.2)
Total			191.9	1126.9 (5.9)**

* Figures in parenthesis indicate productivity in tons ha^{-1}, ** Average productivity

humid to semiarid western ghats) for better land use planning. Altogether, three approaches are commonly utilised for identifying characteristic features of major citrus growing belts and identifying the potential areas. The first approach in that regard was proposed by Planning Commission (Khanna, 1989). According to this approach, whole country is divided into 15 broad agroclimatic regions viz., western himalayan region, eastern himlayan region, lower gangetic plain region, middle gangetic plain region, upper gangetic plain region, transgangetic plain region, eastern plateau and hills region, central plateau and hills region, western plateau and hills region, southern plateau and hills region, eastcoast plains and hills region, west coast plains and ghats, Gujarat plains and hills region, west dry region, and the Island region, on the basis of specific characteristic of prevailing agroecological parameters. The subzones were further delineated on the basis of characteristics of soil, topography, climate, and water resources. On this basis, a comprehensive salient features depicting rainfall, climate, representative soil, and concerned citrus growing state, revealed the diverse adaptibility of citrus. Under the NARP (National Agriculture Research Project), the concept of zone was launched in 1979 based on ecological land classification recognizing various components like soil, climate, topography, vegetation, cropping pattern, rainfall, water resource irrigation facility etc. as major influencing factors.

Table 1.16. Citrus belts of south India

Major citrus growing states	Cultivars	Major belts	Total area (000' ha)	Production (000'tons)
Andhra Pradesh	Sweet orange	Anantapur, Cuddapah, Warangal, Nalgonda, Karimnagar, Prakadham, east Godavari disticts	67.8	1017.4 (15.0)*
	Acid lime	West Godavari, Nellore, Chittoor, Anantapur, Cuddapah, Krishna, Karimnagar, Adilabad, Vizianagar, Nizamabad, Mehabubnagar, Medak, Rangareddy		
	Lemon, Pummelo mandarins	East Godavari, Guntur, Cuddapah (Reddivari palli, Nagavaram, Gundla palli, Kuchivari palli, Montapam palli, Ummadigunta palli, Mydukur, Vem palli, Pulivendla, Nadigedda and Rangapuram), Nellore (Venkatagiri, Maddelamadugu, Nawab peta, Chennur and Manubrolu), Chittoor (Buchi naidukandriga), Kurnool, Ananthapur, Nalgonda, Khamman, east Godavari, Visakapatnam, Srikakulam		
Karnataka	Mandarins	Kodagu, Chikmagalur, Hassan	35.4	318.6 (9.0)
	Sweet orange	Hassan (Chanmareyapatna, Holemorasipur, Alur, Belur, Arkata, Sakaleshpur), Coorg (Virajpet, Sonwarpet, Mercara), Sakaleshpur, Belur, Arakalagudu, Devdurga, Kustthagi, Sindhanur, Manvi, Lingsgur, Koppal, Gangavathi, Gulbarga, Shorapur, Shahpur, Sagar, Honnali, Channagiri, Belgaum, Gokak, Athani, Bailhongal, Soundatti, Virajpet, Shringeri, Shimoga (Bhadravathi, Shimoga, Shikaripur, Honnali, Channagiri), Chikmagalur (Koppa, Sringeeri, Mudgere, Narasimharaj, Epur, Kodur), Bijapur (Badamil, Bagalkot, Bagewadi, Baligi, Hungund, Indi, Jamkhandi, Mudhol, Muddebinal, Sindgi), Bellary (Hadagali, Hospet, Kudl, Hegaribommanahally, Sandoor, Siruguppa, Harapanahally), Raichur (Lingasugar, Gangvathi, Koppal, Kustagi, Sindanur, Yelberga, Deodurg, Manvi), Gulberga (Afzalpur, Chitapur)		
Tamil Nadu	Sweet orange	Dindigul Anna, Nilgiris, Palani, Yercaud, Shevroy, (Asambur, Muluvi, Semmanathan, Kommakadu, Manjakuttail, Kottachedu, Chengadu, Yeacaud, Keeraikadu, Pottukkadu, Kakkambadi, Pathur etc.), Salem, Dharampuri, Coimbatore, Periyar, Trichy	12.5	179.6 (14.4)
	Lime and lemon	Chennai, South Arcot, N.A. Ambedkar, Kukal, Shevoroy T.V., Sambuvaruar, Dhamapur, Puddukotai, Thanjavar, Nagai Qudie Milleth, Mudurai, Ramnad, Kamarajar, Pasumpon, Thevar, Chidambarrnar, Thirunelveli		
Total			115.7	1515.6 (13.1)**

*Figures in parenthesis indicate productivity in tons ha^{-1}, ** Average productivity

Table 1.17. Citrus growing soil series of India

Soil Series	Citrus belt covered	Major soil groups
Chinwali	Pauri, Garwal, Almora, Pithoragarh, Dehradun	Ochrept Orthent, Aquept, Fluvent, Aqualf, Udalf, Ustalf
Kalpa	Palampur, Kangra, Dhaula kuan, Sirmour	Ustorthent, Udorthent, Udalf, Udult,
Hanrgram	Pedong, Kurseong	Vertic Eutrochrept, Haplustalf, Ustochrept, Dystrochrept, Udalf, Udult
Lahangaon	Jorhat, Tinsukia	Aquic Udifluvent, Udorthent, Udochrept, Aquent, Ustalf, Paleudalf, Ochraqualf, Eutrochrept, Udipsamment, Fluvaquent, Udifluvent.
Gemotali	Siang (Arunachal Pradesh)	Typic Hapludalf, Hapluquent, Haplaquept, Haplustalf
Dialong	West district (Manipur)	Ultic Hapludalf, Udic Halpudalf, Typic Paleudalf, Paleudult/ Hapludult
Phullen	Mizoram	Lithic Udorthent, Typic Dystrochrept, Paleudult, Hapludult
Gemotali	Siang (Aurnachal Pradesh)	Typic Paleudalf, Haplustalf, Haplustult, Rhodoustalf, Haplaquoll
Jirania	West district (Tripura)	Ultic Hapludalf, Udic Halpudalf, Typic Paleudalf, Typic Paleudult/ Hapludult
Selsekgiri	East Khasi hills, Tura, Garo hills (Meghalaya)	Umbric Dystrochrept, Hapludult, Paleudult, Dystrochrept, Palehumult, Lithic Udorthent, Haplaquoll, Typic Hapludalf, Udorthent, Palehumult
Fatehpur	Ludhiana,Ferozpur,Hoshiarpur	Typic Ustipsamment, Aquept, Orthent, Ustochrept,Ustalf
Haldi	Nainital, Almora, Pithoragarh	Typic Hapludoll, Typic Eutrochrept/Dystrochrept, Ustoll, Haplaquoll
Nabha	Patiala	Udic Ustochrept, Psamment,Orthent,Ochrept,
Sadhu	Patiala	Vertic Ustochrept, Ustalf,Aquept, Ochrept, Psamment
Itwah	Varanasi, Allahabad	Aeric Ochraqualf Orthent, Ochrept, Aquent, Aquept, Ustochrept.
Bijapur	Fatehpur	Udic Ustochrept, Orthent, Aqualf
Basiaram	Azamgarh	Typic Eutrochrept, Ustochrept, Orthent, Aqualf
Bhubaneswar	Puri	Typic Haplustult, Haplustalf, Ochrept
Chinnaloni	Cuddapah, Nellore	Paralithic Vertic Ustropept
Patancheru	Medak, Anantapur, Chittor	Udic Rhodustalf, Paleustalf, Haplustalf
Masitawali	Hanumangarh, Sriganganagar	Typic Torrifluvent, Torripsamment, Orthent, Aquept, Tropept
Chomu	Jaipur, Sriganganagar	Typic Ustipsamment, Ustorthent, Ustochrept
Adesar	Kachchha, Kheda	Typic Paleargid, Typic/Fluventic Camborthid
Sarol	Indore, Hoshangabad	Typic Haplustert/Pellustert,Ustochrept
Kamliakheri	Indore, Chhindawara	Lithic Vertic Chromustert, Vertic Ustochrept
Linga	Nagpur	Udic Haplustert, Vertic Ustochrept, Haplustert/Pellustert
Jambha	Amravati, Akola, Wardha	Typic Haplustert, Pellustert, Vertic Ustochrept, Typic/Lithic Ustorthent, Typic Haplustalf
Hingona	Jalgaon, Aurangabad	Typic Haplustert/Pellustert, Ustorthent, Ustochrept
Palathurai	Coimbatore	Typic Haplustalf, Plinthustult, Rhodoustalf
Kalathur	Thanjauvr, Dindigul, Madurai	Haplustert, Pellustert, Udorthent, Pellustert, Vertic Ustochrept
Kagalgomb	Gulberga, Hassan, Shimoga, Coorg	Ustalf, Tropept, Fluvent, Paleustalf, Rhodoustalf
Raichur	Raichur, Bijapur	Typic Haplustert/Pellustert, Ustorthent,Ustochrept
Jamakhandi	Bijapur, Bellur,	Paleustalf, Rhodustalf, Paleustult, Ustochrept
Hungund	Bijapur, Dharwad	Typic Pellustert/Haplustert, Vertic Ustochrept, Ustorthent
Jumkhandi	Belgaum	Typic Paleustalf

Source : Based on Benchmark Soils of India by Murthy *et al.* (1982)

According to second approach, the country is divided into 126 agroclimatic zones under National Agriculture Research Project (Ghosh, 1991). Later on, three more zones for Andaman and Nicobar islands and two zones for Pondicherry were added. The third approach was the concept of agroecological zone, based on which country is divided into 21 agroecological regions using criteria for delineation as : i. growing period as an integrated criterion of effective rainfall, ii. soil groups enjoy precedent over topography, and iii. delineated boundries, adjusted to district boundries and in number of regions as minimal as possible (Sehgal *et al.*, 1990). Vaswani (1990) also described various agro-climatic features of 15 zones in India on the basis of rainfall and dominant soil types (Table 1.18).

Table 1.18. Salient agroclimatic features of 15 zones of India

Zone No.	Rainfall range in subzone (mm)	Climate	Soil types	States represented
I	165-2000	Cold arid to humid	Brown hill, Alluvial (Recent), Mountain, Meadow, Skeletal	Jammu and Kashmir, Himachal Pradesh and Uttar Pradesh
II	1840-3528	Per humid to humid	Alluvial, Red loamy, Red sandy, brown hill soil	Assam, West Bengal, Manipur, Meghalaya, Nagaland, Andhra Pradesh, Tripura and Mizoram
III	1302-1607	Moist subhumid to dry subhumid	Red and yellow delatic alluvium, red loamy	West Bengal
IV	1211-1470	Moist subhumid to dry subhumid	Alluvial	Uttar Pradesh, Bihar
V	721-979	Dry subhumid to semi arid	Alluvial	Uttar Pradesh
VI	360-390	Extreme arid to dry subhumid	Alluvial (recent) calcareous desert	Punjab, Haryana, Delhi, Rajasthan
VII	1271-1436	Moist subhumid to dry subhumid	Red sandy, red and yellow	Maharashtra, Madhya Pradesh, Orissa, West Bengal
VIII	490-1570	Semiarid (drier half) to dry subhumid	Mixed red and black, red and yellow, medium black, alluvial	Madhya Pradesh, Rajashtan, Uttar Pradesh
IX	602-1040	Semiarid	Medium black, deep black	Maharashtra, Madhya Pradesh, Rajasthan
X	677-1001	Semiarid	Medium black, deep black, red sandy, red loamy	Andhra Pradesh, Karnataka, Tamil Nadu
XI	780-1287	Semiarid to dry subhumid	Deltaic alluvium, red loamy, coastal alluvium	Orissa, Andhra Pradesh, Tamil Nadu, Pondicherry
XII	2226-3640	Dry subhumid per humid	Laterite, red loamy, coastal alluvium	Tamil Nadu, Kerala, Goa, Karantaka, Maharashtra
XIII	340-1793	Arid to dry subhumid	Deep black, coastal alluvium, medium black	Gujarat
XIV	395	Arid to extremely arid	Desert, grey brown	Rajasthan
XV	1500-3086	Humid		Andaman & Nicobar Islands

Source : Vaswani (1990)

Soil and climate are the two most important criteria for delineating the success or failure of citrus under a given agroclimate (Srivastava and Kohli, 1997a; Srivastava and Kohli, 1999). In India, many efforts have been made in the past to classify agroclimatic zones to prepare a vibrant land use planning for various crops from time to time excluding fruit crops, especially citrus. All the three approaches identifying the citrus growing agroclimatic zones/agroecological regions (Table 1.19) covering citrus growing areas of the country, indicated that citriculture in India is adapted under varying soil, physiography (land forms), climate, and growing conditions. The later two approaches of agroclimatic zones (Ghosh, 1991) and agroecological regions (Sehgal *et al.*, 1990) appeared more explanatory to the these variations. The suitability of soil and climate may be concurrently earmarked for exploring the newer potential areas so that the acreage under citrus cultivation can be scientifically expanded without facing early citrus decline problem due to unsuitable soil and climate. To attain this objective, a perspective land use planning is highly imperative to be prepared for major commercial citrus cultivars being grown in the country.

The most productive citrus belt occurs in the agroecological region of Deccan (Telangana) plateau and eastern ghats, hot semiarid ecoregion with red and black soil and growing period of 90-150 days (Srivastava and Kohli, 1999); followed by Deccan plateau, hot semiarid and subhumid ecoregion with shallow and medium (inclusion of deep soils) red and black soils and growing period of 90-180 days; and eastern ghat, TN uplands and Deccan (Karnataka) plateau, hot semiarid ecoregion with red loamy soils having growing period of 90-150 days (Table 1.20). These productive belts indicated that best citrus production is possible under semiarid to subhumid tropical climate having red and black soils A thorough analysis of agroclimate under citrus growing areas has been made based on the agroclimatic zones of National Agricultural Research Project of India.

Adaptability of Citrus in West and Northwest India

The region comprises seven states viz., Delhi, Haryana, Punjab, Himachal Pradesh, Jammu and Kashmir, Uttar Pradesh, Uttaranchal, and the union territory of Chandigarh (23°52'N to 37°45' N latitudes and 72°40' to 84°38' E longitudes). Physiographically, the region may be divided into the himalayas (trans, greater, lesser, and outer himalayas); Indogangetic plain (trans, upper, middle, and lower gangetic plain); and southern high lands, scarps, and plateaus (Rajasthan uplands, Bundelkhand, and Vindhyan ranges). Climatic conditions exhibit a great diversification, predominantly subtropical, semiarid, and monsoonic type in Indogangetic plain. It varies from hot and subhumid tropical in southern highlands to temperate, cold, alpine, and glacial in the north and eastern high mountains of himalayan region. The rainfall in the himalayan region varies from less than 100 mm to more than 300 mm, thereby, qualifies for aridic, ustic, and udic moisture regimes. In Indogangetic plain, rainfall ranges from 300 to 1200 mm and qualify for

Table 1.19. Agroclimatic variation in major citrus growing belts of India

Agroclimatic regions (Khanna 1989)	Agroclimatic zones (NARP) (Ghosh, 1991)	Agroecological regions (Sehgal *et al.*, 1990)
Northwest India		
Western himalayan region, Trans gangetic plain region, Western dry region, and Upper gangetic plain region	Submountane and low sub tropical zone of Dhaulakuan (H.P.), Sub-montane undulating zones of Kandil (Punjab), Undulating plain zone of Gurudaspur (Punjab), Central zone of Ludhiana (Punjab), Western Punjab, Plain zone of Faridkot (Punjab), Irrigated north western zone of Bhatinda (Punjab), Plain zone of Sriganganagar (Rajasthan), Gujarat heavy rainfall zone of Navasari, Bhabar, and Tarai zone of Pantnagar, U.P.	Northern plain and cntral highlands including Aravalli, hot semiarid ecoregion, with alluvium derived soils and growth period of 90-150 days And Western himalayas, warm subhumid (to humid with inclusion of perhumid) ecoregion with brown forest and podzolic soils and growth period of 180-210 days.
Northeast India		
Middle gangetic plain region, Eastern Himalayan region, and Lower gangetic plain region	North bank plain zone of north Lakhimpur (Assam), Barak valley zone of Karimganj (Assam), Hill zone of Assam and West Bengal, North-east alluvial plain zone of Bihar, Subtropical hill zone of NEH region	Bengal and Assam plain, hot subhimid (moist) to humid with inclusion of perhumid ecoregion with alluvium derived soils and growth period of 120+ days And Eastern himalayas, warm perhumid eco-region with brown and red hill soils and growth period of 210+ days And Northeastern hills warm per humid ecoregion with red and lateritic soils and growth period of 210+ days
Central India		
Central plateau/ hill region, Western plateau, and Hill region	Eastern and central Vidarbha zone, Central Narmada valley zone of Pawad-Kheda (M.P.), Satpura plateau zone of Chhindwada (M.P.), and Central Maharashtra plateau of Aurangabad	Deccan plateau, hot semiarid ecoregion with shallow and medium (with inclusion of deep) black soils and growth period of 90-140 days And Central highland (Malwa, Bundelkhand, eastern Satpura), hot subhumid ecoregion, with black and red soils and growth period of 150-180 (to 210) days
South India		
Southern plateau and Hill region	North Telegana zone of Jagtrial (A.P.), scarce rainfall zone of Rayalseema (A.P.), Northern dry zone of Bijapur, Central dry zone, Southern dry zone, Southern transition zone of Karnataka, Southern zone of A.P. (Tirupati), Hill zone of Karnataka	Deccan (Telangana) plateau and eastern ghats, hot semiarid ecoregion with red and black soils and growth period of 90-150 days And Eastern ghats, TN uplands and Deccan (Karnataka) plateau, hot semiarid ecoregion with red loamy soils and growth period of 90-150 days

Table 1.20. Important citrus producing agroecological regions of India (based on Agro-ecological regions of India prepared by Sehgal *et al.* (1990)

Sr. no.	Agroecological region	Area (000'ha)	Production (000'tons)
1.	Deccan (Telangana) plateau and eastern ghats, hot semiarid ecoregion with red and black soils and growth period of 90-15 days	26.2	372.0
2.	Deccan plateau, hot semiarid ecoregion with shallow and medium (with inclusion of deep black soils and growth period of 90-150 days		
	And		
	Central highland (Malwa, Budelkhand, and eastern Satpura) hot sub-humid ecoregion with red and black soils and growth period of 150-180 days (to 210) days	189.8	1094.6
3.	Eastern ghats, TN uplands and Deccan (Karnataka) plateau, hot semi-arid ecoregion with red loamy soils and growth period of 90-150 days	69.6	620.4
4.	Northern plain and central highlands including Aravallis, hot semiarid ecoregion with alluvium derived soils and growth period of 90-150 days	41.0	337.8
5.	Northern plain, hot subhumid (dry) ecoregion with alluvium derived soils and growth period of 150-180 days		
	And		
6.	Eastern plain, hot subhumid (moist) ecoregion with allumium derived soils and growth period of 180-210 days	21.3	180.5
7.	Bengal and Assam plain, hot subhumid (moist) to humid with inclusion of per-humid ecoregion with alluvium derived soils and growth period of 120+ days		
	And		
	Eastern himalayas, warm perhumid ecoregion with brown and red hill soils and growth period of 210+ days		
	And		
	Northeastern hills warm per humid ecoregion with red and lateritic soils and growth period of 210+ days	60.5	284.4

Source : Srivastava and Kohli (1999)

aridic and ustic moisture regimes. The mean annual temperature of the northern region varies from less than 1°C to more than 25°C, and they qualify for cryic, mesic, thermic, and hyperthermic temperature regimes (Singh *et al.*, 2000). Under this region, citrus is grown in Gujarat, Rajasthan, Haryana, Punjab, and Himachal Pradesh. Analysis of climatic features of four citrus growing centres viz., Sherpur Pacca (Hoshiarpur), Ludhiana and Abohar of Punjab, and Delhi exhibited no significant difference except rainfall, which varied from 274.1 mm at Ludhiana to as high as 1309 mm at Sherpur Pacca. (Table 1.21). This is the reason why kinnow mandarin is more or less equally successful at all these places. Climate and soil under citrus growing areas in India have been studied according to agroclimatic zone specific research (Saxena, 1991a; 1991b; 1991c; Atar Singh, 1991a; 1991b; 1991c; Gangopadhyay, 1991a; Arun Varma, 1991).

Table 1.21. Variation in meteorological features at four kinnow mandarin growing centres of north India

Month	Mean monthly maximum, minimum temperature (^{0}C) and diurnal variation (^{0}C)				Mean monthly relative humidity (per cent) and total rainfall (mm)			
	Sherpur Pacca	Delhi	Abohar	Ludhiana	Sherpur Pacca	Delhi	Abohar	Ludhiana
January	19.4;4.4	204;5.8	18.9;4.5	18.6;4.3	77.4	58.5	80.8	80.4
	*(15.8)	(14.6)	(14.4)	(14.3)	**(70.0)	(11.2)	(7.6)	(11.0)
February	20.9;5.6	21.9;8.6	20.8;6.5	21.0;3.4	65.5	47.8	73.0	60.0
	(15.3)	(13.2)	(14.3)	(14.3)	(50.6)	(48.1)	(24.9)	(15.7)
March	28.0;11.2	30.5;15.9	27.8;12.6	27.8;12.7	50.1	40.0	67.8	58.7
	(16.7)	(14.6)	(15.2)	(15.1)	(10.4)	(7.6)	(16.3)	(13.1)
April	35.6;17.1	37.3;22.7	36.6;18.5	36.4;18.9	32.1	24.6	56.5	62.6
	(18.5)	(14.6)	(18.1)	(17.4)	(4.0)	(0.0)	(0.8)	(0.0)
May	39.0;21.9	39.9;27.2	39.8;23.2	35.6;22.4	23.4	27.2	45.1	42.6
	(17.9)	(12.7)	(16.6)	(13.2)	(13.4)	(13.4)	(8.0)	(2.3)
June	39.2;23.3	37.9;27.3	39.1;25.3	39.4;22.9	37.6	44.9	54.6	50.7
	(15.8)	(10.6)	(13.8)	(16.5)	(92.3)	(52.0)	(78.4)	(62.1)
July	34.6;23.4	34.2;24.7	36.1;26.2	35.5;24.3	68.6	71.9	68.9	67.7
	(11.2)	(9.5)	(9.9)	(11.2)	(513.2)	(387.6)	(102.3)	(74.1)
August	33.1;23.5	33.4;26.2	35.0;26.3	30.5;17.9	75.3	73.5	71.7	65.8
	(9.6)	(7.2)	(8.7)	(12.6)	(407.6)	(224.8)	(107.4)	(26.6)
September	35.4;21.5	33.7;24.8	34.3;23.6	34.2;22.4	65.9	65.7	70.5	66.8
	(13.8)	(8.9)	(10.7)	(11.8)	(108.1)	(157.3)	(39.9)	(67.5)
October	33.2;13.9	32.8;20.0	34.3;16.2	34.1;16.5	51.7	51.4	60.5	60.9
	(19.2)	(12.8)	(18.0)	(17.6)	(0.0)	(39.4)	(0.0)	(0.0)
November	27.5;5.7	27.9;10.3	27.6;9.0	27.7;8.4	55.3	44.6	64.0	70.9
	(21.8)	(17.6)	(18.6)	(19.3)	(0.0)	(0.0)	(0.0)	(0.0)
December	20.5;4.1	22.4;8.3	21.3;5.6	21.1;5.4	72.0	54.7	77.2	75.4
	(16.4)	(14.1)	(15.7)	(15.7)	(39.4)	(4.4)	(0.3)	(1.7)

* Diurnal variation, ** Total rainfall

Source : Singh and Singh (1979)

Citrus in Gujarat

Citrus in Gujarat is commercially grown in Navsari and Kheda districts which agro-climatically belong to middle Gujarat region.

Agroclimate : Middle Gujarat Zone

Climate: The rainfall in the zone varies from 800-1000 mm per annum (Saxena, 1991a).

Soils: The soils of this zone fall into three categories viz., sandy loam, clay loam, and clay. The soils of almost the entire Kheda district are alluvial in origin with sandy loam to clay loam texture. While, rest of the northeastern part is clay loam (Saxena, 1991a). Physicochemical properties of soil types representing three soil series of citrus growing Navsari district of Gujarat indicated Ustochrept and Halaquept soils (Table 1.22). Inceptisols (Vertic Ustochrept, Typic Ustochrept, Lithic Ustochrept, and Fluventic Ustochrept) and Entisols (Lithic Ustorthent and Typic Ustorthent) covering 51 per cent and 36 per cent of total geographical area, respectively, constitute major soil types (Shyampura, 2000).

Table 1.22. Characteristics of some of soil series of Navsari

Soil properties	Soil series		
	Waghai (Ustochrept)	Jalalpur (Ustochrept)	Danti (Halaquept)
Coarse sand (%)	9.5	0.9	10.3
Fine sand (%)	26.4	6.5	20.3
Silt (%)	22.6	33.9	22.7
Clay (%)	40.8	57.6	45.8
$CaCO_3$ (g kg^{-1})	1.3	46.9	75.6
Water at saturation (mg kg^{-1})	0.480	0.702	0.642
pH (1:2.5)	6.2	7.2	8.2
EC (dS m^{-1})	0.14	0.51	12.82
CEC [cmol (p^+) kg^{-1}]	34.9	50.2	46.1
Exchangeable sodium (% age)	1.2	1.7	19.1
Organic C (g kg^{-1})	37.0	6.3	9.0
Available P (mg kg^{-1})	44.7	16.2	13.3

Source : Patel and Trivedi (1994)

Citrus in Uttaranchal

Citrus in Uttaranchal is concentrated mainly in hill districts. Citrus cultivation has got a wide scope in hills of Garhwal and its cultivation is mainly confined to some pockets of river valley i.e. Alakananda, Pinder, Nandakini, Mandakini, Bhilanga etc. Some local strains of mandarins and sweet oranges viz., Srinagar santra, hill orange, and malta are commonly grown in these valleys under the calcareous soils. In Kumaon and Garhwal hills, sweet orange varieties viz., malta common, mosambi, Washington navel, and blood red performed well. Citrus cultivars like hill orange and kinnow mandarin have been reported to adapt well under U.P. hills (Vinay Shankar and Sinha 1987). Physiographically, this region is divided into trans, greater, lesser, and outer himalayas (Singh *et al.,* 2000), but citrus is concentreated in lesser himalayas only.

Agroclimate : Hill Zone

Great potential exists for citrus expansion in UP hills (Singh *et al.*, 1976) which is now in Uttaranchal state. The foot hill region of Garhwal himalayas occupies a distinct characteristics in the geographical, environmental, and socioeconomical scenario of India. This region is situated between the himalaya and the Indogangetic plains. The hill region, on the basis of altitude variation, can be categorized as subtropical (250-1200 m), subtemperate (1200-1700 m), temperate (1700-3500 m), and alpine zone (3500 m and above). Citrus is grown up to an altitude of 1200 m above mean sea level (Katyal and Sabharwal, 1964). Two main physical features of this zone are the himalayan and the sub-himalayan tracts forming a part of the southern outer spurts of the himalaya, comprising Almora, Uttar Kashi, Chamoli, Pauri, Pithoragarh districts, parts of Nainital, and Dehradun districts. The subhimalayan tract, almost parallel to the southern most boundary of the himalayan, with elevation ranging from 600 to 300 m, represents the submountainous land (Saxena, 1991b).

Climate : Singh *et al.* (1985) described the seasonal changes in vegetative growth of two cultivars of lemon (*Citrus limon* Burn.) in relation to weather parameters including soil temperature in Nainital. The climate of the region is quite different from other parts of the state due to geographical differences. The valleys are hot in summer and much cold in winter. The average temperature in valleys varies from 3 to 30 °C, while the higher hills are covered with snow even in summers. The annual rainfall ranges from 670 to 2500 mm. About 70 to 80 per cent of the total annual precipitation is received during June to September. Winter rains are scanty, accounting for only 10 per cent of the annual precipitation. In hill region, the climate, in general, is humid temperate, but variations exist which largely depend on altitude. The mean monthly temperature varies with the altitude. It ranges from 11°C (December) to 30.2°C (August) at about 300 m, 9.1°C (January) to 22°C (June and July) at 1600 m, and 6.6°C (January) to 22.2°C (August) at 2500 m. The temperature decreases with the increase in altitude. Annual maximum relative humidity varies from 60-70 per cent in the northern hills and 30-40 per cent in the southwestern dry areas of Tarai (Saxena,1991b). The annual mean temperature, rainfall, and potential evapotranspiration in Nogina, Bijnor, Nazibabad, Kotdwar, and Lansdown areas of Dehradun valley have been recorded to vary from 21.9 to 24.7°C, 885.0 to 1734.5 mm, and 1149.5 to 1474.6 mm, respectively (Patel *et al.*, 1999).

The mineralogical make up of hill soils of Ramganga catchment (Kumaon himalayas) developed on quartzite and biotite schists, indicated that mica is most dominant clay mineral followed by chlorite, vermicullite, and smectite. Whereas, kaolinite, allophane, quartz, K-feldspar, Na- feldspar, and Ca-feldspar only in small quantity (Parmar and Agnihotri, 1997). Bisht and Kusumlata (1993) described the meteorological features of Garhwal himalaya which indicated that mean maximum temperature ranges from 12.6°C (December) to 25.6°C (May) and mean minimum temperature from 3.7°C (December) to 16.6°C (June).The mean humidity varies from 35.7 to 75.7 per cent during April and February, respectively. The mean monthly rainfall ranges between 29 mm (July) and 539 mm (August) with average annual rainfall of 1510 mm. The year is divisible into rainy (June – September), winter (October – February), and summer (March

– May) seasons. Climate of Doon valley is subtropical monsoonic characterised by long winter and humid monsoon seasons. The mean annual rainfall is 2050 mm, the greater part of which is received from July to mid-September. Winter rains, though small, are not insignificant. The mean annual temperature is 21.1°C. The mean winter and summer temperatures are 15.3°C and 28.0°C, respectively (Sharma *et al.*, 1989).

Soils : The hill soils have developed from rocks with biotite, schists, and phyllitic material under cool and moist climate with rainfall generally exceeding 1500 mm. The soils are shallow, gravelly impregnated with unweathered fragments of parent rocks, occurring within a few centimeters at elevated spots and deeper in valleys or depressions. The soils are brown to grayish brown and dark grey in colour, generally noncalcareous, and neutral to slightly acidic in reaction. Moderate to highly acidic soils are found at higher elevations, whereas rainfall is high and strong enough to leach down bases from the soil minerals under temperate climatic conditions. The major soil groups are brown forest soils, submontane, mountain meadow soils, skeletal soils, and red loam soils (Saxena,1991b).

Pal (1995) described the properties of Inceptisol of Tehri Garhwal as pH 5.2, silty clay loam texture, EC 0.53 dSm^{-1}, organic C 8.5 g kg^{-1}, exchangeable Al 25.1 mg kg^{-1}, exchangeable Fe 37.8 mg kg^{-1}, total Al_2O_3 8.1 per cent, and total Fe_2O_3 6.2 per cent. Hill soils of Almora and Tehri Garhwal are characterised as coarse loamy, hyperthermic family of Fluventic Ustochrept at Sakana, Ranikhet, Almora; coarse loamy, mixed hyperthermic family of Typic Ustorthent at Amgadhera, Ranikhet, Almora; fine loamy, mixed hyperthermic family of Typic Ustochrept at Kunidar, Ranikhet, Almora; coarse, loamy mixed,hyperthermic shallow family of Typic Ustorthent at Kauhana, Ranikhet, Almora; loamy skeletal, mixed, hyperthermic family of Typic Ustorthent at Dhential khal, Ranikhet, Almora; coarse loamy, mixed hyperthermic family of Lithic Ustorthent at Manoli, Ranikhet, Almora; coarse loamy, mixed hyperthermic dry family of Typic Ustorthent at Bagarra, Ranikhet, Almora; fine loamy, mixed, hyperthermic family of Typic Ustochrept at Siroli, Ranikhet, Almora; fine loamy, mixed, thermic family of Typic Tropudalf at Chaubatia Garden, Ranikhet, Almora; coarse loamy, mixed hyperthermic family of Udic-Fluventic Ustochrept at valley of Kaphra slope, Ranikhet, Almora; coarse loamy, mixed, hyperthermic family of Udic-Fluventic Ustochrept at middle valley slope of Kaphara, Ranikhet, Almora; sandy-skeletal, mixed hyperthermic family of Lithic Ustorthent at top valley slope of Talli Mirai, Ranikhet, Almora; fine loamy, mesic, mixed family of Fluventic Dystrochrept at sloping plain of Ranichauri, Tehri Garhwal; fine loamy, mesic, mixed, family of Dystric Fluventic Eutrochrept at eastern side of Ranichauri, Tehri Garhwal; fine loamy, mesic, mixed family of Dystric Fluventic Fragiudalf at Ranichauri (top of ridge), Tehri Garhwal; clayey over loamy-skeletal, mesic, mixed family of Aqultic Hapludalf at west side of Ranichauri, Tehri Garhwal; and fine loamy, mesic, mixed family of Cumulic Haplumbrept at Ranichauri, Tehri Garhwal (Suman Kumar and Sharma, 1987).

Soils of Tarai region of Nainital district according to Srivastava *et al.* (1986) are classified as coarse loamy Typic Ustorthent (pH 7.0-7.4, organic C 0.25-0.74 per cent, $CaCO_3$ 2.0 per cent, CEC 4.3-8.0 me 100 g^{-1}, sand 65.8-84.9 per cent, silt 11.7-22.1 per cent, and clay 3.8-12.3 per cent), coarse loamy Typic Ustochrept (pH 7.1-7.5, organic C 0.13-0.95 per cent, $CaCO_3$ 1.8-2.5 per cent, CEC 7.9-22.9 me 100 g^{-1}, sand 67.2-90.8 per cent, silt 7.1-24.3 per cent, and clay 5.6-11.4 per cent), coarse loamy, Typic Haplustoll (pH 7.5-8.2, organic C 0.38-1.44 per cent, $CaCO_3$ 3.0-3.8 per cent, CEC 15.5-19.1 me 100 g^{-1}, sand 24.3-51.3 per cent, silt 38.4-59.4 per cent, and clay 9.6-17.1 per cent), and fine silty Aquic Haplustoll (pH 8.2-8.5, organic C 0.59-1.37 per cent, $CaCO_3$ 7.7-16.6 per cent, CEC 9.8-17.7 me 100 g^{-1}, sand 10.2-22.7 per cent, silt 48.6-58.5 per cent, and clay 29.5-34.3 per cent).

According to Singhal and Sharma (1985), soils of Doon valley are classified as coarse, loamy, thermic Udic Haplustoll with high Ca:Mg ratio (pH 5.8 – 6.2, organic C 0.14 – 2.8 per cent, CEC 3.0 – 14.0 me 100 g^{-1}, base saturation 38.3 – 76.8 per cent, sand 60.2 – 92.9 per cent, silt 1.0 – 20.7 per cent, and clay 5.6 – 16.1 per cent); loamy, thermic, Udic Agriustoll (organic C 0.39 – 1.45 per cent, CEC 6.7 – 10.3 me 100 g^{-1}, base saturation 65.5 to 94.6 per cent, sand 38.7 – 63.2 per cent, silt 17.9 – 28.7 per cent, and clay 18.8 – 32.9 per cent); loamy, thermic, Udic Haplustalf with absence of $CaCO_3$, and presence of argillic horizon (organic C 0.27 – 1.55 per cent, CEC 5.5 - 8.2 me 100g^{-1}, base saturation 42.8 – 63.7 per cent, sand 59.4 – 65.2 per cent, silt 13.9 – 20.2 per cent, and 14.6 – 21.5 per cent); coarse, loamy, thermic Fluventic Haplumbrept (organic C 1.32 – 3.88 per cent, CEC 3.1 – 16.2 me 100g^{-1}, 20.5 – 46.9 per cent, sand 62.0-80.0 per cent, silt 8.2 – 21.7 per cent, and clay 10.3 – 20.3 per cent); and skeletal, sandy, thermic Lithic Haplustoll (organic C 0.15 – 2.64 per cent, CEC 2.0 – 9.5 me 100 g^{-1}, base saturation 65.0 – 75.4 per cent, sand 71.4 – 95.2 per cent, silt 1.1 – 13.7 per cent, and clay 2.6 – 11.8 per cent). Soils of Doon valley showed variation in the organic matter from 2.0-5.7 per cent, soluble salts 0.04-0.11 per cent, and texture silty clay loam to gravelly loam with illite and kaolinite based clay mineralogy (Sharma *et al.*, 1989).

Singh and Singh (1983a) described the physicochemical characteristics of Kumaon hill soils comprising locations viz., Reoni (pH 5.8, clay 33.3 per cent, organic C 3.2 per cent, exchange acidity 15.4 me 100 g^{-1}, exchangeable Al 0.30 me 100 g^{-1}, Ca saturation 6.5 per cent, and base saturation 33.0 per cent), Kalakhet (pH 5.7, clay 12.5 per cent, organic C 2.1 per cent, exchange acidity 10.9 me 100 g^{-1}, exchangeable Al 0.14 me 100 g^{-1}, CEC 16.1 me 100 g^{-1}, Ca saturation 6.2 per cent, and base saturation 32.2 per cent), Genthia (pH 5.8, clay 10.4 per cent, organic C 2.3 per cent, exchange acidity 9.5 me 100 g^{-1}, exchangeable Al 0.24 me 100 g^{-1}, CEC 16.4 me 100 g^{-1}, Ca saturation 24.5 per cent, and base saturation 41.9 per cent), Kanalichhina (pH 5.4, clay 26.8 per cent, organic C 1.9 per cent, exchange acidity 19.6 me 100 g^{-1}, exchangeable Al 0.34 me 100 g^{-1}, CEC 30.0 me 100 g^{-1}, Ca saturation 20.8 per cent, and base saturation 34.5 per cent).

The plain soils are alluvial in nature as they are deposited by the rivers and rivulets. These are loam to sandy loam in texture having low organic matter and nitrogen content, low to medium available phosphorus, and potassium. The hill soils on the other hand, are residual and acidic in nature (formed by weathering of the rocks) under varying rainfall and diverse natural vegetation. These soils have low water holding capacity. These are medium in organic matter and total inorganic minerals, but deficient in available P_2O_5 and high in available K_2O. Soils in this region have structural/argillic horizon and are classified as Vertic/Dystric/Typic Fluventic Eustrudept, Typic Udorthent, Typic Haplustalf, and Typic Entic Hapludoll (Singh *et al.*, 2000).

Citrus in Uttar Pradesh

Citrus in Uttar Pradesh is agroclimatically grown in western plain zone, eastern plain zone, and central plain zone.

Agroclimate : Western Plain Zone

Saharanpur, Muzaffarnagar, Meerut, Ghaziabad and Bulandshahar located between Ganga and the Yamuna in the west constitute the part of this zone.

Climate : There is considerable variation in the rainfall in this zone, and some of the very dry areas of the state lie in this zone. The rainfall goes on decreasing from northeast to southwest direction. Saharanpur district receives maximum rainfall (800 to 1200 mm), and Bulandshahar, the minimum. In parts of Meerut, Ghaziabad and Bulandshahar districts, the annual rainfall is less than 700 mm (Saxena 1991b).

Soils : The soils are alluvial, which have developed from the alluvium deposited by the Ganga and the Yamuna rivers and their tributaries. The soil reaction is neutral to slightly alkaline. The organic matter status of the soils throughout the region is low. The content of phosphorus is medium except that in Muzaffarnagar district where it is low. The potash content in Saharanpur district is high and medium to low in remaining districts. The western region faces an acute problem of rising water table and waterlogging due to seepage and percolation from canals, and excessive irrigation, resulting in soil salinity (Saxena,1991b).

Agroclimate : Eastern Plain Zone

Citrus in eastern part of Uttar Pradesh is grown mainly in plain zones, which comprises cultivars like acid lime and kinnow mandarin. Alluvial soils dominant in this region exhibit varying stages of pedogenic development varying from Ac to A-Bt-C, calcareous and mostly deep to very deep. The soils of active floodplains are Typic Ustifluvent/Typic Ustipsamment and Aeric Fluvaquent. Soils also belong to Fluventic Haplustept/Aeric Haplquept (Singh *et al.*, 2000).

Agroclimate : Central Plain Zone

Citrus in this zone is cultivated at Varanasi, Allahabad, Mathura, Agra, and Sultanpur districts.

Climate : The mean maximum and minimum temperature vary considerably at various locations in the zone. In January, the mean temperature ranges from 22.1° to 23.7°C, whereas the respective minimum value lies between 8.9°C and 9.1°C. Maximum temperature being 42.3°C, while the minimum value is 25.3°C. The maximum mean relative humidity varies from 85 to 89 per cent during the month of August, whereas the respective minimum relative humidity value stands between 30 and 46 per cent in the month of April. Average annual rainfall ranges between 885 mm and 1160 mm, with July and August are the months of maximum rainfall in this zone (Saxena, 1991b). Allahabad is situated at an elevation of 98 m from the sea level at 25.57° north latitude and 81.50° east longitude. The average maximum and minimum temperatures are 32.4° and 19.8°C, respectively. The average rainfall is 1013.4 mm, with highest concentration during July to September months (Gauri Shankar, 1978). Bundelkhand region of Uttar Pradesh represents a transition zone of tropical dry subhumid type of climate. The mean annual soil temperature is 26°C with mean maximum and minimum summer soil temperatures of 30.4°C and 18.7°C, respectively. The mean annual rainfall is 1032 mm which amounts to 79 per cent of potential evaporation of the area (Walia and Rao, 1996).

Climatic features of Jhansi, Uttar Pradesh (Table 1.23) are characterised by maximum of 23°, 42°, and 33°C during winter, summer, and monsoon season, respectively. The corresponding minimum temperatures are 5°, 26°, and 24°C with an annual rainfall of 958 mm on the basis of the 10 years of average.

Table 1.23 Monthwise climatological characteristics at Jhansi, Uttar Pradesh (1997 – 99)

Month of year	Temperature (°C)		Relative humidity (%)		Rainfall (mm)	Bright sunshine (hrs day^{-1})
	Max.	Min.	Max.	Min.		
Jan.	21.7	5.0	96	46	-	7.9
Feb.	26.0	9.2	91	33	1.8	8.9
Mar.	29.9	13.0	90	32	22.8	9.5
Apr.	37.3	21.8	69	25	13.4	9.8
May.	41.3	25.7	55	26	17.0	10.2
Jun.	39.1	27.0	64	38	61.6	8.0
Jul.	33.8	25.8	87	65	297.5	4.5
Aug.	31.9	24.9	93	73	281.0	4.0
Sept.	33.3	24.3	91	62	113.9	7.6
Oct.	33.8	17.6	90	52	54.1	8.0
Nov.	27.4	12.7	92	47	14.6	7.0
Dec.	22.1	8.0	93	67	77.1	2.7

Source : Indian Grassland and Fodder Res. Institute, Jhansi, Uttar Pradesh

Soils : The major areas of this zone have got alluvial soils, while Vindhyan soils occur only in the southern tract of Allahabad. These soils have scarcity of available water. Vindhyan soils found in low land area, respond well to phosphate, potash, and micronutrient application. The Lakhimpur-Khiri district of the zone has got tarai soils in the northern part. The alluvial soils are highly productive and constitute one of the most fertile belts of the country (Saxena, 1991b). The soils at Allahabad are deep, fertile, alluvial in nature, and texturally sandy loam having average pH of 7.5 (Gauri Shankar, 1978). Soil variability from across different landscapes in Etawah district showed that soils are mostly Ustipsamments. In recent alluvium plains, soils are in the initial stages of development and are mostly Haplustepts. In the old alluvium plains, soils are more stabilized and comparatively well developed (Verma *et al.*, 2001).

According to Pandey *et al.* (2000), soils of central Uttar Pradesh are classified as Typic Ustifluvents (pH 7.7, EC 0.42 dSm^{-1}, $CaCO_3$ 19.3 g kg^{-1}, organic C 4.8 g kg^{-1}, CEC 8.8 cmol(p^+) kg^{-1}, and clay 9.9 per cent), Aquic Haplustalfs (pH 7.8, EC 0.21 dSm^{-1}, $CaCO_3$ 8.0 per cent, organic C 4.3 g kg^{-1}, CEC 9.0 cmol(p^+) kg^{-1}, and clay 20.8 per cent), Udic Haplustalfs (pH 7.6, EC 0.25 dSm^{-1}, $CaCO_3$ 8.6 per cent, organic C 4.7 g kg^{-1}, CEC 8.4 cmol(p^+) kg^{-1}, and clay 15.2 per cent), Typic Haplaquepts (pH 7.8, EC 0.29 dSm^{-1}, $CaCO_3$ 11.2 g kg^{-1}, organic C 4.2 g kg^{-1}, CEC 11.2 cmol(p^+) kg^{-1}, and clay 17.8 per cent), Fluventic Ustochrepts (pH 7.5, EC 0.22 dSm^{-1}, $CaCO_3$ 11.7 g kg^{-1}, organic carbon 4.1 g kg^{-1}, CEC 11.0 cmol(p^+) kg^{-1}, and clay 14.3 per cent), and Vertic Ustochrepts (pH 7.5, EC 0.24 dSm^{-1}, $CaCO_3$ 10.4 g kg^{-1}, organic carbon 4.3 g kg^{-1}, CEC 11 .1 cmol(p^+) kg^{-1}, and clay 26.7 per cent).

Soils of Bundelkhand region of Uttar Pradesh representing Banda district are physiographically delineated as hill, subdued plateau, flat topped hill, gently sloping plateau, monadnock, and foothill slopes. The soils are formed on sandstone, shale, granite, and colluvium. The soils are deep to very deep, excessively to well drained, reddish brown to red, mild acidic, low to medium in CEC, medium to high in organic carbon with wide textural variations depending upon parent material and physiography. These soils are taxonomically classifed as fine micaceous, hyperthermic, deep family of Typic Ustochrept; coarse loamy, kaolinitic, hyperthermic, deep family of Dystric Ustochrept; fine micaceous, hyperthermic, deep family of Ultic Haplustalf; fine loamy, kaolinitic, hyperthermic, deep family of Ultic Haplustalf; and fine, micaceous, hyperthermic, deep family of Udic Ustochrept (Walia and Rao, 1996). According to Hari Ram and Dwivedi (1994), soils of central alluvial tract of Uttar Pradesh are classified as Typic Ustifluvents (pH 7.0-8.5, organic matter 3.0-7.0 g kg^{-1}, $CaCO_3$ 5.0-24.0 g kg^{-1}, sand 45.0-74.9 per cent, and silt plus clay 24.1-53.3 per cent), Aquic Haplustalfs (pH 7.6-8.4, organic matter 3.0-6.0 g kg^{-1}, $CaCO_3$ 4.0-15.0 g kg^{-1}, sand 39.6-62.8 per cent, and silt plus clay 33.5-60.2 per cent), Udic Haplustalfs (pH 7.9-8.7, organic matter 4.0-7.0 g kg^{-1}, $CaCO_3$ 4.0-15.0 g kg^{-1}, sand 45.9-70.7 per cent, and silt plus clay 27.7-51.9 per cent), Typic

Haplaquepts (pH 7.7-9.4, organic matter 2.0-5.0 g kg^{-1}, $CaCO_3$ traces - 13.0 g kg^{-1}, sand 49.5-57.6 per cent, and silt plus clay 39.9-48.6 per cent), Fluventic Ustochrepts (pH 5.0-8.0, organic matter 5.0-8.0 g kg^{-1}, $CaCO_3$ traces - 25.0 g kg^{-1}, sand 44.6-67.1 per cent, and silt plus clay 30.0-54.7 per cent), and Vertic Ustochrepts (pH 7.1-8.7, organic matter 3.0-6.0 g kg^{-1}, $CaCO_3$ traces - 19.0 g kg^{-1}, sand 50.5-56.5 per cent, and silt plus clay 42.3-46.4 per cent)

Soils of Varanasi district are predominantly classified as Typic Ustifluvents (pH 7.4 – 8.0, sand 9.7-26.7 per cent, silt 58.8-67.6 per cent, and clay 14.5-34.9 per cent), Vertic Ustochrepts (pH 7.9 – 8.5, sand 12.8-31.3 per cent, silt 48.2-64.8 per cent, and clay 16.9-24.8 per cent), Fluventic Ustochrepts (pH 8.0 – 8.5, sand 13.8-30.5 per cent, silt 55.1-60.4 per cent, and clay 13.2-29.7 per cent), Aquic Paleustalfs (pH 7.4 – 7.8, sand 11.8-13.5 per cent, silt 46.7-48.2 per cent, and clay 39.2-41.5 per cent), Udic Haplustalfs (pH 7.9 – 8.4, sand 22.5-61.4 per cent, silt 27.4-60.1 per cent, and clay 9.3-25.3 per cent), and Aeric Ochraqualfs (pH 7.9 – 8.0, sand 7.5-14.6 per cent, silt 56.8-68.9 per cent, and clay 20.9-30.7 per cent) according to Singh *et al.* (1989).

Singh and Tripathi (1983b) characterised the fertility status of citrus growing soils of Agra region comprising Agra (available Fe 7.6-10.8 mg kg^{-1}, available Mn 15.6-26.0 mg kg^{-1}, available Cu 1.4-2.6 mg kg^{-1}, and available Zn 1.3-2.8 mg kg^{-1}), Aligarh (available Fe 5.2-7.6 mg kg^{-1}, available Mn 16.9-23.3 mg kg^{-1}, available Cu 1.2-2.2 mg kg^{-1}, and available Zn 1.8-2.8 mg kg^{-1}), and Mathura (available Fe 4.5-6.4 mg kg^{-1}, available Mn 12.7-23.2 mg kg^{-1}, available Cu 1.4-2.3 mg kg^{-1}, and available Zn 1.6-3.3 mg kg^{-1}) with overall variation in other soil properties viz., pH 7.8-8.3, organic C 0.16-0.64 per cent, $CaCO_3$ traces-3.0 per cent, and clay from 8.0-20.0 per cent. These soils have total S content ranging from 97.5 to 187.5 mg kg^{-1} with a mean of 156.0 mg kg^{-1} and 142.4 mg kg^{-1} in surface (0-30 cm) and subsurface soil (30-60 cm), respectively. The organic S varied from 20.8 to 60.0 mg kg^{-1} with a mean of 43.8 mg kg^{-1} in surface and 36.5 mg kg^{-1} in subsurface. While, SO_4-S ranged from 10.0 to 32.5 mg kg^{-1} and from 10.9 to 23.0 mg kg^{-1}, with corresponding mean contents being 22.7 mg kg^{-1} and 16.0 mg kg^{-1} (Singh and Sharma, 1983).

Citrus in Rajasthan

Citrus is agroclimatically grown in irrigated northwestern plain zone (Zone IB), flood prone eastern plain zone (Zone III B), subhumid southern plain and Aravali zone (Zone IV A), and south eastern humid plain zone (Zone V). The Inceptisols (Typic Ustochrept) are dominantly observed covering 32.4 per cent followed by Entisols (Typic Torripsamment, Lithic Ustorthent, and Ustic Torripsamment), Aridisols (Typic Camborthid), and Vertisols (Typic Haplustert/Pellustert) covering 28.5, 16.3, and 5.0 per cent of total geographical area, respectively (Shyampura, 2000).

Agroclimate : Irrigated Northwestern Plain Zone (Zone I B)

Climate : The climate of the zone is similar to semidesert region, which is characterised by severe drought accompanied by high wind velocity and low relative humidity. The range of average relative humidity is 60-70 per cent. The winters are very cold, and temperature falls below the freezing point. The summers are extremely hot and the intensity of light is very high, with temperature touching the mark of 49°C. Often, dust storms occur in this zone. The average rainfall of Sriganganagar district is 25.4 mm (Atar Singh, 1991a).

Soils : The major citrus growing soil types recognised in this zone belong to alluvial Serozem soils, which are greyish to light grey, very deep, dominantly medium textured (sandy loam to loam), and well drained, distributed in parts of Sriganganagar district. Various soil physicochemical properties of a representative soil type from Sriganganagar area showed an alkaline soil pH and sandy loam texture (Table 1.24).

Table 1.24 Soil properties at Sriganganagar area of Rajasthan

Properties	Soil depth (cm)				
	0-15	15-30	30-45	45-60	60-75
pH	8.1	8.1	8.2	8.2	8.1
EC (dS m^{-1})	0.22	0.20	0.21	0.26	0.22
Bulk density (Mg m^{-3})	1.48	1.54	1.57	1.58	1.58
Particle density (Mg m^{-3})	2.62	2.62	2.65	2.63	2.63
Field capacity (mm^{-3})	0.257	0.267	0.271	0.249	0.240
Sand (%)	83.0	79.3	75.7	74.5	74.8
Silt (%)	6.5	7.5	10.0	8.5	10.0
Clay (%)	9.5	10.5	13.5	14.5	13.5
Texture	sl	sl	sl	sl	sl

sl stands for sandy loam

Source : Verma and Ram Deo (1993)

Agroclimate : Flood prone Eastern Plain Zone (Zone III-B)

Soils are mainly alluvial, recent alluvium, and calcareous. On account of recurring floods and waterlogging conditions, a large part of both Alwar and Bharatpur district is affected by salinity and sodicity.

Agroclimate : Subhumid Southern Plain and Aravali Zone (Zone IV-A)

Climate : The rainfall ranges from about 500 mm in the west and northwest to about 700 mm in the southeast and about 900 mm in the southwest. The mean daily minimum temperature at Udaipur ranges from 24.2°C in January to 38.6°C in May. Likewise, the mean daily minimum temperature ranges from 7.8°C in January to 25.3°C in June (Atar Singh, 1991a).

Soils : The soils are lithosols at the foothills in Udaipur, Bhilwara, and Chittorgarh district and old alluvium away from the hills (Atar Singh, 1991a).

Agroclimate : South Eastern Humid Plain Zone (Zone V)

Jhalawar district stands at the edge of Malwa plateau with an area of low hills and shallow plains. Citrus in Jhalawad district, especially Nagpur mandarin is concentrated prominently in areas like Shyam Pura, Khanpuriya, Mandavar, Asnawar, Haripura, Bakani, Bachpaharh, Misuranli, Kotda, Bhawanimandi, Nandpura, and Sunail

Climate : The climate of Bundi district is moderate, while Kota and Jhalawar have dry climate. In general, the weather starts warming up from the month of March until June. The daily mean maximum temperature ranges from 24.5°C in January to 42.6°C May. The rainfall ranges between 690 mm in northwest and 1000 mm in south-east. The average number of rainy days in the zone are 40 per year. The mean monthly potential evapotranspiration varies from 58.3 mm in December to 225 mm in May (Atar Singh, 1991a).

Soils : The predominant soils of the zone are black soils of alluvial origin. The soils vary in depth from shallow to very deep with kankar and lime casted gravel at varying depths. The texture is generally clay loam to clay. These soils are moderate to slowly permeable and generally nonsaline and alkaline. In some pockets, soil salinity has developed due to careless handling of irrigation water. In poorly drained soils, the problem of salinity and sodicity is encountered (Atar Singh, 1991a).

Citrus in Haryana

In Haryana, citrus is grown in various agroclimatic zones which mainly comprise eastern and western zone. The predominant soil types such as Typic Haplustept, Typic Torripsamment, Typic Ustipsamment, Typic Camborthid, Typic Ustifluvent, Fluventic Eutrochrept, Lithic Ustorthent, and Udic Calciustept are of common occurrence (Singh *et al.*, 2000).

Agroclimate : Eastern Zone

Eastern zone includes the districts of Ambala, Kurukshetra, Karnal, parts of districts of Jind, Sonepat, Rohtak, Gurgoan, and Faridabad. It covers about 40 per cent area of the state, which comprises Gurgaon, Rohtak tract, central plain, and hill tract.

Climate : Eastern zone of Haryana is also called as wet zone. It receives more rains in northern and less in southern parts. The mean annual rainfall varies from 500 mm to 1000 mm. More than 70 per cent rains are received during July to September. Remaining 10-15 per cent rains are received in winter and 10 per cent in summer. Normal rainy days are more than 30 year^{-1}. Intensity of monsoon rainfall varies from 20-30 mm day^{-1}, and in winter cyclonic rainfall varies from 8-14 mm day^{-1} from southwest to northeast direction. Summer monsoon rainfall comes with light drizzles and is easily absorbed into the soils. Temperatures vary greatly in this zone. The May and June months are the hottest months and January and February are the coldest. Mean temperature more than 20°C prevails during 8 to 10 months of the year. During May and June, maximum temperature rises above 40°C and hot dry winds in the area is a common feature. In January, normal mean minimum temperature is 6-8°C. Frost for one/two days may also occur during winter months (Atar Singh, 1991b).

Soils : Soils of this zone are of varying characters. The soils are either without calcium carbonate or calcium carbonate concentrated at lower horizons. The soil texture varies from sandy loam to loam. The water table is high along side of Yamuna river and in the depression between Aravali hills. Soil salinity has developed in certain places due to poor drainage. The tract around Rohtak is subject to flooding and large areas suffer from flood annually. In Karnal and Kurukshetra districts, the landscape is almost flat. High water table, seepage, and irrigation from canals and the flood overflows, have contributed much towards the extension of soil salinity in Rohtak tract. In northern part of Ambala district, where annual rainfall is above 1000 mm and natural vegetation is thick, the soils are well drained. Subsoil water is deep and soil texture varies from loamy sand to sandy loam without any clear horizonation (Atar Singh, 1991b) .

Agroclimate : Western Zone

The western zone includes the districts of Sirsa, Hissar, Bhiwani, some parts of Jind, Rohtak, and Gurgaon.

Climate : The southwest monsoon brings rains from July to September contributing about 80 to 85 per cent of the total annual rainfall. From October to mid-April, the weather remains almost dry from October to mid April, except for occasional light showers during these months. Later on, the weather remains quite dry till mid June with high temperature. A major part of southwest Haryana is arid receiving less than 400 mm rainfall in the district of Bhiwani and Jhajjar tehsils of Rohtak district. The districts of Gurgoan and Mahendragarh fall under semiarid climate, with annual rainfall ranging from 400 to 500 mm. The lowest rainfall, less than 300 mm year^{-1} is received in Loharu subdivision of Bhiwani district. Not only the quantum of rainfall is low in this region, but its variability is also very high.

An average minimum temperature of 5-6°C is reported in month of December and January. The high temperature (35°C) exists during mid-September to mid-October. Total yearly US open pan evaporation average is 2592 mm with maximum evaporation rate of 14 mm day^{-1} in the month of June. An average potential evaportranspiration of 5.3 mm day^{-1} is observed from July to October and 2.7 mm day^{-1} from November to February (Atar Singh, 1991b).

Soils : The southwestern Haryana has flat land interspersed with undulating topography (Aravali hills). The uplands characterised by the presence of sand dunes or hilly topography alternated with flat valleys are quite common with active wind erosion. The soils of the belt include Aridisols and Entisols. The soils are light textured, sandy and loamy sand. In Jhajjar subdivision and in some parts of Bhiwani district, the soils are sandy loam. In general, the surface soils are light in texture which become increasingly heavier in lower horizons. The bulk density and particle density of the soils vary from 1.33 to 1.53 g cm^{-3} and 2.47 to 2.63 g cm^{-3}, respectively. Available water holding capacity of these soils varies from 100 mm m^{-1} for sandy soils to 175 mm m^{-1} for sandy loam bordering to loam soils in texture. A great variation is found in soil profile with respect to depth and intensity of calcium carbonate concretion. The concretion layer is located in patches at 60 to 140 cm depth below the soil surface. Due to presence of shallow calcium carbonate concretion layers, the growth is adversely affected, especially in winter season or during drought. The soils are deficient in organic carbon and available nitrogen, and medium to high in phosphorus and potassium (Atar Singh, 1991b). Physicochemical characteristics of kinnow mandarin orchards at Bahauddin (Sirsa) and Kirad (Hisar) showed the variation in soil texture from sandy loam to clay loam, pH 7.6 to 8.0, EC 0.25-0.42 dS m^{-1}, organic carbon 0.15 to 0.37 per cent, and $CaCO_3$ 0.67 to 8.75 per cent (Table 1.25).

Table 1.25. Physicochemical properties of kinnow mandarin orchards in Haryana

Profile depth (cm)	Texture	pH	EC (dS m^{-1})	Organic C (%)	$CaCO_3$ (%)
		Bahudd, Sirsa			
15	Loam	7.7	0.31	0.33	0.67
30	Clay loam	7.7	0.32	0.27	0.75
60	Clay loam	7.8	0.34	0.25	0.75
90	Clay loam	7.7	0.36	0.22	0.75
120	Clay loam	7.7	0.37	0.22	0.75
150	Clay loam	7.8	0.36	0.15	1.00
		Kirad, Hisar			
15	Sandy loam	7.6	0.26	0.37	1.00
30	Sandy loam	7.7	0.25	0.30	1.25
60	Sandy loam	7.6	0.30	0.30	4.25
90	Sandy loam	7.8	0.34	0.15	6.50
120	Loam	7.8	0.35	0.15	7.56
150	Loam	7.8	0.35	0.15	8.75

Source : Singh *et al.* (1997)

Citrus in Punjab

Citrus in Punjab is grown agroclimatically in undulating plain zone, central plain zone, western plain zone, and western zone. Evaluation of various sweet orange, mandarin, and grapefruit cultivars showed largest fruits of Washington navel followed by valencia late, while juice content and total soluble solids recorded highest in mosambi, and Jaffa orange, (Mathur and Godara, 1990). Major soil types comprising Typic Haplustept, Udic Calciustept, Typic Ustipsamment, Typic Ustifluvent, Ustic Torripsamment, Fluventic Eutrochrept, and Natric Eutrochrept have been observed (Singh *et al.*, 2000).

Agroclimate : Undulating Plain Zone

This narrow and transitional zone runs parallel to the submontane undulating region, which includes Gurdaspur excluding flood plain areas, Hoshiarpur I, western parts of Hoshiarpur II, Nawanshagar block of Jalandhar, Machiwara block of Ludhiana, Bassi Pathana, Rajpura, and Ghanaur blocks except flood plain area of Patiala district.

Climate : The annual amount of the rainfall in the zone is 800 to 900 mm. The pattern of the distribution of rainfall is also very close to zone-I (submontane undulating zone). The only difference here is that there is no month when the amount of rainfall exceeds 300 mm. During the period of June to September, 13 weeks can be termed as humid to wet, 1 week iner-humid, and 3 weeks arid to dry. The climate of foot hills area of Gurdaspur is subhumid and warm humid (submoist). Annual rainfall varies from 1100 to 1300 mm and 70 per cent of rainfall is received during the monsoon months of July through September. The mean annual temperature is 23.2^{0}C with eight months having mean monthly temperature above 20^{0}C (Jassal *et al.*, 2000).

Soils : A number of research workers (Sharma *et al.*, 1986; 1993; Jassal *et al.*, 2000) have characterised the soils of Gurdaspur, northwestern part of submontane belt of Siwalik hills as Typic Udorthents, Typic Hapludalfs, Dystric Eutrochrepts, and Typic Eutrochrepts having pH of 5.9-8.0, $CaCO_3$ nil-0.80 per cent, and clay 13.0-29.4 per cent with higher K than Na and Mg than Ca, except calcareous soils.

On the basis of soil texture, this zone is divided into two subzones (Atar Singh, 1991c) as mentioned below :

Undulating plain region-north and south : The northern part of the zone consists of areas of Gurdaspur district, and southern part covers the areas of Ropar and Patiala district. The soils of this subzone are medium to heavy in texture. The soils have severe alkali problem in Kalanaur and moderate in Dinanagar and Gurdaspur blocks.

Undulating plain region-central : The area of Hoshiarpur district falling in the undulating plain region along with upland areas of Macchiwara constitutes this subzone. The soils are light medium in texture.

Agroclimate : Central Plain Zone

Whole Amritsar district except the Bhikhiwind, Patti, Valtocha blocks, flood plains of river Ravi and Beas, Batala, Fategarh Churian, Dera Baba Nanak, western part of Hargobindpur block of Gurdaspur district (except the flood plain parts, in the extreme north of Dera Baba Nanak, and southern fringes of Hargobindpur block), entire district of Kapurthala (except the flood plain of Beas), Jalandhar, and Ludhiana district (except Nawanshahar, eastern Banga, Macbhiwara block, and flood plain areas of Satluj river), Dharamkot block of Firozepur district (except flood plain areas of river Satluj), Mahal Kalan, Malkotla, Dhuri, Bhawaingarh, eastern parts of Barnals, Sangrur, and Sunam blocks of Sangrur district, entire Sirhind, Nabha, Patiala, Bhunerheri, and Samana blocks of Patiala district fall under this zone (Atar Singh, 1991c).

The general character of the land is homogeneous with very gentle slope of the land. The zone is intercepted transversally by the flood plains of the Satluj and Beas. The northern and southern limits are also marked by the flood plains of the Rabi and the Ghaggar, respectively. The average altitude of this plain is 230 to 260 m above mean sea level. The slope of the land decreases gradually from northeast to southwest, where it diminishes to less than a meter per kilometer. The depth of water table varies from 2 to 20 meter. In blocks of Amritsar district, except Raya, Jañdials, Khadur Sahib, Dera Baba Nanak, and Batala of Gurdaspur district, the strata available from 5 meter to 65 meter depth contains medium to fine sand. The quality of the water is good in this zone, except in parts of Sangrur district, and some pockets of other district where it is marginal to good. The main problem is the presence of residual sodium carbonates.

Climate : The mean maximum temperature recorded during the first fortnight of June is 42°C in the southern half and 41°C in the northern half. Whereas, the mean minimum temperature recorded during the month of January varies from 4 °C in northern parts to 7°C in the southern parts. The mean annual rainfall varies from 200 to 300 mm, in the east to about 500 mm towards the western limits. In the southern half, 3 months of the rainy season receive a rainfall between 200 and 300 mm and one month between 50 and 100 mm, and less than 50 mm in one month. In the rainy season period, June to September in the southern half, 12 weeks can be classified as humid to wet, 2 weeks intermediate dry to intermediate humid, whereas 3 weeks are arid to dry. In the northern side, 13 weeks of the season are classified as humid to wet, 1 week interhumid, and 3 weeks as arid to dry (Atar Singh, 1991c).

Soils : On the basis of the texture of the soil, this zone is subdivided into two subzones :

Central plain region-north : The subzone covers the Amritsar district excluding Bhikhiwind, Pati Valtoha block, Dera Baba Nanak, Fatehgarh Churian, Batala, and western part of Hargobindpur block of Gurdaspur district.

Central plain region-south : The subzone extends from the eastern flood plain of river Beas up to southern limit of Patiala district. The soils of this subregion are light to medium in texture. Mild to severe alkali problem exists in the area of Kapurthala, Patiala, and Sangrur districts. A small pocket consisting of Sirhind, Rajpura, Ghanaur, and Bhunerheri blocks has, however, medium to heavy textured soils and is very close to the southern area of the undulating plain region (Atar Singh, 1991c).

Soils representative of four major agroclimatic regions of Punjab are characterised as Ustifluvents, Ustipsamments, and Ustochrepts (pH 8.2, EC 0.25 dS m^{-1},organic C 4.8 g kg^{-1}, $CaCO_3$ 3.8 g kg^{-1}, sand 48.9 per cent, silt 35.2 per cent, and clay 17.5 per cent) on an average basis in Gurdaspur, Hoshiarpur, and Ropar districts under undulating plain region; Ustochrepts and Haplustalfs (pH 8.7, EC 0.33 dSm^{-1}, organic C 5.2 g kg^{-1}, $CaCO_3$, 8.9 g kg^{-1}, sand 39.2 per cent, silt 39.1 per cent, and clay 21.7 per cent) in

Amritsar, Jalandhar, Ludhiana, and Patiala under central plain region; Ustochrepts, Camborthids, Calciorthids, and Torripsamments (pH 8.8, EC 0.36 dSm^{-1}, organic C 3.6 g kg^{-1}, $CaCO_3$ 16.9 g kg^{-1}, sand 63.1 per cent, silt 21.9 per cent, and clay 15.0 per cent) in Sangrur, Faridkot, Ferozepur, and Bhatinda under western plain region; and Ustifluvents, Ustipsamments, and Haplaquents (pH 8.3, EC 0.29 dSm^{-1}, organic C 4.7 g kg^{-1}, $CaCO_3$ 16.3 g kg^{-1}, sand 63.3 per cent, silt 23.0 per cent, and clay 13.7 per cent) in Kapurthala and Ludhiana under flood plain region (Randhawa and Singh, 1997).

The soils of Punjab in northwest India have developed on alluvial deposits, reworked by aeolian action at some places. False colour composites generated from bands 2,3, and 4 were used to delineate and evaluate soil-landscape relationship of the soils belonging to Entisol order. Entisols mostly occurred on convex shoulder slopes of hills, active flood plains, recent flood plains, and sand dunes in the ustic and aridic zone. At the suborder level, these are classified as Orthents, Fluvents, and Psamments. Orthents are represented by Typic Udorthents on the convex shoulder slopes of hills in the upper Kandi zone; Typic and Aquic Ustorthents in the lower Kandi zone and flood plains in the central plain zone; and by Ustic Torriorthents in the interdunal areas of the aridic zone. Fluvents represented by Typic and Aquic Ustifluvents occurred along the rivers and their tributaries in the ustic region. Psamments are represented by Typic Ustipsamments on the sand dunes and Aquic Ustipsamments in the active flood plains of river in the central ustic zone and by Ustic Torripsamments in the aridic zone (Sidhu *et al.*, 1994).

The soils of Ferozepur and Faridkot districts are loamy sand to loam in texture. The pH, EC, $CaCO_3$, organic carbon, available P, and K_2O contents in soils of Ferozepur district varied from 7.8 to 9.3, 0.2 to 1.1 dSm^{-1}, nil to 7.0 per cent, 0.9 to 10.5 g kg^{-1}, 5.5 to 67.2 kg ha^{-1}, and 150 to 750 kg ha^{-1} with their mean values of 8.4, 0.3 dS m^{-1}, 2.1 per cent, 4.3 g kg^{-1} , 14.7 kg ha^{-1}, and 360 kg ha^{-1} , respectively. In soils of Faridkot district, these characteristics such as pH ranged from 7.8 to 9.4, EC 0.2 to 1.1 dS m^{-1}, $CaCO_3$ 0.2 to 5.8 per cent, organic carbon 0.6 to 25.7 g kg^{-1}, available P 5.0 to 18.7 kg ha^{-1}, and available K 202.0 to 750.0 kg ha^{-1} with their mean contents of 8.6, 0.3 dSm^{-1}, 1.6 per cent, 3.4 g kg^{-1}, 7.8 kg ha^{-1}, and 236 kg ha^{-1}, respectively (Singh and Nayyar,1999).

Agroclimate : Western Plain Zone

The zone lies between the central flat plain on the east and the plain with sand dunes in the extreme west. The zone covers Patti, Bhikhiwind, and Valtoha blocks of Amritsar district; Zira, Firozepur, and Ghall Khurd blocks of Firozepur district; Faridkot, Moga I, Moga II, Nihal Singh Wala, and Bhaga Purana blocks of Faridkot district; Rampura Phul, Mansa, eastern few parts of Sangrur, and Sunam blocks, and entire Lehrangaga block of Sangrur.

It is a transitional region where flat topography merges gradually into a sand dune dotted land surface at an altitude of 200 to 250 m above sea level. There is no stream or river worth mentioning which passes through this region. The northern half is free from any topographical feature, whereas towards the western margins in the southern half, sand dunes are quite frequent feature of the landscape. The water table in the zone varies from 2 to 10 m below the ground level. The quality of groundwater is good to marginal for irrigation.

Climate : As the latitudinal extent is not much, there are not much temperature variation from south to the north. The annual amount of rainfall, however, varies between 400 and 500 mm. During the monsoon period (June to September), 8 weeks are classified as humid to wet, 2 weeks inter-dry to inter-humid, and 7 weeks as arid to dry, indicating the weekly index value as a mark of duration of moisture regime.

Soils : Two subzones have been recognised on the basis of soil texture :

Western plain region-north: The subzone stretches up to the southern and south eastern limits of Moga I and Bagha Purana blocks of Faridkot district. The soils in this sub-zone are sandy loam and loam in texture (Table 1.26). Serious alkali problem with waterlogging exists in Firozepur, Faridkot, and Ghall Khurd blocks.

Western plain region-south : The subzone extends from Moga II in the north to Budhlada block in the extreme south. The soils are coarse to medium in texture. The eastern parts of this subzone are infested with mild to serious alkali problem (Atar Singh, 1991c).

Table 1.26. Fertility characteristics of soils of Ferozepur and Faridkot districts, Punjab

Soil characteristics	Ferozpur		Faridkot	
	Range	Mean	Range	Mean
Soil texture	ls – l	-	ls – l	-
pH	7.8-9.3	8.4	7.8-9.4	8.6
EC (dS m^{-1})	0.2-1.1	0.3	0.2-1.1	0.3
Organic C (kg ha^{-1})	0.9-10.5	4.3	0.6-25.7	3.4
Available P (kg ha^{-1})	5.5-67.2	14.7	5.0-187.5	7.8
Available K_2O (kg ha^{-1})	150-750	360	202-750	236
$CaCO_3$(%)	Nil-7.0	2.1	0.2-5.8	1.6
Available B (mg kg^{-1})	0.22-2.40	0.92	20.20-3.85	1.53

ls, loamy sand; l, loamy

Source : Singh and Nayyar (1999)

Considering the critical values of soil available B as less than 0.5 mg kg^{-1} and 0.5 to 1.0 mg kg^{-1} for deficient and low status, respectively (Singh *et al.,* 1980), the available B is rated to be deficient in 13 per cent and low in 50 per cent soils of Ferozepur district. While, in Faridkot district, a 16 per cent and 28 per cent soils have been rated as deficient and low, respectively, in available B. The remaining 37 per cent and 66 per cent of the soils of Ferozepur and Faridkot districts, respectively, had available B in the high level category. Singh *et al.* (1980) reported that hot water extractable B ranging from 0.20 to 7.20 mg kg^{-1} (average 2.60 mg kg^{-1}) in selected surface samples of coarse textured alluvium derived soils of Jalandhar, 6 per cent and 21 per cent soil samples were deficient and low, respectively.

Agroclimate : Western Zone

The western zone covers the districts of Bhatinda, Faridkot, and Ferozepur. The flat topography of the upper central plain zone of the state gradually merges into sand dune dotted land surface of these districts. As one moves from northeast to the southwest of these districts, the size and frequency of sand dunes generally increases. Sand dunes and sand bars have formed due to strong and dessicating southwesterly winds, which are deficient in moisture and transport clouds of sand from within and adjoining state of Rajasthan. The depth of water table varies from less than a meter to 30 meter below the ground level.

Climate : The climate of the region is relatively more dry and warm than the rest of the Punjab. It is marked by cold winter in December and January and very dry and hot weather from April to June. The mean air temperature during the year, generally varies from 12° to 32°C with temperature as high as 45°C in May and June, and as low as 0°C in December and January. The soil temperature in the active root zone from April to August exceeds the optimum temperature by 2° to 12°C. In the premonsoon summer months, relative humidity of area is very low (35 to 50 per cent). High atmospheric evaporative demand of 10 to 18 mm day^{-1} during premonsoon period, 5 to 12 mm day^{-1} during monsoon period and 2 to 6 mm day^{-1} during postmonsoon period are commonly observed. The annual precipitation varies from 200 to 500 mm, with about 80 per cent rainfall being received during monsoon period only. Other notable features of rainfall are high degree of uncertainty and uneven distribution (Atar Singh, 1991c).

Soils : The soils of the southwest Punjab are formed from alluvial deposits, highly calcareous, and are rich in dispersed lime. Calcium carbonate concretions are common in the subhorizons. The soils are mostly sandy to sandy loam in texture with pockets of fine textured soils. Soil profiles are generally well drained, except where highly impermeable layers of clay and/or fine calcium carbonate occurs in the subsoil. The surface horizons of the soils are generally light coloured and are poor in organic matter. The pH of these soils varies from 8.2 to 10.5 and EC from 0.2 to 5.0 mmhos cm^{-1}. The soils are generally deficient in N,P,K, and Zn. The soils of Dharmkot, Zira, and Guruharshhai blocks of Ferozepur district are low, medium, and high in N, P, and K, respectively. The soils of Jalalabad and Fazilka blocks of Ferozepur, Mukatsar, Malut, and Nihal Singh Wala have status as above, but are medium in K (Atar Singh, 1991c). Studies on a typical sweet orange growing soil profile at Bhatinda by Mann and Sidhu (1983) showed that all the micronutrients, except Cu are deficient due to highly alkaline soil pH ranging from 8.4 to 8.8 (Table 1.27).

Table 1.27. Physiochemical properties of sweet orange orchard at Bhatinda, Punjab

Depth (cm)	pH (1:2)	EC (dSm^{-1})	$CaCO_3$ (%)	Organic carbon (%)	Avail. P	Avail. K	DTPA extractable micronutrients ($mg\ kg^{-1}$)			
							Zn	Cu	Mn	Fe
0-15	8.8	0.20	3.7	0.45	16	400	0.36	0.70	3.70	3.30
15-30	8.7	0.30	2.2	0.40	14	350	0.25	0.62	3.00	2.30
30-60	8.7	0.20	2.2	0.40	13	340	0.14	0.60	2.20	1.80
60-90	8.5	0.20	2.5	0.41	8	310	0.16	0.55	2.40	2.60
90-120	8.5	0.20	3.7	0.38	5	305	0.10	0.53	2.60	2.60
120-150	8.4	0.40	1.5	0.36	5	300	0.09	0.54	1.90	2.50
150-180	8.4	0.30	1.8	0.34	4	305	0.09	0.55	2.00	2.50

Source : Mann and Sidhu (1983)

Citrus in Himachal Pradesh

Agroclimatically kinnow mandarin cultivation in Himachal Pradesh is distributed to two major zones viz., submontane subtropical low hill and midhill subhumid zone (Table 1.28). Major soil types, namely, Typic Udorthent, Dystric Eutrudept, Typic Cryorthent, Lithic Cryorthent, and Lithic Udorthent are very common (Singh *et al.*, 2000).

Agroclimate : Submontane and Low Hills Subtropical Zone

Climate : This zone comprises Una, Bilaspur, Hamirpur district, parts of Sirmour, Kangra, Solan, and Chamba districts. Average rainfall is 1100 mm (1100-2800 mm) of which 80 per cent is received during July to September (Gangopadhyay, 1991a). Kangra district falls under low to midhills subhumid climate with elevation ranging from 650 to 1840 m above mean sea level (Table 1.28). The mean annual rainfall varies from 1000 to 2000 mm (average 1700 mm) and monthly temperature from 10° to 27°C (mean 20°C). About 70 per cent of the total rainfall is received during monsoon period (June to September).

Table 1.28. Agroclimatic zones of kinnow mandarin growing areas in Himachal Pradesh

Areas	Elevation (m)	Soil types	Rainfall (mm)	Temperature (°C)	
				Min.	Max.
Submontane and subtropical low hill zone (Dhaulakuan, Una, Bilaspur, Hamirpur, Sirmour, Kangra, Solan, and Chamba)	650	pH 6.5-8.0, shallow, light textured, and loamy sand to sandy loam	1100	5	25
Midhill subhumid zone (Palampur, Kangra, Rampur, Mardi, Solan, Kullu, Chamba, Sirmour, Seabag Bilaspur, and Kandaghat)	651-1840	Deficient in N, P, poor water holding capacity, and responds to liming	2800	-5	25

Source : Gangopadhyay (1991a)

Soils: The rock systems belong to the middle miocene to lower pleistocene age. These are characterised by sandstones, siltstones, conglomerates, shales etc. Granitic sandstone, the weathered product of the granitic rocks found in the Dhulandhar range, are also very common. Soils of the zone are shallow, light textured, and low in fertility. Citrus growing soils of Himachal Pradesh are taxonomically classified as Typic Udorthemis, Typic Paleudalfs, Typic Udorthents, and Fluventic Eutrochrepts, out of which Typic Udorthents have high available Fe (25.4 – 30.8 mg kg^{-1}), Mn (26.8-52.5 mg kg^{-1}), Cu (1.4 – 4.2 mg kg^{-1}), and Zn (1.4–2.8 mg kg^{-1}) up to 1.50 m of soil depth (Table 1.29).

Soils of Kangra and Sirmour districts of Himachal Pradesh have been characterised by Kaistha and Gupta (1994) which revealed the occurrence of Typic Udorthents, Typic Eutrochrepts, and Typic Udifluvents with soil pH varying between 5.7 and 9.8. High soil pH in few soils was attributed to presence of buried sodic soil and their development on calcareous alluvium. Clay mineralogy consisted of dominance of mica, chlorite, and kaolinite. Distribution of micronutrient in representative soil profiles at Bhota, Theog, Rishikesh, Nagwain, Naggar, Palampur, Kangra, Saproon, Nauni, and Dhaulakuan areas of Himachal Pradesh showed available Cu from 0.4 to 4.8 mg kg^{-1}, Mn 4.4 to 50.4 mg kg^{-1}, Zn 0.1 to 2.1 mg kg^{-1}, and Fe from 6.2 to 40.5 mg kg^{-1} in soils developed from three types of parent materials viz., shales, schists, slates, and phyllites; shales and dolomitic limestones; and standstones and conglomerates under three agroecological regions viz., humid temperate, wet temperate, and humid subtropical (Tripathi *et al.*, 1994).

Table 1.29. Important characteristics and available micronutrient status of the soils of Himachal Pradesh

Horizon	Depth (cm)	pH (1:2.5)	Org. C (g kg^{-1})	Clay (g kg^{-1})	$CaCO_3$ (g kg^{-1})	Available micronutrients (g kg^{-1})			
						Mn	Zn	Fe	Cu
			Kangra (Typic Udorthemis)						
A1	0-20	6.9	9.4	162	0.0	50.9	0.4	40.5	0.7
IIA3	20-53	6.9	4.2	216	0.0	48.8	0.2	33.2	0.8
IIC	53-110	6.8	2.3	234	2.0	50.4	0.2	17.0	0.5
			Palampur (Typic Paleudalfs)						
A1	0-05	6.2	14.1	260	0.0	45.8	0.3	40.2	1.0
A3	05-25	6.1	12.4	262	0.0	29.4	0.3	23.0	0.9
B21t	25-51	5.9	8.2	326	0.0	25.8	0.3	13.0	0.8
B22t	51-73	5.7	4.1	332	0.0	45.0	0.3	10.5	1.2
B23t	73-150	5.7	3.6	390	0.0	19.2	0.3	9.5	0.7
			Nauni (Typic Udorthents)						
A1	0-15	6.7	7.7	240	10.0	52.5	2.8	30.8	4.2
C1	15-35	6.2	3.6	262	10.0	45.1	1.5	29.6	2.3
C2	35-75	6.0	0.5	280	12.0	26.8	1.4	25.4	1.6
C3	75-150	5.8	0.5	254	13.2	28.0	1.6	28.1	1.4
			Dhaulakuan (Fluventic Eutrochrepts)						
A1	0-15	7.3	5.7	180	10.0	34.2	0.8	32.0	1.5
A3	15-30	7.1	3.0	188	3.5	30.8	0.5	22.2	1.4
B	30-70	7.1	3.3	230	5.0	29.0	0.4	22.6	1.1
IIC1	70-90	7.0	2.6	162	5.0	33.8	0.5	23.4	1.1
IIC2	90-130	6.9	3.0	82	4.0	33.0	0.4	21.6	1.2

Source : Tripathi *et al.* (1994)

Fertility status of soils of Hamirpur district (400 – 500m altitude), Himachal Pradesh representing Bijhari (organic C 0.07 – 1.27 per cent, available N 101 – 584 kg ha^{-1}, available P traces-47.7 kg ha^{-1}, and available K 60- 450 kg ha^{-1}), Bhoranj (organic C 0.15 – 0.97 per cent, available N 107 – 591 kg ha^{-1}, available P traces-52.1 kg ha^{-1}, and available K 82- 450 kg ha^{-1}), Hamirpur (organic C 0.15 – 1.20 per cent, available N 107 – 572 kg ha^{-1}, available P traces-67.0 kg ha^{-1}, and available K 105- 450 kg ha^{-1}),Nadaun (organic C 0.07 – 1.05 per cent, available N 101 – 522 kg ha^{-1}, available P traces-57.2 kg ha^{-1}, and available K 82- 450 kg ha^{-1}), and Sujanpur (organic C 0.22 – 1.12 per cent, available N 151 – 610 kg ha^{-1}, available P traces-47.7 kg ha^{-1}, and available K 109-450 kg ha^{-1}) has been described by Sharma *et al.* (1993). These soils are derived from parent material consisting of sandstone, clays, conglomerates, and boulders.

Agroclimate : Midhills Subhumid Zone

This zone comprises of Palampur and Kangra tehsils of Kangra district, Rampur tehsil of Shimla district, parts of Mandi, Solan, Kullu, Chamba, Bilaspur, and Sirmour districts.

Climate : The elevation of the zone varies from 651 to 1800 m above mean sea level. The rainfall in the zone is as high as 300 mm, while in remaining parts, it is as low as 800 to 1500 mm. The total annual rainfall in citrus growing locations, namely, Nurpur, Palampur, Kangra, Hamirpur, and Una has been registered as 1523.3 mm, 2681.3, mm, 2132.7, mm 1372.4 mm, and 1017.2 mm, respectively. While, average annual temperature varied from 12.2° to 31.1 °C, 10.3° to 26.7°C, 12.0° to 30.6°C, 11.6° to 29.3°C, and 12.4° to 32.0°C in Nurpur, Palampur, Kangra, Hamirpur, and Una, respectively (Table 1.30).

Table 1.30. Monthly variation in rainfall and temperature in major citrus growing areas of Himachal Pradesh

Months	Nurpur		Palampur		Kangra		Hamirpur		Una	
	Rainfall (mm)	Temp. (°C)	Rainfall (mm)	Temp. (°C)	Rainfall (mm)	Temp. (°C)	Rainfall (mm)	Temp. (°C)	Rainfall (mm)	Temp. (°C)
Jan.	81.0	12.2	115.6	10.3	78.5	12.0	66.8	11.6	51.3	12.4
Feb.	75.9	14.8	112.3	12.2	73.1	14.6	62.7	13.9	49.5	15.2
Mar.	70.1	19.6	103.4	16.4	70.6	19.2	53.1	18.4	39.9	20.1
Apr.	31.5	24.6	57.9	20.9	36.6	24.2	26.7	23.2	19.8	25.3
May	27.4	29.1	63.5	25.1	39.9	28.6	30.0	27.5	17.8	29.9
Jun.	85.3	31.1	190.0	26.7	108.2	30.6	93.0	29.3	68.8	32.0
July	428.7	27.7	809.7	23.6	578.2	27.2	409.2	26.0	310.6	28.6
Aug.	483.6	26.8	868.9	22.9	692.7	26.3	422.1	25.1	273.3	27.6
Sept.	167.9	26.0	256.8	22.3	218.7	25.6	154.2	24.5	140.7	26.8
Oct.	19.3	22.6	35.8	19.7	19.1	22.2	16.0	21.4	14.5	23.1
Nov.	8.1	17.7	12.5	15.8	8.1	17.6	6.6	17.1	3.6	18.0
Dec.	44.5	13.9	54.9	12.4	37.1	13.8	32.0	13.4	27.4	14.0

Source : Sehgal *et al.* (1987)

Soils : The texture of the soil varies from loam to clay loam. The soils are deficient in N and P with poor water holding capacity. Some of the soil types from Palampur, Kangra, Dhaulakuan, and Nauni are presented (Table 1.31). These soils taxonomically belong to Typic Paleudalf, Typic Udorthent, Aquic Eutrochrept, and Fluventic Eutrochrept. Verma and Tripathi (1982) characterised the soils of Indora (Nurpur valley), Himachal Pradesh as pH 7.9, organic C 0.11-0.44 per cent, CEC 10.4-13.8 me 100 g^{-1}, base saturation 87.8-94.5 per cent, sand 40.2-42.7 per cent, and clay 16.8-21.4 per cent.

Table 1.31. Important characteristics of the soils of northwest himalayas, Himachal Pradesh

pH	EC (dS m^{-1})	Org. C (%)	Clay (%)	CEC [cmol (p^+) kg^{-1}]	Exch. $Ca^{2+}+Mg^{2+}$ [cmol (p^+) kg^{-1}]	$CaCO_3$ (%)	Olsen P (mg kg^{-1})	Total Fe_2O_3 (%)	Total Al_2O_3 (%)
Palampur(Typic Paleudalf)									
6.3	0.21	0.81	32.2	14.9	7.9	0.25	55.0	8.6	16.2
Kangra(Typic Udorthent)									
6.9	0.33	0.84	16.2	14.5	8.9	0.80	147.5	5.2	10.1
Nauni (Aquic Eutrochrept)									
7.0	0.35	0.77	14.8	11.4	6.9	1.00	90.0	5.0	10.0
Dhaulakuan (Fluventic Eutrochrept)									
7.4	0.44	0.67	23.5	10.3	6.9	1.00	178.8	5.3	10.2

Source : Upadhyay *et al.* (1993)

The soils of northwest himalayas are located at an altitude varying from 1000 to 1400 m above sea level. The parent materials of these soils include slates, phyllites, quartize, schists, granite, and gneiss. Illite is the most dominant clay mineral followed by kaolinite, along with chlorite, mixed layer silicates, and smectites in small amounts (Kaistha and Gupta, 1992). Citrus growing soils of northwest himalayas at an elevation of 1000-1450 m above mean sea level vary in pH from 5.3 to 6.2, clay 12.5 to 30.0 per cent, silt 27.5 to 45.0 per cent, and sand from 33.3 to 60.0 per cent (Table 1.32).

The phosphorus fixing capacity of mountain soils from northwest himalayas was found to be fairly high, the average being 16.5 cmol kg^{-1}. According to Kaistha *et al.* (1997), it is further influenced by clay, followed by silt, exchangeable, extractable, amorphous, and crystalline forms of Al and Fe. On an average, exchangeable, extractable, amorphous, and crystalline Al were observed as 22.5, 40.3, 419.0, and 645.6 mg kg^{-1}, respectively. While, average extractable, amorphous, and crystalline Fe have been recorded as 91.3, 2736.3, and 2245.6 mg kg^{-1}, respectively (Table 1.33).

Table 1.32. Some properties of the soils of northwest himalayas, Himachal Pradesh

Location	Altitude (m)	pH	Clay (%)	Silt (%)	Sand (%)
Baijnath	1400	6.2	22.5	27.5	50.0
Padhiarkhar	1250	5.7	30.0	45.0	25.0
Joginder Nagar	1450	6.0	12.5	27.5	60.0
Malan	1100	5.3	20.0	40.0	40.0
Palampur	1300	5.5	30.0	32.5	37.5
Rait	1000	5.9	15.0	37.5	47.5
Mean	—	5.7	21.7	35.0	33.3

Source : Kaistha *et al.* (1997)

Table 1.33. Different forms of aluminium and iron in the soils of northwest himalayas, Himachal Pradesh

Location	Aluminium ($mg\ kg^{-1}$)				Iron ($mg\ kg^{-1}$)		
	Exchangeable	Extractable	Amorphous	Crystalline	Extractable	Amorphous	Crystalline
Baijnath	14	25	325	456	58	2337	2454
Padhiarkhar	35	48	413	521	121	3591	2478
Joginder Nagar	16	32	287	392	69	1864	2079
Malan	22	42	489	708	63	3267	2281
Palampur	30	52	654	1216	147	2602	2812
Rait	18	43	346	581	90	2757	1352
Mean	22.5	40.3	419.0	645.6	91.3	2736.3	2245.6

Source : Kaistha *et al.* (1997)

Mineralogy of sand, silt, and clay fractions of six profiles developed under different parent materials and topography from Kangra district (H.P.) has been characterised by Gupta and Tripathi (1996). The light mineral fraction comprised highest of the total sand with quartz, feldspar, and muscovite, and heavy fraction constituting 11.3 to 0.7 per cent with a number of minerals. Plagioclase feldspar is highly weathered. Weathered nature of biotite, augite, and hornblende has also been found. The mineralogical make-up of clays determined by X ray diffraction, diffraction thermal, and infrared analysis indicated the dominance of mica as chlorite, smectite, vermiculite, and mixed gibbsite, while, infra red analysis indicated the presence of amorphous aluminium/iron oxide and allophane. Similarly, clay mineralogy indicated the dominant influence of parent material. Generally, the soil profiles developed under well levelled land and/or in low land topographic situation have well developed textural and structural horizons. The genesis of these soil profiles is expressed by calcification/decalcification, illuviation, and humification. Soil characteristics of few profiles of northwest himalayas are described (Table 1.34).

Table 1.34. Physicochemical properties of soils of northwest himalayas, Himachal Pradesh

Horizon	pH	Organic C (%)	Mechanical composition			Cations				Mobile		Total	
			Sand	Silt	Clay	CEC	Ca^{2+}	Mg^{2+}	K^{+}	Fe_2O_3	Al_2O_3	Fe_2O_3	Al_2O_3
			(%)			[$cmol(p^{+})kg^{-1}$]				(%)			
Nurpur, elevation 545 m, levelled topography, sandstone/pink-grey coloured clays													
A1	6.0	12.2	41.9	26.6	30.5	17.0	3.2	10.5	0.1	0.6	0.6	10.9	8.5
A2	5.9	8.9	47.7	17.6	34.0	10.1	3.7	3.6	0.1	0.8	0.8	11.9	8.3
B1	5.5	3.9	47.7	18.2	34.8	8.4	3.4	3.6	0.1	0.8	0.8	13.2	15.3
B21	5.8	3.3	43.2	12.2	29.1	8.6	4.9	2.2	0.1	0.2	0.2	12.6	14.2
B23	5.9	0.5	43.2	21.2	35.8	7.1	3.1	1.9	0.1	0.2	0.2	12.6	14.2
Nurpur, elevation 485 m, low valley, sandstone/pink-grey coloured clays													
Ap	6.2	6.0	57.6	21.6	20.7	7.0	4.9	0.7	0.2	0.3	0.2	8.7	10.6
A3	6.3	3.2	56.8	20.2	22.9	15.1	4.4	8.2	0.2	0.3	0.2	9.9	10.6
B1	6.2	3.0	56.7	16.2	27.8	13.4	2.6	6.1	0.1	0.3	0.2	9.1	16.0
B21t	6.3	2.4	51.0	18.1	30.8	14.4	11.0	0.6	0.5	0.4	0.4	9.8	18.0
B22t	6.6	3.0	39.6	27.8	32.6	16.8	4.0	8.8	0.9	0.4	0.4	10.0	16.0

Source : Gupta and Tripathi (1996)

Citrus in Jammu and Kashmir

Citrus in Jammu is grown at lower altitudes with climate similar to any high altitude area. Soils of low altitude (300-1200 m), subtropical zone of Jammu region are developed on parent material consisting quartz, granite, slates, and shales with illitic as dominant clay mineral, pH 5.5-7.9, organic C 2.7-14.7 g kg^{-1}, $CaCO_3$ 1.0-37.0 g kg^{-1}, and clay plus silt 34.2-78.1 per cent. These soils belong to Entisol and Inceptisol (Gupta and Lattoo, 1999). In Kashmir region, the soils of summits and side/reposed slopes showed limited profile development having mostly A and C horizons and are classified as loamy skeletal Lithic Udorthents. The other soils found on similar physiography are fragmental and loamy, skeletal Lithic Cryorthents and patches of fine loamy, Typic Dystrochrepts.

In the upper piedmont plains, soils are relatively developed and show the presence of structural B horizon (cambic). These are classified as fine Dystric Eutrochrepts. In narrow valleys, soils are less developed due to both erosion and depositional activities and are classified as coarse loamy Typic Udifluvents. In the lower piedmont plain, soils are comparatively well developed with the presence of structural B (cambic) as well as clay cutans (argillic horizon). The soils have been classified as Typic Hapludalfs. Other soils found on this physiography are fine loamy calcareous Typic Eutrochrepts and fine Dystric Eutrochrepts. Soils of fluvial valleys are calcareous in nature and classified as fine silty calcareous Typic Eutrochrepts. Associated soils of this physiography are fine loamy calcareous Typic Udifluvents and fine loamy calcareous Fluaquents. Soils of terraces belong to fine loamy Fluventic Eutrochrepts. A majority of soils of Kashmir region according to Mahapatra *et al.* (2001) belong to Entisols (60 per cent) followed by Inceptisols (32 per cent), Alfisols (7 per cent), and small extent of Mollisols (1 per cent).

Commercial Cultivation of Kinnow Mandarin

Kinnow, a hybrid between king mandarin (*Citrus deliciosa* Lour) and willow leaf (*Citrus nobilis* Tanaka), also known as mediterranean mandarin, was developed by H.B. Frost in 1915 and released in 1935. During 1959, this cultivar was introduced in Punjab and since then, due to its better adaptability, spread into entire north and northwest states.

Climate : Kinnow mandarin requires semiarid, subtropical climate for attaining acceptable quality of fruits (Singh, 1989). Such climate is characterised by sharply contrasting warm cool temperature, with chilling temperature, for economical cropping, and adequate quality of fruits due to significant influence of preharvest temperature on acid metabolism and total soluble solid development (Singh, 1989). The cultivation of kinnow mandarin has been found very successful at Allahabad, while the cultivation of other cultivars such as Nagpur, Coorg, and khasi mandarins is not so successful (Gauri Shankar, 1978). The cultivation of kinnow mandarin in Himachal Pradesh is confined to elevation ranging from 350 to 2975 m above mean sea level. Such a wide variation in altitude which varies from temperate in higher hills to subtropical in lower hills and valley areas, is likely to induce a strong influence on yield and quality of Kinnow mandarin (Azad and Chauhan, 1989). Most of the areas obtain about 300-500 chilling hours (temperature < 10°C) during midwinter and maximum temperature seldom goes beyond 40°C during summer. The cool and dry weather characterised by cool nights and sunny warm days during October-November extremely favours the development of high quality kinnow mandarin. Kinnow mandarin has been observed to give total soluble solids of 18.2 per cent at Palampur (Jolka and Awasthi, 1980) and 15.5 per cent at Dhaulakuan (Bhullar, 1978) in Himachal Pradesh in comparison to 11.0 per cent at Abohar in Punjab (Jawanda *et al.*, 1973) and 8 per cent at Coorg in Karnataka (Srivastava and Bopaiah, 1978) showing that kinnow produced in cool and humid climate of Himachal Pradesh has far excellent quality.

Kinnow mandarin has a wide adaptability. The districts like Hoshiarpur and Firozpur districts of Punjab emerging from main centres of production indicated that kinnow performed well under divergent agroclimatic conditions. Success of kinnow mandarin in Punjab is due to presence of extremely uniform relative humidity (65.3-72.6 per cent) during rainy to winter season compared to summer (31.0-52.1 per cent). Whereas, diurnal variation in temperature is equally distributed (15.7°-17.4°C) during winter and summer season compared to rainy season (9.8°-11.9°C). This is in sharp contrast to agroclimatic conditions of hot subhumid tropical climate of central India, where Nagpur mandarin is exclusively grown. The relative humidity in this region on annual basis ranges between 55.1 per cent and 92.8 per cent. The maximum and minimum temperature in citrus belts of Punjab varies as 20.5°-27.7°C and 4.1°-10.3°C, respectively, during harvest period (November–January) compared to 30.4°-32.3°C and 13.4°-15.2°C during *Ambia* fruit harvest period in central India. Similar inferences could also be drawn from its spread to arid as well as submontane areas of Haryana.

The success of its cultivation in Rajasthan on one hand, parts of Himachal Pradesh and Jammu and Kashmir on the other hand, also supports similar conclusions. Various reports about its success in Karnataka, Akola, and Nagpur region of Maharashtra further lend support for its wide adaptability to an extent, which lacked in any of the other citrus varieties in India like khasi mandarin of northeastern region, Nagpur mandarin of central India (Fig. 1.5), and Coorg mandarin of south India (Singh, 1989; Uppal *et al.*, 1989). Earlier studies carried out at Coorg (Srivastava and Bopaiah, 1978) showed that kinnow mandarin adapted well to the warm humid climatic conditions of Coorg with reference to yield and quality. No significant difference was observed when yield and quality parameters were compared with Coorg mandarin. According to Singh (1989) on the contrary, kinnow mandarin cultivated at Tirupati (A.P.) has not been able to even produce flowers, with the result, no fruiting was observed.

Soils : Kinnow mandarin growing soil zones of Himachal Pradesh fall under two major categories:

Fig. 1.5. An orchard of kinnow mandarin in black clay soils of central India.

- Submountainous and subhumid uplands of Chamba, Kangra, Solan, Hamirpur, Sirmour, and Bilaspur districts. Soils are shallow, light textured, loamy sand to sandy loam with pH between 6.5 and 8.0.
- Midhills and sub-humid parts of Chamba, Kangra, Mandi, Solan, Shimla, and Sirmour districts having soils loamy in texture, acidic in pH, deficient in N, P, poor water holding capacity, and responsive to liming (Gangopadhyay, 1991a).

The survey of fertility status of mandarin growing soils of Ferozpur district (Punjab) by Dhillon and Dhatt (1988) revealed minimum organic carbon (0.18 per cent), available P (5.0 mg kg^{-1}), and K (13.3 mg kg^{-1}). Dhatt (1989) observed mandarin soils of Hoshiarpur, parts of Jalandhar, Faridkot, Ferozpur, and Bhantinda being sandy to loam in texture, pH 7.0 to 9.0, low in available N, P, and Zn, medium in K and $CaCO_3$ concentrated in subsurface, occurrence of lime concretions up to 75 cm of soil surface or within the feeding zone, and electrical conductivity > 0.5 mmhos cm^{-1} (Kanwar and Randhawa, 1960; Kanwar *et al.*, 1965), apart from inadequate drainage (Singh and Jawanda, 1963; Kanwar *et al.*, 1965) as the major soil problems.

Kinnow mandarian growing soils of Himachal Pradesh are generally coarse textured comprising loamy sand, sandy loam, and occasionally loam to sandy clay loam embedded with pebbles and stones (Azad and Chauhan, 1989). Raina (1988) observed that soils are medium to high in available K, sufficient in Cu and Fe, deficient in Mn and Zn. Kinnow mandarin soils of Nagpur-Indora area (H.P.) have been found low in available N, low to medium in available P and K, optimum in Ca, Mg, and Mn and high in Cu and Fe. While, available Zn registered low to high status (Sharma and Mahajan, 1990).

Potential Areas for Kinnow Mandarin

The areas earmarked potentially suitable for expanding kinnow mandarin industry in subtemperate zone consist of lower hills of Uttar Pradesh and Jammu and Kashmir; submontanous and foot hills of Uttar Pradesh, and Jammu and Kashmir, central plains of Punjab, Uttar Pradesh, and Haryana, semiarid and arid parts of Punjab, Haryana, and

Fig.1.6. Perspective land use planning for citrus based on agroclimatic variation

Rajasthan (Fig. 1.6). Apart from above areas, kinnow mandarin can be successfully cultivated in hill region of Uttaranchal, comprising of areas, namely, Uttarkashi, Chamouli, Pauri-Garhwal, Tehri-Garhwal, Dehradun, Pithoragarh, Almora, and hill region of Nainital.

Adaptability of citrus in Central India

Citrus in central India is cultivated in two major states like Madhya Pradesh and Maharashtra.

Citrus in Madhya Pradesh

Nagpur mandarin belts in Madhya Pradesh are confined to four major agroclimatic zones viz., central Narmada valley zone, Satpura plateau zone, Malwa plateau zone, and Nimar valley zone (Sonwalkar, 1965). Soil types of Madhya Pradesh belong to Typic Ustochrept, Typic Haplustert, Lithic Ustorthent, Vertic Ustochrept, Chromic Haplustert, and Typic Ustorthent covering 16.1 per cent, 14.9 per cent, 13.0 per cent, 11.6 per cent, 5.6 per cent, and 5.2 per cent of total geographical area, respectively (Gajbhiye *et al.*, 2000).

Agroclimate : Central Narmada Valley Zone

It comprises of Hoshangabad, Seoni, Malwa, Sohapur, Narsingpur, and Budhni and Barelli tehsils of Sehore, and Raisen districts.

Climate : The annual rainfall of the region is 1000-1200 mm.

Soils : The soils are mainly deep black with clay ranging from 40.0 to 60.0 per cent and are deficient in N and Zn. Soil-landform relationship in basaltic terrain in north deccan plateau of Satpura range of Madhya Pradesh comprising areas viz., Khargone, Hoshangabad, Betul, Seoni, Jabalpur, Mandla, and Shahdol districts of Madhya Pradesh showed that depths of soils at hill and hill ranges are shallow (less than 50 cm) or moderately deep (50-75 cm). While, at plateau and plains/flood plains, the soils are deep (more than 100 cm). In hilly terrain, young soils are Lithic/Typic Ustorthents, Ustochrepts; Vertic Ustochrepts on plateaus, Typic Haplustalfs in plains, and Typic Haplusterts in valleys (Tamgadge *et al.*, 1996). These soils show a large variation in their properties, like pH (5.6-8.0), organic carbon (0.11-1.61 per cent), sand (5.0-57.0 per cent), silt (11.0 – 36.0 per cent), clay (28.0-65.0 per cent), CEC (22.6-58.4 cmol (p^+) kg^{-1}), and base saturation (70.0-99.0 per cent) according to Tamgadge *et al.* (1999).

Agroclimate : Satpura Plateau Zone

It comprises citrus growing areas of Chhindwara and Betul districts.

Climate : The region receives an average annual rainfall of 1100 mm (700-1400 mm). The rainfall is higher (1400 m) in the hill regions of Tamia, Harrai etc. and the major rainfall is received during monsoon season from June to September with a peak in July and August (Table 1.35). However, there is an unpredicatable variation in the distribution of rainfall during monsoon season, especially during winter, which affects the productivity adversely.

Soils : The soils of this zone vary from light reddish brown to black clay loam, with very low to high water retention capacity. About 45 per cent of soils are light reddish brown, mostly gravely, 41 per cent of them are yellow soils with silty loam texture, and about 14 per cent soils are black or clay loam (Saxena,1991c).

Agroclimate : Malwa Plateau Zone

It comprises Mandsaur, Rajgarh, Ujjain, Indore, Dewas, Dhar (Dhar, Badnawar, and Sardarpur tehsils), Shajapur, and Ratlam. The Malwa plateau zone lies between 23° 30′ and 70° 10′E. The altitude from mean sea level ranges from 450-675 m at Shajapur and Mandsaur.

Climate : The average rainfall varies from 800 to 1200 mm in the zone. More than 90 per cent of the rainfall comes through southwest monsoon from June to late August (Saxena, 1991c).

Soils : The area has mainly the medium black soils, which occupies the largest area. However, deep black and shallow soils are also common. In general, the soils are deep to very deep with 35 to 45 per cent montmorillonitic clay and are calcareous in nature. The soils are usually dark brown in colour, although, patches of light coloured and red soils are commonly met with. The Malwa table land is occasionally intercepted by intrusive baselbic dykes which function as water divide. The major soil series are Pancderia (Lithic Ustorthents), Kamlia Kheri (Typic Ustochrepts), Baloda and Sarol (Typic Haplusterts), and Antratia, and Malakhedia (Fluventic Ustochrepts). Under broad classification, these soils are grouped into Vertisols and associated soils. These soils occupy nearly level to very gently sloping landscape. The soils are mostly heavy with sandy clay loam to clay texture. According to land capability classification, the soils are mostly Class II and Class III with slight to moderate limitations of soils and topography, which need special soil management practice like land shaping, bunding etc. (Saxena, 1991c).

Table 1.35. Climatic features of Chhindwara region of Madhya Pradesh (1997-2000)

Months	Rainfall (mm)	Temperature (°C)		Relative humidity (%)
		Max.	Min.	
Jan.	17.5	23.3	12.4	69.9
Feb.	32.3	26.0	15.1	69.4
Mar.	13.2	31.5	18.7	54.3
Apr.	5.3	37.5	24.4	55.9
May	21.1	38.0	27.6	60.3
Jun.	138.8	32.6	24.9	73.05
Jul.	248.5	27.7	23.1	82.9
Aug.	100.6	27.9	23.0	83.7
Sept.	264.5	27.8	22.2	83.6
Oct.	127.8	27.6	21.3	76.0
Nov.	26.1	24.1	16.6	71.1
Dec.	-	22.8	11.6	70.3

Source : Zonal Agricultural Research Station, Jawaharlal Nehru Krishi Vishwavidyalaya, Chhindwara, Madhya Pradesh

Agroclimate : Nimar Valley Zone

It includes Nimar (east and west), Hoshangabad (Harda tehsil), Dhar (Manawar tehsil), and Khandwa.

Climate : The annual rainfall of the region is 1000-1200 mm. Perhaps, shallow depth of the soil accelerated the overland flow of water from undulating lands and early withdrawal of monsoon, exposes the crop to temporary water deficit stress.

Soils : The soils are mainly deep black, with clay ranging from 40.0 to 60.0 per cent, and are deficient in N and Zn. The various forms of K in soil exist in equilibrium with one another and depletion of one form is replenished by other form as usual (Chandel *et al.,* 1976). Soil samples from 37 profiles covering seven established series of Madhya Pradesh viz., Umarthana I, Umarthana II, Umarthana III, Koniadong, Doriapura, Peeplyamoti I, and Peeplyamoti II have been investigated for the distribution of different forms of K. Positive and significant relationship between exchangeable, 1 N HNO_3 soluble, and total K indicated the existence of dynamic equilibrium between these forms of K. Clay and silt plus clay showed significant positive correlations with exchangeable, nonexchangeable (1N HNO_3 soluble), and total K. Higher amount of 1 N HNO_3 soluble K in these black soils is attributed to higher silt plus clay content (Chaudhary and Pareek, 1976). The approach of fertilizer use for targetted yield taking into account the crop needs and nutrients present in soil has been found quite effective in these soils (Velayutham *et al.,* 1985).

Citrus in Maharashtra

Citrus in Maharashtra is confined to five major agroclimatic zones viz., western Maharashtra plain zone, western Maharashtra scarcity zone, central Maharashtra plateau zone, central Vidarbha zone, and eastern Vidarbha zone. Dominant soil types commonly observed according to Gajbhiyae *et al.* (2000) are: Lithic Ustorthent (32.7 per cent), Typic Haplustert (26.3 per cent), Typic Ustochrept (17.0 per cent), and Vertic Ustropept (11.0 per cent).

Agroclimate : Western Maharashtra Plain Zone

The western Maharashtra plain zone is a wider strip running parallel to eastern side of submontane zone of Maharashtra and extends towards east up to the boundary of Maharashtra dry zone. This zone includes areas in western mosambi (sweet orange) growing tehsils of Dhule, Ahmednagar, Sangli, and Satara.

Climate : The western Maharashtra plain zone in general, has well distributed rainfall ranging between 700 to 1250 mm annually. The maximum rainfall is received from southwest monsoon during June to October. The rainfall pattern of the zone is characterised by unimodal distribution. The period of water availability ranges from 120 to 150 days with some stress during September-October. The maximum and minimum temperatures of 40°C and 5°C are observed during the months of April-May and December, respectively (Atar Singh, 1991d). Climatic features at Rahuri showed the variation in maximum and minimum temperature from 29.2° to 37.9°C and from 17.7° to 27.5°C with year round high humidity of 70.7 to 76.2 per cent (Table 1.36).

Soils : The topography, in general, is plain in the zone. The soils of this zone are predominantly greyish black with varying texture and depth ranging from 25 to 100 cm. The medium deep soils are dark brown in colour. The soils are clay loam in texture, with moderately alkaline pH ranging from 7.4 to 8.4. The lower layers of these soils have a murrum strata of varying colours. The clay percentage is around 40 per cent. These soils are medium in available N, P, and K, and well supplied with calcium carbonate. The deep black soils, however, pose a problem of waterlogging when excessively irrigated (Atar Singh, 1991d).

Table 1.36. Climatic features of citrus growing area of Rahuri, Maharashtra (1998-2000).

Month	Rainfall (mm)	Temperature (^oC)		Humidity (%)
		Max.	Min.	
Jan.	—	31.6	17.7	71.4
Feb.	—	32.9	26.5	71.9
Mar.	30.0	35.5	19.1	71.8
Apr.	25.1	37.9	26.9	71.6
May	126.6	37.4	27.0	73.9
Jun.	12.6	36.2	27.5	72.4
Jul.	96.9	34.4	26.1	72.0
Aug.	153.7	34.8	24.0	70.7
Sep.	45.9	33.4	24.0	71.7
Oct.	82.7	33.7	23.4	77.2
Nov.	273.6	32.3	22.5	76.2
Dec.	—	29.2	20.0	75.6

Source : Directorate of Research, Mahatma Phule Krishi Vidyapeeth, Rahuri, Maharashtra

Agroclimate : Western Maharashtra Scarcity Zone

The region includes mosambi growing areas of Ahmednagar, parts of Satara, Sangli, Pune, Kolhapur, Dhule, Jalgaon, Nasik districts, and parts of Aurangabad and Jalna districts.

Climate : The drought prone area suffers from the twin problem of low productivity and high instability, as a result of inadequate and undependable rainfall. The annual precepitation is less than 750 mm on an average. Two peaks of rainfall are observed, first during June/July and second during September, resulting in bimodal pattern of rainfall distribution. Besides, the dry spells of varying durations extending from 2 to 10 weeks at a stretch are experienced between July and August. The water availability period worked out on the basis of precipitation and potential evapotranspiration values varies from 60 days to 140 days. Besides, the weather aberrations like delayed onset of monsoon or early cessation of monsoon are frequently observed resulting in shortening the water availability period. This adversely affects the flowering during *Mrig* season and consequently the fruit set. Thus, it becomes difficult to obtain even a single crop in a year, unless there is ample water available to supplement irrigation. Because of the high temperature and high wind velocity, the potential evapotranspiration value is more than 1800 mm, resulting in a deficit of significant quantum of irrigation water. In the years of low rainfall, the climate, therefore, further crosses over to arid type (deficiency more than 65.7 per cent). Maximum temperature on weekly average basis exceeds 41°C during late April and early May. The minimum temperature is noticed between 14°C and 15°C during December (Atar Singh, 1991d).

Soils : The general topography of the scarity zone is rolling with slopes between 1 and 2 per cent. The infiltration rate is about 6 to 7 mm hr^{-1}, whereas the rainfall intensity varies from 60 to 70 mm hr^{-1} in the month of September. The soils are Vertisol with base saturated and have free calcium carbonate content varying from 4.0 to 20.0 per cent. They have montmorillonite clay, which swells when wet and shrinks on drying producing deep cracks. Soils are poor in N (less than 0.5 per cent N), low to medium in P (8 to 30 kg available P_2O_5 ha^{-1}), and well supplied with K (more than 350 kg available

K_2O ha^{-1}). Vertisols have high values of moisture content at 33 Kpa and 1500 Kpa (48 and 25 per cent moisture, respectively). The zone suffers from very low rainfall (less than 750 mm in about 45 days) with uncertainity and ill distribution with low productivity (Atar Singh, 1991d).

Agroclimate : Central Maharashtra Plateau Zone

The zone comprises the parts of Aurangabad, Jalna, Beed, Osmanabad districts, major parts of Parbhani, and Nanded districts excluding southern parts of Washim and Mangrulpir talukas, whole of Amravati (Morshi, Warud, Dariapur, Chandur, Paratwada, Nangaon etc.), eastern parts of Chandrapur, Darwha and Pusad talukas of Yavatmal district, whole of Jalgaon, Shirpur taluka of Dhule, south Solapur, and Akkalkot taluka of Solapur district.

Climate : Specific features of the zone consist of 700-900 mm annual precipitation, 41°C maximum temperature, and 21°C minimum temperature. Most of the rainfall in this region is received during monsoon (June-September) with cold dry postmonsoon (October-February) and hot dry premonsoon (March-November) periods. In general, the evaporative demands and thermal properties of atmosphere are higher in premonsoon, intermediate in monsoon, and lower in postmonsoon periods. In monsoon (June-September), more than 75 per cent of the total annual rainfall is received in all the districts of assured rainfall zone and rest of the rainfall is received in pre- and postmonsoon periods (Atar Singh, 1991d). The variation of maximum and minimum temperature at Achalpur, Amravati is observed from 26.2° to 40.0°C and from 12.4° to 26.1°C, respectively, with total rainfall of 1035.8 mm (Table 1.37). Analysis of 12 years of meteorological data in sweet orange, cultivar mosambi growing areas of Parbhani revealed variation in maximum temperature from 28.2 to 40.7°C and minimum from 10.0 to 24.7°C, respectively, with maximum relative humidity of 45.1 to 80.2 per cent and minimum relative humidity of 18.0 to 62.0 per cent. The average total annual rainfall is 937.7 mm (Table 1.38).

Table 1.37. Climatological features of Nagpur mandarin growing area of Achalpur, Amravati (1997-2000)

Month	Temperature (°C)		Rain-fall (mm)
	Max.	Min.	
Jan.	28.3	13.6	15.2
Feb.	33.5	13.2	23.0
Mar.	37.2	18.7	10.8
Apr.	38.8	21.4	9.3
May	40.0	25.9	24.8
June	28.3	26.1	125.0
July	26.2	22.6	205.0
Aug.	33.0	21.9	240.9
Sept.	30.2	20.4	27.2
Oct.	28.4	18.1	75.6
Nov.	27.8	14.2	46.4
Dec.	26.2	12.4	32.6

Source : Agricultural Research Station, Achalpur, Dr Panjabrao Deshmukh Krishi Vidyapeeth, Akola, Maharashtra

Soils : The topography is rolling in few districts viz., Aurangabad, Buldhana, Beed, Osmanabad, Jalna, and mostly plain in other districts. The pH ranges from 7.0 to 7.5. Soils are derived from deccan trap and range from black to red colours. The major soils types are Vertisols, Entisols, and Inceptisols. Yelvikar *et al.* (1996) reported that 77 per cent vertic soils in Beed district of Maharashtra are deficient in exchangeable Fe. However, values of exchangeable Fe are comparatively higher in Typic Hablusterts due to high clay (53.7-68.4 per cent), low $CaCO_3$ (23.0-132.0 g kg^{-1}), and pH (8.1-8.3) compared to other soil classes such as Vertic Ustochrepts, Typic Ustorthents or Lithic Ustorthents having pH 7.6-8.9, clay 17.1-52.6 per cent, and $CaCO_3$ 60.0-172.5 g kg^{-1}.

Table 1.38. Meteorological characteristics of mosambi growing areas of Parbhani, Marathawada region of Maharashtra (Average of 12 years, 1989-2000)

Months	Temperature (°C)		Relative humidity (%)		Rain-fall (mm)
	Max.	Min.	Max.	Min.	
Jan.	29.7	11.5	74.1	30.0	7.8
Feb.	32.3	12.7	65.2	28.0	10.9
Mar.	36.4	17.4	51.4	20.0	12.2
Apr.	39.4	20.2	47.0	18.0	8.0
May	40.7	24.2	45.1	22.0	25.8
June	34.9	24.7	75.2	42.0	176.0
July	31.6	21.5	80.2	48.0	217.0
Aug.	30.6	21.8	83.0	62.0	223.5
Sept.	31.8	21.7	85.0	56.0	144.0
Oct.	32.7	19.3	78.0	42.0	84.0
Nov.	28.5	14.1	79.1	38.0	14.8
Dec.	28.2	10.0	75.1	31.0	16.7

Source : Agrometeorological observatory, Marathawada Agricultural University, Parbhani, Maharashtra

Agroclimate : Central Vidarbha Zone

Central Vidarbha zone includes entire Wardha district, major parts of Nagpur, Yavatmal, Aurangabad,Parbhani, Nanded, and Jalna. The citrus growing areas in the region are given below (Table 1.39).

The basaltic hill ranges in the northern part of Amravati and Nagpur with an elevation ranging between 600 to 1000 m above mean sea level comprising foothills and steep scarp slopes of Satpuras give rise to many small rivulets. The landform is highly dissected by streams due to high gradients accelerating the runoffs. The western part comprising the basaltic plateau in between 300 and 600 m, is characterised by scarp slope transversed by subparallel drainage lines. The plateau runs northwest to southeast, and forms a water divide between Tapti and Godavari

Table 1.39. Prominent citrus belts of central Vidarbha zone of Maharashtra

District	Locations
Wardha	Wardha, Dewli, Selu, Arvi, Karanja, Hinganghat, Samudrapur
Yavatmal	Yavatmal, Darwa, Pusad, Umerkhed, Digras, Bhindi
Nagpur	Nagpur, Hingna, Katol, Narked, Saoner, Kalmeshwar, Ramtek, Parseoni, Umred.
Aurangabad	Pachod, Karmad, Adul, Dabarwadi, Chittepimplegaon, Pimleraja, Danitabad, Paithan
Jalna	Ambad, Wadigodari, Tirthpuri, Gavrayee, Bhadrapur, Chickengaon, Bazarvaigaon, Nagzari, Wadpimpalgaon, Madpimpalgaon, Parner, Kawadgaon, Golapangri, Bharatkheda
Parbhani	Asola, Zhari, (Parbhani proper), Tarborgaon (Manvat), Satva (Dhinglepimplegaon) Churhawa, Churawa, Yarandeshwar, Pimplebhatta, Alegaon (Purna), Pethpimplegaon (Palam)
Nanded	Mudkhed, Bhokar, Himayatnagar, Limbgaon, Dhanora, Ardhapur, Vishnupuri, Jam, Bori, Takli, Maralak, Pokharni, Sayal, Naleshwar, Malegaon, Kondha, Selgaon, Pedgaon
Hingoli	Purna, Ambachundi, Proper Basmat

drainage systems. It exihibits an abrupt fall into Wainganga basin in the east. The tract is formed of horizontal sheets of lava with characteristic spheroidal weathering, marked with round boulders. In most of the places, the plateau surfaces are flanked by mesas and buttes giving a rolling appearance. The eroded pediment occupying the toe slope of scarp faces, ranging in elevation between 300 m and 450 m above mean sea level are strewn with boulders and rolled down from scarp faces. Clay mineralogy of some Vertisol soil series of Maharashtra showed smectite, which constituted the major phyllosilicate in fine clays (>90 per cent) with small proportion (< 10 per cent) of kaolinite. Coarse clay mineral is composed of smectite, kaolinite, quartz, illite, feldspar, chlorite, and vermicullite (Pharande and Sonar, 1997).

The region forms a part of the peninsular India, and presents varied rock formations from the oldest precambrian to recent alluvium. The formations exposed in order of antiquity are older precambrian, cuddapah, vindhyan, the gondwana, lameta group, deccan trap, and recent deposits. The granite, hornblende, and biotite gneiss with basic intrusives form the oldest rocks. Over these ancient rocks, lie a few basins of later sediments covered by extensive sheets of lava flow comprising the deccan trap formation. The cuddapah system includes crystalline limestone, occupying a small area in the southern part of the Sironcha tehsil of Gadchiroli district, while the limestone, dolomitic limestone, purple shale, and sandstone of vindhyan system occupies a vast area in the district of Yavatmal and Chandrapur. Subsequent movements in the crust have given rise to faults, in which sediments of gondwana age comprising sandstone and shale have been deposited giving rise to the important Pench Kanhan valley and the Wardha valley (Hirekerur, 1983).

Climate : The climate is tropical monsoon type. The annual rainfall varies from 750 to 1600 mm, distributed over 60 and 70 days. From middle of June to end of August is the period of heavy rainfall. September rains are occasional. Few showers associated with cyclonic storms are also received during January and February. The mean annual temperature varies from 25° to 27° C. High temperatures of 45° C or more are witnessed during May, while low temperatures of 8° to 10°C are recorded in the months of December and January. The isohytal map of the region further reveals that the rainfall in the area ranges between 750 mm and 1600 mm. Low rainfall zone with rainfall between 750 mm and 900 mm located in the western and south western parts constitutes the semiarid part. While, the central part with rainfall between 1000 and 1200 mm is dry subhumid. The eastern, south eastern, and the northern tip of Amravati district with high rainfall between 1200 mm and 1600 mm forms the subhumid to humid part (Hirekerur, 1983).

The average annual total rainfall is 1130 mm which is received in 59 rainy days. Out of this, 1000 mm is received in 48 days during rainy season. The month July is the wettest month receiving average of 323 mm of rain in 16 days. The mean maximum temperatures are 33°C in winter season and 38°C in summer season. The mean minimum temperatures are 16°C and 26°C for winter, and summer season, respectively. The average daily relative humidity is 72 per cent for rainy season, 53 per cent for winter, and 35 per cent for summer season. The respective monthly average evaporation rates are 156 mm, 115 mm, and 420 mm for the rainy, winter, and summer season, respectively (Atar Singh, 1991d). Agrometeorological features during 1996-2000 at Nagpur indicate an average rainfall of 778.8-1125.6 mm, minimum and maximum temperature as 13.9°-27.4° C and 30.4°-42.3° C with a relative humidity of 55.1-92.8 per cent (Table 1.40). The striking feature of Nagpur mandarin growing areas under subhumid tropical climate is the occurrence of year round high diurnal variation in temperature during winter compared to either summer or rainy season with mean value of 14.9°C, 17.1°C, and 10.5°C during summer, winter, and rainy season, respectively.

Table 1.40. Agrometeorological features at Nagpur, Maharashtra (India)

Season	Tempera-ture (°C) Max.	Min.	Diurnal variation (%)	Humi-dity (%)	Rain-fall (mm)
1996-97					
Summer	42.3	27.4	14.9	64.1	324.0
Winter	32.0	14.2	17.8	75.2	203.0
Rainy	32.7	23.0	9.7	90.4	598.6
1997-98					
Summer	40.2	26.0	14.2	59.8	110.9
Winter	30.7	15.2	15.5	74.6	209.8
Rainy	32.8	22.4	10.4	89.3	585.0
1998-99					
Summer	41.1	25.7	15.4	64.1	236.7
Winter	30.4	15.4	17.0	73.1	236.7
Rainy	34.3	22.7	11.6	91.9	520.5
1999-2000					
Summer	41.8	27.2	14.6	55.1	244.8
Winter	31.8	14.8	17.0	78.5	23.8
Rainy	32.9	22.7	10.2	92.8	807.3
2000-01					
Summer	42.3	26.8	15.5	65.6	132.8
Winter	32.3	13.9	18.4	72.5	117.1
Rainy	33.8	23.1	10.7	91.5	528.9

Source : Agrometeorological Observatory, NRC for Citrus, Nagpur, Maharashtra, India

Soils : Soils of the zone are derived from basalt rocks, black in colour and have varying depths depending upon their physiography. These soils are medium to heavy in texture, fairly high in clay content, alkaline in reaction, high lime reserves with high base saturation of the exchange complex. The soils occupying low lying areas are rich and fertile, but are prone to deterioration, if not managed properly under irrigation. Vertisols are predominant in the zone having montmorillonitic mineralogy. Swelling after wetting and shrinkage following drying yield deep and wide cracks, is a common feature. Patil and Sonar (1994) reported variation in DTPA extractable Fe, Mn, Zn, and Cu in these soils from 6.1 to 26.0 mg kg^{-1}, 14.6 to 63.0 mg kg^{-1}, 0.58 to 1.70 mg kg^{-1}, and 1.7 to 6.1 mg kg^{-1}, respectively, based on investigation of 20 representative soil series of Maharashtra. Except Zn, none of the micronutrients was found deficient.

Various properties in shrink-swell soils at Talegaon-Karanja plateau of Wardha district showed variation in sand from 40.0 to 53.0 per cent in Typic Ustorthents, 13.0 to 21.0 per cent in Vertic Ustochrepts, and from 24.0 to 28.0 per cent in Typic Haplusterts. The silt to clay ratio in soils varied from 0.33 in cambic horizon of Typic Haplusterts to

0.84 in Bss horizon of Chromic Haplusterts, indicating the uniformity in distribution of silt and clay with a variation of 26.7 per cent and range of 0.59± 0.16 with depth. The bulk density had negative relationship with porosity (r = - 0.79) indicating the increasing trends of bulk density with depth. The soil water held at –33 Kpa and –1.5 Mpa showed increasing trend with depth in Typic Haplusterts and Chromic Haplusterts on foot slope position. The plant available water correlated positively with silt (r = 0.77) and clay (r = 0.78) contents. The soils are slightly alkaline with pH values varying from 7.6 to 8.1 with depth and low salinity levels. The soils are low to medium in organic carbon content varying from 0.8 g kg^{-1} in Cr horizon of Typic Ustorthents to 10 g kg^{-1} in Ap horizon of Chromic Haplusterts. The calcium carbonate content increased with depth and reached to a maximum of 211 g kg^{-1} in Bss2 horizon of Chromic Haplusterts. Calcium and magnesium are the dominant cations at exchangeable site ranging from 31.6 to 62.6 cmol (p^+) kg^{-1}. The CEC varied from 34.0 to 66.6 cmol (p^+) kg^{-1} showing increasing trends in Typic Haplusterts to a depth of 88 cm on footslopes and in Chromic Haplusterts to a depth of 98 cm on middle slopes. The CEC to clay ratio in soils of toposequence had a mean of 1.0 with 21.4 per cent coefficient of variation (Anantwar *et al.*, 2000).

Yadav *et al.* (1998) described the landform – soil type relationship at Bazargaon plateau area on diverse basaltic landforms of Nagpur district on weighted mean basis as : Typic Haplusterts (pH 8.0-8.2, EC 0.12-0.17 dSm^{-1}, organic C 3.0-6.0 g kg^{-1}, $CaCO_3$ 9.0-253.0 g kg^{-1}, CEC 38.3-46.7 cmol(p^+) kg^{-1}, and base saturation 83.7-93.6 per cent) on summit flank and main valley side slope; Vertic Ustochrepts (pH 8.0, EC 0.13-0.15 dSm^{-1}, organic C 3.0-6.0 g kg^{-1}, $CaCO_3$ 115.0-157.0 g kg^{-1}, CEC 48.6-51.9 cmol(p^+) kg^{-1}, and base saturation 91.1-95.6 per cent) on toe slope and main valley side slope; and Typic Ustorthents (pH 6.7-7.3, EC 0.07-0.10 dSm^{-1}, organic C 4.0-5.0 g kg^{-1}, $CaCO_3$ 8.0-12.0 g kg^{-1}, CEC 37.1-40.7 cmol(p^+) kg^{-1}, and base saturation 89.5-92.3 per cent) on shoulder slope and interfluve crest.

Agroclimate : Eastern Vidarbha Zone

The eastern Vidarbha zone consisting of subzone (western Vidarbha) is represented by Buldhana, Akola, Amravati, and six tehsils of Yavatmal district in addition to Umred tehsil of Nagpur.

Climate : The climate of the area is classified as subhumid tropical with isohyte limits of 70 cm on western side and 95 cm on eastern side. Mean normal precipitation of Buldana, Akola, Amravati, and Yavatmal are 80.3, 84.7, 85.4, and 92.4 cm, respectively, with an average rainfall of 845 mm having 47.8 mean number of rainy days. The July month is the wettest, with average precipitation of 239 mm. Mean monthly rainfall for the months of the June, August, and September are 151, 161, and 158 mm, respectively. For this subzone, monsoon season (June-September) accounts for 88.8 per cent of annual rainfall. Winter (October-January) and summer season (February-May) take a share of 10.5 and 5.7 per cent, respectively. The mean maximum temperature is 33°C in summer season. The mean minimum temperatures are 24°C, 16 °C, and 20.6 °C, respectively during summer, winter, and rainy season, respectively. The average daily humidity of 70 per cent for rainy season, 52 per cent for winter, and 34 per cent for summer season are prevalent. The respective monthly average evaporation rates are 162 mm, 120 mm, and 425 mm for winter, rainy, and summer season, respectively. Other characteristic features of monthwise climate at Akola district of Maharashtra are described (Table 1.41).

Table 1.41. Meteorological characteristic of Akola, Maharashtra

Month	Temperature (°C)		Relative humidity (%)		Rainfall (mm)
	Max.	Min.	Max.	Min.	
Jan.	28.7	9.0	63.00	25.00	—
Feb.	32.2	15.1	69.00	31.00	46.2
Mar.	36.6	17.2	43.00	14.00	—
Apr.	41.4	24.9	37.00	17.00	—
May	43.7	28.7	39.00	16.00	—
June	38.5	27.1	65.00	37.00	95.9
July	32.8	24.2	83.00	60.00	164.4
Aug.	31.3	23.8	86.00	68.00	10.2
Sept.	30.7	23.1	89.00	72.00	225.2
Oct.	32.2	20.8	85.00	56.00	80.3
Nov.	29.8	15.4	82.00	44.00	67.8
Dec.	28.8	8.4	70.00	24.00	—

Source : Agrometeorological observatory, Dr. Punjabrao Krishi Vidyapeeth, Akola, Maharashtra, India

Soils : Most of the soils are calcareous, thoroughly base saturated, and fairly well supplied with K, moderate in P, but low in organic matter and N. The pH of soils is alkaline and varies from 7.5 to 8.0. Soils are deeper than 150 cm, clayey in texture, and pose problem of temporary waterlogging. Vertisols predominate in the zone with montmorillonite clay. Deep black soils are poor in infiltration and permeability. Irrigation management on these soils poses several problems. Entisols and Inceptisols are found on rolling topography with relatively low clay content. The per cent of shallow, moderate, and deep soils in the subzone are roughly around 25, 50, and 25, respectively (Atar Singh, 1991d). These Vertisols have developed from a variety of parent materials, namely, grey shale, basalt, granite-granodiorite, granodiorite, and limestone having clay texture all throughout with weakly developed horizon in most of the cases (Table 1.42). Basalt derived Vertisols possess better available and total reserves of Fe, Mn, Cu, and Zn (Table 1.43). Aubert and Pinta (1977) concluded that basalt eruptive rocks such as basalt and gabro usually contained more micronutrient cations than the acid eruptive rocks, metamorphic rocks or sedimentary rocks. Nagpur growing soils of western Vidarbha have shown a wide variation in depth, texture, available water capacity, $CaCO_3$ and exchangeable bases with base saturation varying between 84.3 and 99.7 per cent.

Commercial Cultivation of Nagpur Mandarin

Nagpur mandarin on historical background is claimed to be introduced in 1894 by late Shri. Raghujiraje Bhonsle in central India. The rapid cultivation of Nagpur mandarin, has since then taken place in this region. And world famous cultivar Nagpur mandarin

Table 1.42. Characteristics of Vertisols developed from different parent materials

Horizon	Depth (cm)	Sand	Silt	Clay	pH	EC (dS m^{-1})	CEC [cmol (p$^+$)kg^{-1}]	Org. C (g kg^{-1})	Free CaCO$_3$ (g kg^{-1})
		—— (%) ——							
				Grey shale					
Ap	0-014	13.0	36.6	50.4	7.2	0.10	42.0	7.6	65
A	14-40	11.1	36.2	52.7	7.1	0.25	37.0	4.4	85
AC	40-70	10.0	35.7	54.3	7.1	0.10	38.2	4.3	90
C	70-100$^+$	10.0	35.2	54.8	7.1	0.25	38.2	4.3	93
				Basalt					
Ap	0-14	2.3	29.4	68.3	7.2	0.25	49.5	6.7	155
A	14-30	1.8	27.7	70.5	7.1	0.10	48.2	5.0	130
AB	30-46	1.5	24.4	74.1	7.1	0.10	46.4	3.9	130
AC	46-60	2.1	25.5	72.4	7.1	0.15	48.2	3.8	135
C	60-100$^+$	1.3	25.5	73.2	7.2	0.25	46.3	2.4	150
				Granitic-granodiorite					
Ap	0-10	47.5	7.8	44.7	7.9	1.20	24.4	8.8	100
A	10-40	38.2	13.4	48.4	8.4	1.20	29.4	3.9	115
AC	40-60	36.6	7.0	56.4	8.8	1.10	53.9	1.7	460
C	60-80$^+$	56.8	12.1	31.1	8.8	1.20	32.0	1.2	320
				Granodiorite					
Ap	0-14	58.4	10.8	30.8	8.1	0.40	28.1	11.9	185
A	14-36	33.1	13.0	53.9	8.0	0.75	29.0	8.1	280
AC	36-50	28.5	13.1	58.4	8.0	0.90	30.3	7.4	265
C	50-100$^+$	29.8	8.5	61.7	8.0	0.70	29.4	5.8	195
				Limestone					
Ap	0-10	9.6	16.2	74.2	7.6	0.20	42.0	5.5	195
A	10-28	7.4	16.7	75.9	7.5	0.35	53.9	4.6	190
AC	28-62	6.4	16.4	77.2	7.4	0.25	49.5	4.4	170
C	62-102$^+$	4.7	16.1	76.2	7.2	0.30	54.5	4.1	165

*c- clay, cl- clay loam
Source : Murthy *et al.* (1997)

(Fig. 1.7) has adapted well to soil and climate of central India. The best quality of Nagpur mandarin is produced in Vidarbha region (Nagpur, Amravati, Wardha, Akola, Buldhana, and Yavatmal districts) and adjoining areas of Saunsar and Pandurna (Chhindwara district) of Madhya Pradesh. Anand and Leisram (1963) observed that Nagpur

Fig. 1.7. A typical Nagpur mandarin orchard in a black clay soil type

Table 1.43. Distribution of available and total micronutrient in soils (mg kg^{-1}) developed from different parent materials

Horizon	DTPA-extractable				Total			
	Fe	Mn	Cu	Zn	Fe	Mn	Cu	Zn
				Grey shale				
Ap	0.55	20.8	73	1.85	27.5	33000	750	33.7
A	0.37	15.8	64	1.85	32.5	38000	625	36.2
Ac	0.25	14.8	55	1.91	35.0	36750	625	36.2
C	0.25	12.5	45	1.72	34.0	38600	750	36.2
				Basalt				
Ap	0.68	14.3	157	2.38	42.5	40500	1625	47.5
A	0.58	12.0	115	2.05	37.5	38750	2250	38.7
AB	0.28	13.0	90	1.32	38.0	42500	1375	45.0
Ac	0.25	12.3	73	1.25	42.0	41500	1125	45.0
C	0.25	12.8	60	1.19	45.0	40000	1125	47.5
				Granitic-granodiorite				
Ap	1.05	36.8	150	2.24	17.5	31000	3750	18.7
A	0.38	17.5	145	2.57	15.0	32500	625	18.7
Ac	0.18	9.8	25	1.06	12.8	33750	500	15.0
C	0.18	9.3	13	0.40	10.0	34500	250	15.0
				Granodiorite				
Ap	0.64	24.0	55	5.68	175.5	26500	345	21.2
A	0.55	8.5	17	5.68	30.0	35800	500	24.3
Ac	0.53	8.5	13	5.68	30.0	38750	500	27.5
C	0.38	10.3	17	4.36	20.0	41000	750	36.2
				Limestone				
Ap	0.28	12.0	95	1.39	20.0	30250	1125	27.5
A	0.18	12.0	85	1.45	20.0	32000	1375	33.7
Ac	0.08	11.7	73	1.32	21.5	33000	1125	33.7
C	0.30	11.8	64	1.32	22.5	33500	1125	38.7

Source : Murthy *et al.* (1997)

mandarin has qualitywise good adaptability under Delhi conditions. Jawanda (1969) earlier described the Nagpur mandarin industry including climate and soil. Evaluation of Nagpur mandarin under Kumaon and Garhwal hill, showed poor yield, but excellent quality (Singh and Misra, 1981). Soil and climate of western part of West Bengal covering an area of 1.5 million ha having annual rainfall of 1100-1600 m has also been suggested favourable for growing Nagpur mandarin according to Ghosh and Chattopadhyay (1993).

Climate : The climate of major belts of Nagpur mandarin is basically subhumid tropic. Such climates are characterised by hot and dry premonsoon summer months (March to May), followed by well spread summer monsoon months (June to September). The subsequent short period of October and November receives uncertain and infrequent showers, followed by a fairly dry mild winter (December to February). The mean annual temperature ranges from 24° to 27 °C. The mean summer (April, May, and June) and mean winter (December, January, and February) temperatures vary from 30° to 33°C and from 15° to 22°C, respectively. The mean annual rainfall ranges from 800 to 1500 mm of which 80-90 per cent is received during monsoon months. It represents 40-77 per cent of mean annual potential evapotranspiration. The soil moisture control section remains dry either completely or in parts, for two to eight months in a year, suggesting an ustic moisture regime (Sehgal and Bhattacharjee, 1988). Nagpur mandarin is also cultivated in Satpura plateau zone with an average rainfall of 1100 mm (700-1400 mm) mainly in black soils.

Soils : Nagpur mandarin growing soils vary widely in their surface and subsurface soil properties including fertility status. Naude (1954; 1955) reported that even soils impoverished of nutrients could be made to give high yields of Nagpur mandarin with balanced fertilization and supply of optimum moisture and aeration. Kakde (1956) observed that performance of orange orchards is directly related to physicochemical composition of soils. Bandale *et al.* (1951) observed the content of total Zn in range of 7.9 to 17.6 mg kg^{-1}. However, Jadhav *et al.* (1978) observed values of total Zn between 96 mg kg^{-1} and 164 mg kg^{-1} in citrus orchard soils of Marathwada. Malewar *et al.* (1978) observed DTPA-Zn varying from 0.60 to 1.40 mg kg^{-1} in citrus orchard soils. The soil Zn map of Maharashtra (Malewar, 1989) showed that no systematic survey on available Zn in major citrus growing districts like Nagpur, Amravati, and Wardha districts has been carried out.

The black soils of central India, though look alike, vary when examined through depth. Preliminary studies on physicochemical and fertility status of Nagpur mandarin growing soils indicated that deep Haplustert soils are more fertile compared to either Ustorthent or Ustochrept type of soils. But, the drainage problem is more pronounced in former type of soil (Srivastava *et al.*, 1993). Plessis *et al.* (1975) also observed the good response of citrus to applied K at the soil K level lower than 90-100 mg kg^{-1} exchangeable K. Relationship of leaf K and forms of soil K at different critical growth stages of Nagpur mandarin showed that maximum K availability in soil needs to be ensured before fruit set stage for an effective utilisation of soil applied K. The leaf K content showed significant relationship with exchangeable K and nonexchangeable K at fruit set stage. Nonexchangeable K and exchangeable K governed the uptake of K in leaf at fruit maturity and colour break stage, respectively (Srivastava *et al.*, 1997b). Comparatively high degree of correlation of available K with exchangeable K ($r = 0.842$) than water soluble K ($r=0.572$), showed that soil exchangeable phase played more significant role in regulating the availability of K in black montmorillonitic clay soils.

Soils derived from basalt possess higher distribution of available (DTPA extractable) Fe, Mn, Cu, and Zn compared to soils derived from grey-shale, granitic-granodiorite, granodiorite, and limestone (Table 1.43), which would be an added advantage for sustainable cultivation of Nagpur mandarin. Four major orange growing soils (Musarkhapa,

Karla, Panjra, and Linga series) were studied by Kalbande *et al.* (1983) to characterize the Nagpur mandarin *(Citrus reticulata* Blanco) growing soils. The performance of orange trees was poor in Musarkhapa and Karla soils owing to shallow rooting depth and less available water capacity than in the deep Panjra and Linga soils. Panjra soils had favourable physical condition in subsoil layers, owing to good aggregation and the presence of lime nodules and gravels, which promoted root aeration. The Nagpur mandarin trees, therefore, gave significantly more yield and grew taller in Pajra than the trees grown on Linga soils, which are poorly drained coupled with poor physical conditions. Panjra soils had high soluble calcium to meet the high Ca requirement of the crop (Table 1.44). A comparatively recent study by Jagdish Prasad (1996) suggested that the regular excess bearing in the shallow soils (Typic Ustorthent) exhausted the Nagpur mandarin trees more rapidly than those on the deep soils, and induced the decline in productivity, thereby, the reduced orchard life in such shallow depth soils.

Nagpur mandarin growing soils of this area have been observed to have high available Ca and Mg supply (Srivastava and Singh, 1997b; Kohli *et al.*, 1997), which is

Table 1.44. Physicochemical properties of the Nagpur mandarin growing soils

Depth (cm)	Partical size distribution (%)			$CaCO_3$ (%)	Moisture at 33 kpa (%)	Organic C (%)	pH (1:2)	EC (dSm^{-1})	CEC (me 100 g^{-1})	Pore space (%)	AWC* (mm m^{-1})
	Sand	Silt	Clay								
					Linga series						
0-12	17.4	23.7	58.3	0.6	41.8	0.5	8.1	0.3	46.3	31	4.2
12-28	25.8	18.1	56.2	0.2	41.9	0.5	8.2	0.5	42.3	30	5.7
28-64	22.4	19.4	58.6	0.2	41.6	0.4	8.2	0.3	45.4	30	12.7
64-109	23.4	17.4	59.3	0.5	48.6	0.5	8.1	0.3	44.1	29	18.5
109-137	22.0	28.5	48.5	2.2	40.4	0.3	803	0.4	39.5	23	10.4
					Panjra series						
0-8	24.0	30.2	45.4	10.2	37.4	0.5	8.0	0.4	28.4	36	2.3
8-28	22.6	31.5	45.9	10.8	37.0	0.4	7.1	0.4	29.6	35	5.8
28-66	21.6	30.6	48.7	10.5	38.6	0.4	8.0	0.4	31.1	35	11.5
66-98	24.3	32.0	44.1	9.7	35.7	0.4	8.2	0.5	30.9	40	8.6
98-137	24.6	31.5	40.9	11.5	30.4	0.4	8.3	0.5	25.1	40	8.8
137-150	32.1	38.6	27.5	11.2	22.3	0.2	8.3	0.5	19.9	37	2.6
					Musarkhapa series						
0-12	23.9	29.1	46.2	13.6	27.0	0.9	7.5	0.3	29.9	37	2.5
12-22	22.1	29.5	47.2	12.2	26.1	0.4	7.7	0.4	28.3	32	2.1
22-33	36.8	29.6	29.6	7.9	27.3	0.2	7.5	0.4	21.2	30	2.4
					Karla series						
0-15	20.1	24.2	55.8	Nil	44.1	0.4	7.9	0.4	39.1	37	5.1
15-34	24.1	22.2	53.8	11.2	44.8	0.3	8.0	0.3	35.0	30	6.8
34-50	29.6	22.1	48.3	19.9	32.1	0.3	7.9	0.3	31.9	29	4.9

* AWC stands for available water capacity

Source : Kalbande *et al.* (1983)

claimed to be the prerequisite for not only better fruit quality, but also for the longevity of mandarin orchards. The presence of calcium carbonate nodules of varying dimensions improves the drainage of highly clay soils, which otherwise is poorly drained (Dass *et al.*, 1998; Srivastava *et al.*, 1997; Srivastava and Singh, 2001a), and favourably help to improve the intensity of flowering and fruiting. The studies on soil and climate adaptability of citrus in India further showed that citrus belt accompanying red and black soils with growing period of 90-150 days was most productive and sustainable (Srivastava and Kohli, 1997a; 1997b; Srivastava and Kohli, 1999).

The suitability of Nagpur mandarin based on kind and degree of limitation studied in a variety of soils indicated that Vertisols with slight limitation of drainage, texture, soil reaction, and organic carbon are moderately suitable for Nagpur mandarin. However, soils belonging to Inceptisols are marginally suitable, because these soils suffer from very severe limitations of elevation and calcium carbonate, moderate limitations of soil depth, texture, coarse fragments, soil reaction, and organic carbon. The soils of Entisols were not suitable due to the severe and very severe limitations of elevation, soil depth, texture, and soil reaction which are difficult to be modified. The overall suitability of different soils comparing the yield, suggested that the moderately suitable soils provided the yield of 1075 to 1124 fruits plant^{-1}, the marginally suitable soils gave the yield of 511 to 982 fruits plant^{-1}; whereas unsuitable soils produced yield of about 394 fruits plant^{-1}. The high yield in Vertisols is attributed to good drainage, soil depth, proper duration, and timely alleviation of stress for flower initiation, while low yield in soils indicates poor drainage and improper stress resulted in low intensity of flowering. Quantitative land evaluation further indicated that soil depth, slope, texture, and calcium carbonates are predominant factors which governed the land index (Mohekar and Challa, 1999).

Patil *et al.* (1999a) reported that the Nagpur mandarin soils are characterised by slightly alkaline to moderately alkaline soil pH ranging from 7.5 to 8.1 in surface horizon which increased with depth. The cation exchange capacity values of the soil ranged from 40.88 to 53.17 cmol (p^{+}) kg^{-1} and soils are almost base saturated. In general, Ca dominated the exchange complex of all the profiles, followed by Mg, Na, and K. The free calcium carbonate ranged from 5.6 to 17.8 per cent. The organic carbon content in surface layer ranged from 0.24 to 0.870 per cent and decreased with depth (Table 1.45). Based on the morphological, physical, and chemical characteristics, orchard soils have been classified as Typic Haplusterts, Vertic Ustochrepts, Typic Ustochrept, and Lithic Ustorthents (Patil *et al.,* 1999a). Patil *et al.* (1999b) in another study observed soil type, Lithic Ustorthent as unsuitable for Nagpur mandarin cultivation due to very severe limitation of soil depth.

Soil-site suitability determined quantitatively (statistical approach) suggested that soil depth, and Plant Available Water Capacity (PAWC) are the most important characteristics contributing towards the success of Nagpur mandarin cultivation. Marathe

Table 1.45. Soil physicochemical characteristics of some Nagpur mandarin orchards from western Vidarbha

Horizon	Depth (cm)	Sand	Silt	Clay	$CaCO_3$	Exchangeable bases				BS (%)
		(%)				Ca	Mg	Na	K	
						[cmol (p+) kg^{-1}]				
Ap	0-21	28.14	18.12	53.44	5.56	36.08	5.52	0.49	1.22	91.05
Bw	21-50	29.58	16.28	54.02	6.09	34.36	6.89	0.47	0.49	87.81
Ck1	50-86	34.86	25.54	39.54	13.02	26.18	4.80	0.41	0.31	89.94
Ck2	86-122	44.38	23.26	32.16	13.34	21.30	5.52	0.35	0.25	95.67
Ck3	122-160	49.38	20.25	30.16	21.32	16.24	5.77	0.35	0.35	84.26
Ap	0-21	30.66	22.38	46.46	5.99	26.78	10.87	0.45	1.41	95.55
Bw	21-56	32.02	19.36	48.32	6.62	25.64	13.16	0.45	0.75	93.00
Ck	56+									
Ap	0-22	37.54	11.46	50.72	5.46	31.46	10.39	0.65	2.04	98.34
Ck	22+									
Ap	0-22	28.12	15.26	56.26	11.20	39.26	10.19	0.54	1.21	93.25
Bw	22-64	29.2	12.32	58.35	11.80	35.08	12.92	0.51	0.84	95.12
Bss1	64-99	27.26	12.12	60.42	14.86	38.46	13.24	0.53	0.62	98.40
Ck	99-+									
Ap	0-22	26.62	18.3	55.02	8.57	35.4	3.55	0.66	0.88	97.77
Bw	22-44	30.26	16.28	53.26	10.71	35.86	10.69	0.52	0.79	93.47
Ck	44-75									
Ap	0-23	24.22	22.38	53.23	8.28	34.64	11.36	0.28	0.59	99.46
Bw	23-47	23.28	21.32	55.12	9.66	32.86	13.84	0.28	0.39	97.09
Bss1	47-79	23.50	20.28	56.02	8.86	30.63	13.84	0.28	0.39	91.00
Bss2	79-117	22.22	19.38	58.16	9.54	33.48	12.32	0.26	0.45	90.37
Bss3	117-140	21.34	19.26	59.08	10.69	32.9	14.80	0.26	0.45	92.61
Ap	0-23	15.20	22.10	62.70	17.59	41.46	9.94	1.74	0.75	94.51
Bss1	23-63	15.08	20.46	64.46	13.57	42.86	10.19	1.78	0.65	99.66
Bss2	63-97	13.92	20.43	65.67	15.53	40.02	11.98	1.82	0.69	91.17
Bss3	97-115	14.78	18.98	66.24	16.32	44.26	10.19	2.04	0.67	94.56

AWC and BS stand for available water content and base saturation, respectively

Source : Patil *et al.* (1999a)

et al. (1999) based on texture, depth, drainage properties, and performance of the Nagpur mandarin orchards showed that various soil types can be grouped under 5 categories: i. shallow loamy soils (S1), 1 to 1.5 feet deep underlain by weathered by rock, ii. deep loamy soils (S2) with more than 7 feet deep, iii. shallow black soils (S3), 1 to 1.5 feet deep underlain by weathered rock, iv. deep black soil (S4) up to 6 feet deep underlain by weathered rock, and v. very deep black soils (S5) with more than 6 feet depth. The very deep black soils with depth more than 6 feet is most unsuitable for

Nagpur mandarin cultivation due to poor drainage. Response of 8 soil series belonging to Vertisols, Inceptisols, and Entisols on micronutrient availablity and yield potential of Nagpur mandarin showed that orchards growing on alluvial soils (Wadhona series) produced maximum fruits followed by shallow black soils (Musarkhapa series). The other extremely deep black soils (Karla and Linga series) and shallow loamy soils (Yenva and Pardi series) had DTPA extractable Zn more than 1.0 mg kg^{-1} (Yadav *et al.*, 1997).

There is not much information available regarding nutritional status of Nagpur mandarin orchards of central India. Kharkar *et al.* (1991) found deficient to low range of P, optimum to high range of Mn as well as Fe and excess of Cu in mandarin orchards of Vidarbha region. Detailed survey of 17,799 ha area of central India (Kohli *et al.,* 1997; Srivastava *et al.,* 1997; Srivastava and Singh, 1997b) showed deficient to low level of N, P and Zn in 78.8 per cent, 38.5 per cent, and 41.0 per cent orchards, respectively. Malewar (1989) reviewed the zinc research using soil and plant analysis in Maharashtra and showed that soils derived from basaltic parent material had higher Zn reserve compared to granite-gneiss derived soils. These observations indicated that due to high cation exchange capacity of clay soils having dominantly montmorillonitic mineralogy are nutritionally less problematic. The studies on zinc content of soils collected from different agroclimatic zones of Maharashtra showed that soils collected from assured rainfall are deficient in available Zn (Mahapatra and Kibe, 1972). Malewar and Randhawa (1977) registered relatively higher potential reserve of Zn in calcareous than noncalcareous soils. The soils in western part of West Bengal where the Nagpur mandarin cultivation has shown good promise are red and laterite type of soil having pH between 5.0 and 6.5 (Ghosh and Chattopadhyay, 1993). Assessment of micronutrient status of Nagpur mandarin orchards in Warud tehsil of Amravati district revealed the variation in DTPA-Zn from 1.05 to 4.80 mg kg^{-1}, Mn 3.10 to 13.20 mg kg^{-1}, and Cu from 0.75 to 1.85 mg kg^{-1}, showing adequacy of all the micronutrients except Zn (Patil and Malewar, 1998).

Potential Areas for Nagpur mandarin

Nagpur mandarin has shown a great promise in subtropical zone of Himachal Pradesh, but of late, plantations developed serious decline, perhaps due to physiological disorders (Azad and Chauhan, 1989). One of the reasons for such failure is that during the maturity period of Nagpur mandarin, low temperature might have been the main bottleneck contrary to the kinnow mandarin which developed more deep orange colour during the maturity period (mid-December to mid-January) under the conditions of Himachal Pradesh, Punjab, and Haryana. A good possibility exists for expanding the Nagpur mandarin cultivation in Shajapur, Mandsaur, Rajgarh, Betul, Khandwa, Hoshangabad, Indore, Dhar, Ratlam, Ujjain, Gwalior, and Shivpuri districts of Madhya Pradesh and Jhalawad district of Rajasthan and other areas adjoining the western part of Madhya Pradesh (Fig. 1.6).

Adaptability of Citrus in East and Northeast India

Citrus in this region is grown in states like Bihar, Orissa, eastern Uttar Pradesh, West Bengal, and Assam predominantly. Though, a considerable acreage under citrus exists in other northeastern states comprising Arunachal Pradesh, Meghalaya, Assam, Mizoram, Tripura, Manipur, Nagaland situated between 22° to 22°5′ N latitude and 89° 37′ to 97°30′ E longitude (Abani, 1986) and Sikkim. In this region, rainfall activity increases gradually from the month of March and continues up to the last week of September or first week of October. The premonsoon rainfall occurs mainly due to thunder storm activities. Maximum rainfall is observed during the months of June to September i.e. during southwest monsoon season. The annual average relative humidity in the plain area is 77.5 per cent with a minimum monthly average of 70.0 per cent in the month of March and maximum of 85.0 per cent in the month of August (Dhar *et al.,* 1978; Borthakur, 1986). Northeast India is mainly divided into three meteorological subdivisions, namely, Arunachal Pradesh (average annual rainfall of 2997 mm), Assam and Meghalaya (average annual rainfall of 2497 mm), and Nagaland, Manipur, Mizoram, and Tripura (average annual rainfall of 2314 mm) with synoptic system of rainfall (Memoirs of IMD, 1961; Anonymous, 1980a; Borthakur, 1986).

The topography of the region is mountainous, except the Brahmaputra valley, and the altitude varies from 50 to 3000 m above mean sea level. Citrus in the region is grown under a high range of humidity throughout the year (Fig. 1.8). The Brahmaputra and the Barak are the two principle rivers, that govern the network of drainage system. The entire north-eastern region with humid tropical climate is the largest stretch of acid soils in India, where lands are extremely complex, both geologically and geomorphologically. Physiographically, the entire region is divided into four well differentiated units, namely, i. Eastern himalayan region, ii. Eastern mountain region, iii. Meghalaya Mikir tableland, and iv. Brahmaputra valley (Taher, 1986). These are briefly described :

Fig.1.8. Round the year high humidity in northeastern India is a characteristic feature of citrus cultivation

- Eastern Himalayas: It is the eastern most part of the himalayas and lies within Arunachal Pradesh. These areas have plenty of rivers, rivulets, water reservoirs, and streams which descend from north to south and gravitate to Brahmaputra valley.

- Eastern mountain region: The region encompasses part of Arunachal Pradesh (viz., Tirap, Lohit, and Changlong), eastern part of Assam, whole of Nagaland, Manipur, and Mizoram. The region is dissected by number of important rivers viz., Barak, Doyang, and Tuivai and rivulets. These rivers are typical mountain streams flowing in between rocky mountains, interspersed with channels leading to narrow valleys. Mountain range among other things, embodies Tripura-Cachar plain, Barak, Surma, and Imphal valleys.
- Meghalaya-Mikir tableland: It covers the outlaying Mikir hills, Garo, Khasi, and Jaintia hills of Meghalaya. Geologically, this part of the region is an extension of the Indian peninsular shield.
- Brahamaputra valley: It covers the entire Assam valley lying in the length and breadth of Brahamaputra river. It lies within the girdle formed by the eastern himalayas, Patkai, Naga, Garo, Khasi, Jaintia, and Mikir hills. The valley varies in width from 90 km in upper Assam to about 55 km near Mikir hills about 725 km in length. The northern bank is characterised by tarai to semitarai conditions having inumerable tributaries and proximity with the himalayas. On account of low gradient and high annual sedimentation, Brahmaputra river gives rise to large number of riverine islands that cause periodic flood in Assam state.

According to Geological Survey of India, the region is divided into three distinct macrophysiographic divisions, namely, i. Lofty himalayan mountain, ii. Subdued peninsular block, and iii. Great Brahmaputra alluvial plains. These divisions have been further divided into eight units. These are i. Mountains and hills ii. Piedmont mountain slopes, iii. longitudinal valley, iv. Flood plains, v. River terraces, vi. Slipoff slopes, vii. Sleeves, and viii. Filled river courses. Geologically, the region can be classified into five major morphotectonic domains, such as i. Arunachal himalayas, ii. Mishmi hills, iii. Naga-Lushai mountain range, iv. Meghalaya plateau, the Mikir hills and the insetberges, and v. Brahmaputra and the Barak valley (Abani, 1986; Taher, 1986). The soils of the region are broadly classified into four orders, namely, Inceptisols, Entisols, Alfisols, and Ultisols. Of course, occasional occurrence of Vertisols, particularly in newly formed alluvial belts, is not uncommon. Occurrence of Ultisols having lateritic soils has also been reported from Assam, Nagaland, Mizoram, and Meghalaya. Preliminary soil survey has indicated presence of at least 278 soil series, that are classified, defined, and mapped systematically by the National Bureau of Soil Survey and Land Use Planning, Nagpur. The statewise position of the number of soil series identified so far is : 70 in Arunachal Pradesh , 68 in Assam, 50 in Manipur, 12 in Meghalaya, 28 in Mizoram, 11 in Nagaland, and 27 in Tripura. Soil series so identified have shown high degree of inherent heterogenity due to interaction effect amongst different soil forming factors, which are both ecto- and endodynamomorphic. These include parent material, climate, topography, vegetation, microorganisms, relief, and time.

Most of the soil formation phenomenon in northeastern hills are a result of weathered rocks primarily of shales, and to a lesser extent on slate and veriegated sandstones. High rainfall, temperature, and humidity have created favourable conditions for extensive weathering of rocks with *pari pasu* development of forest flora and fauna. In the process, most of the soil formation has taken place *in situ*, excepting the foothills, lower valleys, and near the water courses and streams, where soils are of colluvial and alluvial origin. Depth of soils on steep slopes in upper part of the hills is generally shallow due to extensive erosion effect. However, since geological erosion precedes soil formation, soil attained stability under vegetative cover. The disturbance of ecological balance resulting from deforestation, jhuming etc. caused widespread soil erosion from upper reaches to foothills and valleys. The soils, thus formed from hill wash are rich in basic nutrients and organic matter. In areas of upper reaches, soils are acidic and poor in nutrients including organic carbon and nitrogen.

Capability of available land resources is an inherent capacity of land to perform for a general land use. While, suitability is a statement of the adaptability of a given area for a specific land use type, two areas of same capability may, therefore, have quite different suitabilities (FAO, 1976). The standard norms of land capability classification, developed in the USDA may be applied to northeastern region with certain modifications, as may be necessary due to peculiar agroclimatic conditions governed by high rainfall, soil type, topography, vegetation etc. prevailing in the region. The region has accordingly been divided into five broad land capability classes :

Class A-I : Slopes of 0.5 per cent, suitable for cultivation without special soil conservation measures.

Class A-II : Slopes between 6 and 50 per cent suitable for cultivation with special soil conservation measures.

Class B-I : Slopes of 51-100 per cent having soil depth of 175 cm or less and found suitable for cultivation of pasture and fodder

Class B-II : Slopes of 51-100 per cent having soil depth of more than 175 cm and suitable for citrus orchards, cash crops, and plantation crops

Class C : Slopes of about 100 per cent suitable for silviculture.

According to land suitability classification, (FAO, 1974b), four types of land suitability classifications are recognised viz., i. Qualitative classifications, ii. Quantitative classifications, iii. Current suitability classification, and iv. Potential suitability classification (Sarkar, 1994). Due to wide variation in soil and climate ranging from tropical to alpine altitude, a great variety of adaptations exists in northeast India (Table 1.46). This region suffers from severe soil erosion due to rugged topography, high rainfall, and shifting cultivation (Biswas and Mukherjee, 1994). In addition, the soils are acidic (Sen *et al.*, 1994; 1997a; 1997b; Nayak *et al.*, 1996a; 1996b). Earlier Thaper (1972) observed that major citrus growing belts of the region have sandy loam to clay loam texture with pH ranging from 4.5 to 6.0, and adequate in organic matter content. Both surface and subsurface soils of the hilly region are highly leached, exhibiting poor base saturation with low cation exchange capacity (Sen *et al.*, 1997c). Soil acidity in general and subsoil acidity in particular, are the major limiting factors for low productivity potential of these soils. Acid soils of the northeastern region have been studied earlier by Mandal *et al.* (1975) and Majumdar (1976). Based on the available meagre information, Govindarajan and Gopala Rao (1978), broadly classified the soils of this region as red loamy soils, (Paleustalfs, Rhodustalfs, and Haplustalfs), red and yellow soils (Haplustults, Ochraqults, and Rhodustults), laterite soils (Plinthaqults, Plinthustults, Plinthudults, and Oxisols), brown hill soils on sandstone and shales (Palehumults, Paleustalfs, Haplaquents, and Tarai soils), and alluvial soils (Haplaquents, Haplaquolls, and Udifluvents).

Table 1.46. Agroclimatic zones of north eastern hilly regions of India

Agroclim-atic zone	Altitude (m)	Mean temperature(^{0}C)		Annual rainfall (mm)	Major areas
		Summer	Winter		
Alpine Zone	3500 and above	15	2	1500-2000	Northern parts of Arunachal Pradesh and Sikkim
Subalpine Zone	1500 - 3500	20	10	2000-1500	Direong, Bomdila, and Shergaon, areas of Arunachal Pradesh, Shillong, Mawphlang area in Meghalaya,Mao, Maram and Ukrul districts of Manipur, Darjeeling district of West Bengal, mid regions of Sikkim and blue mountains in Mizoram, and part of Nagaland.
Subtropical Hill Zone	1000 - 1500	26	12	1600-3000	Khonsa and Basar areas of Arunachal Pradesh, jowai subdivision of Meghalaya, Manchal, and Rongali area in Sikkim,lower parts of Darjeeling district of West Bengal,maximum areas of Mizoram, part of Kohima, and Wokha districts of Nagaland.
Midtropical Zone	200 - 800	28	12	1400-2300	Southern parts of Arunachal Pradesh, southern Jowai, north Cachar districts of Assam, parts of Meghalaya, Manipur west, south Sikkim, jampui hills of Tripura, and Medziphema area of Nagaland
Subtropical Plain Zone	400 -1000	27	10	1200-1800	Manipur valley, Bhagti and Longnak valley of Nagaland, and Umikang area Jaintia hills.
Mildtropical Plain Zone	75 - 200	33	17	1600-2200	Parts of Tirap and Lohit district of Arunachal Pradesh, West Garo hills in Meghalaya, major parts of Tripura, Dimapur area of Nagaland, and northern part of Mizoram.

Source : Sharma (2000)

Pedological and edaphological properties of some benchmark acid soils developed on sedimentary and metamorphic rocks of northeastern region under climatic predominance of humid tropic have been studied by Sen *et al.* (1997d). The soils have acidic pH (4.0 to 5.6) and highly leached, having poor base saturation, and low exchange capacity. They are characterised by high KCl extractable aluminium (18 to 639 mg kg^{-1} soil) throughout the solum. Presence of low activity clay along with subsoil aluminium toxicity rendered them less fertile. Total potential acidity (extractable acidity) is as high as 59.5 to 92.5 per cent of total exchange capacity, while the acidity due to variable charge (pH dependent) ranged between 7.3 and 31.9 cmol(p^+) kg^{-1} contributing to 77 to 97.5 per cent of total acidity (Table 1.47). The most spectacular difference among the soils is in the amount of extractable aluminium, which descriminated the soils of Meghalaya as more weathered in the region. Application of lime would be an important management target for these acidic soils.

The climate is humid subtropical and the annual rainfall varies from the heaviest at Cherrapunjee in Meghalaya (average 11420 mm) to the lowest at Lumding in Assam (average 980 mm). Mean minimun and mean maximum temperature are 18.3° and 29.9°C, respectively. While, mean summer and mean winter temperatures vary from 24.6° to 32.8°C and 9.9° to 24.8°C, respectively. Soil moisture regime is udic, and temperature regime varies from thermic to hyperthermic depending on the elevation. Geology of the region is complex with high spatial variability, and is characterised by sedimentary and metamorphic rocks grading from the most ancient to the recent.

Citrus in Bihar

Citrus cultivation in Bihar is concentrated agroclimatically to plateau region, western undulating zone, and mid central table land zone. Predominant soil types are: Ustochrept, Haplaquept, Paleustalf, Ochraqualf, Ustorthent, Ustifluvent, and Rhodustalf (Sarkar *et al.*, 2000).

Agroclimate : The Plateau Region

Physiographically, citrus in Bihar is confined to plateau region covering Palamu area, Shamastipur, and Bhojpur.

Climate : The state of Bihar is situated in monsoon subtropical zone, characterised by hot summer, wet monsoon, and dry cool winter. On the basis of meteorological parameters, four distinct seasons are identified, namely, summer, monsoon or rainy, postmonsoon, and winter. The summer season prevails from mid March to mid June. The summer is at its peak with average atmospheric temperature ranging between 38°C and 40°C during May and occasionally rising as high to 47°C at places for a period of 7 to 15 days duration, specially in south Bihar alluvial plains. The monsoon or rainy season, extending from June 15 to October 15 receives more than 75 per cent of total annual rainfall. Rainfall is more than 2000 mm in the northwestern part of the plateau

Table 1.47. Physical and chemical properties of the soils of northeastern region of India

Depth (cm)	Clay (%)	Org. C (g kg^{-1})	pH (1:2.5)		Exch. Al (mg kg^{-1})	Exch. acidity (H^++Al^{3+})	CEC	Al sat.	Base sat.	LR (tons ha^{-1})
			H_2O	KCl		[cmol (p^+) kg^{-1}]		(%)		
					Typic Kanhaplohumult (West Siang, Arunachal Pradesh)					
0-13	53.0	27.0	4.6	3.8	198	2.7	13.2	45.1	16	3.6
13-36	66.0	19.6	4.8	3.8	252	3.1	12.3	67.9	18	4.6
36-63	68.0	8.7	5.0	4.5	279	3.4	10.5	77.1	16	5.1
63-100	65.0	6.4	5.4	4.2	297	3.7	9.7	74.8	17	5.4
100-125	47.5	2.9	5.2	4.2	297	3.8	7.1	74.3	18	5.4
					Typic Dystrochrept (Dibrugarh, Assam)					
0-13	29.5	15.0	5.0	3.9	72	1.7	9.6	15.2	37	1.3
13-31	34.5	8.0	4.8	3.9	234	3.2	7.5	47.3	30	4.3
31-70	34.0	6.6	4.9	3.9	198	2.8	6.9	46.0	22	3.6
70-122	44.5	5.7	4.7	4.1	198	2.8	8.1	47.3	18	3.6
122-175	29.0	3.2	5.0	4.1	122	2.0	5.3	41.2	29	2.2
					Typic Dystrochrept (Tamenglong, Manipur)					
0-10	35.0	16.0	4.0	3.6	297	3.5	9.2	39.3	51	5.3
10-28	40.0	12.0	4.0	3.6	369	4.3	8.1	67.2	28	6.7
28-60	40.5	7.0	4.2	3.8	414	4.8	7.9	66.3	26	7.6
60-90	39.0	6.0	4.2	3.8	405	4.6	7.6	65.1	30	7.4
90-125	28.0	3.0	4.2	3.9	351	4.0				
					Typic Kandihumult (West Khasi hills, Meghalaya)					
0-10	28.2	23.0	5.6	4.5	23	0.4	8.8	4.7	43	0.6
10-30	45.1	9.0	5.2	4.3	54	1.0	7.2	15.4	40	1.0
30-72	48.1	6.0	5.3	4.5	18	0.4	5.7	12.7	20	0.3
72-120	56.6	4.0	5.5	4.4	54	0.8	5.0	28.8	25	1.0
120-177	54.9	2.0	5.3	4.2	54	1.0	5.2	29.8	19	1.0
177-210	55.0	2.0	5.3	4.2	81	1.1	(0.6)	4.9	46	-
					Typic Hapludult (Kolasib, Mizoram)					
0-27	37.0	10.0	4.7	3.8	414	4.8	10.6	67.7	19	7.6
27-50	45.0	8.0	4.7	3.8	423	5.0	9.8	75.2	13	7.7
50-110	50.0	6.0	4.9	3.8	450	5.2	10.3	75.4	14	8.2
110+	36.5	4.0	5.0	3.9	288	3.4	(3.4)	7.3	61	-
					Typic Hapludult (Kohima, Nagaland)					
0-15	33.0	14.0	5.5	4.1	18	0.6	11.4	2.5	65	0.3
15-30	44.0	7.0	5.1	3.6	207	2.5	10.0	32.4	47	3.8
30-46	55.5	6.0	5.1	3.6	378	4.4	12.4	45.2	39	6.9
46-71	68.0	5.0	5.2	3.5	558	6.4	14.5	53.6	35	10.2
71-100	67.5	4.0	5.2	3.5	639	7.2	14.9	58.5	38	11.7
100-125	64.5	3.0	5.3	3.5	621	7.0	13.3	57.5	46	11.4
					Typic Paleudult (Bisalgarh, South Tripura)					
0-11	18.5	7.0	4.7	3.9	180	2.2	5.8	48.6	33	3.3
11-38	29.5	6.0	4.5	3.8	306	3.4	6.6	69.7	21	5.6
38-83	35.5	4.0	4.6	3.9	297	3.2	7.0	66.9	22	5.4
83-172	36.0	3.0	4.8	3.9	279	3.2	(3.4)	7.4	69	-

Values in parenthesis indicate the weighted average of the control section (25-100 cm)
CEC and LR stand for cation exchange capacity and lime requirement, respectively
Source : Sen *et al.* (1997d)

region. The annual average precipitation in Chhotanagpur plateau ranges around 1400 to 1500 mm. The July and August are the wettest months with heavy downpour, usually with 330 mm or more rainfall. The postmonsoon season lasts from mid-October to mid-December. The atmospheric temperature gradually falls during this period. The weather is mostly dry. There is about 25 to 100 mm of rains during this period. December is normally the most dry month. The winter season, which sets in during mid-December is at its peak during the month of January when the weather records the lowest temperature of 3°C in the plateau region and south Bihar plains. The winter temperature, however, varies from district to district. The rains received during the postmonsoon, and winter seasons vary between 40 mm and 150 mm in the alluvial plains and plateau regions. By March 5, the atmosphere again gets warmer indicating, thereby, the end of winter season (Rastogi, 1991a).

Soils: The plateau region soils have altogether different problems. They are light to coarse textured with high permeability, which cause them to have low to very low water holding capacity. Their strong acidic reactions also require special ameliorative methodologies.

Agroclimate : Western Plateau Zone

Agroclimatically, the citrus growing belt in Bihar belongs to western plateau zone. Though, this zone comprises the whole of plateau, Gumla, Lohar dega, Chatra sub-division of Hazaribagh, hilly southern part of Gaya, Aurangabad, and Rohtas districts.

Climate : The climate in the zone ranges from subhumid to subtropical with an average annual rainfall of 1324 mm of which 92 per cent is received during April-October through southwest monsoon. Approximately 72 per cent of the rainfall is received between July and September, with July and August being the highest rainfall months. Rainfall in the zone is unevenly distributed. However, areas receiving rainfall more than 900 mm, between 900 and 1290 mm, and more than 1200 mm are well classified. The mean winter and summer temperature are 17.3°C and 31.7°C, respectively. The May month is the hottest month. The humidity ranges between 20 per cent and 70 per cent in the zone (Rastogi, 1991a). The climate at Hazaribagh is characterised by rainfall of 1000 to 1700 mm (1300 mm average) with 85-108 rainy days. The major propertion of rainfall is received during July-August (Bhattacharya *et al.,* 1996).

Soils : The general physiography of this zone varies from place to place, which consists of plains, levelled fields, and undulating hilly and slopy lands. The height from mean sea level varies from 222 m at Daltonganj to more than 1140 m at various places of Netarhat plateau. The soils are generally poor in available P and K. The soils of the hill slopes and hill tops (except that of flats) are very shallow, gravelly, moderately acid to neutral in reaction, and very poor in fertility status. The broad soil association groups of this zone are hill and forest soils of steep slope; yellow reddish, medium deep light textured catenary soils; and upland grey yellow, grey heavy soils on sedimentary and allied rocks (Rastogi, 1991a). Soils at Hazaribagh district in Chhotanagpur plateau are predominantly classified as Lithic Ustorthents, Udic Rhodustalfs, Udic Haplustalfs, and Vertic Ochraqualfs (Singh *et al.*, 1996). Soils at Damodar catchment of Hazaribagh, Bihar are sandy loam type derived from schistose biotite gneiss with 45 – 100 cm depth. The surface soil has granular structure with soft consistence underlain by comparatively hard subsurface with iron nodules in plenty. However, soils are well drained and possess good aeration (Bhattacharya *et al.*, 1966).

Citrus in Orissa

Citrus growing areas in Orissa are confined to western undulating zone and mid central table land zone. Citrus industry of Orissa can be divided into two distinct citrus belts viz., one belt covering Angul, Keonjhar, and Deogarh districts, and other belt comprising Gajpati, Ganjam, and Rayagada districts.

Agroclimate : Northeastern Coastal Plain Zone

It mainly covers three districts viz., Balasore, Jaipur, Keonjhar/Anandpur, Dhenkanal, and Ganjam.

Climate: The mean maximum and minimum temperature of the zone varies between 33.6°C and 14.3°C, respectively. The highest mean monthly relative humidity is 83 per cent with a minimum of 63 per cent in December. Other districts, namely, Balaipal, Balasore, Nilgiri, and Basudevpur have the highest rainfall of 1600 mm, whereas Remuna and Soro receive the lowest mean annual rainfall of 1200 mm.

Soils: The main geological formations that occur in the zone are alluvium deposit of marine and riverine origin. The rocks are sandstones, quartzites, limestones, and granites. The soils of the zone (Gangopadhyay, 1991d) are grouped into 3 broad categories viz., red laterite (Alfisols and Oxisols), alluvial (Entisols), and saline coastal sandy soils (Natraquents and Ustipsamments). Adhikari *et al.* (1997) described the soils of Bolangir (pH 7.3-8.4, organic C 2.6-3.1 g kg^{-1}, sand 81.8 per cent, silt 6.5 per cent, and clay 11.7 per cent), Dhenkanal (pH 6.7-8.6, organic C 2.6-5.1 g kg^{-1}, sand 74.8 per cent, silt 7.0 per cent, and clay 18.2 per cent), Ganjam (pH 6.3-8.3, organic C 2.6-2.8 g kg^{-1}, sand 81.8 per cent, silt 6.0 per cent, and clay 12.2 per cent), Mayurbhanj (pH 6.6-7.6, organic C 10.3-13.8 g kg^{-1}, sand 37.8 per cent, silt 50.0 per cent, and clay 12.2 per cent), Puri (pH 6.0-7.6, organic C 3.6-11.3 g kg^{-1}, sand 79.8 per cent, silt 8.0 per cent, and clay 12.2 per cent), and Sambalpur (pH 6.1-7.9, organic C 2.3-6.0 g kg^{-1}, sand 80.8 per cent, silt 7.8 per cent, and clay 12.2 per cent) districts of Orissa.

Agroclimate : East and South Eastern Coastal Zone

Climate: The climate is subtropical, hot and humid with mean maximum summer temperature of 39°C and mean minimum winter temperature of 11.5°C. The average rainfall in the zone is 1340 mm, which decreases from northeast to southwest. About 74 per cent of the annual rainfall is received during June to September (Gangopadhyay, 1991d).

Soils: The soils of this zone belong to coastal saline and sandy soils, alluvial soils, lateritic soils, black soils, and red lateritic soils (Gangopadhyay, 1991d).

Agroclimate : Western Undulating Zone

Western undulating zone lies between 19°3′-21°55′ N latitude and 82°20′-88°47′E longitude. The zone is divided into two physiographic regions:

- Plain lands at an elevation of 215 to 450 m to more than 900 m above mean sea level with numerous barren hillocks.
- Hilly region at an elevation from 450 m to more than 900 m above mean sea level. The entire hill region is covered with forests and contain mineral deposits of magnesium, graphite, and bauxite. The zone comprises citrus growing Dabagaon block of Koraput district.

Climate : The climate of the zone is hot, moist, and subhumid characterised by a hot dry summer, uneven rainfall in monsoon, and extreme cold in winter. Three distinct seasons occur : rainy (July-October), winter(November-February), and summer (March-May). The maximum temperature ranges between 20° and 45 °C. The driest months are March and April. Average annual rainfall of the zone varies from 960 to 1617 mm. The hilly region of Rampur enjoys a rainfall of 2615 mm. About 90 per cent rainfall is received by southwest monsoon during middle of June to October. Due to erratic and uneven distribution of rainfall, the zone suffers from frequent drought (Gangopadhyay, 1991d).

Soils : The major soil types found in the zone can be grouped under red soil, red and yellow soil, red and black soil, forest red and yellow soils, and alluvial soil covering 44.5 per cent, 27.4 per cent, 9.0 per cent, 12.8 per cent, and 6.3 per cent area of the zone, respectively. (Gangopadhyay, 1991d).

Agroclimate : Mid Central Table Land Zone

The zone is located in the middle of the central table land at an average elevation of 300 m above mean sea level, which rises from east to west.

Climate : The climate of the zone is hot and dry subhumid. The mean maximum summer and minimum winter temperature are 38.7°C and 14°C, respectively. The annual rainfall in the zone ranges from 1300 mm at Angul to 700 mm at Palahara with an average of 1421 mm. Angul is located in the rain shadow areas. The month of June is as the premonsoon period, the months of February, March, and April each receives a rainfall less than 50 mm, while, the month of May receives 50-100 mm. During the postmonsoon period, the month of October receives 50 to 100 mm and November through January, each receiving less than 50 mm rainfall. Climate-water balance studies indicated this zone as a medium moisture deficit zone (Gangopadhyay, 1991d).

Soils : Soils of Dhenkanal and Hindol are light textured, lateritic, and taxonomically classified as Rhodustalfs, Paleustalfs, Haplustalfs, Haplustults and Paleustults. The area adjoining Keonjhar district in north Pallahara, Talcher, Kankabad, Kishorenagar, and Athmalik consists of medium textured red loam Alfisols. The area of Angul, parts of Chandipara, Baunupal, Kishorenagar, and Talcher consist of mixed red and black soil, besides alluvial soils (Gangopadhyay, 1991d). Studies on three red soil profiles belonging to Andiramund series (fine, mixed, hyperthermic family of Typic Paleustalfs), Kuhurdigan series (fine, mixed, hyperthermic family of Ustochrepts), and Palkagera series (fine, mixed, hyperthermic faimly of Typic Ustochrept) of Koraput district of Orissa showed a variation in soil pH of 4.8 to 5.0 in soil-water suspension, and 3.6-4.8 in 1 M KCl solution. The lower soil pH determined in 1 M KCl solution compared to water is due to presence of considerable amount of reserve acidity. The EC of the soils ranged from 0.03 to 0.46 dS m^{-1}, organic carbon from 1.5 to 17.8 g kg^{-1}, cation exchange capacity 3.7 to 14.7 cmol (p^+) kg^{-1}, clay from 26.1 to 48.7 per cent, and texture from loam to clay (Dolui and Sarkar, 2001).

Citrus in West Bengal

Citrus in West Bengal is grown in two major agroclimatic zones viz., hill zone and new alluvial zone. Sharma *et al.* (1985) investigated the prospect of cultivating sweet oranges in Bankura, Birbhum, Purulia, and western Midnapur and found cultivars such as queen, mosambi, jaffa, sathgudi, Washington navel, and valencia as highly promising. Major soil types are: Haplaquept, Ustochrept, Ustifluvent, Orchraqualf, Paleustalf, and Ustorthent (Sarkar *et al.*, 2000).

Agroclimate : Hill Zone

The zone covers Darjeeling, Sadar, Kalimpong, and Kurseong.

Climate : The rainfall varies appreciably from 2500 to 3000 mm, of which 80 per cent is received during June to September. Rainfall is not certain from November to March. Due to rapid run off on hill slopes and undulated land, only a small part of water is untilized effectively. In the high hills above 2000 m, the intensity of sunlight is low, particularly in the monsoon and winter months. Snowfall during January to March is also very common in areas above 2200 m altitude. The average maximum and minimum temperature round the year is recorded as 22°C and 2°C, respectively. The temperature in this zone also varies markedly due to altitude. Even in a small area, the hill tops and foot hills show considerable differences in temperature. The relative humidity varies from 70 to 80 per cent depending on the locality and season of the year. At highest altitude, humidity is more often causing accumulation of fog and reducing the impact of light intensity (Gangopadhyay, 1991b).

Soils : The soils of hill zone are mainly gneiss, schist, phyllite, quartzite, greywake, and dolomite. These soils are generally of high fertility status, crop performance is poor due to low soil depth, and high acidity. High aluminium toxicity has often been observed due to the presence of chloritized 2 : 1 lattice minerals in these soils. The precipitation and fixation of hydroxy aluminium is a common pedogenic process occurring in these soils. The soils are mostly light textured and contain high organic matter. The CEC varies from 10.0 to 20.0 me 100 g^{-1} with very low base saturation. Dominant minerals are mostly mica and chlorite with mixed layer clays. The soils contain high amount of undecomposed organic matter having high C:N ratio (Gangopadhyay, 1991b). Acid hill soils of West Bengal including Kalimpong, Siliguri, and Rayganj have been observed having soil pH 4.2-5.2, CEC 14.4-25.4 cmol (p^+) kg^{-1}, and available P 11.7-54.9 mg kg^{-1} (Debnath *et al.*, 2001).

The soils of Darjeeling hill viz., Monteviot series (pH 5.2, organic matter 2.4 per cent, CEC 25.2 cmol kg^{-1}, exchangeable K 0.20 cmol kg^{-1}, exchangeable Ca + Mg 3.2 cmol kg^{-1}, sand 54.2 per cent, silt 25.8 per cent, and clay 20.0 per cent), Thurbo series (pH 4.9, organic matter 2.1 per cent, CEC 13.6 cmol kg^{-1}, exchangeable K 0.15 cmol kg^{-1}, exchangeable Ca + Mg 2.0 cmol kg^{-1}, sand 73.2 per cent, silt 13.1 per cent, and clay 12.7 per cent), and Teesta valley series (pH 5.1, organic matter 2.5 per cent, CEC 12.7 cmol kg^{-1}, exchangeable K 0.21 cmol kg^{-1}, exchangeable Ca + Mg 2.5 cmol kg^{-1}, sand 62.8 per cent, silt 23.1 per cent, and clay 14.1 per cent) are classified as Umbric Dystrochrepts, Typic Udorthents, and Entic Haplumbrepts, respectively. The minerology of Darjeeling hill soils (pH 5.0, organic C 1.31 per cent, EC 0.15 dSm^{-1}, clay 9.5 per cent, silt 28.4 per cent, and sand 62.1 per cent) is predominantly illitic (Adhikari *et al.*, 1986; Lahiri *et al.*, 1995).

Distribution and nature of organic matter in hill soils of west Bengal at various altitudes in eastern himalayan region by Lahiri and Chakravarty (1995) showed that the CEC of humic and fulvic acids of the high altitude soils is considerably higher than that of low altitude soils. The increase in CEC can be explained by the fact that the present acid groups are of low molecular weight with more aliphatic side chains. The CEC of fulvic acid is higher than that of the corresponding humic acid, which may be due to the larger amount of $COOH^-$ and phenolic OH^- groups in the fulvic acid. In the elementary composition, the percentage of carbon in humic acid is greater than that in the corresponding fulvic acid. These features are considered indicative of the simpler structure of the fulvic acids (Kononova, 1966). The percentage of carbon in the humic acids is low in the high altitude soils to that in the lower altitudes. The organic carbon in these low altitude acid soils increases on maturing, which is indicative of the increasing degree of condensation of the aromatic rings (Kononova, 1975).

The soils of mandarin orchards in Darjeeling district are acidic and structurally poor. The sand content is high with low silt and clay content. The soils have limited holding power for cationic nutrients and deficient in Ca and Mg. The limiting soil properties need rectification for the improvement in productivity, particularly in those locations where these properties have been determined (Table 1.48). The nutrient analysis of

Table 1.48. Soil properties limiting the productivity of Darjeeling mandarin in different locations of Darjeeling, West Bengal.

Location	Properties				
	Sand	Silt	Clay	Nutrient holding capacity	Deficiency
Bhalaukhop (Kalimpong I)	High	Moderate	Very low	Critically low	Ca and Mg
Ecchy (Kalimpong I)	Very high	Very low	Low	Critically low	Ca and Mg
Sakyong (Kalimpong II)	High	Moderate	Low	Critically low	Ca and Mg
Singringtam (Takda)	High	Very low	Low	Low	Ca and Mg
Soureni (Mirik)	High	Moderate	Low	Low	Ca and Mg
Pahalgaon (Mirik)	High	Very low	Low	Low	Ca and Mg
Ambutia (Kurseong)	Very high	Very low	Low	Critically low	Ca and Mg

Source : Mukhopadhyay (1987)

soils show the deficiency of different fertility elements in different locations (Table 1.49). Improvement of the availability of these elements is needed for raising the productivity of the soils particularly in those locations, where enough data base is available.

Table 1.49. Available nutrients in soils at different locations of Darjeeling, West Bengal

Location	Available Nutrients (mg kg^{-1})					C/N ratio
	N	P	K	Zn	Cu	
Bhalukhop (Kalimpong I)	0.0154	9.19	6.57	20.50	7.40	10.45
Ecchy (Kalimpong I)	0.1410	8.35	9.46	1.50	8.40	12.03
Singringtam (Takda)	0.0205	8.35	9.40	30.00	6.80	7.35
Soureni (Mirik)	0.0241	31.74	9.02	32.66	5.60	11.53
Pahalgaon (Mirik)	0.0199	35.05	9.90	32.00	8.00	11.53
Ambutia (Kurseong)	0.1725	18.10	9.40	40.50	16.30	10.81

Source : Mukhopadhyay (1987)

Agroclimate : New Alluvial Zone

It covers Malda, Murshidabad, Hoogly, and Nadia.

Climate : The climate of the zone is subtropical with an average rainfall of 1456.5 mm ranging from 1195 to 1692 mm. The maximum rainfall, more than 80 per cent is received from southwest monsoon, during the rainy months of June to September. While, the range of the minimum temperature of the area is 9.0 to 26.8°C with a maximum temperature of 20.4 to 39.4°C. The maximum relative humidity of the area is 53 and 86 per cent, respectively, in March and August with corresponding minimum relative humidity of 34 per cent and 84 per cent. The area as a whole has humid and warm climate, except having a short spell of winter. The duration of sunshine ranges from 8.5 to 10.5 hours day^{-1} except during the cloudy monsoon having sunshine hours of 4.5 to 5.5 day^{-1} (Gangopadhyay, 1991b).

Soils : The soils vary from loam to heavy clay in texture possessing high water retention capacity, good porosity, and a higher permeability for the surface soils. The soils under the agroclimatic zone are of recent alluvium in nature and is generally deficient in N (0.02-0.06 per cent), low in available P_2O_5 (0.03-0.15 per cent), and K_2O (0.1-1.0 per cent). The pH of the soil ranges from 6.0 to 7.8 having high Ca status. Most of the soils (Gangopadhyay, 1991b) are also low in organic matter (0.3-0.6 per cent).

Citrus in Sikkim

Citrus cultivar, popularly known as Sikkim mandarin is grown widely under high altitude conditions. Soil types belong to great groups, namely, Haplumbrept, Udorthent, Cryorthent, Hapludoll, and Dystrochrept (Sarkar *et al.*, 2000)

Climate : Sikkim is a hill state in the eastern himalayas with altitudes ranging from 300 to 8500 m above mean sea level. The state receives heavy rainfall (2000 - 5000 mm) through southwest monsoon, the middle hills receiving the major portion with low temperature (8° – 15°C) throughout the year. Meteorological features at Pakyong indicated variation in maximum and minimum temperature as 16.1° to 26.5°C and 6.9°-20.5°C, respectively, with total rainfall of 2930.6 mm (Table 1.50)

Soils : Soils of Sikkim have developed mostly on granite and gneiss parent materials, and are by and large, acidic in nature. Intensive leaching of bases under high rainfall (2500-3500 mm) have contributed towards the development of acidity in the soils. The management of acid soils depends on relative contribution of exchangeable and pH dependent acidity of soils for sustained citrus production. However, information on the different forms of acidity in the soils developed on side slopes is scanty. Soils are loamy in texture, moderately deep to deep and somewhat excessively drained. Soil moisture and temperature regimes are perudic and thermic, respectively. Some soils of Sikkim have been reported to be sufficient in available micronutrients (Lahiri and Chakravarty, 1989), low in exchangeable K (Patiram and Upadhyay, 1997), and N (Upadhyay and Patiram, 1996a).

Cultivated soils of five different altitude zones of Sikkim viz., subalpine (>2700 m), temperate (2000-2700 m), midhill temperate (1500-2000 m), subtropical (500-1500 m), and tropical (300-500 m) zone have been studied by Yashoda Avasthe and Avasthe (1995), which showed that soil pH and exchangeable Ca^{2+} are much lower at higher altitudes compared to soils at lower altitudes (Table 1.51). The total

Table 1.50. Meteorological characteristics at Pakyong, Sikkim (1997-2000)

Month	Temperature (°C)		Relative humidity (%)		Rainfall (mm)	Sunshine (hrs day^{-1})
	Max.	Min.	Max.	Min.		
Jan.	16.1	6.9	87.4	49.0	19.8	3.6
Feb.	16.5	7.8	86.8	54.4	85.5	2.2
Mar.	20.8	11.1	86.8	48.4	145.4	4.6
Apr.	23.9	13.4	86.3	49.4	238.6	3.2
May	26.5	18.0	90.0	54.0	328.9	4.2
Jun.	26.5	18.8	91.1	66.5	616.3	-
Jul.	26.2	20.5	90.1	73.5	451.7	-
Aug.	26.2	19.4	91.8	73.7	499.8	-
Sept.	26.4	19.0	92.1	65.9	356.0	2.5
Oct.	25.0	15.0	91.3	56.8	153.3	6.5
Nov.	22.3	11.9	91.5	45.8	2.0	4.5
Dec.	18.7	8.3	91.3	43.7	33.3	2.6

Source : Directorate of Agriculture, Govt. of Sikkim, Sikkim

Table 1.51. Important properties of the soils (0-15 cm) of Sikkim

Properties	Soil properties at various altitudes				
	> 2700 m	2000-2700 m	1500 – 2000 m	500 - 1500 m	300 - 500 m
pH (1:2)	4.4-5.3	4.3-5.5	4.8-6.1	5.0-6.4	5.7-6.4
CEC [cmol(p^+) kg^{-1}]	7.3-18.0	4.8-15.4	6.3-11.0	6.8-16.0	4.1-14.0
Org. C (g kg^{-1})	12.3-44.1	29.4-56.1	10.6-33.6	19.3-27.6	3.6-48.0
$CaCO_3$ (g kg^{-1})	3.0-16.0	6.0-22.0	2.0-25.0	6.0-38.0	7.0-18.0
Sand (%)	75.0-83.0	72.0-98.0	55.0-81.0	58.0-84.0	68.0-81.
Silt (%)	7.0-15.0	7.0-16.0	8.0-20.0	8.0-29.0	9.0-28.0
Clay (%)	7.0-17.0	4.0-15.0	7.0-24.0	6.0-18.0	4.0-12.0
EC (dS m^{-1})	0.1-0.3	0.1-0.2	0.1-0.2	0.01-0.2	0.1-0.9
			Exchangeable cations [cmol (p^+) kg^{-1}]		
Ca^{2+}	1.5-5.0	1.5-4.8	1.1-5.9	1.0-12.8	3.4-10.6
Mg^{2+}	0.2-2.5	0.1-3.9	0.1-2.9	0.1-5.5	1.0-4.8

Source : Yashoda Avasthe and Avasthe (1995)

Zn, Cu, Mn, Fe, B, and Mo content of the surface soils varied widely (Table 1.52). About 94 and 85 per cent of the soils belonging to Umbric Dystrochrept, Typic Hapludoll, Typic Dystrochrept, and Typic Haplumbrept could be rated as deficient in available B and Mo, respectively (Table 1.53) with CEC varying between 4.1 and 12.8 cmol (p^+) kg^{-1}.

Total micronutrients and DTPA extractable Cu and Mn, and hot water extractable B showed a significant negative correlation with altitude. Whereas, DTPA-Zn and Fe correlated positively with altitude (Yashoda Avasthe and Avasthe, 1995). The highest Aluminium saturation when expressed as per cent of effective cation exchange capacity (Table 1.53) was observed in Typic Haplumbrept (33-49 per cent), Typic Dystrochrept (26-47 per cent), followed by Typic Dystrochrept (18-29 per cent), and Umbric

Table 1.52. Available micronutrient content in soils of Sikkim

Altitude zones (m)	Depth (cm)	Micronutrients (mg kg^{-1})					
		Zn	Cu	Mn	Fe	Mo	B
> 2700	0-15	2.9-7.1	2.9-5.6	2.0-46.3	120.0-330.0	0.02-0.04	0.05-0.3
	15-30	1.2-6.5	1.8-4.7	1.0-45.0	80.0-237.0	0.02-0.04	0.02-0.6
2000-2700	0-15	2.8-6.2	1.7-6.1	11.0-37.5	85.0-367.0	0.02-0.06	0.05-0.4
	15-30	1.7-4.1	1.5-5.2	8.7-38.0	66.0-244.0	0.01-0.05	0.05-0.4
1500-2000	0-15	1.2-7.1	3.3-9.8	12.5-105.0	51.0-280.0	0.01-0.04	0.09-0.8
	15-30	1.5-3.4	1.9-8.1	4.2-75.0	42.0-230.0	0.01-0.05	0.01-0.8
500-1500	0-15	1.8-4.8	2.7-6.6	9.5-73.8	47.0-173.0	0.01-0.06	0.05-0.5
	15-30	0.8-5.3	1.5-5.3	5.1-65.0	31.0-222.0	0.01-0.04	0.06-0.4
300-500	0-15	2.7-5.6	4.3-6.1	3.5-28.2	59.0-90.0	0.02-0.10	0.11-0.2
	15-30	1.0-2.5	1.9-4.6	2.5-11.0	31.0-62.0	0.01-0.10	0.06-0.2

Source : Yashoda Avasthe and Avasthe (1995)

Table 1.53. Characteristics of soils on steep slope under high altitude of Sikkim

Horizon	Depth (cm)	pH Water	pH KCl	Sand (%)	Silt (%)	Clay (%)	Org. C (%)	CEC	Acidity [cmol (p^+) kg^{-1}] Exchangeable	Acidity [cmol (p^+) kg^{-1}] pH dependent	(%) contribution of Al to total acidity
				Brang, South Sikkim, Coarse loamy Ubric Dystrochrept							
Al	0-20	4.3	4.0	41.0	65.5	14.5	2.66	10.6	2.80	7.40	70
B1	20-36	4.5	4.1	29.2	55.7	16.5	1.80	8.1	2.30	3.90	71
B21	36-64	4.8	4.3	15.0	46.0	14.0	0.93	6.2	1.00	4.50	80
B22	64-87	5.0	4.4	19.0	46.8	12.2	0.86	5.6	0.42	2.58	83
C	87-100+	5.1	4.6	59.2	32.7	8.1	0.17	4.1	0.32	1.91	85
				Elichu, North Sikkim, Coarse loamy Typic Hapludoll							
A11	0-29	4.4	4.0	18.4	67.8	13.8	3.71	12.5	3.40	9.40	66
A12	29-56	4.7	4.2	21.2	61.9	16.9	2.90	9.8	1.60	5.70	75
B1	56-75	4.7	4.3	33.6	50.8	15.6	2.30	8.2	1.30	5.00	78
B2	75-100+	4.8	4.3	41.0	45.6	13.4	2.20	5.0	1.00	3.40	82
				Payong, South Sikkim, Coarse loamy Typic Dystrochrept							
A11	0-13	4.5	4.1	31.0	53.0	16.0	4.80	12.8	5.90	20.30	75
A12	13-31	4.6	4.1	36.6	46.0	17.4	1.71	10.2	4.80	14.00	78
B1	31-57	4.9	4.3	40.2	46.3	13.5	1.67	8.5	2.81	7.40	82
B2	75-100+	5.0	4.5	31.0	52.9	16.1	1.00	5.7	1.40	5.50	83
				Lingtha, East Sikkim, Fine loamy Typic Haplumbrept							
A11	0-18	4.2	3.9	18.2	67.6	14.2	2.7	7.7	3.40	11.90	75
A12	18-35	4.4	4.1	27.4	56.5	16.1	2.3	6.7	2.20	8.50	75
B1	35-55	4.6	4.2	36.6	45.0	18.4	5.0	5.4	1.63	6.40	78
B2	55-82	5.0	4.4	47.4	37.6	15.0	1.8	4.1	0.94	3.70	81

Source : Das *et al.* (1996)

Dystrochrept (4-25 per cent). Soils of Sikkim are likely to cause considerable damage to sensitive citrus cultivars taking into consideration of 10 per cent Al saturation of as threshold value for identifying toxicity due to Al. The soils of Sikkim, therefore, need to be limed judiciously to lower the Al level to a safer limit for sustained crop production (For more details Chapter 5 may be referred).

The soil pH varied from 5.0 to 6.9, organic carbon 6.4 to 30.9 g kg^{-1}, and exchangeable Ca and Mg 3.82 to 22.93 cmol (p^+) kg^{-1} soil with sandy loam to loam texture. The amount of organic carbon and exchangeable Ca^{2+}+Mg^{2+} decreased with increase in soil depth (Table 1.54). Whereas, clay and pH remained always constant. The high amount of organic carbon in the mandarin orchards is the result of the recycling of nutrients through organic manure. The total soil concentration of Zn, Cu, Mn, and Fe varied widely from 40 to 280 mg kg^{-1}, 27 to 210 mg kg^{-1}, 237 to 1872 mg kg^{-1}, and 2.0 to 5.6 per cent, respectively (Table 1.55), with exchangeable bases, effective CEC, and per cent Al saturation varying from 1.4 to 5.5 cmol (p^+) kg^{-1}, 2.1 to 10.2 cmol (p^+) kg^{-1}, and 4-49, respectively.

Table 1.54. Range and mean of some important soil properties and micronutrient cation reserves of mandarin orchards

Soil properties	Soil depth (cm)					
	0-20		20-40		40-60	
	Range	Mean	Range	Mean	Range	Mean
Clay (%)	7.6-16.8	11.6	7.6-16.2	11.5	7.9-16.9	11.6
Org. C (g kg^{-1})	12.9-30.9	23.3	6.7-30.6	6.2	6.4-30.4	19.9
pH	5.0-6.9	5.7	5.0-6.9	5.7	5.0-6.9	5.7
Exch. Ca^{2+}+ Mg^{2+} [cmol (p^+) kg^{-1}]	4.9-22.9	8.8	4.3-13.9	7.3	3.8-14.6	6.6
Total Zn (mg kg^{-1})	47-210	94	43-283	99	42-202	94
Available Zn (mg kg^{-1})	2.9-11.8	5.1	2.3-11.1	4.5	2.1-10.0	4.1
Total Cu (mg kg^{-1})	27-206	86	29-210	79	36-187	77
Available Cu (mg kg^{-1})	0.4-5.0	0.8	0.3-2.9	0.8	0.4-2.7	0.8
Total Mn (mg kg^{-1})	252-1822	761	237-1872	776	276-1786	762
Available Mn (mg kg^{-1})	3.8-132.1	42.4	4.5-141.7	40.3	4.1-118.9	37.6
Total Fe (%)	2.0-5.0	3.6	2.2-5.6	3.5	2.0-5.0	3.5
Available Fe (mg kg^{-1})	6.8-110.6	50.1	11.2-112.8	51.6	9.2-138.9	9.5
Mandarin leaf micronutrient content (mg kg^{-1})						
Zn	19.0-50.0	27.4	-	-	-	-
Cu	5.5-24.0	12.3	-	-	-	-
Mn	16.5-80.5	26.4	-	-	-	-
Fe	100-370	160	-	-	-	-

Source : Patiram *et al.* (2000)

Available P and various fractions of soil P in mandarin soils of Sikkim decreased up to 40 cm depth and correlated positively with organic carbon, pH, exchangeable Ca^{2+} and Mg^{2+} (Patiram and Upadhyaya, 2000). The nonexchangeable K content followed the trend, Inceptisol > Mollisol > Entisol > Alfisol (Table 1.56). Thus, during K stress condition, Inceptisol is expected to release K from its nonexchangeable reserve, most effectively amongst four soils followed by Mollisol and Entisol. Presence of illitic clay minerals acts as a buffering agent for soil K. Whenever, there is a depletion in soil solution K, illitic minerals are likely to release K from specific sites (nonexchangeable K reserve). Soils with high clay content (Inceptisol, Mollisol, and Entisols) required higher level of exchangeable K than a sandy soil (Alfisol) to reach the same K concentration in soil solution (Table 1.56). At nearly identical levels of exchangeable K (Ultisol and Entisol), a lighter soil has been found to release K at a faster rate into the soil solution, leading to the higher apparent diffusion coefficient value than a heavier soil (Raychaudhuri and Sanyal, 1999a).

Acid soils of Sikkim representing Tadong, Penlong, Rorathang, Aritar, Saramasa, Rumtek, Pangthang, and Merchak areas of Sikkim (pH 4.4 – 5.8, organic matter 2.7 –

5.0 per cent, and available P 12.0 – 27.0 mg kg^{-1}), Hee Bazar, Pakyong Outok, Bermiok, Darap, and Dentam areas of West Sikkim (pH 4.9 – 5.8, organic matter 2.7 – 4.1 per cent, and available P 3.0 – 17.0 mg kg^{-1}), Phedong, Namli, Mangan, Dikchu, and Gyathang areas of North Sikkim (pH 4.4 – 5.1, organic matter 2.3 – 4.1 per cent, and available P 8.0 – 58.0 mg kg^{-1}), and Ralang, Ruruk, Maniram, Lingi Nandugaon , Mellibazar, Sodem, and Namthang areas of South Sikkim (pH 5.0 – 6.4, organic matter 0.6 – 4.1 per cent, and available P 3.0 – 41.0 mg kg^{-1}), have shown considerable variation in various soil properties (Gupta and Srivastava, 1998). Bhutia *et al.* (1986) earlier reported high status of available N and K, and low to medium status of available P in soils of Sikkim.

Table 1.55. Aluminium saturation of soil expressed as per cent of effective CEC

Depth (cm)	Exchangeable bases [cmol (p$^+$) kg^{-1}]	Effective CEC (Ca+Mg+Na +K+Al) [cmol (p$^+$) kg^{-1}]	% saturation with Al
Brang, South Sikkim, Coarse loamy Umbric Dystrochrept			
0-20	5.5	7.3	25
20-36	4.8	6.4	25
36-64	3.5	3.8	9
64-87	3.4	3.6	6
87-100+	2.5	2.6	4
Payong, South Sikkim, Coarse loamy Typic Dystrochrept			
0-29	5.1	7.2	29
29-56	4.5	5.5	18
56-75	3.9	4.7	18
75-100	2.9	3.7	22
Payong, South Sikkim, Coarse loamy Typic Dystrochrept			
0-13	5.4	10.2	47
13-31	4.1	7.6	46
31-57	3.9	6.1	36
57-100+	3.0	4.1	26
Lingtha, East Sikkim, Fine loamy Typic Haplumbrept			
0-18	2.9	5.9	49
18-35	2.4	4.2	43
35-55	1.8	2.7	33
55-82	1.4	2.1	33

Source : Das *et al.* (1996)

Properties of acid soils of Sikkim under subtropical zone are: pH 5.0 – 6.4, EC 0.06 – 0.22 dSm^{-1}, CEC 6.3 – 11.0 cmol (p$^+$) kg^{-1}, organic C 1.1 – 3.4 per cent, $CaCO_3$ equivalent 0.2 – 2.5 per cent, sand 55.0 – 81.0 per cent, silt 8.0 – 27.0 per cent, clay 7.0 – 24.0 per cent, exchangeable Ca 1.1 – 5.9 cmol (p$^+$) kg^{-1}, exchangeable Mg 0.1 – 2.9 cmol (p$^+$) kg^{-1}, exchangeable K 215.0 – 475.0 mg kg^{-1}, available N 50.0 – 150.0 mg kg^{-1}, and Bray-P 14.0 – 89.0 mg kg^{-1} at 1500 - 2000 m altitude; pH 5.0 – 6.4, EC 0.01 – 0.22 dSm^{-1}, CEC 6.8 – 16.0 cmol (p$^+$) kg^{-1}, organic C 1.9 – 2.8 per cent, $CaCO_3$ equivalent 0.6 – 3.8 per cent, sand 58.0 – 84.0 per cent, silt 8.0 – 29.0 per cent, clay 6.0 – 18.0 per cent, exchangeable Ca 1.0 – 12.8 cmol (p$^+$) kg^{-1}, exchangeable Mg 0.1 – 5.5 cmol (p$^+$) kg^{-1}, exchangeable K 55.0 – 435.0 mg kg^{-1}, available N 50.0 – 120.0 mg kg^{-1}, and Bray-P 5.6 – 89.0 mg kg^{-1} at 500-1500 m altitude; pH 5.7 – 6.4, EC 0.13 – 0.90 dSm^{-1}, CEC 4.1 – 14.0 cmol (p$^+$) kg^{-1}, organic C 0.4 – 4.80 per cent, $CaCO_3$ equivalent 0.7 – 1.8 per cent, sand 68.0 – 81.0 per cent, silt 9.0 – 28.0 per cent, clay 4.0 – 12.0 per cent, exchangeable Ca 3.4 – 10.6 cmol (p$^+$) kg^{-1}, exchangeable Mg 1.0 – 4.8 cmol (p$^+$) kg^{-1}, exchangeable K 45.0 – 175.0 mg kg^{-1}, available N 40.0 – 80.0 mg kg^{-1}, and Bray-P 18.0 – 175.0 mg kg^{-1} at

Table 1.56. Important physicochemical characteristics of soils of West Bengal and Sikkim

Properties	Location			
	Kalyani (Entisol)	Susunia (Alfisol)	Bakhali (Inceptisol)	Sikkim (Mollisol)
pH(1:2.5)	7.2	5.7	8.0	4.7
EC(dSm^{-1})	0.52	0.11	6.5	0.27
Water holding capacity (%)	58.5	44.4	61.1	85.9
CEC [cmol(p^+) kg^{-1}]	24.5	10.6	31.5	26.2
Organic carbon (g kg^{-1})	6.6	3.1	5.2	38.2
Sand(%)	34.9	43.2	8.2	23.9
Silt(%)	17.0	15.0	30.0	23.0
Clay(%)	48.1	41.8	61.8	53.1
Texture	Clay loam	Sandy clay	Clay	Clay
Base saturation (%)	72.7	41.3	82.3	40.9
Exchangeable cations [cmol(p^+) kg^{-1}]				
Ca^{2+}	11.2	3.04	12.8	6.9
Mg^{2+}	5.51	0.72	7.14	2.93
Na^+	0.84	0.34	3.65	0.34
Forms of K [cmol(p^+) kg^{-1}]				
Water soluble –K	0.03	0.26	0.58	0.13
Exchangeable –K	0.23	0.28	2.34	0.56
Nonexchangeable –K	3.21	0.92	6.67	3.66

Source : Raychaudhuri and Sanyal (1999a)

300-500 m altitude. The total Zn recorded maximum of 181.5 mg kg^{-1}, 147.8 mg kg^{-1}, and 151.7 mg kg^{-1} in 0-15 cm layer of upper, middle, and lower ridges, respectively, at an altitude of 500 –1500 m compared to higher altitudes (Pradhan *et al.*, 1996). Analysis of soils representing citrus belts of Bermiok, central Pandam, Nazitam, Namthang, Sang, and Sumin at an altitude of 600 - 1500 m showed soil pH 5.1 – 6.9, organic matter 2.6 – 8.4 per cent, available N 116.0 – 340.0 mg kg^{-1}, available P 10.4 – 195.0 mg kg^{-1}, and available K 80.0 – 495.0 mg kg^{-1} (Upadhyay and Patiram, 1996a).

Citrus in Arunachal Pradesh

It is one of the seven sister states of northeast India, where citrus is grown in midhills. Gupta and Dwivedi (1994) described the agrotechniques for growing citrus in Arunachal Pradesh.

Climate : The reverine plain of the Lohit district of Arunachal Pradesh forms a part of the Brahmaputra valley, and is located between 27°45′-28°0′ N latitudes and 96°0′-96°21′E longitudes. The area is surrounded by the Mishmi hills on north and east, while the river Lohit and its tributaries on south and west. The area presents more or less a concave relief. The slope varies from 1 to 3 per cent from northeast to southeast direction. The climate is humid subtropical with mean annual temperature of 23°C. The mean summer and winter temperatures are 27.7°C and 18.7°C, respectively, with mean annual rainfall of 2560 mm. The area, therefore, qualifies for hyperthermic temperature regime.

Table 1.57. Meterological characteristics o Gori, Basar, Arunachal Pradesh

Month	Temperature (°C)		Relative humidity (%)		Rain-fall (mm)	Rain day
	Min.	Max.	Min.	Max.		
Jan.	2.3	17.8	60.9	86.3	103.0	10
Feb.	3.7	16.6	61.4	73.7	42.0	13
Mar.	6.7	18.9	52.7	65.4	77.0	09
Apr.	15.5	22.2	55.6	70.4	51.5	04
May	16.0	26.8	54.5	63.2	171.5	12
Jun.	17.6	27.1	62.6	67.7	540.0	21
Jul.	19.0	29.2	57.5	67.6	441.5	24
Aug.	18.5	28.3	60.9	68.9	339.0	20
Sept.	17.1	28.0	59.2	69.8	390.5	18
Oct.	16.5	25.0	47.8	63.5	60.0	07
Nov.	10.9	23.6	59.4	67.3	98.5	07
Dec.	6.4	19.5	43.9	61.2	00.0	00

The total precipitation at Gori, Basar has been observed as 2315 mm in 145 rainy days and maximum rainfall is received during June, July, August, and September (339-540 mm). Winter rainfall during October to January is 262 mm spread in 24 rainy days. Maximum temperature varied from 9.7 °C in January to 25.7°C in July. The minimum temperature varied from 2.3° to 19.0°C during morning hours. The annual relative humidity varies from 61-86 per cent in the morning and 44 to 61 per cent in the afternoon (Table 1.57).

Soils : The soils are stratified, coarse textured, and slightly acidic to neutral with high organic carbon content. Despite, low cation exchange capacity, all the soils have high base saturation. The terrace and piedmont soils exhibit A-B-C horizons, and are classified as Inceptisols (Table 1.58). While, soils of other landforms lack B horizon and, therefore, key out as Entisols (Walia and Chamuah, 1993). The dominant soil types at great group level, according to Vadivelu *et al.* (2000), are Udorthent, Haplumbert, Dystrochrept, rock outcrops, and Eutrochrept covering an area of 33.4 per cent, 16.1 per cent, 14.5 per cent, 12. 6 per cent, and 6.0 per cent of total geographical area, respectively. While, at subgroup level, the major soil types are Lithic Udorthent (17.5 per cent), Typic Udorthent (15.9 per cent), and Paleic Haplumbert (9.7 per cent).

Characterisation of some hill soils of Lohit district of Arunachal Pradesh revealed the occurrence of Lithic Haplumbrept, Typic Haplumbrept, Entic Haplumbrept, Umbric Dystrochrept, and Entic Haplumbrept with pH 4.9-5.3, CEC 4.1-10.6 cmol (p^+) kg^{-1}, sand 36.5-72.0 per cent, silt 17.9-48.7 per cent, clay 9.6-26.9 per cent, and base saturation 21-54 per cent (Walia and Chamuah, 1996). Walia and Chamuah (1997) later characterised and classified the acid hill soils of Tirap district, which showed the occurrence of soil types viz., Typic Udorthent, Ultic Hapludult, Umbric Dystrochrept, Fluventic colluvium, and Typic Haplaquept developed on sandstone, shale, clay, colluvium, and alluvium having base saturation of 9-95 per cent. The dominant rock types present in these areas are sandstone, shales, and clay (Chamuah *et al.*, 1989).

Table 1.58. Characteristics of the some soils of Arunachal Pradesh

Depth (cm)	Particle size distribution (%)			Sand/ silt ratio	Org. C (g kg^{-1})	pH	CEC	Ca^{2+}	Mg^{2+}	Na^+	K^+	BS (%)
	Sand	Silt	Clay				[cmol (p$^+$) kg^{-1}]					
				Tafargam Series: Typic Udifluvent								
0-14	63.0	34.6	2.4	1.82	11.0	6.6	3.7	1.9	0.4	0.3	0.3	81
14-32	39.5	42.3	18.2	0.19	6.4	6.0	5.6	2.3	1.2	0.5	0.2	76
32-47	75.0	21.1	3.4	3.60	2.7	6.7	3.0	1.5	0.1	0.4	0.2	76
47-80	70.0	13.7	16.3	6.10	3.9	6.4	5.0	2.9	0.3	0.4	0.2	78
				Mallua Series: Typic Eutrochrept								
0-15	59.5	30.9	9.6	1.92	16.7	5.7	6.1	1.9	1.5	0.2	0.3	64
15-35	64.0	27.6	8.4	2.31	10.5	5.6	3.4	1.1	0.5	0.2	0.3	67
35-60	58.5	29.5	12.0	1.98	8.8	5.6	3.2	1.0	0.4	0.2	0.3	60
60-85	69.0	25.7	5.3	2.68	7.6	5.6	2.8	1.5	0.1	0.2	0.2	71
85-105	62.5	36.1	1.4	0.57	5.0	5.8	2.2	1.4	0.1	0.2	0.2	86
				Alubari Series : Aeric Fluvaquent								
0-14	45.5	41.1	13.4	1.10	39.3	6.8	12.2	8.0	0.5	0.3	0.2	74
14-30	29.5	56.1	14.4	0.52	32.0	7.0	8.0	6.0	0.6	0.3	0.2	88
30-60	52.5	31.7	15.8	1.65	25.0	6.9	5.8	4.4	0.1	0.2	0.2	84
60-80	60.5	28.5	11.0	2.12	12.8	7.0	3.8	2.9	0.1	0.2	0.1	86
80-100	43.0	49.5	7.2	0.86	14.0	6.9	4.1	2.9	0.2	0.3	0.1	85
				Kheram Series : Typic Udifluvent								
0-12	60.0	34.7	5.3	1.73	31.0	6.5	8.7	4.3	4.0	0.2	0.0	83
12-35	59.5	35.7	4.8	1.66	15.2	6.6	4.0	2.6	1.6	0.2	0.2	77
35-65	80.0	14.2	5.8	5.63	8.5	6.6	4.4	1.6	1.5	0.2	0.1	91
65-100	87.5	5.8	6.7	15.0	4.0	6.8	2.7	1.4	1.0	0.2	0.1	96
				Maruah Series : Dystric Eutrochrept								
0-14	54.4	39.7	5.8	1.37	32.3	6.1	7.8	3.9	2.0	0.2	0.1	82
14-35	51.0	29.3	19.7	1.74	7.0	6.1	4.9	1.6	1.2	0.2	0.1	69
35-65	39.5	47.5	13.0	0.83	5.6	6.0	3.8	1.0	0.9	0.2	0.2	68
65-95	40.5	47.5	12.0	0.85	5.2	5.9	3.0	0.9	0.5	0.2	0.2	60
95-105	60.0	36.0	4.0	1.66	3.0	5.9	1.9	0.5	0.2	0.2	0.3	68
				Kamliamg Series : Typic Udipsamment								
0-12	78.5	10.6	10.9	8.25	6.2	6.9	2.1	1.3	0.3	0.3	0.1	95
12-100	95.0	2.8	2.2	33.90	8.3	7.6	0.8	0.4	0.2	0.1	0.1	100

BS stands for base saturation
Source : Walia and Chamuah (1993)

Citrus in Assam

A number of species of citrus is claimed to have originated in Assam. Citrus in Assam is cultivated at an altitude of 1600-1900 m (Dutta, 1954; 1959; Singh 1963b). It is believed by the Khasis that originally mandarins came from Siam (Thailand) in the past and settled in Assam. It is considered that the Khasis and Synthengs are a remnant of the first Mongolian overflow into India, who established themselves in their present habitat. The occurrence of major soil groups in Assam included recent riverine, old riverine, old mountain valley, nonlaterized, and laterized soils (Table 1.59). Vadivelu *et al.* (2000) described the soils of Assam as Dystrochrept, Fluvaquent, Haplaquent, Haplustalf, and Paleudalf covering an area of 23.6 per cent, 23.3 per cent, 12.5 per cent, 11.0 per cent, and 9.9 per cent of total geographical area, respectively.

Agroclimate : North Bank Plain Zone

It comprises citrus growing areas namely, Lakhimpur, Sonitpur, and Darrang.

Table 1.59. Major soil groups of the Assam

Major soil group	Soil type	Soil feature	Area
Recent reverine	Alluvial (Entisols)	Sandy loam, Silty loam	Sibsagar Kamrup, and Golpara
Old riverine	Alluvial (Inceptisols)	Sandy loam to loam. silty clay loam, silty clay and clay	Dibrugarh, Jorhat, Sibsagar, and part of Golaghat, Lakhimpur, Sonilpur, Darrang,Nalbari,Kokhrajhar, Barpeta, Dhubri, Nagaaon, Kamprup, and Goalpara.
Old mountain valley	Alluvial (Alfisols)	Heavy /low permeability	Lakhimpur, Sonipur , Darrang, Nalbari, Barpeta, Kokhrajhar, Nagaon, part of Karbi, Anglong, Dibrugarh, Jorhat, Golaghat, Cachar, and Karimganj.
Nonlaterites	Red soil (Ultisols)	(•) Ferrugenous red soil loamy having granular structure	Karbi Anglong, North Cachar, Kamrup, Goalpara, Cachar, and Karimganj.
		(•) Ferrugenous gravelley soil, small gravels.	
Laterites	Red soil (Ultisols)	(•) High land laterized	Low lying areas, Cachar and some pockets of Kamprup, and Goalpara
		(•) Ground level laterized red soil	
		(•) Pearl beel soils	

Source : Arun Varma (1991)

Physiographically, this zone is divided into five types viz., foot hill belt, old flood plains (upper terraces), old flood plains (lower terraces), recent flood plain (riverine belts), marshes and swamps (unsuitable for citrus).

Climate : The average rainfall of the zone is around 2650 mm. The rain water availability period is of 16 to 22 weeks from June to September months. Lowest rainfall (16-30 mm) is received in December and January months. The foot hill tract and the upland middle plains of the zone are marked out as important belts for the fruit crop like citrus.

Soils : Soils of north bank plain zone is developed on alluvium derived from adjacent himalayas by river Brahmaputra and its tributaries. The soil types identified are old alluvial soil (Alfisols) along the foot hills, old alluvial soil (Inceptisols), old flood plains, new alluvial soil (Entisols), and recent flood plains. Major soil problems of the the zone are moderate to high acidity, phosphate fixation, slow water retention capacity, deficiency of B, Mn, and toxicity of Fe (Arun Varma, 1991). Earlier, on the basis of variation in soil fertility in Assam, Bora (1975) classified soils into five classes, namely, recent riverine alluvium, old riverine alluvium, laterized red, and nonlaterized red and old alluvium. As much as 89 per cent soils are rated to have high organic carbon content. A 50 per cent, 34 per cent, and 11 per cent soils are rated as low, medium, and high, respectively, in available P. While, a 43 per cent soils are poor in available K.

Agroclimate : Upper Brahmaputra Valley Zone

It covers entire eastern district beyond central Brahmaputra valley and hills zone, and is situated in the south western side of the state. A number of districts viz., Tinsukia, Dibrugarh, Sibsagar, Jorhat, and Golaghat of this zone are separated from the north bank plain zone by river Brahmaputra.

Climate : Intensity of rainfall varies between east to west and to some extent north to south. Golaghat in the west receives 1778 mm of rains, while Dibrugarh in east receives 2347 mm. High land areas receive relatively less rain (1200mm). Eastern side is relatively cooler than western. The July and August (36°C to 38°C) months are the hottest months, while December and January are the coolest months (5°C). Relative humidity in summer does not go beyond 80 per cent. However, it reaches to 60 per cent in winter. Sunshine hours during winter (6 hrs per day) are more in comparison to summer (Arun Varma, 1991).

Soils : The soils of this zone consist of most immature alluvium in the Char areas, mature Ultisols in piedmont, highland, and hill areas in the southern part. The soils of recent riverine alluvial soils of Char areas fall under Entisol order. The soils on the alluvial plains are immature and vary in stages of pedogenic development. The soils of foot hills and uplands in the adjoining flat valleys are built up from alluvial materials. The soils are mostly mature with small pockets of immature soil. The soils of eastern Karbi Anglong and the extreme north eastern part are grouped under acidic soils. The soil structure ranges from granular to subgranular with friable and sticky consistency (Arun Varma, 1991). Poor drainage and over moist soil conditions coupled with presence of water table very close to the surface are major soil constraints in Tinsukia citrus belts (Ghosh, 1985).

Agroclimate : Central Brahmaputra Valley Zone

This comprises the central area of state surrounded by Karbi Anglong in the east, south Kamrup in the west, and the river Brahmaputra in the north covering the district of Nagaon.

Climate : In the eastern humid piedmont, rainfall is more than 2000 mm. In the alluvial plains of Shillongani and Morigaon, it is between 1200 and 2000 mm and very low (less than 1200 mm) in Jojai, Kak, and Lumding.

Soils : This zone has variety of soils from most recent immature Entisols in Char areas to old mature Alfisols in piedmont highlands and hill areas. Most recent immature alluvium are generally unstable on river sides, but stable when away from the river. Soils on the alluvial plains are immature and are at various stages of piedmont development (Arun Varma,1991).

Karmakar and Rao (1999a) described the soils of lower Brahmaputra valley zone of Assam as Ruptic-Ultic Dystrochrepts (pH 4.8-5.7, free Fe_2O_3 2.1-3.2 per cent, free Al_2O_3 0.9-1.2 per cent, sand 28.4-35.5 per cent, silt 27.9-34.9 per cent, clay 31.8-41.1 per cent, and CEC 18.3-44.0 cmol (p^+) kg^{-1}), Aquic Udipsamments (pH 5.6-6.2, free

Fe_2O_3 0.2-0.7 per cent, free Al_2O_3 0.1-0.5 per cent, sand 68.7-86.2 per cent, silt 1.8-8.7 per cent, clay 3.6-51.7 per cent, and CEC 1.9-10.7 cmol (p^+) kg^{-1}), Umbric Dystrochrepts (pH 5.1-5.9, free Fe_2O_3 0.4-0.9 per cent, free Al_2O_3 0.4-0.8 per cent, sand 35.2-85.2 per cent, silt 6.9-42.3 per cent, clay 7.929.0 per cent, and CEC 5.1-22.4 cmol (p^+) kg^{-1}), Oxyaquic Udifluvents (pH 5.6-6.3, free Fe_2O_3 0.3-1.3 per cent, free Al_2O_3 0.2-0.7 per cent, sand 28.7-93.7 per cent, silt 1.5-35.9 per cent, clay 4.8-35.4 per cent, and CEC 1.3-14.4 cmol (p^+) kg^{-1}), and Typic Hapludult (pH 4.8-5.2, free Fe_2O_3 3.3-5.4 per cent, free Al_2O_3 0.9-1.7 per cent, sand 23.0-43.5 per cent, silt 25.9-26.7 per cent, clay 30.6-47.5 per cent, and CEC 15.9-19.5 cmol (p^+) kg^{-1}). The fine sand mineralogy of these soils is dominated by quartz (41.0-61.3 per cent), orthoclase (27.6-39.2 per cent), plagioclase (1.8-6.3 per cent), and muscovite (9.3-25.8 per cent) in the light fraction, and zircon (19.0-34.3 per cent), biotite (23.2-29.3 per cent), chlorite (19.6-31.4 per cent), hypersthene (0-8 per cent), and kyanite (2.2-7.2 per cent) in the heavy fraction (Karmakar and Rao, 1999b). Major portion of the soil organic carbon remains unhumified in the surface horizon. The formation of metal - humus complex is observed with humic and fulvic acid fractions. But, humic acid fraction has been found to form more complex with both Fe and Al (Karmakar and Rao, 1999c).

Agroclimate : Hill Zone of Assam

The hill zone lies between 24°54′ N latitude and 92°52′ E longitude and comprises 66 per cent area of Karbi Anglong and 34 per cent area of North Cachar hills. North Cachar hills are steep high. Karbi Anglong hills are undulating and less high. In this hill zone, land is classified as plain, hills with mild slopes, and hills with steep slopes.

Climate : The annual rainfall of the zone varies from 1050 to 2000 mm (Table 1.60). The annual average rainfall is 1209.0 mm in Karbi Anglong and 1712.5 mm in North Cachar hills. The average maximum temperature varies from 20° to 26°C with occasional higher temperature of 34.4°C . The average minimum temperature varies between 14°C and 19°C with occasional fall to as low as 6°C. The average minimum temperature for Karbi Anglong varies from 21.9° to 34.8°C and minimum temperature from 6° to 25.7 °C. The comparison of two major citrus growing belts has shown the much higher distribution of rainfall in North Cachar compared to Karbi Anglong hills (Arun Varma, 1991).

Soils : Soils, generally are of laterite on slopes and red loam in valleys. In the adjoining plains, soils are old alluvium. Soils of the zone are divided into ferruginous and red soils. The ferruginous soils are rich in iron oxides. Citrus is grown in two different climatic conditions of the hill districts of Assam, which by and large, are favourable for mandarin (*Citrus reticulata*) grown widely in this region. The soils are deep, moderately well drained,

Table 1.60. Average rainfall distribution (mm) pattern between citrus growing Karbi Anglong and North Cachar (NC) hills of Assam

Months	Karbi Anglong (Diphu)	NC Hills (Haflong)
December-February	175.0	69.7
March-May	600.0	920.3
June-September	1050.0	1994.5
October-November	175.0	260.6
Total	2000.0	3245.1

good structured, and have weakly developed argillic horizon with favourable pH (4.9-6.5). In the high temperature (mean annual temperature 23.4°C) and low rainfall (mean annual rainfall 1247 mm) zone, the soils in the northwestern slope (Ultic Paleustalf) at Labak Tisu have high organic matter (0.52-3.36 per cent), moderately to slightly acidic pH (5.4 to 6.5), and comparatively moderate base saturation (59 to 76 per cent), indicating better fertility status than organic matter (0.31 – 1.29 per cent), pH (5.1 – 5.8), and base saturation (50-64 per cent) of soils on the southern slope. In both the areas, the citrus plants showed good canopy development and growth, the difference, if any, is not noticeable due to mixed cultivation of different age groups in the same area. In the other climatic situation (mean annual temperature 20.1° and mean annual rainfall 1761.3-1892.0 mm), the soils are more leached with lower pH and depleted bases. In this area, high organic matter maintains good N status, which helps in proper canopy development, growth, and productivity of citrus orchards. Within this zone, the soils located at Jatinga in the northwestern slope (Typic Haplohumult) have shown better properties in addition to growth and yield of citrus (Table 1.61). However, it is evident that to get high production and even to maintain the present status of yield in productive orchards, proper management practices like regular addition of phosphatic fertilizers, liming in strongly acidic tracts, and adequate soil conservation measures in more slopy areas need to adopted, beside maintaining proper spacing and avoiding undesirable intercropping.

As regards the fertility status, the soils in general are high in available K, high to medium in organic carbon, and medium to low in available P. Profile at Langperpen has shown high available K status all throughout. While at Jatinga, high organic carbon content is noticed up to 100 cm depth. The influence of profile depth on the fertility status can be judged from the observation that the lower horizons are mostly medium in available K and low in organic carbon and available P (Chakravarty and Barua, 1983). The amount of exchangeable Mg is generally higher than exchangeable Ca, which is followed by exchangeable K and Na. Higher amount of exchangeable Mg than Ca in soils of Assam including the hilly region have been reported earlier by Raychaudhuri (1963) and Chakravarty *et al.* (1978). The cation exchange capacity (CEC) of the citrus growing soils has been found to be rather low varying from 6.2 to 14.8 me 100^{-1} g (Table 1.60). This is apparently due to nature of the soil material, developed through progressive stages of weathering under the climatic conditions of Assam. The low activity clay with low surface charge may be responsible for low CEC of these soils. Wadia (1966) earlier reported the dominance of kaolinite in the hill soils of Assam. According to another study (Sen *et al.*, 1997c), the soils of Assam have been classified as Typic Dystrochrept, Typic Hapludalf, and Typic Haplohumult (Table 1.62) with base saturation between 12 per cent and 57 per cent.

Some chemical and electrochemical properties of six acid soils of Assam representing Alfisols, Inceptisols, and Ultisols showed that the CEC by sum of cations and NH_4OAc methods, gave much higher values than effective CEC. Total potential

Table 1.61. Some physicochemical characteristics of the citrus growing pedons of Assam.

Horizon	Depth (cm)	Organic matter (%)	pH	Exchangeable Al (me 100g^{-1})	Free Fe_2O_3 (me 100g^{-1})	Exchangeable cations (me 100 g^{-1})					BS (%)
						Ca	Mg	K	Na	H	
					Ultic Paleustalf (Labak Tisu)						
A_1	0-5	3.36	6.55	0.66	0.93	4.7	2.6	0.76	-	2.55	76
A_2	5-40	1.66	6.35	1.11	1.49	2.1	2.5	0.32	-	2.08	70
B_{21t}	40-67	1.14	5.85	0.83	2.53	3.0	3.3	0.29	-	3.01	69
B_{22t}	67-95	0.72	5.60	1.11	2.82	3.1	3.2	0.26	0.04	3.71	64
B_{3t}	95-180	0.52	5.40	1.16	2.85	2.6	3.5	0.26	-	4.40	59
					Ultic Paleustalf (Langperpen)						
A_1	0-6	1.29	5.80	0.99	1.18	1.5	3.0	0.74	-	3.24	62
A_2	6-40	0.88	5.30	1.11	1.89	1.5	2.1	0.41	-	3.94	50
B_1	40-94	0.41	5.50	0.49	1.92	2.5	2.9	0.39	-	3.94	59
B_{21t}	94-115	0.33	5.15	0.38	3.17	3.3	3.5	0.43	-	4.64	61
B_{22t}	115-152	0.31	5.40	0.61	3.07	0.38	3.3	3.4	0.52	4.17	64
					Typic Haplohumult (Jatinga)						
A_1	0-6	2.76	5.40	3.36	1.67	2.6	1.7	0.51	0.30	7.88	39
A_2	6-35	2.13	5.05	2.44	1.67	0.8	0.8	0.32	0.32	12.52	15
B_{21t}	35-100	1.88	5.20	3.33	2.39	0.5	1.0	0.19	0.08	9.97	15
B_{31t}	100-225	0.87	6.30	2.77	1.68	1.0	1.0	0.23	0.17	7.65	24
B_{32t}	225-250	0.61	5.75	2.48	1.67	0.6	0.8	0.19	0.28	6.26	23
					Typic Hapludult (Haflong)						
A_1	0-6	1.27	5.60	2.44	1.99	0.9	1.2	0.35	0.02	5.33	32
A_2	6-30	0.58	5.25	2.62	2.10	0.8	0.7	0.50	0.13	6.49	25
B_{21t}	30-80	0.58	5.10	2.44	1.49	0.5	0.9	0.32	0.02	7.19	19
B_{31t}	80-115	0.47	4.90	2.70	2.64	0.4	0.9	0.15	0.10	6.72	19
B_{32t}	115-152	0.38	5.15	1.94	2.99	0.4	0.5	0.23	0.08	5.56	18
					Typic Paleudult (Kakenedong)						
A_1	0-4	4.60	5.10	1.72	0.64	0.6	1.0	0.78	-	9.51	20
A_2	4-18	2.53	5.15	2.55	4.99	0.2	0.3	0.14	-	9.04	7
B_{2t}	18-75	0.98	5.40	3.22	5.53	0.2	0.6	0.14	-	7.65	11
B_{31t}	75-114	0.62	5.60	2.33	5.49	0.2	0.6	0.07	-	5.56	13
B_{32t}	114-150	0.33	6.00	2.05	5.14	0.2	0.6	0.11	-	5.56	14
	115-280	0.32	5.95	1.88	4s.78	0.3	0.4	0.20	-	5.33	14

BS stands for base saturation

Source : Chakarvarty and Barua (1983)

acidity constituted 34 to 38 per cent to the total exchange capacity (extractable acidity plus cations). Acidity due to variable charge (pH dependent) contributed 86 to 95 per cent of total acidity. While, exchangeable acidity contributed 5 to 14 per cent only. The variable charge characteristics correlated positively with free oxide (r = 0.66) and clay content (r = 0.74) of the soils. The clay minerals in association with free sesquioxides and organic matter contributed towards the development of variable charge (Sen *et al.*, 1997c).

Table 1.62. Physicochemical properties of some soils of Assam

Soil	Depth (cm)	Texture	Soil subgroup	pH (1:2.5)	Clay (%)	Org. C (g kg^{-1})	Free sesquioxides (%)	B S (%)
			Kariapani (Dibrugarh)					
1A	0-13	cl	Typic Dystrochrept	5.0	29.5	15.0	2.4	38
1B	13-31	cl		4.8	34.5	8.0	3.1	30
			Laisong (North Cachar Hills)					
2A	0-22	s	Typic Dystrochrept	4.8	51.5	35.0	5.2	22
2B	22-55	sic		4.8	48.4	16.5	5.9	12
			Silonijan (Karbi Anglong)					
3A	0-23	scl	Typic Hapludalf	5.5	28.0	9.8	2.8	47
3B	23-50	cl		5.2	42.0	5.9	4.8	49
			Langtin (North Cachar Hills)					
4A	0-18	sl	Typic Hapludalf	6.2	19.3	10.4	3.4	57
4B	18-47	scl		5.4	26.5	4.2	4.3	52
			Mahur (North Cachar Hills)					
5A	0-23	sic	Typic Hapludalf	5.0	49.0	17.4	5.5	16
5B	23-44	c		5.1	61.0	11.2	7.3	13
			Jorhat (Kamrup)					
6A	0-14	c	Typic Haplohumult	5.0	44.3	22.0	6.8	34
6B	14-33	c		4.8	56.6	15.0	7.9	24

A, surface soil, B, subsurface soil; BS, Base saturation; s, c, cl, sic, sl, and scl stand for sandy, clay, clay loam, silty clay, sandy loam, and sandy clay loam, respectively

Source : Sen *et al.* (1997c)

Citrus in Meghalaya

Citrus in Meghalaya is characterised by tall and lanky upright growth (Fig. 1.9)

Climate : The mean annual rainfall is 2232 mm, with more than 90 per cent occurring during May to October. The relative humidity remains between 70 per cent and 85 per cent during most of the period, but reduces to as low as 55 per cent during April, and attains peak value of 94 per cent during August to October. Analysis of rainfall data (1957 – 87) of Barapani located near Shillong,

Fig. 1.9 Tall, upright, and lanky appearance of khasi mandarin seedlings are the features of citrus plantation in Meghalaya similar to other northeastern states

Meghalaya indicated that rainfall trend remained unchanged during the last 31 years. The percentage of total months falling under normal, abnormal, and drought months during the period comes to be 66.93, 7.54, and 25.53, respectively (Satapathy, 1991). Temperature varies from 1 to 33°C. The evaporation rate during March-April remains very high in the range of 4.1 to 9.0 mm day^{-1}. Bright sunshine hours vary from 9 to 11 during the months of November to April, and remains in the range of 2 to 8 hours during May to October. During the period, from February to April, the wind velocity is quite prominent and remains in the range of 4.2 to 12.9 km hr^{-1} (Satapathy, 1996). Monthwise variation in climatic characteristics at Barapani showed maximum and minimum temperatures of 18.0° - 29.1°C and 6.7° – 19.5°C, respectively, with total annual rainfall of 1951.7 mm (Table 1.63).

Table 1.63. Monthly variation in climate at Barapani, Meghalaya

Months	Temperature (°C)		Rain-fall (mm)	Sunshine hours (hr day^{-1})	Relative humidity (%)	Wind velocity (km hr^{-1})
	Max.	Min.				
Jan.	18.0	6.8	20.8	5.86	74	3.08
Feb.	21.8	8.2	19.0	6.18	56	3.83
Mar.	26.1	13.5	46.5	6.44	53	3.38
Apr.	29.1	16.7	16.1	7.11	47	6.62
May	27.4	16.8	23.5	6.12	73	5.28
Jun.	27.2	17.8	374.0	3.58	78	3.06
Jul.	26.9	19.5	280.4	2.05	84	2.98
Aug.	27.3	19.0	272.9	3.20	81	2.71
Sept.	26.8	18.0	324.1	3.18	80	2.06
Oct.	25.5	14.7	403.4	5.34	75	2.36
Nov.	23.7	10.8	00.0	8.18	68	2.49
Dec.	21.6	6.7	00.0	7.77	69	2.56

Source : Satapathy (1996)

Soils : Soils at Barapani comprise very deep yellowish-red soils with very thick B horizon from weathered metamorphic rocks. The land forms are old and susceptible from moderate to severe erosion leading to formation of gullies. Ths soils taxonomically belong to clay, mixed, thermic family of Typic Paleudalf (Satapathy, 1996). The amount of total P and organic P is higher in Entisols than in the adjoining Alfisols, as a result of higher amount of organic matter and deposition of P, enriched surface soil by water erosion, from the latter group of soils to the former. Therefore, the contribution of organic P to total P is higher in Entisols than Alfisols (36 and 30 per cent, respectively), and both forms of P increased with an increase in elevation. Among the inorganic forms of P irrespective of soil types, the magnitude is in the order : red-P>Fe-P>Al-P>Ca-P. Inorganic forms of soil P such as Al-P and Fe-P are almost double in the Entisols as compared to adjoining Alfisols (Patiram and Munna Ram, 1993).

The Entisols had a higher amount of amorphous Al and Fe that favoured the formation of Al-P and Fe-P. Since, Ca-P is the most soluble inorganic form, it has the tendency to revert to less soluble Fe-P and Al-P in acid soils (Chang and Jackson, 1957). The higher amount of reductant soluble P (red.-P) in Alfisols (29 per cent) than the Entisols (25 per cent) indicated that former soils are more weathered than the latter.

Glimpses of khasi mandarin Tura area of west Garo hills, Meghalaya

A healthy khasi mandarin orchard in acidic red soil (Typic Paleudalf)

Extent of bearing in khai hills

Vertical cut of a yellow red soil (Typic Kanhapludult)

Another view of excellent adaptability of khasi mandarin in the area

Biotite gneiss rock type predominant in the area

Khasi mandarin cultivation on steep slopes

The significant positive relationship of red.-P to available P is due to dissolution of the coating of hydrated iron oxides around the soil particles and release of reductant soluble-P to more soluble Fe-P.

Fertility status of some Alfisols from three districts of Meghalaya (Table 1.64) has been studied by Lyngdoh and Shukla (1993) under various altitudes ranging from 800 to 1900 m above mean sea level. In general, organic carbon content is quite high in the soil samples collected from higher altitude. The soils are generally low in available P. Only a small portion (about 8 per cent) from East Khasi hills may be considered deficient in available K (Table 1.65). Singh *et al.* (1997) observed increase in organic carbon content from 4.9-8.1 g kg^{-1} to 9.8-52.0 g kg^{-1} and decrease in DTPA-Zn from 0.4-9.8 to 0.6-6.4 mg kg^{-1} with increase in altitude from 100-700 m to more than 1500 m representing Ribhoi, Jaintia hills, East Khasi hills, and West Khasi hills of Meghalaya, India.

Table 1.64. Important characteristics of the soils at low and high altitudes

Characteristics	Altitude (m)	
	800-1000	1000-1900
pH	4.1-4.8	4.1-5.5
Org.C (g kg^{-1})	10.2-45.9 (19.0)	10.0-51.6 (21.2)
Available P (mg kg^{-1})	Tr-5.7 (0.84)	Tr 23.4 (2.09)
Available K (mg kg^{-1})	43-375 (137.7)	32-325 (110.7)

Figures in parenthesis denote the average value

Source : Lyngdoh and Shukla (1993)

Among the four micronutrient cations, deficiency of Zn, Cu, and Mn is expected in quite a few soils. According to Vadivelu *et al.* (2000) the dominant soils of Meghalaya are Dystrochrept covering an area of 25.0 per cent, followed by Kandihumult (23.7 per cent), and Haplumbrept (5.6 per cent). None of the soil samples is deficient in available Fe. At higher altitudes, the soils have comparatively less available Cu and Zn than those of soils of lower altitudes, whereas the contents of Fe and Mn are higher in high altitude soils. The high altitude soils have developed in cold, humid temperate climate under mixed deciduous forests. While, low altitude soils have developed under warm, humid, subtropical climate with mixed semi-evergreen and evergreen forest. These contrasting features may account for difference in available P content at high and low altitude (Singh and Datta, 1987).

Considering 12 mg P kg^{-1} soil (Bray, 1948) as critical limit (Singh *et al.*, 1989), almost all the soils in Meghalaya are rated low in available P. The average value of NH_4OAc extractable K in the soils of Jaintia hills is low (103.8 mg kg^{-1}) when compared with those of East Khasi hills (128.9 mg kg^{-1}) and West Khasi hills (145.5 mg kg^{-1}). On the basis of critical value of K (45 mg kg^{-1}) as suggested by Jones (1979), only a small portion (about 8 per cent) from East Khasi hills are deficient in available K. It is observed that the soil samples collected from lower altitude, generally have higher available K content compared to the soils from higher altitudes. This variation is attributed to the development of soils on sandstone parent material and high rainfall at higher altitudes. On the basis of availability range of micronutrients and considering 0.07 mg kg^{-1} Cu soil as the critical limit (Singh *et al.*, 1990b), 24 per cent samples from East Khasi hills, 38

Table 1.65. Important characteristics of the soils in Meghalaya

Location	Texture	pH	Org.C (g kg^{-1})	Available P (mg kg^{-1})	Available K (mg kg^{-1})
			East Khasi Hills		
Lad Nongkrem(1)	l	4.1	18.9	1.0	75
Mylliem(1)	cl	5.5	13.3	0.4	190
Mawphlanmg(3)	sl,l	4.6-4.6	22.9-25.0	Tr-0.4	71-171
Umroi(2)	scl	4.5-4.7	20.2-24.4	Tr-2.8	268-375
Nongsdier(2)	sl	4.3-4.6	19.0-19.3	Tr	32-117
Umsning(4)	sl.scl	4.1-4.8	10.2-16.8	Tr-5.7	43-57
Nongpoh(3)	sl,cl	4.5-4.7	13.5-25.5	Tr-0.4	58-83
Saiden(2)	sl	4.5-4.7	16.6-19.5	Tr-0.4	225-293
Bhoilmbong (7)	sl,cl,scl	4.2-4.7	15.1-45.9	Tr-1.6	53-268
			West Khasi Hills		
Nongstoin(3)	sl,scl	4.4-4.7	17.1-28.3	Tr-7.4	47-325
Jakrem(4)	sl,scl	4.4-4.7	17.1-51.6	Tr-23.4	65-262
Mawkyrwat(2)	sl	4.4-4.6	11.2-23.8	Tr-4.5	62-193
Pongkung(4)	sl,scl	4.0-4.3	10.0-39.3	Tr-8.0	65-137
			Jaintia Hills		
Thadmusem(5)	scl	4.4-4.6	15.6-20.4	1.6-6.9	100-141
Saitsama(5)	scl,cl	4.4-4.6	15.7-26.7	0.4-3.9	71-120
Nongkhroh(5)	l,sl,scl	4.6-4.7	10.6-19.8	0.4	17-177
Bamkamar(4)	scl,sl	4.5-4.9	14.4-26.2	0.4-3.3	73-146
Rymphum(8)	sl,scl	4.1-5.3	14.1-35.7	0.4-1.6	48-141

l, scl and sl stand for loam, silty clay loam, and silty loam, respectively.
Figures in parenthesis denote the number of soil samples.
Source : Lyngdoh and Shukla (1993)

per cent from West Khasi hills, and 15 per cent from Jaintia hills are rated deficient in available Cu.

Stepwise regression analysis indicated that silt alone accounted for 21 per cent variation in DTPA-Mn content. Inclusion of clay only marginally raised the value to 27 per cent, and further inclusion of available P slightly improved the prediction to 33 per cent. However, inclusion of other parameters such as organic C, CEC, EC, sand, pH, and available K, improved the prediction of available Mn only to 40 per cent. Available Fe status of soils from East Khasi hills, West Khasi hills, and Jaintia hills ranged from 21 to 77 mg kg^{-1}, 44 to 73 mg kg^{-1}, and 25 to 72 mg kg^{-1}, respectively (Table 1.66). Considering 4.5 mg kg^{-1} DTPA-Fe as the critical limit (Lindsay and Norvell, 1978), none of the soils was found deficient. Nair and Chamuah (1993) reported the dynamics of exchangeable Al^{3+} in dominant soil series of East Khasi hills district of Meghalaya state and the management implications of high exchangeable Al^{3+} in these soils. Soils of southern region of the district, with very high rainfall and shallow soils on steep slopes, showed high exchangeable Al^{3+} in the surface as well as in the subsoil due to rapid leaching down of bases and residual enrichment of Al^{3+} in exchange complex. Exchangeable Al^{3+}

Table 1.66. DTPA extractable micronutrients in soils of Meghalaya

Districts	No. of samples	Micronutrients (mg kg^{-1})			
		Cu	Mn	Fe	Zn
East Khasi hills	25	0.2-3.6	0.8-21.2	21-77	0.03-2.65
West Khasi hills	13	0.14-3.64	1.0-14.5	44-73	0.05-3.20
Jaintia hills	27	0.06-2.94	1.3-20.7	24-72	0.15-1.70

Source : Shukla and Lyngdoh (1990).

often exceeded the critical limit of 60 per cent in many soil series (Table 1.67). The management strategies for these soils include cultivation of Al^{3+} tolerant varieties and liming to neutralize the exchangeable Al^{3+} or to keep its level within safer limits. Most of the current recommendations for correcting acid soils depend on pH and exchangeable aluminium (Oates and Kamprath, 1983). Liming to raise the pH to neutrality is not recommended, as it will adversely affect crop performance in soils dominated by low activity clay minerals. Surface soils generally have lower Al^{3+} saturation due to complexing of Al by organic materials (Schnitzer and Skinner, 1963), or by recycling of basic cations by plant roots and enrichment in surface through litter fall. Higher percentage of Al^{3+}

Table 1.67. Nature of soil acidity in some representative pedons of East Khasi hills of Meghalaya

Depth (cm)	pH	CEC	Exch. Al^{3+}	Exch. H^+	Al^{3+} saturation (%)	Lime requirement (t ha^{-1})
		[cmol(p^+) kg^{-1}]				
		Cheerapunjee (Typic Udorthent)				
0-10	5.1	2.29	0.75	0.26	44	3.17
10-45	5.1	4.50	2.70	0.90	80	
45-200	5.1	6.41	4.30	0.90	81	
		Mawphlang (Humaqueptic Fluvaquent)				
0-15 .	4.3	3.47	0.65	0.32	28	1.26
15-70	4.2	2.57	0.94	0.17	40	
70-100	4.0	2.85	0.94	0.46	49	
		Shillong (Typic Dystrochrept)				
0-35	5.2	4.21	1.75	0.23	47	
35-82	5.3	4.34	1.46	0.33	41	2.88
82-136	5.6	3.59	0.82	0.15	27	
136-190	5.6	2.94	0.70	0.12	27	
		Sohryngkham (Umbric Dystrochrept)				
0-16	5.4	7.25	2.63	0.27	40	
16-20	5.3	3.72	2.00	0.23	60	3.83
20-40	5.2	2.54	1.40	0.30	67	
		Nongpoh (Typic Kandihumult)				
0-15	4.9	3.85	1.27	0.23	39	
15-42	4.8	3.20	1.42	0.21	51	2.19
42-80	4.8	2.43	1.27	0.31	65	
80-145	4.8	1.51	0.60	0.17	51	
145-250	4.8	1.42	0.37	0.17	38	
		Laitdingsai (Typic Kanhapludult)				
0-25	5.3	2.63	1.29	0.21	57	
25-45	5.4	1.93	0.58	0.19	40	2.13
45-100	5.6	1.09	0.23	0.01	22	

Source : Nair and Chamuah (1993)

saturation in soils of southern parts of the district is indicative of the accelerated removal of bases and residual enrichment of colloidal complex with Al^{3+} under intensive leaching conditions.

It is well known that pH per se has no direct effect on plant growth. Low fertility of acid soils may be due to one of the following factors : Al^{3+} toxicity, Ca and Mg deficiency, and Mn toxicity., The concentration of Al^{3+} is 1 mg kg^{-1} at Al^{3+} saturation of the exchange complex equivalent to 60 per cent. The Al in soil solution rises sharply beyond this limit (Nye *et al.*,1961). Sanchez (1976) described that the primary effect of Al^{3+} is direct injury to root system. Root development is restricted and becomes thicker, and stubby with signs of dead spots. The Al^{3+} ions tend to accumulate in roots and impede the uptake and translocation of Ca and P to the top. The percentage of exchangeable Al^{3+} in the soils of the northern parts of the district seldom exceed critical level of 60 per cent saturation with a few exceptions in subsoil layers. But, in soils of southern region, Al^{3+} saturation exceeds 60 per cent even in surface layers and in all subsoil layers. Management of these soils should, therefore, include liming to neutralize exchangeable Al^{3+} and cropping with Al^{3+} tolerant species (Sanchez and Salinas, 1981).

The doses of liming are calculated for soils based on exchangeable Al^{3+} in surface soils following the equation (Kamprath, 1970) : tons $CaCO_3$ ha^{-1} = 1.65 x exch. Al^{3+} [cmol(p^+)kg^{-1}]. The values of lime requirement in acidic soils of Meghalaya vary from 0.1 to 10.9 tons ha^{-1} (Table 1.68). Application of lime to neutralise the exchangeable Al^{3+} has been found effective for soils of Sikkim (Pradhan and Khera, 1976) and Meghalaya (Prasad *et al.*,1981). The soils of the area are dominated by low activity clay and, hence, liming should be aimed at lowering Al levels to safer limits and to supply Ca^{2+} and Mg^{2+} to promote root movement to lower layers.

The soil characteristics under major land uses in Alfisols in Meghalaya have been studied by Das *et al.* (1997). Land use pattern using horticultural fruit crops including citrus ameliorated acid Alfisols by reducing Al toxicity followed by dairy farming, while

Table 1.68. Exchange properties and lime requirement

Soil	Sum of bases	Exch. acidity	Extr. acidity	pH dependent acidity	Per cent Al sat. of ECEC	LR (tons ha^{-1})
		[cmol (p^+) kg^{-1}]				
1A	3.6	1.7	12.5	10.76	18	1.3
1B	2.2	3.5	12.7	9.15	55	4.8
2A	4.0	5.1	20.2	15.15	54	7.4
2B	2.1	6.6	18.6	12.02	71	9.9
3A	2.9	0.4	5.9	5.52	30	0.1
3B	3.9	2.6	9.1	6.52	33	3.2
4A	5.3	0.2	4.3	4.11	-	-
4B	4.0	0.3	6.6	6.32	-	-
5A	2.4	6.4	19.1	12.71	71	9.7
5B	1.8	7.2	17.7	10.46	78	10.9
6A	3.7	1.8	17.9	16.10	25	2.0
6B	2.5	3.1	18.8	15.70	51	4.3

LR stands for lime requirement

A and B stand for surface and subsurface, respectively

Source : Sen *et al.* (1997c)

Table 1.69. Some characteristics of the soils at different altitudes in Meghalaya

Location (district)	Altitude (m)	Soil pH	Org C. (g kg^{-1})	$CaCO_3$ (g kg^{-1})	Exch. Ca [cmol (p^+) kg^{-1}]	Clay (%)
Ribhoi	100-700	4.7-6.0	8.1-4.9	2-28	2.5-8.25	10.0-49.0
Jaintia hills	701-1000	5.2-5.9	20.7-31.9	2-10	4.0-8.0	22.8-32.0
East Khasi hills	1001-1500	5.0-6.0	8.7-48.0	4-13	2.25-8.75	10-40
West Khasi hills	>1500	5.1-5.7	9.8-52.0	5-8	2.32-4.75	20-40

Source : Singh *et al.* (1997a)

organic matter buildup recorded highest in dairy farming (1.63 per cent), followed by agroforestry (1.6 per cent). A negative correlation between pH and extractable Al^{3+} (r=-0.39 and -0.33) suggested that higher value of soil pH, particularly at the surface, is mainly attributed to the disappearance of Al^{3+}, partly or totally from the exchange sites through the formation of Al-organic matter complex. The role of organic matter in modifying Al solubility is also hypothesized as being that of a sink (Helyer and Conyars, 1987). The activity of exchangeable Al^{3+} depends on the size of the organic matter sink relative to the mineral Al sources.

Soil analysis indicated that soils of Khasi and Jaintia hills are acidic in reaction with pH ranging from 4.7 to 6.0 and fairly rich in organic carbon varying from 8.1 to 52.0 g kg^{-1}. The $CaCO_3$ ranged from 0.20 to 2.80 per cent and clay content varied from 10.0 to 49.0 per cent. Exchangeable calcium varied from 2.25 to 8.75 cmol(p^+) kg^{-1} (Table 1.69). Amongst micronutrients, Zn is predominantly deficient nutrient (Table 1.70) in West Khasi hills (39 per cent), followed by East Khasi hills (35 per cent), Jaintia hills (26 per cent), and Ribhoi (22 per cent). Singh *et al.* (1998) observed the variation in soil fertility status as 0.32 – 2.99 per cent organic C, 1.2 – 37.0 mg kg^{-1} available P, 0.11 – 1.02 me100g^{-1} available K, 0.09 – 0.43 me 100g^{-1} available K, 0.20 – 7.00 me 100g^{-1} available Ca, 0.10 – 4.50 me 100g^{-1} available Mg, 0.6 – 14.0 mg kg^{-1} available Zn, 0.1 – 8.0 mg kg^{-1} available Cu, 0.5 – 68.3 mg kg^{-1} available Mn, and 4.0 – 63.0 mg kg^{-1} available Fe, in khasi mandarin orchards with 60 – 100 per cent orchards having soil pH less than 5.0 representing khasi hills of southern slope, nothern slope, central plateau, and Tura of Garo hills of Meghalaya, India.

Table 1.70. Distribution of DTPA-$CaCl_2$ extractable Cu, Zn, Mn, and Fe in soils in relation to varying altitudes in Meghalaya

Location	DTPA-$CaCl_2$ extractable micronutrients (mg kg^{-1})					
	Cu		Zn		Mn	Fe
	Range	%samples deficient	Range	%samples deficient	Range	Range
Ribhoi	0.6-0.3 (1.66)*	22	0.4-9.8 (6.5)	22	1.2-52.0 (30.5)	26-332 (147.8)
Jaintia hills	0.6-0.3 (1.22)	19	0.6-8.2 (6.1)	26	2.6-35.4 (26.4)	44-368 (160.8)
Eastk Khasi hills	0.4-2.6 (1.26)	15	0.4-0.8 (3.4)	35	2.2-24.8 (10.5)	56-422 (204.0)
West Khasi hills	0.5-1.9 (1.20)	12	0.6-6.4 (2.20)	39	1.6-17.8 (8.90)	68-336 (121.8)

* Figures in parenthesis indicate the mean value

Source : Singh *et al.* (1997b)

Citrus in Tripura

Amongst northeastern states, a considerable proportion of citrus is grown in Tripura, the second largest citrus producing state after Assam. The major concentration of citrus is here confined to Jampui hills, earlier known as Jampui-Tlang and Lushai hills. History of citrus cultivation in Jampui hills dates back to 1911 when Raja Hrangunga the then chief settled at Phuldungsei village from western Mizoram. The orchard concentration starts from lowest point of Hmanpui to the southern most highest point of Phuldungsei. The first citrus orchard is here believed to be established by Shri Rodingluaia at Bangla village of Jampui hills. As old as 87 year old plantation can be seen at Tlangsang village near Church established by Pastor Hkdohnyuna Missionary during 1917-1923. A model citrus orchard can be seen at Navincherra.

Climate : Tripura is situated between 22°56′ and 24° 32′ N latitude to 90°10′ and 92°21′ E longitude at an altitude varying between 280 m and 790 m above mean sea level. Most of the rain is received during April to September. Climate has a profound influence on the formation of soils. Climatically, Tripura is humid and subtropical. The mean annual precipitation is around 2200 mm and its distribution is quite uneven. Monsoon starts generally during the month of May and rains till September. However, some light showers are received in the remaining months also. Sometimes, this region also enjoys premonsoon showers during the month of March and April. The maximum and minimum temperature during winter are 27° and 13°C and 35° and 24°C, respectively, during summer. Sometimes, temperature rises up to 40°C. The maximum and minimum relative humidity (at 1000 hrs) are 85 and 57 per cent during the month of July and January, respectively. The average temperature at Agartala, Subrum, and Dharamnagar have been registered as 23.9°, 23.6 °, and 24.6°C with total annual rainfall of 3203.0 mm, 2678.9 mm, and 2427.4 mm, respectively (Table 1.71).

Table 1.71. Characteristics of climate in various citrus growing areas of Tripura

Month	Agartala		Subrum		Dharamnagar	
	Temp. (°C)	Rainfall (mm)	Temp. (°C)	Rainfall (mm)	Temp. (°C)	Rainfall (mm)
Jan.	15.0	0	18.7	2.6	17.5	18.8
Feb.	20.2	55.7	20.2	31.9	19.8	30.2
Mar.	25.5	174.4	23.5	110.3	24.2	110.2
Apr.	27.7	174.4	26.8	206.8	26.8	233.6
May	27.4	455.4	29.4	302.7	27.6	326.2
Jun.	28.2	513.1	28.0	474.6	27.8	510.0
Jul.	28.0	758.8	29.0	660.4	28.2	391.0
Aug.	28.0	491.1	27.1	416.4	28.2	332.2
Sept.	27.5	418.0	27.6	178.5	28.1	245.2
Oct.	26.4	229.5	25.5	236.8	26.4	191.8
Nov.	22.9	102.0	20.8	46.7	22.3	33.6
Dec.	18.9	0	14.4	11.2	18.8	4.6
Total	--	3203.0	--	2678.9	--	2427.4
Avg	23.9	--	23.6	-	24.6	-

Source : Directorate of Horticulture and Soil Conservation, Govt. of Tripura, Tripura

Soils : Analysis of as many as 20000 soil samples by Laskar (1979) showed that a majority of soils are medium in organic carbon content. Two representative soil profiles at Bhangmun and Siblong areas of Jampui hills showed comparatively higher concentration of clay and lower soil pH in the subsurface with low CEC. The base saturation of 49-57 per cent at Bhangmun and 28-43 per cent at Siblong was observed (Table 1.72). The distribution of soil types (Vadivelu *et al.*, 2000) has been observed in order of Dystrochrept

Table 1.72. Physicochemical soil properties in khasi mandarin growing areas of Jampui hills of Tripura

Horizon	Sand (%)	Silt (%)	Clay (%)	pH		OC (%)	Extractable bases [cmol (p^+) kg^{-1}]				CEC	BS (%)
				H_2O	KCl		Ca	Mg	Na	K		
Bhangmun												
A1	36.6	41.5	21.9	5.2	4.3	1.8	3.5	1.6	0.2	0.2	9.7	57
B21	34.6	35.5	29.9	4.8	4.0	0.9	2.0	1.6	0.2	0.3	7.8	52
B22	31.8	34.3	33.9	4.8	3.9	0.6	2.2	1.6	0.2	0.3	7.9	54
B23	27.2	35.9	36.9	4.9	4.0	0.6	2.0	1.3	0.2	0.4	8.0	49
BC	29.2	33.9	36.9	5.0	4.0	0.5	2.0	1.7	0.2	0.4	8.2	49
Siblong												
A11	32.0	44.1	23.9	5.3	4.4	2.3	2.0	1.5	0.2	0.3	9.1	43
A12	26.2	41.9	31.9	5.0	4.1	1.1	2.6	1.0	0.2	0.3	11.0	37
A13	26.2	44.9	28.9	4.8	4.0	1.0	1.6	0.8	0 2	0.3	9.1	31
AC1	25.2	49.9	24.9	4.7	4.0	0.9	1.4	0.7	0.2	0.2	9.1	31
AC2	29.6	47.5	22.9	4.9	4.1	0.8	1.4	0.6	0.2	0.2	7.1	30
C	26.0	41.1	32.9	5.0	4.1	0.5	2.6	0.6	0.1	0.1	12.0	28

OC, Organic carbon; CEC, Cation exchange capacity; BS, Base saturation
Source : Bhattacharya *et al.* (1996)

(61.9 per cent), followed by Epiaquept (12.6 per cent), Udorthent (8.0 per cent), and Haplumbrept (2.2 per cent). Based on mean values, available/total ratio for Fe though undergoing not much variation, is exceedingly low (3×10^{-4}) in upland soils as compared with that (1×10^{-4}) in lowland soils (Dutta and Munna Ram, 1993). This indicated that Ultisols in uplands are of ferruginous origin that contained high total Fe. Whereas, lowland alluvium has appreciably high Fe availability. Taking 0.5, 0.2, 4.5, and 2.0 mg kg^{-1} available Zn, Cu, Fe, and Mn, respectively, as critical limit (Follet and Lindsay, 1970), all the soils are adequate in these micronutrient cations (Table 1.73). Similar values of micronutrient availability has been earlier reported (Dutta and Gupta, 1984; Gupta and Srivastava, 1990) in soils of northeast India. Grewal *et al.* (1969) and Sharma *et al.* (1988) suggested critical limit of Mo as 0.15 mg kg^{-1} in acid soils having pH less than 6.0.

Citrus in Manipur

Manipur is the eastern most state of the country, having nearly 91 per cent area as hilly.

Climate : The mean temperature ranges from 15.8° to 22.1°C. The intensity of rainfall decreases southward. The average rainfall has been recorded as 2003 mm (Sarkar, 1994). The average temperature has been recorded as 20.4°, 24.2°, and 20.5°C at

Table 1.73. Distribution of micronutrients (mg kg^{-1}) in the soils of Tripura

Soils types	Series	Depth (cm)	Zn T	Zn A	Cu T	Cu A	Fe T x 10^{-2}	Fe A	Mn T	Mn A	B T	B A	Mo T	Mo A
							Upland							
Lalchara	Typic	0-25	227	2.3	31	0.2	201	4.3	327	33.6	97	2.7	1.3	0.12
	Dystrochrult	25-46	241	3.0	15	0.3	225	1.8	277	27.7	83	0.9	1.3	0.15
Kathlichara	Typic	0-18	62	2.1	37	0.3	264	14.0	481	18.9	83	0.4	2.6	0.16
	Paleochrult	18-36	295	2.7	33	0.2	249	5.3	354	11.4	92	0.5	1.3	0.09
Sarshima	Typic	0-10	151	4.0	40	1.7	309	16.4	404	18.8	71	0.9	1.9	0.18
	Paleochrult	10-23	206	3.5	37	0.6	316	3.8	437	16.8	110	1.1	6.6	0.23
Mean	-	-	197	2.9	32.2	0.6	259	7.6	380	21.2	89	1.1	2.5	0.16
SEm(±)	-	-	28	0.2	3.1	0.2	16	2.1	26	2.8	5	0.3	0.8	0.02
							Lowland							
Bisramganj	Typic Aeric	0-15	161	2.3	24	1.2	242	18.5	207	26.9	106	0.4	1.6	0.08
	Ochraquept	15-51	116	2.2	33	0.5	184	12.5	173	12.4	102	1.5	1.2	0.13
Baluchara	Typic	0-13	450	2.5	37	2.0	321	61.8	548	33.6	161	0.3	3.9	0.16
	Paleudept	13-36	239	3.2	20	1.8	190	23.1	526	23.6	182	0.3	1.3	0.15
Batchara	Typic	0-18	99	4.0	29	1.7	90	23.9	625	18.1	18.1	1.5	1.3	0.12
	Paleochrult	18-48	293	3.2	24	0.9	56	6.6	543	16.3	144	0.4	7.9	0.18
Vidyanagar	Typic	0-10	326	3.1	22	2.2	82	15.5	300	33.6	113	0.7	1.3	0.07
	Haplaquent	10-28	276	3.3	11	0.9	94	13.0	332	27.6	117	0.9	1.9	0.11
Mean	-	-	245	3.0	25	1.4	157	21.9	407	24.0	138	0.8	2.5	0.13
SEm(±)	-	-	42	0.2	2.9	0.2	33	6.1	62	2.8	11	0.2	0.8	0.01

T and N stand for total and available nutrients, respectively

Source : Dutta and Munna Ram (1993)

citrus areas of Thanlon, Jiribam, and Imphal, respectively, with total annual rainfall of 2652.9 mm, 2088.7 mm, and 1353.0 mm (Table 1.74).

Table 1.74. Characteristics of climate in citrus growing areas of Manipur

Month	Thanlon Temp. (°C)	Thanlon Rainfall (mm)	Jiribam Temp. (°C)	Jiribam Rainfall (mm)	Imphal Temp. (°C)	Imphal Rainfall (mm)
Jan.	13.6	33.9	16.5	12.0	12.4	14.3
Feb.	16.5	32.3	18.0	0.0	14.7	34.2
Mar.	20.3	131.7	20.0	10.0	18.7	61.1
Apr.	23.3	217.3	26.0	135.0	21.6	93.3
May	21.8	413.8	27.0	618.0	26.8	107.3
Jun.	24.0	518.9	30.0	383.0	24.6	316.9
Jul.	23.5	356.7	29.5	314.0	25.1	225.0
Aug.	24.3	418.8	26.5	90.2	24.7	209.3
Sept.	23.9	292.4	28.0	251.2	25.0	115.5
Oct.	21.8	204.1	24.5	231.2	21.9	134.2
Nov.	15.8	9.5	23.5	9.1	17.1	25.7
Dec.	15.5	23.5	24.0	35.0	13.4	16.2
Total	-	2652.9	-	2088.7	-	1353.0
Mean	20.4	-	24.2	-	20.5	-

Source : Directorate of Horticulture and Soil Conservation, Govt. of Tripura, Tripura

Soils : The soils are derived from sedimentary type of rocks and possess clay texture. An analysis of a soil type showed the dominance of kaolinite mineralogy. Water soluble, exchangeable, and nonexchangeable forms of K decreased with increasing doses of lime. Different forms of Al, except the amorphous and crystalline forms, decreased with liming. The K content from a representative soil types of Manipur (Table 1.75) to estimate various forms of Al, especially extractable and organically bound, increased with increasing lime levels (Raychaudhuri and Sanyal, 1999a). These studies suggested that the availability of soil potassium of an organic matter rich acid hill Ultisol decreased with increasing levels of lime application.

Table 1.75. Important characteristics of a soil type (Ultisol) from Manipur

pH	4.8	**Exchangeable cations [cmol (p^+) kg^{-1}]**	
CEC [cmol (p^+) kg^{-1}]	10.8	Na^+	0.45
Sand (%)	49.3	K^+	0.45
Silt (%)	16.3	Ca^{2+} + Mg^{2+}	4.1
Clay (%)	34.4	Al^{3+}	1.6
Texture	scl	**Semiquantitative clay minerological make up (%)**	
Water holding capacity (kg kg^{-1})	0.36	Kaolinite	60
Organic C (g kg^{-1})	18.6	Illite	12
		Smectite	20

scl stands for sandy clay loam
Source : Raychaudhuri and Sanyal (1999b)

Information on the taxonomic classification of these soils is rather inadequate. Considering the morphological and soil characteristics, these soils are classified into the order Ultisols. Unfortunately, identification of argillic horizon, the major diagnostic criterion for Ultisols is difficult in the field (Eswaran, 1972; Beinroth, 1982). This has created confusion for their proper placement in Soil Taxonomy. However, recent introduction of 'Kandi' concept in Soil Taxonomy (Soil Survey Staff, 1992) eliminated the need for the presence of argillic horizon in the classification of low activity clay (LAC) soils, which has facilitated better classification of these soils. With this concept, morphology and soil characteristics of four representative pedons formed on sedimentary parent material in hilly terrain of Manipur showed that these soils are classified as Typic Kanhaplohumult and Typic Kanhapludult (Sen *et al.*, 1994). Vadivelu *et al.* (2000) described the distribution of soil types as Dystrochrept covering an area of 29.3 per cent with Typic Dystrochrept (19.4 per ent) followed by Typic Udorthent (23.1 per cent), Typic Haplohumult (12.3 per cent), Typic Hapludult (3.9 per cent), and Typic Kanhapludult (6.9 per cent).

The total potential acidity is high in the Entisols, Inceptisols, and Ultisols of Manipur. Exchange acidity contributed only 2.2 to 21.5 per cent. Whereas, pH-dependent acidity contributed 78.5 to 97.8 per cent of total potential acidity. All types of soil acidity are relatively high in the hill than alluvial soils due to variation in physicochemical properties of the soils. High amount of ammonium acetate (pH 4.8) extractable Al indicated the presence of appreciable quantity of soluble $Al(OH)_3$ and hydroxy Al polymers in soils. The major controlling soil factor on exchange acidity is exchangeable Al^{3+} and pH-dependent acidity influenced by organic matter, nonexchangeable Al, and free iron oxides. Soils are taxonomically classified as Typic Udorthent, Typic Haplumbrept, Typic Kanhapludult, Typic Palehumult, Aeric Fluvaquent, Typic Haplaquept, and Typic Dystrochrept (Table 1.76).

Table 1.76. Comparison of soil characteristics of hill versus alluvial soils of Manipur

Sample no.	Depth (cm)	Soils (Subgroup)	pH	Org. C (g kg^{-1})	CEC [cmol (p$^+$) kg^{-1}]	BS (%)	Clay (%)	Free iron oxide (g kg^{-1})
				Hill soils				
1A	0-15	Typic Udorthent	5.1	24	15.5	31	52.8	59
1B	15-30		4.8	19	14.1	19	54.4	61
2A	0-15	Typic Haplumbrept	4.9	29	9.9	38	35.5	41
2B	15-30		4.8	22	8.9	22	39.4	44
3A	0-15	Typic Kanhapludult	5.3	23	10.8	47	32.8	57
3B	15-30		5.1	11	10.3	35	55.3	64
4A	0-15	Typic Palehumult	5.2	16	14.9	49	46.5	30
4B	15-30		5.1	12	14.6	36	52.3	44
				Alluvial soils				
5A	0-15	Aeric Fluvaquent	4.8	18	11.0	47	30.5	19
5B	15-30		5.2	20	14.7	51	37.5	28
6A	0-15	Typic Haplaquept	5.2	24	9.2	44	39.0	29
6B	15-30		5.8	9	8.8	65	54.0	30
7A	0-15	Typic Dystrochrept	5.5	9	8.0	71	27.0	26
7B	15-30		5.4	6	6.2	52	31.6	25

C, clay; cl, clay loam; sicl, silty clay loam; sil, silty loam
Source : Nayak *et al.* (1996a)

Information on micronutrient status of the soils of Manipur is almost nonexistent. Recent studies (Sen *et al.*, 1993; 1994; Nayak *et al.*, 1996a; 1996b) indicated that the soils are acidic to various degrees, with high content of exchangeable aluminium, poor base saturation, and low cation exchange capacity. Micronutrient deficiency is of wide concern for managing acid soils of the region. Studies by Sen *et al.* (1997b) showed that Typic Hapludalf had highest amount of variable Zn (1.4 mg kg^{-1}), followed by Typic Haplaquept (1.2 mg kg-^{1}), and Typic Hapludult and Typic Palehumult (0.6 mg kg^{-1}). While, most deficient soil types have been observed as Typic and Mollic Haplaquept (Table 1.77).

Valley soils of Manipur representing Swambung (pH 5.3 – 6.1, organic C 1.2 – 3.7 per cent, Bray-P 3.9 – 17.1 mg kg^{-1}, exchangeable K 0.13 – 1.20 me 100g^{-1}, exchangeable Ca 1.7 – 10.1 me 100 g^{-1}, and exchangeable Mg 1.8 – 6.8 me 100g^{-1}), Saikul (pH 4.8 – 5.9, organic C 0.78 – 1.81 per cent, Bray-P 6.6 – 9.2 mg kg^{-1}, exchangeable K 0.23 – 0.89 me 100g^{-1}, exchangeable Ca 0.6 – 7.8 me 100 g^{-1}, and exchangeable Mg 1.9 – 6.7

Table 1.77. Soil properties and availability of micronutrient in dominant soil pedons of Manipur

Depth (cm)	Clay (%)	Org. C (g kg^{-1})	pH (1:2.5)	CEC	Exch. Al	Available micronutrient (mg kg^{-1}) Zn	Fe	Mn	Cu
				— [cmol (p^+) kg^{-1}] —					
				Churachandpur, Typic Hapludult (Hill)					
0-17	32.2	25.0	4.9	13.8	2.4	0.6	99.4	190.6	1.4
17-35	42.4	8.0	4.9	11.0	4.5	0.2	20.8	26.8	0.6
				Bishnupur, Typic Palehumult (Hill)					
0-16	18.8	17.0	5.2	15.5	2.3	0.6	176.0	119.0	7.2
16-23	27.2	16.0	4.5	10.4	3.0	0.3	10.8	13.2	2.4
				Imphal, Typic Hapludalf (Hill)					
0-21	36.0	31.6	6.0	22.4	0.0	1.4	45.4	144.2	3.0
21-48	46.0	5.1	5.5	12.2	0.8	0.4	13.2	33.6	0.6
				Chandel, Typic Kanhapludult (Hill)					
0-15	32.8	22.8	5.3	10.8	0.6	0.2	61.6	156.2	1.0
15-30	63.5	7.3	5.1	10.1	3.8	0.1	18.2	83.8	0.4
				Thoubal, Typic Haplaquept (Valley)					
0-15	39.0	24.1	5.2	9.2	0.3	0.2	261.4	34.6	2.8
15-30	54.4	8.9	5.8	8.7	0.0	0.1	51.4	28.0	1.2
				Senapati, Typic Haplaquept (Narrow Valley)					
0-15	42.4	22.9	5.2	10.8	0.4	1.2	97.0	212.8	3.0
15-30	21.7	9.1	6.1	8.4	0.0	0.2	36.0	208.4	1.4
				Thoubal, Mollic Haplaquept (Valley)					
0-13	54.5	25.2	7.0	19.4	0.0	0.2	197.4	116.6	5.8
13-25	61.5	15.7	7.4	19.5	0.0	0.1	42.2	68.4	4.6

Source : Sen *et al.* (1997b)

me $100g^{-1}$), Bishenpur (pH 5.0 – 6.3, organic C 0.12 – 3.4 per cent, Bray-P 3.9 – 10.5 mg kg^{-1}, exchangeable K 0.18 – 1.1 me $100g^{-1}$, exchangeable Ca 2.6 – 12.1 me 100 g^{-1}, and exchangeable Mg 1.7 – 4.6 me $100g^{-1}$), Moirang (pH 5.4 – 5.8, organic C 0.72 – 1.0 per cent, Bray-P 3.6 – 10.8 mg kg^{-1}, exchangeable K 0.15 – 1.33 me $100g^{-1}$, exchangeable Ca 5.2 – 5.6 me 100 g^{-1}, and exchangeable Mg 1.8 – 2.9 me $100g^{-1}$), Thoubal (pH 5.4 – 6.6, organic C 0.24 – 1.26 per cent, Bray-P 1.3 – 23.1 mg kg^{-1}, exchangeable K 0.18 – 0.56 me $100g^{-1}$, exchangeable Ca 2.2 – 11.8 me 100 g^{-1}, and exchangeable Mg 2.5 – 9.5 me $100g^{-1}$), Kakching (pH 5.4 – 6.0, organic C 0.99 – 2.16 per cent, Bray-P 2.6 – 11.8 mg kg^{-1}, exchangeable K 0.13 – 1.48 me $100g^{-1}$, exchangeable Ca 2.1 – 9.7 me 100 g^{-1}, and exchangeable Mg 1.4 – 9.5 me $100g^{-1}$), Wangoi (pH 5.3 – 6.0, organic C 0.51 – 2.04 per cent, Bray-P 2.62 – 9.18 mg kg^{-1}, exchangeable K 0.13 – 0.58 me $100g^{-1}$, exchangeable Ca 2.9 – 10.4 me 100 g^{-1}, and exchangeable Mg 2.9 – 8.7 me $100g^{-1}$), and Hourangsabal (pH 4.9 – 6.0, organic C 0.57 – 1.32 per cent, Bray-P 1.3 – 10.5 mg kg^{-1}, exchangeable K 0.14 – 0.48 me $100g^{-1}$, exchangeable Ca 2.7 – 8.9 me 100 g^{-1}, and exchangeable Mg 2.7 – 8.6 me $100g^{-1}$) are highly acidic in nature, low in available P, Ca, and Mg according to Kailash Kumar and Rao (1990).

Another study by Kailash Kumar *et al.* (1996) in acid soils of Manipur representing Entisols, Inceptisols, and Ultisols from mid altitude of 800-1800 m revealed following soil characteristics: pH 4.6-6.3, organic carbon 1.4-5.1 per cent, base saturation 11.9-98.7 per cent, Al saturation to ECEC 54-56 per cent, clay 1.6 to 33.4 per cent, neutral salt

CEC 3.8-27.7 cmol(p^+) kg^{-1}, pH dependent CEC 1.0-6.8 cmol(p^+) kg^{-1}, pH dependent CEC due to organic carbon 0.69-4.95 cmol(p^+) kg^{-1}, pH dependent CEC due to Al/Fe 0.08-2.65 cmol(p^+) kg^{-1}, ECEC 3.1-22.8 cmol(p^+) kg^{-1}, exchangeable Ca+Mg 1.3-21.6 cmol(p^+) kg^{-1}, exchangeable Al 0.0-3.0 cmol(p^+) kg^{-1}, and extractrable Al 0.6-3.5 cmol(p^+) kg^{-1}. Various soil P fractions in these soils (Kailash Kumar and Raychaudhuri, 1996) have been observed as total P 345-1993 mg kg^{-1}, inorganic P 220-1533 mg kg^{-1}, organic P 67-501 mg kg^{-1}, active P 59-798 mg kg^{-1}, saloid-P 1-9 mg kg^{-1}, Al-P 4-330 mg kg^{-1}, Fe-P 32-428 mg kg^{-1}, occluded P 3-138 mg kg^{-1}, and Bray-P 3.0 – 26.7 mg kg^{-1}.

Citrus in Mizoram

Citriculture in Mizoram is by and large, the high altitude citrus cultivation adapted on most undulating topography (Fig. 1.10).

Climate : Most of the citrus is concentrated in the hills with some citrus in valley lands. The temperature in these hills fluctuates between 18° to 26°C in summer and 7° to 14° C in winter. Whereas, in the valleys and lower hills, it ranges from 22° to 31°C in summer and 15° to 18° in winter with perhumid climate receiving around 2900 mm annual rainfall on an average. The relative humidity throughout the year is very high. Analysis of climatic features at Aizawl showed an average temperature of 20.1°C (15.7°-22.0°C) with a total rainfall of 2263.8 mm (Table 1.78).

Soils : The organic matter content in soils of high hills has been observed higher than the soils of low hills. Soils of Mizoram hills are the product, mostly of slow diagenetic changes of the parent material (Singh and Datta, 1983), comprising micaceous schist, not particularly rich in iron content. Hill soils of Mizoram representing various altitudes (250- 1700 m) showed a distinct horizonation with clear differentiation and highly leached surface horizon, strongly acidic in reaction (4.1-5.1), low in soluble salts (EC 0.14 x 10^{-4} – 0.036 x 10^{-3} mmhos cm^{-1}), free of $CaCO_3$, CEC in range of 5.6 – 13.0 me $100g^{-1}$, distribution of exchangeable ions in the order of H + Al > Ca + Mg > K + Na, and base saturation of less than 50 per cent (Singh and Datta, 1989).

Table 1.78. Climatic features of Aizawl, Mizoram

Month	Temp (°C)	Rainfall (mm)
Jan.	15.7	19.4
Feb.	17.0	20.6
Mar.	20.5	70.8
Apr.	22.0	172.0
May	21.8	338.8
Jun.	22.0	413.7
Jul.	22.1	340.5
Aug.	22.1	314.2
Sept.	21.8	313.0
Oct.	21.3	194.9
Nov.	18.9	50.0
Dec.	16.5	15.9
Total	-	2263.8
Average	20.1	-

Source : Directorate of Horticulture and Soil Conservation, Govt. of Mizoram, Mizoram

Due to high precipitation round the year, the entire profile remains fully charged with moisture. Under the prevailing strong acidic conditions due to hydrolytic type of weathering, this moisture charged with H^+ and in intimate contact with the soil adsorptive surface, has caused an effect of slow ageing like in a H-clay system. This process has resulted high accumulation of exchangeable Al in soil. Soils of Mizoram at high altitude tend towards Spodosol, but not true Spodosol formation having variation in soil pH between 4.0 and 5.5 with exchangeable Al^{+} as high as 5.5 me $100g^{-1}$ in subsurface due to presence of other pH dependent charge constituents like humic matter and hydroxy Fe and Al (Singh *et al.*, 1986). The soils are further characterised by high subsoil acidity (Fig. 1.11). The main rock types are sandstone and shales of Barail and Disang series containing dominantly micaceous primary minerals (Table 1.79).

Fig.1.10. Khasi mandarin orchard on slopy land in one of the valley of Mizoram

Mineralogically, Mizo hills are composed of schists, granite-gneiss, and phyllite of Archean era, overlain by tertiary sediments containing sandstone shales and conglomerates. Clay minerals, vermicullite, chlorite, and kaolinite showed a parallel trend and also comparative enrichment at lower altitude of below 800 m. Soils are very deep and well drained with base saturation of about 50 per cent in surface layer and less than 36 per cent in the subsurface. The soils at 1150 m altitude have been classified as fine, sandy loam, mixed hyperthermic family of Typic Hapludults. The soils of low altitude (250 m) are acidic, olive brown in colour, very deep, and moderately well drained with base saturation of nearly 35 per cent in all the layers. The soils are classified as fine loamy, mixed, hyperthermic family of Umbric Dystrochrepts (Singh *et al.*, 1991)

Fig.1.11. A typical acid soil profile at Aizawl, Mizoram with highly acidic subsurface

Table 1.79. Physicochemical characteristics of the soils of Mizoram

Horizon	Depth (cm)	pH	Org. C (%)	Total exchangeable acidity [cmol (p$^+$) kg^{-1}]	Sand (%)	Silt (%)	Clay (%)	Exch. cations [cmol (p$^+$) kg^{-1}] Ca	Mg	K	Na	Base saturation (%)
1150 m altitude at west slope of 10-20% (Developed on siltstone)												
A11	0-10	4.6	2.4	3.9	55.8	24.6	19.9	1.0	0.9	1.0	0.8	48.4
A12	10-25	4.5	1.9	4.8	52.7	26.0	21.3	3.1	1.6	0.3	0.8	54.1
B21	25-42	1.3	0.8	5.2	50.4	25.2	24.4	1.5	0.8	0.3	0.7	35.6
B22t	42-77	4.5	0.7	6.1	49.0	20.6	30.4	1.2	0.8	0.3	0.8	34.7
B23t	77-185+	4.2	0.4	5.5	51.0	20.2	28.8	1.1	0.4	0.4	0.9	31.6
800 m altitude at north-east slope of 15-20% (Developed on siltstone)												
A1	0-9	4.8	18	3.7	27.1	42.1	30.8	1.9	1.0	0.7	0.9	54.9
B2t	9-15	4.8	1.0	4.8	20.1	38.4	41.5	2.5	1.4	0.8	0.8	53.0
B22t	15-43	4.4	0.7	5.7	29.0	31.2	39.8	0.6	0.4	0.3	0.8	27.6
B23t	43-90	4.4	0.4	5.8	32.8	30.5	36.7	0.6	0.4	0.3	0.7	26.2
C1	90-190+	4.9	0.3	6.9	28.5	42.0	29.5	0.9	0.3	0.2	0.3	19.0
550 m altitude at southwest slope of 10-15% (Developed on siltstone)												
A1	0-19	4.9	1.5	3.9	40.2	37.4	22.4	1.2	1.6	0.6	0.3	49.1
B21t	19-39	4.2	1.0	3.6	37.6	34.0	28.4	1.9	1.4	0.3	0.4	62.9
B22t	39-70	4.1	0.8	4.3	32.6	29.7	37.7	1.1	0.6	0.3	0.3	33.5
B23t	70-104	4.1	0.6	5.1	32.6	30.5	36.9	1.0	0.8	0.2	0.5	30.8
B24t	104-190+	4.5	0.4	4.5	36.5	29.5	34.0	1.0	0.6	0.2	0.5	34.4
250 m altitude at north west slope of 20% (Developed on sandstone)												
Ap	0-23	5.1	1.4	3.0	37.2	36.0	26.8	1.9	1.3	0.7	0.5	59.8
B1	23-41	4.8	0.8	5.0	27.3	27.3	27.0	1.9	1.8	0.5	0.4	47.5
B2	41-80	4.5	0.5	6.4	26.0	26.0	20.0	1.6	1.6	0.4	0.4	38.5
C1	80-201+	4.3	0.3	6.1	23.5	23.5	15.2	1.4	2.0	0.3	0.4	40.0

Source : Singh *et al.* (1991)

Citrus in Nagaland

Nagaland is one of those northeastern states having comparatively less acreage under citrus and grown as mixed crop, by and large.

Climate : The climate of citrus growing areas in Nagaland is predominantly humid subtropical with a mean annual rainfall of 2000 mm. The major proportion of rainfall is received during June to October. The maximum and minimum temperature vary from 18.2° – 28.8°C and from 7.2°-18.1°C, respectively (Table 1.80) with 52-86 per cent relative humidity.

Soils : The Soil Survey report of Nagaland (Anonymons, 1975a) showed that soils are acidic having soil pH 4.5 to 6.8. Bora (1975) related soils of Tuensang and Kohima districts as low in available K. Soils of Nagaland based on physiography have been characterised by Zende (1987). The soils of high hills are classified as loamy skeletal Typic Udorthents, loamy skeletal Lithic Udorthents, and loamy skeletal Typic Dystrochrepts at narrow ridges (less than 200 m wide); loamy skeletal Typic Hapludults, fine loamy Typic Hapludults, and fine loamy Typic Paleudults on broad ridges (more than 200 m wide); loamy skeletal Typic Dystrochrepts, loamy skeletal Lithic Udorthents on escarpment slopes (more than 100 per cent); fine loamy, Typic Hapludults, fine loamy Typic Dystrochrepts on very steep slopes (40-100 per cent), fine loamy Typic Paleudults, fine loamy Hapludults on steep side slopes (16 to 40 per cent); and fine loamy Typic Paleudults, fine loamy Typic Hapludults, loamy skeletal Typic Palehumults on straight slopes.

Table 1.80. Monthwise variation in meteorological features at Medziphema, Nagaland (1996-1999)

Months	Temperature (°C)		Rainfall	Relative humidity
	Max.	Min	(mm)	(%)
Jan.	20.1	8.1	31.2	68
Feb.	18.2	8.9	18.1	62
Mar.	24.1	14.2	38.1	52
Apr.	28.8	17.2	10.4	58
May	28.0	16.8	21.4	74
Jun.	26.2	14.6	281.2	73
Jul.	24.0	15.2	198.0	86
Aug.	26.8	17.1	304.6	84
Sept.	27.2	18.1	382.6	80
Oct.	24.1	15.8	316.4	74
Nov.	20.1	8.4	80.2	69
Dec.	19.6	7.2	74.0	71

Source : Regional Centre, ICAR Res. Complex for NEH Region, Medziphema, Meghalaya

While, the soils of medium hills are characterised as : loamy skeletal Typic Dystrochrepts, loamy skeletal Typic Udorthents, fine loamy, Typic Hapludults at narrow ridges (less than 200 m wide); loamy skeletal Typic Hapludults, fine loamy Typic Dystrochrepts, fine loamy Typic Paleudults at broad ridges (more than 200 m wide); loamy skeletal Lithic Udorthents, loamy skeletal Typic Dystrochrepts at escarpment slopes (more than 100 per cent); loamy skeletal Typic Hapludults, fine loamy Typic Paleudults, loamy skeletal Typic Udorthents at very steep side slopes (40-100 per cent); fine loamy Typic Paleudults, fine loamy Typic Hapludults, fine loamy Typic Dystrochrepts at steep side slopes (16-40 per cent); fine loamy Typic Hapludults, coarse loamy Typic Dystrochrepts, clayey Typic Palehumults at low hills (100-300 m height above ground); fine loamy Typic Paleudults, coarse loamy Typic Hapludults, fine loamy Typic Dystrochrepts at very low hills (30 to 100 m height above ground); sandy Typic Udifluvents, fine loamy Typic Hapludults, coarse loamy Aeric Paleaqults at narrow valley (less than 500 m wide); fine loamy Typic Hapludults, sandy Typic Udifluvents, coarse loamy Typic Paleudults at broad valley (more than 500 m wide); loamy skeletal Typic Udifluvents, fine loamy Typic Dystrochrepts, loamy skeletal Typic Dystrochrepts at lower river terrace; fine loamy Typic Dystrochrepts, coarse loamy Typic Udorthents at middle river terrace; and fine loamy Typic Hapludults, fine loamy Typic Dystrochrepts at upper river terrace.

The physicochemical properties of some selected soil types are fine loamy Typic Hapludults (pH 5.4-5.7, sand 38-65 per cent, silt 25-30 per cent, clay 10-32 per cent, and base saturation 18.3-23.1 per cent), loamy skeletal Typic Udorthents (pH 5.5-5.7, organic C 0.31-1.53 per cent, sand 67-76 per cent, silt 7-15 per cent, clay 18-25 per cent, and base saturation 40.0-44.9 per cent), loamy skeletal Typic Hapludults (pH 5.5-5.7, organic C 0.50-2.00 per cent, sand 17-50 per cent, silt 34-55 per cent, clay 15-35 per cent, and base saturation 20.0-24.3 per cent), fine loamy Typic Paleudults (pH 4.9-5.1, organic C 0.81-2.05 per cent, sand 24-42 per cent, silt 25-43 per cent, clay 24-44 per cent, and base saturation 20.0-24.3 per cent), and sandy Typic Udifluvents (pH 5.5-5.7, organic C 0.50-1.50 per cent, sand 40-82 per cent, silt 5-42 per cent, clay 4-18 per cent, and base saturation 45.5-57.1 per cent).

According to other studies (Ghosh and Ghosh, 1982; Patiram *et al.*, 1991), available Zn, Cu, and Mn status in soils of Nagaland have been reported to vary from 2.1 to 4.0 mg kg^{-1}, 0.2 to 2.2 mg kg^{-1}, and 11.4 to 33.6 mg kg^{-1}, respectively, with corresponding values of total content as 62 to 450 mg kg^{-1}, 11 to 40 mg kg^{-1}, and 173 – 625 mg kg^{-1}.

Table 1.81. Rainfall distribution pattern in various citrus belts of northeast India

District	Rainfall (mm)	
	Annual	Seasonal (June-September)
Shillong	4142.1	2742.8
Goalpara	2801.3	1944.6
Kamrup	2125.4	1391.8
Lakhimpur	2929.7	1838.5
North Cachar	3293.9	2041.5
United Mikkir & North Cachar hills	3071.6	1853.4
Garo hills	2735.0	1909.0
Jaintia hills	6344.9	4749.6
Naga hills	2377.3	1581.8
Lushai hills	2821.9	1927.6
Tripura	2100.7	1325.2

Source : Singh (1994)

Commercial Cultivation of Khasi Mandarin

Khasi mandarin is one of the commercial citrus cultivars mainly grown in northeastern states of India under varied soil and climate.

Climate : The cultivation of famous khasi mandarin cultivar is mostly confined to valley lands, submountane tracts with limited area in hills up to an elevation of 1500 m and rainfall from 125 cm to 1250 cm (Ghosh, 1978). The cultivation of khasi mandarin is equally successful in Mawsynram and Cherrapunjee areas of Meghalaya (Table 1.81) located at an elevation of 1313 to 1401 m above mean sea level are the wettest places in the world with an average rainfall high as 12500 mm under Khasi-Jaintia hill range (Singh, 1994). In some of the citrus growing areas like Cherrapunjee, Mawsynram, and Jawai in Meghalaya, highest one day rainfall has been recorded as 1036, 990, and 1019 mm, respectively (Dhar *et al.*, 1978; Anonymous, 1980). But now, due to continuous heavy rainfall over the years coupled with absence of any effective soil conservation measures, citrus orchards in these areas have either declined irreversibly or in a severe state of decline. In Mizoram, certain orchards are situated at 70-80 per cent slope. Such steep slope leads to the heavy loss of fertile soil by strong erosive rains every year, with the result, regular fertilization apart from conservation practices, are required to maintain the normal productivity level of these orchards.

Soils : Soil of the plains are alluvial type, having sandy loam texture, while in the hills, they are mostly lateritic with soil pH varying from 4.5 to 6.0 (Ghosh, 1978). Bhattacharya and Dutta (1952) earlier reported that citrus grows well in soil pH range of 5.5 to 6.5 under Assam conditions. Ramamurthy and Desai (1946) during first survey of citrus orchards of northeast India recorded iron toxicity and manganese deficiency. Later, B, Zn, and Ca deficiencies have been recorded in varying proportions (Choudhary, 1954; Dutta, 1959; Ghosh *et al.*, 1982; Ghosh and Singh, 1993).

Potential Areas for Khasi Mandarin

The khasi mandarin can be successfully expanded (Fig 1.6) to the newer potential areas, namely, Mawlai, Umiam, Shella areas of khasi hills of Shillong, Dawki area of

Meghalaya, Kamrup, Dibrugarh, Lakhimpur, Tinsukiya, and North Cachar area of Assam, Mockakchung and Tueusang areas of Nagaland, Jampui hills of Tripura, Thangel area of Manipur, West and East Siang districts in Arunachal Pradesh, and East and West districts of Sikkim.

Adaptability of Citrus in South India

Citrus in south India is grown on a large scale as a rainfed crop in the regions of Coorg, Wynad tract of Malabar (Kerala), Kulkul valley (Nilgiris), the Palnis, Shevroy, and Yercaud hills at an altitude of 300-1700 m (Singh, 1969). South India citrus industry comprises three major states, Andhra Pradesh, Karnataka, and Tamil Nadu. Suitability of predominant soil types of India, such as red stone (Alfisols), black soils (Vertisols), alluvial soils (Entisols), and acid laterite soils (Ultisols) for cultivation of acid lime with regard to their physicochemical properties like pH, EC, texture, nutrient, and water availability showed the poorest performance in black soil (Vertisol) due to poor growth coupled with appearance of Fe deficiency, low level of N, K, Zn, Fe, and Mn. In alluvial soil (pH 7.2), the growth has been observed as moderate, but the potassium is low. Whereas, in lateritic soils (pH 4.8), the growth is poor due to low leaf level of N, P, and K. Poor root growth is also observed due to presence of free aluminium toxic to roots (Edward Raja, 1999).

Citrus in Andhra Pradesh

Geographically, acid lime is concentrated in coastal and Telangana region, whereas sathgudi sweet orange is grown in Rayalaseema region(Table 1.82). Krishnan (2000) described predominant soil types as: Ustopept, Haplustalf, Paleustalf, Ustorthent, Ustifluvent, Rhodustalf, Camborthid, and Paleargid.

Table 1.82. Geographical distribution of citrus in Andhra Pradesh

Geographical region	Areas covered
Coastal	East Godavari, West Godavari, Guntur, and Nellore
Rayalaseema	Kurnool, Cuddapah, Ananthpur, and Chittoor
Telangana	Karimnagar, Nalgonda, and Warangal

Source : Rastogi (1991b)

Climate : The agroclimates of the areas, where citrus is grown in Andhra Pradesh have been summarised (Table 1.83). Climatic features of sweet orange growing areas of Anantapur consist of mean annual precipitation as 542.8 mm, with peak monthly rainfall of 124.1 mm during July. Mean maximum and minimum temperatures are observed as 39.3^{0}C (May) and 16.2^{0}C (January), respectively (Table 1.84). Climatic features at Tirupati indicated variation in maximum and minimum temperature from 28.6^o to 38.9°C and from 16.9^o to 26.8°C, respectively, with the total rainfall of 1057.8 mm (Table 1.85).

Soils : A variety of soils are covered under above six citrus growing agroclimatic regions.

Table 1.83. Characteristic features of various agroclimatic zones

Citrus under agroclimatic zones	Annual rainfall (mm)	Temperature (°C)		Soil type		
		Max.	Min.	Red	Black	Others
Krishna-Godavari Zone	800-1100	29-38	16-34	35	25	40
Southern Zone	700-1050	31-39	13-25	90	02	08
Northern Telangana Zone	900-1150	30-42	13-27	55	20	25
Southern Telangana Zone	700-900	28-37	16-26	97	01	02
Scarce Rainfall Zone of Rayalaseema	500-750	32-40	17-27	40	40	20
High Altitude and Tribal area	Over 1400	24-38	12-25	-	-	100

* Expressed in proportionate percentage

Source : Rastogi (1991b)

Agroclimate : Krishna Godavari Zone

The important soil groups of the zone are deltaic alluvium (association of Entisols and Vertisols), red soils with clay base (Alfisols), and black soils which are heavy and deep to very deep (Vertisols). Red loamy soils are also deep to very deep (Alfisols), coastal sands (Entisols), and saline soils (association of Aridisols, Alfisols, and Inceptisols) also occur to a significant proportion. Alluvial soils occur in the deltas, while in the uplands, red and black soils are found. This zone has peculiar agroclimatic conditions characterised by heavy rains during late September-October and often cylones during November (Rastogi, 1991b).

Agroclimate : Southern Zone

The predominant soil group is that of red loamy soils which are shallow to moderately deep (association of Alfisols and Entisols) followed by red earths with loamy subsoils i.e. Chalkas (association of Inceptisols and Alfisols). Very small patches of black soils, which are light and moderately deep to deep (association of Vertisols and Inceptisols) are also present.

Agroclimate : Northern Telangana Zone

Red soils are predominant in the zone. They include: i. the Chalkas i.e. red earths, with loamy subsoils, ii. red sandy soils i.e. Dubbas and Chalkas (association of Entisols, Inceptisols, and Alfisols), and iii. deep to very deep red loamy soils (Alfisols). The second important soil group is that of soils having an association of Vertisols and Inceptisols, and heavy, deep to very deep black soils (Vertisols). A small patch of acidic laterite soils (Oxisols) occurs in the southwestern corner of the zone (Rastogi, 1991b).

Agroclimate : Southern Telangana Zone

It is a predominantly red soil tract having red earths, with loamy subsoils i.e. Chalkas and red sandy soils i.e. Dubbas and Chalkas. Small areas along the western boundary have light and moderately deep to deep black soils (association of Vertisols and Inceptisols).

Table 1.84. A decade of mean monthly meterological data (1988-1997) at Anantapur, Andhra Pradesh

Month	Temperature (°C)		Rainfall (mm)	Relative humidity (%)		Sunshine (hrs day^{-1})	Wind velocity (km hr^{-1})	Evaporation (mm day^{-1})
	Max.	Min.		Max.	Min .			
Jan.	30.8	16.2	2.4	74	31	9.6	8.0	6.4
Feb.	34.2	18.4	0.1	64	23	10.4	8.0	8.8
Mar.	37.7	22.0	7.3	55	19	10.5	8.8	11.0
Apr.	39.2	25.3	18.6	51	20	10.1	9.2	11.9
May	39.3	25.8	46.1	57	25	9.4	12.7	12.0
June	35.8	24.0	66.4	68	36	7.6	17.2	10.2
July	33.4	23.1	124.1	75	44	5.2	18.1	9.0
Aug.	32.7	22.6	102.5	78	47	5.3	16.8	8.2
Sept.	32.7	22.7	24.5	78	45	6.9	10.1	7.5
Oct.	31.9	21.6	104.7	79	47	7.5	6.0	5.8
Nov.	30.5	19.0	35.3	81	47	7.6	6.7	5.3
Dec.	29.4	16.7	10.8	80	42	8.3	7.7	5.2

Source : Subbi Reddy *et al.* (1999)

Agroclimate : Scarce Rainfall Zone of Rayalseema

The important soil groups occurring in the zone are i. red earths with loamy subsoils, i.e. Chalkas, ii. red earths with clayey subsoils (association of Alfisols and Inceptisols), iii. red sandy soils i.e. Dubbas and Chalkas, iv. black soils which are light and moderately deep to deep (association of Vertisols and Inceptisols), and v. black soils which are heavy and deep to very deep (Vertisols). Anantapur is the driest district (544 mm rainfall), in which red soils predominate. The mean monthly maximum and minimum temperature are 29.4°C and 39.3°C, respectively. While, average monthly rainfall varies from 10 to 124.1 mm and sunshine from 7.6 to 10.5 hours day^{-1} (Table 1.84). Climatological features at Tirupati showed the variation in maximum and minimum temperature from 29.6° to 38.9°C and from 16.9° to 26.8°C, respectively. While, maximum and minimum relative humidity varied from 60.6 to 86.4 per cent and from 28.3 to 62.9 per cent, respectively, with total annual rainfall of 1057.8 mm (Table 1.85). The soils are shallow and have low fertility, besides low moisture holding capacity (Rastogi, 1991b). The sweet orange growing soils of Anantapur district of Andhra Pradesh (Table 1.86) showed that the soils are well drained, gravely sandy clay loam, neutral to moderately alkaline in reaction, and are highly responsive to

Table 1.85. Monthly variation in climatological features of Tirupati, Andhra Pradesh (1996-2000)

Month	Tempera-ture (°C)		Humidity (%)		Rain-fall (mm)
	Max.	Min.	Max.	Min.	
Jan.	29.6	16.9	84.4	49.8	5.12
Feb.	31.9	18.7	78.2	42.6	8.5
Mar.	36.2	19.8	68.9	31.3	0.3
Apr.	37.9	24.7	68.3	28.3	24.8
May	38.9	26.8	66.3	34.7	51.8
June	36.5	26.4	60.6	39.4	96.6
July	34.8	25.8	65.9	45.2	67.3
Aug.	33.3	25.6	69.4	51.6	130.9
Sept.	32.1	24.7	73.2	55.7	142.7
Oct.	33.1	22.3	77.6	62.7	167.9
Nov.	29.7	20.4	80.2	62.9	201.6
Dec.	28.6	18.8	86.4	62.2	160.3

Source : Citrus improvement project, S.V. College, Tirupati, Andhra Pradesh

Table 1.86. Properties of sweet orange (Sathgudi) growing Anantapur Soils

Soil depth (cm)	Particle size distribution (%)			Tex-ture	Gravel (%)	pH (1:2.5)	Organic carbon (%)	CEC [cmol (p^+) kg^{-1}]	Fertility status (kg ha^{-1})		Available Zn (mg kg^{-1})
	Sand	Silt	Clay						P_2O_5	K_2O	
					Anantapur soils						
0-7	60.6	19.7	19.7	sl	62	7.5	0.7	11.0	18	339	0.60
7-27	60.4	22.9	16.7	sl	26	7.3	0.62	14.9			
					Muchukota soils						
0-13	38.7	30.5	30.8	cl	50	8.2	0.58	26.8	24	338	0.68
13-41	36.6	27.1	36.3	cl	52	8.3	0.50	31.5			
					Putlur soils						
0-11	84.6	8.2	7.2	ls	20	7.1	0.28	5.7	15	339	0.98
11-30	54.3	7.8	37.9	sc	55	6.5	0.61	22.5			
30-63	58.2	8.5	33.3	scl	30	6.5	0.39	20.1			
63-96	71.7	6.2	22.1	scl	50	6.3	0.23	20.0			

sl, cl, scl, ls, and sc sand for sandy loam, clay loam, loamy sand, sandy clay, and sandy clay loam, respectively.
Source : Subbi Reddy *et al.* (1999)

agronomic management. Plants show sickly appearance in only two soil types viz., Anantapur (Loamy skeletal, Lithic Ustorthents) and Mutchukota soil types (Clayey skeletal, Paralithic, Ustropepts), which are deficient in zinc. On the other hand, Puttur soil type (Loamy skeletal Typic Paleustalfs) which is comparatively deeper, nearly adequate in zinc, support better plant growth and higher yield. The longevity of the orchard is found more in Putlur soils compared to shallow soils of Anantapur and Mutchkota (Subbi Reddy *et al.*, 1999).

Dutta *et al.* (1999) described the major soils of Anantapur, Andhra Pradesh as fine loamy, mixed, isohyperthermic family of Typic Haplustalfs (pH 6.6-7.8, organic C 0.10-0.56 g kg^{-1}, sand 60.3-73.0 per cent, silt 8.1-12.4 per cent, and clay 14.6-31.6 per cent); fine loamy, mixed isohyperthermic family of Typic Rhodustalfs (pH 5.7-7.1, organic C 0.09-0.32 g kg^{-1}, sand 48.4-72.1 per cent, silt 10.3-12.7 per cent, and clay 16.2-39.2 per cent), and fine loamy, mixed isohyperthermic family of Typic Paleustalfs (pH 6.3-.1, organic C 0.10-0.53 g kg^{-1}, sand 47.9-74.6 per cent, silt 10.0-21.1 per cent, and clay 15.4-36.3 per cent). Rao *et al.* (1985) characterised the Cuddapah basin soils as Vertic Ustochrept, Typic Ustochrept, Typic Eutrochrept, Typic Hapludalf, Typic Haplustalf, and Typic Haplustert.

Agroclimate : High Altitude and Tribal Zone

Soils are shallow red and lateritic and require major emphasis on slopy land management (Refer Chapter 4 for more details) with acid lime on a substantial scale.

Citrus in Karnataka

It is one of the south Indian states, where acid lime and lemon are grown on a large scale after Andhra Pradesh and Himachal Pradesh in an area of 13.1 thousand ha with a total production of 1.39 lakh tons (Anonymous, 1988a). The major soil types are Ustropept, Paleustalf, Ustorthent, Rhodustalf, Haplustalf, Ustifluvent, Camborthid, Calciorthid, Kandiustult, Haplohumult, and Kandihumult (Krishnan, 2000).

Climate : Citrus in Karnataka is grown under 10 agroclimatic zones, under predominantly on red, black sandy loam and clay soils at an elevation ranging from 300 to 900 m above mean sea level having an averge rainfall of 465 to 4694 mm (Table 1.87).

Table 1.87. Rainfall, elevation, and soil types covered under citrus with different agroclimatic zones in Karnataka

Sr. no.	Region	Rainfall (mm)	Elevation (m mean sea level)	Soil description
1.	Northeast Transition Zone	830 - 919	200-300 in major areas	Shallow to medium black clay soils in major areas, and red lateritic soils in remaining areas.
2.	Northeast Dry Zone	633 - 807	82-150 in all talukas	Deep to very deep black clay soils in major areas, and shallow to medium black soils in minor pockets.
3.	Northern Dry Zone	465 - 786	150-250 in 26 talukas and 225-275 in remaining talukas	Black clay medium and deep in major areas, and sandy loam in remaining areas.
4.	Central Dry	456 - 717 and 800-900	150-200 in major areas in remaining talukas 225-300	Red loam in major areas and shallow to deep black soils in remaining areas.
5.	Eastern Dry Zone	679 - 889	225-300 in major areas and 225-450 in remaining areas.	Red sandy loam in major areas and clay, lateritic soils in remaining areas.
6.	Southern Dry Zone	671 - 887	260-300 in major areas and 150-260 in remaining areas.	Red sandy loam in major area and blacks soils in remaining areas
7.	Southern Transition Zone	611 - 1054	250-300 in major areas, partly 300-500 and partly 150-250.	Red sandy loam in major areas and red loamy soils in remaining areas.
8.	Northern Transition Zone	619 - 1303	250-300 in major area and 150-250 in remaining areas.	Shallow to medium black clay soils and red sandy loamy soils in nearly equal proportion.
9.	Hilly Zone	904 - 3695	250-300 in major areas in 4 talukas, 300-500 in 6, and 4 talukas at 120-270.	Red clay and loamy soils in major areas.
10.	Coastal Zone	3011 - 4694 and 450 - 800	Less than 90 major areas in remaining areas.	Red lateritic and coastal alluvium.

Source : Rastogi (1991c)

Soils : About 16 per cent of the net cultivated area in Hassan district, Karnataka is acidic. Murli *et al.* (1974) observed the mineralogy of red soils of Mysore as kaolinitic, slightly acidic, low in iron oxides, organic matter, and cation exchange capacity. The soils belong to Alfisol, Inceptisol, and Entisol orders, with the suborders viz., Ustalfs, Tropepts, and Fluvents. In Haşsan district, 16 per cent of the total cultivated area is acidic. Majority of the soil is sandy loam with low cation exchange capacity. The nature of acidity in these soils is pH dependent. Extent of calcium saturation ranged from 5.3 to 26.4 per cent. While, aluminium saturation ranged from 11.6 to 39.4 per cent (Table 1.88). Lime addition based on exchangeable calcium method correlated well with soil physicochemical properties (Ananthanarayana and Ravikumar, 1997).

Table 1.88. Characteristic of soils of Hassan district of Karnataka

Character	Hilly zone		Southern transitional zone			
	Sakaleshpur (75)	Belur (75)	Alur (50)	Hassan (25)	Arakalgud (20)	Holenarasipur (5)
pH	4.3-5.8	3.0-3.7	4.0-5.9	4.9-5.9	4.9-5.9	4.9-5.9
Organic matter (g kg^{-1})	4.1-47.7	4.5-33.6	3.2-35.1	1.3-25.8	5.1-25.3	2.6-6.7
Texture	sl-sc	sl-sc	sl-sc	sl-sc	sl-sc	sl-sc
	Soil exchange properties [cmol (p^+) kg^{-1}]					
ECE	1.5-7.5	1.3-5.5	1.4-5.0	1.4-4.3	2.0-4.3	1.4-2.8
Aluminium (pH 4.9)	0.2-1.7	0.3-1.4	0.3-1.5	0.3-1.3	0.3-1.0	0.5-2.8
Calcium (%)	1.6-6.1	1.2-8.5	2.8-10.1	2.8-7.2	2.6-5.7	2.8-6.4
Aluminium saturation %)	1.6-39.4	12.3-28.5	11.9-29.5	9.5-32.5	10.1-28.4	7.8-30.4
Calcium saturation (%)	6.5-26.4	5.3-19.7	8.2-21.4	11.3-24.4	13.2-25.1	8.4-14.3

Numerical values in parentheses indicate the number of samples analysed

Source : Ananthanarayana and Ravikumar (1997)

In Shimoga district, 33 per cent of the total cultivated area is acidic. Total potential acidity contributed to major portion of cationic exchange capacity, which comprises pH dependent acidity and exchange acidity. The values ranged from 0.3 to 12.3 and 0.6 to 11.2 cmol(p^+) kg^{-1} in southern transitional and hilly zone, respectively. Aluminium saturation increased with increasing total potential acidity, whereas calcium saturation decreased. The pH dependent acidity contributed more than 90 per cent of the potential acidity (Ananthanarayana and Hanumantharaju, 1994). According to Yuan (1963), low base saturation is mostly associated with soils having low pH (5.3 to 5.8). Besides exchangeable H^+, Al^{3+}, Fe^{3+}, Mn^{2+}, and some other hydrolytic ions on soil exchange complex also contribute towards acidity. As expected, low calcium saturation in these soils is due to acidic parent material (granite and laterite) and leaching loss of exchangeable bases under high rainfall. Weak acids formed during organic matter decomposition in the soil is also one of the causes for soil acidity (Table 1.89).

Table 1.89. Characteristics of the soils of Shimoga district of Karnataka

Location	Number of samples	Clay (%)	Texture	pH	EC (dS m^{-1})	Organic matter (g kg^{-1})	CEC	ECEC	Exch. Al^{3+}	Exch. Ca^{2+}
							[cmol (p^+) kg^{-1}]			
					Southern transitional zone					
Bhadravathi	31	8.8-18.8	ls-scl	4.0-6.2	0.04-0.57	5.4-17.3	4.3-21.0	3.3-9.4	0.4-1.2	1.7-5.5
Channagiri	5	17.2-20.2	ls-scl	4.9-6.0	0.08-0.52	6.7-10.9	5.4-18.8	5.3-8.0	0.7-1.6	1.9-4.8
Shimoga	30	9.8-16.8	ls-scl	4.6-6.2	0.09-0.41	6.9-12.9	6.9-17.4	5.4-9.6	0.6-1.2	2.1-5.6
Shikaripura	30	10.6-19.6	ls-scl	4.1-6.0	0.08-0.30	6.7-21.2	8.0-25.6	3.0-6.3	0.6-2.5	2.2-8.9
					Hilly zone					
Hosanagara	28	9.4-15.4	ls-scl	4.3-6.1	0.02-0.56	8.3-26.6	5.6-21.6	4.6-12.4	0.6-2.4	1.9-5.4
Sagara	25	9.4-15.4	ls-scl	5.0-6.1	0.02-0.18	7.8-14.7	8.9-26.4	3.9-7.9	0.6-1.2	1.9-4.8
Soraba	26	9.5-14.4	ls-scl	5.0-6.1	0.07-0.33	5.4-18.2	11.1-27.9	5.2-11.5	0.6-1.6	3.7-9.5
Thirthahally	30	8.2-16.2	ls-scl	4.3-6.2	0.02-0.49	6.5-22.5	6.6-22.8	3.1-6.9	0.6-1.8	1.3-4.7

ls, loamy sand; scl, sandy clay loam

Source : Ananthanarayana and Hanumantharaju (1994)

Citrus in Tamil Nadu

Acid limes and sweet oranges are mainly grown in western and southern zone of the state, while mandarin is grown in the hill regions. Citrus belt of Dindigul, Anna, Salem, Triunelveli, Trichy, Madurai, Tiruchirapalli, Salem, and Kamarajar is occupied at an elevation below 800 m above mean sea level. While, sweet orange is cultivated at above 1800 m from mean sea level at Nilgiris and argoclimatically distributed to western zone, southern zone, and high altitude hilly zone (Table 1.90). According to Krishnan (2000), the major soil types are: Ustropept, Rhodustalf, Haplustalf, Kandiustult, Haplustert, Ustorthent, Ustifluvent, and Ustipsamment.

Table 1.90. Citrus growing agroclimatic zones of Tamil Nadu

Zone	Delineations
Western Zone	Periyar and Coimbatore, Tiruchengode taluka of Salem, Karur taluka of Trichirapalli, and northern part of Mudrai.
Southern Zone	Ramanathapuran and Tirunelveli district, Dindigul, Natham, Melur, Tirumangalam, south Mudrai, and Pudukkottai
High Altitude and Hilly Zone	Nilgiris, Shevroys, Elagiri-Javachi, Kollimalai, Pachdi Malai, Annamalais, Palnis, and Podhezai Malai.

Source : Gangopadhyay (1991c)

Agroclimate : Western Zone

Climate : The climate in the zone ranges from semiarid to subhumid with frequent occurrence of drought. Four distinct seasons occur : southwest monsoon (June-September), northeast monsoon (October-December), winter (January-February), and summer (March to May). The cool months of the year are November and December, and the hot months are March, April, and May. The maximum temperature of the zone ranges from 28° to 34°C, and the minimum from 20.9° to 24.3°C. The maximum temperature is experienced during the months of March, April, and May which gets reduced gradually, and reaches the minimum during the months of December and January. Being an interior region, the diurnal variation in temperature is large, particularly in the dry and hot seasons. The areas adjoining to the western ghats, such as Thandikudi, Virupakshi of lower Palanis, Bodimettu, and Cumban valley of Uthamapalayam talukas experience a minimum of 10° -15°C during the cooler months.

Soils : The soils of this zone are divided into six groups. They are: i. red non-calcareous, ii. red calcareous, iii. black calcareous, iv. alluvial, colluvial, mixed soils, and associations, v. forest soil, and vi. problem soils. Among the different types, the red non-calcareous soil is the most predominant one, which comprises Irugur, Vannapatti, and Pinchanur series. The soils of Irugur series represent 37 per cent of the total area of the zone. It is distributed widely in Periyar and Coimbatore districts, and to a certain extent in Karur, Usilampatti, Nilakottai, and Palani Talukas. Vennapati series is another noncalcareous soil series, which occupies major area in Karur taluka of Trichy district. The Vylogem is another important soil series, although its distribution is less, but it is a problem creating soil series. Development of soil crust is a common feature associated with soil types of this series, besides subject to erosion. They are found at Nilakottai, Usilampatti, and Karar. Palaviduthi series occupies vast area in Nilakottail and Usilamapatti talukas, and to a small extent in Palanil taluka. The second predominant soil type is red calcareous. It comprises Palladam, Tulukanur, Palathurai, and Anaipur series. Palladam series is found to occur mostly in Coimbatore district. Tulukanur series occupies major areas of Periyar district and Karur taluka. Sankararaj *et al.* (1983) exclusively suggested the various properties of soils of Salem, Tamil Nadu which are by and large, characterised by acidic soil pH (90 per cent), low to medium available N, P, and K (50-60 per cent).

In black calcareous soil type, Peelamedu series is an important one. It is distributed in almost all talukas of this zone. Alluvial, colluvial, and soil associations are distributed in Nilakottai, Usilampatti, and Karur. Peelamedu, Anaipur, and Pelathurai soil series are also found in this zone (Gangopadhyay 1991c).

Agroclimate : Southern Zone

The zone mainly comprises the southern plain and foot hills of the western ghats ranging to 500 m above mean sea level. The topography is undulating and tapered towards east coast. The north and south areas such as Dindigul and Madurai are covered under this zone.

Climate : The mean monthly maximum temperature ranges from 30° to 37°C and mean minimum temperature from 21° to 26°C. The high temperature that prevails increases the evaporative demand of the crop. The annual potential evapotranspiration is 163 mm. The precipitation exceeds potential evapotranspiration (PET) only during October and November, and PET is higher than precipitation during the rest of the year. Major portion of this zone lies on the rain shadow of western ghats. The climate is semiarid tropic. The mean annual rainfall is 883 mm and northeast monsoon contributes maximum of 514 mm, southwest monsoon 213 mm, and the transitional dry period 156 mm. The northeast monsoon rains are more dependable and cropping is centered around northeast monsoon rains (Gangopadhyay, 1991c). Climatic features of commercially well known acid lime belt (Periyakulam) of Madurai district showed the variation in maximum and minimum temperature of 21.5° - 32.8°C and 10.6° - 19.0°C, respectively, with a total annual rainfall of 1208.8 mm (Table 1.91).

Table 1.91. Meteorological features of Periyakulam, Madurai, Tamil Nadu

Month	Temperature (°C)		Humidity (%)		Rainfall
	Max.	Min.	Max.	Min.	(mm)
Jan.	28.0	12.0	87.0	66.0	--
Feb.	29.9	10.6	81.2	52.0	2.5
Mar.	32.8	14.6	80.3	51.4	--
Apr.	32.5	17.3	79.0	51.0	101.0
May	31.0	18.9	79.6	53.5	73.5
June	24.8	19.0	86.0	83.0	442.4
July	21.5	17.0	98.0	90.2	285.4
Aug.	24.0	18.0	88.0	80.0	166.7
Sept.	25.0	16.2	83.0	67.2	108.0
Oct.	26.2	16.0	78.0	60.1	80.0
Nov.	27.0	17.1	78.5	73.0	94.8
Dec.	23.0	14.0	82.2	52.0	50.8

Source : Horticulture Research Institute, Periyakulam, Tamil Nadu

Soils : The major soil groups of the southern zone consist of deep to medium black soil (Vertisol), red soils varying in texture and depth, red loamy soil, laterite soil, river alluvium, and saline coastal alluvium.

Agroclimate : High Altitude and Hilly Zone

The Palni hill region of western ghat is situated in Maduri, Dindigul, and Quiad-e-Millath districts of Tamil Nadu. The areas such as Kodaikanal, Yercaud, Shevroy hills, and Vijayanagaram (Ooty) are covered under this zone.

Climate : Number of rainy days from March through December is 69 with total rainfall varying from 1173 to 1237 mm. Temperature in the upper Palni is cool and fairly uniform throughout the year. January and Febraury nights are cool reaching to the low of 8°-9°C with frost frequently forming in the valleys. Maximum temperature ranges from 15° to 23°C and minimum temperature from 2° to 15°C. The summer months in the lower Palni is somewhat hotter, the temperature even goes as high as 30°C and winter temperature goes below 8°C in certain months. Maximum temperature ranges from 18° to 31°C, while minimum temperature ranges from 8° to 18°C.

Soils : There are 3 distinct soil types, namely, laterite, dark brown soils, and red soils. The soils spread uniformly through Nilgiris, Shevroys, Elagiri Jevadhi, Kothi Mallai, Annamalais, Palnis, and Padhigai Malai. Most of the citrus cultivation is in lateritic soil. Cultivation in the zone is constrained by soil erosion, longer crop duration improvished soils (Gangopadhyay, 1991c).

Commercial Cultivation of Coorg Mandarin

Coorg mandarin is claimed to be one of the oldest citrus cultivations in India (Rice, 1878). O'Connor (1870) once expressed in his earlier writings about the existence of flourishing orange gardens in Coorg during his visit in 1815. Richtor (1870) later said "the oranges are celebrated fruit and as common as plantains". The mandarin is not indigenous to Coorg, but have come from China and Cochin China, which is the original home of this species (Batchelor and Webber, 1948). In south India, largely comprising of Karnataka, Tamil Nadu, and Kerala, citrus is grown in warm humid region with tropical climate in Coorg, Wynad, Nilgiris, Palni, and Shivroy hills. Adverse soil conditions, such as high water table and inadequate drainage brought about early decline of citrus in south India, reported as early by Naik (1948). Coorg mandarin is successfully grown in submontane region of Wynad in Kerala (Devadas *et al.*, 1988).

Climate : Coorg mandarin is concentrated in the central zone of Coorg district, having annual rainfall of 1250-2500 mm at an altitude of 2000-3000 feet above mean sea level, and temperature varying between 17° to 33°C (Varadarajan and Subramanian, 1953; Aiyappa and Srivastava, 1965; Aiyappa *et al.*, 1965). The Coorg mandarin cultivation is mainly concentrated in Hassan, Kodagu, and Chikmagalur districts. The features of variation in climate are given below:

Location	Agroclimatic zone	Average rainfall (mm)	Elevation (m)
Hassan			
Alur, Belur, and Arakalgudu	Southern transition zone	611-1054	250-300
Coorg			
Virajpet, Somvarpet, Madikeri, and Markara	Northern dry zone	671-887	250-300
Chimagalur			
Mudigere, Koppa, Sringeri, Narasimharajepur, and Chikmagalur	Hilly zone	904-3695	250-450

Source : Rastogi (1991c)

The above areas are covered under the agroecological region of eastern ghats, TN uplands and Deccan (Karnataka) plateau, hot semiarid ecoregion with red loamy soils and growth period of 90-150 days. Coorg mandarin in Malnad (then Mysore state) is grown under climate having 1250-3000 mm of annual rainfall at an altitude range of 500-1700 m above mean sea level as rainfed crop (Anonymous, 1972).

Soils : The soils under Coorg mandarin have been formed by the weathering of the metamorphic rocks. Seven groups have been distinguished in the rocks of Coorg by Holland (1897). They are clay slate or argillaceous schist intruded by a granitic core in the centre, biotite gneiss, and charnokite in the east, hornblende, schist, and quartz, haematite, and schist in the north. Syenite, ferruginous laterite, and feldspar appear sporadically. The land area is composed mostly of hilly slopes with the table lands in between. Due to variation in the natural vegetation, mean temperature, rainfall, and altitude within short distance, the soils in Coorg are highly heterogeneous. According to

Rastogi (1991c), the soils of Coorg area are red sandy loams (Alfisols), red clay, and loamy soils (Alfisols and Inceptisols). Mukherjee (1949) reported Zn and Mn deficiencies in Coorg soils. Govindarao and Reddy (1957) reported deficiency of Mn, Mg, Fe, B, and Cu in Coorg mandarin growing soils. Exposure to continuous drought period of 5 to 6 months has been noticed by Naik (1948) and Ramkrishnan (1954a). Under this situation, rainless period with soil moisture deficit induces dormancy needed for flowering and fruiting (Chadha and Singh, 1989).

Climate and soil under Wynad, Kerala characterised by humid tropical climate, has shown good promise for citrus cultivation in the past. But, now most of the orchards are either replaced by other plantation crops or in severe state of decline. A majority of soils have developed over weathered granite-gneiss having pH 5.5 – 6.6, organic C 15.0 –34.0 g kg^{-1} , sand 30.1 – 49.0 per cent, silt 38.0 –45.3 per cent, KCl extractable Al 0.08-0.42 cmol(p^+) kg^{-1}, and base saturation 42.6 – 72.6 per cent Mineralogically, the soils contain opaque minerals (40 per cent), followed by pyroxene (15 per cent), hornblende (12 per cent), tourmaline (9 per cent), biotite (8 per cent), muscovite (6 per cent), chlorite (3 per cent), and zircon (3 per cent). These soils are comparatively younger and possess higher amount of nutrient reserves (Lekha *et al.*, 1998).

Citrus in Andaman and Nicobar Islands

Amongst different citrus cultivars, acid lime/lemon is grown in Andaman and Nicobar island on a small scale.

Climate : Agroclimatically, Andaman and Nicobar islands is divided into northern zone, central zone, and southern zone, identified based on the variation in topography and soil. These three agroclimatic zones belong to hot, perhumid ecoregion having red loamy and sandy soils with growing period more than 210 days (Sehgal *et al.*, 1990). Citrus cultivation is adapted in hill region of central zone, which covers around 225 ha area with a total production of 563 metric tons. The citrus growing areas of hill slopes of Andaman and Nicobar Islands represent tropical rain belt. Cultivated area in the Andaman and Nicobar Islands is very limited. In this tropical rain belt, soil management is the single most important factor that determines their perpetual crop productivity. The climate of islands is typically equatorial. The difference between the annual maximum and minimum temperature is rather small. The peak temperature is experienced during dry month when evapotranspiration losses are highest. The maximum daliy sunshine hours (8 to 10 hrs) are also experienced during this period, but owing to moisture stress, this advantage is not fully reaped by vegetation, particularly the drought sensitive species. Temperature between May and December is moderated by rains. There is no moisture stress during this peroid, but the cloudy weather restricts sunshine to 3 - 8 hours. The islands receive an average rainfall of 3000 mm distributed unevenly throughout the year (Table 1.92).

Table 1.92. Characteristics of climatic variation at Andaman and Nicobar islands

Month	Rainfall[1] (mm)	Temperature (^{0}C)		Relative humidity[3] (%)		Wind speed[4] (km hr^{-1})	ET[5] (mm)	Sunshine[5] (hr. day^{-1})
		Min.	Max.	Max.	Min.			
Jan.	42.7	22.1	29.8	71	75	7.2	156.2	8.42
Feb.	24.0	21.8	30.5	70	73	5.7	138.9	9.73
Mar.	9.7	22.5	31.6	68	72	5.3	191.3	9.45
Apr.	68.3	23.9	32.6	68	73	5.6	180.0	8.75
May	397.2	23.9	31.1	79	82	11.9	138.3	6.12
June	497.5	23.8	29.4	83	84	18.4	122.1	3.61
July	459.7	23.5	29.1	85	86	15.8	113.5	3.06
Aug.	442.3	23.6	29.0	84	86	19.1	120.5	3.84
Sept.	464.1	23.1	29.0	83	86	12.4	112.5	4.92
Oct.	305.9	23.0	29.5	81	86	8.1	98.9	4.66
Nov.	220.2	23.1	29.8	78	82	6.9	97.2	6.90
Dec.	155.2	22.8	29.3	73	78	8.1	119.0	7.88
Total	3086.8	-	-	-	-	-	1588.2	-
Avg.	-	23.1	30.1	77	80	10.4	-	6.16

[1] Average of 38 years i.e. from 1949 to 1986; [2] Average of 26 years i.e. from 1961 to 1986
[3] Average of 21 years i.e. from 1966 to 1986; [4] Average of 12 years i.e. from 1975 to 1986
[5] Data for 1986 only
Source : Singh and Gajja (1987)

Soils : A study of soils of these islands by Singh *et al.* (1988) provided the information regarding soil characteristics, fertility status, and management of soils, mostly in South Andaman (Table 1.93). Hill slopes are mostly used for raising citrus orchards and spices, but the productivity is low. Five pedons of the hill slope soils of Middle Andaman Islands have been characterised and classified taxonomically. These are situated on a slightly sloping land terrain with undulating surfaces. Soils are strongly to extremely acidic and fairly high in organic matter. The value of CEC varied from 10.6 to 23.2 cmol(p^+)kg^{-1} and calcium dominated the exchange complex along with exchangeable Al^{3+}. Soils are classified under two orders: namely, Entisols and Inceptisols (Table 1.93). Soils of middle Andaman Islands are classified as Fluventic Eutrochrepts, Aquic Eutrochrepts, Aeric Tropaquepts, and Typic Udifluvents with surface organic matter content of 0.06-0.37 g kg^{-1}, CEC of 8.4-23.0 cmol (p^+) kg^{-1} , and loam to sandy clay texture. Sand mineralogy indicated the dominance of quartz in sandstone parent material (Sahu and Nirmalya Bala, 1995).

Ganeshamurthy *et al.* (1989) classified hill slope soils of Andaman and Nicobar Islands as Orthent (pH 4.9-5.3, organic C 0.51-1.53 per cent, silt 3-8 per cent, clay 4-42 per cent, and exchangeable Ca + Mg 1.6-3.6 cmol (p^+) kg^{-1}), Ochrept (pH 6.0-6.4, organic C 0.48-1.92 per cent, silt 10-42 per cent, clay 18-76 per cent, and exchangeable Ca + Mg 4.8-16.2 cmol (p^+) kg^{-1}), and Ustalf (pH 5.0-8.5, organic C 0.26-1.15 per cent, silt 25-52 per cent, clay 39-77 per cent, and exchangeable Ca + Mg 17-37 cmol (p^+) kg^{-1}). Sarkar *et al.* (2000) described the soil types as: Troporthent, Eutropept, Humitropept, Tropaquept, Tropudalf, and Hydraquept.

Table 1.93. Physicochemical characteristics of the soils of Andaman and Nicobar island

Depth (cm)	Particle size distribution Sand	Silt	Clay	pH (1:2)	EC (dSm^{-1})	Org.C ($kg\ kg^{-1}$)	Exch. H^+	Exch. Al^{3+}	Total exch. acidity	Free oxides (%)
	——(%)——						——[cmol (p^+) kg^{-1}]——			
					Typic Udifluvent					
0-17	75.9	13.7	10.4	5.3	0.05	0.016	0.6	5.9	6.5	3.3
17-45	61.9	16.7	21.4	4..7	0.04	0.008	0.4	6.3	6.7	
45-60	77.9	9.7	12.4	4.1	0.24	0.006	0.3	6.8	7.1	
60-110	57.9	15.7	26.4	4.7	0.09	0.005	1.6	4.6	5.9	
					Typic Ustifluvent					
0-15	47.9	27.7	24.4	4.7	0.07	0.014	0.5	2.1	2.6	4.6
15-40	61.9	13.7	24.4	4.6	0.06	0.012	0.3	4.3	4.6	
40-90	51.9	17.7	30.4	4.5	0.04	0.008	0.8	9.0	9.8	
90-130	51.9	33.7	14.4	4.7	0.04	0.016	0.3	2.0	2.3	
					Typic Udifluvent					
0-20	59.3	17.7	33.0	4.6	0.08	0.022	0.4	2.0	2.6	4.0
20-50	51.3	27.7	21.0	4.5	0.04	0.013	2.2	9.1	11.3	
50-70	69.3	8.7	22.0	4.5	0.08	0.006	1.6	9.4	11.0	
70-105	37.3	26.7	36.0	4.7	0.03	0.004	0.4	2.2	2.6	
105-145	61.3	1.7	37.0	4.8	0.04	0.002	1.0	2.7	3.7	
					Typic Udifluvent					
0-15	77.3	15.7	7.0	5.2	0.05	0.016	0.3	-	0.3	4.2
15-65	49.3	20.7	31.0	4.8	0.04	0.005	0.6	3.0	3.6	
65-95	63.3	9.7	27.0	5.0	0.05	0.003	0.9	0.4	1.3	
95-115	51.3	19.7	29.0	5.2	0.05	0.013	0.3	-	0.3	
115-145	54.3	22.7	23.0	5.6	0.06	0.015	0.2	-	0.2	
					Fluventic Dystrochrept					
0-15	52.2	26.0	21.8	5.3	0.08	0.0.19	0.3	-	0.3	5.6
15-40	42.2	24.0	33.8	4.7	0.04	0.009	0.6	3.9	4.5	
40-95	60.2	20.0	19.8	5.2	0.06	0.008	0.8	0.3	1.1	
95-130	57.3	19.7	23.0	4.8	0.04	0.004	0.3	-	0.3	

Source : Nirmalya Bala and Sahu (1993)

Commercial Cultivation of Sweet Orange

The well marked belts of sweet orange occur in the country. To name few of them, agroclimatically are: Agra, Aligarh, and Mathura regions having plain alluvial soils of Uttar Pradesh, Marathwada region (Nanded, Parbhani, Jalna, and Aurangabad) of central Maharashtra (Fig. 1.12), and Ahmednagar, Pune and Nasik area of western Maharashtra, and sathgudi growing area of Andhra Pradesh (Anantapur, Kodur, Cuddapah, and Tirupati).

Climate : According to Naik (1948) production of sweet oranges (Sathgudi) seems to be largely favoured by dry or arid conditions, prevailing during the good part of the year on the plains of south India as in Cuddapah, Kurnool, and Chittoor districts of Rayalseema and in the northern Ciracars of Andhra Pradesh. Under humid environment, and on higher altitude of Wynad, lower Palni, and Coorg, the fruits do not develop colour properly and turn insipid, though juicy.

Sweet oranges are adapted to arid tropics and subtropics. Contrary to the mandarin, sweet orange has poor adaptability under warm humid conditions. The Kodur sathgudi is grown agroclimatically in central dry zone which has poor colour development under tropics, but developed excellent deep orange colour when grown under subtropical climate (Chadha and Singh, 1989). Scarce rainfall zone of Rayalseema of Andhra Pradesh consisting of sweet orange growing areas of Kurnool, Anantapur, part of Cuddapah, and Prakasham has average rainfall of 500-750 mm, temperature maximum as 32°-40°C, and minimum as 17°-27°C (Rastogi, 1991b). In Uttar Pradesh, sweet orange is grown in alluvial soils in south western, semiarid zone comprising of areas namely, Agra, Aligarh, and Mathura regions having average rainfall of 765-776 mm (Saxena, 1991b). Agroclimatically, sathgudi sweet orange growing areas of Andhra Pradesh are similar to the subhumid tropical climate of Vidarbha region, and adjoining areas of Satpura plateau of Madhya Pradesh where famous Nagpur mandarin is cultivated on a large scale. The precise reasons for existence of such distinct belts of different citrus cultivars are yet not known and require further to be elucidated in more details. However, comparison of climatic features across sweet orange belts (Table 1.83) revealed that sweet oranges prefer more hot and dry climate than mandarin.

Fig. 1.12. Sweet orange, cultivar mosambi grown in Marathawada region of Maharashtra

Ghosh and Srivastava (2001) suggested that best growth of sweet oranges is obtained at temperature of 16°-20°C. It can tolerate maximum temperature of 32°-40°C and minimum of 17°-27°C while analyzing the climatic variation across sweet orange belts of the country.

Soils : Important soil types occurring in scarce rainfall zone of Rayalseema of Andhra Pradesh are red earths with loamy subsoils (Entisols), red earths with clayey subsoils (association of Alfisols and Inceptisols), red sandy, and black clay soils (Vertisols) of varying depths (Rastogi, 1991b). The soils of sweet orange growing areas of Uttar Pradesh are largely alluvial in nature, where large scale Zn deficiency is commonly observed (Singh and Tripathi, 1985).

Fig. 1.13. A mature sweet orange, cultivar Mosambi tree of Marathwada region of Maharashtra

Khanduja (1969) observed a large scale deficiency of N, P, Fe, Mn, and Zn in mosambi orange orchards of Maharashtra. The sweet orange growing areas of Maharashtra have black soils derived from basaltic parent material (Fig. 1.13). According to Malewar *et al.* (1979), 31.0 per cent sweet orange growing soils are deficient in available Zn in Aurangabad division. The available Zn content of Marathwada soils has been observed in the order of shallow black > deep black > medium black > forest shallow > loamy soils (Malewar and Randhawa, 1978). Malewar *et al.* (1978) observed lower available Zn (0.60-1.40 mg kg^{-1}) in citrus orchard soils than noncitrus orchard soils (0.28 - 9.88 mg kg^{-1}). Large scale Zn deficiency is noticed in sweet orange orchards of Marathwada region (Malewar *et al.,* 1978). The studies on sathgudi sweet orange grown in Alfisol, Inceptisol, and Vertisol type of soils in Rayalseema region of Andhra Pradesh showed best quality of sweet orange in Inceptisol type of soils (Vijayasankar Reddy *et al.*, 1992).

Potential Areas for Sweet Orange

Good potential of sweet orange exists in the areas of alluvial plains of Agra region, lower foot hills of Uttaranchal, Rajasthan (Jaipur, Bhilwara, Udaipur and Ajmer), Madhya Pradesh (Morena, Gwalior, Patia, Hamirpur, and Lalitpur), Gujarat (Gandhinagar, Himatnagar, and Surendragarh), Maharashtra (Kolhapur, Sangli, Solapur, Satara, and Pune), Andhra Pradesh (Nizamabad and Adilabad), and Karnataka (Raichur, Belgaum, and Dharwad) having more or less similar agroclimate. Heavy rainfall and high humidity regions, like Konkan, Kerala, Karnataka, Tamil Nadu, West Bengal, Assam, Meghalaya, Mizoram, Manipur, and Sikkim are not ideally suitable for the cultivation of sweet oranges. In these regions, due to high humidity, the trees showed irregular flowering at different periods of year, and fruits did not develop proper colour and ripening, which later became insipid in taste (Randhawa and Srivastava, 1986).

Commercial Cultivation of Acid Lime

Acid lime is the hardiest of all citrus cultivars and adapts well under a wide variety of climate and soil (Fig. 1.14), since flowering and fruiting is not a problem in acid lime.

Guillen Paiz (1975) in his review on prospects of lime in agricultural diversification described the climatic (temperature and rainfall) and soil requirements for commercial plantations as fresh fruit, canned juice or essential oil production. Principal acid lime producing regions of Andhra Pradesh are Godavaries (Palakol) and Nellore (Venkatgin) (Singh, 1969; Chadha and Singh, 1989). In Andhra Pradesh, acid lime is commercially cultivated agroclimatically in Krishna-Godavari zone, which has annual rainfall of 800-1100 mm and maximum and minimum temperature as 29°-38°C and 16°-34°C, respectively. The important soil groups of this zone are deltaic alluvium (association of Entisols and Vertisols), red soils with clay base (Alfisols), coastal sands (Entisols), and Vertisols of varying depths (Rastogi, 1991b). Acid lime in Tamil Nadu is grown agroclimatically in western zone (semiarid to subhumid) consisting of Periyar, Dindigal, Coimbatore, Salem, Karur taluka of Trichulapally, and northern part of Madurai having rainfall of 692-811 mm received in 45 rainy days. The soils of this zone consist of red noncalcareous/ calcareous types, alluvial, and black soils (Gangopadhyay, 1991c).

Hilly tract of Naubutak district surrounded by Pauri in west, Almora in north, Pithoragarh in east, and Tarai, Bhabar part of Nainital district in south is also a commercial acid lime belt in Uttaranchal. A vast area under acid lime is cultivated up to 1200 m elevation. Annual precipitation of the area varies from 1250 to 1500, mm with temperature from –5°C in winter to 30°C in summer. Soils of this zone are generally formed from biotite, phyolyte, granite, schist, and at some places from calcareous rocks forming various textural classes. Available N is medium in Ramgarh, Okhalkanda, Dhari, and Betalghat areas, while available P is low in all the blocks. The available K is low in Okhalkanda and Dhari blocks, and medium in Ramgarh, Bhimtal, and Betalghat areas, which are near to low range (Divakar *et al.*, 1989).

Fig. 1.14. An orchard of acid lime in black clay soil of Vidarbha region of Maharashtra

Nellore district of Andhra Pradesh lies at 14°05'N latitude -79°08'E longitude, where acid lime is commercially grown on a large scale. The area receives mean annual rainfall of about 843.7 mm, more than 60 per cent of which is received during June to September. The mean annual temperature is 29.2°C with mean summer temperature of 31.7°C, but goes up to more than 40°C during April and May. The mean annual winter temperature is 25.3°C, but goes down to 16°C during December and January. The soil temperature and moisture regimes for this area are hyperthermic and ustic, respectively.

Six representative profiles of Somasila and Teluguganga project area in Nellore district located on the gently rolling plains showed the presence of hard and compact laterite or ferruginous layer. The soils are light in texture, with low water holding capacity, slightly acidic to neutral pH, low organic matter, and CEC. No vertical movement of total Fe within the profile is observed. But, substantial loss of Ca, Mg, and K from the soil is noticed. These soils are relatively infertile, because of the paradoxical situation existing in the semiarid monsoonic climate that do not permit sufficient weathering to release nutrients and are subjected to serious erosion of high intensity in the east coast of Andhra Pradesh (Table 1.94). The soils have been classified belonging to order of Alfisols, Inceptisols, and Entisols (Bhaskar and Subbiah, 1995). Lack of distinct horizonation is the general characteristic of laterite soils as earliar observed by Kellogg (1950).

Higher values of coarse sand in the intermediate layers of individual profile is due to the separated iron oxides forming nodules or pseudoaggregates and is one of the

Table 1.94. Physicochemical characteristics of soils of Nellore district of Andhra Pradesh

Depth (cm)	pH (1:2)	Particle size distribution (%)			Org. C (%)	Exchangeable cations [cmol (p^+)kg^{-1}]				Base saturation (%)
		Sand	Silt	Clay		Ca	Mg	K	Na	
				Allimadugu loam (Typic Ustorthent)						
0-9	6.5	67.0	10.9	21.9	0.40	1.4	1.5	0.08	0.3	75.4
9-21	6.3	64.3	12.8	22.8	0.34	3.7	1.8	0.07	0.3	74.3
21-38	6.5	67.6	12.1	20.2	0.26	3.2	3.0	0.14	0.3	74.4
				Veenkateshwarapalem red clay loam (Udic Rhodudalf)						
0-15	17.9	57.9	16.1	26.1	23.1	26.1	1.4	32.7	2.1	3.6
15-31	15.6	51.3	12.6	36.1	34.6	36.1	1.3	34.1	4.2	4.1
31-53	15.4	46.4	12.1	41.5	32.0	41.5	1.4	37.6	3.2	3.8
63-68	21.9	13.7	37.7	31.6	30.8	31.6	1.5	28.0	1.0	1.5
68-92	18.1	42.8	24.3	32.9	28.6	32.9	1.5	31.1	0.7	1.8
				Venkatachalam sandy loam (Udic Ustochrept)						
0-10	27.9	70.0	4.4	15.5	54.1	14.5	1.5	23.7	0.9	18.2
10-22	66.0	73.2	9.3	20.5	46.6	20.5	1.6	34.7	1.0	7.5
22-41	69.4	49.0	16.3	34.7	36.1	34.7	1.5	35.8	1.4	3.0
				Saidapuram loam (Udic Ustochrept)						
0-11	13.0	60.3	18.3	21.4	40.0	21.4	1.6	27.0	1.1	3.3
11-26	64.2	60.9	15.3	23.8	37.9	23.8	1.6	31.0	1.2	4.0
26-35	64.1	65.9	9.3	25.0	41.9	25.0	1.5	35.5	2.3	7.1
				Nellore sandy loam (Typic Ustorthent)						
0-13	46.1	68.6	10.7	21.0	44.2	21.0	1.6	30.2	1.2	6.4
13-46	72.1	58.3	13.6	28.1	31.7	28.1	1.6	33.7	1.3	2.0
				Nellore sandy loam (Typic Usorthent)						
0-10	22.5	83.3	5.4	11.2	54.4	11.2	1.7	28.3	1.2	5.3
10-40	47.6	74.4	12.4	13.3	56.3	13.3	1.6	30.0	1.2	5.8

Source : Bhaskar and Subbiah (1995)

essential characteristics of laterite soils under low rainfall conditions (Jessup, 1960). The low CEC values in these soils reflect the presence of high percentage of mica, like minerals throughout the profile (Ahmed *et al.*, 1968a, 1968b). The exchangeable complex is mostly dominated by Ca^{2+}, Mg^{2+}, Na^{+}, and K^{+}. The very low K status in these soils is due to slow weathering of coarse size mica and K fixation on the other hand. The increasing CEC of the soils in relation to the clay content of sesquioxidic nature and amorphous oxides of Fe and Al is ascribed to the coatings of these oxides on clay fractions.

The $Al_2O_{3/}Fe_2O_3$ content in general, increased with depth and clay content in all soils. The high content of Fe and accumulation in B horizon of the profiles is related with the form in which Fe is weathered, translocated, and deposited in a particular particle size. It is possible that Fe in association with clay could move in solid phase as postulated eariler. The data emphasize the lack of vertical movement of total Fe within the profiles. All the soils showed that considerable weathering of Ca, Mg, and K bearing minerals has taken place. The very low status of total CaO, MgO, and K_2O in the soils is due to low weathering of coarse size mica or presence of small amounts of feldspars as a source of bases.

Potential Areas for Acid Lime

A good potential for future commercial cultivation of acid lime exists in Gujarat, Maharashtra, Andhra Pradesh, Karnataka, and Tamil Nadu. Cultivation of acid lime and lemon may be successfully expanded in south western semiarid zone of Uttar Pradesh comprising areas like Aligarh, Etah, Mainpuri, Mathura, and Agra in agroecological region of northern plain and central highlands including Aravallis, hot semiarid eco-regions with alluvium derived soils having growth period of 90-150 days.

1.8.1.2. Citrus in Azerbaijan

History of citrus cultivation in Azerbaijan has been outlined by Kuliev (1985). A number of soil types of Azerbaijan have been characterised :

Leached Mountain Chernozems : These soils are confined to the medium mountain zone, plateau of Slovyanskaya plain of the Kedabek Rayon in the Lesser Caucasus, having moderately cold and semihumid climate. The soils are deeply leached and have dark humus norizon with clay loam texture formed on diluvial parent material. The upper horizons have up to 7.40 per cent humus, 0.14 per cent total N, and a C:N ratio of 10.4. The total exchangeable cations range from 24.1-41.7 me $100g^{-1}$ of soil dominated by Ca (Table 1.95)

Chestnut Soils: These soils occur in Maraz area in the Shemakha upland of the Greater Caucasus having dry steppe climate. The soils have formed on diluvial calcareous loam deposits having clay loam texture. The humus content is 2.32 per cent with 0.20

Table 1.95. Features of various soil types of Azerbaijan

Soil type	Depth (cm)	Soil pH	Ca (%)	Mg (%)	Na (%)	H (%)	Humus (%)	Total nitrogen (%)	Fraction of diameter (mm) <0.001 (%)	Fraction of diameter (mm) < 0.01 (%)	Total absorbed (me 100 g^{-1})
Leached mountain chernozems	0-14	6.9	87.6	12.4	None	None	7.40	0.41	40.6	62.2	40.5
	14-30	7.0	93.4	6.6	None	None	4.18	0.25	36.6	59.6	41.1
	30-62	7.2	98.1	1.9	None	None	2.37	0.17	44.1	67.7	41.7
	62-86	7.3	95.2	4.8	None	None	1.42	0.12	38.7	65.2	29.0
	86-120	7.1	90.9	9.1	None	None	0.41	0.09	36.5	61.4	24.1
Chestnut	0-20	8.1	74.4	14.6	11,0	None	2.32	0.20	18.9	51.0	21.9
	20-36	8.2	67.1	23.1	9,8	None	1.05	0.11	20.7	51.3	21.6
	36-53	8.1	73.3	20.5	6,2	None	1.01	0.10	21.1	54.4	21.0
	53-76	7.8	72.9	19.9	7,2	None	0.87	0.17	19.7	55.6	19.6
	76-102	8.3	63.8	30.0	6,2	None	0.26	-	23.2	59.3	21.0
Gray-brown	0-20	8.3	79.4	8.7	11,9	None	1.49	0.13	22.2	53.7	25.2
	20-50	8.4	82.4	8.4	9,2	None	1.00	0.11	21.3	42.1	27.3
	50-57	9.2	75.1	11.6	13,3	None	0.75	0.08	28.4	52.9	24.1
	57-90	9.4	74.4	13.0	12,6	None	0.45	0.06	23.6	55.5	24.6
	90-100	9.2	84.2	4.9	12,9	None	0.04	0.01	18.7	44.8	22.2
Weakly podzolic yellow earths	0-6	6.2	66.4	31.8	None	2.8	5.16	0.29	14.4	49.3	27.8
	6-16	6.0	62.2	35.8	None	2.0	4.20	0.26	19.0	53.2	32.5
	16-42	5.7	63.7	34.2	None	2.1	2.95	020	28.4	63.8	34.6
	42-62	5.7	64.8	32.8	None	2.4	1.94	0.17	35.5	67.7	33.7
	62-100	5.5	63.0	34.8	None	2.2	0.44	0.05	17.5	35.7	30.6

Source : Aliyev *et al.* (1981)

per cent total N and the C:N ratio of 6.7.

Gray-brown Soils: The soils occur on the Apsheron Peninsula in a dry semidesert climate with hot dry summers. The soils have formed on diluvial-proluvial calcareous and gypsiferous clayey parent material. Soils vary in texture from loam to clay loam. The C:N ratio is 6.7 with humus and total N content of 1.49 per cent and 0.13 per cent, respectively. The reaction is weakly alkaline to alkaline. The total amount of exchangeable cations varied between 22.2 me $100g^{-1}$ and 27.3 me $100g^{-1}$ of soil rich in Ca and presence of an appreciable amount of absorbed Na.

Weakly Podzolic Yellow Earth Soils: These soils are common in the piedmont plain belt of the Lenkoran zone where they developed under Jerusalem sage-oak-hornbeam type of vegetation. Diluvial deposits of the yellow earth weathering crust are the parent material with a shallow profile high in humus content in the 0 - 16 cm layer, which decreased sharply with the depth. Other features comprised of a low N content, wide C:N ratio in the upper humus layer, which decreased sharply in the lower horizon. These features are coupled with an elevated total amount of exchangeable cations, predominantly rich in Ca and Mg among adsorbed cations, exchangeable H, and a weakly acid reaction of surface and subsurface horizon. The soil profile is lithogenic with signs of leaching of sesquioxides and fine particles from the upper to the compacted and argillic illuvial horizons. The soils include Mn-Fe concretions, an isolated whitish pale-yellow podzolized horizon etc.

1.8.1.3 Citrus in Bangladesh

Climate : Two mandarin growing districts, namely, Moulavibazar and Sylhet experience very high annual rainfall (4500-5000 mm). Nazimuddin and Ahmad (1975) reported that sweet oranges are more adaptable than mandarins in Bangladesh

Soils : Most of the citrus plantations exist on erosive soil due to slopy lands. Low soil pH (4.5 –5.5) further induced many deficiencies like Ca, Mg, and Zn. Waterlogging due to inundation of water and poor drainage further aggravated the nutrient deficiency. Soils with flood coatings are mainly classified as Aeric Hapluquepts, Typic Haplaquepts, Typic and Aeric Haplquents, Aquic Haplorthents, Typic Ustochrepts, and Typic Dystrochrepts (Table 1.96). According to Karim *et al.* (1973), soils of series Gopalpur, Kaunia, Tejgaon and acid sulphate of Bangladesh are classified as Aquic Eutrochrepts, Aeric Haplaquepts, Typic Dystrochrepts, and Thiomic Fluvisols (Table 1.97).

Table 1.96. Analytical soil profile data for Dhamri of Bangladesh

Horizon	Depth (cm)	pH	Sand	Silt	Clay	Exchangeable cations (me 100^{-1}g)				
			———	(%)	———	Ca	Mg	K	Na	H
Ap1g	0-12.5	5.7	33	35	32	10.1	2.1	0.20	0.24	2.2
Ap2g	12.5-18	6.6	34	38	28	-	-	-	-	-
B21g	18-33	6.8	23	44	33	9.3	2.3	0.23	0.27	1.1
B22g	33-46	7.0	35	40	25	-	-	-	-	-
C1g	46-58	7.2	68	21	11	-	-	-	-	-
C2g	58-66	7.3	86	6	8	-	-	-	-	-
C3	66-102	7.4	53	29	15	-	-	-	-	-

Source : Brammer (1971)

Table 1.97. Characteristics of different soils of Bangladesh

Area	Soil series	Soil parent material	Drainage	Textural class	Clay (%)	CEC (me 100^{-1} g)	pH	Adsorbed P (mg kg^{-1})
Chanchora, Jessore	Gopalpur	Recent Gangetic alluvium	Imperrectly drained	Sandy clay loam	26	12.5	7.8	13.5
Assadpur, Rangpur	Kaunia	Recent Teesta alluvium	Poorly drained	Sandy loam	17	6.6	5.5	55.0
Porbata, Dhaka	Tejgaon	Pleistocene terrace deposits	Moderately well drained	Sandy clay loam	21	11.4	4.8	32.6

Source : Karim *et al.* (1973)

1.8.1.4 Citrus in Bhutan

Climate and Soils : Climate in the major citrus belts is hot and humid with temperature range of 15° to 30°C and average rainfall from 2000 to 4000 mm. Both climate and soil are ideally suited for loose skin mandarins (Ghosh and Singh, 1993). A number of orchards have been developed on very steep slopes (40-50 per cent) under tropical to subtropical foothills using bench terraces.

1.8.1.5 Citrus in China and Taiwan

Fossil excavations in Yunnan and Szechwan plateau revealed that many Rutaceae pollens were preserved from the quarternary glaciation, and considered to be blessed with a number of citrus species growing wildly, the origin of which dates back to 10th century B.C. evident from the books of poem "Si Chin" and book of Chou entitled "Chou Shu". About 2000 years ago, the great poet, Chu Yuan, in his famous writing "Chu Zhong", highly praised the oranges growing in Hupeh at that time. About 900 years ago, Han Yen-Chih wrote "Chu Loh" which was published in 1178. These publications enumerated 27 superior varieties grown in Wenchou and Chekiang, together with various cultural practices, claimed to be earliest monograph of citriculture in the world. Hu (1956) recorded that citrus fruits are being cultivated in southern China over 4000 years ago. At present, more than 1000 counties are producing some kind of citrus (Pieniazek, 1967; Wen Cai, 1981).

Luh (1965) and Yen (1966) described the citrus culture in Taiwan. Citrus cultivation in Taiwan started in 17th century following large scale migration of people from coastal areas. Southern China is the most important citrus producing area, including Taiwan and the provinces of Kwangtung, Fukien, and Kwangsi. The second most important areas are the upper Yangtze river district in western and southwest China, including Szechwan, where citrus was first grown commercially, besides Kweichow and Yunnan provinces. The third major producing area is in the lower Yangtze river basin of central China, including orchards in the provinces of Chekiang, Kiangsi- Kiangsu, Hupeh, Anhwei, and Hunan. There are also a few small orchards in northwest China in the provinces of Shensi and Chamdo. These citrus growing areas are mainly distributed between 20-30° N and below 700-1000 m altitude (Wen-Cai, 1981). In Yunnan province of China, citrus tree stands above sea level of 17-1800 m in northeastern, 18-1900 m in central, and 1800-2400 m in western part (Ding *et al.*, 1990). Number of studies have described the area and production scenario of citrus in China (Shen, 1989; 1991) which showed that Guangdong and Sichuan are the major citrus producing provinces. Shen and Li (1997) recently highlighted the present position of citrus research and strategies for further development in China.

Climate : Based on ecological conditions, the citrus producing areas have been delineated into six districts and five subdistricts (Shen, 1989). These are described as follows:

- Hilly and plain district in southern China growing sweet oranges and mandarins: This is divided into two subdistricts viz., coastal hill and plain growing Liucheng, Xinhuicheng, Ponkan, and Tonkan; and central and northern hilly subdistrict growing sweet oranges.
- Coastal lower hill district in Fujian and Zhejiang provinces and five ridges of south which grow sweet oranges and mandarins.
- Sichuan basin district producing mandarins: Lower hill subdistrict along the upper reaches of the Changjiang river, and the middle and lower reaches of Minjiang, Tuojiang, Jinshajiang, and Jialingjiang rivers practising cultivation of sweet orange cultivar Jincheng. The subdistrict of the lower hills in the middle and upper reaches of Minjiang, Tuojiang, and Jialingjiang adapt the cultivation of sweet oranges and mandarins. In the hilly subdistrict on the edge of Sichuan basin and the lower elevations of the mountains of Sichuan basin, famous satsuma mandarin is grown.
- The district consists of middle and lower mountains and dry and hot valleys in the Yunnan-Guizhou plateau growing a variety of citrus species.
- Districts on the edge of subtropics grow a variety of citrus.

Ye and Lin (1991) identified climatically most suitable summer orange growing zones in Fujian province of China. The climatic features of most suitable zones are : mean yearly temperature of 20°-22°C, mean January temperature of 10°-13°C, cumulative day degree of above 10° of 6500-7700°C, and an absolute minimum temperature of -4°C. Area with mean yearly temperature of 18°-20°C, a mean January temperature of 0°-10°C, cumulative day degrees above 10°C of 6000-6500°C, and an absolute minimum temperature of -5°C, may also successfully grow the summer oranges provided careful selection of cultivars is made. Rongchun county is famous for the concentrated production of *Citrus reticulata* cv Pongan. Many years of practice demonstrated that this cultivar prefers an altitude below 1300 m above mean sea level, optimum mean annual temperature below 19.2°C, mean January temperature below 10.3 °C with accumulated temperature above 10°C of below 6000°C (Chen, 1998).

Southern China has a hot, humid, rainy summer climate, and frost free rather dry winter. Temperature in the summer is sufficiently high to partly defoliate trees. Climatic hazards are heavy in form of rains, typhoons, and hot weather. Western and southwestern China, the upper Yangtze river district including Szechwan province, also has heavy summer rainfall. But, winters are colder in southern China with light frost and snow in some seasons. The lower Yangtze river basin of central China, including the provinces of Chekiang, Kiangsi, Anhwei, Hunan, and Hupeh, has less rainfall than southern China with frost and light snow in some winter seasons. Climatic data containing maximum and minimum temperature, and rainfall for the five major citrus producing regions of

Table 1.98. Average climatic data for the major citrus areas of China and Taiwan

Province	District	Mean max. temperature (°C)	Month	Mean min. temperature (°C)	Month	Annual rainfall (mm)
		Southern China				
Kwangtung	Canton	37.0	July	13.3	December	1822
Kwangtung	Swatow	36.6	August	14.8	December	1572
Kwangsi	Wuchow	32.2	June	18.9	January	1092
Fukien	Amoy	35.5	August	15.9	December	1017
Fukien	Foochow	37.4	August	13.3	December	867
		Western China				
Szechwan	Chengtu	28.8	October	16.2	January	800
Sekang	Tatsienlu (Kang-Ting)	32.2	July	17.8	January	1017
Szechwan	Nan-Chung	31.1	July	15.5	January	1265
Kweichow	(unspecified	30.0	July	15.9	January	1125
Yunnan	K'un Ming	26.2	June	17.8	January	1025
		Central China				
Chekiang	Hangchow	31.5	June	12.2	January	1207
Kiangsi	Nanchang	31.8	July	15.2	January	1605
Hupeh	Ichang	31.1	August	15.5	January	1077
Hunan	Ch'ang-Sha	31.8	July	15.9	December	1390
		North-western China				
Shensi	Han-Chung (Nan-Cheng)	31.8	July	12.2	January	725

Source : Dept. of Hort., College of Agriculture, National Taiwan University, Taipei, Taiwan

China are presented (Table 1.98). Red mountainous soils present at Jiangxi, the Hongguang selected from local mandarin (*Citrus reticulata*), cultivar Nanfeng Guangju is well adaptable under wet weather conditions (Cheng and Zeng, 1991). Most Chinese citrus growing areas have sufficient rainfall to make irrigation unnecessary. Wang and Gong (1995) described the climate of Jiangxi province located at an altitude of 66 to 132 m above mean sea level as subtropical with mean annual temperature of 18.6°C, annual rainfall of 1360 mm, and sunshine of 1705 hours.

Table 1.99. Meteorological variations at Henggang, Nanchang county of China

Month	Mean monthly temp. (°C)	Mean monthly daily temp. (°C)	Mean daily precipitation (mm)
	1981		
Oct.	16.50	6.6	5.7
Nov 1-10	10.80	3.4	1.2
	1982		
Oct.	21.30	7.6	4.3
Nov 1-10	17.20	5.4	6.2
	1983		
Oct.	19.70	6.1	6.8
Nov 1-10	16.50	7.0	4.1

Source : Li *et al.* (1990)

The daily temperature range during 1981-1983 at Nan Chang county varied from 3.4° to 7.6°C with a average monthly temperature from 16.5° to 21.3 °C and daily average precipitation of 1.2 to 6.8 mm (Table 1.99). Guangdong province located in southern China, has yearly average temperature range of 20°-24°C. The yearly accumulated temperature of more than 10°C is around 6000-8500°C with an annual rainfall of 1400 mm, concentrated during April to September (Gan, 1990). Rainfall pattern in citrus areas of Taiwan is characterised by two distinct peaks, one in summer and another in winter. Most of the rains is received as monsoon rains. The summers are long and warm, while the winters are short and cool. The mean monthly summer maximum temperature in southern Taiwan is 28-20°C. High humidity prevails for the most of the year which averages 70-80 per cent.

Soils : Based on variation in soil types, land forms, the citrus growing areas have been divided into five districts, namely, i. paddy ii. alluvial, iii. red hill soil with gentle slopes, iv. hilly purple and yellow soil hill land, and v. coastal mud flat and beach district (Shen, 1989). These are briefly described below:

- In the paddy areas, mounds of soil are built up in the field before planting the citrus trees. The mounds are expanded in the first and second year after planting, and then connected to each other to form high beds between which deep ditches are filled with water. In these areas, hilling up and water requirement are important components of orchard management.
- There are some citrus orchards along the banks of rivers known as alluvial district which covers Fujian, Hunan, Jiangxi etc. Mud washed down in floods serves as rich source of nutrients, with the result, trees are very tall and vigorous.
- District of red hill soil having gentle slope is quite large and blessed with suitable climate for growing citrus. But, at the same time, poor and thin, strongly acid soil has compact structure.
- Purple and yellow soils in hill districts are rich in P and K under warm, humid, and foggy winter, which are quite suitable for citrus.
- In the coastal mud flat and beach districts, cultivation of citrus is gradually picking up. Due to presence of salinity and alkalinity, Fe deficiency induced chlorosis is a common feature.

Soils of citrus orchards of Fujian province have been observed to have soil pH less than 5.5 with nutritional disorders in form of Ca and P, and excess of B and Cu (Li *et al.*, 1988). Citrus growing soils of Zhejiang area from sea deposits are clay in nature, pH 7.9, clay 27.6 per cent, organic matter 1.38 per cent, $CaCO_3$ 14.4 per cent, free iron oxide 1.27 per cent, amorphous iron oxide 0.19 per cent, and DTPA-Fe 11.5 mg kg^{-1}. The red soil at Guangzhou is loam in texture having pH 6.2, total C 1.82 per cent, available N 100.6 mg kg^{-1}, P 19.6 mg kg^{-1}, and K 77.0 mg kg^{-1} (Huang and Yan, 1990). Reddish clay loam soils are common to the citrus areas of central and southern China having Zn as one of the major soil fertility constraints (Outjang, 1993; Quyang, 1993; Xie *et al.*, 1993). Earlier, Sato (1969) described the Mo deficiency induced by excess of Ni in soil derived from surpentine containing rock in Japan.

While, such soils are fertile, heavy, and poorly drained, which create disease problem, and result in orchard with a short commercial life. In other areas of China, grey, sandy loam, alluvial soils are widely used for citrus. The fertile, well drained soils, usually near rivers, make up the best citrus areas of Szechwan, Fukien, Kwangtung, and

Taiwan. A large part of the southern citrus orchards are on heavy soils near the sea, where, the water fills the drainage ditches between rows. Such groves have a short life because of the high water table, and are usually protected from typhoons by windbreaks of thorny bomboo. Citrus soils of Hunan have average soil pH 5.6, effective P content 70.4 mg kg^{-1}, and total P content 0.16 per cent (Chen *et al.*, 1990b). Red soils at Jiangxi contain 0.78 per cent organic matter, 0.063 per cent N, 0.055 per cent P, and 1.67 per cent K (Sheng, 1990). While, soils at Zhejiang are clay developed from sea deposits having pH 7.9, clay 27.6 per cent, organic matter 1.38 per cent, total $CaCO_3$ 10.4 per cent, free Fe oxide 1.27 per cent, amorphous Fe oxide 0.19 per cent, and DTPA Fe 11.5 mg g^{-1} soil. Zhou *et al.* (1996) reported large Mn, Zn, and Fe deficiency on alkaline Hongju growing soils at Sichuan.

The topography of land devoted to citrus culture varies widely in different sections of China. The terraced hillside is the common type of topography utilized in the Kiang-Ting district of Szechwan province, and in the provinces of Hainan Island, Kiangsu, Kweichow, Hunan, Hupeh, and Shensi. Level land is the common type of topography in citrus orchards in the Nan-Chung district of Szechwan Province, and in the provinces of Kwangtung, Fukien, Kiangsi, and Chekiang. Combinations of these two types of topography are found in the provinces of Kwangsi-Chuang and Anhwei. The average soil quality of citrus orchards developed on argillaceous red, sandstone red soils, paddy soils, and meadow soils showed some increase in soil quality when evaluated over time and space. The average relative soil quality index of soil increased by 5.0 and 11.2 in red soil and meadow soil, respectively (Table 1.100). However, the relative soil quality index in citrus

Table 1.100. Some properties of the major soils in Jiangxi province of China

Soil type	Horizon	Depth (cm)	pH (H_2O)	Organic (%)	Total N (%)	CEC [cmol (p^+) kg^{-1}]	Base sat. (%)
Argillaceous red soils	A	0-12	6.5	1.60	0.10	11.5	85.5
(Haplic Acrisols)	AB	12-20	7.1	1.20	0.07	10.7	88.2
	B	20-50	7.6	0.80	0.03	10.1	90.5
	C	50-103	7.6	0.40	0.02	7.7	92.2
Sandstone red soils	A	0-7	4.9	1.21	0.05	8.2	28.3
(Aric Anthrosols)	B	7-27	5.0	0.65	0.04	7.3	19.1
	C	27-52	5.3	0.40	0.03	10.2	15.7
Paddy soils	A	0-14	5.3	3.50	0.16	20.5	28.5
(Eutri Fluvisols)	B	14-58	5.4	1.98	0.12	12.0	30.1
	C	58-93	5.6	1.01	0.08	10.3	35.2
Meadow soils	A	0-11	7.5	1.05	0.10	15.3	84.7
(Umbric Fluvisols)	Br	11-69	8.1	0.45	0.50	10.4	98.8
	C	69-110	8.2	0.12	0.21	3.6	100
Meadow soils	Ap_1	0-10	6.2	2.71	0.15	10.7	31.7
(Umbrihumus)	Ap_2	10-20	6.5	1.80	0.09	10.6	37.9
	Bg	20-55	7.0	0.65	0.04	9.6	45.2
	BC	55-75	6.2	0.38	0.02	10.8	16.2

Source : Wang and Gong (1998)

orchards established on Umbrihumus meadow soil, decreased by 3.4. (Wang and Gong, 1998).

According to Chang *et al.* (1994), the major nutritional problems encountered by citrus producers in Taiwan are high soil N content and low soil K, Mg, Zn, and B contents. Hsieh (1990) described nutrient deficiencies inTaiwan in form of P, K, and Zn in addition to excess Fe and Mn in citrus orchards.

Sichuan basin actually represents citrus growing area of southwest China. It is surrounded by continuous mountain ranges because the area is outside the mountain shelter that protects the crop from cold waves from the north. No citrus is grown on the high plateaus further west beyond 103°E longitude, where the winters are cold with dry summers. Other provinces estward beyond the famous Wu-Shan Gorge of Yangtze river are also unprotected by mountains and freezes occur once every six or seven years. Topographically, Sichuan basin can be divided into three districts: the eastern hilly lands, the western Chengtu plain, and the western plateaus. Most of the citrus is confined to eastern hilly land, whereas western plateau zone is not suitable for citrus because of low winter temperature. Sichuan basin is classified as subtropical zone. Along the Yangtze river, the year round temperature is mild with a small fluctuation in diurnal temperature. The mean annual rainfall is from 18.1° to 19.3°C in citrus producing area. Absolute maxima and minima vart from 39.4° to 42.6°C and –1.2° to -2.6°C , respectively. Rainfall averages 1000 mm with a mean annual relative humidity of 80 per cent as in Fong-gi and De-Chang counties.

Soil of Sichuan basis are classified as a purple coloured soils, since most of soils are derived from weathering of the purple coloured slate rock or sandstones. The soils are rich in Fe, P, K but poor in N and organic carbon with soil pH of 6 – 7. Soils of western plateau are often too dry to grow citrus fruit. The Chengtu plain consists of alluvial soil with a high water table which is harmful to citrus roots. The eastern plateau hilly lands consist of well drained, purple coloured loam soil on the separate broad hills, favourable for citrus growing which suggested that areas planted under citrus becomes colder after the trees attain 15 to 20 years of age. The large leaf area of the older trees causes a more rapid loss of radiation heat at night and prevents direct heating of the soil by the sun during the day. Thus, a warm area in the desert may become progressively colder with trees attaining in size.

1.8.1.6 Citrus in Cyprus

The major citrus areas of Cyprus are Famagusta, Morphou, Lefka, and Episkopi.

Climate : The climate of the island is mild and moderately humid. Average annual rainfall ranges from 300 mm at Morphou in the north to 450 mm at Limassol in the south, nearly all of it occur between October and May. Mean maximum temperatures at Morphou are 23°C in January and 34°C in July. Mean minimum temperatures are 16°C in January and 25°C in July. Despite low summer rainfall, humidity is high enough for warm nights, resulting in shamouti oranges that are still green or only breaking in color when the fruit is ready for harvest in November. Average rainfall in most areas is over 40 mm a year, except on the Adzharian black sea coast, where it may exceed 100 mm. Average temperatures are 27°C in July and 15°C in January. Coastal areas are humid in the summer. The annual minimum rainfall occurs in May and June when trees are in bloom. Spring droughts in some seasons cause blossoms to shed. Climatic hazards include frosts and winds that are strong enough to defoliate trees. Temperature drop to 7°-8°C in most districts (Reuther *et al.*, 1967).

Soils : Soils are generally light textured as red sandy loam to loam type and acidic in pH.

1.8.1.7 Citrus in Indonesia

Citrus cultivation in Indonesia gained expansion since 1950. Mixed citrus orchards have been popular for several centuries (Thrower, 1959).

Climate : Indonesia is the largest country of the equatorial region. Citrus is grown on a wide range of pedoclimatic conditions which range from tidal acid swamps in hot and humid Pontianak (southwest Kalimantan) to volcanic wet high lands of Sumatra and Java having distinctive xerophytic landscape (Aubert, 1990). Citrus producing areas are centered around Java followed by Sumatra, Kalimantan, and Sulawasi Iran Jaya. Soils are by and large, deficient in N, Fe, Mn, Zn, and Cu (Lamsayun, 1980)

1.8.1.8 Citrus in Iran

Two major regions contribute to about equally to Iran's citrus production. The vast area of southern Iran may be compared with the California-Arizona desert citrus areas, while the coastal region along the Caspian sea resembles northern Florida (Opitz, 1974). Sweet orange (Bumi) is the major cultivar grown in the north. Washington navels have been introduced in the Shahi area. Best quality mandarins are produced at Jahrom. Sweet lime is also grown as seedling. Citrus cultivation in Iran is concentrated in the Caspian belt, inland southern belt, and persian gulf (Bar-Akiva, 1965; Asif, 1977).

Climate : Citrus districts of southern Iran receive scanty rainfall and endure long hot dry summers short and sometimes cold winters. An exception is the limited citrus grown in isolation along the Persian Gulf. On the other hand, the narrow citrus belt of the Caspian littoral annually averages about 1000 mm of fairly well distributed rainfall. The Caspian littoral is a humid region moderately warm, but with a shorter summer and cold winter. The average rainfall ranges between 1000 mm and 1500 mm. The climate of south Iran experiences extreme heat in summer and frosty nights in winter. This is the reason, the establishment of citrus orchards compared to valleys helps to protect from frost hazards (Singh, 1969).

Soils : Soils utilised for citrus cultivatin in southern Iran are generally rocky and well drained. As with most desert soils, the reaction is alkaline and excessive salts may limit the growth and production. Climbing the steep slopes and distributed irregularly on alluvial deposits, next to the Caspian seas, there are many citrus orchards mostly seedling in origin which mature early.

The majority of the soils is lighter in texture, especially at Nowshar and Babol-Shahi of Ramsar in north Iran, and Daseroon and Kaseroon areas of south Iran. But, heavier soils are also used for citrus plantation such as Shahsaver area. Many orchards are planted in gravely or in a rock soils in south Iran. Citrus planting on slopes reduces the chance of injury from frost hazard. Some scattered plantations have also taken place in desert area, which are coming up nicely. Foroughi *et al.* (1974) reported B deficiency in Iran. Bamdad *et al.* (1977) and Rao (1993) reported the prevalence of Zn deficiency in 59 per cent orchards followed by Mn, and Cu in 17 per cent and 14 per cent orchards, respectively, in Jiroff valley of Iran.

1.8.1.9 Citrus in Iraq

A majority of citrus orchards is concentrated in Baghdad, Karbela, and Diyala provinces.

Climate : Suitability of Dez irrigation project in the Khuzestan plain of southwestern Iraq has been assessed for commercial planting which had less possibility of frost hazard than Brownsville, but greater than Yuma (Neild *et al.*, 1970). Citrus growing areas in Iraq possess a typical mediterranean climate, a subtropical and semiarid where rainfall is received during winter (December-March), while in the summer and autumn (June-November), weather remains dry in most part of the year. Sehgal and Bhattacharjee (1988) described climate of northeast Iraq is characterised by mean annual air temperature and rainfall of 18°C and 765 mm, respectively. The soil moisture control section remains dry for 4 to 5 months suggesting the presence of xeric soil moisture regime.

Soils : Intermontane valley plain soils have been classified as Xererts represented by Malawais and Sirwan soils developed from calcareous alluvium having limestone, dolomite, and marls. Soils of northeast Iraq have been classified as Palexeroll Chromustert, and Aquentic Chromxerert (Sehgal and Bhattacharjee, 1988) with soil pH varying between 7.5-8.0 (Table 1.101).

Table 1.101. Physical and chemical properties of Vertisol types in northeast Iraq

Horizon	Depth (cm)	Particle size (%) Silt	Clay	Bulk density ($g\ cc^{-1}$)	AWC (mmm^{-1})	pH	Org. C (%)	$CaCO_3$ (%)	CEC (me 100 g^{-1})	ESP (me100 g^{-1})
	Malwals silty clay (Palexerollic Chromustert), moderately well drained									
Surface	0-38	50.0	39.5	1.55	179	7.6	1.3	11.4	42.2	2.3
Subsoil	47-107	53.8	28.8	1.80	-	8.0	-	11.2	42.0	2.3
Substratum	61-166	40.0	47.0	-	-	8.0	-	18.6	36.1	3.7
	Sirwan silty clay (Aquentic Chromoxerert), imperfectly drained									
Surface	0-22	-	-	-	180	-	-	-	-	-
Subsoil	22-70	51.0	44.0-	-	-	7.5	0.96	25.0	40.8	0.6
Substratum	70-120+	49.0	44.0	-	-	7.7	0.61	30.2	36.4	0.7

AWC and ESP stand for available water capacity and exchangeable sodium percentage, respectively
Source : Sehgal and Bhattacharjee (1988)

1.8.1.10 Citrus in Israel

In the last two decades, new grapefruit gowing regions have been developed in the hot and dry inland valleys. While, the south is better suited for lemons and Washington navel oranges (Chalutz and Roessler, 1986). Orchards are primarily concentrated along the mediterranean coastal plain. Other citrus growing areas are Yerreel, Biesan, and the Jordan valley, the northwestern Negev, and the north central Negev (Milella, 1966; McCarty, 1967).

Climate : The citrus districts of Israel, a relatively small country, evince surprising differences and possess a climate ranging from mediterranean sea on the coast to semiarid in the interior (Turrell *et al.*, 1964). The main interacting factors are the influences of the mediterranean sea which predominate for only few kilometers from the coast and large bodies of desert in the neighbouring citrus growing countries, the influence of which becomes increasingly detectable with larger the distance from the sea. Such influence has been even more marked for districts sheltered from marine winds by hills and mountain ranges. Evaporation is higher, the more arid the climate. The dew point is reached in July on the coast almost every night, but only 10-15 nights per month 15-30 km inland. A clear decreasing trend of winter rains exists from the north and south from a maximum of about 700 mm near the Lebanese border to a minimum of about 150 mm in the northern Negev, contributing towards the aridity of climate (Table 1.102).

Table 1.102. Climate characteristics of diverse representative citrus districts of Israel

Climatic parameters	Western Yezreel (Interior semiarid)	Carmel shore coast(Coastal)	Magen Northen Negev (Intermediate semiarid)
Mean temp. May (°C)	22.6	20.5	22.3
Mean temp. July (°C)	27.1	25.7	26.2
Total evaporation of May-Aug. (mm)	940	799	1043
Mean humidity of May-Aug.(%)	59.3	65.9	61.4
Rainfall (1977-78) winter (mm)	628	597	134
Rainfall (1978-79) winter (mm)	349	385	142

Source : Monselise (1981)

Table 1.103. Climatic characteristics at southern coast of Israel

Average evaporation from class A pan (mm day^{-1})							
Apr.	May	Jun.	Jul.	Aug.	Sept.	Oct.	
5.1	6.8	7.8	7.7	7.5	6.4	4.7	
Average monthly rainfall during 1967-72 (mm)							
Sept.	Oct.	Nov.	Dec.	Jan.	Feb.	Mar.	Apr.
3.1	19.9	51.6	118.6	124.1	42.6	43.1	49.0

Source : Yagev (1977)

The influence of climate is most felt during May to August when peel growth occurs and acidity increases to peak before decreasing towards maturity (Reuther *et al.*, 1969).

Climatic characteristics of citrus growing areas of southern coastal plain of Israel (Table 1.103) showed average evaporation varying between 4.7 mm and 7.8 mm day^{-1} during April to October and average monthly rainfall from 3.1 mm in September to as high as 124.1 mm in January (Yagev, 1977). The average annual rainfall at citrus growing areas of northern Negev is 230 mm, all of which falls during winter months of November to April. The mean annual minimum and maximum temperatures are 22°C and 32°C, respectively. The mean relative humidity in summer is 55-60 per cent. The evaporation from US Weather Bureau Class A Pan varies from 7 mm day^{-1} during April-September to 9 mm day^{-1} during July-August (Levy and Shalhevet, 1990a). In the coastal areas, humidity is relatively high throughout the year. A 80 per cent of the rainfall occurs from November through February. Average annual rainfall at Rehovot is 550 mm. Mean maximum temperatures at Rehovot range from 24°C in January to 33°C in July with corresponding minimum temperature range of 17°-25°C. The climate is moderate mediterranean with high humidity and annual rainfall of 400-700 mm (Levy, 1999).

Soils : Citrus soil analysis up to 150 cm depth (Table 1.104) in southern coastal plain of Israel showed variation in pH from 8.1 to 8.4 with SAR between 4.0 and 9.6 (Yagev, 1977). Soils at upper Bet Dagan area range from sandy loam in central coastal plain orchards with pH range of 7.0-7.5 to heavier alluvial soils in the northern coastal plains and upper Gallilee with pH range of 7.5 – 8.5 (Shaked and Ashkenazy, 1984). Levy (1984) described citrus soils at Gilat as Typic Xerorthent (loessial Serozem) with a uniform profile of 180 cm having bulk density 1.45 g cm^{-3} (Table 1.105). The soils of coastal plain are red sandy loam soils, which are well drained.

Table 1.104. Citrus soil analysis in southern coastal plain of Israel

Depth (cm)	0-30	30-60	60-90	90-120	120-150
Saturation per cent	68.5	67.5	74.5	86.0	87.5
B (mg kg^{-1})	-	-	-	1.56	2.08
pH	8.1	8.2	8.1	8.3	8.4
EC (mmhos cm^{-1})	0.55	0.75	0.95	1.0	1.0
Cl (me l^{-1})	1.2	1.8	3.2	4.6	5.4
Na (me l^{-1})	4.4	6.2	8.5	9.1	10.0
Ca + Mg (me l^{-1})	2.28	2.38	2.46	2.30	2.08
SAR	4.0	5.7	7.6	8.5	9.6

Source : Yagev (1977)

A soil pH of 6.5-7.0 and sometimes presence of impervious horizon having high reducible Fe is observed. The soils of Hulla, Yesrell, Jordon, and Biesan valleys which are below the sea level have developed from alluvial type of parent material with moderate calcareousness (Levy, 1999). The soil at Negev is classified as Loessial Serozem (Typic Xerorthent) with the uniform texture up to a depth of 180 cm. The soil has average bulk density 1.45 Mg m^{-3}, field capacity 18 per cent, wilting point 7 per cent on dry weight basis, and cation exchange capacity 0.14 mol kg^{-1} (Levy and Shalhevet, 1990a). Soils are extremely variable in the coastal citrus zones, where soils are 80-150 cm deep, well drained, basaltic Protgromosol (65 per cent clay) on basaltic rocks, having soil pH 7.6 and $CaCO_3$ 7.6-12.1 per cent. Although, most coastal soils are sandy loam and some are considerably heavier soils. Heavy clay loam soils are found in the Jordan valley near Lake Tiberias. Some arid soils of middle east part of Israel are typical of Orthids (Camborthids and Gypsiorthids) with old desert gravel plains, while young plains are considered as Entisols (Torripsamments or Torriorthents) according to a number of studies with regard to characterisation of Israel soils (Dan and Raz, 1971; Dan, 1979; Dan *et al.*, 1981; Dan and Yaalon , 1980). Dan *et al.* (1982) classified the soils in arid part of Negev and Sinai as Torriorthent and Gypsiorthids (Table 1.106). Horesh *et al.* (1986) reported lime induced Fe chlorosis in Beth Guvrin calcareous soil.

Table 1.105. Citrus soil analysis at Gilat, Israel

Depth	Na	Ca	Mg	Cl	SAR	ECe
(cm)	——(mole m^{-3})——					(dSm^{-3})
0-30	8.1	7.4	4.2	6.1	3.4	1.3
30-60	8.3	5.6	4.4	6.1	3.9	1.1
60-90	9.2	6.4	3.5	7.1	4.1	1.2
90-120	12.4	5.9	3.1	8.9	5.9	1.5
120-150	14.1	6.9	4.4	11.8	5.9	1.9

Source : Levy (1984)

The analysis of soil samples collected from upper 90 cm from citrus orchards of upper Gallilee near the Hulla valley (GAL), Nahalat David (ND), Gilat (GIL), and Bet Dagan (BD) showed four different soil types which are composed of clay with 33 per cent $CaCO_3$, silty clay loam with 23 per cent $CaCO_3$, sandy loam loess with 15 per cent $CaCO_3$, and loamy sand with 1 per cent $CaCO_3$ (Table 1.107). Eisenberg *et al.* (1982) described eight major soil types, namely, sandy Regosols, red loamy Regosols, cumulic light brown loam, loessial light brown clay loam, quartzic dark brown clay loam, natric grumic dark brown clay Regosols, and clayey saline Regosols (Table 1.108).

1.8.1.11. Citrus in Japan

After Taji Mamori, about 1260 years ago, a new citrus seedling tree was derived from seed brought from China. Commercial citriculture was established in Kumamoto, Wakayama, and Shizuoka during 16th century. The citrus cultivars such as Satsuma mandarin and natsudaidai are the two major commercial cultivars grown in Japan. Satsuma mandarin appeared in Kagoshima about 300 years ago, and its expansion increased during last 100 years. The other cultivars include Hassaku, Washington navel, iyo, sanbokan, and lemon with 90 per cent of satsuma mandarin on *Poncirus trifoliata* rootstock. Citrus growing areas of Japan extend from south Tokyo, at about 35.5° N

Table 1.106. Physicochemical properties of some soil profiles in southern Israel and Sinai

Horizon	Depth (cm)	pH	Mechanical composition (%)			Gypsum $CaSO_4$ (%)	Carbon-ates $CaCO_3$ (%)	CEC (me $100g^{-1}$)	Exchangeable cations (me $100g^{-1}$)	
			Sand	Silt	Clay				Na	K
En Yahav, Coarse Desert Alluvium or Typic Torriorthent										
A	0-1	7.6	66.4	27.2	6.4	0.01	33.6	6.20	0.95	1.0
B	1-4	7.6	71.0	22.4	6.6	0.17	39.5	7.17	0.80	1.0
Clcs	4-15	7.6	96.2	3.8	-	0.075	51.0	2.72	0.65	0.38
C21	15-30	7.5	95.5	3.3	1.2	0.83	45.1	2.39	0.70	0.45
C22	30-50	7.5	97.1	1.3	1.6	0.23	44.2	2.17	0.80	0.47
C23	50-75	7.4	94.7	4.7	0.6	0.01	42.5	2.39	0.85	0.51
Southeast Hazeva, Petrogypsic Regosol or Cambic Gypsiorthid										
A	0-1	7.5	47.0	43.7	9.3	0.9	27.4	8.04	3.90	0.92
B21sa	1-5	7.3	50.0	43.6	6.4	0.2	33.6	9.88	1.71	0.83
B22cs	5-12	7.8	57.8	23.7	18.4**	29.2	36.0	3.93	3.25*	0.38
B3sa	12-20	7.4	73.2	12.5	4.3	8.1	23.9	4.61	9.32*	0.39
C11sacs	20-39	7.1	67.4	13.8	18.8**	20.7	14.7	4.46	21.65*	0.31
C12	39-52	7.6	92.5	6.1	1.5	8.0	14.2	3.55	8.09*	0.29
C13cs	52-100	7.7	74.8	11.0	14.3**	9.9	8.8	3.87	4.56*	0.29
C14	100-117	7.6	94.4	4.8	0.8	7.7	13.8	2.29	4.89*	0.28
C2	117-135	7.7	95.6	2.2	2.25	3.1	13.9	2.39	10.30*	0.24
Southeast Qual'at en Nakhl, Deep Regosol or Typic Gypsiorthid										
A11	0-3	7.8	52.5	36.6	11.4	0.07	-	10.59	2.9*	0.88
A12	3-9	7.4	67.5	25.9	6.0	0.078	-	8.36	-*	0.45
B1	9-19	7.3	73.9	20.6	5.6	0.72	-	8.94	2.9*	0.41
B2cs	19-35	7.3	66.5	21.8	11.7**	13.35	-	10.43	-*	0.33
B3sacs	35-47	7.2	4.8	35.9	22.2**	9.30	-	9.17	-*	0.27
C11cs	47-75	7.7	-	-	-	42.96	-	4.41	3.1*	Tr
C11cs	75-100	7.7	-	-	-	29.36	-	4.68	-*	Tr
C12cs	100-110	7.7	-	-	-	32.55	-	4.69	6.6*	Tr
Near Bir-eth Thamada in Central Sinai, Coarse Desert Alluvium or Typic Torriorthent										
A	0-10	7.9	88.2	9.9	1.9	0.004	24.1	4.20	1.04	0.49
Ac	10-24	7.8	90.9	7.2	1.9	0.001	21.6	3.94	0.99	0.45
C11	24-40	8.0	95.8	2.2	2.0	0.23	34.4	2.19	0.60	0.19
C12	40-56	8.1	95.4	3.3	1.2	0.06	37.0	2.37	0.60	0.20
C13	56-110	8.2	95.7	3.0	1.4	0.27	42.9	2.28	0.71	0.15
South Bir-eth Thamada in Central Sinai, Young Regosolic or Cambic Gypsiorthid										
A	0-4	7.8	56.1	22.8	21.1	0.16	28.9	14.06	4.64	0.96
B2sa	4-15	7.6	70.2	20.9	9.0	1.35	31.1	10.15	2.63	0.42
B3cs	15-22	7.6	74.4	7.8	17.8	18.8	24.8	8.98	2.20	0.24
C11sacs	22-34	7.5	64.5	12.8	22.8	21.4	20.1	10.38	7.31	0.27
C12cs	34-78	7.8	88.0	2.6	9.4	4.90	40.0	4.88	1.55	0.16
C13	78-130	7.7	84.6	7.4	8.0	2.78	47.6	7.64	1.55	0.15
C2	130-150	8.0	93.8	2.7	3.6	0.29	60.7	41.2	1.55	0.06

*The exchangeable sodium values are meaningless due to the high salt content

** The high clay values include a large portion of gypsum

Source : Dan *et al.* (1982)

latitude to the island of Kyushu. Major citrus growing prefectures in order of importance are Shizuoka, Ehime, Wakayama, Hiroshima, Kumamoto, Kanagawa, Saga, Tokushima, and Fukuoka (Hodgson, 1967; Singh, 1966). Other citrus species such as lemon and sweet orange have been introduced into Japan from USA since 1859. It was possible to grow satsumas, but too cold for natsudaidais. Good growth, fruit yield, and quality are obtained with the satsuma cultivars viz., Tachima-wase, Okitsu-wase, Yonezama, Okitsu no. 6, Sliverhill, and Hayashi in western coastal region of Shimane prefecture of Japan (Blondel, 1970; Kuranaka *et al.*, 1972).

Climate : The climate in citrus growing areas of Japan is temperate in general. Spurling (1969) reported that rainfall of citrus growing regions ranges from 1200 – 1300 mm with supplementary irrigation in Wakayama and Eihme. Nakagawa (1969) analysed the climatic conditions in the major areas of citrus production in the world with reference to satsuma mandarin. Citrus areas in Japan have both features of subtropical and subfrigid climate. In citrus growing zones, the annual mean air temperature is higher than 14.5°C, and the extreme mean air temperature is usually higher than –7°C. The citriculture in Japan is limited by low temperature, which facilitate the cultivation of early maturing and cold resistant satsuma mandarin harvested before the onset of low temperature (Kozaki, 1981). The average annual temperature and rainfall in major citrus producing prefectures ranged from 15.2 to 17.0°C and 1400 – 2400 mm, respectively (Nito, 1999).

Table 1.107. Characteristics of different citrus growing soil types in Israel

Parameters	Soil types at			
	GAL	BD	ND	GIL
Particle size distribution (%)				
Sand	13.8	83.9	59.8	69.8
Silt	38.8	6.4	18.8	12.8
Clay	47.4	9.7	21.4	17.4
$CaCO_3$	33.2	0.1	23.2	15.4
Texture	Clay	Loamy sand	Silty clay	Sandy loam
pH	7.9	8.0	7.9	8.0
EC (mmho cm^{-1})	1.0	1.6	1.6	1.4
SAR	0.98	1.78	1.26	1.37
Saturated paste (me 100 g^{-1})				
Cl^-	1.90	7.06	1.03	0.78
K^+	0.30	0.19	0.67	0.40
Na^+	1.70	4.00	2.30	1.85
Ca^{2+}+Mg^{2+}	5.90	10.00	6.60	3.60
Exchangeable cations (me 100 g^{-1})				
Ca^{2+}	28.07	5.26	14.49	7.70
Na^+	0.38	0.28	0.28	0.18
K^+	1.35	0.17	1.18	0.82
Mg^{2+}	2.70	1.04	3.35	1.60
CEC	32.50	6.75	19.30	10.30
% distribution of CEC				
Ca^{2+}	86.36	77.92	75.07	74.80
Na^+	1.17	4.75	1.45	1.75
K^+	4.15	2.50	6.10	7.90
Mg^{2+}	8.30	15.50	17.30	15.50
Elements (mg kg^{-1})				
NH_4^-N	9.02	5.86	6.76	5.67
NO_3^--N	21.70	13.67	3.51	8.22
P	50.00	5.70	154.20	6.50
B	0.02	0.10	0.38	0.24

Source : Sagee *et al.* (1992).

The coastal citrus growing areas are characterised by fairly high humidity throughout the year, considerable cloudiness, and excellent summer growing weather. Annual rainfall in selected prefectures has been observed as 1925 mm, 2200 mm, 1775 mm, 1375 mm, 1350 mm, and 1900 mm at Kanagawa, Shizuoka, Wakayama, Hiroshima, Ehime, and Nagasaki, respectively. The corresponding mean maximum temperatures at Okitsu in the Shizuoka prefecture are 20°C in January and 32°C in July. The corresponding mean

Table 1.108. Physicochemical characteristics of various soil types in Israel.

Horizon	Depth (cm)	Texture	Clay (%)	CEC (me 100 g^{-1})	Exchangable cations (me 100 g^{-1})			
					Ca	Mg	K	Na
Soil type, Sandy Regosols (Typic Quartzipsamment)								
A	0-25	Loamy sand	4.0	4.6	3.2	1.2	0.14	0.11
C	25-50+	Sand	3.0	3.1	1.9	1.0	0.08	0.10
Soil type, Red loamy Regosols, Husmas (Typic Xerorthent)								
A1	0-16	Sandy loam	11.0	11.5	9.9	1.1	0.3	0.2
A3	16-38	Sandy loam	12.0	10.7	9.3	1.0	0.2	0.2
C11ca	38-63	Loamy sand	7.2	5.2	4.2	0.7	0.1	0.2
C12ca	63-110	Loamy sand	7.3	5.9	4.6	1.0	0.1	0.2
C13ca	110-130	Loamy sand	10.8	10.7	9.4	1.0	0.1	0.2
B21cab	130-170	Loamy sand	16.0	8.5	7.2	1.0	0.1	0.2
B22cab	170-200	Sandy loam	18.2	6.1	4.9	0.8	0.1	0.3
Soil type, Natric grumic dark brown clay (Vertic Natrargid)								
A	0-7	Loam clay loam	26.9	21.5	14.0	6.2	0.9	0.4
B1	7-23	Clay loam	28.5	20.5	13.0	6.9	0.5	0.5
B21ca	23-55	Silty clay loam	38.8	26.2	11.9	12.5	0.5	1.3
B22ca	55-110	Silty clay	43.0	27.0	6.4	16.7	0.6	3.3
B23ca	110-150	Clay	44.0	26.7	2.9	18.3	0.6	4.9
Bb31ca	150-185	Clay loam	37.0	23.4	2.6	15.6	0.6	4.6
Bb32ca	185-200	Clay loam	28.5	19.9	2.9	13.1	0.5	3.4
Bb33ca	200-230	Loam	25.0	13.0	2.5	11.1	0.5	2.9
BCb	230-270	Sandy loam	18.0	13.5	3.6	6.9	0.4	2.6
Soil type, Clayey saline Regosol (Vertic Terriorthent)								
A	0-1	Clay	50.0	29.9	6.8	13.5	0.9	8.7
C1	1-20	Clay	50.0	28.3	8.4	13.0	0.9	6.0
C21	20-40	Clay	49.0	27.8	3.6	13.8	0.7	9.7
C22ca	40-60	Clay	47.0	24.4	1.8	15.5	0.7	6.4
C23ca	60-80	Clay	49.0	27.8	4.5	16.1	0.8	6.4
C23ca	80-110	Clay	48.0	25.0	3.8	14.3	0.8	6.1
C24	110-155	Clay	50.0	28.0	5.9	14.6	0.9	6.6
Bb21ca	155-195	Clay	52.0	30.7	7.6	14.6	0.9	7.6
Bb22	195-220	Clay	54.0	32.0	8.3	14.8	0.9	8.0
Bb22	220-250	Clay	56.0	33.7	9.3	15.0	0.9	8.5

Source : Eisenberg *et al.* (1982)

minimum temperatures are 13° and 27°C. At Tadono in the Wakayama prefecture, mean maximum temperatures are 20°C in January and 34°C in July with corresponding mean minimum of 13°C and 27°C, respectively. Temperatures in these two prefectures are representative of the wide range of Japanese citrus growing areas. Navel orange in Japan is mainly grown in areas with a mean annual temperature of more than 15.5 °C and a minimum winter temperature of -5°C before harvest, the temperature is no lower than -2°C (Sun, 1998).

Soils : Citrus in Japan is mostly cultivated on mountainous areas where soil pH is very low coupled with poor soil fertility and heavy erosion as reported earlier by many workers (Kanki *et al.*, 1968; Kanki and Imamura, 1968). A majority of citrus growing soils are sandy to clay of granitic origin. Soils in large part of citrus growing areas consist of volcanic ash and weathering matter of andesite (Nito, 1999). Comparative study of soil types in relation to growth of satsuma mandarin trees by Yamada (1964) in Kangawa prefecture showed best response in Sanuki andesite soil type followed by Izumi sandstone, granitic, biotite, andesite, and diluvial soils in that order due to high cation exchange capacity and available Mn. To avoid frost, citrus production is restricted to within about 25 – 30 km of coast line and most of the orchards are established on slopes. A 41 per cent of the orchards are on slopes of more than 15 per cent and 4 per cent of the orchards on slopes greater than 30 per cent. This ensures good drainage. But, makes it necessary to adapt terracing in the orchards. These terraces are often 5 to 6 ft high and only 6 to 8 feet wide (Spurling, 1969). According to Ono *et al.* (1986), performance of 19 year old Sugiyama satsumas trees in a sloping, south facing orchard showed superior growth of trees on moderate slope (17-19°) to that of trees on steep slope (27-32°). Boron deficiency has long been reported as hard fruit disease (Morris, 1938) under these conditions.

Nearly, all citrus is grown on relatively steep hills and bench terraced mountain slopes near the sea. Citrus growing soils are mostly acidic, heavy silt loams, varying in color as yellow, red, and brown, grey and dark grey. The geology of the citrus belts in Japan differs widely from paleozoic strata to alluvium. The nature of soil varies from sandy soil of granitic origin to clay. The soils in large part, consist of volcanic ash and the weathering matter of andesite. The natural fertility in most of the soils in the citrus zone is low as the topographic and climatic conditions cause not only considerable soil erosion and leaching, but also the development of acidic soils (Kozaki, 1981) with Mo toxicity in surpentine soils (Otsuka and Takahashi, 1962).

Morisada and Ohusumi (1993) described the soils of Bandai area in Japan as brown forest soils (Eutrochrepts or Dystrochrepts) mudflow, black soils (Andept), and Podzolic soils (Spodosols) with highly acidic soil pH (Table 1.109). Number of workers (Ogata, 1962; Iwamoto *et al.*, 1969; Kanki *et al.*, 1968; Kanki and Imamura, 1969) have reported very low soil pH (4.0 in KCl extractable) having high water soluble Mn (10 mg kg^{-1}) leading to abnormal defoliation. Such a low soil pH is further lowered due to heavy application of nitrogenous fertilisers (Yuda, 1977; Okada, 1981) which have induced a large defoliation of leaves on account of Fe and Mn toxicity. In southern Kyushu, about 55 per cent of citrus orchards are covered with soils derived from volcanic ash which are characterised by their poor base saturation, high humus content, active alumina (coefficient of P_2O_5) absorption, and relatively favourable physical properties. Soil reactions of horizons in a representative soil profile from Miyazaki perfecture are somewhat acidic (Table 1.110), high in cation exchange capacity, high coefficient of P_2O_5 absorption, poor in exchangeable cations and available P_2O_5 (Wada *et al.*, 1981). Takatsuji and Ishihara (1980) reported N, P, K, and Mg deficiency.

Table 1.109. Soil physicochemical characteristics of soils of Bandai area of Japan

Horizon	Clay	Silt	Fine sand	Coarse sand	pH	Exchange acidity	CEC	Ex. Ca	Ex. Mg	Ex. K
	(%)				(H_2O)	[cmol (p^+) kg^{-1}]				
	< 2	2-20	20-200	> 200 µm						
					Pine					
A	13.2	11.9	41.3	33.6	5.5	1.2	30.8	11.1	1.6	0.5
B	10.0	14.5	48.0	27.5	5.4	10.3	12.2	2.2	0.5	0.2
C1	2.5	10.5	41.7	45.3	6.0	0.7	1.8	0.4	0.1	0.1
C2	2.9	7.0	33.6	56.5	5.9	0.6	0.5	0.1	0.0	0.1
					Alder					
HA	8.6	9.1	36.2	46.1	6.2	0.7	37.0	16.8	3.2	1.5
B1	5.1	8.8	36.9	49.2	5.5	2.9	4.8	1.1	0.2	0.6
B2	4.3	10.4	35.5	49.8	6.3	0.9	2.0	1.1	0.1	0.5
B3	5.4	9.5	37.7	47.4	6.2	1.3	3.0	1.2	0.1	0.4
					Birch					
A	14.4	12.3	44.0	29.3	5.1	14.0	21.2	5.3	1.1	0.5
B1	13.2	11.2	35.8	39.8	5.8	2.7	8.8	2.7	0.7	0.8
B2	14.6	11.2	30.0	44.2	6.0	0.4	10.6	1.2	0.4	0.5
C	3.8	2.8	30.4	63.0	6.0	1.0	2.3	30.4	0.1	0.6
					Beech					
A1	27.7	27.1	22.4	22.8	4.8	14.3	24.1	0.3	0.2	0.3
A2	27.5	23.9	22.4	26.2	4.9	6.4	24.4	0.2	0.2	0.2
B1	24.8	34.5	34.5	6.2	5.0	2.9	21.7	0.1	0.1	0.1
B2	23.8	35.0	35.1	6.1	5.0	7.8	14.9	0.3	0.2	0.1

Source : Morisada and Ohusumi (1993)

Table 1.110. Chemical properties of a type of Miyazaki perfecture of Japan

Soil horizon	Depth (cm)	pH	Total N (%)	Humus (%)	Base ex. capacity	Exchangeable cations Ca	Mg	Available P_2O_5 (mg kg^{-1})
					[cmol (p^+) kg^{-1}]			
A11&A12	0-60	4.8	0.35	9.03	24.0	4.3	0.4	2.8
IIB21	60-75	4.5	0.20	1.22	14.9	0.4	0.3	3.3
IIB22	75-100	4.5	0.10	0.38	8.3	0.6	0.6	3.1

Source : Wada *et al.* (1981)

1.8.1.12. Citrus in Lebanon

The areas of Lebanon suitable for citrus expansion are limited, and all orchards are located on the narrow, eroded coastal plain. The acreage under orange, lemon, tangerine, and grapefruit has gradually expanded in the districts of Tripoli and Akkar in north Lebanon and Sidon (Saida) and Tyre district in south Lebanon.

Climate : The bulk of rainfall in Lebanon is distributed between October and May with 150 to 200 mm as mean annual rainfall, each month typical from December through February. Little rain occurs during the summer months of May to November. Average annual rainfall ranges from 575 mm in south Lebanon to 875 mm at Beirut. Mean maximum temperatures at Beirut range from 23°C in January to 33°C in July with corresponding mean minimum temperature of 19°C to 25°C. Humidity averages between 60 and 80 per cent from November to March and between 65 and 40 per cent from April to October. Severe frosts occur at interval of a decade or more, damaging trees and fruits. The mild, humid climate results in mature shamouti oranges in November with peel that is still green or only breaking in color.

Soils : Citrus orchards planted on terraces and sloping grounds are provided with an elaborate stone or concrete drainage ditches to carry off water and prevent erosion during the period of heavy rainfall. Italian cypresses is used as windbreak to protect orchards in exposed areas. Calcareous nature of soil and poor drainage are the major soil constraints in addition to shallowness of soil. This has resulted in reduced availability of N, Mg, Na, and B. Soils at Adlound showed variation in the sand, silt, and clay from 32.4 to 36.2 per cent, 13.6 to 20.0 per cent, and 46.5 to 54.0 per cent, respectively, with soil pH of 7.5 – 7.2 and EC of 1.20-1.35 dSm^{-1} (Khalidy *et al.*, 1969b).

1.8.1.13. Citrus in Malaysia

Climate : Rainfall is abundant in tropical Malaya with year round high humidity. Average rainfall at Johore Bahru is about 95 mm and well distributed throughout the season.

Soils : The soils in the tropical lowlands of Sarawak region belong to Podzols (Orthox). The other soil physicochemical properties of three representative soil profiles (Table 1.111) are depicted. Soils of peninsular Malayasia have been classified as Plinthia Orthoxic Tropudult (Wong,1984), pH 4.9-5.7, sand 64-71 per cent, silt 10-16 per cent, clay 18-21 per cent, exchangeable Ca, 0.54-1.00 cmol (p^+) kg^{-1}, Mg 0.24-0.43 cmol(p^+) kg^{-1}, and exchangeable K 2.7-4.7 cmol (p^+) kg^{-1}, in addition to low fertility status.

1.8.1.14 Citrus in Nepal

Nepal is one of the citrus growing Asian countries, characterised by high altitude citriculture.

Citrus Growing Belts : In the hills of Nepal, mandarin (*Citrus reticulata* Blanco) is cultivated in the kitchen gardens as well as in form of commercial orchards at an altitude from 650 to 1400 m above mean sea level. Mandarin is successfully grown in 55 out of the 75 districts of Nepal and the cultivated area under citrus is increasing every year in the western development region (NCDP, 1989; Chitrakar, 1990). Among the tropical/subtropical fruit

Fig. 1.15. Physiographic variation under which citrus is grown in Nepal

Table 1.111. Chemical analysis of the fine earth fraction of Sarawak podzols and grey-white podzolic soils of Malaysia

Horizon	Depth (cm)	Sand	Silt	Clay	pH (1:2.5)	Total N (%)	Org. C (%)	CEC	Ca	Mg	K	Na	Al
		(%)						(me 100 g^{-1})					
A1	0-7	72.4	17.5	5.8	4.2	0.03	2.50	6.51	0.63	1.55	0.10	0.06	1.31
A1.2	7-9	78.1	13.3	5.6	4.2	0.04	0.47	1.47	0.40	0.97	0.05	0.02	0.22
A2	9-12	75.3	19.6	3.7	5.0	0.01	0.16	0.63	0.38	0.79	0.04	0.04	0.11
A3	12-15	77.0	18.7	6.1	4.7	0.03	0.65	1.89	0.34	0.73	0.04	0.27	1.30
B1.1	15-18	73.7	16.8	7.6	4.5	0.06	1.38	4.52	0.38	0.77	0.04	0.27	2.10
B1.2	18-24	66.9	16.2	10.0	4.8	0.07	3.36	18.59	0.68	0.73	0.04	0.38	5.80
B2	24-28	63.3	16.8	12.5	5.1	0.06	1.61	11.24	0.18	1.03	0.04	0.38	4.92
B3	28-50	61.7	17.0	18.7	5.1	0.03	0.44	3.78	0.37	0.78	0.04	0.27	1.51
C	60-67	54.3	15.4	31.2	5.4	0.02	0.08	7.25	0.67	1.14	0.08	0.04	6.90
O	0-2	42.2	9.5	6.1	3.0	1.02	25.82	14.52	0.54	0.02	0.25	0.40	0.62
A1	2-5	86.4	5.5	4.7	3.3	0.16	4.78	6.4	0.42	0.02	0.17	0.39	0.61
A1.2	5-9	91.4	4.6	4.8	4.2	0.02	0.46	3.0	0.30	0.01	0.05	0.33	0.15
A2	9-13	90.7	8.6	3.0	4.7	0.004	0.14	0.5	0.42	0.02	0.05	0.33	0.10
B1	13-18	88.7	11.5	8.3	3.6	0.05	4.52	4.7	0.19	0.02	0.06	0.36	3.82
B1	13-18	88.7	11.5	8.3	3.6	0.05	4.52	4.7	0.19	0.02	0.06	0.36	3.82
B1.2	18-22	71.5	9.5	15.7	3.9	0.006	2.64	7.5	0.12	0.02	0.06	0.37	4.53
B2	22-33	72.4	10.5	17.2	4.3	0.01	0.57	1.0	0.19	0.02	0.06	0.33	2.41
B3	33-44	70.9	12.0	18.7	4.3	0.003	0.18	0.5	0.19	0.01	0.06	0.33	2.55
C	44-68	78.2	14.3	9.5	4.1	0.003	0.05	0.5	0.42	0.04	0.07	0.33	1.70
CII	68-76	46.4	19.4	35.1	4.2	0.01	0.10	4.3	0.19	0.01	0.24	0.41	2.84
A1	0-3	47.0	24.1	27.6	5.0	0.09	0.89	4.41	0.34	Tr	0.11	0.02	4.0
A2	3-15	44.6	28.4	26.9	4.6	0.04	0.11	2.94	Tr	0.24	0.19	0.10	4.4
B1.1	15-30	30.1	25.2	47.0	4.7	0.07	0.05	5.99	0.18	0.18	0.11	0.01	6.0
B1.2	30-60	22.4	31.7	46.8	4.5	0.09	0.03	7.77	0.06	0.24	0.14	0.01	7.15
B2.1	62-69	13.7	44.3	42.1	4.6	0.11	0.05	7.67	0.06	0.06	0.16	0.01	8.85
B2.2	90-98	13.5	46.8	37.5	4.7	0.10	0.05	6.62	0.12	Tr	0.16	0.01	6.85
B3	124-136	37.9	35.1	27.1	4.6	0.07	0.03	4.41	0.12	0.06	0.16	0.02	4.65
C	160-172	35.9	32.4	31.9	4.5	0.08	0.05	4.41	0.16	0.12	0.18	0.07	4.40
C/D	at 180	32.0	39.2	29.7	4.3	0.12	0.08	5.57	Tr	0.30	0.19	0.02	5.50
D	at 250	30.9	43.9	25.9	4.3	0.12	0.08	4.20	0.30	0.12	0.20	2.24	4.51

Source : Andriesse (1968/69)

species, citrus and mango have been given top priority. The total area of citrus is 17,026 ha, out of which productive area is 10,034 ha with 10 tons ha^{-1} as average productivity (Table 1.112). Citrus in Nepal is grown in areas, that include mainly Kaski, Gulmi, Dhankuta, Tanahu, Gorkha, Lamjung, Dailekh, Palpa, Syangja, Ilam, Sankhuwasabha, and Tehrathum districts (Table 1.113). Out of total area under citrus, 58 per cent, 21 per cent, 17 per cent, and 4 per cent are covered under mandarin, sweet orange, acid lime, and other citrus cultivars, respectively.

Physiographic Setting and Climate : Nepal is situated in between 22.22°N-30.27°N latitude and 80.4°E-88.12°E longitude. The altitude variation is characterised by citrus in mountains, and ridges interspersed by deep valleys and low lands. The country has five distinct physiographic regions : Terai, siwalik range, mid hills, high hills, and the himalayas (Table 1.114). The country is further divided into four agroecological zones viz., tropical (60-1000 m), subtropical (1,000-1,500 m), warm temperate (1,500-2,000 m), cool temperate (2,000-3,000 m), and alpine climatic regions (Table 1.115). Various citrus cultivars are adapted to these physiographic regions (Fig. 1.15) such as pummelo in tarai areas, pummelo, sweet orange, lime, and lemon to siwaliks and mandarins to midhills.

Table 1.112. Area, production and yield of citrus in Nepal during the last five years

Types	Total area (ha)	Productive area (ha)	Production (mt)	Yield (mt ha^{-1})
1993-94	13.5	7.9	76.5	9.68
1994-95	14.6	8.5	83.4	9.82
1995-96	15.2	9.0	88.6	9.87
1996-97	15.9	9.3	92.9	9.97
1997-98	17.0	10.0	100.0	10.00

Source : Fruit Development Division, Nepal Agricultural Research Council, Nepal

Table 1.113. Major citrus growing areas of Nepal representing various agroclimatic ranges

District	Area	Cultivar/ variety
Dhankuta	Khku, Budhimorang, Syaule and Dhankuta	Mandarin
Bhojpur	Chhinamarkhu, Sangpang, Pyuli	Mandarin, Sweet orange
Sankhuwasabha	Chanipur, Chabhue, Mangtewa	Mandarin
Sindhuli	Nakajoli, Ratonchura	Sweet orange
Ramechap	Lakhapur, Bhaluwajor	Mandarin
Kavre	Khopsi, Sankhu, Sharada	Mandarin
Sindhupalchok	Golangkhola	Mandarin
Kaski	Bhalam, Hyangia, Batulechaur	Mandarin
Syangia	Bahumara, Phoksing, Dopahara	Mandarin
Baglung	Kalika, Bhimpokhari	Mandarin
Dailekh	Dulla	Mandarin
Dadeldhura	Bagarkot	Mandarin
Baitadi	Jhalalgaun	Mandarin
Jajarkot	Sima	Mandarin
Gorkha	Bhorletar, Shishaghat, Dhimiregaun, Syant	Mandarin
Lamjung	Bakrang, Ashrang, Ghairung, Tanglichok, Manakamana, Gaikhar	Mandarin

Source : Fruit Development Division, Nepal Agricultura Research Council, Nepal

Citrus Soils - Nature and Properties : Pummelo growing soils of Tarahara (Anonymous, 1994a) are highly acidic (soil pH 5.7-5.9), good in organic carbon (1.56-1.84 per cent), high in total N (0.060-0.089 per cent), available P (186-288 kg ha^{-1}), and K (75-96 kg ha^{-1}). Soil textural analysis in citrus growing districts viz., Kathmandu, Kavri, Kaski, Lamjung, Ramachap, and Gorkha showed that 55 per cent of the soils are loam to loamy sand and silty loam followed by 28 per cent sandy soils, and 17 per cent clay with available water capacity between 16.8 per cent and 18.5 per cent (Anonymous, 1994b). A system analysis of soil fertility in relation to mandarin orchards in the midhills of western Nepal indicated that bearing trees (Fig. 1.16) are better cared than young seedlings. Though, there has been no consistent manuring practice in mandarin (Suntala)

Table 1.114. Characteristics of various physiographic regions of Nepal

Features	Physiographic region				
	Terai	Siwalik	Mid-hill	High-hill	Himalayas
Elevation (meter)	60-330	200-1500	800-1240, relief 1500 m with isolated peaks to 2700 m	1000-5000, relief 3000m from valley floor to ridges	2000-8848
Climate	Tropical	Tropical and subtropical	Subtropical, warm temperate (but tropical in lower valleys, cool temp. on high ridges)	Warm to cool temperate Alpine	Alpine to Arctic (Snow 6-12 months.)
Moisture regime	Humid in FE-MW, humid in W,C and E	Subhumid in most of the area, humid in W, C, E and Dun valleys	Humid and semihumid above 2500 m	Subhumid to per humid	Semiarid
Rainfall intensity	High	High	Medium	Low	Low

FE, Far eastern; MW, Mid-western; W, Western; C, Central; E, Eastern development region

orchards (Joshi *et al.*, 1995). Gupta *et al.* (1989) reported the deficiency of B, Mg, Cu, Ca, and Zn in potential mandarin growing pockets of Dhankuta district. Lohar *et al.* (1995) observed poor soil fertility, as one of the major constraints for low citrus productivity at Lohi (550 m), Bachegaon (750 m), and Bandipur (1000 m) areas of Nepal.

Most of citrus growing soils are acidic in nature, which covers 49 per cent of the total geographical area. Soils which have less than pH of 7.0 are said to be acidic. Brady (1992) classified soil reaction (pH) into six categories : very strongly acid (4.5-5.0), strongly acid (5.1-5.5), medium acid (5.6-6.0), slightly acid (6.1-6.5), neutral (6.6-7.3),

Table 1.115. Characteristics of agroecological zones of Nepal

Features	Agroecological zones			
	Tropical	Subtropical	Warm temp.	Cool temp.
Elevation range (m above mean sea level)	60-1000	1000-1500	1500-2000	2000-3000
Daily temp. fluctuations				
Minimum (°C)	7-24 Dec.-Jan.	2-17 Dec.-Jan.	9-10 Dec.-Jan.	15-8 Dec.-Jan.
Maximum (°C)	24-44 Jun.-July	13-17 Jun.-July	12-21 Jun.-July	8-15 Jun.-July
Mean annual (°C)	20-24	17-20	15-17	10-12
Annual rainfall (mm)				
Eastern part	1300	2800	900	700
Western part	600	1000	140	110
Physiography covered	Some part of mid-hills and Siwalik and all parts of Terai	Some part of high hills and many parts of mid hills and Siwalik	Some part of mid hills and many parts of high hills	Some parts of mid hills and high hills and all parts of Himalayas
Citrus varieties	Pummelo, lime, lemon,	Suntala (Mandarin), and Junar (Sweet orange)		

Fig. 1.16. Suntala (*Citrus reticulata* Blanco) cultivation in one of the valleys of Nepal

and slightly alkaline (7.4-7.8). In addition to soil pH, available Al, Fe, and Mn present in soil affect the crop yield at their toxic levels. Concerns over soil acidification have featured prominently in recent years (Shah *et al.*, 1991). One of the major contributing factors to decline in soil fertility in Nepal, is the increasing menance of soil acidification (Tripathi and Shah, 1986; Turton *et al.*, 1996), resulting in an overall decline in productivity of citrus orchards adapting citrus based farming systems.

Causes of Soil Acidity : Five major causes of development of soil acidity in Nepal have been identified:

Parent Material of Soil : The dominant bedrock is sandstone, siltstone, and quartzite, all of which acidify the soil.

Loss of Calcium and Magnesium due to Excess Rain : High rainfall within a short period of time during monsoon period (May-Sept.), solubilizes Ca and Mg nutrients, which are susceptible to leaching beyond the root zone. In Nepal, rainy season starts from May and lasts up to November. Leaching as well as surface run-off of Ca and Mg occur along with downward movement of soil particles. Thus, losses of Ca and Mg are high in the hills of Nepal. High leaching rates of basic cations (Ca, Mg, and K) have been reported by Shah (1995) and Shah *et al.* (1991) in Nepal. At higher altitudes of Nepal, soils are darker with a high organic matter content. They are usually more acidic due to high rainfall, which removes bases from the soil. With the oxidation of organic matter, acidity becomes a stronger limiting factor.

Continuous Use of Acidic Fertilizers : Ammonium sulphate and urea are the most commonly used fertilizers in Nepal and both of these fertilizers acidify the soils. Although, the addition of N fertilizers in Nepal is not as high as in the developed countries, the increasing use of acid producing fertilizers slowly increases soil acidity (Schreier *et al.*, 1995). However, in high and mid hills, farmyard manure and compost are still being the major sources of plant nutrients. Although, in the accessible areas and intensively cultivated river basin/low hills, the use of chemical fertilizers is increasing due to reduced availability of organic manures (Tripathi, 1996). Acidification affects the physical, chemical, and biological properties of soil, thereby, soil becomes hard and needs more labour to cultivate the land.

Crop Removal of Calcium and Magnesium : Soil Ca and Mg experience a depleting pattern with each crop, the menace of which becomes increasingly more acute under no addition of dolomite.

Presence of Large Scale Pine Plantation : In recent years, the planting of pine trees has become a national tradition in Nepal. As the forests become more pine-dominated, the forest litter collected by farmers during the dry season is dominated by pine. Pine and pine litters have a tendency to acidify the soils (Schreier *et al.*, 1995).

Comparison of Khet (Irrigated Low Land) and Bari (Upland land Rainfed Hills) Lands : Lands in the mid hills of Nepal are grouped into two categories based on irrigation availability : upland soils, mainly sloping terraces with rainfed crop (Bari) and lowland plain terraces with irrigation system (Khet). In the bari land, losses in soil fertility are attributed to depletion of soil organic matter (Maskey *et al.*, 2001).

Various soil properties showed higher fertility of bari than khet land (Table 1.116). The physicochemical properties at various land use patterns and physiographic position indicated most favourable physico-chemical properties under alley fruit tree cropping (Table 1.117).

Table 1.116. Soil properties of khet and bari lands in the western hills of Nepal

Soil property	Khet land (n = 249)	Bari land (n= 340)	SEm (±)
pH	5.5	5.6	0.05
Organic carbon (%)	2.6	3.2	0.09
C/N ratio	12.8	12.8	0.41
Total N(%)	0.2	0.3	0.01
Available P(mg kg^{-1})	122.9	174.3	16.13
Exchangeable K (cmol kg^{-1})	0.3	0.4	0.03

Source : Turton *et al.* (1996)

Fertility Constraint in Red and Nonred Soils : Red soils are prevalent in the river basin, lower hills of eastern, and western parts of Nepal (Fig. 1.17). In the western hills, red soils are dominant in tars, which comprise 2.3 per cent of the total agricultural land in the research command area having 11 hill districts of ARS Lumle Research Command Area (Gurung, 1994). Vaidya *et al.* (1995) reported that red and black soils are the common soil types in western hills. Black soils are regarded as fertile soils, since they contain high organic matter, available P, and moisture content compared to red soils. The highest priority area of research in the eastern hills is on the red soils and these soils are highly prone to

Table 1.117. Physicochemical properties of the soils of Chyandanda site, Nagorkot, Karve Palan chowk

Treatments	Soil texture	pH (H_2O)	OM (%)	N (%)	P (mg Kg^{-1})	K	Ca	Mg	CEC
						(cmol kg^{-1})			
1. Alley cropping- fruit trees	Silty loam	4.9	1.4	0.1	7.3	0.1	1.0	0.6	8.4
2. Hill side ditch	Silty loam	4.8	1.5	0.1	7.0	0.1	0.9	0.4	7.9
3. Strip cropping	Silty loam	4.8	1.2	0.1	6.5	0.1	0.9	0.4	4.7

OM stands for organic matter
Source : Maskey and Joshy (1991)

erosion (Sherchan and Baniya, 1991), because of poor infiltration rates and physical properties due to high silt fraction (Carson, 1985).

Schreier *et al.* (1995) compared soil pH and chemical properties of red soils and nonred soils of Jhikhu Khola, Karve Palan Chowk district, which revealed that red soils have slightly higher pH and significantly higher exchangeable cations such as K, Mg, and Ca (Table 1.118). In contrast, base saturation of red soil is lower due to the difference in cation exchange capacity and soil texture between these soils. It can be noted

Table 1.118. Properties of red soil and black soil in the western hills of Nepal

Soil properties	Red soil (n= 77)	Black soil (n= 159)
pH	5.6 ± 0.01	5.7 ± 0.04
Organic C (%)	2.9 ± 0.11	3.1 ± 0.08
Total N (%)	0.2 ± 0.01	0.3 ± 0.01
Available P (mg kg^{-1})	102.0 ± 19.30	138.0 ± 18.59
Exchangeable K (cmol kg^{-1})	0.5 ± 0.05	0.4 ± 0.03

Source : Vaidya *et al.* (1995)

that available P and percentage base saturation are higher for the nonred soils (Table 1.119). Although, the nonred soils have higher available P, the lower pH made P poorly available to crops, and thus, the rate by which the soil supplies P to the plant is insufficient to meet the seasonal requirements, or even the daily requirement. Shah *et al.* (1991) pointed out a problem of P fixation with Fe and Al oxides in red soils in Jikhu Khola watershed. Tripathi and Harding (2001) reported occurrence of large scale B deficiency in mandarin growing Bhorletar, Shishaghat, Dhimire, and Syaut areas of mid hill Lamjung and Bakrang, Ashrang, Ghairung, Tanglichok, Manakamana, and Gaikhur of Gorkha district.

Fig. 1.17. Suntala, *Citrus reticulata Blanco* a seedling mandarin grown in a red clay loam soil type in Nepal

Table 1.119. Differences in soil fertility between red and nonred soils at Jhikhu Khola, Karve Palanchok

Variables	Red soil (mean, n=90)	Black soil (mean, n=110)
pH (H_2O)	4.9	48
Available P (mg kg^{-1})	60.8	137.0
Org. C (%)	1.0	1.0
Cation exchange capacity [cmol(p^+) kg^{-1}]	13.1	8.9
Exch. K (cmol kg^{-1})	0.4	0.2
Exch. Mg (cmol kg^{-1})	1.8	0.1
Exch. Ca (cmol kg^{-1})	4.0	3.5
Base saturation (%)	46.8	55.7
K: Base saturation	0.8	0.4
Mg: Base saturation	3.8	1.9
Ca: Base saturation	8.5	6.3

Source : Schreier *et al.* (1995)

1.8.1.15 Citrus in Pakistan

The most extensive orchards in west Pakistan are in the Rawalpindi, Lahore, Multan, Sargodha, and Bhagawalpur divisions (Ahmed and Ghafoor, 1962-64). The Bhawal tehsil in Minawali district and Peshawar valley of northwest province are very well known. The annual rainfall in these areas ranges from 500 to 700 mm, relative humidity 50 to 54 per cent, highest maximum and minimum temperature as 42° and 12°C, respectively (Singh, 1969). The climate of west Pakistan varies from subtropical in the south to temperate in the north (Ahmed and Mazhar, 1962-64).

Soils are deficient in available Mn and Zn (Ibrahim and Ali, 1970). Soils are sandy clay loam to clay loam in Shapur, sandy loam to clay loam in Bhalwal Sargodha, loam to clay in Sahiwal, and sandy clay loam to clay loam in Sillanwali. A majority of citrus soils are classified as hyperthermic Typic Haplargid having predominantly low organic matter low available P, and adequate K. Khan (1969) described the suitable soil and climate for lemon growing in west Pakistan. A number of studies described soil and climate for sweet orange cultivation in Peshawar and Dera Ismail Khan divisions (Said and Inayatullah, 1964) and Sind (Jagirdar and Maniyar, 1962-64).

Haq *et al.* (1995) described the various physicochemical properties of sweet orange orchard soils of Northwest Frontier province and Malakand division showing majority of soils deficient in available K (Table 1.120) with mean soil pH 7.0-8.2, EC 0.10-0.25 dsm^{-1}, $CaCO_3$ 2.9-16.1 per cent, organic C 0.32-1.3 per cent, available P 4.5-46.5 mg kg^{-1}, and available K 75.0-165.0 mg kg^{-1}. Among micronutrients in Northwest Frontier province, available Fe, Mn, Zn, and Cu varied from 4.3 to 67.0 mg kg^{-1} (15.0 mg kg^{-1} mean), 4.0 to 35.0 mg kg^{-1} (15.0 mg kg^{-1} mena), 0.25 to 4.92 mg kg^{-1} (1.42 mg kg^{-1} mean), and 0.31 to 1.3 mg kg^{-1} (0.84 mg kg^{-1} mean), respectively. While, in Malakand division, available Fe, Mn, Cu, Zn, and B showed the variation of 4.5-48.6 mg kg^{-1} (13.8 mg kg^{-1} mean,), 7.6-35.2 mg kg^{-1} (19.2 mg kg^{-1} mean), 0.37 – 4.3 mg kg^{-1} (1.3 mg kg^{-1} mean), and 0.35-1.3 mg kg^{-1} (0.74 mg kg^{-1} mean), respectively, with 50 per cent soils deficient in available Zn and 95 per cent soils deficient in available B.

1.8.1.16 Citrus in Philippines

Extensive planting of mandarins has been taken up in Batangas region of Luzon. This has encouraged cultivation of valencia orange in Davao region of Mindnao (De Leon, 1955). Major commercial areas are located in southern Tagalog, Bicol, east and west Visayas, and Mindarao. There are also some commercial orchards in Icocos, Cagaya, Valley, and central Luzon. Similarly, good plantation also exists in Basilan island of just off Zamboanga (Wallihan, 1964; Gonzales, 1990).

Climate : The mean annual rainfall in northern part of Luzon is over 2000 mm with a pattern showing two distinct seasons: a predominantly dry season from November

Table 1.120. Physicochemical properties of sweet orange orchard soil types (mean of three depths viz., 0-15, 15-45, and 45-90 cm) in Northwest Frontier province and Malakand division of Pakistan

Location	* Northwest Frontier province						Location	** Malakand division					
	PH	EC (dSm^{-1})	$CaCO_3$ (%)	Org. C (%)	Avail.P (mg kg^{-1})	Avail. K (mg kg^{-1})		pH	EC (dSm^{-1})	$CaCO_3$ (%)	Org. C (%)	Avail. P (mg kg^{-1})	Avail. K (mg kg^{-1})
Shah Kali	8.2	0.14	4.2	0.61	19.7	98.6	Chungai, Swat	7.8	0.20	12.0	0.18	6.9	77.0
Sherpao	8.2	0.10	5.7	1.16	22.7	75.0	Barikot	7.7	0.17	12.2	0.85	9.1	95.5
Berlin Tangi	8.0	0.14	4.3	0.49	24.0	96.3	Nut Mira	7.7	0.20	11.0	1.0	4.5	80.0
Ziam	7.8	0.19	11.2	0.88	20.2	174.3	Colrra	7.8	0.16	10.6	0.68	10.5	69.0
Ghazi	7.8	0.20	2.9	0.39	34.0	60.3	Dedawar	7.8	0.18	13.7	1.3	9.6	96.0
Shah Mansoor	8.0	-	3.8	0.33	35.6	119.0	Malk Aabad	7.8	0.10	13.0	0.68	7.8	86.0
Baja (Swabi)	8.0	0.12	4.7	0.47	25.9	138.8	Nawagai	8.2	0.25	7.2	1.38	8.3	165.0
Dagai	7.8	0.10	2.9	0.46	21.4	122.3	Maina Kand	8.0	0.15	8.1	0.96	9.5	100.7
Shahbaz Garhi	7.7	0.19	3.7	0.41	28.3	119.3	Kot Kand	7.7	0.24	4.5	0.88	16.2	105.0
Chamagali	7.9	0.15	9.5	0.61	29.9	131.6	Dheri Dad	7.8	0.21	3.7	0.92	13.0	125.0
Kharkai	7.9	0.14	3.3	0.63	32.8	111.7	Palai, Swat	7.8	0.15	9.1	0.78	8.2	90.0
Palai	7.8	0.16	8.2	0.47	26.3	93.8	Bazdera	7.8	0.16	9.8	0.58	6.7	89.0
Khankali	8.0	0.18	16.1	0.51	20.3	153.0	Hajidabad	7.7	0.14	4.7	0.85	10.3	91.5
Kot (Dargai)	7.8	0.21	3.3	0.57	24.2	126.1	M. Barangola	7.7	0.14	5.0	0.80	13.3	105.5
Dedawar Shamzai	7.8	0.18	11.6	1.03	28.7	97.7	Ranai	7.7	0.10	5.0	0.93	12.3	95.6
Malikabad	7.9	0.15	13.3	0.35	24.1	77.0	Gandigar	7.9	0.18	8.5	0.68	11.5	107.0
Nawagai	8.2	0.18	8.1	0.87	27.3	136.0	Khal	7.7	0.10	4.5	0.82	7.7	92.0
Barikot	8.2	0.21	9.2	0.57	29.8	125.5	Akhagram	7.0	0.12	7.2	0.68	12.2	115.0
Fazal Aabad	8.2	0.19	11.2	0.53	23.3	164.6	Rabbat	7.8	0.16	5.5	1.11	11.0	108.0
Khazana	8.0	0.13	5.4	0.85	41.0	168.3	Khan Abad	7.8	0.14	3.7	0.84	13.7	125.0
Gosam (Munda)	8.0	0.18	11.2	0.84	37.9	103.6							
Pindi Kas	8.0	0.16	5.1	0.57	46.5	116.3							
Manzare	7.9	0.10	4.2	0.88	26.9	91.7							
Kadhi	7.7	0.16	4.7	0.43	25.8	100.7							
Wudi Gram	7.5	0.14	4.4	0.63	32.3	104.0							
Shahzadi	7.7	0.18	4.4	0.37	27.8	79.7							
Shabzadi Bala	7.8	0.15	9.8	0.48	30.5	116.3							
Manukhawar	8.0	0.20	3.8	0.63	16.9	122.3							
Khwar Kote	8.0	0.10	4.5	0.68	20.1	107.0							
Talash	7.9	0.10	4.7	0.35	19.5	67.3							
Chakdara	8.0	-	4.3	0.32	28.8	79.0							

* Sand – silty clay loam texture ; ** Sandy loam- silty clay loam texture
Source : Haq *et al.* (1995).

to April and a wet season from May to October.

Soils : The analysis of representative soil types at Quezon have been observed to have 66 per cent clay, 21 per cent silt, 13 per cent sand, 19.2 cmol (p^+) kg^{-1} cation exchange capacity, 1.5 cmol (p^+) kg^{-1} exchangeable Ca, and 0.1 cmol (p^+) kg^{-1} exchangeable K belonging to Ultisols (Palis *et al.*, 1994).

1.8.1.17 Citrus in Sri Lanka

Cultivation of citrus in Sri Lanka dates back to over a thousand and five hundred years, which is described in King Buddhadasa's medical treatise *Sarratha Sangrahaya* (Richards, 1958). Citrus in Sri Lanka is cultivated in two main areas consisting of Bibile-Moneragala belt of lower Uva province as rainfed and Vavuniya-Omentai districts of northern part of island under irrigation (Ghosh and Singh, 1993). Monaragala, Karungella, and Badulla districts are the leading centres of citrus production.

Climate : The climate of Sri Lanka is generally tropical and in some areas, the annual rainfall exceeds 1200 mm. The island slopes upwards on all sides of the coast towards the central Massif and Nuwara Eliya. During the monsoon season, from May to July, the west side of the island is wet, but from November to January, the east side becomes wet with the northeast wind blow. Average annual rainfall in the capital city of Colombo is 2000 mm. Mean temperatures at Colombo are 30°C in January and 31°C in July. Other orchards may be found in the northern part of the island in the Vavuniya-Omantai district, where the dry season extends over seven months with irrigated citrus. Climatic features at some of the citrus growing areas of Peradeniya and Maha Illuppallama showed average monthly rainfall between 35-371 mm and 20-302 mm, respectively (Table 1.121) with relative humidity of 80-89 per cent and 76–93 per cent.

The total amount of rainfall distribution during 1976-95 at Bandarawa is observed between 100 mm and 2400 mm. While average monthly sunshine hours varies from 3.2 hours day^{-1} in December to 6.3 hours day^{-1} in June. The average annual relative humidity varies between 61 per cent in August and 83 per cent in December. The difference of maximum and minimum temperature on average monthly basis varies from 7.8^{0}C in December to as high as 11.9^{0}C in February-March having maximum 20 rainy days in November and minimum 5 rainy days in June.

Citrus growing Monaragala district experiences a bimodal rainfall pattern with

Table 1.121. Climatic features of some citrus growing areas of Sri Lanka

Months	Location										
	Peradeniya					Maha Illuppallama					
	Rainfall (mm)	Relative humidity (%)	Temperature (°C)		Sunshine hours (hr day^{-1})	Rainfall (mm)	Relative humidity (%)	Temperature (°C)		Sunshine hours (hr day^{-1})
			Max.	Min.				Max.	Min.	
Maha season (September – February)										
Sept	163	83	29.1	-	4.2	73	80	32.3	24.4	7.2
Oct.	371	81	29.4	-	4.2	282	81	31.9	23.7	6.6
Nov.	257	83	29.8	-	6.8	302	85	31.2	22.6	6.7
Dec.	123	82	-	-	-	167	91	30.1	22.1	5.3
Jan.	58	80	28.8	18.7	6.6	82	93	29.0	21.0	5.6
Feb.	46	80	30.3	20.5	7.5	69	91	30.8	22.2	7.4
Yala season (March - August)										
Mar.	35	80	32.0	19.3	8.1	48	86	33.5	21.8	9.2
Apr.	159	84	29.9	21.5	5.4	145	82	32.5	24.3	6.8
May	140	89	29.3	21.9	5.4	107	81	32.0	24.6	7.9
June	159	84	28.5	20.8	6.4	20	81	31.4	24.2	7.8
July	183	83	28.6	21.7	5.9	34	79	32.0	24.3	8.4
Aug.	105	82	29.1	21.2	7.4	38	76	32.3	24.3	9.1

Source : Anonymous (1999)

annual rainfall of 1610 mm. Assured rainfall during October to December of the Maha season is well above the 100 mm threshold. Even though, the average rainfall during the rest of the season (January-February) is above the 100 mm, but its variability is considered high compared to the early part of the season. The minor rainy season, the Yala, lasts for only two months. The variability of rainfall during this season is substantially higher than that of the Maha season. During the Yala season, daily pan evaporation values exceed 4 mm day^{-1}. The minimum and maximum temperature in the area is around 20^{0}C and 35^{0}C, respectively. However, during the Maha season, maximum temperature fluctuates around 30^{0}C. The relative humidity during both day and night time is well above 70 per cent, except during inter-season dry period where day time humidity approaches 50 per cent. Agroecological features of citrus growing areas of Sri Lanka are further described (Table 1.122).

Soils : Most of the citrus in Sri Lanka is grown in reddish brown earths, followed by alluvial soils, and red-yellow podzols. Physiographically citrus orchards are concentrated on undulating land. Mendel (1963) earlier attributed poor yield of citrus orchards to physiographic and management condition. Site selection for commerical citrus orchards should have at least 6 feet deep soils and water table 4 feet below the surface during the wet months should preferably be selected for establishing commercial citrus orchards in Sri Lanka. Mostly, the soils are reddish brown and well drained having an adequate fertility that are usually considered suitable for citrus growing in Sri Lanka. But, generally, they lack the soil depth required for successful cultivation.

The trees are subjected to considerable water stress during the dry season, resulting in the death of feeder roots, wilting, and dieback of branches. During the wet season, these shallow soils are quickly saturated, some develop a temporary or sometimes perched ground water table reaching almost to the surface. In the low lying areas, soils may be flooded for lengthy periods. The alluvial soils are in many cases, deep and adequately drained. And if, no high groundwater table develops during the wet season, they are very much suitable for citrus growing. Many healthy and old trees are found in such alluvial soils in Liangolla-Syambalanduwa area and

Fig. 1.18 Cultivation of Bibile sweet orange on slopy land in Sri Lanka

Table 1.122. Agroecological characterization of citrus growing areas in Sri Lanka

Areas	75% expectancy value of annual rainfall (mm)	Climatic features								Major Soil groups
		75% expectancy of dryness for various months								
		J	F	M	My	Ju	Jl	A	S	
Amparai	> 875	J1/2	F	Y2M	My1/2	Ju	Jl	A		Noncalcic brown soils, reddish brown earths, old alluvial soils, low humic gley soils, and Regosols
Mahalllupallama Anuradhapura, Batticaloa, Gampola, Polonnaruwa, Vavuniya	> 750	J	Fy2	M	My	Ju	Jl	A	Y/2S	Reddish brown earths and low humic gley soils
Badulla, Moneragala	> 1375	*	F1/2	Y2M	My1/2	Ju	Jl	A	S	Red yellow podzolic soils and mountain Regosols
Bibile, Siyambalanduwa	> 1125	*	F1/2	M	My1/2	Ju	Jl	A	1/2S	Reddish brown earths, immature brown loams, and low humic gley soils
Colombo, Kalutara, Lunngala	> 1500	J	F	1/2M	*	*	*	A	*	Red yellow podzolic soils with soft and hard laterites
Galle	> 1500	J	F	1/2M	*	*	*	A	*	Red yellow podzolic soils with hard laterites, bog and half bog soils
Bentota	> 500	Jy/2	F	M	My	Ju	Jl	A	1/2S	Reddish brown earths with high amount of gravel in subsoil, low humic gley soils and solodized solonetz
Jaffna, Puttalam	> 575	J1/2	F	M	My	Ju	Jl	A	1/2S	Solodized solonetz, Solonchaks and Grumsols (Jaffna), Red yellow Latosols and Regosols (Pattalam)
Kandy	> 1250	J1/2	F	1/2M	*	*	*	A1/2	*	Reddish brown latosolic soils, Immature brown loams, and Red yellow podzolic soils
Kurunegala	> 1000	J	F	1/2M	*	*	Jl	A	1/2S	Red yellow podzolic soils with strongly mottled subsoil, low humic grey soils, Red-yellow podzolic soils with soft and hard laterite, and Regosols on old red and yellow soils
Uda Walawe	> 1050	J1/2	F	1/2M	MY1/2	Ju	Jl	A	1/2S	Reddish brown earths and low humic gley soils
Matale	> 1250	J1/2	F	1/2M	*	*	*	A	1/2S	Reddish brown latosolic soils, Immature brown loams and Red yellow podzolic soils
Nuwara Eliya	> 1375	J1/2	F	1/2M	*	*	*	*	*	Red yellow podzolic soils with dark B horizon and red yellow podzolic soils with prominent A_1 horizon
Ratnapura	> 2500	J1/2	F	*	*	*	*	*	*	Red-yellow podzolic soils and Red-yellow podzolic soils with semi-prominent A_1 horizon

J, F, M, My, Ju, Jl, A, and S stand for January, February, March, May, June, July, August, and September.
Source : Wijesinghe (1979)

on the Monaragala-Pottuvil road (Reuther *et al.*, 1967). A number of workers (Soyza, 1955; Panabokke, 1958; 1959; Moormann and Panabokke, 1961; Alwis and Panabokke,1972) have earlier described the soils of Sri Lanka. According to Ghosh and Singh (1993), citrus soils of Sri Lanka are shallow due to slopy lands. Heavy leaching of nutrients has been estimated in reddish brown permeable soils. The soils are well drained and deep in Liangolla-Syambalanduwa areas. Soils are poorly drained due to high water table especially deep alluvial soils.

Reddish and Brown Earths : This is the most widespread soil great group (Rhodustalfs and Haplustalfs) in Sri Lanka, which occupies the largest area compared to all other soils. Soils belonging to this great group are mainly developed on residuim and colluvium derived from intermediate and basic Precambrian rocks that have sufficient ferromagnesian minerals. The reddish brown earths are mainly confined to dry zone. They occur in a catenary sequence with other drainage members on the undulating mantled plain of the lower planation surfaces of the dry zone. They occupy the crests and the well drained upper and midslopes of the undulating landscape. A small extent is found in the transitional planation surface of the intermediate zone where they occur on rolling to hill terrain. Very often, the upper part of the profile is developed on semirecent colluvial materials and the lower part on much older, and somewhat transported residual materials. The two parts are often separated by a stone line or erosion pavement. Illuviation of clay or lixiviation appears to be a dominant process in the development of the reddish brown earths as evident from the occurrence of cutans on ped surfaces and on pore walls.

The colour of the surface soil is characteristically reddish brown when dry, changing to dark reddish brown when moist. The colour of the subsoil is distinctly redder than the surface soil. The normal uneroded soil profile consists of sandy loam to a sandy clay subsoil argillic a horizon. Most commonly, a subsoil horizon with a high proportion of quartz gravel is present. The depth to this gravel horizon is quite variable. The gravel itself is made up of subangular quartz gravel, angular feldspar gravel, and Fe-Mn nodules in different proportions. The profile sequence is usually A1, B1t, B2t, B3t, C. The surface soil structure is weak to moderate, coarse, and subangular blocky. The soils are extremely hard when dry, friable to firm when moist, and sticky when wet. The base saturation in the subsoil is in the range 60-80 per cent with soil reaction slightly acid to neutral.

The soil moisture relationships are characterised by a low water holding capacity with a rapid release of soil moisture at tensions lower than one atmosphere. According to the studies of Panabokke (1959), the clay minerals of these soils are mainly kaolinite, followed by illite and some smectite occurring in the lower B horizon and upper C horizon. Presence of significant quantities of smectite in the upper horizons of the hydromorphic associates of the reddish brown earth and the presence of 25 to 30 per cent illite in the surface horizons are the characteristic mineralogical feature. The presence of weatherable minerals in the sand fraction indicated an intermediate stage of weathering of these soils (Alwis and Panabokke, 1972). Its closest analogues are found in similar climate of east and west Africa, especially where the soils are derived from the basement complex of rocks. Their equivalents in Asia have been described as red mediterranean soils and red brown earths in Thailand.

Low Humic Gley Soils : Next to the reddish brown earths, low humid gley soils are the most extensive soil great group in Sri Lanka. The low humid gley soils are found throughout the lowlands of Sri Lanka, usually in the low lying ground above alluvial or colluvial parent materials. In the dry and intermediate zones of the country, they occur as the lower members of the drainage catena in association with either the reddish brown earths or the noncalcic brown soils. They are not found in the hilly terrains or in the regions of mountainous relief. These are essentially hydromorphic soils developed on the local colluvium that has built up during long periods of landscape formation at the footslopes of the undulating landscape. The dominant factor that governs the expression of the soils is the periodically high ground water level. This may be true ground water or a water table that develops on an impermeable stratum during the rainy season.

Soils are characterised by wetness or gleying throughout the profile, or gleying immediately below the surface horizon. They have a textural B or argillic horizon, which frequently shows conspicuous mottling. The colour of the surface soil is dark greyish brown to dark brown, and that of the subsoil is greyish to yellowish brown with distinct mottles and gleying. The normal profile consists of a sandy loam to sandy clay loam surface horizon underlain by a sandy clay to clay subsoil horizon. Calcium carbonate concretions, some times referred to as a kankar, are found in the lower depths of the profile in the drier environments. The depths of occurrence of these carbonate concretions correlate well with the degree of leaching or the dryness of the present climate.

In a well developed undisturbed profile, a horizon sequence is A1, A2, Btg, C. So far, two broad categories of low humid gley soils are found in the dry zone. The first and more widespread category consists of deep sandy clay loam underlain by sandy clay and with abundant calcium carbonate concretions below 1 meter depth. The second category is usually of a much lighter texture and with very little or no carbonate concretions. The former is associated with the reddish brown earths, and the latter with the noncalcic brown soils. The surface soil structure is subangular blocky to massive. Structural forms are more visible in the subsoils when the profile dries out to some depth during the dry season. These soils are extremely hard when dry and sticky when wet.

Brown Loams : These are immature brown loam soils (Ustropept) of the dry zone are mainly developed on residuum and colluvium of the micaceous bintenne gneisses. They occur on the dissected middle level planation surface as well as on the transitional area between the intermediate planation surface and the middle planation surface. They also appear in scattered locations with a marked relief and also on rock knowb plains. In north Bibile, however, these soils are found on moderate slopes in places where the reddish brown earths have been entirely eroded after intensive cultivation several centuries ago.

Red Yellow Podzolic Soils : The central concept of the red yellow podzolic soils is not as yet sharply defined to that of the reddish brown earths. The red yellow podzolic soils are the nodal soils of the wet zone of Sri Lanka, both in the lowlands as well as in the central highlands. Soils belonging to this soil great group are still referred as lateritic soils, lateritic loams, etc., usually with a qualifying adjective such as red, yellow, reddish to yellow, gravelly etc., to differentiate individual memebers. The earlier term lateritic does not necessarily indicate the presence of laterite, but more the earlier criterion of the low silicasesquioxides ratio of less than 2.0, which differentiated the lateritic from the nonlateritic soils. The term red yellow podzolic is, however, the accepted terminology for these soils, which are now recognized as the more widespread and dominant soils of the humid tropics of Asia. Besides the nodal group, a number of subgroups, namely, strongly mottled subsoil, soft or hard laterite, prominent or semiprominent A1 horizon, and dark B horizon have been prominently recognized.

1.8.1.18 Citrus in Syria

Citrus orchards in Syria are confined primarily along the mediterranean coast. The Syrian climate is subtropical mediterranean. Rainfall varies from 875 mm annually along the coast to 100 mm in the desert, and falls mostly between November and April. Average temperature ranges about 28°C in summer and 17°C in winter. Knorr and Vaughn (1964a) described world citrus problems with special reference to Syria.

1.8.1.19 Citrus in Thailand

A French messionary, Delaloublies mentioned about Citriculture in Thailand as early as 1680, spread to eastern, central, and northern provinces (Aubert, 1990; Sethpakdee, 1997). Major commercial areas of citrus production are coastal region and mid north inland region growing mandarins, limes, and pummelos. Sweet oranges are grown successfully in drier northern and northeastern provinces. The tidal coastal area of Nakorn Patoin of Bangok is well known for pummelos (Anchern, 1965). Northeast Thailand is the largest part of the country. Korat plateau is one such region where citrus is grown. The region is divided into Sakon, Nakhon, and Korat basins by the Phu Phan mountains which range from the northwest to the southeast of the region (Craig and Pisone, 1988).

Important areas of citrus cultivation are Chiengmai, Ping river valley, and Nan valley in the north; Petchboon in central Thailand, Roiet and Lamchi valley in the plateau; Nakorn, Pathom, and Donburi in the central west; Chonburi to Chandhaburi in the southest, and Haadyai and Songkla-Pattoni near the southern boundary line along Malaysia (Wahlberg, 1967).

Climate : A Climatic variation at Bangkhen (Bangkok) is characterised by mean maximum and minimum temperature of 30.9°-35.7°C and 19.3°-24.7°C, respectively, with a total annual rainfall of 1352 mm (Table 1.123). According to Vorasoot (1988), the climate of northeast Thailand is classified as tropical savanna with three distinct seasons, namely, the rainy season, winter season or cool season, and summer. Rainfall is erratic and dominated by both the southwest monsoon and tropical cyclones originating over Indian Ocean. The average annual rainfall varies from less than 1000 mm in the rain shadow in the west to over 2300 mm in the northeast along the Mekong river.

Table 1.123. Climatic variation at citrus growing region of Thailand

Month	Temperature (°C)		Rainfall (mm)
	Min.	Max.	
Jan.	19.3	33.4	5
Feb.	21.4	33.5	18
Mar.	23.1	34.7	29
Apr.	24.5	35.7	61
May	24.7	34.9	157
June	24.6	33.4	161
July	24.3	32.7	198
Aug.	24.4	32.4	213
Sept.	24.1	31.9	328
Oct.	24.0	31.4	297
Nov.	22.5	30.9	60
Dec.	19.5	31.0	4

Source : Reuther (1973b)

Soils : A majority of citrus orchard soils have been observed deficient in available Mn, Mg, and Zn (McCall, 1965; Spurling, 1969). The system of Citriculture common in Thailand, Malaya, and Indonesia in which trees are grown on the tidal floodplain soils by planting on narrow ridges builtup, some 1-1.5 meters above tide level (Spurling, 1969). A total of nearly 35 soil types have been identified and a majority of them belong to Ustifluvents, Tropaquepts, Paleaqults or Paleustults (Table 1.124). Soils are also derived from weathered ferrolysis and possess sandy texture with low clay content and organic matter (Brinkman *et al.*, 1977), imparting low bufferring and cation exchange capacity. Ragland and Boonpuckdee (1987) observed an average CEC of 24 different soil types as 18 mmmol c kg^{-1} based on analysis of soil samples from 0-15 cm depth. These soils are characterised by very low clay content (< 5 per cent), low organic carbon (< 1 per cent), and very low CEC [< 5 cmol (p^+) kg^{-1}], besides soil pH < 5.5. Virtually, all the soils in northeast Thailand have derived from sandstone, shale or siltstone, and low in fertility in terms of available P, K, Ca, and Mg (Ragland *et al.*, 1984). Watson (1994) described that soils of northeast plateau of Thailand and classified into approximately ten great groups and 20 soil series based on difference in colour, texture, cation exchange capacity, depth, etc. Most of them are alike in terms of having sandy surface horizon with not more than 6 per cent clay and 0.3 per cent organic carbon compared with low acitivity kaolinitic mineralogy. Pronounced Zn deficiency was noticed with certain soil types such as various latosols and sandy soils of the Korat plateau (McCall, 1965).

Table 1.124. Some soil properties of Thailand soils

Location	Soil pH	Organic C (%)	Particle size distribution (%)			Classification
			Sand	Silt	Clay	
Udon Thani	6.4	0.6	65	33	2	Paleustult
Khonkaen	4.7	0.6	67	25	8	Paleustult
Chaiyaphum	6.2	0.1	79	19	2	Qartzipsamment
Mahasarakham	6.4	0.4	84	12	4	Qartzipsamment

Source : O'Donnell *et al.* (1994)

1.8.1.20 Citrus in Turkey

Satsuma mandarin is the predominant cultivar amongst various citrus species and introduced in Turkey from Russia via Batum to Riza province. Over 90 per cent of production is obtained from the mediterranean sea coastal areas. Production areas along the aegean sea and black sea make up most of the balance. The Mersin-Adana district is the most concentrated area of plantings. Nearly, all orange production is in the mediterranean province of Icel (the Mersin area), Hatay (Iskenderun), Adana, and Antalya. Lemon production is centered in the Icel and Antalya provinces. The Aegean region of Turkey is the most important belt that contributes as much as 35-40 per cent of total citrus production (Burke, 1965; Anonymous, 1994c).

Climate : Mandarin type is well adapted to climatic conditions of Aegean region. The mediterranean citrus districts have a harsh climate with hot and dry summers. Citrus areas on the black sea at Ordu, Giresun, and Rize have cool and wet climate. In the mediterranean districts, nearly all the rain falls within the six months period from November to May. Annual rainfall averages about 575 mm at Mersin, 750 mm at Iskendrun, and 1000 mm at Antalya. Mean maximum temperatures at Antalya range from 22°C in January to 13°C in July, with corresponding mean minimum temperature ranging from 16° to 27°C. Summer temperature of over 38°C is common in the mediterranean and aegean coastal districts. All coastal areas are subject to some frost damage. In the humid black sea area near Rize, average annual rainfall is 96 mm distributed throughout the every month of the year. Mean maximum temperatures range from 19°C in January to 29°C in July, with corresponding mean minimum ranging from 15° to 25°C. The black sea areas and the Izmir district along the Aegean sea are subject to serious frost damage (Reuther *et al.*, 1967). Variation in minimum and maximum temperature at Adana has been recorded as 4.6°-22.4°C and 1.42°-35.0°C with an annual rainfall of 625 mm (Table 1.125). Areas of central Anatolya are characterised by arid climate with an average annual rainfall of 390 mm, mean annual air temperature of 11.1°C, and the mean annual soil temperature of 15.1°C (Seyyid and Recep, 2000)

Table 1.125. Climatic variation at citrus growing Adana area of Turkey

Months	Temperature (°C)		Rainfall (mm)
	Min.	Max.	
Jan.	4.6	14.2	102
Feb.	5.6	15.7	104
Mar.	7.6	19.0	66
Apr.	11.0	23.3	46
May	14.9	28.1	51
June	18.9	31.2	21
July	21.9	33.9	4
Aug.	22.4	35.0	5
Sept.	18.9	33.0	15
Oct.	14.6	28.9	39
Nov.	10.4	22.9	58
Dec.	6.6	16.8	114

Source : Reuther (1973b).

Soils : A majority of the soils are acidic in nature and suffer from deficiency of N, P, and K. The roots of satsuma mandarin are fairly widespread and as much as 90 per cent of feeder roots are distributed at 0-90 cm (Anonymous, 1988c). Ercivan (1974) reported Zn deficiency in Izmir region of Turkey. While, B toxicity, killing 10-20 per cent sweet orange and mandarin trees on trifoliate orange in Izmir region of Turkey has been reported by Mahmood (1971a). Soils in these affected orchards had B concentration up to 5 mg kg^{-1} and 395 ppm in leaves due to percolation of sea water (Mahmood, 1971b). Ozbek (1969) observed more acute Zn deficiency on soils with low pH, $CaCO_3$, and light texture. While, Fe deficiency appeared only on soils of high pH, $CaCO_3$, and moderately heavy texture. Boron and molybdenum deficiency are also invariably observed. Physical and chemical characterisation of citrus growing soils of Antalya, Alanya, Aksu, Dortyol, Iskenderun, Mersin, and Finike regions further showed mean variation in soil pH 7.7 – 8.4, $CaCO_3$ 0.17-78.1 per cent, organic matter 0.99 – 1.74 per cent, cation exchange capacity 9.2 – 31.6 cmol kg^{-1}, and soluble salts 0.02 – 0.90 per cent with soil texture from sandy loam to clay (Table 1.126).

Table 1.126. Physical and chemical characteristics of soils of different citrus growing regions of Turkey

Region	pH (1:2.5)	$CaCO_3$ (%)	Organic matter(%)	CEC (me $100g^{-1}$)	Soluble salts(%)	Texture
Antalya	8.42	78.07	1.72	11.85	0.02	Loam
Alanya	8.36	6.43	0.99	9.19	0.04	Loam – Silty loam
Aksu	8.36	27.52	1.70	24.40	0.05	Sandy clay – Clay loam
Dortyol	7.80	0.25	1.74	10.92	0.02	Sandy loam- Loamy sand
Iskenderun	8.15	31.79	1.69	31.60	0.10	Clay loam- Clay
Mersin	8.12	25.65	1.33	31.13	0.90	Clay
Finike	8.23	13.39	1.12	24.56	0.04	Loam

Source : Ozbek (1969)

Soils of central Antalya have developed on the conglomerate material with solum depth ranging between 22 cm and 100 cm with moderately to weakly developed structure in Ap and Bw horizons. Soils are taxonomically classified as Vertic Cambid and Fluventic Cambid (Table 1.127). A number of sweet orange and mandarin growing soils in Izmir region having available > 5mg kg^{-1} and up to 395 ppm in the leaf sample have shown tree collapse on account of B toxicity (Mahmood, 1971b). Citrus soils in the mediterranean part of Turkey are by and large, deficient in Mg, Mn, and Zn. Most centres of citrus production except Silifke, Limon lu, Erdemli, and Iskenderun have Na excess. In the western subregion, excess P and Mg deficiency are more frequent (Tuzcu *et al.*, 1981; 1986). Nutritional survey of mandarin orchards in eastern mediterranean coastal region of Turkey showed a number of soil fertility constraints like deficiency of N, P, and K (Ibrjkci, 1994). In southern Turkey, Cakmak *et al.* (1995) demonstrated a major role of high light intensity in the development of low temperature induced Zn deficiency bound leaf chlorosis in field grown citrus species viz., grapefruit cv Star Ruby and mandarin cv Fairchild. Similarly in Izmir region, Kovanci *et al.* (1978; 1985b) reported the occurrence of large scale Zn deficiency.

Table 1.127. Physical and chemical properties of soils of central Antalya, Turkey

Horizon	Depth (cm)	pH (1:1)	$CaCO_3$ (%)	Particle size distribution (%)			CEC	Exchangeable cations		
				Sand	Silt	Clay		Ca+Mg	K	Na
							[cmol (p^+) kg^{-1}]			
				Vertic Cambid						
Ap	0-22	7.8	10.2	18.8	45.2	36.0	34.3	33.1	0.46	0.65
A	22-46	7.8	12.6	23.1	42.6	34.3	34.8	33.8	0.40	0.60
Bw1	46-74	7.6	14.6	12.0	39.6	48.5	41.3	40.7	0.38	0.20
Bw2	74-100	7.9	14.8	13.1	40.2	46.7	38.9	38.3	0.20	0.36
C1	100-126	8.0	16.4	31.2	34.0	34.8	32.8	31.5	0.92	0.38
C2	126-150	7.6	18.6	29.9	37.5	33.5	22.3	21.4	0.61	0.20
				Fluventic Cambid						
Ap	0-18	7.7	8.7	36.5	30.0	33.4	22.0	21.9	0.20	0.48
A	18-32	7.9	9.5	34.5	31.6	34.8	22.8	22.4	0.20	0.20
Bw	32-60	7.8	12.5	30.5	30.0	39.4	22.8	21.3	0.66	0.85
C1	60-90	7.9	14.8	25.2	40.0	34.7	21.7	20.3	1.04	0.34
C2	90-124	8.0	15.9	38.5	26.6	35.8	13.0	11.8	1.00	0.20
				Fluventic Cambid						
Ap	0-15	7.5	11.6	23.6	44.2	32.2	26.4	25.4	0.60	0.86
Bw	15-22	7.8	12.0	27.3	38.0	34.6	21.3	20.3	0.60	0.38
C1	22-34	7.6	14.5	23.1	48.2	28.7	24.3	23.7	0.38	0.21
C2	34-56	7.8	10.8	32.1	42.3	25.6	12.0	10.5	0.92	0.60

Source : Seyyid and Recep (2000)

1.8.2 Citrus in Europe

Major citrus growing countries in Europe are France, Greece, Italy, Portugal, Spain, and Yugoslovia (Table 1.128).

1.8.2.1 Citrus in France

Citrus in France is mainly concentrated in Corsica island where modern Citriculture started in 1960 (Cassin *et al.*, 1969a). Vanniere (1990) described the cultivation of grapefruit, Star Ruby in Corsica Island with reference to terrain, soil, and climate.

Climate : There are three main climatic seasons in France. Cool and dry weather prevails from May to September. Weather remains mostly dry until December. Hot and humid weather occurs from January to April. Hurricane (tropical depression) occurs during January to April. Annual rainfall of more than 1500 mm occurs in most citrus growing areas, with an exception of sites located to the west and southwest of the island, a less than 300-500 m altitude (Pruvost *et al.*, 1997). Citrus is widely grown on Corsica Islands of France. Analysis of six years (1960-1965) of meteorological features at San Giuliano showed average monthly temperature of 16.3, total temperature of 333°C above 12°C and an annual rainfall of 819 mm (Table 1.129).

Average annual rainfall at Corsica is about 575 mm with a moderately dry climate. Mean maximum temperatures are 20°C in January and 33°C in July, and mean minimums are 16°C and 26°C, respectively. Climatic hazards are frost and wind.

Table 1.128. Citrus growing belts across the Europe

Name of the country	Citrus growing areas
France 48°00′ N lat.; 2°00′ E long.	Southern area (Bassin Martin, Bassin Plat, Saint-Louis, Etang-Sale-les-Sables, Etnag-Sale-les-Hauts, Montvert-Les-Hauts, Ravine des Cabris, Petite-Ie, Bel-Air, Berive, Plaine des Gregues Le Tampon), Western area (Bois de Nefles, Grand Fond, Tan Rouge, Villele, Trois-Bassins, Le Guillaume), Eastern area (Sainte-Anne, Pont Payet, Piton Sainte-Rose, Beaufonds, Chemin de Ceinture), Norther area (La Bretagne, Beaudonds Sainte-Marie, Sainte-Suzanne), Cirque de Cilaos (Palmiste Rouge, Mare Seche, Bras sec), and the entry of Cirque de Salazie (Petit Trou), Corsica island.
Georgia 41°40′ N lat.; 45°00′ E long.	Adzharia, Colchis plains (foothills and plains), Batuni, Poti, Sukhumi, Makharadze, Kolkhida, Tbilisi
Greece 38°00′ N lat.; 23°00′ E long.	Ipiros (Arta-Preveza), Thessalia (Volos), central Greece (Mesolongion Etoloakarnania), Peloponnesus (Corinth, and Heraklion Amalias, Pirgos, Patrai, Nav plion, Sparta, Achaia, Illia, Laconia Agrolis Kalamai), Crete (chania), Chios, Kiparissia, Rodhos, Rhods, Arta, Astypalea, Akheloos, Solonigios, Kerkira, Khan, Rethimnon, Iraklion, Ay-Nikolaos-kos, Corfu,Kefallinia
Italy 42°00′ N lat.; 12°00′ E long.	Fondi, Terraoina, Sperlonga, Scanzano, Metaponto, Massafra, Montabano along Agri and Sinni river basin, Sorrento, Salerno, Amalfi, Pontecoganga, Bellizzi irpino, Battipaglia paestum, Diamante, Belverdere, Crotone, Marina di catanzaro, Soverato marino, Polistena, Taurianova, Santa di caulonia, Villa san giovanni, Messina, Reggio calabria, Melito, Barcellona, Santa teresa di riva, Taormina, Santa Agata, Francavilla di sicilia, Cefalu, Bagheria, Palermo, Carini, Adranco, Aci castello paterno, Catania, Ramacca, Lentini, Francofonte, Floridia, Noto, Vittoria
Portugal 40°00′ N lat.; 13°00′ E long.	Ribtejo, Alto Douro, Braga, Oeste, Sotavento, Algarvio, Barlavento, Beja-Vidigaeira, Coimbra, Outra-Banda, Baixo-Douro, Santarem, Dalmau, Prata, Coroa de rei, Vernia, Demi sanguine, Sanguinelli, Verde doce de espana
Spain 40°00′ N lat.; 5°00′ E long.	Valencia, Alicante (La Murada, R. Bonanza, Crevillente, Los Montesinos) , Castellon de la plana, Almeria (Antas, Vera, Cuevas de Almanzora), Cordoba, Malaga, Murcia (Alhama, Librilla, Totana), Seville, Castile, Tarragona, Codiz, Huelva, Granada, Balecares, Canarias, Otras.
Yogoslovia 43°30′ N lat.; 20°00′ E long.	Adriatic sea island, Dubrovnik, Bar, Ulcint, Montenegro

Soil : Citrus in Corsica is generally grown on alluvial soils. Old fluvial deposits with acid reaction are widely used than calcareous fluvial deposits. These soils are sandy clayey or sandy loam in texture (Table 1.130). Comparison of soils located between trees against the soil under trees receiving fertilizer, former soils showed more acidic (pH 4.6 against 5.5), less rich in exchangeable Ca (1.2 against 2.4 me $100g^{-1}$), and exchangeable Mg (0.3 against 0.7 me $100\ g^{-1}$).

Favreau (1978) described the soils of eastern Corsican plain and their suitability for cultivation of various types of citrus species. Salinity is the major problem affecting the mediterranean citrus area. Saline stress affects the productivity and the survival

Table1.129. Climatic characteristics of San Giuliano, France (1960-1965)

Month	Monthly Temp. (°C)	Rainfall (mm)	Total temp. > (12°C)
Jan.	9.3	64.3	108.5
Feb.	9.3	101.4	98.0
Mar.	11.2	82.2	49.6
Apr.	14.2	79.4	42.0
May	17.4	29.6	142.6
June	21.5	14.2	261.0
July	24.5	32.8	362.7
Aug.	24.6	18.4	365.8
Sept.	21.8	39.0	270.0
Oct.	18.5	116.0	176.7
Nov.	13.8	122.2	30.0
Dec.	10.3	119.4	77.5
Mean	16.3	819*	333

* Total rainfall

Source : Cassin *et al.* (1969a)

of citrus (Mass and Hoffman, 1977). Alkaline pH due to calcareous soils is an another serious concern for the mediterranean basin citrus industry. It is associated with debilitating iron chlorosis, that is particularly prevalent in semiarid regions, where irrigation water contains high carbonates. Crop tolerance to iron deficiency has proved to be genetically controlled (Brown and Wann, 1982). Cassin *et al.* (1977) described citrus soils of Corsica Island as sandy to clayey, fairly rich in fine silt (2-20m), and characterised by slightly acidic pH, low buffering capacity due to presence of illitic and kaolinitic clays with satisfactory level of exchangeable Ca, K, and Mg (Table 1.131). The exchangeable Ca has a positive effect on the yield of Troyer citrange, *Poncirus trifoliata*, and sour orange rootstock, whereas, the exchangeable K has negative effect (Cassin *et al.*, 1977). Citrus growing soils at Corsica have been described as brown soils, red Palesols, and old deposits of schists origin having characteristic features of absence of $CaCO_3$, pH slight to highly acidic, illitic-kaolinitic mineralogy, and relatively high in clay (14-30 per cent) with fairly high cation exchange capacity (Cassin, 1968; Cassin *et al.*, 1969b; Lacoeuilhe *et al.*, 1968; Cassin *et al.*, 1977). Cassin *et al.* (1981) observed 140-380 mg kg^{-1} exchangeable Al, 30-700 mg kg^{-1} active Mn, and 34-109 mg kg^{-1} exchangeable Mn, showing leaf burn symptoms due to B and Mo deficiency, which reduced following

Table 1.130. Physicochemical properties of common citrus soil types in Corsica

Soil properties	Depth (cm)	
	0-20	20-40
pH	5.0-6.5	5.0-6.0
Silt (%)	22-32	21-32
Clay (%)	11-16	11-17
Exch. Mg (me 100 g^{-1})	< 1.0	< 1.4
Total-N (%)	0.8-1.3	0.7-1.1
Truog-P_2O_5 (%)	0.04-0.09	0.02-0.04
Exch. K (me 100 g^{-1})	0.3-0.7	0.1-0.3

Source : Cassin *et al.* (1969a)

Table 1.131. Vertical composition of citrus soil at Corsica Island of France

Properties	Depth (0-20 cm)		Depth (20-40 cm)	
	Average	Extremes	Average	Extremes
Clay* < 2 μ (%)	22.0-0	19-25	23.7	20-27
Fine silt 2-20μ (%)	24.81	19-29	24.4	19-31
Coarse silt 20-50 μ (%)	17.32	15-19	17.4	16-22
Fine sand 50-200 μ (%)	23.21	20-29	22.4	17-28
Coarse sand 200-2000 μ (%)	12.20	10-15	11.7	7-16
Exch.Ca (me100g^{-1})	6.87	6.2-8.8	6.01	5.6-7.6
Exch. K (me 100 g^{-1})	0.42	0.35-0.50	0.35	0.29-0.41
Exch Mg. (me 100 g^{-1})	1.50	1.30-1.70	1.52	1.40-2.20
Truog- P (mg kg^{-1})	16.70	8-54	8.13	6-15
Total N (%)	0.10	0.06-0.14	0.09	0.06-0.15
pH (H_2O)	6.93	6.4-7.3	6.51	6.2-6.9
pH (KCl)	5.72	5.3-6.1	5.27	5.1-5.5
Exch. cap. (me 100 g^{-1})	15.71	10.6-27.0	17.0	11.0-27.0

* Mainly illites and kaolinites.

Source : Cassin *et al.* (1977)

the application of $CaCO_3$ than $CaSO_4$. Soils of Corsica Island have shown large scale deficiency of Mo (Juste, 1978; Triboi, 1978; Cassin *et al.*, 1982).

1.8.2.2 Citrus in Georgia

Citrus in Georgia is confined predominantly on along the black sea coast. Adzharia is the major citrus growing region of Georgia with a satsuma mandarin on trifoliate rootstock as a major commercial cultivar. A major citrus is grown on the slopes in the subtropical zone of Western Gerogia (Kinkladze, 1987). Chanuvadze (1989) used agroclimatic information (soil and weather) for suitability of various citrus cultivars such as mandarin, sweet orange, and lemon regionwise in Georgian SSR. Lemons have been found to be highly suitable under warmest region.

Climate : Pogorelov (1970) described the geographic distribution and climatic characteristics of Abhazia citrus growing regions. Citrus production has been economical in warmer regions. The climate of Georgia is characterised by an average annual rainfall of 1000 mm, except Adzharia and black sea coast which has rainfall of as high as 2500 mm. Average temperatures are 27^{0}C in July and 15^{0}C in January. Coastal citrus growing areas are humid in summer. Climatically, Mtirala and Rikotskiy (Region I and II) have the features in common with other mountain regions of Western Georgia (III) and Krasnodar krayon on the southern slope of the greater Caucasus (IV). The almost continuous drizzle and fog make this belt of Western Georgia, similar to the so called rain forest of the high mountains of India, Vietnam, Burma, and other regions. Multilocation work on frost protection by using nonwoven or bamboo mats for improved snow retention of citrus showed best response of prostrate training for lemon, orange, and mandarin trees, low growing cultivars, and dense planting (Ramishvili *et al.*, 1988).

Citrus (Sweet oranges, mandarins, and lemons) is grown in the USSR on the eastern shores of the black sea which is considered to be the northern limit of their cultivation. Although, the spring, summer, and autumn temperatures are favourable to citrus growing, the winter temperatures are low, and once every 10-15 years, the trees are killed by frost in January or February. The studies on changes in leaf water potential during the winter and active growth phase, leaf water deficit and leaf water holding capacity suggested that the water regime of the different citrus species is a reliable index for their resistance to extreme growing and overwintering conditions (Goncharova *et al.*, 1989).

Soils : Most of the citrus cultivation is confined to subtropical area of Western Georgia on slopy lands as steep as 30^0 in the thermal belts. Hill side thermal belt soils are acidic, red, and yellow laterite types. These soils are well drained, but subjected to severe erosion. Citrus is, therefore, grown in trenches to avoid the trees from frost during winter (Reuther *et al.*, 1967). Soils of Western Georgia are classified as transitional to red earths and are called as red brown soils having extremely acidic soil pH (Sabashvili,

Table 1.132. Climatic characteristics of the mountainous region at an elevation of 700-2000m above mean sea level of Western Georgia

Climatic index	Mtirala (I)	Rikotskiy mountain pass (II)	Mountainous Abkazia (III)	Krasnaya Polyana (IV)
Av. Temp. in warmest months	17.0	17.5-18.5	15.0-20.0	19.5
Av. temp. in coolest months	-1.0 - 3.0	-0.5 - 2.5	-0.1 - 4.0	-0.1
Sum of temp. > 10^0C	2000-3000	2500	2000-3000	2000-2900
Av. annual temp. (^{0}C)	8.5	8.0-8.5	6-10	8-10
Av. frost free duration (Days)	170-200	180-200	160-230	210
Stable snow cover (Months)	4	2-3	2.5-4	4

Source : Romashkevich (1978)

1967). The climate of the region is temperate, moist, and excessively moist (Table 1.132).

The catenas representing upper, middle, and lower part of slope in mandarin growing subtropical part (Abkhazia) of Georgia reflected the following differences in the present modern dynamics of soil formation : in gleyey yellow earths, processes of gley formation are clearly expressed by the presence of perched water table at a depth of 130-140 cm, subsurface water movement down slope, and by an increase in the number of bluish grey mottles. Under these conditions, yellow earth formation is associated with seasonal excess wetness. Waterlogging decreased with an increase in thickness of the parent material and signs of a gley process (bluish mottles and marble appearance) diminish. Resultantly, the soils gradually acquire a yellower colour owing to an increase in their iron hydroxide content. The more intense the drainage, the redder the soil and colour (rubification) due to the transformation of iron hydroxides to oxides. The second catena is probably of similar age to the first, but it has greater natural drainage. All reduced iron is in oxides and hydroxides from which gives the soil a reddish colour. Part of the ferrous iron, on oxidizing, precipitates from solution, forming siliceous iron concentrations as weathering cursts in the red earth soils. Based on morphological structure, the concretions can be divided into two groups : fine, relatively soft, and presumably younger concretions characteristics of present day soil forming conditions. Formation of iron concretions in yellow earths make them more closely related to red earths. Unlike the yellow earths, signs of gleying are uncharacteristic of red earths, and are conspicuous only under conditions of water stagnation. Characteristics of the mineral composition of the soils and vertical (with depth) and horizontal (across the slope) changes in its composition showed a distinct difference due to atmospheric moisture and lateral subsurface flow.

On comparison of the bulk composition of the yellow earth and the red earth, the following features have been noted : Clay content (< 0.001 mm) in the soil profile 2 varied from 31 to 42 per cent and from 50 to 68 per cent due to more intense weathering in other soil profile (Table 1.133). In yellow earths, the SiO_2 content is high (especially in the low horizon due to frequent soil cultivation), decreasing uniformly down the profile, the content of Al_2O_3 and Fe_2O_3 is the same throughout the profile. In red earths, there is a large amount of sesquioxide and a clearly expressed tendency for the sesquoxide content to increase with depth. The main chemical feature of these soil profiles is the predominance of MgO over CaO except for the upper soil horizons, which must be considered originating from their parent material rather than any biogeochemical characteristics. In the yellow earth, the loss of CaO and MgO did not reach these critical values, while in red earth, the limit of CaO and MgO is reached as 0.30 and 0.60 per cent, respectively, as a result of which their secondary or biogenic accumulation became apparent. The intensity of weathering is more expressed in the red earth than in the gleyey yellow earth. The yellow earth retains a ferrsiallitic composition in the clay part, while red earth is clearly ferrallitic.

Red colour of soils showed no correlation with total iron oxide content. The yellow earths contained 48-81 per cent of the total amount of free oxides and up to 13-33 per cent of weakly crystallized forms of oxides. While red earths, with similar amounts of free oxides, contain up to 40-50 per cent of the total amount of strongly crystallized forms of Fe compounds. Red earths had higher humus than yellow earths, besides exchangeable Mg content in red earths being three times greater to that in yellow earths with exchangeable Ca having an eluvial distribution in both the soils (Table 1.131). The basic changes in the Al_2O_3 content are due to redistribution of clay within the catena as a result of lessivage as well as relative increase in the clay content owing to loss of SiO_2. A similar phenomenon occurred with CaO and MgO, which imparted a concurrent difference in fruit yield pattern of both, the red earth and yellow earth catenas with former having more favourable features.

According to clay formation and removal i.e. precipitation of substance and particles, the completely developed rubbly soil crusts of the mountains of Western Georgia have three types of profile : i. Metamorphic, ii. Eluvial metamorphic, and iii. Eluvial-illuvial metamorphic. The formation of the first type of profile is determined by youthfulness of soil formation, when the eluvial - illuvial removal and redeposition of soluble substances and dispersed particles against the background of argillation are not yet manifest or are inhibited by continuous denudation. The formation of second type of profile is determined by the superposition of vertical and lateral removal of substances and particles (resulting from longer periods of soil formation or less intense denudation) on the active argillation and liberation of substances in the profiles during weathering of the frontal lateral removal of soluble substances and dispersed particles, there are no signs of their illuvation in the profile. The formation of the third type of profile is determined, as in the second by more prolonged soil formation and/or relief conditions favouring the precipitation of substances in the profiles. Romashkevich (1978) studied the crust formation in mountain soils of Western Georgia and classified them as ferrallitic or ferrallitized depending upon three

Table 1.133. Some indices of the chemical composition and physicochemical properties of yellow and red earth soils of Western Georgia

Depth (cm)	pH (Water)	Hydrolytic acidity	Exchange activity	Available Al	Exchangeable cations Ca	Exchangeable cations Mg	Humus	Total N	P_2O_5	K_2O
		(me 100^{-1} g)					%		(mg kg^{-1})	
				Yellow Earths						
				Gvandra, upper part of the slope						
0-19	5.1	8.40	3.50	3.00	1.40	0.60	4.38	0.30	15.50	26.0
19-33	5.1	8.66	12.60	6.60	1.40	1.0	2.76	0.20	9.25	16.0
43-58	5.0	8.46	14.20	7.60	9.0	1.50	1.80	0.11	7.50	11.5
68-135	5.7	3.09	6.20	0.80	12.10	0.50	0.36	0.09	8.75	11.0
				Gvandra, middle part of the slope						
0-22	4.8	18.20	10.20	9.10	3.00	0.30	3.30	0.24	18.0	9.0
22-41	4.7	16.57	8.40	7.60	2.80	0.30	3.90	0.11	16.25	10.5
41-59	4.7	16.37	9.8	7.60	2.90	0.05	2.26	0.14	6.75	11.0
59-138	4.8	16.10	15.2	7.90	5.50	1.5	1.76	0.12	6.25	11.0
				Gvandra, lower part of the slope						
0-13	4.5	31.37	16.3	11.1	1.7	0.7	3.60	0.24	14.25	10.0
13-34	4.7	24.21	15.6	7.4	2.40	2.3	1.02	0.15	8.00	11.0
34-63	4.8	20.72	16.5	12.0	2.20	1.8	0.90	0.12	8.00	11.0
63-84	4.9	19.60	15.3	5.3	2.0	0.5	0.78	0.11	8.00	6.0
84-135	4.9	18.09	14.2	5.7	2.6	0.6	0.35	0.10	8.75	6.0
				Red Earths						
				Kokhora, upper part of the slope						
0-13	4.9	12.69	4.5	4.3	0.5	2.5	6.60	0.20	10.50	10.50
13-38	5.1	12.25	5.0	4.8	0.5	1.7	0.40	0.11	9.25	8.00
38-58	5.3	11.72	5.0	4.3	1.2	2.5	0.84	0.10	6.25	6.00
58-80	5.3	11.45	5.0	3.8	1.4	1.8	0.46	0.10	7.51	5.5
80-123	5.2	11.0	4.0	3.5	6.2	2.0	0.72	0.10	11.75	5.3
123-153	5.4	10.55	4.2	4.1	11.4	4.1	0.40	0.10	7.50	5.5
				Kokhora, middle part of the slope						
0-22	4.9	11.02	9.2	7.6	0.18	2.4	4.68	0.20	10.25	6.00
22-32	4.8	9.97	3.6	3.0	0.20	1.5	4.38	0.20	15.00	6.00
32-53	5.0	7.24	2.0	1.5	0.10	0.9	1.80	0.10	7.50	5.50
53-98	5.0	8.75	2.8	2.6	0.20	1.3	0.78	0.10	8.75	6.00
98-128	5.2	8.92	3.5	3.4	1.3	2.4	0.72	0.10	5.50	7.00
128-153	5.3	6.27	2.5	2.4	1.8	3.1	0.78	0.10	9.25	15.50
				Kokhora, lower part of the slope						
0-15	5.5	6.98	1.8	1.6	1.1	9.3	6.00	0.3	10.25	33.0
15-39	5.1	7.0	2.0	1.0	0.5	9.0	3.18	0.41	31.75	6.00
39-58	5.7	4.3	1.2	0.8	1.0	8.5	1.68	0.1	9.25	11.50
58-92	5.7	3.94	2.6	1.8	0.8	6.6	1.44	0.1	7.50	15.00
92-142	5.9	2.19	1.6	1.0	2.2	9.6	0.46	0.1	6.25	15.00

Source : Gvaliya and Zonn (1990)

criteria viz., presence of free Fe_2O_3 and Al_2O_3 in some mineral form in the weathering products, molecular SiO_2 : Al_2O_3 ratio in the clay (up to 2 ferrallitic composition and 2.2-5.0 ferrallitized), and ratio of clay minerals (crusts are ferrallitic when Kaolinite group predominates or equal in content to the group of mixed layer minerals and ferrallitized following predominance of mixed layer minerals over kaolinite. These criteria are used for Oxisols of Brazil (Lepsh and Buol, 1974), Natal soils in South Africa (Le Roux, 1973), and southwest Japan (Nagatsuka, 1975).

A number of other attempts have been made to characterise the soils of Western Georgia from many angles (Gradusov and Urushadze,1968; Nakaidze, 1970; Urushadze and Mkheidze, 1971; Machavariani, 1971; Bobrovitskiy, 1971; Zyrin *et al.*, 1973; Urushadze and Gradusov, 1976; Nakaidze and Archvadze, 1977). Tsanava *et al.* (1979) observed highest coefficient of N utilization as 45 per cent by mandarin cultivar, Kovano-vase in alluvial clay soil compared to 36 per cent in red soil. Talakvadze (1977) reported Zn deficiency in Georgia. Blondel and Brun (1973) described the soils under citrus grown in Georgia in addition to various cultural practices. Response of K fertilization at the rate of 100 g tree^{-1} showed an increase in mandarin fruit yield by 14 per cent in podzol soil type compared to 49 per cent in alluvial soil type of Georgia (Kharebava, 1971).

Glonti (1973) described changes in properties of red soil as a result of application of various N sources in Georgia. Cherkezishvili (1987) suggested a differential N rate depending upon soil type and agroclimate as 350, 400, and 250 kg ha^{-1} for Sukhumi, Makharadze, and Poti regions, respectively. Beridze (1986b) reported that an annual application of 210 g Ca tree^{-1} along with basal NP, 100 g K_2O, and Zn 150 mg tree^{-1} Zn increased mean yield of satsuma mandarin by 35 per cent (from 13.5 tons ha^{-1} in control to 18.2 tons ha^{-1}) in a 8-12 year old orchard on old tea soil.

1.8.2.3 Citrus in Greece

Citrus culture in Greece according to Theophratus is known from the period of Alexander, the Great (Webber, 1969). Greece is situated in the eastern part of mediterranean basin having mediterranean climate. Citrus orchards are more or less scattered throughout Greece. The principal growing areas are: the Arta-Preveza district of Ipiros; the Volos district of Thessalia; the Mesolongion district of Central Greece, the Corinth, Amalias, Pirgos, Patrai, Navplion, Sparta, and Kalamai districts of the Peloponnesusm and the Chaina district of Crete (Burke, 1966).

Climate : There are widespread variations in rainfall among citrus growing areas. Rainy season in most areas occurs during October to May. Average annual rainfall ranges from 550 mm in Crete to 1075 mm in the Arta district of Ipiros. Greecian summers are generally hot and dry. Mean maximum temperatures in the Khania district of Crete range from 23°C in January to 32°C in July, with corresponding mean minimums ranging from 17° to 25°C. Mean maximums at Arta range from 21°C in January to 35°C in July with corresponding mean minimum temperature of 15° to 25°C (Reuther *et al.*, 1967). The climate of citrus growing regions has geen described by Argyriou and Mourikis (1981). Comparison of rainfall at two citrus growing areas viz., Arta and Chaina showed higher rainfall at Arta (1080 mm) than Chaina (556 mm) coupled with higher variation in mean monthly maximum and minimum temperature (Table 1.134). The annual rainfall in most part of the country is approximately 1000 mm including some mountainous region, east, and south region. The mean minimum temperature in northern citrus growing areas, Preveza and Arta is 7°C in January, and 18°C in July with corresponding mean maximum of 13°C. While, in southern citrus growing areas, at Chaina, the mean minimum temperature is the same, with mean maximum temperature of 16°C in January and 28°C in July.

Table 1.134. Climatic variation at some citrus growing areas of Greece

Months	Arta			Chaina		
	Temperature (°C)		Rain-fall (mm)	Temperature (°C)		Rainfall (mm)
	Min.	Max.		Min.	Max.	
Jan.	3.9	12.8	145	7.6	15.4	109
Feb.	4.4	14.4	125	7.9	15.2	84
Mar.	6.7	17.8	104	8.9	17.8	46
Apr.	9.5	21.7	76	11.1	21.1	28
May	13.3	26.2	66	13.9	23.9	15
June	16.7	30.0	28	17.2	27.8	5
July	18.9	33.3	10	19.4	30.0	3
Aug.	18.9	33.9	10	19.4	30.6	0
Sept.	16.7	30.6	43	17.2	27.8	15
Oct.	12.8	24.4	147	14.4	24.4	89
Nov.	8.3	18.3	152	12.2	21.7	66
Dec.	5.6	14.4	174	9.5	17.8	96

Source : Reuther (1973b)

Soils : The citrus growing soils, by and large, are sandy loam in texture. Soil fertility of citrus soils at Messara valley of southern Crete showed 0.10 per cent N, 6.5 and 3.5 mg kg^{-1} $NaHCO_3$ extractable P, and 0.62 and 0.53 me 100 g^{-1} exchangeable K at 0-30 and 30-60 cm depth, respectively. The $CaCO_3$ content was registered as 35.2 and 35.5 per cent at 0-30 and 30-60 cm depth, respectively (Androulakis *et al.*, 1992). Dermetriades *et al.* (1963) earlier reported B toxicity in citrus growing soils as major fertility constraint. Nychas and Kosmas (1984) described Vertisols of Greece having montmorillonitic mineralogy (Table 1.135).

Table 1.135. Physical and chemical properties of the Greece soils

Sample depth (cm)	pH (1:1)	Silt (%)	Clay (%)	Organic C (%)	$CaCO_3$ (%)	CEC (me 100 g^{-1})	Extractable cations (me 100 g^{-1})			
							Ca	Mg	K	Na
Very fine, montmorillonitic, thermic (Typic Chromoxerert)										
0-30	8.0	16.4	67.4	1.06	0	57.4	31.10	24.89	0.36	0.70
30-70	8.1	14.8	70.4	0.90	0	55.5	29.00	23.00	0.12	0.74
70-120	8.0	13.4	73.4	0.89	2.0	60.3	31.50	30.85	0.20	2.70
120-150	8.2	16.2	71.4	0.73	2.4	56.7	27.00	31.25	0.34	3.20
Very fine, montmorillonitic, thermic (Typic Pelloxerert)										
0-15	8.0	16.7	60.1	0.93	0	68.4	28.65	41.00	0.36	1.40
15-42	8.4	12.8	66.4	1.25	0	70.0	24.75	44.20	0.22	2.50
42-99	8.7	11.4	71.8	0.77	2.8	59.6	20.15	40.10	0.20	5.20
99-141	8.1	9.8	75.0	0.63	13.6	56.0	21.00	36.33	0.24	7.20
Very fine, montmorillonitic, thermic (Typic Chromoxerert)										
0-25	7.1	22.0	61.3	0.82	7.3	32.4	28.20	5.64	1.56	0.70
25-50	7.1	25.8	56.2	0.78	8.2	34.0	29.50	6.04	0.68	0.90
50-87	7.0	21.0	67.3	0.92	7.3	35.2	31.25	4.08	0.48	0.90
87-130	6.8	23.4	59.3	0.56	2.6	36.0	29.80	6.67	0.20	0.60

Source : Nychas and Kosmas (1984).

1.8.2.4 Citrus in Italy

Citrus species, citron (*Citrus medica*, L.) was known in Italy as early as 310 B.C. during the Roman period (Webber *et al.*, 1967). The other citrus species such as lemon and sour orange (*Citrus aurantium* L.) were introduced by Arabs during 1000-1200, the commercial cultivation of which started around about 1830 (Russo, 1955). Though, introduction of sweet orange in Italy is not clearly defined, but claimed to be introduced during 1400-1600 as described by Ferrari (1646). The most important citrus producing areas of Italy are in Sicily and the southern part of the peninsula. Although, orchards in Sicily are scattered more or less over the island. They are nearly limited to the coastal areas and valleys extending further inlands. The greatest density of distribution is one, the east coast (Giacomo, 1987). Few researchers (Crescini, 1965; Crescimanno, 1966) compared the citrus orchards of the past with the present status at Benaco and suggested relevant remedies to be adapted. Valicenti (1962) and Di Martino (1962) described the citrus industry of Matera province. Most of the orchards in southern Italy are in warm areas of low altitude in a belt skirting the coast from Fondi to the Gulf of Taranto. This district includes provinces such as Lazio, Campania, Calabria, Puglia, and Basilicata. In northern Italy, the Ligurian section, famous in the history of the industry during the 14th and 15th centuries, occupied a narrow coastal belt extending from Genoa to St. Remo and the French boundary. Other researchers in the past have described the soil and climate under citrus in Italy (Burke, 1962a; Spina, 1975; Bangulo, 1977; Vilicenti, 1977). Cametti (1987) described the geological, pedological, hydrological, climatological, and cultural features of the Piana de Sibari, Calabria with special reference to irrigation.

Climate : According to Autori (1957), Rosini (1965), and Pinna (1977), climatewise Italy can be divided into ten regions, namely, Alpin, Padana, North Adriatic, South-central, Adriatic, Appenninica Riviera region, North Tyrrhenean, Ionian, South Tyrrhnean, and Sardinia on the basis of rainfall, minimum, maximum temperature, and intensity of rainfall (Table 1.136). Mean monthly maximum and minimum temperature at Messina varied from 13.9° to 30.0°C and from 8.9° to 22.8°C with total annual rainfall of 947 mm. While, at Taranto, a variation of 12.8°-31.6°C and 5.9°-21.1°C has been registered, as maximum and minimum temperatures, respectively (Table 1.137) with 360 mm rainfall. The regions such as Ionian, South Tyrrhnean, and Sardinia having average minimum temperature above 10°C and low rainfall are most important from citrus cultivation point of view. These regions have been described by Russo (1981) more analytically. Southern Italinan peninsula as Sicily has a mild climate having summer temperature over 34°C (Reuther *et al.*, 1967).

Review, outlining the climatic features of main Italian citrus regions indicated differential cultivarwise requirement. The areas viz., Sicily, Calabria, Basilicata, and Puglia have proved to be suitable for specific Citrus species (Spina *et al.*, 1980). The season of heaviest rainfall is from September to March. Average annual rainfall for selected citrus

Table 1.136. Climatic variation in citrus growing regions of Italy

Regions	Description
Ionian region	It is represented by Puglia, Brasilicata, Calabria, and Sicily. A very cold climate exists in Pollino, Sila, Aspromonte, Peloritani, Nebrodi, Madonie, and Etna mountains along the Ionian arc of Puglia and Lucania, rainfall is 600 mm and 800 mm in Sicily on the low eastern and southeastern sides of Mt.Etna, respectively, where citrus is grown. Under this climatic region, heaviest rainfall of 2000 mm is received in Calabria. This zone is further subdivided into 5 zones:
	Zone 1: It is represented by three provinces namely Calabria, Brasilicata, and Pugulia, to some extent Metaponto, Sibari, Corigliano Calabro, and southern coast of Catanzaro and Reggio Calabria
	Zone 2: It is represented by east and southeast Etna, Alcantara valley, and coast of Messina area suited for lemon production.
	Zone 3: It is represented by south of Catania in the province of Syracuse and Ragusa and coastal areas, characterised by mild winter and warm and dry summers.
	Zone 4: It is represented by the southern coast of Sicily from Gela to Marsala having mild winters, warm and dry summer. Citrus is concentrated in the valleys of Verdura and Magazzolo rivers.
	Zone 5: It is represented by southwest side of Mt.Etna, Catania plain, interiors of Catania, Syracuse, Ragusa provinces, and Centuripe area (Enna), characterised by warm summer and cold winter.
South Tyrrhenean region	This region is further subdivided into zone 6 and zone 7 having citrus culture.
	Zone 6: It is represented by coastal area of Campania, plain of Gioia Tauro, and coastal area of Calabria and Amalfi. Diamante in Scalea along the Tyrrhenean coast of Cosanza province area where only citron is cultivated.
	Zone 7: It is represented by Tyrrhenean coastal areas of Messina, Palermo, and Tarapani provinces, and more suitable for lemons and mandarins than sweet oranges.
Sardinia Region	It covers the east, west coast, and inland area of Campidano plain

Source : Russo (1981)

districts has been recorded as 975 mm, 350 mm, 600 mm, 625 mm, and 850 mm at Messina (Sicily), Taranto, Catania, Palermo, and Naples, respectively. Rain is received in every month of the year in the Liguirian district of northern Italy. Average annual rainfall at Genoa is 1150 mm. Bosco *et al.* (1977) described citrus growing areas of Sicily as mediterranean type with an average monthly temperature never below 10°C (subtropical area), a hot and dry climate with very little rain (Table 1.138). Germana (1992) reported annual rainfall as 500 mm and most of them is received between November to January (Table 1.139).

Table 1.137. Climatic variation at citrus growing Messina and Taranto areas of Italy

Months of Year	Messina			Taranto		
	Temperature (°C)		Rainfall (mm)	Temperature(°C)		Rainfall (mm)
	Min.	Max.		Min.	Max.	
Jan.	9.4	13.9	160	5.9	12.8	41
Feb.	8.9	15.0	109	6.3	13.9	23
Mar.	10.0	16.7	89	7.2	15.3	33
Apr.	12.2	18.9	45	10.0	15.2	20
May	15.0	22.8	51	14.4	24.4	25
June	17.8	27.2	13	18.9	28.9	15
July	22.2	30.0	15	21.1	31.6	10
Aug.	22.8	30.0	20	21.1	31.6	18
Sept.	20.6	27.8	21	18.3	28.3	25
Oct.	16.7	23.3	150	14.4	22.8	56
Nov.	13.3	20.0	137	10.6	17.2	46
Dec.	10.6	15.6	137	7.2	14.4	48

Source : Reuther (1973b)

The temperature of all citrus areas is greatly modified by proximity to the warm waters of the mediterranean sea. The climate of Sicily is particularly mild. Mean maximum temperatures range from 21°C in January to 32°C in July with corresponding mean minimums of 18°C and 27°C. Mean temperatures in the southern part of the Italian peninsula are slightly lower than those of Sicily. In the northern Ligurian, mean maximum at Genoa range from 19°C in January to 31°C in July, with mean minimum of 15°C to 26°C during these months. Calabrese and Marco (1981) described climate of Bagheria area of western Sicily as having an average annual rainfall of about 550 mm (September to June) with average summer temperature of 24°-26°C.

Table 1.138. Maximum and minimum temperatures and rainfall during May-October period at Sicily, Italy.

Month	Period	Temperature (°C)				Rainfall (mm)	
		Maximum		Minimum			
		1975	1976	1975	1976	1975	1976
May	I decade	21.1	20.4	12.5	12.9	2.4	----
	II decade	22.0	20.8	13.7	14.0	23.6	----
	III decade	23.9	22.5	16.9	15.2	----	----
June	I decade	22.3	23.6	15.1	15.1	----	5.6
	II decade	25.7	26.4	17.7	17.5	1.0	----
	III decade	27.3	27.9	18.3	17.0	----	12.2
July	I decade	27.2	28.5	19.0	17.7	----	4.4
	II decade	30.0	30.4	20.6	19.0	----	----
	III decade	29.5	28.5	20.8	18.0	----	----
August	I decade	30.0	29.2	20.5	17.4	----	----
	II decade	30.0	29.5	22.0	17.8	----	----
	III decade	27.1	27.0	19.4	17.5	45.8	39.2
September	I decade	27.8	25.6	20.1	16.3	----	3.4
	II decade	29.7	27.8	21.2	17.0	8.4	----
	III decade	26.2	26.7	18.7	13.8	12.6	----
August	I decade	25.2	27.0	18.4	18.1	2.6	10.2
	II decade	23.0	22.3	16.2	13.7	33.6	45.6
	III decade	20.4	23.2	13.2	14.0	1.2	87.6

Source : Bosco *et al.* (1977)

Table 1.139. Climatic trend in the two years period (1990-92) at eastern Sicily, Italy

Month	Temperature (°C)		Relative humidity*(%)		Rainfall (mm)	
	1990	1991	1990	1991	1990	1991
Jan.	10.3	8.2	82.0	64.0	55.6	63.4
Feb.	12.1	8.7	68.8	39.2	3.8	82.6
Mar.	12.7	12.3	4.8	71.4	4.6	72.2
Apr.	14.2	12.5	68.0	65.5	28.2	15.0
May	17.6	15.4	66.0	60.3	31.2	33.8
June	22.0	21.6	60.2	55.6	1.0	27.6
July	23.2	23.9	60.0	59.5	7.2	0.0
Aug.	21.5	25.4	61.0	62.3	42.8	1.4
Sept.	24.0	22.2	63.6	73.3	23.2	19.8
Oct.	21.2	19.2	72.6	77.2	52.6	45.6
Nov.	15.0	12.2	64.2	75.5	106.4	23.4
Dec.	9.0	7.4	7.3	76.0	133.8	112.6
Total					487.2	497.4

* Mean between maximum and minimum values

Source : Germana (1992)

Soils : Soil types in all areas vary from heavy to light and sands to loam in texture. Soils on valley floors tend to be heavier than those of hillside orchards. Summer lemons are produced in parts of Sicily by the verdellie practice, which involves drying orchards to the point of wilt in June and July, and afterwards forcing an out of season bloom by irrigation and fertilization. This practice results in a larger lemon crop for April to August harvest, with reduced winter and spring crops. The higher returns for summer lemons encourage growers having terraced orchards with light soil to force summer lemons. But, such practice weakens the trees. Analysis of soil samples from valencia orchards representing Messina, Catania, and Siracusa areas of eastern Sicily (Table 1.140) showed that these soils have pH 6.8-7.8, clay 12.5-26.5 per cent, silt 10.0-26.0 per cent, sand 47.5-79.5 per cent, active $CaCO_3$ 7.5-8.5 per cent, and organic matter 8.3-9.4 per cent. Soils at west central Sardinia are classified as Aquic Haploxeralf characterised by an AB profile, in which horizon A (approx. 50 cm depth) showed a sandy bound texture, field capacity soil moisture content of 23.5 per cent (volume), and a permanent wilting point of 8.8 per cent and wilting point of 24.7 per cent (Anonymous, 1975b). Raciti and Salerno (1974a) described the various mineral deficiencies and excesses in Sicily with B toxicity more prominent (Raciti and Salerno, 1974b). Potassium deficiency in Sicily and N and Mg deficiencies in Calabria have been commonly observed (Pennisi, 1975). Miyagawa satsuma is grown on the slopes of Mount Etna of Catania province at 350 m altitude above mean sea level. Soils in these areas are of volcanic origin, with subacid pH 6.5 and very low active lime content (Continella and Gentile, 1996). Majorana (1960) reported Cu deficiency in sweet orange orchards of Catania province for the first time. Clementine growing soils at Uta, south of Sardinia are noncalcareous alluvial soils rich in skeleton grain of 42 per cent with 60 per cent sand, 16 per cent silt, 24 per cent clay, and 0.09 per cent total N (Lovicu *et al.*, 1996). Low soil organic matter in commercial citrus belts of Italy is a new emerging problem (Calabrese, 1992).

Table 1.140. Some characteristics of soils from valencia orchards of eastern Sicily, Italy

Soil Characteristics	Locations		
	Catania	Siracusa	Messina
pH	7.8	7.8	6.8
Clay(%)	18.0-26.5	26.5	10.5
Silt(%)	22.0	26.0	10.0
Sand(%)	60.0	47.5	79.5
Active $CaCO_3$ (%)	8.5	7.5	-
Organic matter (%)	8.3	8.5	9.4

Source : San Lio *et al.* (1988).

Calabrese and Marco (1981) observed soils at Bagheria of western Italy have sand, silt, and clay content varying from 43.2 to 77.5 per cent, 3.6 to 21.9 per cent, and 5.9 to 34.8 per cent, respectively. Whereas, Bosco *et al.* (1981) described soils near Palermo (western Sicily) as typical red mediterranean soil, 1 m deep with sandy clay loam texture (27 per cent clay, 21 per cent silt, and 52 per cent sand), available water capacity 11 per cent on dry weight basis, and 50 per cent porosity with bulk density of 1.2 g m^{-3}. Citrus soils at eastern Sicily are characterised by a uniform, loamy, fertile soil having 1.19 per cent N 37.7 mg kg^{-1} assimilable P_2O_5, and 119 mg kg^{-1} exchangeable K_2O (Intrigliolo *et al.*, 1992). Cicala and Catarina (1992) observed citrus soils in the area of Catania plain as loamy-sandy soils, alluvial in nature with pH 7.5, and $CaCO_3$ 14.5 per cent. Some physicochemical characteristics of eastern Sicily soils showed variation in hydraulic conductivity from 2.01 to 12.18 mm hr^{-1} (Table 1.141). Physicochemical

Table 1.141. Main properties of the soils in the eastern Sicily

	Bulk density (g cc^{-1})	Ca^{2+} (%)	ESP	K^+/Mg^{2+}	Hydraulic conductivity (mm hr^{-1})
Average	1.19	1.20	0.85	0.112	6.11
Range	1.12-1.31	1.09-1.31	0.32-2.77	0.022-0.21	2.01-12.18

Source : Sardo (1992)

characteristics in soils of Metaponto showed higher pH and conducibility of clay soil than sandy loam soil (Table 1.142).

Soils at Corigliana area have observed pH 8.8, EC 241 mg cm^{-1}, sand 55 per cent, silt 24.2 per cent, clay 20.8 per cent, total limestone 2.36 per cent, and active lime 1.75 per cent (Recupero-Reforgiato and Russo, 1988). Basile *et al.* (1984a) described the soils of Patermo province of Catania having pH 8.2, clay 28.4 per cent, silt 25.8 per cent, and sand 45.8 per cent (Table 1.143). While in another study, Basile *et al.* (1984b) observed a wide difference in soil characteristics at Catania and Sircausa provinces of Italy (Table 1.144). Lio *et al.* (1984) described the citrus soils of Catania provinces having pH 5.8-8.0, organic matter 5.9-6.5 per cent, and texture varying between sandy and clay loam (Table 1.145). Bonifacio *et al.* (1997) described the soil characteristics of Valle Erro province of northern Italy and classified them as Lithic Usorthent, Lithic Ustochrept, Typic Haplustalf, Typic Ochraqualf, and Typic Haplustoll (Table 1.146).

Table 1.142. Soil characteristics at Metaponto, Italy

Soil property	Loam	Clay
pH	8.1	8.3
Exchangeable K (me 100 g^{-1})	1.96	1.19
Olsen-P (mg kg^{-1})	7.73	23.86
Active-Ca (%)	1.72	8.74
Total Ca (%)	5.08	15.24
Clay (%)	34.7	45.7
Silt (%)	20.4	47.8
Sand (%)	44.9	6.5

Source : Palazzo *et al.* (1992)

Table 1.143. Physicochemical characteristics of citrus growing

Soil type	Per cent constituents			Organic matter (%)	pH
	Clay	Silt	Sand		
Particle size	(<0.002)	(0.002-0.05)	(0.05-2.0)	-	-
	28.4	25.8	45.8	3.7	8.2

Source : Basile *et al.* (1984a)

Table 1.144. Physical characteristics of various citrus soil types at Catania and Siracusa provinces of Italy

Soil proeprty	Particles size	Palagonia	Grammichele	Catania	Lentini	Villasumundo
Clay (%)	<0.002 μm	36.13	12.00	20.07	12.04	20.07
Silt (%)	0.002-0.05 μm	20.08	0.04	14.06	0.04	1.47
Sand (%)	0.05-2 μm	43.79	87.96	65.87	87.92	78.46
pH	–	7.23	7.68	7.72	8.10	7.79

Source : Basile *et al.* (1984b)

Table 1.145. Citrus soil types at Catania, Italy

Characteristics	Soil type I	Soil type II	Soil type III	Soil type IV
pH	5.8	8	7.9	5.9
Clay (%)	10.5	10.5	30.0	10.5
Silt (%)	2.0	16.0	24.0	6.0
Sand (%)	87.5	73.5	46.0	83.5
Textural class	Sandy	Sandy loam	Clay loam	Sandy
Organic matter (%)	6.5	6.0	6.0	5.9

Source : Lio *et al.* (1984)

Table 1.146. Some physicochemical characteristics of soils in northwestern Italy

Horizon	Depth (cm)	Sand (%)	Silt (%)	Clay (%)	pH	Organic C (%)	CEC [cmol (p^+) kg^{-1}]	Ca saturation (%)
				Lithic Ustorthent (Mesic)				
A	0-2	64.5	23.5	12.0	7.1	2.32	12.6	21.7
C	2-12	57.5	26.1	16.4	7.0	1.48	18.2	17.2
R	12+	-	-	-				
				Lithic Ustochrept				
A	0-8	53.7	31.2	15.1	6.5	2.85	20.9	13.2
Bw	8-12	48.7	32.9	18.4	6.4	2.96	23.4	18.4
C	12-18	41.0	36.9	22.1	6.7	1.78	19.8	16.2
R	18+	-	-	-				
				Typic Haplustalf				
A1	0-25	25.1	46.1	28.8	6.5	4.58	21.4	31.3
A2	25-40	25.4	47.8	26.8	6.4	2.18	20.7	14.5
Bt1	40-60	27.9	31.7	40.4	7.0	1.14	21.1	12.3
Bt2	60-90	14.3	30.5	55.2	6.6	0.79	43.2	5.1
B	90-140	18.8	25.8	55.4	6.7	0.33	45.1	11.0
C	140-165	ND *	ND	ND				
R	165+	-	-	-				
				Typic Ochraqualf				
Ah	0-5	29.3	39.3	31.4	6.8	6.78	42.6	67.5
B1	5-25	25.4	39.9	34.7	6.7	3.22	38.9	45.0
B2	25-50	22.0	44.9	33.1	6.5	4.30	35.4	65.5
Bt	50-65+	16.7	39.0	44.3	6.9	0.89	30.0	33.5
				Typic Haplustoll				
Ah	0-25	15.6	29.3	55.1	4.7	14.34	31.3	30.0
Bw	25-50	21.9	49.4	28.7	5.3	3.51	18.6	27.6
C	50-75	34.6	29.8	36.6	6.6	0.79	18.9	10.2
R	75+	-	-	-				

* Not determined

Source : Bonifacio *et. al.* (1997)

1.8.2.5 Citrus in Portugal

Citrus is one of the important perennial fruit crops of Portugal and holds a significant acreage. There are four major citrus districts viz., Moura, Elvas, Lisbon, and Setubal.

Climate : The climate is hot and dry with poor soil in southeren Portugal compared to cooler climate having better soil fertility in north Portugal (Cameron, 1953). Coastal citrus growing areas are moderately dry. Interior valleys are somewhat humid. Rainfall is mostly distributed between October and March. Average annual rainfall is 440 mm at Faro and 690 mm at Santarem. Mean maximum temperatures at Faro are 22°C in January and 31°C in July, with corresponding mean minimum temperature of 18°C and 24°C. Frost and wind hazards are low in most areas (Reuther, 1973b).

Soils : Soils in citrus growing areas range from medium to coarse. All Portugese citrus is irrigated, raised as a monocrop, and sometimes mixed with other tree or vine crops. Some orchards are terraced. Some citrus is grown at lower elevations of Madeira island having predominantly on Andosols derived from balsaltic rock and pyroclastic material, and considered mostly as Dystrandepts and occasionally Placandepts (Ricardo *et al.*, 1984). Madeira *et al.* (1994) described these soils (Table 1.147) and classified them as Typic Dystrandepts and Oxic Dystrandepts.

1.8.2.6 Citrus in Spain

Cultivation of citrus is located primarily in the eastern coastal zones (Levant) and in the south (Andalucian) of the peninsula with nearly all the orchards close to the shoreline or in the river valleys. The citrus area of the Levant is comprised of the provinces such as Castellan, Valencia, Alicante, and Murcia. In this area, rainfall diminishes from the north to south with a mean annual rainfall of 426 mm in Castellan and 294 mm in Murcia.

Describing the citrus industry of Spain, Rivero (1981) considered four important milestones: i. introduction of citrus during 9th-10th century by Arabs, ii. introduction of citrus into America by Spain during expedition in 1493 by Colon, iii. large spread of citrus plantings due to introduction of sour orange as a rootstock, and iv. outbreak of tristeza in 1957, to pave the way for change in Citriculture. Rubel (1962) described the citrus industry of Spain with the help of soil and climate characteristics. According to Burke (1961; 1967), orange trees were introduced by Arabs into Andalucia in the 9th century and spread from there into Murcia, Valencia, and other parts of Spain. Spanish citrus orchards are centered in three areas, with the greater part of production along the mediterranean coast. Some citrus is also grown in the provinces of Almeria, Malaga, and Cordoba. The main zones for citrus culture in southeastern Spain are found in the high, middle (Murcia), and lower valleys (Alicante) of the Segura river system; Guadalentin valley (Murcia), Almanzora valley, and Campo de Dalias (Almeria).

Climate : Climate is basically semiarid. Rainfall in Spanish citrus growing areas is distributed mostly from September to April. Average annual rainfall in selected provinces has been recorded as 375 mm, 300 mm, 215 mm, 440 mm, and 575 mm at Valencia, Alicante, Almeria, Malaga, and Seville, respectively. Out of season, rainfall is not unusual in some districts. Summer thunder showers are occasionally heavy. Mean maximum

Table 1.147. Features of some soil types at Madeira Island of Portugal

Horizon	Depth (cm)	pH H$_2$O	pH KCl	Organic C (g kg^{-1})	Ca	Mg	K	Na	Ext. Al. (KCl)	ECEC
					Exchangeable bases [Cmol (p$^+$) kg^{-1})]					
				Typic Dystrandept						
Ah	0-22	5.4	4.6	117	1.17	0.59	0.12	0.31	1.17	3.96
Bw$_1$	22-53	5.5	4.9	89	1.14	0.21	0.05	0.27	0.40	2.07
Bw$_2$	53-97	5.4	4.9	80	0.43	0.12	0.04	0.26	0.26	1.11
2C	120-150	5.2	3.9	6	0.73	1.05	0.06	2.13	0.36	4.33
				Typic Dystrandept						
Ah	0-20	5.3	4.4	112	2.13	1.79	0.58	0.36	1.24	6.10
Bw	20-45	5.4	4.7	68	0.61	0.43	0.05	0.32	1.50	2.91
				Typic Dystrandept						
Ah$_1$	0-15	5.4	4.3	208	3.83	2.36	0.23	0.59	2.57	9.58
Ah$_2$	15-35	5.5	4.5	166	0.62	0.36	0.06	0.34	1.29	2.67
Ah$_3$	35-60	5.5	4.7	155	0.42	0.22	0.04	0.28	0.91	1.87
C	> 60									
				Typic Dystrandept						
Ah$_1$	0-10	5.2	4.3	176	5.26	6.61	0.36	0.76	1.48	14.47
Ah$_2$	10-35	5.4	4.5	101	1.28	1.50	0.04	0.37	1.66	4.90
Bw$_1$	35-50	5.4	4.6	85	0.93	1.30	0.03	0.33	1.11	3.70
Bw$_2$	50-75	5.6	4.8	63	1.04	0.98	0.02	0.28	0.4	2.86
Bw$_3$	75-105	5.7	4.9	47	1.00	1.41	0.01	0.28	0.36	3.06
C	105-130	5.9	5.2	52	0.86	1.51	0.02	0.31	0.12	2.85
				Oxic Dystrandept						
Ah$_1$	0-10	5.5	4.5	212	13.58	7.60	0.95	0.99	0.87	24.09
Ah$_2$	10-35	5.4	4.6	70	0.80	0.60	0.15	0.35	1.10	3.00
Bw$_1$	35-63	5.4	4.6	64	0.52	0.28	0.15	0.30	0.91	2.16
Bw$_2$	70-110	5.5	4.8	61	0.63	0.34	0.10	0.14	0.71	2.16
Ahb(?)	113-150	5.5	5.0	75	0.39	0.30	0.14	0.39	0.15	1.37
				Oxic Dystrandept						
Ah	0-25	4.5	3.9	121	0.13	0.21	0.06	0.28	6.07	6.75
Bw$_1$	25-51	5.4	4.1	76	0.07	0.07	0.02	0.15	4.12	4.43
Bw$_2$	51-72	5.5	4.5	70	0.19	0.04	0.02	0.20	1.50	1.95
Bw$_3$	72-108	5.4	4.7	51	0.07	0.02	0.01	0.13	0.90	1.13
C$_1$	108-145	5.5	4.7	50	0.16	0.01	0.02	0.20	0.56	0.95
C$_2$	145-170	5.5	4.8	52	0.06	0.01	0.02	0.14	0.40	0.63
				Oxic Dystrandept						
Ah$_1$	0-11	5.0	4.4	54	0.06	0.09	0.02	0.17	2.72	3.06
Ah$_2$	11-23	5.1	4.3	52	0.06	0.12	0.01	0.17	3.02	3.38
Bw$_1$	23-63	5.0	4.3	53	0.07	0.10	0.01	0.18	3.65	4.01
Bw$_2$	63-86	4.9	4.2	51	0.08	0.12	0.01	0.19	1.70	2.10
Bw/c	86-113	5.0	4.6	44	0.07	0.03	0.01	0.17	1.07	1.35
C	113-133	5.0	4.7	34	0.04	0.02	0.01	0.17	1.27	1.51

ECEC stands stands for effective cation exchange capacity
Source : Madeira *et al.* (1994)

temperatures in the province of Valencia are 22°C in January and 31°C in July with minimum of 15°C and 26°C. Mean maximum temperatures in the summer are slightly higher in the citrusgrowing province of south Valencia, reaching a mean maximum of 36°C in July in Seville Province. Gomez De Barreda *et al.* (1984) described the climate at

Moncada, characterised by mean annual rainfall of 231 mm at Almeria to 576 mm at Tortosa (Tarragona) with a dry season from June to August and rainy period between October and November.

Table 1.148. Climatic variation at citrus growing areas of Valencia, Alicante, Spain

Months of Year	Valencia			Alicante		
	Temperature (°C)		Rainfall (mm)	Temperature(°C)		Rainfall (mm)
	Min.	Max.		Min.	Max.	
Jan.	5.0	14.4	34	5.0	16.1	25
Feb.	6.1	15.6	31	5.9	16.6	25
Mar.	8.3	17.2	39	7.8	18.9	23
Apr.	10.6	19.4	39	10.0	21.1	28
May	13.3	22.8	42	12.8	23.9	23
June	17.2	25.6	21	16.1	27.2	13
July	20.0	28.3	12	18.9	30.0	6
Aug.	20.6	28.3	9	19.4	30.6	10
Sept.	17.8	26.7	75	17.8	28.4	46
Oct.	13.9	22.8	84	13.9	25.0	31
Nov.	8.9	18.3	50	7.2	19.4	51
Dec.	5.5	10.6	48	6.8	16.6	33

Source : Reuther (1973b)

The climate of citrus zones in Murcia region of Spain is semiarid according to Thornthwaite classification. Medium annual pluvimerty is about 300 mm with medium temperature of 18°C. Summer maxima of +40°C and minima in winter rarely under 0°C. The potential evapotranspiration fluctuates around 900 mm year^{-1}, which represents an annual deficit of moisture of about 600 mm, originating a medium index of –40°C. These features define the region as a semiarid close to aridity (Sanchez and Artes, 1981). The climate of southeastern Spain is characterised by mean yearly rainfall of 250-300 mm, mean evaporation 1000-1500 mm, mean temperature 17-18 °C, and mean relative humidity 65-75 per cent, representative of semiarid zone. The climate of southward region is more arid (Carpena *et al.*, 1969). Reuther (1973b) described the variation in maximum and minimum temperatures and rainfall at Valencia and Alicante aeras of Spain (Table 1.148).

The Andalucian zone is distinguished by two region: the coastal, made up of the provinces, namely, Almeria, Malaga, Cadiz Huelva, and the valley of the Guadilquivir, which embraces Cordoba and Seville. In the former, the rainfall is very scanty in the estern most part of the zone (230 mm) and somewhat higher in the remainder (450-500 mm). In the valley of Guadilquivir, rainfall diminishes from east to west and varies between 630 and 560 mm year^{-1}.

Soils : Soil types are red calcareous (Xerochrepts) and grayish limestone (Xerorthents) predominating Xerofluvents, and Xeralfs. Sandy soils (Xeropsamments) occur infrequently. Hernando (1969), while studying the citrus growing soils of Valencia province suggested five distinct soil types. These are Regosol (Orthents), Xerorendzina (Orthids), Mediterranean soil (Xeralfs), Regosol (Psamments), Pseudogley (Aquents), and Alluvial soil (Fluvents). Soil type, Regosol (Psamments) has two subtype, namely, beach having soft limestone fragments and colluvial (having high soil pH and $CaCO_3$). Soil type, Regosol (Orthents) and Regosol (Psamments) subtype beach have been observed highly favourable for citrus cultivation, recording fruit yield more than 35 tons ha^{-1} on an average, despite having $CaCO_3$ as high as 37.8-37.9 per cent and soil pH 7.92-

Table 1.149. Chemical characteristics of various citrus growing soil types of Valencia, Spain

Soil type	pH	Exchangeable cations (me 100 g^{-1})				$CaCO_3$	Organic	Texture
		Ca	Mg	P	K	(%)	matter (%)	
Regosoil (Orthents)	7.92	496.7	25.0	12.3	35.1	37.9	2.18	Sandy loam
Xerorendzina (Orthids)	7.90	456.1	21.9	27.5	37.1	16.4	1.84	Loamy sand
Mediterranean Soil (Xeralf)	7.88	429.6	21.4	26.5	33.9	12.1	1.76	Sandy loam
Regosol (Psamments)								Sandy
*Beach	8.10	376.2	8.5	17.2	8.9	37.8	1.15	Sandy loam
*Colluvial	7.92	292.2	9.6	37.5	15.7	10.4	1.13	
Pseudogley (Aquents)	7.84	547.7	31.0	23.2	45.9	29.9	2.79	Sandy loam
Alluvial (Fluvents)	7.90	468.5	22.5	22.9	32.3	25.8	2.12	Sandy loam

Source : Hernado (1969)

8.10 (Table 1.149). Soils of southeastern Spain are derived mainly from calcareous parent material and calcareous throughout ($CaCO_3$ 5.0 to as high as 72.5 per cent). A majority of soils belong to calcic Browns and Sierozems having sandy loam texture (Comas, 1966; Carpena *et al.*, 1969). Pennisi (1974) reported deficiency of N, P, K, Ca, and Mg in varying proportions in the citrus orchards of Sicily, Calabria, and Brasilicata, Spain.

A majority of citrus growing soils at Murcia are deficient in Mn and Zn (Llorente *et al.*, 1973), Fe (Hellin *et al.*, 1988), and excess of B problem (Leon *et al.*, 1983; 1984). In a survey of over 200 valencia orchards of eastern region, it has been observed that over 92 per cent orchards are sufficient in N status. But, over 50 per cent orchards showed low level of P and 90 per cent deficient in K (Hernando, 1978). Comas and Cros (1984) described the citrus growing areas in Spain with emphasis on Mg and various micronutrients deficiency. Hellin and Alcarez (1980) reported Mn deficiency in lemon orchards of cultivar Cerna. According to Gomez De Barreda *et al.* (1984), citrus soils at Moncada (Valencia) are characterised as calcareous ($CaCO_3$ 34 per cent), alkaline pH (8.4), sandy loam texture, low organic matter (1.1 per cent), high available P (38 mg kg^{-1}), and K (380 mg kg^{-1}). At Valencia, the citrus growing soils have been observed as sandy loam, pH 8.2, $CaCO_3$ 29 per cent, and active Ca 6.8 per cent (Legaz *et al.*, 1992b).

The soils of Vega valley with very fine texture and frequent presence of impermeable horizons prevent the normal growth of lemon trees. On the contrary, drainage is perfect in the calcareous soils and in the semidesert soils. Most of the soils have large amount of calcium carbonate and low organic matter content with soil pH reaching up to 8.5 (Carpena-Artes and Alcarez, 1990). Major soil type is Typic Xerorthent, 50-90 cm deep loam soil profile. Spanish citrus soils vary from sandy light loams to heavy brown loams, and have a pH value between 6 and 8. Most of the heavy soils are near the mediterranean coast or on low ground. Lighter soils are towards inland at higher elevations. Soils at

Villalonga (Valencia) are clay loam in texture, pH 8.1, carbonates 12.2 per cent, and active lime 4.2 per cent (Bono *et al.*, 1988).

Physicochemical studies of soils up to 60 cm depth from Washington navel orchard at Moncada, Valencia showed that soils have pH 8.4-8.5, sandy loam in texture, organic matter 0.6-1.1 per cent, available P 28-31 mg kg^{-1}, available K, 260-372 mg kg^{-1} and active lime 1.0-1.3 per cent (Table 1.150). Hernando (1969) reported Mn deficiency in citrus orchards of Valencia.

Table 1.150. Physical and chemical characteristics of clementine mandarin soils Moncada, Spain

Soil properties	Depth (cm)	
	0-30	30-60
pH	8.4	8.5
Organic matter(%)	1.1	0.6
Phosphorus (mg kg^{-1})	38	21
Potassium ((mg kg^{-1})	372	260
Calcium (%)	3.4	2.8
Active lime (%)	1.3	1.0
Conductivity (dS m^{-1})	0.22	0.21
Texture	Sandy loam	Sandy loam

Source : De Barreda *et al.* (1988)

Another report by Castel and Buj (1992) indicated sandy loam to sandy clay loam nature of clementine growing soils at Moncada (Table 1.151) whose effective depth is limited to 50-60 cm by a layer of partially cemented $CaCO_3$ (Petrocalcic horizon). Pomares *et al.* (1981) described the salustiana and Washington navel growing soils at Moncada, Valencia as reddish brown, sandy loam, and shallow, less than 1 m deep having $CaCO_3$ 6.7-7.2 per cent, pH 8.4-8.5, organic matter 1.2-1.5 per cent, Olsen-P 29.0-55.0 mg kg^{-1}, NH_4OAc-K 233.0-285.0 mg kg^{-1}, and NH_4OAc-Mg 219.0 – 266.0 mg kg^{-1}. Clementine soils of Moncada are red calcareous in majority, with low organic matter and nitrogen, optimum in P and K having soil moisture constraint in form of limited available water content (Table 1.152). Soils under Clementia de Nules (*Citrus clementina* Hort. ex Tan.) grafted on Carrizo citrange *(Citrus sinensis* Osb. x *Poncirus trifoliata* Raf.) at Moncada are sandy loam to sandy clay loam, with a petrocalcic horizon at 50-60 cm depth. Soils have available water capacity as 125 mg kg^{-1} and bulk density 1.43 to 1.55 g cm^{-3}. The soils are rich in K (250 ppm), poor in organic matter (0.7 per cent) and Olsen-P (30 mg kg^{-1}) with 0.5 dS m^{-1} EC (Castel and Ginestar, 1996).

Table 1.151. Physicochemical properties of citrus growing soils in Murcia region of Spain

Parameter	Depth (cm)	
	0-30	30-60
pH	8.6	8.6
Coarse elements (%)	10	18
Sand (%)	71	59
Silt (%)	17	20
Clay (%)	12	21
Bulk density (Mgm^{-3})	1.43	1.55
Field capacity (% vol).	29	26
Wilting point (% vol)	15	15
Total carbonates (%)	8.6	10.0
Active lime (%)	1.4	1.9
Organic matter (%)	0.7	0.5
Total nitrogen (%)	0.1	0.05
Potassium (mg kg^{-1})	260	245
Olsen-P (mg kg^{-1})	32	24
ECe (dSm^{-1})	0.5	0.5

Source : Castel and Buj (1992)

In Murcia region, the outcrops of Kenper appear to be as numerous and they are

Table 1.152. Physical and chemcial characteristics of clementine mandarin soils Moncada, Spain

Parameters	Depth (cm)	
	0-15	15-30
Sand (%)	64.4	68.4
Silt (%)	17.6	17.6
Clay (%)	15.0	15.0
Texture	Sandy loam	Sandy loam
pH (H_2O)	8.0	8.0
Calcium carbonate (%)	17.7	17.6
Organic matter (%)	1.0	0.5
Total N (%)	0.09	0.06
Olsen P (mg kg^{-1})	48.0	42.0
Exchangeable K (mg kg^{-1})	246	234

Source : Legaz *et al.* (1992a)

disseminated around its surface. Upper valley of the Segura river, sector from Archena to Djor as well as the following one to Blanca Abaran and Cieza, the outcrops of the Kenper are very frequent. In other valleys of rivers which are affluents to the Segura, of which are distinguished from those of rivers Argos and Quipar in central western Murcia region. Soils formed on gypseous iridescent marl of the Kenper or Trias layer in the Murcia zone in the southeast of Spain, constituents within semiarid climate of the region, an ecological means generally suitable for cultivation with irrigation of citrus, especially orange and lemon trees.

The lithological material indicated mixed with other adjacent parent materials of a different nature, such as fragile limestone on grey miocene marl, provides also a soil of anthropic character, suitable for citrus cultivation under irrigation. The presence of oligoelements has been determined in these marls, on account of their influence. On complete development of citrus species grown under ecological circumstances which are defined by Kenper soil, semiarid climate and nonsaline irrigation water from the river Segura. The characters of soils showed moderately carbonated, limestone, and gypseous. Texturally, soils are sandy loam with some enclaves where clay is dominating having illitic and monmorillonitic mineralogy. Soils have been classified as Gypsiorthids mixed with Arents (Table 1.153). These soils have good aptitude for Citriculture provided terracing and irrigation facilities are expanded properly.

1.8.2.7 Citrus in Yugoslavia and Croatia

Citrus is grown near Dubrovnik, as early as 1600, and a small marginal citrus industry has existed in Yugoslavia ever since. All citrus in Yugoslavia is grown near the Adriatic sea on Islands and along the coast mostly from split south. There are production centres near Dubrovnik and Bar and at Ulcinj in the extreme south.

Climate : Along the Adriatic coast, the climate has hot and dry summers and generally mild winters. Cultural hazards include drought, wind, and frost. Citrus orchards in Yugoslavia are usually small and scattered. Tabain and Miljkovic (1978) divided the citrus growing areas into 5 agroclimatic regions. Variation in monthly and annual air temperature showed a large variation (Table 1.154). Citrus on the Dalmation coast in Croatia is cultivated further north (up to 44.32°N latitude) than in other mediterranean regions (Tabain and Miljkovic, 1978).

Table 1.153. Physicochemical properties of citrus growing soils in Murcia region of Spain

Depth (cm)	Mechanical composition Sand (%)	Silt (%)	Clay (%)	$CaCO_3$ Total (%)	Active (%)	Org. matter (%)	Total N (%)	Available nutrients (mg kg^{-1}) P	K	Gypsum (%)	EC (dS m^{-1})
38°08′ 50″ N lat.; 2°22′ 10″E long, 110 m altitude											
0-24	28.8	39.3	23.5	16.5	11.30	0.47	0.030	2	0.43	0.70	2.30
24-46	38.5	-	-	16.0	9.97	0.34	0.020	2	0.72	1.14	2.45
46-90	51.6	38.0	19.8	22.5	9.97	0.31	0.023	1	1.02	0.68	2.50
38°08′50″ N lat.; 2°20′35″ E long, 142 m altitude											
0-30	25.9	50.2	22.7	35.0	9.90	0.38	0.030	1	0.46	0.80	2.30
30-78	36.0	-	-	14.5	12.08	0.31	0.016	1	0.16	1.11	2.35
78-118	53.1	-	-	11.0	10.50	0.40	0.020	1	0.15	1.14	2.20
38°08′50″ N lat.; 2°18′40″ E long., 340 m altitude											
0-14	80.0	11.0	16.4	14.0	9.06	1.31	0.07	8	0.22	0.80	2.18
14-28	72.5	7.0	16.7	6.0	-	0.88	0.055	6	0.31	0.80	2.18
28-46	45.6	7.6	10.6	3.0	-	0.57	0.051	6	0.36	0.80	2.13
46-113	37.8	11.8	9.9	4.5	-	0.47	0.035	6	0.54	0.65	2.14

Source : Sanchez and Artes (1981)

Soils: As early as Skoric *et al.* (1963) provided a detailed study about the soils of Yugoslavia. In the present classification of the soils of Yugoslavia (Nejgebauer *et al.*, 1963), the podzol is designated as soil type. It has been inserted into the class of soils with an A-B-C sequence of horizons together with the brown podzolic and lessive soils. Of the lower systematic units, the following subtypes are mentioned: iron podzols, humus-iron podzols, and gleyed podzols. A subdivision into varieties has been made according

Table 1.154. Mean monthly and annual air temperatures (°C) and rainfall (mm) at Zalesina, Gorsk; Kotar; Gerski Kotar Slzeme, Zagrebacka Gora; Topusko, Kordun; and Stubica, Hrvatsko Zagorje regions of Croatia, now an independent country

Months	Zalesina Temp.	Rainfall	Sljeme Temp.	Rainfall	Topusko Temp.	Rainfall	Stubica Temp.	Rainfall
Jan.	-2.0	188	-2.3	96	-0.1	93	-0.5	100
Feb.	-1.8	156	-2.1	68	1.0	73	1.1	61
Mar.	1.3	121	0.9	69	5.2	64	5.3	53
Apr.	6.3	153	57	91	10.6	85	10.8	74
May	11.2	160	10.2	127	15.1	89	15.3	102
June	14.9	145	13.4	167	18.6	104	18.8	131
July	16.6	127	15.5	141	20.2	90	20.6	93
Aug.	16.1	123	15.4	112	19.3	64	19.4	96
Sept.	12.4	193	12.2	100	15.6	86	15.5	88
Oct.	7.9	207	7.2	105	10.7	107	10.5	90
Nov.	3.3	205	2.3	121	5.9	117	5.7	101
Dec.	0.7	221	0.0	100	2.6	103	2.4	90
Annual	7.2*	1999**	6.5*	1297**	10.4*	1075**	10.4*	1079**

* Mean, ** Total

Source : Racz (1968/69)

to the parent material. Thus, in humus-iron and iron-podzols, varieties are distinguished on quartz sandstones and on other silicate parent rocks. The lowest systematic units are forms, and in podzols, they are subdivided according to the distinctiveness and depth of the A2 subhorizon (Racz, 1968/69).

1.8.3 Citrus in North and Central America

Citrus in north and central America consists of Honduras, Jamaica, Trinidad, Mexico, USA, Cuba, Puerto Rico, and Costa Rica. (Table 1.155)

Table 1.155. Citrus growing belts across North America

Name of the country	Citrus growing areas
Costa Rica 10°00′ N lat.; 85°20′ W long.	Pacific valley, Heredia, Turrialba, Orotina, Baudrit
Cuba 22°00′ N lat.; 80°00′ W long.	Guane, CarTomas, Ceiba Arimao, Moron, Sola, Vilorio, Contramaestre, Ciego De Avila, Jaguey, Isle of Youth, Troncoso, Havana
Honduras 14°00′N lat.; 87°00′ E. long.	Corozal, Orange walk, Maskalls bank, Cayo, Stann creek, Alta vista, Middlessex, Melinda, Pomona, Sula valley, Bajo Aguan,Guinope, Valle de Angeles, Signatepeque
Jamaica 18°00′N lat.; 77°00′ W long.	Anchovy, St. James, Westmoreland, St. Elizabeth, Gibraltar, St. Ann. Brown's town, Oracabessa, St. Mary, Paradise, Portland, St. Thomas, St. Andrew, Bog walk, St. Catherine, Linstead, Trout hall, Maypen, Clarendon, Manchester, Mandeville
Mexico 19°26′ N lat.; 99°01′ W long.	Brownsville, Monterrey, Matamoros, Montemorelas, Linares, Carmen, Cindad, victoria, Llera, Tompico, Cindad valles, Tan cannuitz, Tamazunchale, Rioverde, Tuxpan, El Alto, Tenila Colima, Mazanillo, Tecoman, Wruapan, Apatizingan, Tepalcatepec, Jalapa, Vera Cruz, Cordoba, Michoacan, Tamaulipas, Nuevo Leon, San Luis, Potosi, Guerrero, Oaxaca
Puerto Rico 18°15′ N lat.; 66°20′ W long.	Mayaquez, Isabela, Barcelonta, Barranquitas, Toro Neqro, Carozal, Cidra, Rio Grande
Trinidad 10°30′ N lat.; 10°20′ W long.	St. Clara, St. Augustine, St. George, St. Antoine, Piarco, Cumuto, El Carmen, Sangre Grande,Chayuramas, Matura, St. David, Gunapo, Talparo, Caroni, Rio Claro, Maygo, Moruga, Patrick, San Francique (Erin pt.)
United States of America 24°00′ – 47°00′ N lat.; 120°16′- 70°00′ W long.	Miami, Florida (Lake Polk, Highland, coastal Brevard, St. Lucie, Martin, Palmbeach, Indian river, Dade, Fort pierce, Lake hamilton, Avon park, Homestead, Hendry, Hardee, DeSoto, Hillsborough, Manatee, Orange, Collier, Osceola, Pasco, Charlotte, Okeechobee, Lee, Glades, Sarasota, Seminole, Volusia, Marion, Broward, Hernando, Pinellas, Citrus, Smuter, Putnam), Texas (Raymondville, Edinburg Mission, Alamo, Weslaco, Harlingen, Brownsville, Montealto, Rio Grande valley, Camaron,Hidalgo, Wallacy), Coastal ventura county, California (Central California –Tulare, Kern, Fresno, Clovis Porterville, Arvin; Southern California - Orange, Ventura, Los Angeles, Riverside, San Bernardino, Santa barbara, San diego, Indio, Brawley; Northern California - Glen, Butte, Tehama, Sacramento, Orlando) Arizona (Lower colorado river valley, Desert mesas around Yuma, Welton- Mohawk area, Salt river valley around Phoenix), Louisiana, Albana, Georgia, Hawaii, Mississippi, Pacific valley, Heredia, Orotina, Baudrit

Besides USA, Carribean basin countries contribute significantly towards world basket of citrus. Carribean Island is that part of Atlantic ocean surrounded by Central America, West Indies, and South America. Moratalla (1968) described the climate and soil aspects of *Criollo* lemons in Guatemala. Citrus in Carribean Islands is grown in a number of countries comprising Antigua and Barbados, Bahamas, Bbelise, Columbia, Costa Rica, Cuba, Dominica, Dominician Republic, El-Salvador, French Guiana, Grenada, Guadeloupe, Guyana, Haiti, Honduras, Jamaica, Martinique, Mexico, Montserrat, Nicaragua, Panama, Puerto Rico, Saint Lucia, Saint Vincent and Grenadines, Surinam, Trinidad and Tobago, and Venezuela. The Carribean coastal valleys have wet tropical climate, with some locations at very low elevation (Albrigo and Menini, 1984), where citrus is grown without irrigation.

1.8.3.1 Citrus in Costa Rica

Costa Rica has relatively dry valleys at lower elevation on the Pacific ocean. Most of the areas are below 50 m elevation and unsuitable for citrus cultivation. While, central valleys are at higher elevation within a volcanic mountain range (Saenz, 1966; Albrigo and Menini, 1984).

Climate : The climate of tropical region is subequatorial. According to available data (Saenz, 1966), the mean annual amount of precipitation is 2400-2500 mm year^{1}. It increases with elevation and in some places, it is as high as 3900 mm. Precipitation falls unevenly throughout the year, but usually exceeds evaporation. The mean annual temperature varies from 18.9^{o} to 20.8 ^{o}C, decreasing with the elevation by approximately 0.6^{o}-0.8 ^{o}C every 100 m. Relative humidity ranges from 64 to 87 per cent. The humidity is highest in October and lowest in June. The primary plant associations have been replaced everywhere by secondary forest (Gomez *et al.*, 1981).

Soils : Soils of Costa Rican lowlands, El Valle Central (Albarado and Boul, 1975) contain amorphous compounds, allophanes of varying compositions, halloysite and gibbsite. However, as in other volcanic regions, many aspects of supergene mineral formation remain obscure. The soils investigated by Gomez *et al.* (1981) in tropical region of Costa Rica are divided into 4 groups, depending on their position in the relief. Soils at San Pedro de Poas, San Isidro, and Sabana Redonda Poacinto occupy flat surface at an elevation of 1400-1600 m above mean sea level. The mean annual precipitation is more than 3000 mm, sometimes as much as 3900 mm. The soils have developed on pyroclastic material that has scarcely been altered or not altered at all by ancient hypergenesis under weak denudation conditions. Soils at San Pedro de Poas and Sabanilla are confined to about 15° slopes at elevations of about 1100-1275 m above sea level. The mean annual precipitation here is 2500 mm. The soils have developed on deluvium with pyroclastic material and a substantial quantity of clay particles. The parent material contains interlayers of different textures, including gravel and pebbles about 10 mm in diameter or more. Soils at foothills of the central Cordilleras are confined to the foot of slopes at elevation up to 950 m above sea level. The mean annual precipitation is 2300 mm. Soils at Orosi region occur at the foot of 20° slopes at an elevation of 1051 m above sea level. The mean annual precipitation is 2200 mm. Bornemisza *et al.* (1985) reported Mg deficiency at Turrialba, Zn deficiency at Orotina, and Mn deficiency at Orotina and Baudrit. Soil analysis from Atlantic zone of Costa Rica indicated nutritional deficiencies in form of N, P, K, Ca (Alvardo *et al.*, 1994), Zn, and Mn (Araya *et al.*, 1994).

1.8.3.2 Citrus in Cuba

Citriculture in Cuba has been introduced by Spaniards at the end of 15th century. Most of the Cuban citrus industry did not begin developing until the late 1960s (Muraro *et al.*, 1996). The two citrus growing regions which have the greatest impact on the export market are Jaguey Grande and Isle of Youth. Gras Guerra (1987) described a comprehensive account of the soil and climate under Cuban citrus industry.

Climate : The Contramaestre region is the northern part of the Sierra Maestra mountains in the eastern Cuba, at an altitude of 100 to 150 m above mean sea level. It has a humid tropical climate having mean maximum temperatures of 32°C in August (the hottest month) and 28°C in February (the coldest month), with corresponding mean minimum temperature of 21°C and 15.2°C and annual rainfall of 1450 mm (Sanchez-Garcia and Fernandez, 1981).

Based on principal component and cluster analysis involving 18 climatic variables and 10 localities, classified the citrus producing regions of Cuba into 3 main groups and 5 main groups, respectively, taking soil type into account (Lima *et al.*, 1988). Lobaina *et al.* (1988) described the climatic features of Guantamamo, Cuba with reference to air, temperature, rainfall, evaporation, and relative humidity for over 10 years. The minimum and maximum temperature vary between 15°C and 20°C and 31°C and 36°C, respectively. Evaporation exceeds rainfall, and values for relative humidity are generally high (72-74 per cent on an average). The citrus growing areas in Cuba belong to tropical climate with only grapefruit being native to this climate (Bello, 1996). Due to geographical location of Cuban Island within the Caribbean region, having modified tropical climate, and where the relative humidity and air temperature remain high all the year, apart from characteristic rainfall pattern, several vegetative flush periods take place year round in most of the cultivars (Otero *et al.*, 1996).

Soils : A thorough description about the soils of Cuba has been focussed comprehensively earlier by Bennett and Allison (1928). Nipe clay and serpentinite soils at Cuba are highly deficient in CaO and rich in Fe_2O_3 (Table 1.156). Hernandez *et al.* (1975) described the soils at south Havana as red ferrallitic. Potassium has been found one of the important fertility constraints in leached quartzitic yellow ferrallitic soil at Pinar del Rio which has been confirmed through the significant response of K fertilization (60 kg ha^{-1}) on various fruit quality parameters (Gonzales *et al.*, 1987).

Table 1.156. Chemical composition (%) of Nipe clays and serpentinites of Cuba

Depth (cm)	SiO_2	Al_2O_3	Fe_2O_3	TiO_2	MgO	CaO	K_2O	P_2O_5	MnO	SiO_2:Al_2O_3
0-66	3.28	18.46	63.04	0.80	0.33	0.12	0.06	0.03	0.42	0.18
66-100	2.25	11.13	69.56	0.26	0.48	Traces	0.08	Traces	0.28	0.20
100-400	1.83	12.36	71.12	0.80	0.64	0.01	0.02	Traces	0.38	0.15
400-500	1.55	14.66	68.10	0.80	0.60	0.15	0.05	Traces	0.47	0.11
500+	41.93	2.00	7.84	0.05	34.02	1.50	0.08	Traces	0.12	20.97

Source : Beinroth (1982)

1.8.3.3 Citrus in Honduras

Citrus industry of Honduras with reference to soil and climate has been described by Burke (1956a).

Climate : The climate under citrus is predominantly subtropical, where flowering is regulated by temperature as well as rainfall. Average annual rainfall at four of the major citrus producing areas has been recorded as 2515 mm, 2520 mm, 2125 mm, and 2460 mm at Alta Vista, Middlesex, Melinda, and Pomona. Mean minimum temperature of Honduras ranges from about 25^{0}C in January to 27^{0}C in July. While mean maximum temperature ranges from 31^{0}C in January to 33^{0}C in July. The average annual humidity ranges from 77 to 90 per cent (Reuther *et al.*, 1967).

Soils : Number of citrus growing regions are located at a very low elevation (Ton and Alfonso, 1980). Citrus plantations have been undertaken on virgin sandy loam soils after removing forest. Presence of large scale Mg deficiency is a common feature (Weir, 1971).

1.8.3.4 Citrus in Jamaica

It is one of the Carribean Islands, where citrus is grown primarily as a rainfed crop under year round high humidity.

Climate : Jamaica has a warm, humid, and tropical climate with an annual rainfall ranging from 1375 to 1800 mm. Average maximum and minimum humidity of the year is 78 per cent and over 66 per cent, respectively. In most districts, at least 50 mm of rainfall is observed each month. Temperatures are remarkably uniform, with mean minimum and maximum ranging from 23^0 to 32°C in January and from 26° to 34°C in July. In winterless climate of Jamaica, the blooming of citrus is controlled by rainfall that normally occurs after heavy rains breaking the fall drought of December to February (Reuther *et al.*, 1967). Variation in mean monthly maximum and minimum temperatures has been observed as 30.5°-33.1°C and 16.0 – 21.3 °C, respectively, with annual rainfall (Table 1.157) of 800 mm according to Reuther (1973 b).

Soils : Most of the Jamaican citrus orchards are of the plantation type. Such orchards are planted on rolling land in areas of heavy rainfall, where irrigation is unnecessary. More recent orchards in Jamaica, comprising not over 10 per cent of citrus orchards, are established on ground level in dry areas with rather light soil. They are cultivated with tractors in a manner similar to California citrus orchards. The clean cultivation technique is practised only in areas with less than 50 mm of annual rainfall.

Table 1.157. Climatic variation at citrus growing region of Kingston (St. Andrew), Jamaica

Months	Temperature (°C)		Rain-fall
	Min.	Max.	(mm)
Jan.	16.9	30.5	23
Feb.	17.0	30.7	15
Mar.	17.5	31.7	23
April	18.8	31.8	30
May	20.2	31.9	102
June	20.9	32.7	89
July	20.8	32.9	38
Aug.	21.0	33.1	91
Sept.	21.0	31.9	99
Oct.	21.3	31.9	180
Nov.	20.7	31.5	74
Dec.	19.7	31.3	36

Source : Reuther (1973b)

1.8.3.5 Citrus in Mexico

Citrus has been introduced by Spanish explorer in Veracruz area of Mexico (Ramirez-Daiz, 1983). While, commercial Citriculture started in 1896, and later spread to various areas. Growing of lemons and limes is concentrated in the southern states of pacific coast, while the cultivation of sweet oranges, mandarins, and grapefruits in the states along the Gulf of Mexico (Burke, 1926b; Calabrese, 1976; Cardenas Teresa, 1986).

Climate : Climatically, Mexico has large areas of dry, hot desert, high cool desert, hot dry tropics, and wet tropics (Albrigo and Menini, 1984). Rainfall in citrus producing districts varies from area to area falling mostly between June and October. Highest temperatures normally occur in the spring and early summer. The rainfall pattern at some of the premier citrus growing areas has been observed as 375, 850, 700, 825, 500, and 1375 mm at Montemo relos, Atotonilco el Alto, Colima and Apatzingan, Linares and Ciudad Victoria, Rio Verde, and Jalapa, respectively (Burk, 1962b; Reuther *et al.*, 1967). Reuther (1973b) described the climatic variation in terms of mean monthly minimum and maximum temperature along with rainfall at citrus growing areas of Vera Cruz and General Teran (Table 1.158).

Table 1.158. Climatic variation at Vera Cruz and General Teran areas of Mexico

Months	Vera Cruz			General Teran		
	Tempera-ture (°C)		Rain-fall	Temperature (°C)		Rain-fall
	Min.	Max.	(mm)	Min.	Max.	(mm)
Jan.	18.9	24.7	25	3.8	17.9	11
Feb.	19.3	25.4	16	7.5	23.5	10
Mar.	20.2	26.5	8	9.5	26.1	18
Apr.	22.2	28.2	30	14.3	30.3	23
May	24.0	29.8	69	19.5	34.6	84
June	24.2	30.4	244	21.3	35.7	83
July	23.6	30.6	312	22.2	36.9	7
Aug.	23.6	30.7	338	21.7	36.3	18
Sept.	23.1	30.2	346	19.7	33.0	64
Oct.	22.7	29.4	162	17.0	30.0	39
Nov.	20.6	27.0	75	9.9	22.1	24
Dec.	19.7	25.6	31	8.5	20.0	11

Source : Reuther (1973b)

Soils : The variation in soil type

in citrus growing areas of Mexico has been thoroughly described by Cardenas Teresa (1986). The varying elevation of land has given Mexico, three broad climatic regions. i. Tierra Caliente (hot land) from sea level to 3000 feet with dry winters and wet summers, ii. Tierra Templada (temperature land), from 3000 to 6000 feet free from extreme heat or cold, and iii. Tierra Fira (cold land) above the Tierra Templada. All the commercial citrus is grown in the Tierra Templada and the Tierra Caliente. Mostly, the soils are clayey with excellent fertility. Lopez-Valdovinos *et al.* (1984) described mexican lime growing soils at Colima as noncalcareous loamy sand with acidic soil pH. Soils are taxonomically classified as Camborthids, Gypsiorthids, and Salorthids. Most of the gypsum in these soils is allogenic i.e. transported from other locations either by ground water or as dust, and autogenic i.e presence of distinct layer of maximum accumulation (Table 1.159). While, Orozco-Romero and Sepulveda-Torres (1981) described Mexican lime growing soil Colima as deep sandy loam having 83 per cent sand, 6 per cent clay, 9 per cent silt, pH 8.3, $CaCO_3$ 2.5 per cent, EC 0.75 dSm^{-1}, Ca 375 mg kg^{-1}, Mg 80 mg kg^{-1}, and organic matter 0.70 per cent. Soils in Tularosa basin of New Mexico with a shallow water table (52 cm) developed no distinct salic or gypsic horizons. Soils with moderately deep water table (96 cm) developed gypsic horizons and soils with deep water table (112 cm) developed salic horizons (Table 1.160).

1.8.3.6 Citrus in Puerto Rico

Citrus is claimed to be introduced into Puerto Rico by Spanish settlers prior to World War II. Cedeno – Maldonado *et al.* (1990) observed that pummelos are adapted to the soil and climate conditions prevailing in the central mountainous region of Puerto Rico.

Climate : The mean annual precipitation is highest at site 11 in the Luquillo Mountains and at Los Guineos in the Corlillera Centra, reaching 4500 mm and 2300 mm, respectively. The lower elevations in the central mountains and along the Manati, Barranquitas, Cidra, and Jayuya areas of northcentral coast, the rainfall decreases to between 1650 mm and 2000 mm annually. Matanzas and Coto areas receive somewhat less rain (~1500 mm) due to displacement of the orographic barrier to the south in western Puerto Rico. Mean annual air temperatures are subject to less local variation than rainfall, but characteristically temperature decreases with increasing altitudes. At the lower elevations up to to 200 m above sea level, mean annual temperature is about 25°C at Mayaguez, Isabela, and Manati. Intermediate altitudes (Jayuya and Cidra) have temperatures between 23°C and 24.5°C. The highest locations such as Los Guinos and Picacho series are also the coolest with mean values of 21°C and 20°C, respectively. The variation in meteorological data is due to the migration of the equator trough. This also accounts for annual temperature cycle having a range of less than 5°C, and diurnal variation is greater, which reach 10°C or more (Table 1.161).

Table 1.159. Physicochemical properties of the soils of Tularosa basin in New Mexico

Depth (cm)	Sand	Silt	Clay	EC	Na	Ca	Mg	Fe	Zn	$CaCO_3$	$CaSO_4$
	(%)			(Me l^{-1})				(mg kg^{-1})		(%)	
						Soil 10					
0-15	68.9	17.9	13.2	75.1	675.6	101.6	132.4	2.3	0.5	50.4	3.4
15-30	47.0	22.4	30.6	24.0	13.7	42.5	66.1	3.4	0.4	9.9	0.2
30-45	42.3	24.1	32.6	22.0	156.0	41.0	66.0	4.3	0.4	12.0	0.2
45-60	44.5	24.9	30.6	15.6	91.7	30.9	39.1	5.0	0.6	12.2	0.5
60-75	48.4	27.6	24.0	11.5	55.7	36.2	38.5	4.4	0.2	11.9	0.4
75-90	48.4	27.6	24.0	6.6	58.1	58.9	20.5	5.2	0.1	8.2	0.2
90-105	60.0	23.5	16.5	6.9	15.0	34.5	24.9	5.5	0.1	16.8	6.5
105-120	67.4	21.8	20.8	8.4	23.0	31.4	20.4	2.8	0.1	16.8	25.6
120-135	55.6	20.5	23.9	9.1	24.8	38.6	22.1	2.3	0.1	18.8	24.7
135-150	74.0	12.1	13.9	-	14.0	22.1	195	2.0	0.1	3.3	34.5
						Soil 11					
0-15	60.6	26.4	13.0	24.7	158.3	27.9	57.3	2.7	0.2	9.6	25.6
15-30	29.1	61.0	9.9	19.0	141.7	34.6	52.1	0.5	0.1	3.4	41.7
30-45	45.3	49.2	5.5	13.3	91.8	30.2	33.1	0.7	0.1	3.1	40.6
45-60	46.2	45.2	8.6	9.7	33.0	27.2	25.0	0.6	0.2	3.2	42.8
60-75	39.8	43.2	17.0	7.0	23.6	25.9	19.9	0.7	0.1	7.2	45.1
75-90	47.6	31.4	21.0	7.4	26.4	25.5	23.9	0.7	0.1	8.6	46.2
90-105	49.1	24.6	26.3	8.1	28.5	22.5	26.9	0.4	0.1	8.4	45.6
105-120	74.6	14.7	10.7	6.6	27.2	23.1	28.9	2.0	0.2	3.2	25.2
120-135	75.2	13.2	11.6	6.4	23.9	25.7	29.8	2.0	0.1	3.2	27.9
135-150	74.0	12.1	13.9	5.5	13.9	22.1	19.4	2.0	0.1	3.3	34.5
						Soil 20					
0-15	39.6	37.8	22.6	17.9	145.7	32.9	59.7	0.9	0.2	8.7	39.5
10-20	35.1	48.9	16.0	60.9	477.3	49.4	153.8	1.4	0.1	6.2	44.0
20-30	37.5	47.1	15.4	59.0	523.2	57.9	153.7	1.4	0.1	5.9	43.4
30-40	39.1	49.6	11.3	83.9	592.6	57.6	142.1	0.7	0.1	4.1	43.9
40-50	31.2	59.7	9.1	85.0	729.4	45.7	143.9	0.4	0.1	4.0	43.4
50-60	34.7	58.5	6.8	49.2	379.9	58.8	152.9	0.4	0.1	3.2	43.9
60-70	33.9	57.2	8.9	71.9	680.1	50.4	144.4	0.9	0.1	2.9	44.5
70-80	51.4	40.9	7.7	39.8	360.3	51.3	134.4	1.8	0.1	2.9	41.7
80-90	58.0	35.5	6.5	39.7	329.6	45.3	129.2	1.6	0.1	2.9	40.6
90-100	57.0	35.4	7.6	26.3	235.8	37.2	80.7	1.9	0.1	2.2	37.8
100-110	58.7	32.4	8.9	23.7	168.0	33.2	66.3	2.8	0.1	2.8	36.9
110-120	60.7	30.4	8.9	23.9	223.0	34.5	78.8	3.2	0.1	3.5	38.4
120-130	57.6	34.9	7.5	23.3	220.0	29.3	60.58	2.4	0.5	3.5	38.4
130-140	60.3	31.2	8.5	24.8	229.3	29.3	75.9	3.6	0.1	6.2	41.1
140-150	60.4	29.0	10.6	27.3	317.7	38.4	98.6	3.7	0.1	2.0	40.0

Source : Nash *et al.* (1994)

Soils : Puerto Rico, the smallest and east most Island of the Greater Antilles, exhibits and intriguing pedologic diversity. All the orders established in soil taxonomy are found except Aridisol. More than 150 soil series have been identified which laid the basis for citrus fertilization (Kingman, 1915; Beinroth, 1982). Baldwin *et al.* (1938) classified the soils of Puerto Rico into great groups of lateritc soils (Nipe), reddish brown laterite soils (Bayamon, Catalina, Matanzas), yellowish brown lateritic soils (Coto), and red yellow podzolic soils (Los Guineos). Later, Beinroth (1982) characterised the soils of Puerto Rico. The Nipe series meets criteria for Typic Acrorthox. Matanzas and Bayamon series as Eutrorthox with base saturation exceeding 35 per cent throughout the profile. The Catalina series is a Tropeptic Haplorthox owing to moderate structure in the major part of the oxic horizon. Pina series is transitional between Ultisol and an Oxisol and classified as Ultic Haplorthox. Coto series is considered as Tropeptic Eutrustox. The Los Guineos, Carreras, Daguey, and Torres series have argillic horizon with base saturation less than 35 per cent. These are classified as Epiaquic Orthoxic Tropohumult, Typic Tropohumult, Orthoxic Tropohumult, and Orthoxic Tropudult. Soils of Picacho series are classified as

Table 1.160. Soil physicochemical characteristics at Colima, Mexico

Physicochemical characteristics	Depth (cm)		
	0-30	30-60	60-90
pH	6.4	6.6	6.8
Organic matter (%)	0.44	0.27	0.24
$CaCO_3$(%)	0.0	0.0	0.0
Saturation paste characteristics			
EC (mmohs cm^{-1})	2.54	1.72	1.29
Ca (Meq l^{-1})	13.05	11.32	8.12
Mg (Meq l^{-1})	6.68	5.27	5.29
K (Meq l^{-1})	3.57	1.30	2.80
Na (Meq l^{-1})	4.02	3.49	3.01
NO_3 (mg kg^{-1})	51.00	61.00	53.00
Cl (mg kg^{-1})	649.00	244.00	412.00
B (mg kg^{-1})	0.19	0.19	0.20
Exchangeable cations (me 100 g $^{-1}$)			
Ca	3.52	3.11	2.44
Mg	0.97	0.92	0.71
K	0.46	0.22	0.30
Na	0.48	0.52	0.46
CEC	5.9	5.0	4.0
Olsen-P (mg kg^{-1})	55.0	35.0	37.0
Texture	Loamy sand	Loamy sand	Sand

Source : Lopez-Valdovinos *et al.* (1984)

Table 1.161. Climatic data at few selected areas of Puerto Rico

Location	Jan.	Feb	Mar.	Apr.	May	Jun.	Jul.	Aug.	Sept.	Oct.	Nov.	Dec.	Annual
Mean precipitation (cm)													
Mayaguez	5.08	5.16	9.35	12.50	20.93	22.38	26.52	27.91	27.41	23.52	14.88	6.55	202.18
Isabela	8.59	7.59	7.21	9.22	16.74	12.37	10.13	14.24	14.68	14.68	17.12	11.20	143.81
Barceloneta	14.07	9.88	7.65	9.07	15.32	9.63	12.62	13.64	13.56	13.21	17.83	13.44	149.91
Barramqiotas	9.30	8.41	6.20	8.76	14.55	12.83	10.08	14.50	15.27	22.73	12.34	15.60	150.57
Toro Negro	10.24	11.86	11.07	18.60	32.66	17.93	17.20	25.40	34.14	38.51	22.96	11.94	252.53
Corozal	13.11	10.29	10.26	14.91	20.63	14.00	19.05	21.23	18.87	18.42	22.05	17.07	199.87
Cidra	10.59	7.87	5.54	11.71	17.91	15.52	18.42	18.59	20.68	17.96	14.25	11.10	170.13
Rio Grande Upper	27.97	24.43	18.95	24.43	39.65	36.30	32.94	37.11	38.81	36.47	38.61	31.88	385.55
Mean temperature (°C)													
Mayaguez	23.7	23.7	23.9	24.6	25.5	26.2	26.1	26.3	26.4	26.2	25.3	24.4	25.2
Isabela	23.6	23.9	24.0	24.6	25.6	26.1	26.6	26.6	26.5	26.2	25.5	24.3	25.5
Barramqiotas	19.6	19.7	20.7	21.5	22.1	22.9	23.2	23.2	23.0	22.8	21.6	20.3	21.7
Corozal	22.5	22.4	23.0	23.9	25.1	25.6	25.7	25.8	25.7	25.5	24.6	23.3	24.4

Source : Beinroth (1982)

Table 1.162. Particle size distribution of selected horizons of eleven soils types of Puerto Rico

Horizon	Depth (cm)	Sand (2-0.05 mm)	Silt (0.05-0.002 mm)	Clay (< 0.002 mm)
		Oxisols Bayamon (Tropeptic Eutrorthox)		
A1	0-20	56.2	4.1	39.7
B21	20-40	43.2	2.5	54.3
		Catalina (Tropeptic Eutrustox)		
Ap	0-16	1.5	15.5	83.0
B21	16-35	0.1	11.8	88.1
		Matanzas (Tropeptic Eutrorthox)		
A1	0-35	8.1	7.1	84.6
B21	35-70	3.3	5.4	91.3
		Nipe (Typic Acrothox)		
A1	0-25	9.8	35.2	55.0
B21	40-70	3.0	13.0	84.0
		Pina (Ultic Haplorthox)		
Ap	0-20	89.3	1.6	9.1
B21	34-60	70.9	1.2	27.9
		Carreras (Typic Tropohumult)		
Ap	0-14	15.2	23.0	61.8
B21t	14-28	12.0	23.1	64.9
		Daguey (Orthoxic Tropohumult)		
A1	0-10	6.4	37.5	66.1
B21t	10-30	3.0	20.0	77.0
		Los Guineos (Epiaquic Orthoxic Tropohumult)		
A1	0-8	6.3	20.7	73.0
B21	8-25	3.9	16.7	79.4
		Torres (Oxthoxic Tropudult)		
Ap	0-10	11.0	23.8	65.2
B21t	10-30	6.5	15.8	78.2
		Picacho (Oxic Humitropept)		
A	0-18	51.0	16.7	32.3
B21	18-35	36.0	18.9	45.1

Source : Beinroth (1982)

Oxic Humitropepts having base saturation less than 50 per cent and 12 kg organic carbon in the surface cubic meter (Table 1.162).

1.8.3.7 Citrus in Trinidad

Citrus growing conditions have been described by Maliphant (1966) in Caribbean territories, Blazquez (1966) in Jamaica and by Tai and Storey (1968) in Trinidad.

Climate : Citrus in Trinidad is grown under tropical climate. Rainfall averages from 1500 to 2500 mm year^{-1}, falling mostly between June and November. Mean maximum temperatures average 32°C in January and 33°C in July. Mean minimum temperatures average 26°C in January and 27°C in July. Most citrus orchards are near the rain forests with extensive drainage systems. Reuther (1973b) described the variation in climate at St. George (Table 1.163), characterised by 19.9-22.1°C and 30.2-32.0°C as mean monthly minimum and maximum temperature, respectively, with a total annual rainfall of 1601 mm.

Soils : The soil types of Trinidad have been classified as Piarco sandy clay loam, cocal fine sand, Mayaro sandy loam, and Monstserrat clay (Table 1.164). The various physicochemical properties revealed variation in pH from 4.7 to 6.9 with 38-85 per cent sand, 6-20 per cent silt, 9-44 per cent clay, and CEC 5.9-35.0 me 100 g^{-1} according to Chenery (1952), Watson *et al.* (1958), and Brown and Bally (1970). Deficiency of Mg, Zn, and Mn has been observed as a common feature (Weir, 1965a; 1965b; 1969a; 1971). The other details of these soil types are further elaborated below:

Table 1.163. Climatic variation at citrus growing areas of Trinidad

Months	Temperature (°C)		Rainfall (mm)
	Min.	Max.	
Jan.	20.1	30.2	69
Feb.	19.7	30.3	38
Mar.	19.9	30.7	46
Apr.	20.7	31.5	47
May	21.4	32.0	88
June	21.7	30.8	194
July	21.7	30.8	222
Aug.	21.9	30.7	242
Sept.	22.1	31.0	185
Oct.	22.0	31.5	169
Nov.	21.7	31.2	180
Dec.	20.9	31.3	121

Source : Reuther (1973b)

Piarco Sandy Clay Loam : The soil is derived from micaceous (muscovite) schists and believed to be predominantly kaolinitic in mineralogy (Ahmad and Jones,1969). In the surface layer, kaolinite derived from the weathering of mica comprises about 78 per cent of the clay fraction. Other minerals present are mica (13 per cent) and montmorillonite (9 per cent). The surface layer is loose, fine sandy loam which is dark greyish brown and becomes whitish grey upon exposure. The B horizon is grey with yellow and red mottling which increases with depth. Iron concretions are present. Below this, a gleyed horizon is also observed with some red and yellow mottling. The soil is acidic and low in fertility level.

Cocal Fine Sand.: The soil is developed on beach sand. Under forest, the mineral surface layer is very dark brown beneath a layer of litter, when exposed. The mineral layer is yellowish brown under moist condition and whitish under dry condition, typical colour for siliceous sands. Drainage is free to excessive. The profile has low soil fertility level.

Table 1.164. Physical and chemical properties (0-20 cm) of some soil types of Trinidad

Soil type	pH	CEC (me 100g^{-1})	Particle size distribution (%)			Fe_2O_3 (%)
			Sand	Silt	Clay	
		Trinidad				
Cocal fine sand	6.9	4.8	85	6	9	2.29
Piarco sandy clay loam	5.1	6.5	49	20	29	0.59
Mayaro sandy loam	4.7	5.9	63	15	10	1.01
Montserrat clay	6.7	35.0	38	20	44	29.50
		St. Vincent				
Montreal sandy loam	6.0	14.7	76	13	10	2.01
Akers sandy clay loam	5.8	10.9	77	7	20	2.10
Bellevue sandy loam	6.3	6.1	74	11	17	1.42
Soufrierecindery gravelly loam sand	6.3	6.3	81	11	10	1.05

Source : Chaenery (1952); Watson *et al.* (1958); Brown and Bally (1970)

Mayaro Sandy Loam : The soil is derived from soft miocene sandstone and sandstone conglomerate. The soil profile lacks horizon development and has a uniform yellowish colour. Drainage is free to excessive. The soil is acid, only moderately supplied with nutrients, and subject to severe desiccation.

Monstserrat Clay : The parent material is calcareous upper miocene glauconitic sandstone. The minerals in the clay fraction of the surface layer are montmorillonite (47 per cent), mica (17 per cent), and kaolinite (15 per cent). Goethite (13 per cent), quartz, and amorphous materials (20 per cent) are also present (Ahmad *et al.*, 1968a; 1968b; Davies *et al.*, 1971).

Ahmad and Jackson (1960) described (Table 1.165) the valencia growing soils of Trinidad viz., Cumuto, El Carmen, Piarco, San Antoine, Sangre Grande, and Chaguaramas (Table 1.163). Cumuto soils are fine sandy clay loam with a deep profile having soil pH 5.2 – 5.3, CEC 7.7 – 8.9 me $100g^{-1}$ and base saturation 26.1 – 32.8 per cent. El Carmen soils are known as fine sandy clay loam to silty caly loam according to Chenery (1952) which comparatively compact having imperfect drainage. Occurrence of Zn deficiency is usually observed. Soils are highly acidic (pH 4.8 –5.2) besides low CEC (9.8 – 10.5 me 100 g^{-1}) and base saturation (25.4 – 41.5 per cent). There is not much difference between exchangeable Ca^{2+} (1.05 – 1.67 me $100g^{-1}$). Piarco soils belong to Piarco

Table 1.165. Physicochemical analysis of citrus growing soils in Trinidad

Sampling depth (cm)	pH	Exchangeable cations (me 100 g^{+})				CEC (me 100 g^{-1})	Base saturation(%)
		K^+	Na^+	Ca^{2+}	Mg^{2+}		
				Cumuto			
0-15	5.2	0.02	0.35	1.51	0.44	8.9	26.1
15-30	5.3	0.05	0.46	1.50	0.52	7.7	32.8
30-45	5.2	-	-	-	-	-	-
				El Carmen			
0-15	5.2	0.13	1.15	1.67	1.14	9.8	41.5
15-30	4.9	0.17	0.75	1.05	1.08	10.0	30.1
30-45	4.8	0.13	0.41	1.09	1.10	10.5	25.4
Piarco							
0-15	6.7	0.04	0.48	2.26	0.17	6.7	43.9
15-30	6.3	0.21	0.57	2.12	0.08	3.0	98.3
30-45	5.7	-	-	-	-	-	-
				San Antoine			
0-15	5.3	0.03	0.48	0.77	0.30	6.5	24.4
15-30	5.4	0.04	0.50	0.77	0.57	6.4	24.6
30-45	5.4	0.11	0.53	0.76	0.28	6.5	25.9
				Sangre Grande			
0-15	4.9	0.05	0.61	1.69	1.00	8.7	38.4
15-30	5.4	0.07	0.70	2.23	1.30	14.4	29.8
				Chaguaramas			
0-15	5.1	0.07	0.31	0.41	0.19	5.7	17.3
15-30	5.1	0.07	0.43	0.73	0.25	3.9	37.4

Source : Ahmad and Jackson (1965).

series which are texturally sandy loam and have planosol development. These soils are naturally waterlogged and dessicated during wet and dry season, respectively. Citrus in such soils is cultivated on cambered beds having good external drainage. Prevalence of Mg and Mn is commonly observed. Soils of San Antoine are fine sandy clay loam with a depth of 2m overlain by detrital material from the northern range. Chemically, soils have a pH range of 5.3 to 5.4, CEC 6.4 – 6.5 me 100 g^{-1} and base saturation of 24.4 to 25.9 per cent with exchangeable Ca^{2+} (0.76 – 0.77 me 100 g^{-1}) as dominant cation, followed by exchangeable Na^{+} (0.48 – 0.53 me 100 g^{-1}), exchangeable Mg^{2+} (0.27 – 0.30 me 100 g^{-1}) and exchangeable K^{+} (0.03 – 0.11 me100g^{-1}). Occurrence of deficiency symptoms of Mg, Mn and Zn is a common feature.

Sangre Grande Soils according to Chenery (1952) are slity clay in texture predominantly, but some soils are clayey in texture. Soils remain waterlogged for most part of year. Citrus in such soils is grown on cambered beds having good external drainage. Deficiency of Mg and Zn is of common occurrence. Soils vary a great deal in pH (4.9 – 5.4), exchangeable K^{+} (0.05 – 0.07 me 100g^{-1}), Na^{+} (0.61 – 0.70 me 100g^{-1}), Ca^{2+} (1.69 – 2.23 me 100g^{-1}), Mg^{2+} (1.00 – 1.30 me 100 g^{-1}), CEC (8.7 – 14.4 me 100 g^{-1}), and base saturation (29.8 – 38.4 per cent). Soils of Chaguaramas are most shallow in depth but coarse textured. Chemically, soils have pH around 5.1, very low CEC 3.9 – 5.7 me 100 g^{-1} and base saturation 17.3 – 37.4 per cent.

1.8.3.8 Citrus in United States of America

Citrus was brought to the western hemisphere by Christopher Columbus in 1493 during his second voyage to the new world to establish a colony on Haiti (Webber, 1967). Citrus planting in the United States extends from approximately 26°N latitude (Miami, Florida, and Brownsville of Texas) to 40°N latitude (Oroville district of California). They are chiefly at comparatively lower altitude ranging from slightly above sea level up to 400 m. Although, in California, cultivation of citrus extends from 55 m below sea level (in the southern part of the state) to some 800 m above sea level (in the interior valley). The climates represented by Puerto Cartez, San Pedro Sula, Aracataca, Manaus, Altamira, and Belam are all rather typical of a very common type in lowland coastal plains, and valleys of tropical America with a monson type rainfall distribution (Bellia, 1964; Reuther, 1973a). Yee (1974) described the soil and climate requirement for pummelo in Hawaii. The potential impact of change in climate on citrus production in USA has been discussed by Rosenzweig *et al.* (1996). A majority of citrus growing soils belong to a wide range of taxonomically defined soil orders viz., Entisols, Alfisols, Ultisols, Spodosols, and Histosols. Cameron *et al.* (1970) described the fruit quality of various citrus cultivars in relation to climate in ojai valley of Ventura county.

Arizona Citrus Belt

Citrus belt of Arizona is older to California citrus. Majority of citrus in Arizona consists of grapefruits. Citrus production in Arizona is concentrated to Salt River valley and Yuma Mesa. The Salt River valley area consists of five separate districts situated in the warmer zones near the low mountains which irregularly border valley. Studying the basic requirements for establishing citrus in Arizona, Hilgeman and Van Horn (1954) suggested that three factors must be considered in the selection of orchard site, winter temperature, quantity and quality of water, and soil characteristics. Usually in the Yuma Mesa areas, the warmer zones are near the edge of Mesa or on the top of the ridge of the Mesa. In the Salt River valley, small areas near the mountains, northwest of Glendale area is generally warm. Other warm zones are isolated from small areas near the Camel Back mountain, south mountain district, and the northeast portion of the Mesa district. Observation in the Phoenix and Mesa.

Climate : Citrus belts of Arizona have dry climate with dry desert atmosphere (Rodney, 1964; 1969; Finch, 1977; Buross, 1978). Citrus orchards in Arizona are situated mainly in three districts: lower Colorado river valley and desert mesas around Yuma; Wellton-Mohawk area; and Salt river valley, centering around Phoenix. Citrus orchards are established in the desert area about eight miles south of Yuma area on high, level mesasatan altitude of 45 m above sea level (Johnson and Martinet, 1981). Average annual rainfall is about 90-100 mm, with low humidity. Temperatures vary from 38° to 42°C during the hot summers. Winters are mild, but damaging frosts do occur. Citrus orchards in the Salt river valley near Phoenix are grown at an altitude of about 350 m in sandy and clay loam soils. The climate is generally cooler than that of the Yuma district, and humidity is higher. Summer temperature over 38°C is common. An annual rainfall of 175-200 mm is supplemented by about ten irrigations year^{-1}, averaging between 50 and 100 mm per irrigation (Reuther *et al.*, 1967; Hilgeman *et al.*, 1969). Salt accumulation and Fe chlorosis are the problems in certain orchards. The Wellton Messa, an area extending from about 35 to 40 miles east of the Yuma Messa, has climate similar to that of the Yuma district.

Soils : In Arizona and other arid regions where flood irrigation is practised citrus sites should be selected on lands liable to be graed properly (Hilgeman and Rodney, 1961). Soils of Yuma district of Arizona consist of deep fine sand and silt from the Colorado river. Sharples and Hilgeman (1969) described the soils in Salt River valley near Tempe having 47 per cent sand, 36 per cent silt, and 17 per cent clay, underlain by clacareous caliche at 75 m. Chemically, soils contain $CaCO_3$ 2.5 – 14.6 per cent, total P 450-480 mg kg^{-1} , K 2.9-4.7 g kg^{-1} with pH 7.9-8.1 and EC 1.20-1.56 dS m^{-1}.

California Citrus Belt

Historically, citrus reached California in 1863. California citrus fruits for home use are grown in coastal areas far from north as San Francisco and in interior valleys to Butte and Tehama counties (Johnston, 1953; Butterfield, 1965; Plat, 1968; Opitz, 1969). Commercial areas, however, are relatively limited, and are roughly classified into coastal, interior valley, and desert areas, each with different climatic conditions. The interior valley area includes three distinct districts: i. the southern or upper Santa Ana river valley district in the western part of Riverside and San Bernardino counties; ii. the central or San Joaquin valley district, including orchards in Tulare, Fresno, and Kern counties; and iii. the northern or Sacramento valley district, including orchards in Butte, Glenn, Tehama, and Sacramento counties (McCarty, 1966; Ebeling, 1979). The California desert region is a unique citrus region similar to citrus areas of Arizona and Israel. It includes orchards in two districts of the south eastern part of the state, both in the Colorado desert area and east of the San Jacinto mountains. The Coachella valley district is in the northwestern part in Riverside county. The Imperial valley district is the southeastern part in Imperial county, extending to the Mexican border.

It is evident that sweet oranges can adapt and produce heavily in areas such as Tulare county which is characterised by very hot dry summers, comparatively cool winters, and many sunshine days. Conversely, in areas such as San Juan Capistrano where the cool summer temperatures prevail, with mild winter, fairly high humidity, and hours of sunlight. These are concerned conditions for high production in all of the commercial orange districts of California. However, two aspects of climate are definitely prejudicial to top performance. These are : i. the degree of exposure to high velocity and dessicating winds which blow in from the desert to the east and north, at periodic intervals during the fall and winter months and ii. frequency of forst (Harding, 1951).

Climate : Humidity in the coastal areas is relatively high, ranging from an annual average of about 72 per cent in San Diego to about 63 per cent in Los Angels. Total annual rainfall averages from about 10 mm in the south (San Diego) up to 20 mm in the north (Sant Barbara and Ventura). Mean annual temperatures range from 23° (San Diego) to 24°C (Los Angeles). Absolute maximum of 42°C and minimum of 28°C are experienced in the coastal areas. The interior valley districts are characterised by medium humidity, averaging 51 to 53 per cent annually, medium hot summers or growing periods, and cold winters as compared to the coastal and desert areas. Mean annual temperatures in the three districts range from 23° to 24°C. Absolute maximum and minimum temperatures for the southern district range from 42° to 43°C and 7° to 8°C, respectively, 43° to 45°C, and 7° to 9°C for the central districts, 44° to 45°C and from 6° to 7°C for northern district. Average annual rainfall in the southern districts ranges from 10 to 20 mm, 9 to 15 mm in the central districts, and from 18 to 28 mm in the northern districts. Rainfall throughout the desert region is light, averaging only about 3 mm annually.

Humidity is generally low, averaging 37 to 39 per cent. Although, in the driest months of April to September, humidity as low as 5 per cent has been recorded. The mean yearly temperature ranges from 27° to 29°C. Absolute maxima have reached 45°C (Imperial) to 47°C (Indio). But, in ordinary years, maximum temperatures do not exceed 44° or 45°C. Actual orchard minimum as low as 5°C has been recorded. The desert receives a greater percentage of sunshine than the coastal or interior valley areas, a factor presumed to be associated with early maturity of the fruit. This area is especially adapted to early maturing grapefruit (Reuther, 1973a).

Soils : More than 300 soil series have been identified and described in California. Out of them, the best known and developed soil families in California is the San Joaquin. In San Bernardino county and western Riverside county except Carona district, members of San Joaquin family are practically the only soils used for raising citrus trees. In Los Angeles, San Diego, Orange, and Tulare counties, this family occupies a leading position in citrus soils originated from the sourrounding granite hills and mountains. Mineralogically, these soils contain 60 per cent feldspar, 30 per cent quartz, and 10 per cent dark minerals. San Joaquin family soils contain a wide range of soil types, ranging from young to old soils with differential productivity potential. Tujunga soil series is the youngst soil types found around Cucamonga in San Bernardino county contain low water holding capacity. Hanford is another famous soil series having higher clay content than Tajunga soil series, with the result, water movement through soil profile is comparatively slower. Greenfield series is another fine citrus soil, with just enough clay accumulated in the subsoil to provide good water holding capacity, but not enough to inhibit root penetration. Ramona series, is another wide spread and much used citrus soils with considerable profile development. The Placentia series, the most mature member of the family occurring extensively in southern California, with well developed profile having compact subsoil. San Joaquin series, the oldest member of the group, is not found in southern California, although, it covers wide areas in San Joaquin valley. In Tulare county, the San Joaquin series is successfully used for citrus cultivation with the presence of hardpan in subsurfaces at places (Donnelly, 1947).

California soils are generally fertile, but range from fine sand to heavy clay. Alluvial valley soils are deep and hillside soils tend to be shallow and heavy. In some districts, the soils in even a 50 acre orchard may range from sand to heavy black loam. Storie and Harradine (1958) have earlier described the citrus soils of California.

Analysis of soil samples (30-60 cm depth) from Grangeville, Cajon, and Hanford (tentative) soil series of California showed that the Grangeville and Cajon soils are derived from granitic alluvium, and the Hanford soil from basic igneous alluvium. The layer silicate minerals present in the Grangeville and Cajon soils are principally biotite weathering products (vermiculite and hydrobiotite). These soils are essentially free of montmorillonite. The Hanford soil contains principally montmorillonite with only traces of vermicullite and

Table 1.166. Soil chemical properties from some citrus growing some soil series of California

Soil depth (cm)	pH (Sat. paste)	Exchangeable cations (me 100 g^{-1})				Total ex. cations (me 100 g^{-1})
		Na	K	Mg	Ca	
				Escondido		
0-15	7.4	0.33	0.21	2.1	7.6	10.2
15-30	7.4	0.40	0.17	1.9	5.2	7.7
30-60	7.5	0.40	0.16	2.1	5.4	8.2
60-90	7.5	0.64	0.14	2.1	5.1	8.1
90-120	7.5	0.74	0.12	2.2	5.2	8.2
				Rancho Santa		
0-15	5.6	1.3	0.36	16.9	17.9	36.5
15-30	6.6	2.4	0.22	18.3	18.5	39.4
30-60	7.3	3.9	0.20	18.1	22.3	44.5
				Carpinteria		
0-15	7.0	0.30	0.22	1.4	7.4	9.3
15-30	6.9	0.45	0.10	2.8	6.8	10.1
30-60	6.4	1.30	0.11	6.6	5.6	13.6
60-90	6.6	2.48	0.13	8.6	5.4	16.6

Source : Page *et al.* (1969)

hydrobiotite. Layer silicates present in the clay fraction (< 2m) of Hanford soil are basically montmorillonite with only traces of trioctahedral vermicullite and hydrobiotite. Dominant layer silicates present in the coarse and fine silt, and coarse clay fractions of Grangeville and Cajon soils also have trioctahedral vermicullite and hydrobiotite. The clay fraction for these soils showed only traces of montmorillonite. A large variation in cation exchange capacity due to difference in mineralogical make up has been noticed with Ca as predominant cation followed by Mg, Na, and K (Table 1.166).

Behaviour of these soils in relation to dynamics of K has been analysed by Page *et al.* (1969). Amounts of K fixed and the exchangeable plus soluble K content of every other 3 cm increment in the soil column to a depth of 30 cm indicated that the Hanford soil fixed little or no K throughout the soil column. Whereas, the Grangeville and Cajon soils fixed rather large amounts of K to depths of 15 and 21 cm, respectively (Table 1.167). The exchangeable plus soluble K levels for the Hanford soil

Table 1.167. Amounts of K fixed and exchangeable plus soluble K for the Hanford, Grangeville, and Cajon soil columns treated with K_2SO_4 and leached

Depth cm	Hanford	Grangeville	Cajon
	Fixed K (me 100 g^{-1})		
0-3	0.1	6.6	1.6
6-9	0.2	6.0	2.8
12-15	0.0	5.2	2.8
18-21	0.3	0.3	1.2
24-27	0.0	0.1	0.5
30-33	0.2	0.0	0.0
	Exchangeable plus soluble K (me 100 g^{-1})		
0-3	0.47	2.25	0.58
6-9	1.22	1.78	0.98
12-15	1.18	0.88	0.94
18-21	1.22	0.05	0.27
24-27	1.43	0.06	0.13
30-33	1.55	0.06	0.06

Source : Page *et al.* (1969)

below 3 cm are rather uniform at 1.2-1.6 me 100 g^{-1}. This soil with no K added contained about 0.2 me $100g^{-1}$ exchangeable plus water soluble K. The Grangeville and Cajon soil columns showed relatively high levels for exchangeable plus soluble K to a depth of 15 cm. At depths below 15 cm, the amount of exchangeable plus water soluble K decreased to levels, which approximate the levels found for these soils under conditions with no K added (Page *et al.*, 1969). The brown to reddish brown soils of Lake Mathews area near Riverside have been formed from granite rocks with shallow to moderate depth and underlain by porous weathered granitic parent material. In this area, five soil series viz., Cajalco (Fine loamy, mixed thermic family of Mollic Haploxeralf), Las Posas (Fine montmorillonitic, thermic family of Typic Rhodoxeralf), Buren (Fine loamy, mixed, thermic family of Haplic Durixeralf), Vista (Coarse loamy, mixed, thermic family of Typic Xerochrept), and Fallbrook (Fine loamy, mixed, thermic family of Typic Haploxeralf) have been identified (Sterling Davis *et al.*, 1969). Fertility constrains such as B toxicity on eastern slopes of coastal range of Fresno county (Browne and Platt, 1966) and Cu deficiency in Ventura county (Lee, 1964; Bradford *et al.*, 1962a; 1962b; 1964) have been commonly observed.

Florida Citrus Belt

Citrus is not native to Florida. Florida has the world's largest citrus in terms of area and production, using relatively infertile soils (Calabrese, 1969). This has been made possible through development of sound nutritional programme including both macro- and micronutrients (Grierson and Reitz, 1981). Geographically, citrus cultivation in Florida is confined to three major regions viz., central Florida (Highlands, Osceola, Poik), southwest (Charlotte, Collier, Glades, Hendry, Lee, Okeechobee), and Indian River (Brevand, Broward, Indian Riverm Martin, Palm Beach, Saint Lucie) (Muraro and Ford, 1990). A large acreage of flatwood soils have been drained, bedded, and planted since 1960 (Sites, 1961). Krome (1968) discussed the suitability of lime growing in Florida. Extreme soil variability as well as topography and rainfall patterns are unique compared to other citrus belts of United States.

Fig. 1.19 Valencia orange on a red sandy soil of Florida

Climate : The portion of the central Florida

ridge used for citrus is located in the central and westcentral part of the Florida peninsula between latitude 27°10′ and 29°40′ N. Fifty years of climatic data from Talahassee and Homstead indicated that there may be a zone from 27° to 29° latitude where the temperate climate of north Florida weakens sufficiently for the subtropical climate of south Florida to exert its influence in the region and two climate types are well balanced. Number of hours below the certain threshold value is considered one of the important indicators of climate (Gerber and Chen, 1986). The topography of the region is level to rolling, with slopes up to 40 per cent. Elevation is from 3 to 100 m above sea level. Most of the area is underlain by calcareous beds whose solution gives rise to sinkholes and depressions. The climate of the region is considered to be subtropical. The mean air temperature is approximately 29°C in July and 12°C in January. The Florida peninsula stretches for 5° latitude covering temperate to subtropical climatic zones. It is not uncommon to find maximum temperature greater than 27°C during winter months of December to February. Thus in winter, Florida becomes a meeting ground between intruding polar air masses and native warm and humid subtropical air masses. As a result, crop survival in each winter depends not only on the temperature of the intruding air masses, but also on the preconditioning from the native subtropical air masses, and finally, the interaction between the two (Gerberand Chen, 1986). The mean annual precipitation is approximately 1350 m. The rainfall distribution is very seasonal. However, with about one half occurs during the four summer months of June through September (Boston and Prine, 1968).

Cooper *et al.* (1969b) described climatic features of Leesburg (Table 1.168). Rainfall in Florida occurs every month in the year. The amount falling in any individual year may varies from 925 to 2100 mm. Average annual rainfall within the Florida citrus belt is about 1290 mm. Orth (1981) described the climate at Dade county as tropical with 1600 mm of rainfall and rainy season continues from May to October. Most precipitation takes place in the summer, during May to September. Rainfall is often insufficient for the needs of citrus trees from October through April. The average daily relative humidity of 72 per cent is fairly constant throughout the year. Highest daily temperature in summer usually ranges from 35° to 37°C and all the time, maximum for the state is 40°C (Jackson, 1991). Climate of southwest Florida provides mild winters with less chance of damaging freezes which creates a longer citrus growing season compared to central Florida (Obreza and Arnold, 1990).

Table 1.168. Climatic features at Leesburg, Florida

Month	Temperature (°C)		Relative humidity (%)		Rain-fall
	Max.	Min.	Max.	Min.	(mm)
Jan. 1967	25.9	19.17	98	58	29
Feb.	26.3	18.04	94	54	65
Mar.	28.5	21.43	98	52	20
Apr.	31.2	22.9	99	47	0.00
May	32.3	48.7	100	47	44
June	33.4	25.9	100	60	180
July	33.8	26.7	100	60	170
Aug.	33.1	26.7	100	60	323
Sept.	31.9	25.5	100	56	64
Oct.	30.4	22.5	99	51	5
Nov.	27.8	19.9	100	50	4
Dec.	27.4	20.3	98	55	61
Jan. 1968	48.8	17.3	95	51	4
Feb.	23.3	15.8	99	46	38

Source : Cooper *et al.* (1969b)

Soils : Efforts are being made to

establish new citrus orchards in newly reclaimed swamp lands (Grierson and Reitz, 1981). However, a majority of soils are infertile (Bradsley, 1987). Most of the soils in the flatwood citrus growing region of Florida are poorly drained due to low land elevation and existence of slowly permeable subsurface layer. This layer can be either argillic or spodic in nature (Watts and Stankey, 1980). Shallow water table exists above this layer during period of consistence rainfall (Obreza and Admire, 1985). Citrus grown on flatwood soils must be planted on raised bed in order to create enough unsaturated soil volume for proper root development (Ford *et al.*, 1985). Soil series namely, Winder, Riviera, Wabasso, Pineda, Chobee, Nettles, EauGallie, and Oldsmar are occupied under Indian River and St. Lucie citrus growing counties of Florida. Out of these, first four soil series have argillic or spodic horizon and are most important from Citriculture point of view.

During the growth and development of the citrus industry in Florida, which began around 1900. Numerous nutritional problems are encountered including deficiencies of N, P, K, Fe, Mg, Zn, Mn, Cu, B, and Mo. (Stewart and Leonard, 1953; Koo *et al.*, 1958; Stewart *et al.*, 1968; Koo and Young, 1970; Phillips, 1980; Koo, 1984; Jones and Smith, 1984). Nutritional survey of Hamlin orange orchards in south central (Hardee, De soto), southwest (Hendry, Collier), and east coast (Indian River, St. Lucie, Martin, and Palm Beach) regions of Florida showed majority of soils possess optimum level of P, Ca, and Mg (Tucker *et al.*, 1990). Calvert *et al.* (1977) described primary soil types at Fort Pierce as Oldsmar sand, a Spodosol and a member of sandy siliceous, hypethermic family of Alfic Aresenic impermeable spodic horizon ranging in depth from 85 to 107 cm below the soil surface with an average depth of 96 cm underlain by the layers of fine sandy clay loam. Hammond *et al.* (1971) described the pedological features of Oldsmar soil types. Most citrus varieties are well established throughout the mediterranean countries by the time of Columbus expeditions. Typically, two thirds of the crop is produced on the sandhills of the central Florida ridge (Anonymous,1980b). The soils on the ridge are primarily sandy Entisols derived from deep beds of marine sediments. Many are dominated by sand size quartz, to depths of more than 2 m having low native soil fertility and low water holding capacity. Soils possess a low CEC (1 to 4 me 100 g^{-1}), arising from their organic matter content (0.5 to 2.5 per cent), and low water holding capacity (0.5 - 1.2 cm water per 15 cm soil). Many are dominated by sand sized quartz to a depth of more than 2 m. They have characteristically low level of native soil fertility, besides low nutrient and water holding capacity. The soils are infertile and strongly acidic throughout the profile. The chief value for citrus is their natural warmth during the winter due to their elevation and rolling topography, which contribute to good drainage, off the hillsides on cold, and still nights.

The major soil problems of Dade county of Florida are the shallowness and rockiness of Rockdale (provisional Lithic Ruptic Alfic Eutrochrept, clayey, mixed hyperthermic) and Rockland (provisional Lithic Udorthent, loamy skeletal, mixed carbonetic, hyperthermic) soils, and the nearness of the water table. Under natural condition, the Rockdale and

Rockland areas contain outcrops of oolitic limestone which has resisted the weathering process. Rockdale is the primary soil on higher elevation. The soils tend to be a fine sandy loam in the southern half. The later contains some red clay and the region is known locally across the Rockdale areas. Wutscher and Perkins (1993) described 6 major soil types from entire citrus belts of Florida as Typic Quartzipsamment, Arenic Haplohumod, Arenic Glossaqualf, Mollic Ochraqualf, Aeric Haplaquod, and Typic Psammaquent. A survey of citrus acreage in Indian River and St. Lucie counties of Florida by Boman (1987) showed that 85 per cent citrus in these counties is grown on 4 major soil types viz., Rivers, Winder, Wabasso, and Pineda. The other soil types viz., Chobee, Nettles, Oldsmar, and Eau Gallie accounted for only 10 per cent citrus on these soil types.

Citrus in Florida is distributed more or less throughout the state, where soils are suitable. Most favourable citrus soils in Florida are acid, usually ranging from pH 5 to 6.5. A few calcareous soils ranging in pH from 7 to 8 are also used for citrus cultivation (Calvert and Reitz, 1966; Smith, 1967). The natural plant association areas known as hammocks, high pinelands, and flatwoods are recognized in the order named as the best lands for citrus culture. The most concentrated areas of citrus production are found in the central Ridge section, which consists largely of high pine soils found in Lake, Polk, Orange, and Highlands counties. The coastal Indian River section includes Breward, St. Lucie, Martin, Palm Beach, and Indian River counties. Almost every county in the state, however, has some small citrus orchards. The major lime production areas is Dade county, relatively free from frost hazards. Citrus growing soils of Florida have been grouped into series on the basis of origin, colour, structure, and other characters. These are further subdivided into classes according to the texture (particle size) of the surface layer. Astatula fine sand (formerly classified as Lakeland free sand) identifies the series Astatula and class fine sand of one of the important soils used for citrus production in central Florida. Such grouping makes possible, the classification of soils, as naturally occurring bodies, and the study of soil characters with regard to their usefulness. Drainage of water is perhaps the most important of these characters, that can be used for grouping soils. Citrus growing soils of United States are classified on the basis of drainage characteristics viz., well drained soils and poorly drained soils (Jackson, 1991):

Well Drained Soils : Well drained soils are found in the area extending from north and south through the central part of the Florida peninsula as far as Highlands county, with lateral extensions into Pasco, Hillsborugh, and Pinellas counties. The main area from Clermount to Lake Placid is often spoken as the Ridge. Isolated areas of well-drained soils are found along both coasts and even on some of the coastal islands. These soils belong principally to the Astatula association, which are the soils, most extensively used for citrus culture.

- *Astatula Fine Sand* : Astatula find sand is the predominant soil for citrus cultivation through the greater portion of the central citrus area of the state. Its general characteristics are: usually sandy throughout the profile, contain a small amount of clay at varying depths, and surface 8-10 cm of brownish grey, fine sand, with the subsoil of yellow fine sand, and water table low most of the time. Although, during rainy season, it may temporarily be high. Soil reaction (pH) under natural conditions ranges from 4.8 to 5.4. But under good cultural practices, citrus orchards are usually maintained at pH 5.5-6.5 or even slightly higher, with very low organic matter (1-2 percent), thereby, producing a low exchange capacity for mineral nutrient elements. Citrus (Ruby Red grapefruit, *Citrus paradisi* on sour orange, *Citrus aurantium* rootstock) growing soils at St. Luice county, Florida is observed on a Pineda fine sand series soil (loamy, siliceous, hyperthermic Arenic Glossaqualfs). These excessively drained soils are managed by wide double row beds raised above the drainage furrows, and presence of clay layer in the subsurface provides a slowly permeable barrier on which the perched water table is formed during the high rainfall (Boman, 1996). The salient features of the soils consist of good water drainage, air drainage, and depth for root development. Soil uniformity, lack of hardpan, and generally low water table allow for deep, well ramified root systems are other advantages. Such an extensive rooting habit enables the tree to overcome to a considerable extent, the handicaps of the low ability of these soils to hold water nutrient elements. Deep and well branched root systems enable trees attain standard size and enjoy comparatively long producitve life.

The constraints of these well drained soils lie in their initial low level of mineral nutrients, organic matter, and their small ability to hold water and the mineral elements. Excessive leaching through the centuries, together with the low base exchange capacity, is responsible for the low nutrient level in the soil. Exchange capacity is based on the relative amounts of clay and organic matter in the soil. Both of these two properties are low in these soils, which also impart the poor ability to hold water too. Presence of clay close to the surface makes the orchard responsive to fertilizer application compared to presence of clay deep down or not at all. But, an extensive development of root system that occupies the soil thoroughly to a considerable depth can overcome these constraints to a remarkable degree.

- *Coastal Soils* : The low hammock lands are found principally along the east and west coast. Although, they occur in isolated areas in the central portion of the state. They are represented by the Myakka, Oldsmar, Immokalee, Riviera, Winder, and other soil series. The first three soils are usually quite acidic and have an acidic hardpan. Riviera and Winder are similar. But, they may occasionally have marl in the profile, making them somewhat less acidic. The organic matter content is high, ranging from 3 to 8 per cent. Usually,the water table is close to the surface and in the natural state of these soils, it may often be above the soil surface. They must be drained and bedded before,they can be used for citrus planting. The important merits of these soils are their high inherent soil fertility, high base exchange capacity, and good water supply. The high exchange capacity permits better utilisation by the trees of the nutrient elements present or added in fertilizer to the acid surface layer. This is reflected in the great vigour of growth, usually exhibited by trees in the early years of the orchard life.

The demerits of these soils include, poor air and water drainage, shallow depth with comparatively small root volume, and occasional alkalinity of the subsoil (Syvertsen *et al.*, 1990). Although, the soils of this type used for citrus planting are usually located geographically where frost hazard is relatively small. But, they are by no means free from frost and their low elevation does not allow for compensating air drainage. Ditching systems are absolutely necessary for these soils that take care of normal problems of water drainage (Smith and Beville, 1964). Shallow depth to water table indicates that roots can only be formed to a shallow depth and thus, only a small volume of soil is available for storing available water. In periods of drought, the moisture supply to root depth is quickly exhausted, so that these soils are drought prone in spite of their high water table. Irrigation water must be applied in small amounts frequently, which makes its application more expensive than on the deep well drained soils. The availability of Cu, Zn, Fe, Mn, and B decreases with increase in soil pH from acidic to basic due to subsoil alkalinity. The other nutrients such as K and Mg must be present in greater amounts before the tree can get an adequate supply, because of the effect of the high calcium content of subsoil. This situation has made it necessary for the grower to supply most of the above micronutrients as foliar sprays, since they are more readily absorbed through the leaves than through the roots in these soils.

- *Flatwood and Marsh Soils* : The flatwood soils and associated swamp soils are found in many areas of the citrus belt as soon as one leaves the elevated central Ridge (Anonymous, 1961). These soils are considered quite marginally suitable for citrus cultivation. The other flatwood sites contain acid and alkaline soil types ranging from those suitable for citrus with proper drainage and irrigation to types that are definitely submarginal, even with adequate drainage. The lines of demarcation between poorly drained soils suitable for citrus and those considered submarginal have not been clearly established (Ford, 1963). The soils underlain with alkaline materials have been in greatest demand for citrus cultivation. This group comprises such recognized soil types viz., Felda, Parkwood, Bradenton, Sunniland, Charolotte, Adamsville, and Pompano. Presence of leached sand in the profile also applies to the alkaline soil types (Ford, 1964a; 1969). Typical soil types in the area include Myakka, Immokalee, Oldsmar, Pineda, and Wabasso. Orchards are planted on single or multiple row beds. Spodic or hardpan layers are sometimes broken up by deep ploughing or use of dragline. Limestone is incroporated into those soils with acid reaction. Besides, these problems, high water levels and accumulation of sulfides due to impediment in form of aeration, has sometimes been quite serious problem.

- *Oolitic Limestone Soils* : One soil series in particular, Rockdale occurs in Dade county and is extensively used for lime orchards. A very shallow layer of clay or sand overlies a very porous limestone formed by cementing together the shells of tiny marine organisms. To provide adequate depth for root development, this oolite layer must be broken up by heavy sacrifying plough. Although, the surface is not many feet above the water table in winter, the roots can penetrate a few feet at most, and trees easily suffer from drought in the dry season. In summer, the water may be above the surface for some time during periods of very heavy rainfall.

Poorly Drained Soils : In poorly drained areas, three types of soils have been identified:

- *Hammocks Soils* : The hammocks are usually low lying areas, rich in organic matter, and originally occupied by a natural growth of oak, hickory, magnolia, and cabbage palmettos in the areas near the coast. Fruits grown on hammock lands has a reputation of superior quality. Most of the early Florida orchards have been established on these soils. Few hammock lands now remain unplanted except in high frost hazard areas.

- *High Pineland Soils* : The natural growth on the high pineland soils are originally used for long leaf pine with scattered red oak and post oak. The soils are in general very sandy, very deep, and very well drained. While, generally good citrus soils are quite lacking in native fertility, and are thus, dependent on plant nutrients added from outside.

- *Flatwood Soils* : These soils are also called as ground water podzols formed from moderately thick bed of sands. Large acreage of flatwood soils have been planted to citrus in the southern part of the state since the 1957-58 freeze following extensive investment in reclamation through drainage works, subsoiling, tilling, and planting on cambered beds. Flatwood soils are low and level, and originally remain covered with pine. Other natural vegetation includes saw palmetto, wire grass, and cypress. Flatwood soils may have a hard pan layer near the surface and many are poorly drained. The potential for expansion of citrus in flatwood soils is nearly unlimited (Todd, 1987). In acid poorly drained soils, the colour and thickness of the surface soil, the amount of white sand below the surface soil, depth and thickness of organic pan layer influence the growth of citrus. In many soils, the organic pan is the important zone for satisfactory root growth. In general, the whiter the sand in the leached zone above the pan and the greater its depth below the surface, poorer is the land for citrus. The better soil types that contain more organic matter produces satisfactory trees on somewhat less root growth (Ford, 1962; Ford and Eno, 1962). On this basis, the acid soils ranked in the order of decreasing desirability for citrus are : Blanton, Scranton, Ona, Leon, Immokalee, Pomello, and St. Lucie fine sands (Ford, 1963).

Soils of the flatwoods have an organic subsoil horizon within about 1 m of the surface. This horizon, composed of amorphous organic material and aluminium, is usually dark coloured and very acidic. Some spodic horizons are strongly cemented (Table 1.169). Soils have light coloured surface layers and clayey subsoils. The clay fraction is dominated by layered lattice silicate clays well saturated with bases such as Ca and Mg (Table 1.170).

Table 1.169. Soil properties for Myakka fine sand, sandy, siliceous, hyperthermic Aeric

Depth (cm)	Particle size (%)			Organic matter (%)	Bulk density (g cc^{-1})
	Sand	Silt	Clay		
0-13	96.1	3.6	0.3	1.9	1.28
13-38	97.1	2.8	0.1	0.5	1.40
38-64	94.0	3.0	3.2	4.9	1.08
64-74	96.1	2.6	1.3	1.5	1.37
74-91	96.8	2.7	0.5	0.5	1.50
Depth (cm)	**CEC (me 100g^{-1})**	**pH**	**Moist. (eq.%)**	**Hydraulic conductivity (cm hr^{-1})**	
0-13	3.6	4.3	3.8	50	
13-38	0.5	4.7	2.2	52	
38-64	13.2	4.3	6.1	3	
64-74	4.6	4.5	2.8	20	
74-91	1.2	4.8	1.6	20	

Source : Buol *et al.* (1973); Calhoun *et al.* (1974)

The soils that are eventually, most extensively planted to citrus on the Ridge are the Quartzipsamments sandy, Entisols having 95 per cent or more inert quartz (Buol *et al.*, 1973; Craddock and Wells, 1973; Calhoun *et al.*, 1974). Soils are loose, incoherent, and have developed from thick beds of marine sediments, having low CEC of 1-4 me 100 g^{-1}, derived primarily from their organic matter content ranging from 0.5 to 2.5 per cent (Anderson and Albrigo, 1972). Presence of organic matter in the surface horizon, is the most noticeable feature that distinguishes surface from subsurface soil. The Quartzipsamments have low water holding capacity from 0.5 to 1.2 cm water 15 cm^{-1} soil. Internally, they are well drained to excessively drained, and strongly acidic throughout the profile. The soil series most commonly planted to citrus are Astatula, Candler, Tavares, and Paola, classified as hyperthermic which by

Table 1.170. Soil properties for Felda sand, loamy, siliceous, hyperthermic Arenic Ochraqualfs

Depth (cm)	Particle size (%)			Organic matter (%)	Bulk density (g cc^{-1})
	Sand	Silt	Clay		
0-15	90.1	7.3	2.6	4.4	1.43
15-33	93.8	3.9	2.3	0.4	1.58
33-64	94.3	3.6	2.1	0.1	1.69
64-104	76.6	4.2	19.2	0.2	1.66
104-152	76.4	11.0	12.6	0.3	--
Depth (cm)	**CEC (me 100g^{-1})**	**pH**	**Moist. (eq %)**	**Hydraulic conductivity (cm hr^{-1})**	
0-15	6.5	5.4	4.6	29	
15-33	1.7	6.2	2.2	60	
33-64	0.9	7.9	1.6	21	
64-104	9.3	7.5	13.9	1	
104-152	5.5	8.1	--	--	

Source : Buol *et al.* (1973); Calhoun *et al.* (1974)

definition are soils that occur in Florida in a subtropical environment capable of supporting citrus (Carlisle and Nesmith, 1972). Specifically, the annual mean temperature of these soils is greater than 22^{0}C at a depth of 50 cm. A typical virgin Astatula fine sand characterised by Calhoun *et al.* (1974) showed that soil has predominantly with low CEC varying between 0.3 me 100 g^{-1} and 1.3 me 100 g^{-1} up to 279 cm soil depth (Table 1.171).

Closely associated with the Quartzipsamments and extensively planted for citrus, are some Paleudults, that are deeply weathered, low base saturated Ultisols (Buol *et al.*, 1973 , Calhoun *et al.*, 1974; Carlisle and Nesmith, 1972). They also have sandy surface horizons, but have loamy to clayey subsurface horizons. Potentially, they are better soils than the Quartzipsamments. But, they are generally located in colder areas on the central Florida Ridge, citrus seldom reaches its full potential on these soils. Examples of Paleudults (Table 1.172) preferred by citrus growers are Eustis, Arredondo, Apopka, and Sparr series characterised by Calhoun *et al.* (1974).

Table 1.171. Soil properties for Astatula fine sand, a hyperthermic, uncoated typic Quartzipsamment

Depth (cm)	Horizon	Particle size (%)			Bulk density (g cc^{-1})	Moisture (%)
		Sand	Silt	Clay		
0-12	A1	96.5	2.0	1.5	1.30	1.57
12-43	C1	96.8	1.7	1.5	1.52	1.04
43-91	C2	97.5	1.2	1.3	1.50	1.40
91-152	C2	97.8	0.8	1.4	1.52	0.95
152-279	C2	97.6	1.2	1.2	1.60	1.08

Depth (cm)	Organic carbon (%)	N (%)	C/N	NH_4O Ac-P (pH 4.8) (mg kg^{-1})	pH 1:1 (1N KCl)	pH 1:1 (H_2O)
0-12	0.36	0.01	27.7	5.0	4.7	5.6
12-43	0.16	0.01	26.7	5.4	5.0	5.7
43-91	0.08	0.01	20.0	5.8	5.0	5.7
91-152	0.04	0.01	20.0	7.1	5.0	5.5
152-279	0.03	0.01	6.0	3.8	5.1	5.4

Depth (cm)	Extractable bases, 1N NH_4OAc (pH 4.8) (me 100 g^{-1})					CEC NH_4OAc (me 100 g^{-1})
	Ca	Mg	Na	K	Sum	
0-12	0.6	0.2	Traces	Traces	0.8	1.3
12-43	0.3	0.2	Traces	Traces	0.5	0.7
43-91	0.2	0.1	Traces	Traces	0.3	0.5
91-152	0.2	0.1	Traces	Traces	0.3	0.3
152-279	0.2	0.1	Traces	Traces	0.3	0.3

Source : Calhoun *et al.* (1974)

The most important characteristic of the soils on the central Florida Ridge is their topography which indirectly provides natural warmth during the winters. Cold damage is always of paramount concern among growers, because all parts of the Ridge are subject to occasional freezing due to extremely low winter temperature. Freezes occur most frequently on still nights as a result of radiant heat losses. As the air collects just above the soil surface, but on rolling land, it flows down the slopes and collects in pockets and lowlands. The hilltop and hillsides remain relatively warm. Occasionally in Florida, freezing temperature is accompanied by high winds. Heat is lost by convection. The terrain in central Florida provides some protection against that type of freeze, due to the presence of numerous fresh water lakes and marshes. These bodies of water act as heat sinks. As masses of cold air move across them, the air is warmed to the considerable benefit of crops on the leeward side of the lakes. Florida citrus growers preferred the orchard sites on the hills overlooking the southern and eastern shores of large lakes.

Table 1.172. Soil properties for Arredondo sand, a loamy, siliceous, hyperthermic Gross Arenic Paleudult

Depth (cm)	Horizon	Particle size(%)			Bulk density (g cc^{-1})	Moisture (%)
		Sand	Silt	Clay		
0-23	A1	93.2	4.4	2.4	1.51	3.1
23-64	A21	93.1	4.5	2.4	1.55	2.0
64-97	A22	95.0	3.2	1.8	1.52	1.5
97-127	A23	95.3	3.5	1.2	1.52	1.1
127-183	Bt	68.8	1.2	30.0	1.71	15.9

Depth (cm)	Organic carbon (%)	N (%)	NH_4OAc-P (pH 4.8) (mg kg^{-1})	pH 1:1 (1N KCl)	pH 1:1 (H_2O)
0-23	0.8	0.05	10.9	5.1	5.8
23-64	0.1	0.01	1.2	4.5	5.5
64-97	0.04	0.01	0.6	4.6	5.4
97-127	0.01	0.01	0.5	4.7	5.4
127-183	0.1	0.02	0.2	4.1	4.8

Depth (cm)	Extractable bases, 1N NH_4OAc (pH 4.8) (me100 g^{-1})					CEC NH_4OAC (me 100 g^{-1})
	Ca	Mg	Na	K	Sum	
0-23	1.2	0.3	Traces	0.1	1.6	3.3
23-64	0.1	Traces	Traces	Traces	0.1	0.8
64-97	0.1	Traces	Traces	Traces	0.1	0.5
97-127	0.1	Traces	Traces	Traces	0.1	0.3
127-183	0.4	0.3	0.1	0.1	0.9	2.6

Source : Calhoun *et al*. (1974)

Texas Citrus Belt

Citrus orchards in Texas extend from the Gulf of Mexico about 75 miles westward along the Rio Grande River and northward slightly above Raymondville. This is a strip of rather level valley land with an elevation ranging from 6 to 45 m above sea level. Soils range from sandy loam to heavy clay, and are fairly fertile.

Climate : Texas has humid climate with 500 to 700 mm of rainfall. The rainfall, however, is usually in the form of thunder storms which are undependable either from the stand point of amount or location. The long growing period coupled with an early season for citrus is sole featrue of Texas citrus. Bailey *et al.* (1963) and Bailey and Maxwell (1963) poionted out that in Rio Grande Valley of Texas, citrus on higher land or Ridges is subjected to warmer temperatures during radiation frosts, but to colder temperatures during advection freezes. Citrus usually blooms in late February and March when temperatures of 32°C are not unusual. Average annual temperature in the Rio Grande Valley is about 28°C. Absolute maxima range from 39° to 41° C and absolute minima from 5° to 7°C (Sauls and Rouse, 1985). Humidity is usually high, averaging about 74 per cent at Brownsville, near the coast. Rainfall is heaviest from May to September, averaging from 500 to 750 mm a year. Tropical storms and the rising of the Rio Grande River occasionally cause floods. There is water shortage in the Rio Grande and rainfall is not dependable (Reuther *et al.*, 1967). Over the last 20 years at Wesclaco (Table 1.173), annual rainfall has risen from 300 mm to more than 1000 mm with an average of 623 mm according to Leyden (1977).

Table 1.173. Montly and annual rainfall with maximum and minimum (20 years mean data) at Wesclaco, Texas

Month	Rainfall (mm)		
	Max.	Min.	Mean
Jan.	178	2	38
Feb.	155	0	30
Mar.	109	0	23
Apr.	106	0	36
May	208	14	67
Jun.	281	3	81
Jul.	238	0	56
Aug.	149	0	48
Sept.	331	9	119
Oct.	198	0	71
Nov.	116	0	37
Dec.	885	0	26
Annual	1006	306	632

Source : Leyden (1977)

Soils : All citrus orchards are irrigated, mostly by flooding. Salinity is a frequent problem, and some soils require drainage due to presence of harpan in the subsurface Williams *et al.* (1977) and Turner (1982) described the soils of Cameron and Wallacy county, respectively. There are orchards operated on a nontillage and permanent sod culture. Hidalgo sandy loam soil (fine loamy, mixed, hyperthermic Typic Calciustolls) is a common occurrence in Texas. The soil is calcareous throughout the profile, with pH in the top 50 cm varying from 8.0 to 8.3 (Swietlik, 1996a).

1.8.4 Citrus in South America

Citrus cultivation in South America is predominently confined to countries like Argentina, Brazil, Chile, Columbia, Ecuador, Peru, Paraguay, Suriname, Uruguay, and Venezuela (Table 1.174). Larue (1971) described soil and climate for citrus growing in South America.

1.8.4.1 Citrus in Argentina

Citrus was introduced in the northwest course in the present province of Salta, Jujuy, Tucuman, and Santiago del Estero, in 1556, when Harnan Mejja de Miraval brought orange and lemon seeds during the trip of Chile and Peru. Later on, better citrus orchards were developed from those ancient plantations (Luchini, 1965; Sartori, 1981). A majority of citrus production emerges from Mesopotamia, northwest, and littoral areas.

There are three principal citrus producing regions in Argentina. The largest citrus is confined to Rivera area known as Mesopotamia region consisting of delta area of Parana and Uruguay rivers, the Entre Rio, Corrientes, and Missiones region. The second area is in the northwest in the Tucuman, Salta, and Jujuy. The third and smallest section is in central zone covering provinces of Cordoba, San Luis, Catamarca, and La Rioja (Chapman, 1954).

Climate : Average annual rainfall pattern at major citrus growing provinces has been observed as 1150, 1025, 925, 675, and 975 mm at Corrientes, Entre Rios, Tucuman, Salta, and Jujuy, respectively. Rainfall is mostly distributed from October through March.

Table 1.174. Citrus belts across south America

Name of the country	Citrus growing areas
Argentina 30°00′ S lat.; 65°00′ W long.	Misiones, Cerrientes, Tucuman, Saita, Jujuy Buenos Aires, Entre rios, Bella vista (Coastal plain), Padilla, Colonia santa rosa, Oran, Tabacal, Urundal, Caimancito, Ladesma, Saladas, Santa lucia, Santa fe rio, Rosario
Brazil 10°00′ S lat.; 55°00′ W long.	Rio de janerio (Araruama, Cabo Frio, Rio Bonito, Casemiro de Abreu, sao Pedro D'Aldeia squarema, Silva Jardim, Casemiro de Abreu); Sao Paulo (Bebedouro, Barretos,Pitangueiras,Colina, Severinia, Olimpia, Monte Azul, Pualista, Limeira, Taquarilingu, Olimpia, Matao, Araraquara, Sano Jose do Rio Preto, Aguai-Mogi-Guacu river valley, Barretos); Bahia (Agreste de Alagoinhas, northern coast, Recancavo, Sergipe (Agrestede Legarto); Minas Gerais (Cerrado, Tringulo Mineiro, Zona da Mata, Frutal, Uberaba); Rio Grande do Sul (Montenegro, Sao Sebastiao do Cai, General Camara, Triunfo, Taquari.
Chile 30°00′ S lat.; 71°00′ W long.	Copiapo, Vallenar, La Serena, Elqui, Ovalle, La Ligua, Quillota, Metropolitan, Melipilla, Santiago, Peumo, Santa Cruz, Esmeralda. O'Higgins, Colchagua
Colombia 5°00′ N lat.; 75°00′ W long.	Cartagna, Palmira, Medellin, Monteria, Barranquilla, Giradot, Cali, Buenaventura, Bogota, Villavicencio, Magdalena river valley,
Ecuador 0°00′ N lat.; 78°00′ W long.	Quito, Patate valley, Salcedo, Pillaro, Andes
Paraguay 23°00′ S lat.; 58°00′ W long.	Asuncion
Peru 10°00′ S lat.; 75°00′ W long.	East Andean slope, Huaral, Piura,
Suriname 5°00′ N lat.; 5°00′ W long.	Paramaribo coastal plain
Uruguay 32°00′ S lat.; 57°00′ W long.	Salto, Paysandu, Rivera, Mela, Cerro largo Montevideo, Canelones, San jose, Federacion, Colonia, Artigas, Cuarai, , Merceder, Treintay Tres, Rocha, Maldonado, Minas, Durazro,Rio Negro, Tacurembo
Venezuela 5°00′ N lat.; 65°00′ W long.	Miranda, (Carabobo), Falcon, Monagas, Sucre, Poterito (Yaracuy), Zulia, Santa Cruz, Dutch island (Curacao), Isla de la Juventud

On the coastal plain at Bella Vista in Corrientes province, mean maximum temperatures are 35°C in January and 26°C in July. In the interior province of Jujuy, where the elevation is much higher, mean maximums are 31°C in January and 25°C in July with corresponding mean minima of 22°C and 15°C, respectively (Reuther *et al.*, 1967). The mean annual rainfall at Concordia has been recorded as 1330 mm and temperature 19°C with frost risk during May to September (Anderson and Benatena, 1996). Variation in climate and soil in citrus growing regions (Mesopotamia, Northwest Littoral, Central, Northeast, and Andean zone) is further described in more detail (Table 1.175).

Soils : The soils are silty, clayey, and sandy with high water table and danger of flooding in the delta region. Along the rivers soils are sandy with inlands heavier soils in northwest region and red soils in Missiones province (Chapman, 1954). Trifoliate rootstock is used on heavy soils, while rough lemon, sweet orange, and cleopatra mandarin rootstocks are used on lighter soils. In the Misiones district, only 25 per cent of the citrus orchards are considered commercial. The shallow, red clay soils are exposed to erosion following the removal of forest and adaption of clean cultivation. Citrus orchards are

Table 1.175. Climate and soil variation in major citrus growing regions of Argentina

Area and location	Climate			Soil type description
	Altitude (m)	Annual temp. (°C)	Rainfall (mm)	
1. Mesopotamia				
Misiones (Semitropical) 25°28′ – 28° 18′ latitude S 53°38′ – 56°03′ longitude W	300-400	20.1	1575	Organic matter rich lateritic soil with contour planting
Corrientes province (Semitropical) 27° 5′ – 30°43′ latitude S 55°37 – 59°41′ longitude W	60-80	21.7	1205	Sandy to clay, deep well drained, poor in organic matter, pH 5-6.5 having lateritic origin
Entre Rios province (Semitropical) 30°10′ – 34°03′ latitude S 57°48′ – 60°47′ longitude W	100-300	18	1250	Sandy to sandy loam texture, low in organic matter, and soil pH 5.0
2. Northwest Jujuy province 21°46′ – 24°37′ latitude S 64°09′ – 67°13′ longitude W	300-400	21.5	550	Sandy alluvial soils in the valleys with pH between 6 and 9 besides soil salinity problem
Salta province 22°24′ – 26°24′ latitude S 62°21′ – 68°34′ longitude W	200-300	20	800	Sandy alluvial soils with soil salinity problem
Tucuman (Subtropical) 26°05′ – 28′ latitude S 64°28′ – 66°13′ longitude W	300-500	19.2	954	Mountainous alluvial soils, sandy loam – clay loam texture, organic matter 2-6%, and pH 6.3-7.4
3. **Littoral (Temperature climate, comprises the western coast and delta of Parana river)** 31°51′–34°50′ latitude S 59°50′–61°50′ longitude W	100-300	10.9	1100	Clay soils rich in organic matter, but heavy clay below 1 or 2 feet, not very suitable for citrus, acidic pH with deficiencies of Ca and P
4. **Central (Subtropical, comprise areas like Santiago del Estero, Cordoba and San Luis**	-	-	-	Good fertile soils
5. **Northeast (Formosa, Chaco)** 25°-28° latitude	100-300	22.4	800	Alluvial clay loam soils, deep with good drainage and fertility
6. **Andean zone (Subtropical desert, covers Catamarca, La Rioja provinces**	300-400	10.2	920	Alluvial soils

Source : Sartori (1981).

planted on rolling ground near forest and no irrigation is practised. In Tucuman province, soils vary from sandy to black loam (Reuther *et al.*, 1967). Aso and Dantur (1970) reported about the N, Cu, Fe, Mg, and Zn deficiency in Tucuman. Soils at Concordia are well drained, deep sandy soils with soil pH 5.0 (Anderson and Benatena, 1996)

1.8.4.2 Citrus in Brazil

The history of citrus cultivation in Brazil dates back to 1540, only 40 years after the country was discovered. The most extensive orchards are established in the states of Sao Paulo and Rio de Janeiro, with smaller areas in Minas Gerais, Rio Grande do Sul, Parana, and Sarta Catarina. Most of the citrus produced in Sao Paulo are located at an elevation higher than 600-700 m above mean sea level (Burke, 1958; Giacometti, 1981). Mota *et al.* (1990) described the soils and climate of Rio Grande do sul region of Brazil for cultivation of sweet oranges, mandarins, and lemons

Climate : The climate varies from hot super humid (equatorial) in the north to mesothermic superhumid (subtropical) in the south and southeast region. There are no climatic limitations for citrus growing, except in certain areas, where the rainfall is high, such as in the humid Amazon basin, or low as in the semiarid northeast and in the south, where frost may occur. The altitude for the cultivated citrus areas varies from 20 to 500 m above mean sea level. The rainfall varies from 1000 to 1800 mm. Most of the rains occur during the winter (March-August) in the northeast. Rainfall distribution is not regular except in the south (Rio Grande do Sul), where it rains every month. However, under normal conditions, irrigation is not necessary. Relative humidity is higher in the north and northeast reaching almost 100 per cent in the winter. The average relative humdity ranges from 75 to 80 per cent. The average annual temperature varies from 19°C in the south to 26° C in the north. Although, the highest and lowest temperatures occur in the south. The further away are the citrus planted from the south and closer to the equator line, the shorter the harvesting time and the greener the fruits. Regardless of the region, flowering occurs mainly in September. In tropical areas, flowering occurs two or more times during the year in response to the rainfall patern (Passos, 1999).

Rainfall in the various citrus areas ranges from 40 to 70 mm or more annually, mostly distributed in the summer from September to April. Average annual rainfall of selected areas has been recorded as 1497, 1172, 1082, 1165, and 1820 mm at Limeira (Sao Paulo), Taquari (Rio Grande do Sul), Universidad Rural (Rio de Janeiro), Belo Horizonte (Minas Gerais), and Salvador (Bahia), respectively. At Limeira, in the coastal valley area of Sao Paulo state, mean maximum temperatures are 32°C in January and 29°C in July with corresponding minimum temperatures of 24°C and 21°C, respectively. Further north in Bahia State, mean maxima in Salvador are 32°C in January and 30°C in July, and mean minima are 28°C and 26°C, respectively. There are considerable temperature variation between day and night in the citrus districts, especially in the dry winter period The bloom cycle of trees is affected both by temperature and rainfall.

The climates of Salvador and Cruz das Almas in Bahia, represent intermediate type between the typical humid, mild and cloudy climate of Canete (Lima, Peru) and the typical humid hot and sunny climate of most of the Carribean and Amazon basin. Monthly mean minimum temperatures in the coolest months at Salvador are around 3°C cooler and around 4° C at Cruz das Almas cooler than Belam. Accumulated climatic data at Cruz das Almas from 1949 to 1977 showed an average annual temperature of 24°C, relative humidity 83 per cent, and annual rainfall of 1214 mm, with more rains from March to August (Soares Filho *et al.*, 1984).

The performance of citrus at Cruz das Almas (Passos and Sorbinho, 1970) suggested the presence of moderate cloudiness and mild temperatures during the April-July period having highest rainfall among other factors contributed to good quality of navel oranges (Passos *et al.*, 1971). Amaro (1984) described the climate of six major citrus growing regions of Brazil. Climate of Minas Gerais in the mountainous region is comparatively not well suited for citrus, and only Cerrado area is suitable for citrus. A favourable climate exists for Citriculture in Rio de Janeiro characterised by annual rainfall of 1000 – 2000 mm with no defined dry season, and average temperature ranging from 21 to 22°C having average minimum and maximum temperature as 18°C and 27°C, respectively. The annual rainfall at Rio Grande do Sul is abundant (1537 mm) and well distributed with average annual temperature of 19.4°C. In this region, low temperature prevents commercial Citriculture above an altitude of 300 m. Citrus in Sao Paulo is cultivated at an altitude ranging from 400 to 600 m above sea level. Pluviometric precipitation averages 1250 mm year^{-1} with dry period during April to August. The average annual temperature is 22°C with maximum of 36°C and minimum above 0°C coupled with occurrence of frost. The climate of Sergipe is hot and humid with the rainfall in winter and dry in summer. Pluviometric precipitation ranges from 1000 to 1500 mm per year with average temperature of 25°C and relative humidity of 80 per cent. The climatic data of Bahia region showed annual rainfall of 1100 to 1500 mm, mainly concentrated during March to September, with average temperature of 24.5°C and relative humidity of 82 per cent.

The state of Sao Paulo is crossed by the Tropic of Capricorn. It has a plateau, where the most frequent altitudes vary from 400 to 600 m above sea level. The predominant climate may be classified as tropical at altitudes, where the temperatures are pleasant. The climate of Sylvio Moreira region is mesothermic with dry winter in which the average temperature of the coldest month is lower than 18°C, and the warmest month is higher than 22°C. Average annual rainfall is 1400 mm and lower than 30 mm in the driest month (Quaggio *et al.*, 1992b). Climatic variations in Cerrado region of Brazil covering states like Goias, Mato, Grosso Minas, Gerais, Piaui, Bahia, Maranhao, and Distrito Federal showed that temperature is not a limiting factor. Mean annual temperature is around 21.3°C (Table 1.176).

The average of minimum that occurs generally in July is 12.6°C, with maximum average temperature of 30.1°C in September. Rains, although, are abundant but not very well distributed having average annual rainfall of 1500 mm and monthly average temperature varying from 42.7 mm in April to 342.4 mm in December. During rainy season, it is common to have two or three weeks without rains, a phenomenon known in the area Veranico. Relative humidity is low in some periods of the year, ranging from 49.2 to 83.0 per cent with lowest values in June, July, August, and September and highest during December, January, February, and March. The potential evapotranspiration in the Cerrado region reaches most severe limit during August and September (Genu and Pinto, 1984). Coelho *et al.* (1984) described the meteorological features of citrus growing states of Manans, Cruz das Almas, Limeria, and Taquari of Brazil (Table 1.177). Climatic variation of Limeira during 1954-1966 showed annual mean temperature 21.2 °C, mean maximum temperature 27.6°C, mean minimum temperature 14.2°C, and annual mean rainfall 1388 mm (Rodriguez and Moreira, 1969).

Table 1.176. Average climatic data, over last 35 years at Formosa, Brazil

Month	Average temp. (°C)	Tempera-true (°C)		Relative humidity (%)	Rainfall (mm)	Evapora-tion (mm)	Solar radiation (cal cm^{-2} dia^{-1})
		Min.	Max.				
Jan	22.0	17.8	27.4	80.2	271.9	73.2	425.0
Feb.	22.1	18.0	27.8	80.8	204.2	63.7	210.1
Mar.	21.9	17.9	27.6	81.5	220.6	67.1	380.9
Apr.	21.5	17.0	27.6	77.3	42.7	75.3	377.9
May	20.1	14.8	27.6	71.0	17.0	97.8	377.9
June	19.0	13.1	26.4	66.0	3.2	113.0	376.8
July	18.9	12.6	26.3	59.4	5.5	141.3	428.3
Aug.	20.7	13.7	28.4	49.6	2.5	188.3	445.1
Sept.	22.8	16.2	30.1	51.7	30.0	189.2	423.2
Oct.	22.9	17.8	29.2	66.0	127.1	138.1	405.5
Nov.	21.6	18.0	27.4	79.3	255.3	75.2	408.4
Dec.	21.9	18.1	26.6	83.0	342.5	60.8	409.5
Annual	21.3	16.2	27.6	70.6	1,572.5	1,285.0	-

Source : Goedert *et al.* (1980)

Soils : Rodriguez (1985) described the importance of soil in orchard productivity with a mention of physical characteristics (water permeability, depth, air, and water), chemical characteristics (reaction and nutrient availability), biological characteristics (organic matter and soil organisms), orchard management, and soils in Sao Paulo most used for citrus. Brazilian citrus growing soils are highly weathered, high in Al, and low both in P and base exchange capacity. Under such conditions, both liming and fertilization are necessary practices for improved productivity. In Santa Cantarina, there are more than 80 soil types. Amaro (1984) described the citrus growing soils in Brazil regionwise. The soils at citrus orchards of Sergipe are mostly sedimentary, sandy, red or yellow on level and gently rolling hillsides, commonly deep well drained, quite acidic, and possess low fertility at an altitude ranging between 40 m and 250 m above sea level. The predominant soils at Bahia region are Latosols related to coastal tablelands classified as dystrophic red yellow Latosols on level to gently undulating with low natural fertility, deep or very deep, and markedly drained. The podzolic soils are moderately and strongly acidic, having low fertility, sandy textured, well drained on a level and gently undulated topography with P and Zn as major fertility constraints (Coelho and Matos, 1991).

Soils in the Rio de Janeiro are Latosolic, very deep with low natural fertility and

Table 1.177. Climatic conditions of Manaus (AM), Cruz das Almas (BA), Limeira (SP) and Taquari (RS) of Brazil

	Jan.	Feb.	Mar.	Apr.	May	June	July	Aug.	Sept.	Oct	Nov.	Dec.	Year
Manaus (AM) (1931-60)													
Av.max.temp. (^{0}C)	30.0	29.9	30.0	29.9	30.8	31.1	31.6	32.7	33.1	32.7	32.0	31.1	31.2
Av.min.temp. (^{0}C)	23.3	23.2	23.3	23.3	23.6	23.4	23.2	23.5	23.9	24.1	24.0	23.7	23.5
Av. temp. (^{0}C)	25.9	25.8	25.8	25.8	26.4	26.6	26.9	27.5	27.9	27.7	27.3	26.7	26.7
Rainfall (mm)	276	277	301	287	193	98	61	41	62	112	165	228	2101
Cruz das Almas (BA) (1949-77)													
Av. max. temp. (^{0}C)	32.6	31.8	32.6	30.5	28.8	27.3	26.5	27.0	28.8	29.8	31.2	31.8	29.9
Av. min. temp. (^{0}C)	20.3	20.6	20.6	20.4	19.0	17.9	16.7	16.8	17.4	18.8	19.8	20.1	19.0
Av. temp. (^{0}C)	25.8	25.8	25.6	24.8	23.8	22.4	21.9	21.7	22.7	24.1	24.8	25.4	24.1
Rainfall (mm)	74.7	88.5	109.1	131.0	149.4	120.2	128.0	80.3	76.7	81.4	99.6	71.1	1210
Limeira (SP) (1962-71)													
Av. max. temp. (^{0}C)	29.3	29.2	29.4	28.1	25.6	24.6	24.7	21.3	28.7	27.8	28.8	28.7	27.2
Av. min. temp. (^{0}C)	17.9	18.2	17.5	14.8	12.4	11.5	10.9	12.1	13.9	14.9	15.8	17.2	14.2
Av. temp. (^{0}C)	22.6	22.6	22.3	20.5	18.1	17.2	16.9	18.8	20.5	20.6	21.5	22.1	20.3
Rainfall (mm)	239.1	193.5	155.1	32.9	44.6	30.4	22.3	26.1	59.8	165.2	121.6	219.7	1310
Taquari (RS) (1964-75)													
Av.max.temp.(^{0}C)	30.5	30.1	27.9	25.1	22.4	19.6	19.8	19.7	22.2	24.4	26.9	29.4	24.8
Av. min. temp. (^{0}C)	19.4	19.6	18.2	14.7	12.3	10.2	10.1	10.3	12.4	13.9	15.7	17.8	14.5
Av. temp. (^{0}C)	24.8	24.6	22.9	20.0	16.7	14.3	13.9	15.2	16.6	18.6	21.2	23.5	19.4
Rainfall (mm)	127.0	108.0	108.0	128.0	151.0	147.0	141.0	153.0	160.0	108.0	101.0	105.0	1537

Source : Coelho *et al.* (1984)

favourable physical characteristics. While, soils in the Rio Grande do Sul are deep, sandy, porous, red yellow, acidic, less fertile, and well drained. Soils of Sao Paulo region are characterised by sandy and sandy clay texture, deficient in P, pH between 4.8 to 5.2, medium in soil fertility, good drainage, depth, and topography from gently to strongly undulating. Soils at Minas Gerais are coarse textured, low fertility, and well drained shallow on slopy lands. Silva *et al.* (1984) described citrus growing soils at Boquim in Sergipe region of Brazil as sandy loam, red yellow podzols with 1.4 per cent organic matter, 3 mg kg^{-1} P_2O_5, 26 mg kg^{-1} K_2O, 1.4 me 100^{-1} g Ca^{2+}+ Mg^{2+}, and pH 5.4. In the west, basaltic soils are predominant, while in the coast, poorer sedimentary soils are more common. Soils at Barra Velha on the north coast of Santa Catarina at 30 m above mean sea level, have been characterised as Haplorthox and at Rio do Oeste, Santa Catarina as Dystrochrept (Koller, 1988). Soils at Cruz das Almas are sandy, clay loam, deep, well drained, moderately acid, low in fertility, and classified as Alic Yellow Latosol (Soares Filho *et al.*, 1984).

Uneven lime distribution due to the difficulty in incorporating high levels of lime in soil, and the impossibility of incorporating top dressed lime into the soil after planting have led to the development of overliming the top soil layers. This condition has resulted in an increased soil pH and decreased nutrient availability, mainly the microelements in

Table 1.178. Physical characteristic of the main soils occupied by citrus in Sao Paulo, Brazil

Soil	Altitude (m)	Relief	Drainage	Superficial texture	% in the state (%)
Podzolized on calcareous sandstone	400 to 600	Heavily undulated and undulated	Moderate	Sandy	8.7
Podzolized on calcareous sandstone	300 to 600	Smoothly undulated and undulated	Good	Sandy	11.0
Dark red Latosol sandy phase	300 to 560	Smoothly undulated	Accentuate	Sandy and Sandy clay	19.7
Ortho dark red Latosol	540 to 700	Smoothly undulated and undulated	Good	Clay and Clay sandy	4.4
Red yellow Latosol sandy phase	500 to 1000	Undulated and smoothly undulated	Good	Sandy and Sandy clay	4.9

Source : Rodriguez *et al.* (1977)

these layers. Overliming has caused B deficiency (Soprano and Koller, 1991). In Brazil, there are at least four different methods for determining lime requirements (LR): base saturation, buffer SMP, exchangable Al, and through neutralization curves (Quaggio, 1983). In southern Brazil, LR is based on the buffer SMP method for pH 6.0. Lime levels recommended by this method for citrus are relatively high, and in some cases reach 21 or 42 t lime ha^{-1}, for incorporation at 0-20 or 0-40 cm depth, respectively (Comissao De Fertildae, 1995). Citrus growing soils at Itajai, Santa Catarina state is podzolic red yellow Hapludults having clay texture (Soprano and Koller, 1996). Except for the swallow soils of some areas like the Cocoa region in the south of Bahia state, the humid soils of the Amazon basin or the clay soils of Sao Paulo, Parana, and in the sea coast of northeast, where coffee and sugarcane are being cultivated, there is an unlimited area where soils are suitable for Citriculture. Although, Brazilian soils have poor fertility, especially in terms of P availability and an acid pH, the soils are generally sandy, deep, well-drained, and with smooth topography (Passos, 1999). Citrus orchards are generally planted on sloping terrain in deep, sandy loam soils, and some orchards are planted on contour. A large proportion of citrus orchards are irrigated, and during periods of dry winter and spring weather, trees may suffer on account of water deficit stress.

The Sao Paulo citrus orchards, amounting to 80 million trees, are located mainly in podzolized soils on calcareous sandstone, dark red Latosol sandy phase, ortho dark red Latosol, and red yellow Latosol sandy phase. These soils are considered to be best for citrus which have most favourable physical and chemical characteristic are shown (Tables 1.178 and 1.179). The citrus plateau of the state of Sao Paulo is undulated or lightly undulated, the altitudes of which varies from 300 to 1000 m above sea level, but more frequently at 400 to 600 m. The citrus nutritional problems in that area are basically related to soil acidity, and deficiency of P, Ca, Mg, Zn, Mn, and B. Experimental results indicated the necessity to increase the pH around 6.0-6.5 with dolomitic limestone.

For more details, Chapter 5 may be consulted.

Table 1.179. Chemical characteristic of Ap horizon of the main soils occupied by citrus in Sao Paulo, Brazils

Soil types	Acidity (pH)	Sum of bases ($mg\ 100\ g^{-1}$) soil	Cation exchange capacity ($e\ mg\ 100\ g^{-1}$)	Base saturation (%)
Podzolized on calcareous sandstone	Acid to neutral	2.3-14.0	3.3-14.0	70-100
Podzolized on calcareous sandstone	Acid to smoothly acid	1.1-4.7	2.8-7.5	50-80
Dark red Latosol sandy phase	Strongly acid to acid	0.4-3.5	2.0-4.8	17-74
Ortho dark red Latosol	Acid	1.6-4.2	8.5-12.3	9-42
Red yellow Latosol sandy phase	Strongly acid to acid	0.3-1.9	2.8-4.8	11-40

Source : Rodriguez *et al.* (1977)

Through a survey carried out in the main citrus areas of the state of Sao Paulo, a knowledge about the nutrition problems of the citrus trees has been generated by Rodriguez and Gallo (1961). The soil analysis indicated variation in pH from neutral to strongly acid (pH 4.3); Ca^{2+} 4.9-0.7 me 100 g^{-1}; Mg^{2+} 1.1-0.1 me 100 g^{-1}; K^{+} 0.5-.01 me 100 g^{-1}; Al^{3+} 2.9-0.0 me 100^{-1}, and PO_4^{3-} 2.9-0.4 me 100 g^{-1}. Bebedouro is the one of the large production regions of Sao Paulo. Predominant soils are pozolized on calcareous sandstone and dark red Latosol sandy phase specially adequate for citrus growing (Table 1.180). Kolier *et al.* (1986) reported deficiency of Mg, Zn, N, and Mn in the state of Rio Grande do Sul.

Citrus growing soils at Monte Azul, Matao, Pirassununga, Olimpia, Botucatu, and Ararequara of Sao Paulo are red yellow Podzols, red yellow Latosols and dark red Latosols having soil pH between 4.0 and 4.9 and base saturation between 7 and 71 per cent (Table 1.181). The pH determined in water (Table 1.182) ranges from 4.0 to 5.6 and

Table 1.180. Soil characteristics of some citrus orchards in Bebedouro region of Sao Paulo, Brazil

Parameters	County						
	Bebedouro	Barretos	Pitangueiras	Colina	Severinia	Olimpia	Monte Azul Paulista
Cultivar	Pera Natal Valencia	Pera Natal Valencia	Natal Pera Valencia	Pera Natal Valencia	Pera Valencia Natal	Pera Natal Valencia	Natal Valencia Pera
Yield*	1.8-5.0	2.9-4.3	3.2-6.0	3.5-6.1	3.5-5.0	2.3-5.6	3.1-5.6
Soil analysis							
pH	5.5-5.9	5.5-6.0	5.1-6.4	5.2-5.4	5.2-5.3	5.4-5.8	5.7-6.1
Organic matter (%)	1.4-1.8	1.4-2.1	1.6-2.5	1.1-1.9	1.2-1.4	1.1-1.5	1.4-2.5
K^{+} (me 100^{-1} ml)	0.12-0.26	0.05-0.09	0.17-0.37	0.26-0.50	0.22-0.29	0.23-0.33	0.29-0.58
Ca^{2+} (me 100^{-1} ml)	1.3-2.4	1.3-2.0	1.1-3.8	1.5-2.6	1.2-1.7	1.8-3.5	1.8-2.7
Mg^{2+} (me 100^{-1} ml)	0.9-1.1	0.8-1.2	0.9-1.0	0.4-0.7	0.3-0.4	0.4-0.6	0.5-1.1
Al^{3+} (me 100^{-1} ml)	0.0-0.3	0.0-1.1	0.0-0.4	0.2-0.4	0.3-0.6	0.0-0.2	0.0-0.2
P (mg kg^{-1})	5-24	7-14	11-48	11-35	34-40	20-45	10-50

* Expressed in box $tree^{-1}$

Source: Caetano *et al.* (1984)

Table 1.181. Distribution of citrus soil types and their physicochemical properties in Sao Paulo, Brazil

Location	Soil depth (cm)	P-resin (mg dm^{-3})	pH $CaCl_2$	Exchangeable cations K^+	Ca^{2+}	Mg^{2+}	CEC	Base saturation	Org. matter
				[cmol (p^+) d m^{-3}]				(%)	
			Eutrophic Red Yellow Podzol						
Monte Azul	0-20	14	5.5	0.31	2.6	1.0	5.8	67	1.5
	20-40	05	5.2	0.18	2.7	0.7	5.6	64	1.0
			Eutrophic Red Yellow Podzol						
Matao	0-20	09	5.6	0.29	2.2	0.9	5.3	64	1.9
	20-40	06	5.2	0.14	1.7	1.1	5.1	57	1.7
			Allic Red Yellow Latosol						
Pirssununga	0-20	45	5.8	0.12	2.6	0.8	5.0	71	1.4
	20-40	45	5.9	0.11	2.5	0.9	4.9	71	1.3
			Dystrophic Dark Red Latosol						
Olimpia	0-20	22	4.5	0.30	1.3	0.4	5.3	35	1.6
	20-40	06	4.0	0.20	1.5	0.2	4.8	19	1.3
			Dystrophic Dark Red Latosol						
Botucatu	0-20	04	4.5	0.07	0.6	0.5	3.8	32	1.7
	20-40	02	4.1	0.04	0.1	0.1	3.0	07	1.1
			Dystrophic Dark Red Latosol						
Araraquara	0-20	17	5.1	0.32	3.1	1.1	6.9	67	2.0
	20-40	02	4.3	0.23	1.4	0.7	5.1	46	1.4

Source : Cantarella *et al.* (1992)

base saturation is < 50 per cent for all soils. The dominant base cations are Ca and Mg, with levels of Mg > Ca for most horizons. Al saturation is high, especially in well drained soils. Horizons with high smectite contents also had clay fractions with relatively high CEC values. Fe-oxide mineralogy is related to hillslope position. Lepidocrocite is dominant in the soils on footslope positions (Typic Plinthaqult and Plinthic Paleaqult) where reductomorphic conditions prevail. Goethite dominates in soils on the better drained shoulder positions (Inceptic Plinthustult and Plinthic Dystropept) in which oxidizing conditions prevail. In these soils, much of the Si released from the weathering of alumino-silicates is leached and Al substitutes for Fe in Fe-oxide minerals.

Sergipe is the another citrus growing region of Brazil, where a large proportion of citrus is grown on a sandy loam Red yellow podzolic soil. The various soil properties revealed organic matter 1.4 per cent, available P_2O_5 3 mg kg^{-1}, K_2O 26 mg kg^{-1}, Ca + Mg, 1.4 e.mg (Silva *et al.,* 1984). The soils at Sylvio Mereira, Brazil is dark red clay textured Latosol with the following chemical properties: organic matter 3.1per cent, pH ($CaCl_2$) 4.0, exchangeable cations in me 100 cm^{-3}, Al 1.5, Ca 0.4, Mg 0.1, and K 0.1, CEC (pH 7.0) 10.1 me 100 cm^{-3}, and base saturation 6 per cent, in addition to high buffering capacity (Quaggio *et al.*, 1992b). Soils at Bebedouro belonged to dark red Latosol, Allic Haplustox (Donadio *et al.*, 1996).

Table 1.182. Physical properties and organic carbon content of the Brazilian soils

Profile	Horizon	Particle size (% of < 2 mm)			pH (1:2.5)	Sum bases	Ext. Al^{3+}	Ext. H^+	CEC sum	CEC* clay
		Sand	Silt	Clay		(cmol kg^{-1})				
Soil on summit position (Ustic Dystropept and Typic Ustorthent)										
P_1	A	75	15	10	5.3	3.8	0.7	4.9	9.4	42.2
	Bw1	64	15	21	4.9	2.6	5.1	2.6	10.3	41.1
	Bw2	67	16	17	5.0	2.7	8.0	1.0	11.7	62.7
P_5	A	71	16	13	4.0	1.3	3.4	4.3	9.0	20.8
	C1	74	7	19	5.1	2.3	9.0	0.2	11.5	53.5
	C2	77	6	17	5.1	2.7	9.8	0.6	13.1	73.4
Soil on shoulder position (Inceptic Plinthustult and Plinthic Dystropept)										
P_4	Ac	64	21	15	5.0	2.3	0.8	5.3	8.4	6.5
	Btv	35	28	37	5.0	2.2	5.3	1.6	9.1	21.5
	BCv	36	37	27	5.0	2.2	7.3	1.4	10.9	36.9
P_6	Ap	53	31	16	5.6	2.1	1.1	3.6	6.8	21.4
	Bwv	47	30	23	5.1	3.4	3.3	2.4	9.1	33.7
	BCv	62	20	18	5.3	4.3	5.0	1.6	10.9	56.3
Soil on footslope position (Typic Plinthaqult and Plinthic Paleaqult)										
P_2	A	81	9	10	5.2	1.7	0.8	3.4	5.9	40.5
	Btv1	67	9	24	5.0	1.7	5.4	2.0	9.1	33.6
	Btv2	62	14	24	5.1	2.7	7.5	0.8	11.0	41.9
	Plinth.	56	27	17	-	-	-	-	-	-
	N/Plinth.	59	14	27	-	-	-	-	-	-
	Fe-sandstone	3	59	38	-	-	-	-	-	-

CEC clay = [(CEC Sum) – (4.5 x % c)] x 100 / % clay

Source : Dos Anjos *et al.* (1995)

A great variation in soils of Cerrado region due to their genesis, parent material, relief, altitude, depth, texture, permeability, and drainage has been observed (Freitas and Silveira, 1977). Soil types in Cerrado region have been observed distributed as red yellow Latosols (41 per cent), dark red Latosols (11 per cent), Latosols (4 per cent), and considerable area under quartziferrous sand (20 per cent). According to Sanchez (1982), Goedert *et al.* (1980), and Genu and Pinto (1984), physicochemical analysis of red yellow Latosols showed an extremly acidic pH, low P content, and Ca^{2+} + Mg^{2+}, medium K, and high Al^{3+} (Table 1.183). Soils contain medium status of organic matter with little activity (Goedertt *et al.*, 1980). Genu and Pinto (1984) described the soils of Cerrado region as deep, well drained with a sandy to very clay texture, much favourable to Citriculture.

Table 1.183. Properties of red yellow Latosol soil type in Cerrado region of Brazil

Depth (cm)	Sand (%)	Silt (%)	Clay (%)	pH H_2O (1:2.5)	Al (me 100 ml^{-1})	P	K^+	Ca^{2+} + Mg^{2+} (me100 ml^{-1})
						(mg kg^{-1})		
0-20	60	09	31	4.38	1.40	1.50	49	1.06
20-40	54	12	34	4.27	1.37	1.07	46	0.40

Source : Genu and Pinto (1984)

1.8.4.3 Citrus in Chile

Cultivation of citrus in Chile is traced to be documented way back in 1646, and introduced by Spanish (Burke, 1967). Most Chilean citrus production is within a hundred miles north and south of the capital city Santiago and Esmeralda region. The Peumo-San Vicente valley is the most important lemon and orange producing area, followed by scattered orchards on the Pacific coast south and west of Santiago, mostly near San Antonio. A number of valleys are known for citrus in Chile such as Pica, Azapa (Region I), Copiapo, Huasco (Region III), Elqui (Region IV), Limari, Petorca, Aconcagua (Region V), Cachapoal and Tinguiririca (Region VI) are famous (Ortuzar, 1996). Joublan *et al.* (1997) described eight citrus agroecological regions and cultivarwise preferences for various soil and climatic conditions.

Climate : There is a vast climatic range in Chile, whose 2,600 mile length stretches from a tropical desert in the north, through the temperate zone, and finally ends in cold, bleak, rocky Islands. In the citrus producing areas around Santiago, summers are cold and dry, and winters are wet. Average annual rainfall at Santiago is 14 mm, distributed mostly from May through September. The mean maximum temperatures of Santiago are 32°C in January and 22°C in July. While, mean minimums are 20°C and 14°C, respectively (Reuther *et al.*, 1967). The maximum and minimum temperature at Melipilla showed a variation from 15.0° to 27.7°C and 3.7° to 9.6°C, respectively (Table 1.184). Appraisal on climatic characteristics of regionwise citrus growing areas showed a wide diversity in diurnal variation in tempartures and rainfall (Table 1.185).

Table 1.184. Mean monthly temperatures at the Melipilla commune in Chile

Temp(°C)	Maximum	Minimum
Jan.	27.7	9.2
Feb.	27.5	9.1
Mar.	24.2	9.6
Apr.	22.3	8.8
May	19.1	8.5
June	16.6	6.6
July	15.0	3.7
Aug.	18.3	6.2
Sept.	18.2	7.4
Oct.	19.6	7.4
Nov.	24.1	7.8
Dec.	26.2	9.5

Source : Razeto and Villavicencio (1996)

Soils: The principal physiographic features of citrus growing areas of Chile consists of long chain of Andes in the east, extending from north to south, the coastal region consisting of various chains of lower mountains and hills some parallel to Andes and others going in an east – west direction. The valley soils of Chile are deep grayish sandy loam type and in coastal areas, these are red clay loam in character(Chapman, 1954).

A majority of soils has been found deficient in Mn and Zn which showed good response of micronutrient application (Razeto and Solas, 1986; Razeto *et al.*, 1988). Benito and Ruiz (1975) described the nutritional constraints in Santiago, O'Higgins, and

Table 1.185. Climatic characteristics of the current and potential citrus areas in Chile

Region	Annual heat units (above 12.5°C)	Chilling hours	Rainfall (mm $year^{-1}$)	Salient features (below 7°C)
Pica, Azapa	1000-1500	350-560	10-25	Good potential for expansion in the Huasco valley, with excellent fruit quality expected
Copiapo, Huasco	920-1400	160-560	90-190	Excellent fruit quality in intermediate to interior areas and great potential for expansion
Elqui	700-1000	400-600	260-350	Good fruit quality and good potential for expansion
Limari, Petorca	800-850	500-900	350-450	Good fruit quality, greater frost hazard in some areas and greater Pateca problems
Aconcagua, Cachapoal, Tinguiririca	800-850	500-900	580-700	Good fruit quality, sometimes thick rind problems and greater Pateca problems

Source : Santibanez (1986); Santibanez and Uribe (1991)

Colchagua provinces of Chile. Veregara *et al.* (1973) showed widespread N, Zn, and Mn deficiency with some indications of P and S deficiency.

1.8.4.4 Citrus in Columbia

Giacometti *et al.* (1966) described various soil and climatic requirements for growing citrus in Valle del Cauca.

Climate: Citrus growing locations such as Cartagena, Palmira, and Medellin belong to tropical climate. Cartagena is located on the northern coastal plain on the Carribean. It has a uniform hot climate throughout the year, having diurnal fluctutations of only 5° to 6°C, and a mean temperature of 27.2°C. Palmira is in the narrow Cauca river valley at about 1000 m elevation between two high Andean ranges. It is uniformly warm throughout the year, averaging between 9° and 13°C in diurnal fluctutation with a mean annual temperature of 23.9°C. Medellin is located in a small Andean valley at about 1400 – 1600 m elevation, with a mild year round climate having diurnal fluctutation of 11°-14°C,

Table 1.186. Climatic variation at citrus growing region of Palmira valley, Columbia

Months	Temperature (°C) Min.	Temperature (°C) Max.	Rainfall (mm)
Jan.	17.9	30.3	77
Feb.	18.2	30.7	67
Mar.	18.4	30.5	86
Apr.	18.4	29.7	133
May	18.3	29.1	117
June	18.0	29.4	67
July	17.5	30.5	28
Aug.	17.7	30.7	35
Sept.	17.7	30.5	55
Oct.	17.9	29.4	146
Nov.	18.1	29.1	112
Dec.	18.1	28.6	87

Source : Reuther (1973b)

and mean annual temperature of 21.4°C (Reuther and Rios-Castano, 1969). Citrus (cultivars viz., ICA-Bolo, ICA-Amaime, ICA-Jumundi, and Oreco nucellar) in Palmira is cultivated at 1006 m above mean sea level with an average rainfall of 956.7 mm, temperature of 23.4 °C, and relative humidity of 72.5 per cent (Rios-Castano and Camacho-Bustos, 1969). Reuther (1973b) also described the climate of citrus growing area of Palmira having mean monthly maximum and minimum temperature as 28.6°-30.7°C and 17.5°-18.4°C, respectively, with annual rainfall of 1010 mm (Table 1.186).

1.8.4.5 Citrus in Paraguay

The climate of Paraguay is midway between tropical and subtropical. Summers are hot and rainy. Winters are generally temperate. Average annual rainfall at the capital city of Asuncion is about 1350 mm. Annual mean temperatures at Asuncion are 31°C in January and 25°C in July.

1.8.4.6 Citrus in Peru

Peru has two principal citrus producing areas, one along the coast at elevations from 240 to 480 m, and another on the east Andean slope at about 900 m. Cultural practices and producing seasons are distinctly different in the two areas. Wolfe (1969) highlighted the problems of Citriculture in Peru.

Climate : The climate of the coastal plain is warm and equable. But, rainfall is almost nonexistent, averaging between 1 and 2 mm year^{-1}. At the Lima area, mean maximum and minimum temperatures in January are 31°C and 25°C. In July, the mean maximum and minimum temperatures are about 22°C and 20°C. The climate of the eastern slope is tropical and humid, but chilling is insufficient to produce normal rind colour of sweet oranges. There is heavy rainfall, ranging from 50 to 120 mm annually. Annual rainfall in citrus growing areas varies between 50 mm and 120 mm. Climatic variation in terms of mean monthly minimum and maximum temperature has been observed as 14.4-20.3°C and 20.0-29.5°C, respectively, with annual rainfall of 39 mm (Table 1.187).

Table 1.187. Climatic variation at citrus growing region of Lima, Peru

Months	Temperature (°C)		Rainfall (mm)
	Min.	Max.	
Jan.	19.6	28.4	< 1
Feb.	20.3	29.3	2
Mar.	20.0	29.5	< 1
Apr.	18.2	28.1	< 1
May	16.1	24.7	5
June	15.2	21.4	9
July	14.9	20.2	2
Aug.	14.4	20.0	7
Sept.	14.8	21.0	6
Oct.	15.5	22.5	2
Nov.	16.4	23.9	2
Dec.	18.0	26.4	1

Source : Reuther (1973b)

Soils : Citrus orchards are planted in the valleys at an elevation under 500 m in sandy loam soil that is ideal for citrus since rainfall is rare. All the orchards are irrigated by gravity water from streams, supplemented by water pumped from wells. The seedy oranges and valencias of the tropical east Andean slope are budded on sour orange rootstock. The east Andean slope is Lima's main source of orange supplies in the months of January, February, and March (Reuther *et al.*, 1967). Soils of Colca valley have been described by Sandor and Furbee (1996) and accordingly, they have been classified into two broad classes, namely, fluvial surfaces and colluvial surfaces, and high plateaus (Table 1.188).

Soils on the flood plains and lowest fluvial terraces show no subsurface horizon development. Soils on the highest of these lower fluvial surfaces have cambic horizons. Soils on the intermediate fluvial surfaces have argillic horizons, and are weakly to moderately developed. Soils on the high fluvial surfaces have greatest development of argillic horizons with common Bt horizon, clay maximum upto 70 per cent developed in

Table 1.188. Classification of soils in Andes of southern Peru

Fluvial surfaces	Colluvial surfaces and high plateau
Channel and floodplain of Rio Colca : Sandy or loamy skeletal, nonacid, Aquic or Typic Ustorthents	**Steeply sloping erosin surface (primarily andesite) :** Clayey skeletal, Typic Durustolls; loamy or clay skeletal, Typic or Agriustolls
Low to intermediate stream terraces and alluvial fans < 5 to about 75 m above Rio Colca : Typic or Cumulic Haplustolls (variable particle size classes viz., sandy, coarse or fine loamy, sandy or loamy skeletal); Endoaquolls having lateral seepage	**Steeply sloping erosion surface that overlie alluvial slopes mainly 20-60 per cent :** Fine loamy or loamy-skeletal, Entic, Typic or Cumulic Haplustolls; Coarse or fine loamy or loamy skeletal, Typic or Pachic Agriustolls; Coarse or fine loamy or loamy skeletal Typic Durustolls; Sandy coarse loam, Typic Ustorthents; Loamy-skeletal nonacid Typic Ustipsamments.
Intermediate stream terraces and aluvial fans about 100 to 125 m above Rio Colca : Coarse or fine loamy, Typic or Cumulic Haplustolls; Fine loamy or fine Typic or Pachic Agriustolls; Aquic Agriustolls on nearly level surface	**Steeply sloping erosion surfaces that overlie by laminated fine sand and silt :** Typic Ustorthents; Sandy or fine loamy, nonacid, Typic Ustipsamments
High stream terraces, alluvial fans, and piedmont surfaces about 150 to 500 m above Rio Colca : Fine loamy or fine Typic or Pachic Agriustolls; Fine loamy or fine Typic Durustolls	**Landslide topography on alluvium and laminated fine sand and silt :** Mollic epipedons
	High plateau (Altiplano) surfaces > 700 to 800 m above Rio Colca : Volcanic rock (mostly andesite), Alluvium, Colluvial glacial deposits.

Source : Sandor and Furbee (1996)

1.8.4.9 Citrus in Venezuela

The most important citrus producing areas of Venezuela are reported in the states of Carabobo, Falcon, Miranda, Monagas, Sucre, Yaracauy, and Zulia (Knorr and Vaughn, 1964b).

Climate : The climate of Venezuela varies greatly with altitude, ranging from tropical to temperate. But, citrus is grown mostly in subhumid tropical areas (Mendt,1989). Garcia Benavides (1971) described the climate under which *Citrus sinensis* is grown in Venezuela. Average annual rainfall of Caracas is slightly over 225 mm. Annual mean temperatures for Caracas are 25°C in January and 26°C in July (Reuther *et al.*, 1967). The climate of Miranda in Carabobo state is characterised by an annual rainfall of 995 mm and average temperature of 22.8°C. The mean temperature in the high valleys of Carabobo-Yaracuy areas is 23.4°C with an annual rainfall of 980 mm (Monteverde *et al.*, 1988). The climate is divided into rainy season (May to November) and dry season (December to April). Climate of Valencia and Criollo Montero growing Montalban area under high valleys of Carabobo – Yaracuy located at an altitude of 600-1400 m is characterised by annual rainfall of 950 mm with a 7 month rain period during May to November and a 5 month drought from December to April. The mean annual temperature is 23.5°C with 17.9°C and 29.1°C as minimum and maximum mean temperature, respectively (Monteverde *et al.,* 1984).

Table 1.190. Soil texture, fertility, and pH at Montalban-Carobobo, Venezuela

Properties	Araguita	Montero		La Majada
Texture	Clay loam	Sandy loam	Sandy loam	Sandy loam
Phosphorus (mg kg^{-1})	25	16	43	36
Potassium (mg kg^{-1})	124	44	48	80
Nitrate (mg kg^{-1})	36	39	67	100
Calcium (mg kg^{-1})	540	480	680	240
Organic matter (%)	2.43	1.62	2.02	1.62
pH	6.0	5.8	5.6	4.9

Source : Monteverde *et al.* (1984)

Soils: Soils at Miranda area of Carabobo state are medium clay with low available P and acidic pH. Monteverde *et al.* (1984) described citrus soils at Montalban as sandy loam to clay loam with soil pH varying between 4.9 to 6.0 (Table 1.190). Such a low soil pH has resulted in Ca deficiency mainly due to K antagonism (Birriel *et al.*, 1984). Soils in high valleys of Carabobo – Yaracuy are clay loam, pH 4.5-5.0 with medium NO_3 content (9-15 mg kg^{-1}), low P (1-2 mg kg^{-1}), low to high K (32-920 mg kg^{-1}), low to medium Ca (120-160 mg kg^{-1}), and low to medium organic carbon (0.6 - 3.5%) according to Monteverde *et al.* (1988). Rivero *et al.* (1998) described the alluvial valley and Savannahs in north and central Venezuela where citurs is extensively grown (Table 1.191). Magnesium and Zn deficiency are commonly observed in citrus orchards of Venezuela (Servicio Para El Agricultor, 1974a; 1974b). Pinto and Leal (1974) reported deficiencies of Ca, P, N, and Zn in Valles Altos of Carabobo, Venezuela. Suboptimum level of B is also one of the major fertility constraints (Anonymous, 1964b; 1966b; Laborem, 1977).

Table 1.191. Soil classification and some properties of Venezuelan soils

Soil order	Location	pH	Organic carbon (g kg^{-1})	Total N (g kg^{-1})	CEC (cmol (p^+) kg^{-1})
Inceptisol	Santa Cruz	7.6	12.5	9.5	6.2
Alfisol	Chaguaramas	5.0	13.3	8.5	2.3
Entisol	Calabozo	4.3	36.6	3.8	2.6
Vertisol	Apure	4.4	22.4	2.0	6.8
Oxisol	Amazonas I	4.3	29.3	1.2	3.6
Spodosol	Amazonas II	4.6	48.8	1.5	3.2

CEC stands for cation exchange capacity

Source : Rivero *et al.* (1998)

1.8.5 Citrus in Africa

Arab traders are believed to have introduced citrus into eastern mediterranean basin, approximately 1000 AD. Crusadors probably spread these citrus fruits to Italy and Spain from who later introduced to north African countries like Morocco, Algeria, and Egypt (Eshuys, 1967). Citrus in Africa is cultivated in counties like Algeria, Morocco, Egypt, Zimbabwe, Ivory Coast, Sierra Leone, Guinea, Senegal, Congo Brazzaville, Ghana, South Africa, Swaziland, Mozambique, Uganda, Ethiopia, Nigeria, Kenya,Tanzania, Zambia, and Malawi (Table 1.192). Burke (1952), Devezla (1968), and Castro (1972) described citrus growing in Mosambique with reference to climate and soil, which highlighted the wide scale occurrence of Zn and Mn deficiency (Mendes, 1972). Soil and climate under citrus have been described by Hodgson (1966a; 1966b) in Libya and by Bigi (1966) in Somalia. Rey (1999) discussed a wide range of climatic conditions for growing citrus in western and central Africa. According to Barry (1996), the climatic diversity in citrus growing areas of southern Africa is divided into 3 major classes:

i. Semitropical areas of low lying eastern sea board (Zimbabwe, Mozambique, northern and eastern Transvaal, Natal)

ii. Higher lying subtropical areas (Nelspruit, Letaba, Zimbabwe, Middle Veld)

iii. Coastal areas of Cape

The eastern areas are all summer rainfall areas, while the western and southern Cape enjoy a mediterranean type climate with winter rainfall. In the eastern Cape, a biomodal rainfall pattern exists with rains mostly occuring in spring and the fall. The southern African region comprises six main climatic zones or production areas which are divided in to 40 subareas. The six climatic zones are : hot humid; hot dry; intermediate; cool inland; cold semicoastal; and semidesert areas (Table 1.193).

Table 1.192. Citrus in Africa

Name of the country	Citrus growing areas
Algeria 30°00′ N lat.; 0°00′ long.	Algiers (Orleansville, Tizi - Ouzou), Oran (Mostaganem, Tlemcen), Constantine (Bone, Setif), Castal zones – Algiers, Oran, Phillipeville, Bone; Inland area - Orleansville, Voltaire
Botsawana 23.00∘′S lat.; 22.00∘′ E long.	
Egypt 25°00′ N lat.; 30°00′ E long.	Cairo, Kalioubiya, Charkieh, Menoufieh, Gharbieh, Bahera, Faiyum, Minya, Giza, Abouxah, Shakshook, Etsa, Aswan, Assiut, Dakahlia, Beheira, Ismailia, Tahrir
Ethiopia 10°00′ N lat.; 40°00′ E long.	Near lake Awasa
Kenya 1°00′ N lat.; 36°00′ E long.	Meru, Thika, Eastern and Coastal region, Kwale, Marsabit, Northern Rift valley
Malawi 12°00′ S lat.; 34°00′ E long.	Lilomgwe , Limbe, Blantyre
Libya 30°00′ N lat.; 15°00′ E long.	Suani, Saida, Zenzour, Ana Siria, Azzahara, Amiria, Azizia, Ben, Gashir, Sghedeia
Morocco 32°00′ N lat.; 5°00′ W long.	Oujda, Tetouan, Ourergha, Rharb, Rabat, Casablanca, Beni mellal, Marrkech, Agadir, Fes, Meknes, Taza, Nador,Al Hoceima, Janger, Quarzuzate, Ksar es Sonk, Jarfaya.
Mozambique 15°02′ S lat.; 40°08′ E long.	Beira, Marques, Lourenco
Nigeria 10°00′ N lat.; 08°00′ E long.	Ibadan, Abeocua
Sierra Leone 9°00′ N lat.; 12°00′ E long.	Coastal areas
South Africa 25°00′ S lat.; 25°00′ E. long.	Citrusdal, Nelspruit, Zebediela, Tzaneen-Letaba, Sundays river area of eastern Cape province, Natal, Salisbury, Cape midland, Cape coastal, Triangle ranch, Western Transvaal (Barbie hall, Pretporia, Johansburg), Western Cape, Eastern Transvaal.
Swaziland 26°23′ S lat.; 31°30′ E. long.	Tambuti, Big Bend, Ngonini
Tanzania 04°00′ S lat.; 34°00′ E long.	Zanzibar island, Pemba, Coastal region, Tanga, Dar es Salaam, Morogoro, Mbeya
Tunisia 33°00′ N lat.; 10° 00′ E long.	Soliman, Menzel Bou Zelta, Grombalia, Beni khaled, Nabeul, Beni Kriar, Tunis, Hammamet
Uganda 1°00′ N lat.; 32°30′ E., long.	Busoga, Iganga, Kamuli, Masaka, Kampala
Zambia 17°30′ S lat.; 24°00′ E long.	Ndola, Lusaka
Zimbabwe 19°00′ N lat.; 30°00′ E long.	Mazoe, Hippo valley, Tambuti, Ngonini, Big Bend, Umtali, Sinoia

Table 1.193. Commercial citrus growing areas of southern Africa

Climatic zones	Citrus growing areas
Hot dry	Limpopo valley, Letsitele, Hoedspruit
Hot humid	S. Mozambique, Onderberg, Swazi lowveld, Pongola, Nkwaleni
Intermediate	Letaba, Levubu, Marble Hall, Nelspruit, Karino, Hazyview, Barbertown, Ngonini, Schagen
Cool inland	Rustenberg, Potgietersrus, Zebediela, Burgersfort, Ngodwana
Cold semi-coastal	ECM, SRV, GRV, Knysna, Boland, Berg river, Breede river, Grabouw, Citrusdal, S. Natal
Semi desert area	

Source : Barry (1996)

Other than several regions in Africa and the mediterranean area, the various localities listed have no months with an average temperature below 12.5°C. However, the pattern is quite different considering the number of months with an average less than 17.5 °C (up to 5°C over the minimum). Especially at the higher elevations in tropical regions (Table 1.194), appreciable number of months with presumably reduced growth activity is a common sight.

The hot production areas are split into i. humid and ii. dry areas. The former being less than 200 km from the sea and the latter up to 600 km. These are low lying locations at less than 600 m above mean sea level, with the humid climate and dry climate at less than 300 m above mean sea level, along the eastern seaboard of southern Africa in southern Zimbabwe, southern Mozambique, northern Transvaal, eastern Transvaal, and Natal. These areas are suitable for the production of high quality grapefruit and valencia oranges, and to a lesser extent midseason oranges, and certain mandarin types such as Minneola. The intermediate production areas are so called due to their being, somewhere between hot, low lying and cool, highlying areas in terms of altitude, temperature, and cultivars suitability. These are at an altitude of 600-900 m above mean sea level and form the high lying areas of the eastern and northern Transvaal and lower elevation of the central Transvaal. These areas are suitable for the production of valencia, the midseason oranges and lemons, and marginally suitable for grapefruit (too cool) and navel oranges (too warm).

The cool, inland production areas are the high lying (above 900 m altitude) areas of central Transvaal and Natal. They are well suited to navel oranges and lemons, the warmer microclimates are suited to valencia oranges, and to a lesser extent, certain mandarins such as clementine, nova, and possibly temples. The cold production areas are semicoastal areas situated in southern latitudes, between 32°30′ and 34°30′ in the eastern, southern and western Cape. High quality navel oranges, satsuma, clementine, Nova mandarins, and lemons are produced in these areas. While, the warmer microclimates are suitable for valencia orange production. The semidesert areas are relatively new in terms of citrus production and characterised by exterme hot summers

Table 1.194. Meteorological variation in some tropical citrus growing countries of Africa

Month	Jan.	Feb.	Mar.	Apr.	May	Jun.	July	Aug.	Sept.	Oct.	Nov.	Dec.
Kindia, lat. 10⁰ - Guinea (Africa)												
MMT	23.8	25.1	26.5	27.2	25.9	24.3	23.4	23.3	24.6	24.8	25	23.9
R	3.4	2.0	35.5	56.8	188.4	280.6	397.3	477.8	348.9	238.9	86.8	40.5
B	-	-	-	-	B	-	-	-	-	-	-	-
M	-	-	-	-	-	-	-	-	-	M	M	M
ET	-	-	-	-	406.1	345	328.6	325.5	354	372	366	344.1
Odienne, lat. 10°, Ivory Coast (Africa)												
MMT	24.9	25.0	28.9	28.3	27.1	25.7	25.0	24.7	24.2	25.2	25.0	23.4
R	3.0	14.0	38.0	76.0	121.0	177	288	387	291	164	53	9
B	-	-	-	-	B	-	-	-	-	-	-	-
M	M	-	-	-	-	-	-	-	-	M	M	M
ET	-	-	-	-	443.3	387	378.2	368.9	342	384.4	366	328.6
Foumbot, Alt 1100 m, lat. 5°.5′, Cameroon (Africa)												
MMT	29	29.7	29.8	27.9	26.4	25.2	24.4	24.5	24.7	25.4	27	28.2
R	9.2	27.6	94.9	137.5	133.2	170.7	223.2	250.1	308.1	246.2	67.5	13.1
B	-	-	-	B	B	-	-	-	-	-	-	-
M	M	-	-	-	-	-	-	-	-	-	M	M
ET	-	-	-	453	421..6	372	359.6	362.7	357	390.6	426	477.4
Grands Fonds, Alt 50 m, lat. 16⁰, Caribbean area (Guadeloupe)												
MMT	23.4	23.5	23.9	25	26	26.7	26.7	26.7	26.6	26	25.3	24.2
R	128.2	59.7	69.9	147.5	111.3	134.9	184.3	172.9	212.8	207.8	198.9	115.5
B	-	-	-	B	B	-	-	-	-	-	-	-
M	M	-	-	-	-	-	-	-	-	M	M	M
ET	-	-	-	366	409.2	417	430.9	430.9	414	409.2	375	353.4
St. Clair, lat. 10′, Caribbean area – Trinidad												
MMT	25.2	25.2	25.3	26.1	26.8	26.2	26.2	26.3	26.6	26.6	26.4	25.8
R	68.6	37.6	45.7	47.2	87.9	194.3	222	241.8	185.2	169.2	180.3	121.4
B	-	-	-	-	-	B	B	-	-	-	-	-
M	M	M	-	-	-	-	-	--	-	-	-	M
ET	384.4	-	-	-	-	402	415.4	418.5	414	427.8	408	403
St Laurent, Alt 4 m, lat. 5°30′, French Guyana (South Africa)												
MMT	25.5	25.6	25.8	26.1	26.1	26	26.4	26.9	27.3	27.2	26.8	26
R	217.5	129.5	148.8	131.5	249	294.4	239	158.3	83.5	58.5	158.8	247.2
B	-	-	-	-	B2	B2	-	-	-	-	B1	B1
Paramaribo, lat. 6⁰, Suriname (Africa)												
MMT	26.1	26.4	26.7	26.9	26.7	26.5	26.8	27.7	28.3	28.3	27.8	26.6
R	214.9	160	195.1	229.1	312.9	304	232.9	160	78	78	119.9	214.9
B	-	-	B2	B2	-	-	-	-	-	-	B1	B1
M	-	-	-	-	M1	M1	M1	-	-	M2	M2	M2
ET	412.3	380.8	430.9	423	430.9	411	434	461.9	465	480.5	450	427.8
Dakar, lat. 15°, Senegal (Africa)												
MMT	21.3	20.4	20.7	21.6	23.5	26.1	27	27	27.5	27.5	26.2	23.4
R	1	2	-	-	3	9	118	244	196	71	4	7
B	-	-	-	-	-	-	B	B	-	-	-	-
M	M	M	-	-	-	-	-	-	-	-	-	-
ET	263.5	212.8	-	-	-	-	440.2	440.2	441	455.7	402	328.6

MMT, Monthly mean temperature (°C); R, Rainfall (mm); B, Blooming; M, Maturity; ET, Effective temperature between mid-bloom and maturity; A, Average; and T, Total

Source : Cassin *et al.* (1969b)

and cold winters with the occurrence of frosts due to advectic conditions. In the cooler Vaalharts areas, navel and valencia oranges are produced, while grapefruit and valencia oranges are produced in the hotter lower orange river area. Due to Zimbabwe's more northerly latitude, ranging from 17° to 22°S, higher altitudes are required to produce similar conditions to those experienced further from the equator. Therefore, hot areas are located at less than 900 m altitude as opposed to 600 m, intermediate areas range from 900 m to 1200 above mean sea level, and cool areas occur above 1200 m (Barry, 1996).

1.8.5.1 Citrus in Algeria

Citrus growing areas of Algeria consist of three districts; Algiers (Orleansville and Trizi-Ouzou), Oran (Mostaganem and Tlemcen), and Constantine (Bone and Setif).

Climate : Algiers, Oran, Phillipeville, and Bone of coastal zone are moderately humid and have an annual rainfall of 375 to 625 mm, falling mostly from September to May. Mean maximum temperatures at Algiers range from 29°C in January to 38°C in July. Mean minimum temperatures are 17°C in January and 25°C in July. The inland citrus areas such as Orleansville and Voltaire are more arid, and have lower humidity. Rainfall occurs mostly between October and March. Average annual rainfall at Orleansville is 14 mm. Mean maximum temperatures at Orleansville range from 27°C in January to 42°C in July. Mean minimum temperatures are 12°C in January and 24°C in July (Reuther *et al.*, 1967). Climate variation in terms of mean monthly minimum and maximum temperature at citrus growing area of indicated 9.5°-21.7°C and 14.0°-29.4°C, respectively, with annual rainfall of 759 mm (Table 1.195).

Soils : Citrus growing soils of Algeria are mostly heavy loam, light to deep red in colour with high clay percentage. A few citrus areas have brown loam soils, and a small coastal lemon area in north of Algiers has light sandy loam soils. Some areas near Blida have a water table only five feet from the surface, and drainage tiles and ditches are extensively used. Orchards with heavy clay soils are often deeply cultivated with subsoilers between each tree row, and between the grove and windbreaks. The deep cultivation results in root cutting. Winter cover crops are used, but supplemented with mineral fertilizers.

Table 1.195. Climatic variation at citrus growing region of Algiers, Algeria

Months	Temperature (°C)		Rainfall (mm)
	Min.	Max.	
Jan.	9.5	14.0	112
Feb.	9.5	16.1	84
Mar.	11.1	17.2	72
Apr.	12.8	20.0	41
May	15.0	22.8	46
June	18.3	25.6	15
July	15.6	28.3	3
Aug.	21.7	29.4	5
Sept.	20.5	27.2	41
Oct.	17.2	23.3	74
Nov.	13.3	18.9	129
Dec.	10.6	15.6	137

Source : Reuther (1973b)

1.8.5.2 Citrus in Botswana

Climate: In an adaption and yield potential studies, the most important factor is the climate. The climate is influenced by environmental factors which determine the crops/cultivars to grow and yield (Hudson, 1975). In commercial citrus production, quality an important aspect of marketing, is determined by climate. Quality comprises flavour which is influenced by balance between carbohydrates and citric acids (Juice composition), rind thickness, rag toughness, and colour. Consumer's preferences are influenced by the orange colour of fruits. This condition is dependent on temperature (Green *et al.*, 1975). The optimum mean daily temperatures for normal growth of citrus is 23°–30°C. Growth is markedly reduced above 38°C and below 13°C. Citrus is susceptible to severe frost. However, the agrometeorology section of the Institute of Tropical and Subtropical Fruits at Nelspruit, Republic of South Africa, has established that citrus can withstand temperature down to 2°C for a short period without any adverse effect on yield.

Citrus, especially sweet orange, cv valencia and navels require shorter days and cooler temperatures in winter in order to initiate flower bud formation. Flowering takes place in spring that would result in production of large fruits. Warm weather and long sunshine hours encourage high yield due to increased fruit size, provided optimum soil moisture condition throughout fruit development. In subtropical regions, where citrus is produced, the best yield obtained ranges from 50-100 kg tree^{-1} (Bredell, 1979). In subtropical areas, days tend to be shorter and cool, fruit maturity takes place when temperatures fall from April onward and the fruit colour is good. However, in the semiarid, tropical and wet tropical areas, the development of fruit colour is poor. In the former, high temperature tends to shorten the period of fruit maturity. In the former, days tend to be long and hot, and the maturity of fruit takes place when temperatures are high. In the latter, high relative humidity is responsible for poor fruit colour (Doorenbos and Kassam, 1979). The average maximum and minimum temperature have been observed varying from 26.0° to 37.2°C and 7.3° to 21.1°C, respectively, at Maun in northwest Botswana (Table 1.196).

Table 1.196. Average maximum and minimum temperatures during last two years at Maun located in north-western part of Botswana

Month	1992		1993	
	Max.	Min.	Max.	Min.
Jan.	34.2	19.9	31.5	19.7
Feb.	37.2	21.1	29.7	19.9
Mar.	34.1	19.7	32.5	18.3
Apr.	33.0	17.0	31.0	16.3
May	29.7	10.2	30.5	12.7
June	26.3	7.3	26.1	8.2
July	26.0	8.0	26.0	10.0
Aug.	27.6	9.1	25.9	9.9
Sept.	34.4	16.7	28.9	11.9
Oct.	35.4	20.8	36.5	21.2
Nov.	33.5	19.8	34.2	20.8
Dec.	34.3	20.1	33.8	20.9

Source : Machacha (1996)

Soils : Most of the cultivation is confined between 40° N and 40° S at 1800 m above sea level in the tropics, and up to 750 m above sea level in subtropics (Doorenbos and Kassam, 1979). Although, citrus trees grow satisfactorily in a wide range of soils, good drainage is a pre-requisite, otherwise root rot becomes a problem (Plessis, 1975). However, in areas where draiange is a problem, rootstocks which tolerate poorly drained soils are used.

1.8.5.3 Citrus in Egypt

The citrus in Egypt is being grown at the time of Alexander, the great (Brown, 1924). Egyptians believe that early Coptic-Arabic literature furnishes evidence that the sour or bitter orange, lemon, and sweet orange were known to Egypt earlier than the 9th and 10th century AD, the generally accepted period of their introduction into the mediterranean by Arab conquerers. The mandarin reached Egypt in modern times. However, having been brought in from Europe in 1830 by Youssef Effendi, a member of the mission sent by Mohmed Ali Pasha, founder of the dynasty ended by an explusion of king Farouk. Citrus is cultivated in every province in the Nile valley with majority of citrus occurring in delta provinces of Kalioubiya, Sharkiya, Menoufiya, Gharbiya, and Beheira in northern Egypt. The focus of the industry is located in upper (southern Egypt) portion of the delta towards north of Cairo with Assiut province having maximum contribution. The citrus growing middle Egypt provinces are Fayoum, Minia, and Giza.

Climate : Upper Egypt, which includes the citrus region of Assiut, has a desert climate with intense, arid heat that contributes to a long growing season. Winters are short and mild, and average annual rainfall is less than a mm, falling mostly from November to February. Mean maximum temperatures in Assiut province range from 26°C in January to 37°C in July. Mean minimum temperatures range from 16°C in January to 27°C in July. Middle Egypt, which extends from Cairo several hundred miles south, is slightly cooler, and has somewhat higher humidity. Average annual rainfall in Giza province is 25 mm, falling mostly between December and March. Mean maximum temperatures for Giza range from 26°C in January to 36°C in July. Mean minimum temperature are 15°C in January and 25°C in July. Lower Egypt, where the bulk of citrus orchards are concentrated, possesses a climate similar to that of the Rio Grande Valley in Texas. Humidity is relatively high in late summer and fall. Rainfall is relatively scant and ranges annually from 25 mm in Kalioubiya province to 175 mm in the Alexandria Governorate. Mean maximum temperatures in Kalioubiya province range from 25°C in January to 36°C in July, and mean minimums during the corresponding months range from 16° to 26°C. These mean temperatures may be considered generally typical for lower Egypt as a whole (Reuther *et al.*, 1967).

Three climatic zones under citrus have so far been identified. Upper Egypt from Luxor to the southern boundary. The chief citrus producing province is Assiut, lies partly in the tropics with tropical desert climate, characterised by very long growing season with intense heat and aridity, and a short and very mild winter. It also covers Coachella valley characterised by warmer winters and hotter summers. Middle Egypt has typically desert subtropical climate and differs from upper Egypt, principally in total amount of heat and mean winter temperature. The growing season is featured by slightly less intense heat and somewhat higher humidity. Lower Egypt where bulk of acreage under citrus is located, consists of an extensive delta of Nile subject to annual flood and overflow of river. With the rise in water level of rivers during late summers which brings about a rise in the water table, imparts increased saturation of soil and evaporation, thereby, high atmospheric humidity.The lower Egypt differs significantly from the rest of the Nile valley in the amount of humidity. The climate of lower Egypt is similar to Lower Rio Grande valley of Texas, USA. Due to the strong desert winds, wind protection is indispensable in Egypt particularly on the edge of the valley and delta. Use of windbreak trees like Casuarina equisetifolia, Casuarina stricta, and Casuarina glauca has been found quite effective (Hodgson,1954).

Soils : Physiographically citrus growing areas of Egypt can be divided into western desert, eastern desert, delta coastal, central delta, upper delta, and upper Egypt (Hodgson, 1937; 1954; Bingham, 1964). The soils in the Nile river valley and delta proper are of alluvial origin. The soils are deep, fertile, silt, and clay loams. Citrus cultivation is restricted to soils of light to medium texture, and located at high elevation to avoid the adverse impact of rising water table, preferably more than six feet from the surface. The valley soils are rich and fertile.

In the past, a good account of Egyptian soils has been documented highlighting the general characteristics and geography of the soils (Elgabali *et al.,* 1969a; Elgabali *et al.,* 1969b; Rozanov, 1970). The mediterranean coast of Egypt consist of a succession of marine terraces created by gradual tectonic uplift of the land. Different coastal landscapes are found on these terraces, to wit: i. coastal dune sands on terraces ranging in age from recent to middle Paleolithic, ii. terraces ranging in age from early paleolithic to early pleistocene are covered with loess like rocks, the top of which consist of light sierozems and where surface and internal salt, gypsum, and calcareous paleohydrogenic crusts are common, iii. rocky desert on terraces ranging in age from the middle pleistocene to the miocene and situated over 50 m above sea level, that is, hammadas completely lacking in vegetation and soil cover. The main distinguishing feature of soils of the Egyptian coastal semideserts is their very high carbonate content. Further distinguishing characteristics are the secondary distribution of carbonate in the soils and the formation of hydrogenic calcareous crusts, a phenomenon observed long ago by many Egyptian researchers (Zein El Abedine and Abdalla, 1952; Shata, 1955; 1960; Tadros, 1956; Said, 1962).

It is convenient to define calcareous soils as those containing 5 per cent or more of free $CaCO_3$ by weight in any horizon or throughout the section, i.e. vigorously liberating CO_2 in 10 per cent HCl. The great majority of Egyptian soils fall into this category, the exceptions being the alluvial and gley soils of the Nile valley and the barhan sands of the Libyan deserts. The secondary accumulation of carbonates in soils is an evolutionary process, closely related to the age of the soils, the older the soil that exists under given geomorphological and geochemical conditions, the larger is the secondary accumulation of $CaCO_3$ (Rozanov *et al.,* 1982).

Nutritional survey of orchards of sweet orange cultivars Balady and Sukkary, and mandarin cultivars, Balady at 8 localities of El-Fayoum Governorate showed adverse effect of soil salinity and high Ca content at some localities. However, balady orange trees did best at Abouxah, whereas Sukkary trees did best at Shakshook and Balady mandarin trees yielded best at Etsa (Montasser *et al.*, 1985). The calcareous soils of all types decreased the growth of a majority of citrus types (Assal *et al.*, 1994). These soils have shown a good response of macro-and micronutrient fertilization (El-Fouly, 1985; El-Fouly *et al.*, 1988). Calcium enters the soil of various landscapes in four different ways i. via diluvial, colluvial, or alluvial accumulation of calcareous particles in the course of primary sedimentation, whereby the parent material of the soil is deposited at the same time, the soil is formed; ii. eolian impulverization i.e. impregnation with fine calcareous dust lifted from limestone surface; iii. biological accumulation due to presence of lagoon, marsh, and land mollusks, that is formation and deposition in a thick layer of calcareous mussel shells; and iv. deposition of calcium from groundwater flowing from topographic elevations to depressions. All these modes of calcium deposition can be observed in the Egyptian soils (Rozanov *et al.,* 1982). The $CaCO_3$ content gradually increases with decreasing diameter of the fraction (Table 1.197). However, the highest $CaCO_3$ content is found in the finest fraction, the peak total soil calcium carbonate content occurred in the coarse sand fraction.

Table: 1.197. $CaCO_3$ content of the various fractions of a typical sandy terrace soil of Egypt

Depth	Particle diameter (μm)													
(cm)	> 500		500-297		297-210		210-100		150-105		105-53		< 53	
	1	2	1	2	1	2	1	2	1	2	1	2	1	2
0-10	34.9	0.2	17.9	9.5	27.2	12.5	7.5	15.5	11.9	12.0	0.2	10.0	0.8	37.5
60-70	50.8	1.1	20.9	2.0	7.0	7.8	2.5	15.8	13.9	14.5	1.8	7.5	1.8	12.6
200-210	40.6	0.7	22.9	2.0	11.5	8.0	6.4	12.1	16.7	8.6	0.9	8.5	1.0	10.8
490-500	35.0	2.3	22.8	3.8	16.6	15.6	6.2	18.4	17.0	10.5	1.1	15.0	1.2	34.0
600-610	19.7	4.5	11.6	4.1	5.6	15.7	2.3	24.5	9.0	15.0	0.5	17.0	0.5	41.3

1, Content of fraction in per cent; 2, $CaCO_3$ content of the fraction in per cent

Source : Rozanov *et al.* (1982)

Soils of the Nile valley are a major natural source of Egypt. Wahab (1969) and Wabha (1972) studied the morphology, chemical, and physical characteristics of representative soils in the valley and concluded that textural differences among the soils reflected the differential rate of sedimentation of suspended material carried by the river. Soils are classified as Torrents and Torrifluvents having mineralogy predominantly as monmorillonitic, kaolintic, and illitic (Hamdi, 1959; 1967; Hamdi and Barrada, 1960; Hashad and Mady, 1963; Khadr, 1963). The clay content of these soils ranges from 36 to 49 per cent. The pH values lie between 8.0 and 8.8, having 2.4 per cent calcium carbonate, which increases upto 7 per cent in the soils near desert. The organic matter content ranges between 2.1 per cent and 2.4 per cent in the Ap horizon that decreases with depth (Hanna and Beckmann, 1975). A large portion of citrus industry in Egypt is located in low lying areas close to the mediterranean coast. Conditions of high water table and sea water intrusions, thus, prevail posing potential salinity and boron hazards to citrus industry (Elseewi, 1974). Egyptian soils are fertile and consist of deep alluvial silt and clay loams. The Nile river canal system furnishes water with up to 350 mg l^{-1} of soluble minerals in June and July. Most of the land in that region is below sea level with water tables varying in depth between 90 and 120 cm below the surface.

Table 1.198. Some chemical and physical characteristics of orchard soils (based on average of three depth)

Orchard	pH range	$CaCO_3$-equivalent (%)	ECe (mmhos cm^{-1})	Texture surface/ subsurface
1	7.3-7.5	9.1	3.5	Sandy clay loam/Clay loam
2	7.6-9.1	8.9	2.1	Sandy/Sandy
3	7.2-8.2	6.4	1.8	Sandy clay loam/Loam
4	7.0-7.9	11.3	3.0	Sandy clay/Sandy clay loam
5	8.2-8.3	13.0	1.2	Sandy clay/Sandy clay loam
6	7.8-8.1	11.2	3.7	Sandy clay loam/Sandy clay loam
7	7.6-7.7	4.7	3.5	Sandy clay/Loamy sand
8	7.5-7.9	5.0	1.9	Clay/Clay loam
9	8.2-8.3	13.2	3.3	Sandy clay/Clay
10	7.7-7.8	6.1	2.1	Sandy clay loam/Clay
11	7.3-7.5	9.2	3.1	Sandy clay loam/Sandy clay

Source : Elseewi and Elmalky (1977)

Nile river water is used for irrigation and open drainage is maintained in all the orchards. The orchards are irrigated by sprinkler or permanent basins. Most soils are medium to heavy textured. The average $CaCO_3$-equivalent varies between 4.7 per cent and 14.2 per cent, indicating the calcareous nature of the soils. Consequently, the pH of the soil ranges between 7.0 and 9.1. The average electrical conductivity values (Table 1.198) are below the critical limit of soil salinity (ECe 4 mmhos cm^{-1}), established by the Staff of U.S. Salinity Laboratory for establishing the salt sensitivity of plants. Soils of the orchards are well drained, and salts indigenously found in soil or added from irrigation water are leached beyond the root zone (Elseewi and Elmalky, 1976). Based on analysis of 6000 soil samples collected by Nile valley, large scale Fe deficiency dominant on calcareous soil, Zn deficiency on sandy soils, and Mn deficiency on alluvial soils (Wallace *et al.*, 1975; Elsokkary, 1976; El-Fouly *et al.*, 1984) in addition of Mg deficiency responding significantly to $MgSO_4$ either as soil application or foliar spray (Haggag *et al.*, 1987).

Nile valley alluvial soils are mostly low in salt, lime, organic matter, N, and Zn; high in pH, P, K, and Cu; and variable in Fe and Mn. However, low P at Monoufia, Fe and Mn at Kalioubiya with very low content of Mn in Assiut soils (Table 199). Sandy soils have very low concentration of organic matter and nutrients, except P, which range between low and moderate (Table 1.200). Calcareous soils have comparatively higher pH and EC, and low organic matter and Fe, with very low N and P, irrespective of soil types (Table 1.201). El-Gazzar *et al.* (1975) reported an adequate amount of Fe with marginal Mn status in orchards of Balady, Succari and Washington navel oranges, and balady mandarin. Salem *et al.* (1995) also reported N, Fe, Mn, and Zn deficiency under sandy soil conditions. Under exchangeable K as a common indicator of soil K fertility evaluation, alluvial soils and calcareous soils have been rated rich and low, respectively. But, Nile alluvial soils require annual K fertilization due to wide scale K nutritional problem (Fawzi *et al.,* 1990)

Table 1.199. Soil characteristics of the Nile loamy alluvial soils (1-10 cm depth)

Location	Valley			Delta	
Properties	Assuit	Minia	Kaliobia	Monoufia	Dakahlia
pH	7.6	8.2	8.3	8.3	8.2
EC (mmhos cm^{-1})	0.4	0.4	0.2	0.5	0.5
$CaCO_3$ (%)	3.2	2.6	1.8	1.6	3.0
Org. matter (%)	1.6*	1.0*	1.5*	1.2*	1.4*
N (mg 100 g^{-1})	49**	117**	142*	148*	129**
P (mg 100 g^{-1})	4.3	3.1	3.8	6.4	4.4
K (mg 100 g^{-1})	56	68	49	56	45
Fe (mg 100 g^{-1})	42	21	10*	24	34
Mn (mg 100 g^{-1})	136	28	4**	14*	31
Zn (mg 100 g^{-1})	1.9*	1.5*	1.3*	1.4*	2.1*
Cu (mg 100 g^{-1})	4.9	4.7	4.5	3.6	4.4

* low , ** very low

Source : El-Fouly *et al.* (1984)

Table 1.200. Characteristics of sandy soils (0-30 cm depth)

Location / Properties	West Desert (South Tahrir)	East Desert (Ismailia)	Valley Fringes (Minia)
pH	8.4	8.1	8.3
EC(mmhos cm^{-1})	0.2	0.2	2.4
$CaCO_3$ (%)	1	1	9
Org. matter (%)	0.6**	0.8**	0.3**
N (mg 100 g^{-1})	28**	55**	14**
P (mg 100 g^{-1})	1.5	1.7	1.7
K (mg 100 g^{-1})	5.6**	9.3**	9.3**
Fe (mg 100 g^{-1})	4**	13	5**
Mn (mg 100 g^{-1})	5**	10	3**
Zn (mg 100 g^{-1})	1.7*	1.6*	1.5*
Cu (mg 100 g^{-1})	1.7	3.0	1.0

* low , ** very low

Source : El-Fouly *et al.* (1984)

1.8.5.4 Citrus in Kenya

Citrus in Kenya is cultivated up to an altitude of 1800 m. A majority of citrus is confined to eastern and coastal parts of country. Johnson (1966) described various keys to growing citrus successfully in Kenya.

Climate : Climate in citrus growing areas of Kenya is basically humid tropical with no distinct delineation between summer and winter months. Climate of west Kenya is characterised by annual rainfall of 1200-2200 mm (Wielemaker and Boxem, 1983) and from 1200 mm at the coast to 700 mm in inland received within an altitude range of 10 to 300 m (Boxem *et al.*, 1987).

Soils: Citrus growing soils are acidic in nature which predominate in most of the areas. The cations such as Ca^{2+} , Mg^{2+}, K^{+}, and Na^{+} are leached and Al^{3+} and H^{+} ions occupy the exchange complex. Soils are deficient in nutrients such as N, P, B, Fe, Zn, Cu, Mn, and Mo (Table 1.202). Soils in west Kenya have been classified as Paleudolls and

Hapludolls (Wielemaker and Boxem, 1983) and Haplustox and Quartzipsamment (Boxem *et al.*, 1987).

Table 1.201. Characteristics of calcareous soils (0-30 cm depth) in the Nubaria area (west Desert)

Properties	Location			
	Mariut	North Tahrir	Gianaclis	Nobaseed
pH	8.4	8.4	8.7	9.7
EC (mmhos cm^{-1})	0.2	0.2	0.4	0.5
$CaCO_3$ (%)	25	26	23	11
Org. matter (%)	1*	1*	0.1*	0.8*
N (mg 100 g^{-1})	30*	30*	22*	2**
P (mg 100 g^{-1})	0.4*	0.4*	0.6*	5*
K (mg 100 g^{-1})	52	43	48	16**
Fe (mg 100 g^{-1})	9*	10*	8*	7*
Mn (mg 100 g^{-1})	14	16	12	3**
Zn (mg 100 g^{-1})	3	3	3	1**
Cu (mg 100 g^{-1})	3	2	2	0.5**

* low ** very low

Source : El-Fouly *et al.* (1984)

1.8.5.5 Citrus in Libya

The history of citrus in Libya begins from the period of Turkish occupation (1835 – 1912) and a few old plantation still remains that are known to have been flourishing at the time of the Italian invasion during 1911.

A great majority of citrus in Libya is situated in the province of Tripolitania, and a portion of the Jefera plain that extends along the coast from Sorman to Gharabulli and southward to Azizia. The old centres of production aside from the suburbs of Tripoli include Suk El-Jumaa, Tagiura and Zavia. Among the more developed districts are Suani, Saiad, Zenzour, Ana Siria, Azzahara, Amiria, Azizia, Ben Gashir, and Sghedeida.

Climate : Feature of Libyan climate is characterised by prevalence of cool on shore winter winds, sufficient to induce winter chilling for regular and adequate bloom, development of satisfactory peel colour with enough acidity, necessary for good flavour

Table 1.202. Chemical characteristics of citrus growing soils of Nairobi, Kenya

Properties	A		B		C		D	
	Surface	Subsurface	Surface	Subsurface	Surface	Subsurface	Surface	Subsurface
PH	5.80	6.20	6.10	6.3.0	5.80	6.3.0	5.60	6.30
Na (me %)	0.17	0.12	0.12	0.13	0.15	0.12	0.12	0.10
K(me %)	0.26	0.17	0.19	0.19	0.17	0.12	0.20	0.20
Ca (me %)	3.90	0.60	2.30	0.60	5.00	1.80	3.40	1.00
Mg (me %)	1.40	0.40	0.80	0.60	1.00	0.50	1.20	0.40
Mn (me %)	0.53	0.70	0.46	0.44	0.46	0.48	0.36	0.41
P (mg kg-1)	13.0	10.0	4.00	13.0	4.00	8.00	4.00	6.00
N (%)	0.06	-	0.04	-	0.06	-	0.05	-
C (%)	0.66	-	0.32	-	0.49	-	0.46	-
Cu (mg kg^{-1})	0.50	0.70	0.40	1.20	0.40	0.50	0.70	0.90
Fe (mg kg^{-1})	3.00	3.80	1.50	3.80	1.50	3.00	3.00	1.50
Zn (mg kg^{-1})	7.60	1.90	7.10	1.50	7.10	1.90	4.30	2.00

Source : Kimani (1984).

of fruits. The early long and very hot growing seasons provide a high total amount of heat sufficient to insure early fruit maturity. The presence of high atmospheric humidity of the maritime zone, is reflected in the desirable features of the fruit shape and texture in terms of smoothness and thinness of rind, and juiciness of flesh. In Tripolitania, the desert ghibli remains usually hot and intense dry, sometimes reduces the crop and causes tree damage (Hodgson, 1966a).

Soils : The majority of citrus growing soils are sandy in texture and possess good depth, the primary requirements of citrus in Jefera plain is met, since the soils are well drained coupled with good premeability. The light textured sandy soils have a coarsening effect on fruit texture through roughness and thickness of rind, lowered juice content, and acidity. Washington navels have proved more productive with better quality on the heavier textured loam soil type than sandy soil (Hodgson, 1966b).

1.8.5.6 Citrus in Morocco

Citrus is grown in localised areas with latitudes higher than 29° (Spina and Giudice, 1975). The major citrus production regions are Souss, Ghargb, Oriental, Tadla, Haouz, and Loukkos covering cities like Agadir, Kenitra, Berkane, Beni Mella, Marrakech, and Larache (Chapot, 1964; Wilbeat, 1964; El-Otmani, 1990; El-Otmani *et al.*, 1990).

Climate : There are four principal climatic zones. The desert zone, including Marrakech and the Souss valley, is hot and dry with rainfall mainly distributed between October and April. Average annual rainfall has been recorded as 250 mm and 225 mm at Marrakech and Valley district, respectively. The corresponding mean maximum temperatures at Marrakech range from 25°C in January to 38°C in July, with corresponding mean minimum temperature are 15°C and 25°C. The coastal citrus zones along the Atlantic ocean, including Rabat, Larache, Kenitra, and Casablanca, have a temperate and humid climate. Average annual rainfall is 500 mm at Rabat and 400 mm at Casablanca, falling mostly between October and April. Mean maximum temperatures at Rabat are 24°C in January and 31°C in July, with corresponding mean minimum temperature of 17°C and 24°C, respectively. Climatic conditions are similar for the east Moroccon citrus district, centered around the city of Berkane on the mediterranean sea. Average annual rainfall at Berkane is 200 mm. Mean temperatures in January and July are roughly comparable to those of Marrakech.

The interior districts of the Rharba and Tadla have rather humid climate, with rainfall as high as 900 mm. Average annual rainfall is 575 mm at Mechra Bel Ksiri in the Rharb and 500 mm at Beni Mellal in the Tadla district. Both districts, while influenced by the Atlantic ocean, are out of the fog belt, and Tadla, which is further inland, has the hotter summers. Mean maximum temperatures at Beni Mellal are 24°C in January and 38°C in July. While, mean minimum temperatures are 14°C in January and 24°C in July. The northern citrus areas of Meknes, Fes, Taza, and Ouergha districts are higher in elevation than other Moroccan citrus regions, that have dry hot summers with temperatures up to 43°C. Average annual rainfall at Meknes is 450 mm, distributed mostly from November to May. Mean maximum temperatures at Meknes range from 24°C in January to 37°C in July. While, mean minimums range from 15° to 22°C, respectively. Reuther (1973b) described climatic variation at citrus growing area of Mechra Bel Ksiri (Table 1.203).

Table 1.203. Climatic variation at citrus growing region of Mechra Bel Ksiri, Morocco

Months	Temperature (°C)		Rainfall (mm)
	Min.	Max.	
Jan.	6.7	18.0	93
Feb.	7.4	19.6	85
Mar.	9.5	22.5	72
Apr.	11.1	24.7	52
May	12.6	27.1	29
June	16.1	31.1	8
July	17.9	34.8	0
Aug.	18.5	34.9	1
Sept.	16.6	32.5	10
Oct.	13.7	28.4	52
Nov.	10.2	22.7	78
Dec.	7.7	19.2	97

Source : Reuther (1973b)

Soils : Soils in citrus growing areas of Morocco are highly calcareous with a pH of about 8.0 Salinity and drought are the two major problems in these areas (El-Otmani, 1996). Evaluation of clementines grown in three serozeam soil types viz., dark serozem, typical serozem, and light serozem which showed that trees in dark serozem produced best growth with highest yield and quality. Trees in calcareous light serozem grew poorly, turned chlorotic, and gave poor yields. In a typical serozem, growth and cropping are intermediate (Penkov, 1979b). Sagee *et al.* (1992; 1993) demonstrated that there is a significant effect of soil texture on the response of citrus to $CaCO_3$. A better plant performance in clay soil (14 per cent sand, 39 per cent silt, and 47 per cent clay) with 33 per cent $CaCO_3$ has been observed than in a silty clay soil (60 per cent sand, 21 per cent clay, 19 per cent silt) containing only 23 per cent $CaCO_3$ (El-Otmani, 1996). Moroccan soils are lighter than those of Algeria, mostly red and brown, light and well-drained loams. Desert districts and northern areas along the water course have sandy loam soils.

1.8.5.7 Citrus in Nigeria

Climate: Ibadan has an annual rainfall of 1100-1200 mm with biomodal character of rainfall distribution (Lal, 1976). The growing season from late March to late July is followed by a dry spell of one month (Ministry for Agriculture, 1959; Crossley, 1966).

Soils : Soils on major land forms in central Nigeria are described as alluvial in nature (Okusami *et al.,* 1987) and the changes following the long term management practices (Aina, 1979). Soils of the coastal plain sands, beach sand, and sandstone are coarse textured with textural class ranging from loamy sand in the surface horizons to sandy clay loam in the subsurface horizons (Table 1.204). Soils derived from shale and alluvial parent materials have a surface sandy clay loam and subsurface clay loam texture. The soils are generally moderately acidic with pH values ranging from 5.0 to 5.9 in the

Table 1.204. Some physical and chemical properties of the soils in Nigeria

Depth (cm)	pH (H_2O)	Org. C (%)	Ex.Al. —cmol(p^+) kg^{-1})—	ECEC	BS (%)	Oxalate extractable Fe_2O_3	Al_2O_3	Particles size Sand (%)	Silt	Clay
Ayadeghe										
0-15	5.2	2.53	2.00	9.26	78	ND	ND	46.4	31.4	22.2
15-40	4.8	1.65	3.40	8.23	57	ND	ND	42.4	35.4	22.2
40-75	5.5	0.77	0.40	1.71	77	ND	ND	30.4	29.4	40.2
75-100	5.8	0.50	0.40	9.88	96	ND	ND	42.4	25.4	34.2
100-120	5.9	0.42	0.40	5.27	92	ND	ND	34.4	23.4	32.2
Idu-Uruan										
0-20	5.9	3.20	1.00	5.96	83	5.14	0.53	89.2	6.0	4.8
20-60	5.9	1.00	1.60	3.20	50	5.49	0.26	71.2	4.0	24.8
60-105	5.7	0.93	1.60	4.08	61	4.83	0.47	73.2	0.0	26.8
105-135	5.5	0.54	1.60	2.69	41	3.38	0.52	73.2	2.0	24.8
135-200	5.5	0.39	1.60	2.71	41	1.32	0.52	73.2	2.0	24.8
Okosi-Udu										
0-13	5.2	4.03	1.60	2.29	30	3.20	0.52	81.2	7.4	11.4
13-35	5.5	2.04	1.80	3.53	49	1.94	1.19	79.2	3.4	17.4
35-61	5.5	1.57	2.20	4.22	48	1.60	0.94	73.2	5.4	21.4
61-94	5.5	1.01	2.00	2.51	20	1.37	1.45	69.2	3.4	27.4
94-180	5.7	0.43	1.40	2.93	52	3.43	1.57	73.2	3.4	23.4
Ikot Afanga										
0-16	5.4	3.95	1.00	2.00	50	1.00	0.57	85.2	5.4	9.4
16-50	5.5	1.85	1.20	2.04	41	1.20	0.89	85.2	3.4	11.4
50-100	5.6	1.43	1.20	2.00	40	1.14	1.45	85.2	3.4	11.4
100-157	5.6	0.85	1.00	2.34	57	2.03	1.30	81.2	1.4	17.4
157-200	5.7	0.50	1.00	3.90	74	1.69	1.10	83.2	1.4	15.4
Edebo										
0-8	5.6	2.96	1.40	1.93	27	1.03	1.25	95.2	0.0	4.8
8-22	5.7	1.09	1.20	2.65	54	0.91	1.81	93.2	1.4	5.4
22-58	5.8	0.86	1.00	2.11	53	1.03	2.02	89.2	1.4	9.4
58-115	5.9	0.54	1.00	2.11	53	0.77	0.94	89.2	0.0	10.8
115-170	5.8	0.31	0.80	1.25	36	0.27	1.15	85.2	3.4	11.4
Nto Ndang										
0-14	5.0	5.11	1.80	4.59	61	0.49	1.30	79.2	7.4	13.4
14-28	5.2	2.92	2.00	3.12	36	0.72	1.40	73.2	7.4	19.4
28-105	5.4	0.73	1.40	1.99	30	0.69	1.57	61.2	1.4	31.4
105-136	5.3	0.73	1.60	4.82	67	0.69	0.77	69.2	3.4	27.4
136-192	5.4	0.35	1.40	3.75	63	0.59	1.03	67.2	1.4	31.4
Obotme										
0-15	5.6	3.24	1.40	14.06	90	1.03	1.04	60.6	15.4	24.0
15-44	5.7	1.08	1.60	18.20	91	1.56	2.07	48.6	7.4	42.0
44-80	5.7	0.77	3.00	29.93	90	1.71	2.29	40.6	19.4	40.0

ECEC, Effective cation exchange capacity; BS, Base saturation ; ND, not determined
Source : Ibia and Udo (1993)s

surface. Organic carbon contents which decreased with depth, ranges from 0.31 to 5.11 per cent in the surface.

Asadu (1990) described soil at Nsukka of Enugu state with kaolinite as predominant clay mineral. The soil ranged 4.5-5.0 and cation exchange capacity 45-50 $mmol_c\,kg^{-1}$ of which exchangeable K amounted to 0.9 mmol kg^{-1} with clay content varying between 12 per cent to 40 per cent (Table 1.205). The effective cation exchange capacity (ECEC) is low in the soils formed from coastal plain sands, beach sand, and sandstone materials with values ranging from 1.2 to 0.59 cmol (p^+) kg^{-1} of soil. The soils formed from river alluvium and shale parent materials have values ranging from 1.7 to 29.2 cmol(p^+) kg^{-1}. Base saturation is correspondingly low for most of the soils at N to Ndang giving some inconsistent trends within the profile. The content of oxalate extractable oxides varied considerably among the soils with extractable Fe_2O_3 ranging from 0.27 to 5.19 per cent and extractable Al_2O_3 from 0.26 to 2.26 per cent. Lal (1976) described a soil type at Ibadan having pH 5.9-6.8 , clay increasing from 21.3 to 65.4 per cent, and sand decreasing from 66.9 to 20.2 per cent down the profile with 97 – 98 per cent base saturation (Table 1.206).

Table 1.205. Some soil characteristics at Nsukka of Enugu, Nigeria

Depth (cm)	Particle size distribution (%)			Organic matter (%)
	Sand	Silt	Clay	
0-30	75-80	4-7	12-19	1.5-3.0
30-60	70-75	25-30	25-30	0.8-1.2
60-90	30-70	0-4	30-40	0.5-1.0

Source : Asadu (1990)

Major soils of southwestern Nigeria belonging to soil orders viz., Alfisols, Oxisols, and Ultisols have been observed to have pH 4.4-6.1, organic C 7.3-28.4 g kg^{-1}, sand 54.8-87.5 per cent, silt 5.6-14.4 per cent, and clay 5.7-30.8 per cent (Adetunji, 1994). These soils according to Adetunji (1997) are further classified as Oxic Paleustalf (pH 5.0-6.4, organic C 11.8-15.6 g kg^{-1}, clay 80.0-225.0 g kg^{-1}, ECEC 2.2-6.2 cmol(p^+) kg^{-1}, free Fe_2O_3 6.0-39.0 g kg^{-1}, free Al_2O_3 8.0-28.0 g kg^{-1}, exchangeable Fe 10.1-24.1 mg kg^{-1}, and exchangeable Al 5.9-15.4 mg kg^{-1}), Oxic Tropudalf (pH 5.8-6.0, organic C 10.2-15.5 g kg^{-1}, clay 88.0-164.0 g kg^{-1}, ECEC 1.8-5.6 cmol(p^+) kg^{-1}, free Fe_2O_3 11.0-26.0 g kg^{-1},free Al_2O_3 11.0-20.0 g kg^{-1}, exchangeable Fe 13.1-20.1 mg kg^{-1}, and exchangeable Al 5.6-12.8 mg kg^{-1}), Typic Tropudalf (pH 6.3, organic C 9.8 g kg^{-1}, clay 88.0 g kg^{-1}, ECEC 1.1 cmol(p^+) kg^{-1}, free Fe_2O_3 7.0 g kg^{-1}, free Al_2O_3 9.0 g kg^{-1}, exchangeable Fe

Table 1.206. Physical and chemical characteristics of a soil type at Ibadan, Nigeria

Depth (cm)	Physical characteristics				Chemical characteristics			
	Gravel (%)	Particle size analysis (%)			pH	CEC	Organic carbon	Base saturation
		Sand	Silt	Clay			(%)	(%)
0-5	25	66.9	11.8	21.3	6.5	5.45	1.55	97
5-15	20	57.9	16.8	25.3	6.4	6.99	1.50	98
15-45	50	48.9	13.8	37.3	6.8	5.26	0.76	97
45-65	40	29.9	13.8	56.3	6.4	4.33	0.28	96
65-95	40	29.6	6.1	62.3	6.0	4.06	4.26	96
95-110	33	20.2	14.4	65.4	5.9	4.49	0.18	96

* CEC is in cmol (p^+) kg^{-1}

Source : Lal (1976)

15.4 mg kg^{-1}, and exchangeable Al 5.0 mg kg^{-1}), Arenic Paleustalf (pH 5.5, organic C 13.6 g kg^{-1},clay 180.8 g kg^{-1}, ECEC 4.9 cmol(p^+) kg^{-1}, free Fe_2O_3 30 g kg^{-1}, free Al_2O_3 25.0 g kg^{-1}, exchangeable Fe 18.0 mg kg^{-1}, and exchangeable Al 11.2 mg kg^{-1}), and Aquic Paleudalf (pH 4.8, organic C 14.5 g kg^{-1}, clay 196.0 g kg^{-1}, ECEC 4.2 cmol (p^+) kg^{-1}, free Fe_2O_3 27.0 g kg^{-1}, free Al_2O_3 12.0 g kg^{-1}, exchangeable Fe 22.6 mg kg^{-1}, and exchangeable Al 14.0 mg kg^{-1}).

Soils of Ultisol order are characterised as Oxic Paleudult (pH 5.0, organic C 15.0 g kg^{-1}, clay 285.0 g kg^{-1}, ECEC 6.5 cmol(p^+) kg^{-1}, free Fe_2O_3 35.0 g kg^{-1}, free Al_2O_3 26.0 g kg^{-1}, exchangeable Fe 274 mg kg^{-1}, and exchangeable Al 16.0 mg kg^{-1}), Oxic Paleustult (pH 4.8, organic C 18.6 g kg^{-1}, clay 430 g kg^{-1}, ECEC 7.2 cmol(p^+) kg^{-1}, free Fe_2O_3 342.0 g kg^{-1}, free Al_2O_3 38.8 g kg^{-1}, exchangeable Fe 38.8 mg kg^{-1}, and exchangeable Al 34.5 mg kg^{-1}), Orthoxic Tropudult (pH 4.9, organic C 21.5 g kg^{-1}, clay 405.0 g kg^{-1}, ECEC 6.9 cmol(p^+) kg^{-1}, free Fe_2O_3 36.0 g kg^{-1}, free Al_2O_3 26.0 g kg^{-1}, exchangeable Fe 30.7 mg kg^{-1}, and exchangeable Al 20.2 mg kg^{-1}). While, the soils of Entisol are classified as Typic Tropopsamment (pH 6.1, organic C 10.6 g kg^{-1}, clay 98.0 g kg^{-1}, ECEC 4.0 cmol(p^+) kg^{-1}, free Fe_2O_3 18.0 g kg^{-1}, free Al_2O_3 6.0 g kg^{-1}, exchangeable Fe 128 mg kg^{-1}, and exchangeable Al 5.6 mg kg^{-1}), Typic Ustipsamment (pH 6.0, organic C 11.1 g kg^{-1}, clay 105.0 g kg^{-1}, ECEC 5.0 cmol(p^+) kg^{-1}, free Fe_2O_3 15.0 g kg^{-1}, free Al_2O_3 9.0 g kg^{-1}, exchangeable Fe 9.2 mg kg^{-1}, and exchangeable Al 9.0 mg kg^{-1}), Lithic Ustorthent (pH 5.8, organic C 12.0 g kg^{-1}, clay 117.0 g kg^{-1}, ECEC 3.1 cmol(p^+) kg^{-1}, free Fe_2O_3 8.0 g kg^{-1}, free Al_2O_3 4.0 g kg^{-1}, exchangeable Fe 114.0 mg kg^{-1}, and exchangeable Al 6.4 mg kg^{-1}), and Lithic Troporthent (pH 6.0, organic C 17.0 g kg^{-1},clay 165.0 g kg^{-1}, ECEC 4.1 cmol(p^+) kg^{-1},free Fe_2O_3 24.0 g kg^{-1}, free Al_2O_3 18.0 g kg^{-1}, exchangeable Fe 16.0 mg kg^{-1}, and exchangeable Al 9.2 mg kg^{-1}).

1.8.5.8 Citrus in Sierra Leone

Climate : The climate in the citrus areas is hot tropical with distinct rainy and dry seasons occuring from May through October and November through April, respectively. The rainy season starts in April / May with a peak in August-September in most years and ceases towards December. Rainfall ranges from 3100 to 5100 mm along the Coast and decreases towards the north and east with 2100 to 2500 mm or less. The warm days are observed during the months of January, February, March and April with mean maximum monthly temperatures of 30° to 35°C along the Coast and with mean temperature of 27° to 28°C in July and August. In the interior, the coolest nights are during December, January, and February with mean monthly minimum temperatures of 14°-20°C (Hyward, 1968).

Soils : Majority of the soils are derived from old rocks of igenous and metamorphic types mainly - granite, acid gneiss and numerous schists and sedimentary rocks mainly sandstones, siltstones, mud-stones, alluvium and colluvium. Citrus is grown on two types of upland soils. Haque and Godfrey (1976) observed a large scale deficiency of N,P, K, Ca, Mg, and Zn, besides Al toxicity in varying proprotions in various agroecological zones of Sierra Leone.

Ferrallitic Soils : These are formed on old upland erosion surfaces and steep hills. Seventy to eighty percent of these soils are underlain by hard plinthite. These soils are highly weathered, low in water holding capacity and available plant nutrients.

River Terrace Soils : These are found along river banks in the interior. They are formed to a large extent from gravel-free alluvium and colluvium parent material. Some of these soils become waterlogged during the rainy season. They have more water holding capacity and available plant nutrients as compared with ferrallitic soils (Dijkerman, 1969).

Table 1.207. Chemical properties of various ferrallitic soil types of Sierra Leone

Location	pH (Soil paste)	P (ppm) (Bray I)	Organic C (%)	CEC (me 100 g-1)
Lumley	5.2	16.5	0.58	4.6
Newton	5.3	11.2	0.65	4.4
Kaprion	5.1	3.5	0.65	3.9
Romaka	5.6	6.3	1.42	7.2
Haffia	6.3	19.6	1.04	9.3
Kabala	5.3	20.3	1.04	8.6
Kanikeh	6.3	19.9	0.42	4.0
Kalangba	5.7-6.0	2.1-20.3	1.85-2.73	9.5-11.7
Yonibana	5.2	30.8	1.00	10.1
Njala	5.0-5.3	7.0-8.8	1.66-2.19	12.1-16.5
Bo-Mattar Estate	5.4	37.8	1.54	11.5
Kenema	5.7	23.1	0.96	7.8
Gbandaru-Shimbeck	5.3	3.5	1.12	9.5

Source : Haque (1974); Haque and Godfrey (1976)

1.8.5.9 Citrus in South Africa

South African citrus industry was founded in 1654 in the vicinity of Cape Town (Grobler and Bredell,1981) when citrus trees were planted in Governor's garden from Island of St.Helena. The citrus industry of South Africa is classified into two distinct categories, namely, small fruit areas, comprising the largest part of the citrus producing area and large fruit areas (Plessis, 1980; Plessis and Koen, 1996a). The differences between these two areas are mainly climatic, with the large fruit areas representing the hotter regions eg. Malelane, Komatipoort, Zebediela, Hoedspruit, Nkwaleni, Tshipise, and Citrusdal. The small fruit areas representing cooler areas are Letsitele, Nelspruit, Zebediela, Sunday river valley, Gamtoos, and several others. Citrus is the principal crop in the large scale farming area of Beit Bridge. It has been cultivated under irrigation over 30 years with water extracted from sand to close to the Mzingwane and Limpopo rivers containing high proportion of sodium magnesium bicarbonate and choride in South Africa (Moore, 1962).

Citrus is grown in South Africa under a wide range of soil and climatic conditions. Soils vary from very sandy, acid types of low fertility in the Citrusdal area of the western Cape province to heavier, alkaline and often relatively saline soils of high fertility in the eastern Cape province. Preference is usually given to soils of neutral to slightly acidic reaction of medium texture, good depth, and drainage. These inherent features are more important and can be readily supplemented by means of fertilization.

Climate : The essential components of climate found to be important in determining fruit size include maximum and minimum air temperature, sun shine hours, humidity, and wind (Hilgeman *et al.*, 1959; Da Ponte, 1964; Reuther, 1973a; Fucik and Norwine, 1979; Plessis, 1980). Bain (1958) stated that a cold winter rest period is required for greater flowering and fruit set. The mean of the three coldest months should be 13°C or lower in order to induce dormancy. This permits the build up of nutrient reserves and carbohydrates, for the subsequent flowering and fruit set (Anonymous, 1988b). Averge daily air temperature for Citrusdal during June, July, and August is approximately 13°C. While, it is approximately 15° C over the same time period at Nelspruit. The same climatic factors of dormancy and daily average air temperatures around 13°C during the winter up until September, appear to be responsible for causing late flowering at Citrusdal. At Nelspruit, the absence of true dormancy together with higher daily average air temperatures around 15°C during the winter months appear sufficient for early flowering to occur. Survey for fruit quality of 5 citrus species from across 13 districts of South Africa with a view to define the best climatic conditions revealed high proportion of good quality Washington navel and poor quality grapefruit (Bozalek, 1974).

Citrus growing regions of South Africa are classified into seven major provinces namely, western Cape, eastern Cape, Natal, eastern Transvaal, northern Transvaal, central Transvaal, and western Transvaal (Table 1.208), the climatic features of them have been described by Grobler and Bredell (1981). Newman (1969) suggested that the net energy balance is of importance in determining crop growth and yield. The energy balance of Citrusdal and Nelspruit warrant further study due to differences in altitude and topography of the two areas. Comparison of the energy richness between these two areas may reveal further information about causes of the difference in fruit size. Both, the average maximum and minimum humidity for Citrusdal are lower over the critical fruit growth period It is evident that climatic conditions during the critical growth period are more favourable at Citrusdal than Nelspruit (Plessis, 1996). The diversity in climatic conditions is even greater, which varies from semitropical conditions of summer rainfall and high temperature in the typical Lowveld regions of the

Table 1.208. Climatological features of citrus growing areas of South Africa

Area	Altitude (m)	Rainfall (mm)	Max. temp. (°C)	Min. temp. (°C)
Western Cape	250	450	33	7
Eastern Cape	100-500	600	30	6-7
Natal	100-400	850	30	7
Eastern Transvaal	5-600	1,000	29	6
Northern Transvaal	550	275	33	4
Central Transvaal	1000	500	30	5
Western Transvaal	1000	1,200	26	5

Source : Grobler and Bredell (1981)

eastern Transvaal; subtropical, semiarid conditions in the eastern Cape province to typical mediterranean climate in parts of the western Cape province. Hailstorms are quite common in all summer rainfall areas and cause considerable damage to trees in certain seasons. Wind is also a factor of importance in most areas. More attention is required towards the planting of suitable windbreaks. Frost and cold injury occur occasionally, especially in the eastern Cape midlands, but is usually of minor importance. Broembsen (1988) placed commercial citrus under four climatic groups, namely, warm/hot, intermediate, cool, and cold for the cultivation of grapefruit, orange, lemon, and other citrus cultivars, respectively.

The interior of South Africa is an elevated plateau. Because of frost, citrus culture is precarious below approximately 26° south. The citrus area of the Transvaal, lie in the interior between latitudes 23° and 26° south in the region where the temperature becomes sufficiently high to permit successful citrus growing at altitudes up to about 800-900 m. At higher altitudes, however, there is considerable danger of frost injury. Rainfall in most South African districts occurs principally during the summer months (October through April), that ranges from about 250 mm in the Citrusdal area to 1250 mm or more in part of the northern Transvaal. Tzaneen- Letaba district has an average rainfall of 500 mm year^{-1}. The monthly mean maximum and minimum temperature at Nelspruit in January is 32°C and 25°C, respectively. In July, mean maximum and minimum are 28°C and 17°C, respectively. Nelspruit is fairly representative of the eastern and northern Transvaal climate where most of the citrus is grown. The Citrusdal area has one of the lowest rainfalls of the South African citrus producing districts. The average rainfall at Clanwilliams is only 8 mm annually. The rainfall pattern is the opposite of that in the Transvaal, where winters are dry with wet summers. The highest precipitation in the Citrusdal area occurs in winter from May through August. The climatic norms as developed for navel and valencia oranges grown in South Africa using homogeneous climatic

Table 1.209. Broad climatic norms as per cultivar group for citrus production in southern Africa

Season	Ave. min. temp (°C)	Ave. max. temp (°C)
Navel oranges a. Inland production areas with effective heat units 1600-2950		
Spring (Aug. – Nov.)	9.5-14.5	25.5-28.5
Summer (Dec. – Feb.)	15.0-19.5	27.5-31.5
Autumn (Mar. – May.)	8.5-15.0	25.0-28.0
Winter (Jun. - Jul.)	2.0-8.0	20.5-24.5
Navel oranges b. Coastal production areas with effective heat units of 1600-2200		
Spring (Aug. – Nov.)	8.0-12.0	19.0-29.0
Summer (Dec. – Feb,)	14.5-17.0	24.5-32.0
Autumn (Mar. – May)	8.0-13.5	23.0-27.0
Winter (Jun.-Jul.)	2.5-9.0	14.0-22.5
Valencia oranges a. Inland production areas with effective heat units of 1300-3500		
Spring (Aug. – Nov.)	10.5-16.5	25.0-30.0
Summer (Dec. – Feb.)	16.0-21.5	27.5-33.0
Autumn (Mar. – May)	10.5-17.5	24.5-30.0
Winter (Jun. - Jul.)	2.0-11.5	19.5-26.0
Valencia oranges b. Coastal production areas with effective heat units of 1300-3500		
Spring (Aug. – Nov.)	10.0-13.0	23.5-26.0
Summer (Dec. – Feb.)	15.0-17.0	28.0-32.0
Autumn (Mar. – May)	10.5-13.5	23.0-26.0
Winter (Jun. - Jul.)	6.0-9.5	18.5-22.0

Source : Barry *et al.* (1996)

Table 1.210. Predomiant soils types in various citrus growing areas of South Africa

Major citrus growing areas	Description of soil types
Western Cape	Mostly deep, grey and dark brown, alluvial sands; frequently stratified and underlain by grey clay loam. Gravelly red and yellow sandy clay loam also occur extensively
Eastern Cape	Brown, sometimes structured, sandy clay loam and clay loam soils of alluvial origin predominates. Salinity is a problem and pH values up to 8.5 are not uncommon.
Natal	Brown, reddish and yellowish sandy loam to clay loam with frequently acidic pH
Eastern Transvaal	
i. Western region	Gravelly, sandy greyish brown soils derived from granite occurs dominantly, interspersed with more fertile clayey soils derived from dolerite
ii. Eastern region	Extensive areas of reddish brown, sandy clay, with common occurrence of gravel
Northern Transvaal	Brown, grey and red, ferruginous sandy loam to sandy clay and frequently acidic soil pH
Central Transvaal	Brown to reddish brown sandy loam to sandy clay
Western Transvaal	Dark and red ferruginous sandy loam to sandy clay

Source : Grobler and Bredell (1981)

zones (Dent *et al.*, 1988) on a geographical information system (GIS) have been developed (Table 1.209), which could generate precise data about the suitability of potential areas for future citrus expansion.

Soils : Major soil types (Table 1.210) in citrus growing provinces have been described by Grobler and Bredell (1981). The soils of South African citrus areas range from light sandy loam to heavy clay. The rich and most productive soils are the flat alluvial soils, along the rivers, which are deep and contain an abundance of organic matter. Such are the valleys of the Kat, Mooi, Fish, Gamtoos, and Sundays rivers. In general, the soils are derived from disintegrated granite and closely resemble those of California. In some sections, difficulty is experienced with hardpans and salinity. Soils of the area are light and sandy, having an excellent drainage, but require careful irrigation and a heavy fertilization. De Villiers (1969) observed the soils at Zebediela estate of Transvaal province as sandy loam, very deep, and well drained with pH 5.0 in 1 : 2 soil – water suspension, exchangeable bases 0.9 me Ca, 0.9 me Mg, 0.2 meq K, 0.05 me Na, and 0.05 me 100 g^{-1} Mn at an altitude of 1100 m under warm temperate to subtropical climate. Citrus growing soils at Sovenga are established on low base content loamy sand with pH 6.0 having average depth of 105 cm (Du Sautoy,1992). At Nelspruit, soils are clay loam with pH 6.2, low P level (4.0 mg P kg^{-1} soil of resin extractable P), and exchangeable K (72 mg kg^{-1}) according to Plessis and Koen (1972). Plessis (1977) observed that various soil properties viz., pH varying from 5.6 to 6.9 (mean 6.0), CEC 1.0 to 25.3 me 100 g^{-1} (mean 6.5 me 100 g^{-1}), Mg saturation 14 to 44 per cent (mean 27 per cent), exchangeable Ca 1.0 to 22.2 me 100 g^{-1} (mean 5.0 me 100 g^{-1}), and Mg from

0.27 to 900 me 100 g^{-1} (mean 2.04 me 100 g^{-1}) in citrus soils of Nelspruit. At Citrusdal in the western Cape, soil is very sandy (95 per cent sand) with pH 6.0, 5.5 in the subsoil, and low in K, Ca, and Mg status (Plessis *et al.*, 1992).

Table 1.211. Soil textural composition at Nelspruit

	Topsoil (0-300 mm)	Subsoil (300 – 600 mm)
Clay (%)	15	30
Silt (%)	6	10
Sand (%)	79	60

Source: Mostert and Hoffman(1996)

The soils at Nelspruit have been observed granite derived, well drained, and sandy loam (Table 1.211). The soils of Palmer navel orange orchards established on Florida rough lemon have been as Calcic Xerosol/Combisol according to FAO classification with clay, silt, and fine sand contents of 15-20 per cent, 12-15 per cent, and 15-20 per cent, respectively, in the A horizon. The B horizon exhibited clay, silt, and fine sand contents of 20-25 per cent, 15-20 per cent, and 25-75 per cent, respectively (Abercrombie and Hoffman, 1996). Soils in Sundays river valley vary texturally from sandy to sandy loam (10-12 per cent clay) stratified alluvium along the river to more structured (30-40 per cent clay) soils on the upper terrraces. Soil pH is normally 7.5-8.5 with irrigation water containing 600-700 mg l^{-1} total dissolved salts and 150-180 mg l^{-1} chlorides (Miller *et al.*, 1996).

Most of the citrus orchards in South Africa are irrigated, usually by means of the full basin surface system, or modifications thereof. Various sprinkler systems have come into limited use in the western Cape province and elsewhere, especially on very sandy soils or on fairly steep slopes. Sources of irrigation water consist chiefly of storage dams, streams, rivers as well as underground supplies. In many parts of the eastern Cape province and in some areas of the northern Transvaal, high salinity due to NaCl and other salts frequently accompained by excess boron, cause considerable damage to citrus orchards, especially during times of drought. Nutrient deficiencies and fertilizer programmes are naturally dependent largely on soil types, climatic conditions, as well as age of trees (Oberholzer *et al.*, 1958). Nitrogen is the most common limiting factor and must usually be applied at or soon after planting time. Virgin alluvial soils of high fertility as in the eastern Cape province have, however, support good tree growth and profitable crops for as long as 20 years without additional nitrogen.

Zinc deficiency is wide spread in most South African citrus orchards, besides Mn and P (Plessis and Prando, 1975). It is normally corrected by foliar spray of zinc oxide. Other micronutrients which periodically need supplementing in certain areas are Cu, Mn, and B. These nutrients often occur at near toxic levels during dry periods, in parts of the northern Transvaal and eastern province of coastal citrus regions. With the exception of the eastern Cape province, P deficiency occurs in most other citrus regions and exerts important influences on fruit quality (Oberholzer *et al.*,1958). In recent years, K deficiency has assumed increasing importance in most citrus orchards, especially where kraal manure is no longer applied. Some important soil properties from citrus growing areas of Nelspruit (Table 1.212) and northeast Cape province are further described (Table 1.213).

Table 1.212. Some citrus soil properties orchards at Nelspreut in South Africa

Elements	Mean	Range
Soil pH	6.0	5.6-6.9
CEC (me 100 g^{-1})	6.5	1.0-25.3
Mg (sat.%)	27	14-44
Ca (me 100 g^{-1})	5.0	1.0-22.2
Mg (me 100 g^{-1})	2.04	0.27-9.00
Ca/Mg	3.35	1.58-7.30

Exchangeable Mg (me 100 g^{-1}) as a % of the CEC

Source : Plessis (1977)

Singer *et al.* (1995) described that in Cape province, most of the sepiolite or polygorskite containing soils that belong to the Hutton, a few to the Fernwood, Glenrosa, and Clovelly soil forms in the South African Classification System, at a level corresponding approximately to the Orthid and Ochrept suborders in the USA Soil Taxonomy or to the Calcic Yermosols and Calcic Xerosols subclasses in the FAO classification system. Commonly, the soils have a medium to coarse sandy texture in the upper horizon, passing in a clear, smooth transition to a loamy, medium to coarse, sandy texture in the lower horizon. In some soils, the texture is medium sandy throughout the profile. The lower horizons are weakly structured, single grained too, but in other cases, have developed a weak angular blocky or a massive hard (cemented) structure. Reflecting their texture, water retention of the soils is low. While, hydraulic conductivity is high. Sepiolite containing soils have slightly higher electrolyte concentration than polygorskite containing soils. Relatively higher Ca^{2+} and SO_4^{2-} concentrations in the saline soils indicate the presence of gypsum. The major salt in the saline soils, however, is NaCl. The exchange capacity of the soils is low (< 10 cmol kg^{-1}) and the exchange complex is dominated by Ca^{2+} and Mg^{2+}. The ESP in some soils reached very high values and commonly increased with soil depth. The exchangeable Ca:Mg ratio is commonly higher than 4. All the soils are poor in organic carbon (<0.5 per cent). The CBD extractable Fe is between 0.5 and 1.0 per cent in the soils containing more clay and < 0.5 per cent in the clay poor soils (Table 1.213).

Magnesium deficiency is fairly wide spread, especially on acid sandy soils such as those occuring in western Cape province and eastern Transvaal (Oberholzer *et al.,*1958). This deficiency is usually aggravated by applications of kraal manure containing high potassium values (Oberholzer *et al.,*1958). Sulfur deficiency as well as toxicity in South Africa has been reported by Stanton and Basson (1970) and accordingly grouped into 5 classes. Soil applications of dolomite, magnesite, and magnesium sulphate are normally effective. Although, responses are generally somewhat slow. Iron chlorosis is fairly common on calcareous and alkaline soils, especially in the eastern Cape province. However, it is seldom serious enough to warrant any treatment.

Table 1.213. Some physical and chemical characteristics of the palygorskite and sepiolite containing soils of Namaqualand district of the northwest Cape province

Depth (cm)	Horizon	Particle size (%)			O.C. (%)	Fe (%)	CEC (cmol kg^{-1})	$CaCO_3$ (%)	pH (H_2O)	EC (mS cm^{-1})	ESP
		Clay	Silt	Sand							
					Sepioite containg soil profiles						
0-5	A	5	8	86	0.4	0.5	8.6	4.35	8.1	0.43	5
5-30	B2ca	6	12	81	0.5	0.5	12.8	10.15	8.0	0.77	9
Module centre		-	-	-	-	-	-	86.7	-	-	-
Module white coating		-	-	-	-	-	-	93.4	-	-	-
0-10	A	3	0	95	0.3	0.1	5.9	1.75	8.0	0.40	3
10-55	B21	4	0	95	0.2	0.0	7.2	5.4	7.9	0.57	3
55-70	C1	3	2	94	0.1	0.1	8.1	6.5	7.8	0.52	5
70-80	C2	2	0	96	0.1	0.5	5.5	1.0	7.9	0.61	9
0-25	A	2	3	86	0.1	0	3.2	1.15	7.7	0.68	3
25-140	C1	3	3	95	0.1	0.1	8.7	0.4	7.4	5.53	25
140-160	C2	7	14	80	0.1	0	25.6	7.35	7.4	13.83	42
0-10	A	6	3	91	0.3	0.4	4.7	0.4	7.6	0.16	2
10-50	B2	5	4	91	0.1	0.2	3.8	0.25	7.5	2.10	37
50-70	C1	9	9	81	0.1	0.2	21.9	1.25	7.4	15.50	49
70-80	C2	7	8	81	0.1	0.3	14.2	5.05	7.6	6.63	22
					Palygorskite containing soil profiles						
0-35	AB	18	2	79	0.2	0.7	5	0.39	7.3	0.33	4
35-60	Cca	15	6	74	0.1	1.0	7.3	20.09	7.4	1.02	10
0-30	A	5	5	91	0.1	0.4	6.2	0.96	8.2	0.40	10
30-70	B21	5	5	89	0.1	0.4	9.7	0.93	7.7	4.93	31
70-100	B22	10	7	81	0.1	0.4	12	2.80	7.2	11.71	32
0-3	A	8	1	93	0.1	0.4	4.5	0	-	-	2
3-30	B21	7	1	90	0.1	0.8	4.5	0	7.8	0.14	2
30-70	B2ca	8	1	89	0.1	0.6	9.0	0.54	7.7	0.52	1
70-100	B3ca	10	1	84	0.1	0.5	9.1	1.28	7.8	0.21	2
0-15	A	8	13	78	0.1	0.5	8.4	2.74	8.0	0.21	1
15-30	B2	9	11	79	0.1	0.5	7.8	3.46	7.9	0.23	2
0-5	AB	12	8	79	0.2	0.8	14.7	2.9	8.1	0.21	3
5.50	Cca	22	4	69	0.2	0.8	15.5	12.63	7.9	0.76	6

Source : Singer *et al.* (1995)

1.8.5.10 Citrus in Swaziland

Commercial citrus cultivation in Swaziland started since 1954 and majority of citrus orchards are concentrated near Tambuti, Ngonini, and Big Bend. Rainfall is about 625 mm $year^{-1}$ in the low Veld and about 950 mm in the middle Veld, falling largely from October through April in both the regions. In the low Veld, where most orchards are located, the monthly mean maximum and minimum temperatures are 34°C and 26°C, respectively, in January. The corresponding July temperatures are 29°C and 18°C. The main temperature ranges for growth are : mimimum 12.5-13°C; optimum 23-34°C; and maximum (limiting growth) 37-39°C. Dodson (1964) described the citrus industry of Swaziland including soil and climate features.

1.8.5.11 Citrus in Tanzania

Climate: The climate of Kodoma region of Tanzania ranges from semiarid in the southwest, with about 600 mm of annual rainfall to subhumid in the northeast having annual rainfall of 900 mm (Payton and Shishira, 1994).

Soils: The soils of Haubi basin of Kodoma region have been classified into quartz gravels and Lithosols (Lithic Troporthents) of the rocky hillslopes; quartz gravel, Regosols (Typic Troporthents); and eroded Chromic Luvisols/Haplic Lixisols (Kandic Haplustalfs) of the upper pediment slopes; eroded deep ferric Lixisols (Kandic Palustalfs and Kanhaplic Rhodustalfs) of the middle to lower pediment slopes; Albic Arenosols with Petroplinthite (Plinthite Quartzipsamments) of the gentle footslopes; Gley soils (Tropaquents) of lower footslope; immature sandy Arenosols (Typic Tropopsamments) of recent sand fans; and buried Vertisols (Payton and Shishira 1994). Rocky hill slopes are dominated by angular quartz gravels upto 35 cm thick overlaying quartzo-feldspathic gneiss interspread with common hard rock outcrops (FAO, 1974b). On the upper pediment upto 30 cm of angular quartz gravel overlies upto 150 cm of light brown, grity sandy loam or sandy clay loam saprolite developed from *in situ* weathering loam of quartzo-feldspathic biotite gneiss (Table 1.214), having large proportion of low activity kaolinite. On very deeply gullied pediment midslopes to the east and northeast of Haubi, the parent rock is often dominated by biotite, schists, and amphibolites. In these cases, red clayey saprolite from which ferric Lixisols have developed, which may extend to the gully floor at as much as 20 m depth.

1.8.5.12 Citrus in Tunisia

The primary citrus producing areas are Cape Bon peninsula near Soliman, Menzel Bou Zelta, Grombalia, and Beni Khaled. Other major citrus plantings are on the southern side of Cape Bon near Hammamet, Nabeul, Beni Kriar, and adjacent to the city of Tunis.

Climate : The Tunisian climate is arid with low humidity and wide extremes in temperatures. Maximum temperature of 45°C has been recorded at Tunis. Rainfall occurs in winter from September to March and averages 200 to 250 mm year^{-1}. Climatic hazards include drought, heat, desert wind storms, and frost. In the cool winters, snow has been reported in Tunisia.

Soils : Tunisian soils are light textured. Salinity is a continuing problem. Some citrus is planted on fine sand. In such orchards, cover crops are grown to prevent wind erosion.

Table 1.214. Physicochemical properties of soil types of east Haubi region of Tanzania

Horizon	Depth (cm)	pH Water	pH KCl	Organic C (%)	Particle size distribution (%) Sand	Silt	Clay	Exch cations	ECEC
								—[cmol (p^+) kg^{-1}] —	
Truncated Saprolite : Rocky Hillslope									
C (Saprolite)	30-150	5.2	4.9	0.18	69	13	18	3.8	3.8
R (Weathered)	150-350	6.0	5.3	0.20	83	9	8	2.7	2.7
Eroded Haplic Lixisol : Upper Pediment									
Bt1	0-50	5.8	5.0	0.06	50	6	44	7.2	7.2
Bt2	50-154	5.2	5.0	0.05	50	8	43	5.1	5.1
C (Saprolite)	150-200	5.6	5.1	0.01	57	9	34	5.4	5.4
R (Weathered)	540-550	5.3	4.8	0.00	80	80	13	2.9	2.9
Eroded Ferric Lixisol : Pediment Midslope									
Bt1	0-120	5.2	4.7	0.19	37	11	52	6.2	6.5
Bt2	120-320	5.2	4.7	0.13	38	10	52	7.8	8.2
C (Saprolite)	320-500	5.1	4.9	0.10	38	11	51	8.2	8.4
C (Saprolite)	700-750	6.3	5.6	0.00	60	6	34	8.4	8.8
Ferric Lixisol : Lower Pediment									
Ap	0-30	5.9	5.1	0.28	36	13	51	12.9	13.2
Albic Arenosol : Footslope									
Ap	0-22	6.6	6.2	0.50	86	8	6	5.0	5.2
E	22-110	6.1	5.3	0.40	75	22	3	2.4	2.6
Bms(Ironst)	110-190	5.6	4.9	0.10	80	9	11	4.5	4.7
BCg	190-350	5.9	4.9	0.10	71	7	22	6.6	6.8
Haplic Arenosol : Stabilised Toeslope Sand Fan									
Ap	0-25	6.3	5.3	0.80	90	5	5	2.2	2.6
C	25-35	4.8	4.3	0.20	83	5	12	3.5	3.8
C	70-100	5.8	5.3	0.00	94	1	5	1.5	1.8
B2Bw	240-350	6.4	5.6	0.20	6	16	78	32.0	32.0
Reddish Raw Sand : Toeslope Sand Fan									
AC	0-25	6.6	5.5	0.20	91	1	8	1.9	1.9

ECEC stands for effective cation exchange capacity
Source: Payton and Shishira (1994)

1.8.5.13 Citrus in Zimbabwe

Citrus growing in Zimbabwe dates back to 1914 when commercial citrus orchards were established in Mazoe valley. Later, plantations were undertaken near Umtali and Sinoia, and recently in Hippo valley. Sweet oranges are grown at an elevations of 120-180 m above mean sea level (Burke, 1958).

Climate : The two major citrus areas are at the Mazoe citrus estate and in Hippo valley. The Mazoe citrus estate located at an elevation of about 1000 m has an annual rainfall of 925 mm. The monthly mean maximum and minimum temperatures are 31°C and 23°C in January. Comparable mean temperatures for July are 28°C and 15°C, respectively. Rainfall in Hippo valley ranges from 500 to 625 mm year^{-1} although, highly erratic and mostly distributed from November through March. In the Hippo valley area, mean maximum temperatures are 34°C in January and 30°C in July. While, mean minimum temperatures are 26°C and 16°C, respectively (Reuther *et al.*, 1967). According to Oberholzer (1969), large part of southern Rhodesia have a semitropical climate. Reuther (1973b) described variation in mean monthly maximum and minimum temperature at citrus growing area of Mazoe as 23.2°-30.5°C and 4.3°-16.8°C, respectively, with annual rainfall of 919 mm (Table 1.215). Hussein and Adey (1994) described the climatic data of Chisumbanje and Tshotsholo having 580-620 mm as mean annual rainfall and 20.6°-22.0°C mean annual temperature with mean maximum temperature between 31.2° and 31.5°C.

Table 1.215. Climatic variation at citrus growing region of Mazoe, Zimbabwe

Months	Temperature (°C)		Rain-fall (mm)
	Min.	Max.	
Jan.	16.8	27.4	230
Feb.	16.5	27.4	202
Mar.	15.6	27.5	128
Apr.	12.5	27.3	28
May	7.9	25.6	7
June	5.1	23.2	3
July	4.3	23.4	< 1
Aug.	5.8	25.2	4
Sept.	9.4	27.5	3
Oct.	13.4	30.5	26
Nov.	15.8	29.2	98
Dec.	16.4	28.0	217

Source : Reuther (1973b)

Soils : Mostly, the soils are coarse textured and have an excellent soil fertility with acidic soil pH. Irrigation is mostly by surface basin method. Quality of irrigation water is good with salinity problem in Hippo valley area (Oberholzer, 1969). A serious B deficiency in citrus growing regions of southern Rhodesia is a common feature (Morris, 1938). The Chisumbanje and Tshotsholo Vertisols have been derived from basalts extruded during the Jurrassic era. The basalt in the northwest areas are predominantly labradorite phenocrysts in a ground mass of augite, magnetite, and feldspar (Stagman, 1978). Chisumbanje Vertisols are located on a flat plain at an elevation of 400 m above mean sea level. The soils are uniform in surface morphology with characteristically well developed loose crumb structure in the surface horizon and classified as fine, montmorillonitic, hyperthermic, Typic Pellustert or Chisumbanje 3B.2 under the Zimbabwe soil classification system (Thomson and Purves, 1978; Nyamapfene, 1991). The Tshotsholo soils have a coarse angular blocky surface structure and are also smectite rich (Table 1.216) with two base saturation (Nyamapfene, 1984).

Table 1.216. Some important soil properties of Vertisols of Zimbabwe

Soil properties	Location	
	Chisumbanje	Tshotsholo
Exchangeable Ca ($mmol_c\ kg^{-1}$)	500	434
Exchangeable Mg ($mmol_c\ kg^{-1}$)	340	250
Exchangeable Na ($mmol_c\ kg^{-1}$)	2	5
Exchangeable K ($mmol_c\ kg^{-1}$)	21	13
Cation exchange capacity ($mmol_c\ kg^{-1}$)	736	718
Clay ($g\ kg^{-1}$)	690	640
Organic carbon ($g\ kg^{-1}$)	9	7
Soil pH	7.5	6.8

Source: Nyamapfene (1984)

1.8.6 Citrus in Australia

Citrus in Australia continent is concentrated in Australia, Cook Island, and New Zealand (Table 1.217).

1.8.6.1 Citrus in Australia

Citrus in Australia was introduced by the first white settlers in 1788 in New South Wales (Bowman, 1955). Later, during 1800s extensive plantings were established further inland due to establishment of irrigation systems along the Murrey and Murrumbidgee river and later in 1828 in seven hills area. A number of researchers (Bowman,1956; Botham, 1963; Burke, 1963; Spurling, 1963; King 1969; Gallash and Ainsworth, 1989) described the citrus culture in Australia with reference to soil and climate. The major citrus production in Australia comes from Riverland district of Victoria and Sunraysia district of New South Wales with minor production along both sides of river Murray from Murray bridge in south Australia to Swan (seven) hill in Victoria. They are also known as lower Murray valley having semiarid climate. Rest of the production emerges from central coast area of New South Wales and from Queensland (Gallasch *et al.*, 1984). Other pockets of production are in the central west of New South Wales around Narromine and Bourke and in the western Australia in addition of Burnett region of Queensland around Gayndah and Munduberra (Cope, 1988). Ngo (1988) suggested areas suitable for growing tropical lemons and limes in the top end of the northern territory comprising Katherine and Darwin regions.

Table 1.217. Major citrus growing belts across Australia

Name of the country	Citrus growing belts
Australia 25°00S lat.; 135°00′ E long.	western Australia, Queensland, South Australia, (Goulbourn valley) Colignan, New South Wales, Central Burnett Victoria, Perth, Adelaide, Melbourne, Coastal area of Sydney, Brisbane, Nembour, Cordwell, Murray, Fremantle, Riverland, Sunraysia
Cook Island 49°35 S lat.; 170°00′ E long.	Rarotonga, Atiu, Mauke
New Zealand 40°00 S lat.; 175°00′ E long.	Kerikeri, Auckland, Bay of Plenty, Gisborne, Kaitaia, Tauranga, Wellington, Christchurch

Australia is basically a dry continent, but a range of climatic zones are available for citrus growing from tropical to arid. Coastal areas are generally mild to humid and although, annual rainfall can exceed 1200 mm. The larger and more important inland citrus areas are hotter and drier, with average rainfall of 650 mm. Coastal areas tend to be undulating. Citrus is cultivated only in selected localities where soil, drainage and water supply are adequately available. Inland citrus areas are generally flatter and plains or wind blown ridges, are usually adjacent to river system. Climate in Gayndah and Mundubbera areas of Queensland located 100 m above mean sea level is characterised by average daily maximum temperature 31.2°C, daily minimum temperature 20.7°C, mean relative humidity 67.1 per cent and annual rainfall of 750 mm. The highest maximum temperature recorded at Gayndah is 42.2°C and lowest of 8.0°C (Chapman, 1963). The climate of Murray river area consists of average daily maximum temperature 27.8°C, daily minimum temperature 18.9°C, daily mean temperature 23.4°C, with maximum and minimum relative humidity of 55.0 per cent and 37.0 per cent, respectively, coupled with 250 mm rainfall.

Climate : Australia is basically a dry continent. The climate in citrus areas of Australia includes humid tropical, humid subtropical, and subarid subtropical. Coastal areas are generally mild to humid and although, annual rainfall can exceed 1200 mm. The larger and more important inland citrus areas are hotter and drier with an average rainfall of 650 mm (Cope, 1988). The climate of lower Murray district of New South Wales consists of mean maximum temperatures, ranging from 15.6°C in July to 32.3°C in February with corresponding mean minimum temperatures of 4.7°C and 17.2°C (February), respectively with an average annual rainfall of 289 mm (Bevington, 1992).

Average annual rainfall record at some citrus producing areas are 850 mm, 525 mm, 625 mm, 1175 mm, 1125 mm, 1650 mm, and 1925 mm at Perth, Adelaide, Melbourne, Sydney, Brisbane, Nembour, and Cordwell, respectively. In the state of Victoria, mean maximum temepratures at Watsonia are 30°C in January and 21°C in July. While mean minimum temperatures are 20°C and 16°C, respectively. In western Australia, mean maximum temperatures at Kalamunda are 32°C in January and 23°C in July, with mean minimum temperatures of 23°C and 17°C, respectively (Reuther *et al.*, 1967). The average annual rainfall at Riverland, south Australia is 264 mm, which is usually received during winter months having hot dry summer and cool winters (Robinson, 1977). At Nambour near the north coast of Queensland, mean maximums are 31°C in January and 26°C in July, with mean minimum temperatures of 25°C and 17°C, respectively, (Reuther *et al.,* 1967). Weather factors have been observed to guide the variation in productivity level in Australia. Of the many factors, mean temperature during November at the time of fruit setting, some two to six weeks after full bloom has been observed to have the strongest influence (Moss, 1969). Temperatures are very similar during November for various citrus growing areas in Australia (Moss, 1970), although some differences are apparent (Table 1.218). Reuther (1973b) described the climatic variation at citrus growing areas of Mildura and Leeton (Table 1.219) in terms of mean monthly and maximum temperature along with annual rainfall.

Soils : In south Australia, citrus is grown on the lighter mallee soils which have sandy surface horizons (often containing free lime) overlying calcareous subsoils varying in texture from loamy sand to clay loam (Table 1.220). The Daney mandarin and Ellendales grown in mallee highland soils have at least 60 cm of sandy to sandy loam texture in the surface with good drainage (Northcote, 1971). Citrus orchards in Riverland and Sunraysia districts are planted on deeper soils of sandy texture and possess a good buffering capacity due to presence of lime and clay in subsurface. Analysis of soil samples collected

Table 1.218. Temperature and evaporation during November (period of fruit setting) for some citrus-growing disticts of Australia

Location	Griffith (M.I.A.)	Dareton (Sunraysia)	Luxton (S. Australia)	Narara (Coastal New South Wales)
Mean minimum (°C)	11.7	12.4	10.5	12.1
Long term average 1970	11.3	13.2	10.0	12.3
Mean maximum (°C)	26.9	27.5	26.5	26.8
Long term average 1970	26.4	27.2	23.6	23.2
Mean temperature (°C)	19.3	20.0	18.5	19.5
Long term average 1970	18.8	20.2	16.8	17.7
Mean monthly (Class A pan)				
Evaporation (mm)	235	279	205	N.A.*
Long term average 1970	219	233	189	N.A.

* Not applicable

Source : Moss (1970)

from these sites (roadside, dripline, and midrow) at Sunlands showed subsoil acidity and high extractable aluminium as soil constraints (Table 1.221) which could affect the performance of citrus especially at extractable Al over 5-10 mg kg^{-1}. Depth of root proliferations is usually restricted to the surface 100 cm, as texture becomes heavier. Commonly, exchangeable sodium percentage from 5 to 10 is found at 60-100 cm portion of root zone and these values do not present any serious sodicity hazard related to effects on soil structure (Richards, 1954). Moss (1976c) described the problem of high acidity in fruit juice at Murrumbidgee irrigation areas. Martin *et al.* (1952; 1961) showed that exchangeable sodium percentage such as above can bring about considerable reduction in citrus growth, particularly in presence of free lime. The soil type at lower Murray district of New South Wales is an alkaline loamy sand with a depth of 75-90 cm (Bevington, 1992). Most of citrus orchards are established on sandy to sandy loam soil type. Because, irrigation water usually contains 300 to 600 mg kg^{-1} of total soluble salts, many Australian citrus soils have a high natural salt content. A chloride content of 0.02 per cent in topsoil is common. Nonirrigated citrus is grown primarily on the coastal plain of New South Wales, in the foothills near Adelaide in south Australia, and in the hills near Perth in western Australia. Halse (1963) reported deficiency of Cu, Zn,

Table 1.219. Climatic variation at citrus growing areas of Mildura and Leeton, Australia

Months	Mildura Temperature (°C)		Rainfall (mm)	Leeton Temperature (°C)		Rainfall (mm)
	Min.	Max.		Min.	Max.	
Jan.	16.1	32.0	19	17.3	31.6	31
Feb.	16.5	32.2	23	17.4	31.5	27
Mar.	14.0	29.1	18	15.0	28.1	34
Apr.	10.3	23.6	14	10.7	22.5	34
May	7.5	19.3	26	7.3	18.2	38
June	5.1	15.8	27	4.7	14.2	39
July	4.7	15.3	23	3.8	13.7	36
Aug.	5.9	17.7	26	4.7	15.7	41
Sept.	7.9	20.0	24	6.7	19.3	32
Oct.	10.5	24.7	25	9.9	23.2	43
Nov.	13.0	28.4	21	13.1	27.3	32
Dec.	15.3	31.2	18	16.0	30.2	27

Source : Reuther (1973b)

and Mn in western Australia. Duncan (1969) described the deficiency of N, P, and B as major nutritional problems in citrus orchards of two districts of New South Wales.

Table 1.220. Chemical analysis of a composite soil profile at south Australia

Depth (cm)	EC (mS cm^{-1})	pH (1:5)	CEC (me 100 g^{-1})	ESP
0-15	0.8	8.4	7.0	9
15-30	1.2	8.5	7.2	11
30-45	1.6	8.6	7.4	50
45-60	2.8	8.3	7.0	12
60-75	2.4	8.6	6.5	14
75-90	2.4	8.4	7.8	19

CEC, Cation exchange capacity; ESP, Exchangeable sodium percentage

Source : Robinson (1977)

Soils at Riverland, south Australia vary in pH from 0.3 to 0.6, ECe 0.8 to 2.4 mS cm^{-1}, cation exchangeable capacity 6.5 to 7.8 me 100 g^{-1}, and exchangeable sodium percentage from 9 to highest of 50 at soil depth of 75 - 90 cm (Table 1.222). Marshall and Hooper (1937) earlier classified these soils as Murray sand (Renmark Type).

Table 1.221. Depthwise description of three soil profiles collected from citrus growing areas of Sunlands, Australia

Depth (cm)	pH	Extractable Al (mg kg^{-1})
Roadside		
0-15	7.4	0.12
15-20	7.9	0.06
30-45	8.1	0.0
45-60	8.1	0.0
Orchard		
0-15	6.7	0.06
15-30	4.7	1.16
30-45	4.5	3.49
45-60	7.1	0.11
Mid-row		
0-15	4.2	8.79
15-30	3.7	19.66
30-45	3.7	34.95

Source : Robinson (1989)

Soils of Gayndah and Mundubbera area along Burnett river range from sandy loam to loam and clay loam, largely deep alluvial, and highly permeable with a soil pH of 5.8-7.0. Depth of soils of Murray river area is 4 to 6 feet having high permeability and virtually free from profile characteristics. Some soils show accumulation of lime in lower horizons, with pH values 6.5 - 8.0 with occurrence of large scale Zn deficiency (Chapman, 1963).

1.8.6.2 Citrus in Cook Islands

The fifteen islands (excluding Niue Island) are divided into two groups, seven in the northern group and eight in the southern. Cultivation of citrus in these islands is claimed to have started during 15th century by visiting Spanish colony. Jollie (1965) described the history of citrus cultivation in Cook Islands.

Climate : Average annual rainfall on the four major citrus producing islands has been recorded as 2075 mm, 2300 mm, 1725 mm, and 1875 mm at Rarotonga, Atiu, Aitutaki, and Mauke, respectively. The mean annual temperature on Rarotonga is 28°C.

Soils : Soils on the Islands vary from light sandy loam near the sea to medium volcanic further inland. Hume *et al.* (1985a; 1985b; 1985c) described the various soil types occurring at Rarotonga. The Pouara soil, which is older, more weathered, and

Table 1.222. Some physical and chemical properties of typical red earth profile (50 per cent kaolinite, 40 per cent illite, and 10 per cent quartz and iron oxides) at New South Wales, Australia

Depth (cm)	pH (1:5)	EC 1:5 (μSm^{-1})	Particle size analysis (%)				Organic carbon (%)	Available N (mg kg^{-1})	Available P (mg kg^{-1})
			< 0.002 mm	0.002-0.02 mm	0.02-0.2 mm	0.2-2.0 mm			
0-0.6	6.5	22.0	19	16	54	6	1.90	78.1	17.2
0.6-3.5	6.3	56.2	23	17	53	6	1.27	29.8	6.6
3.5-18.0	5.8	35.8	27	17	50	9	0.94	23.1	2.6
18.0-14.0	5.6	29.8	33	15	44	9	0.50	12.6	1.9
40.0-48.0	6.0	40.3	44	14	34	5	0.39	11.0	2.2

Source : Greene and Tongway (1989)

more leached than Tikioki soil, has low base saturation, pH, exchangeable Ca and reserve K, and high absorbed S.

1.8.6.3 Citrus in New Zealand

Citrus industry of New Zealand has been focussed earlier by Levitt (1960) and Burke (1963) taking climate and soil into account.

Climate : Climatically, New Zealand is at the very southern limit for citrus cultivation. It is only in the sheltered relatively frost free conditions in eastern areas of the northern Island of New Zealand that allow commercial citrus production (Fletcher, 1959; 1966). North Island is warmer than south Island, and all citrus production is confined to the rather cool, humid subtropical climate of the former. The rainy season occurs during the period of April through October. Annual rainfall at the four main citrus-growing areas has been observed as 1592.5, 1635, 1280, and 972.5 mm at Kerikeri (Bay of Island), Auckland, Bay of Plety, and Gisborne, respectively. There is little difference in temperature among the areas. Fairly typical is Kerikeri where in January, the mean maximum and minimum temperatures are 29°C and 21°C, respectively. Mean maximum and minimum in July are 23°C and 16°C, respectively (Reuther *et al.*, 1967).

Climate of citrus growing areas showed mean minimum and maximum temperature as 5.6°-13.9°C and 15.6°-25.0°C, respectively, with annual rainfall of 1656 mm (Table 1.223). North land region of New Zealand at Kerikeri has

Table 1.223. Climatic variation at citrus growing region of New Zealand

Months	Temperature (°C)		Rainfall (mm)
	Min.	Max.	
Jan.	12.8	24.5	109
Feb.	13.9	25.0	130
Mar.	12.8	23.3	99
Apr.	11.1	21.1	142
May	9.5	18.3	170
June	6.7	16.1	183
July	5.6	15.6	196
Aug.	6.1	16.1	147
Sept.	7.2	17.8	127
Oct.	8.9	19.4	137
Nov.	10.0	21.7	114
Dec.	11.7	23.3	102

Source : Reuther (1973b)

meritime and warm temperate climate with average annual rainfall of 1700 mm with a total heat units of 872°C. Average maximum and minimum temperatures for the warmest month of January are 24.3° - 13.9°C and 14.9°-6.6°C, respectively, for the coolest month of July (Harty *et al., 1*996).

Table 1.224. Selected properties of some soils of New Zealand

Parameters	Hopai	Horotiu	Wairna	Planawatu
pH (H_2O)	5.6	5.1	5.2	5.6
Organic carbon (%)	9.2	7.9	4.2	4.9
ECE (cmol kg^{-1})	34.5	28.2	32.2	20.0
Bull density (Mg m^{-3})	0.85	0.85	0.90	1.12
Clay (%)	58	20	38	32
Clay mineralogy	MKI	A	KVI	ICV

M, K, I, A, C, and V stand for montmorillonite, kaolinite, illite, allophane, chlorite, and vermicullite

Source : Lieffering and Mclay (1995)

Soils: No supplementary irrigation is used except in the dry period at Kerikeri where the light volcanic soils are shallow and low in moisture holding capacity. Soil type at Kerikeri is weathered, gravelly underlain by a clay ironstone horizon (Harty *et al.,* 1996). Lieffering and Mclay (1995) described the soil at Hopai, Horotiu, Wairna, and Planawate having pH 5.1-5.6, organic carbon 4.2-9.2 per cent, and clay content 20-58 per cent (Table 1.224). Graeme *et al.* (1993) classified soils of New Zealand as Umbraqualf, Ustochrept, Dystric Eutrochrept, Haplustalf, and Hapludalf (Table 1.225). Earlier studies by Campbell *et al.* (1977) reported total Fe in clay and fine clay (< 0.2 mm) fractions up to about twice the total Fe in coarse fractors in an allophanic Andesol from New Zealand. Other studies by Hewitt (1989) and Parfitt (1992) described soils of New Zealand belonging to soil orders viz., Andesols, Oxisols, Ultisols, and Inceptisols (Table 1.226). The cations Ca, Mg, K, Na, and the anions NO_3, Cl, SO_4 are lost by leaching from most New Zealand soils every year because rainfall exceeds evapotranspiration, particularly in winter (Edmeades *et al.*, 1985; Parfitt 1990; Sale, 1998).

Table 1.225. Textural composition along with Fe and Mn availability in some soils of New Zealand

Depth	Texture (%)			Fe	Mn
(cm)	Sand	Silt	Clay	(mg g^{-1})	(μ g^{-1})
Tamuka soil (Umbraqualf)					
3-15	44.5	28.0	27.5	22.0	346
40-55	54.2	21.6	24.2	19.0	247
40-55	23.9	44.4	31.7	27.2	247
88-92	79.4	12.5	8.1	17.6	141
131-136	20.2	56.5	23.3	26.0	242
Templeton soil (Ustochrept)					
0-5	54.8	24.6	20.6	20.7	464
10-15	53.0	24.4	22.6	20.6	487
10-15	52.5	25.7	21.8	20.1	462
15-20	52.8	24.7	22.5	21.5	464
20-25	51.7	26.0	22.3	20.5	484
Wakanui (Dystric Eutrochrept)					
0-5	54.9	26.1	19.0	18.9	457
0-10	52.8	26.4	20.8	19.2	444
10-15	52.7	27.0	20.3	19.3	419
15-20	52.0	27.4	20.6	18.4	374
20-25	53.7	26.1	20.2	18.6	383
Timpendean (Haplustalf)					
5-10	33.4	31.8	44.8	45.6	849
12-17	33.7	21.2	45.1	35.5	703
35-40	28.4	17.1	54.5	48.2	210
55-60	34.9	15.5	49.6	46.6	146
90-95	33.2	16.0	50.8	48.1	81.1
Cookson (Hapludalf)					
5-10	20.0	26.1	53.9	126.3	2254
17-22	18.1	24.6	57.3	131.1	2114
35-40	23.3	21.0	55.7	130.1	3816
55-60	31.5	20.9	47.6	122.8	1741
70-75	22.1	30.1	47.8	114.7	1464

Source : Graeme *et al.* (1993)

Table 1.226. Important soil properties of some New Zealand

Order	Series and classification	Depth (cm)	Hori-zon	Texture	pH (H_2O)	Organic C (%)	Silt (%)	Sand (%)	Clay (%)
Allophanic	Otorohanga	0-8	Ap	Silty loam	5.5	12.9	64	20	16
	(Typic Hapludand)	23-47	Bw1	Silty loam	6.4	2.4	47	17	36
	Te Puke	0-10	Ap	Sandy loam	5.8	6.0	28	69	3
	(Typic Udivitrand)	27-45	Bw	Sandy loam	5.8	1.8	44	46	10
Oxidic	Okaihau	0-8	Ap	Silty clay loam	5.7	8.5	39	12	49
	(Anionic Acrudox)	18-41	Bw1	Silty clay loam	5.1	2.5	23	20	57
	Kerikeri	0-8	Ap	Silty loam	5.1	6.5	46	7	47
	(Humic Kandiudox)	17-39	Bw1	Silty clay loam	4.7	2.9	31	7	62
Granular	Hamilton	0-9	Ap	Clay loam	5.7	4.0	51	19	30
	(Typic Haplohumult)	29-46	Bt	Silty loam	5.1	0.8	43	11	46
Brown	Westmere	0-12	Ap	Silty loam	6.4	5.2	55	17	28
	(Mollic Hapludalf)	41-61	Bt1	Silty clay loam	6.0	1.0	49	9	42
Recent7	Karamea	0-7	Ap	Fine sandy loam	5.3	2.6	41	48	11
	(Fluventic Eutrochrept)	19-30	B	Silty loam	6.1	1.2	43	43	14
	Manawatu	0-10	Ap	Silty loam	5.5	3.1	77	14	9
	(Dystric Eutrochrept)	33-60	Bw1	Silty loam	6.1	0.7	76	7	17

Source : Hewitt (1989); Parfitt (1992)

The water soluble Ca^{2+} is the predominant cation in New Zealand soils (Edmeades *et al.*, 1985) and it behaves differently from K^+ or Na^+. The other anions such as Cl^- and NO_3^- are the predominant anions in solution and are both sorbed in a similar manner. Soils used for Citriculture in New Zealand are usually well drained and range from lighter recent soils to the heavier Oxidic soils.

Summary

Continent as well as countrywise analysis of climate and soil under citrus has provided a solid foundation to understand the behaviour of citrus under varying agroclimates, and prepare a blue print of most favourable type of climate and soil leading to maximum quality production. Such conditions act as model requirements which could later be utilized in developing climatic and soil analogues to forecast behaviour of citrus in a specific growing condition elsewhere. The role of altitude and floral response in relation to climate has further established the fact that yield and quality of citrus orcahrds are affected a great deal by these temporal and ecotrophic variables.

A vast diversity in climate and soil occupied by citrus crop worldwide highlights the excellent adoptability of crop. Such modulation of citrus can help adapt a precision Citriculture in due course of time duly backed up by precision techniques like remote sensing and geographical information system.

2 Soil Suitability Criteria

2.1 INTRODUCTION

A good soil is one, that is productive and at the same time, has the characteristics to contribute towards productivity (Lawrence,1958; Hausenbuiller, 1963). Therefore, it must be always an endeavour to relate as many soil properties to its capacity for fruit yield as high as possible. An inadvertent expansion of an acreage under citrus has brought different kinds of soils under cultivation, having varying degrees of suitability for citrus, with the result, the total production has not recorded a proportionate increase. It is common to observe wide difference in productivity level from one soil to another soil type, even under similar management level in an agroclimate. Such a yield variation has attributed varying orchard efficiency due to difference in soil surface and subsurface properties. So far as the soil is concerned, three most important conditions influencing citrus growth and production are: i. chemical environment, means the achievements and maintenance of sufficient nutrients, and the avoidance of excess of salts or other harmful conditions, ii. maintenance of proper moisture conditions to avoid moisture excess or deficit, and iii. soil biological environment to control the detrimental soil organisms.

Azzi (1956) considered three major characteristics, namely, chemical capacity, water balance, and workability while evaluating the effect of soils on yield. The ideal deep, well drained, and sandy loam soils have become increasingly inaccessible due to competition by other fruit crops. The increasing area under fruit crops, therefore, must utilise the cultivable wastelands of low capability (Ghosh, 2000). Fruit crops including citrus in India are grown under a wide range of soil conditions having physical constraints like shallowness and subsurface hardpans, higher water table and/ or poor drainage, calcareousness, salinity, alkalinity or acidity are also wide spread (Iyengar and Joshi, 1991). These constraints lead to genesis of varied nutrient constraints either as a resultant effect of interaction between two or more than two soil factors or as an influence of individual factor. Such constratints need to be wisely managed to get the best fruit yield from given soil types. Therefore, before demarcating an area for citrus plantation, it is important to establish parameters to evaluate the suitability of a given soil type. Such an attempt on soil suitability is a must for successful citrus cultivation.

The deteriorating condition of citrus orchards in India and abroad has been attributed mainly to soil related nutritional factors. Chapman (1961) suggested that no new orchard should be established without a thorough consideration of climate and soil characteristics. There is scarcely an existing orchard, which would not benefit by a thorough appraisal of all the factors affecting orchard productivity, and adapting such changes in management, would lead to better understanding on soil-citrus response relationship. The longevity of a citrus orchard, the investment involved, greatly increased returns, and lowered costs, which can often be achieved, amply justify the time, effort, and costs involved. A land has mainly two types of limitations. Temporary limitations are those characteristics, which have an advance effect on capability. Permanent limitations are those which can not be easily changed, at least by minor land improvements. They include slope angle, soil depth, liability to flooding, and some adaptations of climate. Temporary limitations can be removed or ameliorated by land management like reduced availability of nutrients and miner drainage problems. Land is classified on the basis of permanent limitations. Storie index (Storie, 1954) has often been used for land classification. Land potentiality is described by productivity index (LPI) as : A x B x C x D x E, where A, B, C, D, and E stand for general character of soil profile, texture of horizon, slope of land, site condition (salinity, water quality, drainage), rainfall, and temperature.

A survey of soils in the citrus belt of Punjab by Kanwar *et al.* (1965) showed that out of 9317 acres of planted area, 50 per cent area as suitable, 12 per cent marginally suitable, and another 37 per cent entirely unsuitable for citrus plantation. An ideally a deep, uniform, loamy, moderately well drained on level topography, lime free, and slightly acidic pH is best for citrus, but such ideal soils are not frequently encountered. Actually excellent quality fruit production can be obtained on a wide range of soils, provided production and management operations are such that basic soil handicaps can be efficaciously dealt with. Top citrus production is being achieved on pure sands to clay adobe and peat soils; shallow soils no more than nine inches deep, and underlain by an almost impervious sandy clay layer (coastal Australia); soils underlain at 1.5 ft. by hardpan (California); on extremely heavy clay soil with a water table at 4 ft. (California); other soils with a static water table at 1.5 to 2.0 ft. (Florida); highly laterized soils; and on soils ranging from pH 4.5 to 8.5 (Chapman, 1961).

Soil quality is a concept, which describes soil in terms of its capacity to perform three major functions, those of enhanced productivity, environmental protection, and health (Parr *et al.*,1992). The citrus soils differ from other cultivated soils. The cultivated lands remain fallow for 3 to 6 months every year. This results in depletion of soil organic matter in cultivated lands as very little C is added during the fallow phase, while biological oxidation of existing C continues at the same rate as in cropped lands (Sharma and Singh, 2001). Earlier, Martel and Paul (1974) suggested that degradation of organic matter effectively ceases between 60 to 70 years of cultivation. Substantial decrease in soil organic C and N due to cultivation has been reported by Bauer and Black (1981) and Veroney *et al.* (1981). Larson and Pierce (1991) suggested that minimum data set consisting of soil attributes, namely, nutrient availability, total organic carbon, labile organic carbon, particle size, plant available water capacity, soil structure, soil strength, rooting depth, pH, and electrical conductivity must be considered for suitability appraisal. The quantitative changes occurring over few years must be measured to later use them in evaluating the changes in soil quality (Larson and Pierce, 1991).

Compaction of soils with different contents of organic matter produces general patterns of change in differential porosity: i. a decrease in total porosity occurs upon compaction as a result of a sharp decrease in the volume of large water and air containing pores with the effective diameters of 10 mm or more, which otherwise enable aeration of the soil and uptake of water and ii. reduction in pore volume with effective diameters of less than 0.2 mm containing water that is inaccessible or sparsely accessible to plants. The result of these processes is a degradation of the hydrophysical properties of soil. Thus, at optimum soil bulk density and at a moisture content equal to the minimum moisture capacity, the ratio of the amount of air and productive moisture is close to 1:1 or 1:1.5, which results in favourable conditions for soil productivity (Kuznetsova, 1990). However, in all such extremes, successful citrus production is possible only because of support of special climatic conditions on the one hand, and careful management on the other. While, excellent citrus production is obtained on almost pure sands of Florida, which have never been encountered on soils of such texture in California. The apparent reason for such a differential response is ascribed to rainfall pattern. There is about 1250 mm of rainfall in Florida and associated humidity conditions, whereas, under the intense summer heat and prevailing low humidity of California, it is almost impossible to avoid occasional extreme moisture stresses. Thus, in the latter circumstances, top performance is impossible. The soil singled out as the best or ideal in terms of productivity and represents the standard by which the other soils are judged (Connor, 1954; Ollier *et al.*,1971; Kotur *et al.*, 1982; Dvorak, 1988; Osunade, 1988; Srivstava *et al.*, 1999) provides a dress rehearsal for developing soil suitability guidelines for citrus. An appraisal of soil suitability criteria may, therefore, help to identify suitable soils, and scientifically expand the acreage under citrus, so that big existing gap between the present and potentially possible productivity level may be abridged by avoiding the risk of crop failure, due to soil related constraints.

2.2 SOIL QUALITY

Soil quality indexing, while perhaps originally focussed on indexing and optimization of limited collection of specific attributes, has evolved to assessment of highly generalised, sometimes unspecific overall worth, value or condition of soil. This approach risks certain pitfalls. Soil quality evaluation is one of the highest priority and most difficult problem of modern Soil Science. Lackey (1998) noted , quality of managed natural systems is not an objective scientific attribute. Such quality definitions are contextual, subjective, value ladden, outcome driven, and infinite in possibilities. Singer and Ewing (2000) stated that useful evaluation of soil quality requires agreement about why soil quality is important, how is it defined, how should it be measured, how to respond to measurements, with management, restoration, or conservation practices. Because, determining soil quality requires one or more value judgement which are not easily addressed. These formidable barriers notwithstanding, a key focus of soil quality assays and research has been developed as soil quality assessment tools (Larson and Pierce, 1991; 1994; Arshad and Coen, 1992; Granatstein and Bezdicek,1992; Pierce and Larson, 1993; Gregorich *et al.,* 1994; Turco *et al.,* 1994; Doran and Parkin, 1994b; Romig *et al.*,1995; Warkentin, 1995; Liebig *et al.*, 1996; Halvorson *et al.,* 1996; Harris *et al.,* 1996; Sinclair *et al.,* 1996) Hortensius and Welling, 1996; Anonymous, 1996b). Most of these assessment tools stem from and are based on attempts to define parameters and functions linking crop performance and soil properties. A major concern, however is that none of these objectively and simultaneously consider both the potential positive and negative outcomes of all the indicators employed for all three major consideration of soil management viz., production, sustainability, and environmental impact (Sojka and Upchurch, 1999).

Larson and Pierce (1991) stated in the past, Q (soil quality) is defined in terms of productivity. However, Q is not limited to productivity and such a limited view of soil quality does not serve well in addressing current problems. Thus, they suggested delinking the concept of soil quality from productivity with the rationale that productivity is determined by the efficiency in the use and management of resource inputs whereas, soil quality is related to a set of intrinsic soil properties. Further, they suggested establishing a set of standards for evaluating soil, Serry (1981) proposed the procedures for agroecological considerations based soil quality evaluation. While, Nel and Bennie (1983) proposed critical values, beyond which growth of citrus trees is restricted for citrus orchards of South Africa. Sims *et al.* (1997) proposed a nonpolluted soil criteria for soil quality that they referred to as the clean state of soil. After early definition of soil quality offered by Larson and Pierce (1991) stated that soil quality can be defined as the state of existence of soil relative to a standard, or in terms of a degree of excellence. Soil Science Society of America Adhoc Committee said of soil quality: by encompassing productivity, environmental quality and health as major functions of soil, require that values be placed on specific soil functions as they relate to the overall sustainability of alternate land use decision. Although unstated, the definition presumes that soil quality can be expressed by a unique set of characteristics for every kind of soil. It recognizes the diversity among soils. A soil that has excellent quality for one function or product can have very poor quality for another (Allan *et al.*, 1995).

Mausbach and Tugel (1995) defined soil quality and soil condition separately as: 'Soil Quality' reflects the capacity of a specific kind of soil to function within natural or managed ecosystem boundaries, to sustain plant and animal productivity, maintain or enhance water and air quality, and support human health and habitation. Whereas, 'Soil

Condition' (Health) is the ability of the soil to perform according to its potential. Soil condition changes over time due to human use and management or to unusual natural events. Thus, soil quality must be defined in terms of distinct management and environmental considerations specific to one soil, under explicit circumstances for a given use. The soil quality literature repeatedly emphasizes the need for indexing to encompass the diversity of soil function (Larson and Pierce, 1991; Lal, 1993; Pierce and Larson,1993; Allen *et al.,* 1995). Yet, the indices formulated to date are narrow in scope, mainly emphasizing the soil factors related to plant growth and crop productivity (Sinclair *et al.,* 1996). Multiplicity of definition is rationalized as providing flexibility to accommodate the manifold uses of soil demanded by production, sustainability, environmental, economic, and social imperatives. Appropriately, the soil quality concept has focused increased interest on integrating the soil microbiological assessments into soil evaluation and better understanding the functioning and make up of soil microbial communities (Turco *et al.,* 1994; Kennedy and Smith, 1995; Yakovchenko *et al.,* 1996). Kennedy and Pependick (1995) stated that size and composition of soil microbial populations could be useful indicators of soil quality. The extensive focus of soil quality indices on the microbial ecology and dynamics, is disturbing given that microbiologists acknowledge the critical roles and functions of soil microorganisms are yet to be fully explained (Sojka and Upchurch, 1999).

Wang and Gong (1998) introduced soil quality index (SQI) and its difference (RSQI) to determine the soil quality changes over a period of time. Soil quality index is calculated as SQI = ΣWi. Ii, where Wi are the weights of the indicators and Ii, the marks of the indicator classes. For example, if the slope of a soil is 7^0, it belongs to class II (Table 2.1). As the weight for slope is 13 and the mark for class II is 3, then the SQI_{slope} = 13 x 3 = 39. In this way, SQI for every indicator can be calculated. Therefore, the SQI

Table 2.1. Soil quality indicators and their weights and classes for the evaluation of soil quality in China

Indicators	Weights	I	II	III	IV
Soil depth (cm)	13	> 100	80-100	50-80	< 50
Texture	11	loam	clay or sandy loam	Clay or sand	Gravel
Slope (degrees)	13	0-5	5-10	10-20	> 20
Organic matter (g kg^{-1})	13	> 30	20-30	10-20	< 10
Total N (g kg^{-1})	6	> 2.0	1.25-2.0	0.75-1.25	< 0.75
Total P (g kg^{-1})	6	> 1.5	1.0-1.5	0.5-1.0	< 0.5
Total K (g kg)	5	> 25	17.5-25	10-17.5	< 10
Available N (g kg^{-1})	6	> 200	150-200	100-150	< 100
Available P (g kg^{-1})	6	> 15	10-15	5-10	< 5
Available K (g kg^{-1})	6	> 150	100-150	50-100	< 50
CEC [cmol (p^+) kg^{-1}]	10	> 15	10-15	5-10	< 5
pH	5	5.5-7.0	5.0-5.5	4.5-5.0	< 4.5

Source : Wang and Gong (1998)

value for a soil can be produced by summing up its 12 indicator - SQI values. The maximum value of SQI for the soil is 400 and the minimum value is 100. An optimal soil in any region will have a normalized RSQI of 100, but real soils will have lower values which indicate directly their distance from the optimal soil. By computing RSQI values, soil quality in different regions can be compared even if they are evaluated with different evaluation systems. Similarly, the RSQI could quantify changes in soil quality in a comparable way between two regions.

Characterisation of soils on the basis of relative horizon development and relative profile development is one of the ways to analyse pedological features. Sarkar *et al.* (1997) suggested that larger the rating scale values for a particular horizon, the greater is its pedological development. Meixner and Singh (1981) reported high values in B horizon in older soils. Patil *et al.* (1999) characterised the citrus growing soils on different landforms of Deccan basaltic landscape of Akola district, Maharashtra under semiarid monsoon type of climate with a ustic moisture regime and hyperthermic temperature regime as clayey, mixed hyperthermic family of Lithic Ustorthent; very dark greyish brown to brown in colour, clayey with cambic B horizon, fine montmorillonitic hyperthermic family of Typic Ustochrept; deep very dark greyish brown to dark yellowish brown in colour, fine montmorillonitic, hyperthermic family of Vertic Ustochrept; and deep, very dark greyish brown to dark brown in colour, presence of slickenslides, very fine montmorillonitic hyperthermic family of Typic Haplustert. These soils are observed to have relative horizon development values from 1 for horizons Bssk1/Bssk2 and Bssk2/ Bsssk3 to as high as 14 for horizon Bssk3/Ck. While, relative profile development values varied from 2 for profile Ap/Ck to 15 for profile Bssk2/Ck.

2.3 PHYSICOCHEMICAL PROPERTIES OF SOIL

Steps to restore good physical conditions in the soil may be based only on the objective, up-to-date information on the danger of physical degradation and on the current physical condition of cultivated soils (Berezin *et al.,* 1989). Thus, there is a real need for a theory and methods of evaluating compaction and monitoring physical condition of soils. The development of compaction is often regarded as an increase in soil bulk density, and chemical and physicochemical criteria are often used to measure it. In practice, predictive models of the condition of irrigated soils are based on the salt composition of the irrigation water and the irrigation regime. But, chemical and physicochemical changes taking place during compaction are often quite insufficient to give a satisfactory explanation of catastrophic changes in physical properties. Chemical methods of determining physicochemical characteristics (cation exchange capacity, composition of exchangeable cations, anion exchange, and the like) are in many respects conditional and comparative, and do not always provide the information needed for a quantitative evaluation of phase interaction and physical consequences (Gedroyts, 1955).

The variability of the physical characteristics made it necessary to use the concept of 'equilibrium density'. It is assumed that at some time after mechanical tillage or other treatment, the soil reaches a state that is typical of it under the particular agroclimatic condition. Despite its sketchy theoretical basis, the concept of equilibrium density is widely used in interpreting measurements, even though it does not eliminate all problems in evaluating the physical state from one time measurement of variable characteristics. In traditional agrophysical investigations, several characteristics related to the volume of the stable structural porosity are determined, namely, the noncapillary porosity, interaggregate porosity, aeration porosity, water permeability, yield, and certain others. Koga (1972) described the basic factors of the subsoil affecting productivity of satsuma mandarin orchards in Shikoku district of Japan as a noncapillary porosity and bulk density which alongwith available soil depth can be used as criteria to decide whether a soil can be adapted for satsuma production in addition to methods of soil management.

The estimates of the potential and actual compaction are mutually independent characteristics. The potential compaction limits the technological load in order to prevent the occurrence of actual compaction. The actual compaction is a basis for determining the actual physical state of soil and can be used to plan soil protection measures aimed at restoring or maintaining the soil structure. Realization of the potential compaction depends on the specific conditions of soil use. Under natural conditions, the actual compaction is generally equal to or less than potential compaction. Prolong fallow of a soil earlier classified as compact, may result in the conversion to a normal well aggregated soil. If the actual compaction is greater than the potential compaction, then the soil is experiencing excessive technological stress, whether mechanical, irrigational, chemical, or biological. Optimization of physical properties related to soil structure with an appropriate organizational framework may increase field productivity by a factor of 1.5 to 2.5 without increasing chemicalization and a great decrease in soil tilling operations. (Berezin, 1990)

The deteriorating condition of citrus orchards in India has been attributed mainly to poor physical condition of soil and unbalanced nutrition (Bandale *et al.,* 1951 ; Dhingra and Kanwar, 1963). The above information warrants a thorough evaluation of various soil properties, which are likely to affect the performance of citrus via established guidelines. Citrus growing soils vary widely in their texture, structure, available water capacity, subsoil properties, and soil fertility status. Kakde (1956) observed that performance of orange orchards is directly related to physical characteristics and chemical composition of soils. Studies on soil physicochemical analysis of citrus belt of Punjab revealed that soils from all the series are light textured with loam having low water retention capacity and required frequent irrigation. The pH ranged from 8.1 to 8.5, and EC in the rootzone of soil supporting healthy citrus plants recorded 0.4 mmhos cm^{-1}, and any value above it may lead to chlorosis (Dhingra *et al.,* 1965).

Damigella (1965) observed that soil features viz., pH, deep layer N and P, and to a lesser extent, sand, clay fractions, and total Ca correlated positively with the yield of valencia sweet orange. The deep layer K, silt fraction of both layers, and P content of upper layer correlated negatively with yield. In most cases, the deep layer properties appeared to be more important. Response of soil types (variation in texture) in relation to growth of one year old satsuma mandarin on trifoliate orange indicated best growth in sandy loam with low N fertilization compared to either sand, clay or loam soil, besides having high N requirement (Inoue and Nakayasu, 1982). Other studies by Komamura *et al.* (1982) investigating the behaviour of satsuma mandarin in relation to soil type and depth revealed changes in leaf water potential in various soil types in the order: volcanic ash > clay > gravel.

On the basis of survey conducted by Kanwar and Randhawa (1960), it has been predicted that soil pH should be preferably below 8.5 for successful cultivation of citrus. Nilangekar and Patil (1982) observed poor growth of orchard due to presence of high clay, silt, $CaCO_3$ in subsoil, low organic carbon, available N, P, low exchangeable Mg, high $CaCO_3$, high salinity and ESP in the surface. Singh *et al.* (1967) considered that ideal soil for citrus is as medium light or light loam with slightly heavier subsoil. However, none of the characteristics except EC and $CaCO_3$ have shown bearing on growth of kinnow trees according to Brar *et al.* (1986). Assal *et al.* (1994) observed that calcareous soils of all types decreased the growth of the rootstocks when compared with the normal clay soil. Cleopatra mandarin fared best in calcareous soils, giving generally greater seedling height, growth rate, and dry weight than the other 2 rootstocks. In the clay soil, sour orange performed best. Leaf mineral contents increased, whereas Fe, K, Mn, P, and Zn contents decreased. In general, cleopatra mandarin has been observed as most suitable rootstock for calcareous soils. Koudounas (1994) observed that the amount of clay in the soil correlated negatively with the amount of feeder roots in all soil layers except the top 0-15 cm. Maas (1992) observed a decrease in fruit yield of about 13 per cent for each 1.0 dS m^{-1} increase in electrical conductivity of saturated soil extract once the salinity exceeds a threshold EC of 1.4 dS m^{-1}. The soils having clay content more than 60 per cent beyond 30 cm soil depth are unsuitable for cultivation of Nagpur mandarin under hot subhumid tropical climate of central India (Dass *et al.*, 1998).

Influence of various soil properties on variation in yield of *Citrus reticulata* Blanco, cv Nagpur mandarian in central India indicated that soil properties viz., pH (7.5-7.8) and EC (0.12-0.21 dS m^{-1}) at 0-15 cm and 15-30 cm depth have no effect on fruit yield due to a narrow variation (Table 2.2). Amongst particle size distribution, clay content (42.1-58.6 per cent) correlated more significantly with fruit yield at 0-15 cm depth ($r = -0.784$, p=0.01) than 15-30 cm depth (r=0.548, p=0.05). Fruit yield reduced at the rate of 17.0 tons ha^{-1} (Y=106.21-1.70 clay) and 11.2 tons ha^{-1} (Y=69.10-1.12 clay) per 10 per cent

Table 2.2. Surface (0-15 cm) and sub-surface soil physiochemical properties at various yield levels of mandarin orchards (n=57)

Yield Level (tons ha^{-1})	Soil pH (1:2)	EC (dS m^{-1})	$CaCO_3$ (%)	Particle size (% of < 2 mm)		
				Sand 2.0-0.05	Silt 0.05-0.02	Clay <0.02
0.0-5.1	7.6[a]	0.12[a]	2.3[e]	20.7[a]	20.4[a]	58.6[a]
(4.4)[1]	(7.7)[a]	(0.14)[a]	(2.9)[ef]	(24.2)[a]	(21.2)[a]	(60.8)[a]
5.1-10.1	7.7[a]	0.13[a]	2.6[e]	21.2[a]	21.4[a]	53.2[b]
(8.1)[2]	(7.6)[a]	(0.14)[a]	(3.4)[e]	(24.2)[a]	(22.6)[a]	(58.2)[a]
10.2-15.2	7.6[a]	0.14[a]	4.4[d]	23.1[a]	22.4[a]	58.2[c]
(13.8)[3]	(7.8)[a]	(0.13)[a]	(5.2)[d]	(26.2)[a]	(23.1)[a]	(60.1)[a]
15.3-20.3	7.6[a]	0.14[a]	5.8[bc]	22.4[a]	21.2[a]	52.2[d]
(19.1)[4]	(7.7)[a]	(0.16)[a]	(6.9)[bcd]	(22.9)[a]	(22.4)[a]	(54.2)[b]
20.4-25.4	7.5[a]	0.12[a]	6.2[b]	23.1[a]	20.4[a]	50.2[de]
(23.8)[5]	(7.8)[a]	(0.16)[a]	(7.4)[bc]	(25.2)[a]	(24.2)[a]	(54.1)[b]
25.5-30.5	7.8[a]	0.14[a]	6.9[b]	20.1[a]	23.4[a]	48.2[de]
(26.4)[6]	(7.7)[a]	(0.18)[a]	(8.2)[b]	(24.8)[a]	(25.8)[a]	(50.4)[b]
30.6-35.6	7.5[a]	0.14[a]	8.6[a]	21.4[a]	24.0[a]	44.0[e]
(34.2)[7]	(7.7)[a]	(0.18)[a]	(13.8)[a]	(22.6)[a]	(26.2)[a]	(48.9)[bc]
35.7-40.7	7.5[a]	0.18[a]	9.4[a]	22.2[a]	25.2[a]	42.1[ef]
(38.2)[8]	(7.8)[a]	(0.21)[a]	(14.9)[a]	(26.4)[a]	(26.8)[a]	(44.6)[bcd]
LSD (*p=0.05*)	NS	NS	1.0	NS	NS	3.4
	(NS)	(NS)	(1.2)	(NS)	(NS)	(4.6)

[1] Mean (n=5), [2] Mean (n=10), [3]Mean (n=11), [4]Mean (n=10), [5]Mean (n=10), [6]Mean (n=4), [7]Mean(n=4), [8]Mean(n=3)

* Means in columns followed by different letters are separated according to LSD test (p=<0.05)

** Figures in parenthesis indicate for 15-30 cm depth.

NS, nonsignificant

Source : Srivastava and Singh (2000a)

increase in clay content at 0-15 cm and 15-30 cm depth, respectively. While, sand (20.1-26.4 per cent) and silt (20.4-26.8 per cent) fractions, did not show any corelation with the variation in fruit yield. Fruit yield increased from 4.4 tons ha^{-1} to 38.2 tons ha^{-1} with increase in $CaCO_3$ from 2.3 to 14.9 per cent upto 15-30 cm soil depth. A strong positive correlation between fruit yield and $CaCO_3$ at 0-15 cm (r=0.842, p=0.01) and 15-30 cm soil depth (r=0.709, p= 0.01) showed an increase in yield, respectively, at the rate of 3.40 tons ha^{-1} (Yield = -1.81 + 3.40 $CaCO_3$) and 3.11 tons ha^{-1} (Yield = -2.42 + 3.11 $CaCO_3$) per unit increase in $CaCO_3$.

Elemental analysis of $CaCO_3$ concretions showed mean value of 6-8% Fe, 538.2 mg kg^{-1} Mn, 116.1 mg kg^{-1} Cu, and 51.9 mg kg^{-1} Zn. These values are highly comparable with the values obtained for soil alone as 4.6 mg kg^{-1} Fe, 485.2 mg kg^{-1} Mn, 94.6 mg kg^{-1} Cu, and 41.0 mg kg^{-1} Zn. The $CaCO_3$ concretions during their pedogenesis, act as a strong sink to entrap the available micronutrients under the influence of alternate wetting and drying cycles. Application of heavy dose of farmyard manure coupled with root exudates, and possible recurrence of argillo-pedoturbation (self mulching process) further help release the adsorbed nutrients into the soil available nutrient pool. This maintains the regulated rate of nutrient supply to the plants. The association of $CaCO_3$ concretions with micronutrient containing minerals such as nontronite, suconite, and suponite also aid in maintaining the good supply of nutrients in the rhizoshpere.

Possibly, the gradual disintegration of $CaCO_3$ concretions under these conditions improves the Ca saturation level of soluble and exchangeable phase of soil. Both of these two soil phases are dominated by Ca^{2+} over Mg^{2+}, K^+ or Na^+ and show a significant relationship with yield (Table 2.3). Water soluble (r=0.901 and 0.060) and exchangeable Ca^{2+} (r = 0.916 and 0.902) correlated positively with fruit yield at 0-15 and 15-30 cm soil depth, respectively. Regression analysis later revealed an improvement in fruit yield at the rate of 0.24 tons ha^{-1} (Yield = -9.21+0.24 Water soluble Ca^{2+}) depth at 0-15 cm and 3.2 tons ha^{-1} at 15-30 cm depth (Yield = -24.26 + 0.32 Water soluble Ca^{2+}) per 10 mg l^{-1} increase in water soluble Ca^{2+}. Similarly, unit cmol (p^+) kg^{-1} increase in exchangeable

Table 2.3. Surface (0-15 cm) and sub-surface (15-30 cm) water soluble and exchangeable cations in relation to yield levels of mandarin orchards (n=57)

Yield level (tons ha^{-1})	Water soluble cations (mg l^{-1})				Exchangeable cations [cmol(p^+) kg^{-1}]			
	Ca	Mg	Na	K	Ca	Mg	Na	K
0.0-5.1	78.1[d]	46.4[a]	1.1[a]	2.9[a]	18.1[c]	11.2[a]	2.8[a]	4.1[a]
(4.4)[1]	(81.2)[d]	(48.4)[a]	(1.2)[a]	(3.0)[a]	(21.0)[c]	(11.0)[a]	(3.1)[a]	(4.1)[a]
5.1-10.1	89.2[d]	39.4[a]	1.2[a]	2.9[a]	18.4[c]	10.8[a]	2.9[a]	3.9[a]
(8.1)[2]	(94.1)[d]	(42.4)[a]	(1.4)[a]	(2.8)[a]	(22.0)[c]	(11.4)[a]	(2.6)[a]	(4.2)[a]
10.2-15.2	110.4[c]	48.2[a]	1.6[a]	3.2[a]	19.2[c]	11.2[a]	2.8[a]	4.2[a]
(13.8)[3]	(120.8)[c]	(51.2)[a]	(1.2)[a]	(2.4)[a]	(21.4)[c]	(12.4)[a]	(3.0)[a]	(3.9)[a]
15.3-20.3	112.4[c]	49.2[a]	1.4[a]	2.8[a]	20.4[c]	11.4[a]	2.6[a]	4.2[a]
(9.1)[4]	(122.1)[c]	(54.8)[a]	(1.8)[a]	(2.7)[a]	(22.0)[c]	(12.6)[a]	(2.8)[a]	(4.4)[a]
20.4-25.4	132.1[b]	46.4[a]	1.3[a]	2.6[a]	24.2[b]	10.9[a]	2.7[a]	4.4[a]
(23.8)[5]	(144.6)[b]	(51.2)[a]	(1.7)[a]	(2.9)[a]	(26.8)[b]	(11.2)[a]	(2.8)[a]	(4.6)[a]
25.5-30.5	146.8[b]	42.8[a]	1.2[a]	3.1[a]	28.4[a]	11.2[a]	3.1[a]	4.1[a]
(26.4)[6]	(152.4)[b]	(56.4)[a]	(1.9)[a]	(3.3)[a]	(31.4)[a]	(11.6)[a]	(2.9)[a]	(4.4)[a]
30.6-35.6	161.4[a]	43.1[a]	1.8[a]	3.2[a]	28.0[a]	11.4[a]	2.8[a]	3.9[a]
(34.2)[7]	(181.2)[a]	(52.4)[a]	(2.2)[a]	(3.1)[a]	(32.0)[a]	(12.1)[a]	(3.0)[a]	(4.4)[a]
35.7-40.7	181.4[a]	51.2[a]	1.2[a]	2.8[a]	28.6[a]	11.6[a]	2.9[a]	4.1[a]
(38.2)[8]	(212.4)[a]	(60.4)[a]	(1.6)[a]	(3.2)[a]	(32.4)[a]	(12.0)[a]	(3.2)[a]	(4.4)[a]
LSD	14.1	NS	NS	NS	2.2	NS	NS	NS
(*p=0.05*)	(20.4)	(NS)	(NS)	(NS)	(3.1)	(NS)	(NS)	(NS)

[1] Mean (n=5), [2] Mean (n=10), [3]Mean (n=11), [4]Mean (n=10), [5]Mean (n=10), [6]Mean (n=4), [7]Mean(n=4), [8]Mean(n=3)

* Means in columns followed by different letters are separated according to LSD test (p=<0.05)

** Figures in parenthesis indicate for 15-30 cm depth.

NS, non-significant

Source : Srivastava and Singh (2000a) ; Srivastava and Singh (2001d)

Ca^{2+} at 0-15 cm and 15-30 cm soil depth, respectively, imparted an increase in fruit yield at the rate of 2.92 tons ha^{-1} at 0-15 cm depth (Yield = -44.12 + 2.92 Exchangeable Ca^{2+}) and 2.96 tons ha^{-1} at 15-30 cm depth (Yield = -38.84 + 2.96 Exchangeable Ca^{2+}). Other cations (Mg^{2+}, Na^{+} and K^{+}) showed no correlation with fruit yield.

2.4 SOIL DEPTH REQUIREMENT

The success or failure of citrus is by and large, dependent upon the soil depth. Statistical analysis of data collected in citrus orchards on deep sands in the Citrusdal area (South Africa) showed that coarse sand and clay content are the main factors in determining average tree growth. The depth of occurrence of an impermeable gleyed layer and the presence of gravel also appeared to influence the growth. In Nelspruit area, where citrus trees are grown on three soil types, red sand, red sandy loam, and red clay soil, tree size decreased drastically from 42.0 m^3 on the sand soil to 23.0 m^3 on the clay soil (Nel and Bennie, 1983). Positive correlation between soil depth and yield of satsuma mandarin (Koga and Kawamura, 1970) added strength to the importance of soil depth in growth and yield of citrus. Another study by Tanbara and Kurihara (1963) showed low yields of satsuma mandarin in soils deeper than 80 cm and yield level further had good correlation with available water capacity and noncapillary pores. Takagi *et al.* (1963) reported that trees on more productive soils are deep rooting and better physical properties than soils bearing less productive trees. The exchangeable cation content and the base saturation degree of the subsoil required to be higher in the more productive orchards coupled with much lower than that of less productive soils. Tanbara *et al.* (1964) observed that root and shoot growth of satsuma mandarin reduced with noncapillary pore content of subsoil reading less than 15 per cent. While, Patt *et al.* (1966) suggested a critical limit of 9-10 per cent air space at field capacity for 25-75 cm soil depth.

Canales *et al.* (1974) described that the lemon trees have the problem of adaptability on the shallow stony soils of Vega Baja on the Segura river in Alicante, Spain. In high yielding satsuma orchards, the limits of soil:liquid:air for good fruit yield are suggested as 40-57:20-40:15-37 per cent. The weight:volume ratio in the subsoil of low yielding orchards is high and the root zone shallow, but deep tillage lowered the ratio to the level found in high yielding soils (Tanbara, 1970). Tree size of satsuma mandarin is closely related to the lowest depth at which fine roots occurred. Fine root growth is prevented in some places by a soil layer containing 50 per cent gravel and in others by a highly compacted layer. The bulk density of a soil, an easily measurable parameter has been reported to have the lowest coefficient of variation when compared with other physical properties. The depth at which the layers of bulk density higher than the critical limit occur, determines the productivity potential of the soil. Gupta *et al.* (1995) suggested critical limit of bulk density as 1.62 Mg m^{-3} upto 0.45 m in alluvial soil as a guide for tillage. In order to maintain high orchard productivity on fine or medium textured soil, the bulk density of 1.1-1.3 g cc^{-1} and 40-50 per cent solids in the soil:liquid:gas ratio with atleast 50-60 cm deep soil have been suggested. The lowest limit of subsoil pH and Ca saturation have been suggested as 4.0 and 1.0 per cent, respectively (Kawamura *et al.*, 1970). Identifying the causes for productivity differences across a 50 ha grapefruit (*Citrus paradisi* Macf.) grove in the flatwoods of southwestern Florida, USA, Myhre and Shih (1993) observed that tall high yielding trees grew on the friable Pompano soil (Psammaquents), with good drainage, and a neutral pH upto 90 cm. While, trees grown on oldsmar (Alaquods) soil, low yields are observed due to poor drainage and very acidic pH (4.5-4.8) in the rooting zone. Yields on oldsmar soils could be increased by tile draining at shallower depth and by lime application.

Gandhi (1956) in India observed the unsuitability of shallow soils less than two feet in depth for orchard growing, as encountered in many parts of Deccan. Naik (1948) reported that soils at least eight feet deep, uniform in texture, and having a good subsoil drainage with water table below six feet throughout the year, are best suited for citrus. Hilgeman and Van Horn (1955) and Cahoon *et al.* (1959) observed that most of the roots remain concentrated in top three feet of soil depth. Mendel (1963) suggested suitable depth for citrus under citrus growing dry zone areas of Sri Lanka as at least 5 feet deep without a gravelly layer and water table during the wet season not higher than 3 feet from the surface. Mikhail and El-Zeftawi (1979) in a study on the performance of valencia orange in three different soil types, observed the soil depth of 60 cm provided better concentration of roots than shallow soils, that reflected on the yield. Hernandez *et al.* (1987) observed that in the shallowest and least stony soils, yields are 43 per cent of those obtained in the deepest soils with the same stoniness. In stony (11-25 per cent) deep soils (>150 cm), yields are 58 per cent of those obtained in soils with little or no stones and the same depth. Regression equation showed an increase in fruit yield at the rate of 0.15 ton ha^{-1} per 10 cm increase in soil depth within soil depth range of 25-200 cm in red ferrallitic soil.

El-Gazzar *et al.* (1979) recorded high yield (14-30 tons ha^{-1}) in alluvial soils of valley and deltaic areas compared to yield (6-8 tons ha^{-1}) in sandy and calcareous soils of desert areas of Egypt. In the Doon valley of Uttarakhand, lemons (Eureka round, Eureka long, Baramasi) mandarin (kinnow and Nagpur) and grapefruits (marsh seedless, marsh pink and ruby) have shown good promise under class II land. While, sweet oranges (Malta blood red and mosambi) also proved equally good in denuded lands of Doon valley (Arora and Saraswat, 1983). Prochnow and Boaretto (1995), in the studies on the response of Sicilian lemon on *Citrus volkameriana* to varying levels of gypsum from 0-4 tons ha^{-1} in a dark red Latosol of Brazil, observed that 0-20 cm depth supplied maximum sulphur demand to the crop, which can be considered for an accurate assessment of suitability for cultivation of Sicilian lemon.

Studies by Abercrombie and Plessis (1996) showed that unploughed treatments showed resistance to penetration values of > 2000 Kpa at a soil depth of 200 mm, as measured with a penetrometer. Deep ploughing decreased the mean soil penetration resistance significantly to a depth of at least 500 mm, and still showed values below 2000 Kpa after 3 years of treatment. Strong root regrowth occurred in the 200 to 800 mm soil zone, where the soil has been deep ploughed. Whereas, there are few feeder roots in the control treatments and characterised by soil compaction at shallow soil depths of 200-300 mm. Yield and fruit size showed significant increase over 3 seasons with deep ploughing, compared with controls. It is recommended that a proper soil compaction survey should be done before considering ploughing in existing citrus orchards. Ploughing should be done up to a soil depth of 800 mm on a compacted soil to facilitate favourable rooting volume for regenerated roots. Nunez-Moreno and Valdez-Gascon (1994) observed that a 10 cm water table could be infiltered in 11 days in low yielding areas (48 kg tree^{-1}), while in the high yielding areas (162 kg tree^{-1}), this infiltration occurred in a few hours. A comprehensive review of work done on the citrus soil suitability for citrus, Srivastava and Kohli (1997b) observed that a minimum of 2 feet soil depth, coarse texture with heavier subsurface, well drained, water table below 6 feet throughout the growing period, and free lime not more than 10 per cent are suitable parameters for any of the commercial citrus cultivars grown in India. Jackson and Looney (1999) recommended an ideal soil depth as 5-6 feet for high yield of lemons.

Feeder root distribution pattern is also taken as an index of soil depth requirement for various citrus cultivars, that is governed by the natural surface and subsurface soil properties. Patt *et al.* (1966) found that soil aeration influenced root density and tree productivity, which suggested a critical limit of 9-10 per cent air space at field capacity for 25 to 75 cm soil layers. Yeh (1970) observed maximum distribution of feeder roots to top 10-20 cm depth, about 150 cm from trunk using p^{32} radiotracer technique. Bell *et al.* (1995) working on the Louisiana citrus soil fertility suggested that trace element requirements of satsuma and washington navel trees can not be predicted from testing the top 15 cm soil, especially as soil pH at this depth could be much higher in, these areas. Ono *et al.* (1986) observed an uniform distribution of satsuma mandarin up to a depth of 60 cm on terraces compared to fewer feeder roots on slopes. The root of valencia orange on Mazoe rough lemon rootstock extended 1-2 m beyond the drip line and feeder roots confined sparsely within 1 m of the trunk. The root system is shallow and 90 per cent roots remain confined to top 30 cm of soil (Luttingh and Pyle, 1981). However, state of nutrition, size and yield of trees closely related to the amount of soil explored by the root system (Avilan *et al.*, 1986).

For citrus trees grown in clay loam soil at Griffith in New South Wales, it has been found that the rootzone is largely confined to the top 60 cm depth of soil (West, 1933; 1934). Ford (1954b) reported that although, some roots penetrated to a depth of 5 m,

66 per cent of feeder roots confined to top 75 cm depth. Furr (1955) found effective roots to a depth of 3 m below the surface. Extensive surveys and root distribution studies of drained and undrained flatwood orchards indicated that 1 m of rooting depth is needed for reasonable growth of citrus (Ford, 1969). Mellado and Caballero (1974) reported that maximum efficiency of applied ^{32}P in sandy clay loam soil of Spain confined to 30 cm depth. In 30 year old navel orange trees, the most efficient placement is 3 m from the trunk and 1 m from the trunk in 7-14 year old trees.

Root distribution in 15 years old trees on *Citrus aurantium* rootstock in three serozem soil types showed that roots penetrated to a depth of 180 cm in a dark serozen and to a depth of only 80 cm in a light serozem with intermediate depth in a typical serozem (Penkov *et al.*, 1979a). Castle (1980) and Avilan *et al.* (1987) reported that fibrous roots penetrated to 1.9 m and over half are below 32 cm in 16 year old citrus trees grown in sandy soil of Florida. Sometimes, root distribution is also affected by available air space in soil. Chang *et al.* (1987) demonstrated that nutrient availability in soil up to a depth of 120 cm influenced the various fruit quality parameters. Autkar *et al.*(1988), studying the root distribution of Nagpur mandarin (*Citrus reticulata* Blanco) in Vertisols, suggested that root system radius and depth for 1 year old trees confined to 7.5 - 8.0 cm and 5.0-12.5 cm, respectively, and for 10 years old trees, they expanded to 2-3 m and 80-90 cm. Dhandar and Singh (1989) observed highest root distribution at 120 cm radial distance and 20 cm depth in grapefruit (*Citrus paradisi* Macf.) cv Triumph in alluvial soil. Iyengar and Murthy (1987) suggested that zone of effective nutrient uptake is located in the top 10 cm layer of the soil and within a radial distance of 120 cm for better utilization of applied fertilizers. Kurien *et al.* (1991) using P^{32} radiotracer technique, observed that a majority of feeder roots of acid lime (*Citrus aurantifolia*), cultivar kagzi lime distributed within a radius of 80 cm and a depth of 16-24 cm.

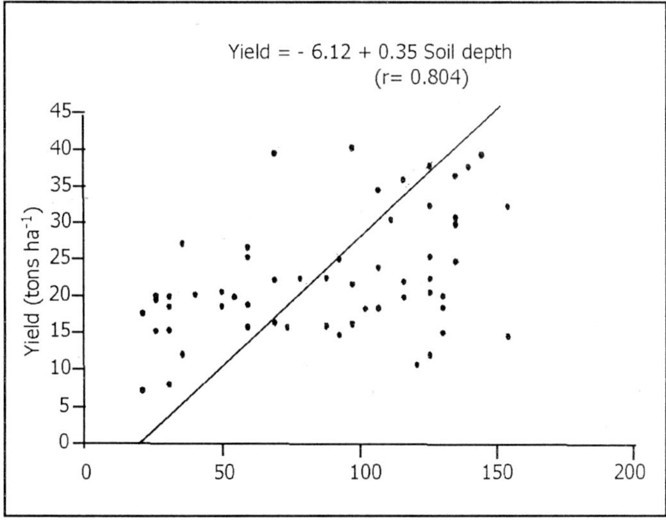

Fig. 2.1. Regression relationship between fruit yield and soil depth in sweet orange orchards in Marathwada region of Maharashtra, India

Studying the impact of soil depth on yield of sweet orange (*Citrus sinensis* Osbeck), cultivar mosambi orchards of Marathwada region of

Maharashtra comprising districts viz., Aurangabad (Jadgaon, Pachod, and Pimpliraja), Jalna (Badnapur, Chickangaon, Ambad, and Kawadgaon), Parbhani (Asola, Parbhani, Sinnapur, Jhari, Parwa, and Jam, 2000), and Nanded (Limgaon and Dhanora), Srivastava and Singh (2000d) observed that the effective soil depth i.e. the soil depth (till the appearance of murrum, disintegrated basalt) in these areas varied from 20 to 160 cm with mean soil depth of 88.5 cm. Similarly, the productivity level of surveyed 59 sweet orange orchards varied from 7.4 to 40.0 tons ha^{-1} with mean productivity of 22.8 tons ha^{-1}. Relationship between soil depth and productivity level of these orchards showed a strong positive correlation (r=0.804) and regression equation, Fruit yield = -6.12 +0.35 Soil depth (x) showed an increase in productivity from 2.6 tons ha^{-1} at soil depth of 25 cm to 46.4 tons ha^{-1} at soil depth of 150 cm (Fig. 2.1). The relationship further indicated that productivity of these orchards increased at the rate of 3.5 tons ha^{-1} per 10 cm increase in soil depth within the above range of soil depth.

2.5 WATER TABLE DEPTH AND DRAINAGE

Effect of water table on depth and concentration of feeder roots (Parker, 1948; Pearson and Goss, 1953; Ford, 1954a; Reitz and Long, 1955; Ford, 1973a) has been summarised as : i. chief threat to orchard health is the fluctuating water table, ii. fluctuating water table accelerated the deleterious effect of high salt concentration, iii. depth of rooting is dependent upon depth of water table, iv. lowering of water table from 75 to 175 cm, doubled the quantity of feeder roots and consequently increased tree size, v. feeder root concentrated more heavily in the 75 to 175 cm zone than 0 to 25 cm zone, and vi. yield of trees remained unaffected, if the peak water table is more than 45 cm below the surface. Number of studies have shown that soil characteristics influenced the tree growth even when the orchard has an excellent drainage with water table below 1.5 m (Ford, 1959; 1963; 1964b). Changes in water table and soil moisture status under valencia and hamlin orange on *Carrizo citrange* showed that the water table under the middle of the beds rose higher following rainfall than at edge. Tree growth on the shoulders of the beds is stunted with excessively high water table, indicating the importance of proper water table management when seepage irrigating multi row citrus beds (Obreza, 1989).

The interaction of salinity and water table is extremely evident, and the worst effect is produced by a combination of the one feet water table with salinity level of 6 mmhos cm^{-1}. The adverse effect of such a shallow water table is more pronounced than salinity (Nadir, 1968; Kanwar and Bhambota, 1969). The growth and fruit yield of two citrus varieties viz., Washington navel orange and Balady mandarin in Egypt decreased with shallowness of water table. Variation in depth of water table influenced the Na, Ca, Mg, K, Cl, and P concentration in leaves and roots (Minessy *et al.,* 1970b). Other studies showed an increase in leaf B content with rising water table level, while nutrients such as Mn, Cu, and Fe contents did not vary much with the depth of water table (El-Azab *et al.*, 1974). Abbadi *et al.* (1974) observed that shallowness of water table induced more die back and reduced the yield of Marsh grapefruit in Spain. Soil analysis indicated that there is no problem of salinity or alkalinity. But, physical analyses showed that per cent clay rose with the shallowness of the water table, in line with characteristics soil moisture curve.

Minessy *et al.* (1970b) showed that raising the water table from 17 to 53 cm for Washington navel sweet orange and from 158 to 89 cm^{-1} for Balady mandarin trees progressively reduced the average total leaf area $year^{-1}$ from 52.4 m^2 and 65.8 m^2, respectively, at the lowest water table level to about 2.5 per cent of these values at the highest water table level. These studies further suggested that yield of Washington navel orange reduced from 16.4 metric tons ha^{-1} at water table depth of 171 cm to zero at water table depth of 78 cm. Under high water table conditions (45-60 cm), successful citrus production can sometimes be achieved by the practice of ridging or mounding, that is planting citrus on the crests of wide ridges, which from trough to apex may be 45-60 cm. Such ridges provide good surface drainage and a roughly triangular or conical soil zone some 90-120 cm above the prevailing water table. In the coastal area of eastern Australia, where highly productive citrus trees are growing on 22 cm of surface soil underlain by an almost impervious clay layers, the reason for good performance is sufficient annual rainfall (45-50 inches) plus supplemental irrigation during dry periods, so that moisture stresses are harmlessly avoided. Fertilizer and other management practices are also adapted to the soil. Many more examples could be cited, but these suffice to illustrate the point that given suitable climatic and good management plus of course, the productive citrus varieties excellent performance can be achieved under a very wide range of soil conditions (Chapman, 1961). Soils supplied with low O_2 level, contained higher concentration of NO_3-N and Ca, and low concentration of N, P, K, Ca, Mg, Zn, Cu, Mn, B, and Fe in sweet orange leaves (Labanauskas *et al.*, 1972b).

Coping with water excesses is a much more serious and difficult problem. Citrus trees have been observed to experience root injury within 3 to 10 days after flooding (Hunziker, 1959; Ford, 1962; Rove and Bradsell, 1973). The situation first is taken up in which a water table exists. It is generally thought that for citrus, a water table which never rises, closer than six feet to the soil surface is satisfactory or if it can be maintained to the levels at 4 to 5 feet. However, if there are fluctuating water tables which at times come to within 2 to 3 feet of the surface and persist for any length of time, successful citrus orchards can not be achieved (Amy *et al.*, 1986). On the other hand, these are some highly productive orchards where a static water table is maintained by proper engineering devices at about 2 feet. Success in these cases is due to the fact that the water table is controlled and stopped fluctuating. In these circumstances, citrus trees developed its root system in the zone above the water table. Under a fluctuating water table, the first problem is to establish the cause. These possible causes comprise of overuse of irrigation water, seepage from canals, and rise from underlying aquifers under hydrostatic pressure. The soils in Immokalee region of Florida are primarily poor to poorly drained with seasonal high water tables. Drainability is the major factor which determines soil suitability for citrus production. Soil suitability for citrus ranked in the order : Flatwoods > Sloughs > Depressions (Obreza *et al.,* 1993).

There are many degrees of plant response to waterlogging which may include any or all of the following symptoms : cessation of shoot elongation, epinastic curvature of leaves and petioles followed by wilting, yellowing, and premature defoliation starting with oldest leaves, blackening and death of roots. Sometimes accompanied by rotting of lower stem region, hypertrophy and formation of adventitious roots along the submerged and exposed stem, less crop and death of plants (Kramer, 1951, Jawanda, 1961). A distinction between excess soil water which is flowing and that which is stagnant is sometimes made, waterlogging being less severe when water is flowing (Cannon, 1925). Water under waterlogged conditions acts as a diffusion barrier to oxygen movement to the roots. It is suggested that the low concentration of oxygen dissolved in water is adequate to prevent root asphyxiation (Yelenosky, 1963; Turner and Patrick, 1968). Waterlogging also implicates plant toxins in root death (Culbert and Ford, 1972). Excess ethylene production (Kawase, 1972), nitrite (Bitcover and Wander, 1950; Chapman and Liebig, 1952), and hydrogen sulphide (Ford and Prevatt, 1958; 1959; Ford, 1965) during waterlogged condition could be lethal to citrus besides root decay (Klotz *et al.,*1966) due to excess depletion of O_2 (Boynton and Compton, 1943; Hunziker, 1959). Culbert and Ford (1973a) showed that O_2 deficiency is relatively harmless to citrus roots during 7 days of flooding, whereas severe root damage occurred when 2.5-3.0 ppm of soluble sulphide at pH 6 is present in the root environment. Rough lemon is known for its tolerance to waterlogging (Hayashi and Wakisaka, 1956) due to tolerance to sulphide injury. Ford (1973b) observed threshold level of H_2S toxic to *Citrus jambhiri* seedlings under waterlogged condition as 2.8 ppm.

It is also possible that flowing water could remove these compounds before these toxins reach dangerous level. A great diversity in reaction of citrus rootstocks to waterlogging has been noticed. The rootstocks such as sour orange and sweet lime have been rated as sensitive to watelogging (Hagin *et al.,* 1965; Ford, 1969), rough lemon, trifoliate, and *Citrus volkameriana* as tolerant rootstocks (Morita, 1955; Bowman, 1956; Salem, 1991), and cleopatra mandarin and sweet orange having intermediate reaction (Ford and Prevatt, 1958; 1959; Ford, 1964b). Schaffer and Moon (1991) observed 100 per cent mortality of budded Tahiti lime (*Citrus latifolia*) after 21 days of flooding, while marcotted trees noticed only 20 per cent. After 35 days of intermittent flooding, tree mortality peaked on sour orange, followed by *Citrus macrophylla.*

Water uptake under waterlogged condition is severely hampered (Kramer, 1951;

Veretennikov, 1964). Labanauskas *et al.* (1966; 1970) found that in citrus seedlings, the amount of N, P, K, Ca, Mg, Cl, Zn, Mn, B, and Fe decreased, and Na increased under conditions of reduced soil oxygen besides growth reduction. Other study by Labanauskas *et al.* (1971) showed that high levels of CO_2 in the soil atmosphere decreased the concentration of N, P, Ca, Mg, and Mn in the tops and in the whole plant increased K and Mg. The top growth is greatly increased with reduced root growth under high CO_2 level in soil. While, earlier Cannon (1925) observed no impact of high CO_2 in the presence O_2 on the root growth of sweet orange. The Chemistry of micronutrients is entirely changed due to waterlogging (Hodgson, 1963). Vu and Yelenosky (1991) observed a direct correlation between leaf photosynthetic rate and stomatal conductance to water vapour using hamlin orange grafted on to sour orange and Rangpur lime rootstocks. Hamlin orange grafted on Rangpur lime rootstock is more tolerant to soil flooding than hamlin grafted on sour orange rootstock.

2.6 ROLE OF HARDPAN

A hardpan is characterised by a hard, stone like horizon that remains hard and impermeable when wet (Raney *et al.,*1955). It is formed by the natural downward movement and cementation of soil particles with organic matter and materials such as calcium carbonate, silica, and iron oxides (Volk, 1954). Hardpans may be formed at any distance from the surface of the soil. Claypans are compact layers, high in clay content that are usually hard when dry. They become sticky and impermeable after wetting (Nikiforoff and Alexander, 1942; Brady, 1978). Hardpans and claypans are common in parts of the San Joaquin valley in USA, arid and semiarid parts of Punjab, Haryana, Cuddapah in south India, Madhya Pradesh, and Maharashtra under subhumid tropical climate (Naik, 1948; Naidu and Rao, 1958; Kanwar and Randhawa, 1960; Chapman, 1961; Kanwar *et al.*, 1965) in addition to many other parts of world citrus. An essential point concerning hardpan, in contrast with clay pan, is that once hardpan is broken up it stays broken up. Clay pan on the other hand, when broken, softens and runs together with first liberal application of water.

Much of the area planted to citrus in the San Joaquin valley is situated in a zone adjacent to the foothills. Soils in this region are derived from old alluvial deposits and it is often underlain by impervious hardpan. Soils in this group, are the San Joaquin, Exeter, and Madera series and are often referred as red hardpan. They are commonly interlaced with mounds giving them the name hogwallow which, impedes rooting and water movement. A perched water table is formed when accumulating water produces a saturated zone above the pan (Amy *et al.*, 1986). Some soils without cemented hardpans have compact subsoil with low permeability. The surface soils absorb water, but the subsurface soil impedes its movement. Soils of the Ducor and Portville series are characterised by the presence of dense clay. Compact soil layers formed immediately below the tillage depth are called ploughpans and are the result of long term cultivation at the same depth, heavy wheel traffic, and/or regular irrigation. They may be partially permeable when wet, but are very hard and nonporous when dry. Tillage pans can cause otherwise well drained soils to become waterlogged after a heavy rainfall or irrigation (Volk, 1954). Even in the seemingly porous sands of the desert valley, formation of clay lenses may restrict water movement appreciably. Silt and other fine particles that blow into the grove settle, and are worked into the soil by tillage and irrigation. These thin bands of clay may be formed anywhere in the soil profile and cause many of the problems associated with hardpans. Although, only true hardpans are completely impenetrable by water at all times. All pans often restrict water movement to such a degree that high water tables are formed after rains or irrigation. Under the presence of compact soil layers along the hillsides, runoff or hillside seepage cause water to accumulate in low lying areas (Burns *et al.,*1976). The behaviour of the hardpan in relation to response of citrus is more comprehensively discussed in subsequent chapter (Chapter 6).

2.7 SOIL PROPERTY NORMS

Optimization of soil properties and condition has recently emerged as a promising field of study. It represents a new stage in managing soil fertility, in which the transition is made from simple improvement of soil properties to regulation of these properties aimed to bringing them into agreement with plant needs in order to achieve maximum yield. The central element of this new concept is the optimum value, best regarded not as a constant, but as having a range of values within which the highest crop yield may be obtained, consistent with the potential crop varieties, climatic condition, and field management. The idea has become firmly established that the optimum value is a rather variable quantity, current models of soil fertility take little or no notice of this fact (Medvedev, 1990). The optimum soil bulk density, as an integral index of the physical conditions governing fertility, is now gaining importance, because agrochemical Soil Science is now changing over, from simple description of the physical conditions governing the fertility of various soil types to finding ways of adjusting them. Models of soil fertility are now being developed, some of them incorporate physical properties, which require a knowledge of their optimum values.

The optimum density values are also needed in order to evaluate the stability of the soil structure of the plough layers for the development of field management techniques and zonal systems of Agriculture, for evaluation of the soil tillage and its compacting effect. The optimum density is defined as the density that results in the highest crop productivity (Bondarev and Medvedev, 1980). But this definition, which without doubt is basically true, does not reveal the essence of the phenomenon. The soil optimum values are further influenced by size composition, temporal variability (refers to variation in the optimal apparent density during crop growth) and ecotrophic variability (produced due to differences in weather condition and soil fertility level). Soils in the high valley of Carababo and Yaracuy have been evaluated for orange production potential in relation to criteria for soil classification confirmed through Duncan test and analysis of variance (Sanchez *et al.*, 1998).

Use of soil testing, specifically for citrus as a guide for fertilizer recommendations has been restricted due to the lack of calibration for P and K in soils and crop response (Obreza, 1990; Hanlon *et al.*, 1995). Shishov and Kapshuk (1984) suggested that reddish brown soils of the Tripolitanian coastal plain of USSR are considered suitable for Citriculture, if the following criteria area met: soil depth more than or up to 60 cm, calcium carbonate less than or up to 8-12 per cent, EC less than or up to 5.0 mmhos cm^{-1}, and gypsum less than or up to 30 per cent. Ko and Kim (1987) observed that high yielding orchards have high mineral content, but low in available P. Available N and P levels are high in orchard soils in which an early cultivar is grown, while Ca and Mg are high in those orchards practising cultivation of mid season cultivars in Cheju county of Korea with average values of soils as pH 5.7, organic matter 8.9 per cent, and exchangeable K, Ca, Mg, and N levels of 1.4, 6.7, 2.3, and 0.2 me 100 g^{-1} soil, respectively. Evaluation of various citrus growing soil series for citrus production in central Taiwan revealed the suitability of 27 soil series and 5 great groups. Orchards established on dark grey colluvial soils registered highest fruit yield of 28.1 tons ha^{-1} and lowest yield of 16.1 tons ha^{-1} on yellow earth soils. In terms of soil series, the soils in grey yellow colluvial soils registered the highest fruit yield of 39.6 tons ha^{-1} and 16.1 tons ha^{-1} in soils of Y & K in the yellow earth colluvial soils (Lay and Wang, 1997). Beridze (1987) observed highest fruit yield of satsuma mandarin (10-12 tons ha^{-1}) in soil having 31-49 mg $100g^{-1}$ exchangeable K, 52-54 mg $100g^{-1}$ Mg, and 197-223 mg 100 g^{-1} Ca with Ca : Mg 2.5-3.0, Mg : K_2O 3.0-4.0, and Ca : K_2O 8.0-10.0 in Western Georgia. Soils maintained above 50-65 per cent of exchangeable Ca and Mg under valencia orange on rough lemon showed no response of Ca or Mg application (Aso and Dantur, 1971).

It is commonly believed, that part of the infertility of serpentine soils is related to high exchangeable Mg percentage (Jacob, 1958). Plants grow well in high Mg and low exchangeable Ca soils (Moser,1933; Hunter, 1948; and Jacob, 1958), it appears that growth of most plants is not significantly influenced as long as the exchangeable Mg does not exceed that of Ca. Relatively few studies have been concerned with exchangeable Mg-percentage greater than Ca. Prince *et al.* (1947) concluded that if the Mg constitutes less than 6 per cent of the exchangeable cations, crops are likely to respond to soil applications of the element. In California where most citrus soils are base saturated some Mg-deficiency has been associated with soils containing about 8 to 15 per cent exchangeable Mg and relatively high K-percentages, namely 6 to 15 per cent (McColloch *et al.,* 1957). Jacoby (1961b) reported that Shamouti oranges seedlings were deficient in Mg when the soil contained about 5 per cent exchangeable Mg. Growth of the citrus seedlings in the Hanford sandy loam soils remained unaffected by exchangeable-Mg percentage of 5 to 11 per cent. About 8 to 15 per cent exchangeable Mg, however, was not adequate for optimum growth in the Altamount loam. Seedling growth in the Hanford soils reduced at the exchangeable Mg exceeding about 50 per cent. Some of the plants growing in the high-Mg and low-Ca soils developed an indefinite leaf edge and intervein chlorosis suggestive of a combination of Mg-Mn and Zn-deficiency. At exchangeable-K percentages of 1 to 4 per cent, the citrus seedlings grew well at Mg-percentages of about 10 to 45 or 50 per cent. When exchangeable Mg exceeds about 50 per cent, growth of the seedlings is significantly reduced (Martin and Page, 1965).

Growth of satsuma mandarin trees is seriously depressed in the presence of 30 mg kg^{-1} Al (ionic concentration about 9 mg kg^{-1}) and 30 mg kg^{-1} Mn. Plants supplied with 30 mg kg^{-1} Mn contained 4000 ppm and 3600 ppm Mn in spring and summer leaves, respectively, compared to 500 and 580 ppm in control leaves (Otsuka and Morizaki, 1969). In Brazil, class limits of soil testing for citrus has been adapted from annual crops (Grupo Paulista, 1988). Earlier, Jorgensen (1978) suggested the soil fertility norms for valencia to be grown in Queensland, Australia. Comparatively recent works by Cantarella *et al.* (1992) and Quaggio *et. al.* (1996) have established the specific calibration for P and K in soils, N content in leaves, and citrus responses to added fertilizers, based on a network of long term field trials (Table 2.4). The results of these works and other previously published in Brazil and summarised by Malavolta (1992) helped to re-evaluate the guidelines for citrus fertilization in Brazil, based on soil testing, leaf analysis, and yield goals (Grupo Paulista, 1994). Koo (1971) obtained high yield and quality of grapefruit and sweet orange at soil exchangeable Mg between 0.4-0.6 me Mg 100 cm^{-3}. Aso and Bustos (1981) working in several citrus regions of Argentina reported the appearance of hidden signs of Mg on soils having exchangeable Mg^{2+} less than 0.8 me 100 g^{-1}. Hidden hunger signs have been also induced due to high K content (K/Mg > 0.4) or Ca (Ca/Mg > 7). Maximum yield of valencia oranges on Rangpur lime rootstock has been obtained when the exchangeable Mg in soil constituted more than 10 per cent of total cation exchange capacity (Quaggio *et al.,* 1992b).

Table 2.4. Soil testing interpretation for citrus in Brazil

Interpretation	P resin * (mg dm^{-3})	Exchangeable (mmol dm^{-3})		Base saturation (%)
		K	Mg	
Very low	< 6	< 0.8	-	< 26
Low	6-12	0.8-1.5	< 4	26-50
Medium	13-30	1.6-3.0	4-8	51-70
High	> 30	> 3.0	> 8	> 70

(*) P extracted by ion exchange resin

Source : Grupo Paulista (1994)

Influence of various soil fertility levels on yield of Nagpur mandarin orchard in central India revealed that fruit yield increased from 4.4 tons ha^{-1} at 82.4 mg kg^{-1} available N 7.6 mg kg^{-1} available P, and 0.58 mg kg^{-1} available Zn to 38.2 tons ha^{-1} at 132.8 mg kg^{-1} available N, 17.4 mg Kg^{-1} available P, and 1.10 mg kg^{-1} available Zn considering 0-

15 cm depth. Fertility variation at 15-30 cm depth had comparatively less magnitude of effect on yield. Yield models based on optimum values of soil fertility have offered a great deal of ease to modulate soil fertility constraints. Fruit yield correlated positively with available N r=0.912, *p* = 0.01 and 0.728, *p*=0.01), P (r=0.842, *p*=0.01 and r= 0.612, *p*=0.05), and Zn (r = 0.882, *p*=0.01 and r=0.589, *p*=0.05) at both 0-15 cm and 15-30 cm soil depth, respectively. Other available nutrients such as K, Fe, Mn, and Cu did not corelate with the fruit yield (Table 2.5). Regression of available N, P, and Zn at 0-15 cm depth on fruit yield indicated an increase in fruit yield at the rate of 0.78 tons ha^{-1} (Yield = -69.20 + 0.78 Available N), 3.10 tons ha^{-1} (Yield = -18.46 + 3.10 Available P), and 66.28 tons ha^{-1} (Yield = -32.42 + 62.88 Available Zn) per mg kg^{-1} increase in Available N, P, and Zn, respectively. Variation in Available N (Yield = -32.46 + 0.54 Available N), P (Yield = -14.20 + 2.28 Available P), and Zn (Yield = -18.76 + 48.21 Available Zn) at 15-30 cm imparted proportionately less magnitude of yield response, at the rate of 0.54 tons ha^{-1}, 2.28 tons ha^{-1}, and 48.21 tons ha^{-1} per mg kg^{-1} increase in available N, P, and Zn, respectively.

Soil test values emerged from various citrus growing corners of the world, which could be used to great effect in evaluating soil fertility constraints of citrus orchards (Table 2.6). Soil fertility evaluation rating for citrus growing soils (Table 2.7) developed by Chapman (1961) showing the soil conditions responsible for development of nutrient deficiency could further prove very helpful in this direction. Smith (1975) suggested that citrus trees do not have a narrow optimal Ca requirement. Sys *et al.* (1993) suggested soil suitability criteria taking into account the characteristics like topography, wetness, physical properties, fertility, salinity, and alkalinity (Table 2.8). Using similar parametric approach, Patil *et al.* (1999b) suggested soil and climatic norms for Nagpur mandarin grown under central Indian conditions (Table 2.9). Comprehensive efforts made over the 10 years or so to develop soil suitability criteria in relation to productivity levels using multivariate quadratic response equation (Srivastava and Singh, 2000a; 2000b; 2000c) have provided useful information with regard to soil fertility evaluation of Nagpur mandarin orchard soils. Nunez-Moren and Valdez-Gascon (1994) observed average values of ECe, water soluble Na, Ca, Mg, and exchangeable sodium percentage as 3.8 dS m^{-1}, 12 me l^{-1}, 16 me q l^{-1}, 3.8 me l^{-1}, and 7.3 me l^{-1} in low yielding orchards (48 kg $tree^{-1}$) compared to 1.1 dS m^{-1}, 6 meq l^{-1}, 4 me l^{-1}, 1.2 me l^{-1}, and 4.6 me l^{-1}, respectively, in high yielding orchards (162 kg $tree^{-1}$).

Summarised soil suitability criteria (Table 2.10) for *Citrus reticulata,* cultivar Nagpur mandarin using multivariate quadratic correlation and regression analysis under sub-humid tropical climate of central India would lead to successful expansion of citrus free of soil related constraints. Such attempts are highly imperative to be undertaken for other commercial cultivars grown in various other parts of India and abroad to have a rationale of soil suitability criteria.

Table 2.5. Surface (0-15 cm) and sub-surface (15-30) soil fertility levels in relation to various yield levels of Nagpur mandarin orchards (n=57)

Yield level	Available nutrients (mg kg^{-1})						
(tons ha^{-1})	N	P	K	Fe	Mn	Cu	Zn
0.0-5.0	82.4^{bc}	7.6^{c}	218.4^{a}	8.2^{a}	12.4^{a}	2.4^{a}	0.58^{d}
(4.4)[1]	$(76.0)^{bc}$	$(7.2)^{c}$	$(222.1)^{a}$	$(8.1)^{a}$	$(10.4)^{a}$	$(2.6)^{a}$	$(0.54)^{c}$
5.0-10.1	85.0^{bc}	7.8^{c}	228.4^{a}	8.6^{a}	12.0^{a}	2.2^{a}	0.52
(8.1)[2]	$(80.1)^{c}$	$(8.1)^{c}$	$(232.1)^{a}$	$(8.0)^{a}$	$(11.4)^{a}$	$(2.1)^{a}$	$(0.50)^{c}$
10.2-15.2	90.4^{b}	12.2^{b}	218.4^{a}	9.2^{a}	14.2^{a}	2.2^{a}	0.68^{c}
(13.8)[3]	$(84.1)^{b}$	$(8.4)^{c}$	$(236.4)^{a}$	$(8.6)^{a}$	$(10.2)^{a}$	$(2.2)^{a}$	$(0.60)^{b}$
15.3-20.3	92.8^{b}	11.4^{b}	204.1^{a}	10.1^{a}	14.8^{a}	2.4^{a}	0.64^{c}
(19.1)[4]	$(88.0)^{b}$	$(10.8)^{b}$	$(220.2)^{a}$	$(9.2)^{a}$	$(10.2)^{a}$	$(2.0)^{a}$	$(0.60)^{b}$
20.4-25.4	100.1^{b}	12.2^{ab}	218.0^{a}	9.6^{a}	14.2^{a}	1.8^{a}	0.82^{bc}
(23.8)[5]	$(90.2)^{b}$	$(10.4)^{b}$	$(208.4)^{a}$	$(8.6)^{a}$	$(11.4)^{a}$	$(2.2)^{a}$	$(0.76)^{a}$
25.5-30.5	118.4^{ab}	15.8^{a}	220.4^{d}	11.8^{a}	16.1^{a}	2.2^{a}	0.94^{b}
(26.4)[6]	$(110.1)^{ab}$	$(13.7)^{a}$	$(220.2)^{a}$	$(9.2)^{a}$	$(14.2)^{a}$	$(2.4)^{a}$	$(0.80)^{a}$
30.6-35.6	125.2^{a}	16.38^{a}	238.4^{a}	12.2^{a}	16.8^{a}	2.3^{a}	1.10^{a}
(34.2)[7]	$(120.1)^{a}$	$(14.8)^{a}$	$(218.0)^{a}$	$(9.2)^{a}$	$(15.2)^{a}$	$(1.8)^{a}$	$(0.86)^{a}$
35.7-40.7	132.8^{a}	17.4^{a}	222.1^{a}	11.4^{a}	16.1^{a}	1.9^{a}	1.10^{a}
(38.2)[8]	$(124.4)^{a}$	$(14.2)^{a}$	$(218.7)^{a}$	$(10.2)^{a}$	$(12.2)^{a}$	$(2.1)^{a}$	$(0.82)^{a}$
LSD (*p=0.05*)	13.4	2.4	NS	NS	NS	NS	0.15
	(10.8)	(1.8)	(NS)	(NS)	(NS)	(NS)	(0.09)

[1] Mean (n=5), [2] Mean (n=10), [3]Mean (n=11), [4]Mean (n=10), [5]Mean (n=10),
[6]Mean (n=4), [7]Mean(n=4), [8]Mean(n=3)

* Means in columns followed by different letters are separated according to LSD test (p=<0.05)

** Figures in parenthesis indicate for 15-30 cm depth.

NS, nonsignificant

Source : Srivastava and Singh (2000a, 2001a) ; Srivastava and Singh (2001d)

2.8. Soil Fertility Evaluation for High Altitude Soils

The critical limit has been evaluated for the soils of Meghalaya under the altitude range of 1000 to 1900m above mean sea level. The organic carbon content below 1.5, 1.5 to 2.25, and above 2.25 per cent are rated as low, medium, and high, respectively, for evaluating N responsiveness. The above limits suggested that the existing limits (<0.5 per cent as low, 0.5- 0.75 per cent as medium, and > 0.75 per cent as high) did not hold good for the soils of higher altitudes (Prasad *et al.,* 1981). Minhas and Bora (1982) suggested various limits of organic C, less than 1.14 per cent as low, 1.14-2.27 per cent medium, and beyond 2.27 per cent as high for high altitude hill soils. The soil fertility rating (Table 2.11) proposed by Mohr *et al.* (1965) has been quite useful in identifying the fertility constraints of horticultural crops including citrus grown in high altitude soils of Nainital (Divakar *et al.,* 1989). Another soil fertility rating (Anonymous, 1994a) used for fertility evaluation of citrus growing soils of Nepal (Table 2.12) can also be used for fertility evaluation of high altitude soils.

Table 2.6. Soil test values for diagnosing the soil fertility constraints for citrus cultivation

Parameters	Indices			Source
	Low	Medium	High	
Soil pH	-	4.0	-	Smith (1957), Rasmussen and Smith (1959)
	< 5.5	5.5-7.5	>7.5	Chapman (1961)
	< 5.5	5.5-7.0	> 7.0	Reitz *et al.*(1972)
	< 8.3	8.3-9.0	>9.0	Brar *et al.* (1986)
	<6.4	6.4-8.5	>8.5	Dhingra and Kanwar (1963)
	< 5.8	5.8-7.5	> 7.5	Plessis and Koen (1996b).
	< 5.5	5.5-7.6	> 7.6	Sys *et al.* (1993)
	< 5.0	5.0-8.0	> 8.0	Jackson and Looney (1999)
	< 4.0	4.0-7.5	> 7.5	Jones and Cree (1954)
Exch. Al (mg kg^{-1})	-	39.0	-	Plessis and Koen (1986).
EC (dS m^{-1})	-	0.50	-	Kanwar and Randhawa (1960), Jones (1980)
	< 1.7	1.7-1.8	> 1.8	Maas and Hoffman (1977), Swietlik (1996b)
	-	1.3	-	Shalhevet (1983)
Salts (g kg^{-1})	-	0.5	-	Di Boixo (1962)
ESP	< 5	5-10	> 10	Robinson (1977)
Water soluble Cl^- (meq l^{-1})	< 10	10-25	> 25	Swietlik (1996b)
$CaCO_3$ (%)	-	16.2	-	Brar *et al.* (1986)
	-	10.0	-	Randhawa *et al* (1966)
Organic carbon (%)	< 0.40	0.40-0.75	> 0.75	Hayes (1960)
Hydrolyzable-N (mg $1000g^{-1}$)	-	30.0-37.0	-	Arsenidze and Chanukvadze (1988)
Total - P (mg kg^{-1})	-	< 350	-	Aldrich and Buchanan (1954)
Available P (mg kg^{-1})	< 6.0	6.0-10.0	> 10.0	Jones (1980)
Olsen - P	< 8.0	8.0-25.0	> 25.0	Plessis (1977), Plessis and Koen (1996b)
Bray - P	< 25.0	25.0-70.0	> 70.0	Plessis (1977)
Resin – P (mg d m^{-3})	-	20	-	Quaggio *et al.* (1996)
Ammonium acetate-P (mg kg^{-1})		27.5	-	Spencer (1963)
BrayP_1(0.03NNH_4F)-P (mg kg^{-1})	-	100		Reitz *et al.* (1972)
BrayP_2(0.03NH_4F plus 0.1N	-	162.5		Reitz *et al.* (1972)
HCl-P (mg kg^{-1})	-			
Exch. P (mmol d m^{-3})	-	2.0		Quaggio *et al.* (1996)
Exch. K (kg $100g^{-1}$)	-	5-20	-	Okada *et al.* (1994)
Available K (mg kg^{-1})	< 60.0	60-150	> 150	Thomas and Peaslee (1973)
Exch. Mg (m eq 100 cm^{-3})	-	0.60	-	Spencer and Wander (1960)
Exch. Mg (mmol d m^{-3})	-	8.0	-	Quaggio *et al.* (1996)
Exch. Mg (mg $100g^{-1}$)	-	15.0-20.0	-	Datuadze (1977)
Exch. Mg (mg 100 g^{-1})	-	21.0	-	Shimizu and Morii (1985)
Ex. Mg (%)	-	5-10	-	Martin and Page (1965)
Exch. Mn (mg kg^{-1})	-	10.0	-	Ogata (1967)
3N $NH_4H_2PO_4$-P (mg kg^{-1})	-	150	-	Ogata (1967)
Exch. Ca/Mg	-	2.2	-	Plessis and Koen (1996b).
Exch. Ca + Mg/K	-	6.0	-	Plessis and Koen (1996b).
Available Zn (mg kg^{-1})	< 0.6	0.6-1.0	> 1.0	Reisenaur (1978), Jones (1980)
	-	1.0	-	Chapman (1960b)
	-	1.05	-	Malewar and Patil (1997)
Available Fe (mg kg^{-1})	-	4.0	-	Jones (1980)
Available Mn (mg kg^{-1})	-	2.8	-	Takkar and Nayyar (1981)
Exch. Mn (mg $100g^{-1}$)	-	21.0	-	Shimizu and Morii (1985)
Available Cu (mg kg^{-1})	-	0.5	-	Jones (1980)
0.5 NHCl (1:10), Cu (mg kg^{-1})	-	6.5, 2.0	-	Ishihara *et al.* (1972), Fiskell and Leonard (1967)
Available B (mg kg^{-1})	<0.50	0.50-0.75	> 0.75	Swietlik (1996a)
Available B (mg kg^{-1})	-	0.5	-	Quyang *et al.* (1984)
Hot water B (mg kg^{-1})	-	0.5	-	Sato *et al.* (1962a), Bingham (1973)
Water soluble B (mg kg^{-1})	-	1.0	-	Elseewi and Elmalky (1977)

Table 2.7. Diagnostic criteria and causes of mineral deficiencies excesses

Elements	Diagnostic criteria – Soil analysis values: Methods of analysis	Low or deficient range	Satis-factory range	Favouring soil conditions
Nitrogen	Solution NO_3 (mg kg^{-1})	< 5.0	> 5.0	**Deficiency** • Sandy soils low in organic matter • Most cropped soils
Phosphorus	Sol. in 0.5 M $NaHCO_3$ (mg kg^{-1})	< 5.0	> 5.0	**Deficiency** • Acid soils, lateritic soils, calcareous soils soils and soils cropped for many years
	Water soluble 1:10 soil water ratio (mg kg^{-1})	< 1.0	> 2.0	**Excess** • Where manures or mixed fertilizer or phos-phate are used heavily over a period of years
Potassium	Exchangeable K (kg ha^{-1}) for 30 cm	< 250.0	> 750.0	**Deficiency** • Acid sandy, peat and muck soil • Heavily cropped leached or eroded soils and leached lateritic soils **Excess** • Prolonged use of manure mixed fertilizers etc.
Calcium	Exchangeable.Ca (Per cent exchange capacity)	< 25.0	75.0 to 90.0	**Deficiency** • Acid soils, sandy soils of humid regions, peat soils, soils from serpentine rocks and alkali soils with ESP 5 per cent. **Excess** • Calcareous and saline soil
Magnesium	Exchangeable Mg in relation to K (meq 100 g^{-1})	0.30 to 0.90	0.1 to 0.3	**Deficiency** • Acid sandy soils having high exchangeable potassium
Sulfur	Morgan's Acetate-acetic acid extract (mg kg^{-1})	< 3.0	--	**Deficiency** • Soils low in organic matter or acid leached soils **Excess** • Alkali and saline soils
Manganese	Soluble in 3N $NH_4H_2PO_4$ (mg kg^{-1})	< 20.0	30.0 to 60.0	**Deficiency** • Neutral and alkaline soil • Acid sandy soils with native low in Mn • Alkaline peat and muck soils **Excess** • Acid soils and mangniferous soils
Copper	0.1 N HCl (mg kg^{-1})	0.9 to 1.6	--	**Deficiency** • Peat and muck soils, leached sandy soils, and calcareous sandy soils
Zinc	Ammonium acetate and Dithizone extract (mg kg^{-1})	< 1.0	2.5	**Deficiency** • Acid sandy soils, alkali soils or soils derived from granites gneisses etc **Excess** • Soils contaminated from mining operations • Accumulation due to prolonged soil or foliar application of zinc compounds

Source: Chapman (1961)

Table 2.8. Soil suitability norms for citrus

Land characteristics	Class, degree of limitation and rating scale.					
	S_1	S_2	S_3		N_1	N_2
	0 1	2	3		4	
	100 95	85	60	40	25	0
Topography						
Slope (%) (1)	0-1	1-2	2-4	4-6	-	>6
(2)	0-2	2-4	4-8	8-16	-	>16
(3)	0-4	4-8	8-16	16-30	30-50	>50
Wetness (w)						
Flooding	Fo	-	-	-	-	Fl+
Drainage	Good, ground water >150 cm	Good, ground water 100-150 cm	Moderate	Imperfect	Poor but drainable	Poor not drainable
Physical soil characteristics						
Texture/ structure	sl, sil, l, sicl, cl,cl, si	scl, ls, lfs	c<60s, sc, s, fs, sics, co	C<60v, C>60s	-	cm, sicm c>60v
Coarse frag. (vol %)	0-3	3-15	15-35	35-55	-	>55
Soil depth (cm)	>200	200-150	150-100	100-75	-	<75
$CaCO_3$ (%)	0-3	3-5	5-10	10-25	-	>25
Gypsum (%)	0-1	1-2	2-3	3-5	-	>5
Soil fertility characteristics (f)						
Apparent CEC [cmol (p^+) kg^{-1} clay]	>16	<16(-)	<16(+)	-	-	-
Base saturation (%)	>35	35-20	<20	-	-	-
Sum of basic cations (cmol (p^+)kg^{-1}	>3.5	3.5-2	<2	-	-	-
pH (H_2O)	6.5-5.8	5.8-5.5	5.5-5.2	5.2-5.0	<5.0	-
	6.5-7.0	7.0-7.6	7.6-8.0	8.0-8.2	-	>8.2
Organic carbon (%)	>1.5	1.5-0.8	<0.8	-	-	-
Salinity and alkalinity						
ECe (dS m^{-1})	0-2	2-3	3-4	4-6	-	>6
ESP (%)	0-4	4-8	8-12	12-15	-	>15

Note : S_1 , S_2, S_3, N_1, and N_2 stands for highly suitable with no significant limitations; moderately suitable having slight limitation; marginally suitable having severe limitation; N, currently not suitable having correctable limitations and permanently not suitable having limitations beyond correction.

Source : Sys *et al.* (1993)

2.9 SOIL TO LEAF NUTRIENT RELATION

The relationship between leaf nutrient content and soil fertility status helps to find out the expected response on the nutrition level of citrus tree as a result of change in fertility due to fertilizer application (Ko and Kim, 1987). A number of studies (Plessis, 1977; El-Gazzar *et al.,* 1977; Koo, 1982) have shown significant correlation between soil

Table 2.9. Soil site suitability criteria for Nagpur mandarin grown in western Vidarbha region of Maharashtra

Climatic and land characteristics	Class, degree of limitation and rating scale					
	S_1		S_2	S_3	N_1	N_2
	0	1	2	3	4	
	Climatic characteristics					
Total precipitation (mm)	> 850	850-750	750-600	500-600	450-500	<450
Mean annual air temp.(^{0}C)	22-25	25-30	30-35	35-40	--	> 40
		19-22	15-19	13-15	--	< 13
Mean relative humidity (%)	60 -0	50 - 60	45 - 50	40 - 45	--	< 40
Topography (Slope %)	0-1	1-2	2-3	3-8	8-15	> 15
Wetness (Drainage)	Well Drained	Well Drained	Mod. well drained	Imper fectly	Poor	Very poor
	Physical soil characteristics					
Texture/clay (%)	l, cl, sl, sicl	scl, l	c, sc, sic	c >70 %	--	sand
Coarse fragments (Vol. %)	< 3	3 - 15	15 - 35	35 -55	--	> 55
Soil depth (cm)	> 150	100– 150	60 - 100	45 - 60	25 – 45	< 25
$CaCO_3$ (%)	0 – 5	5 - 10	10 - 15	15 - 25	25 – 35	> 35
	Soil fertility characteristics					
Apparent CEC [cmol(p^+) kg $^{-1}$ clay]	> 16	> 16	> 16	> 16	--	< 16
Base saturation (%)	> 35	> 35	> 35	> 35	--	< 35
Sum of basic cations [cmol (p^+) kg $^{-1}$ soil]	> 10	>10	--	--	--	>10
pH	6.- 7.5	5.5 - 6.0	5.0 - 5.5	4.8 - 5.0	< 4.8	>8.5
Organic carbon (%)	> 0.8	0.5 - 0.8	0.3 - 0.5	0.2 - 0.3	--	--
	Salinity and alkalinity					
EC (dS m^{-1})	0 – 0.2	0.2 - 0.3	0.3 - 0.4	0.4 - 0.5	0.5 - 1.0	> 1.0

Note **:** S_1 , S_2, S_3, N_1, and N_2 stands for highly suitable with no significant limitations, moderately suitable having slight limitation, marginally suitable having severe limitation, N, currently not suitable having correctable limitations, and permanently not suitable having limitations beyond correction.
Symbols such as l, cl, sl, sicl, scl, c, sc, and sic stand for loam, clay loam, sandy loam, silty clay loam,sandy clay loam, clay, sandy clay, and silty clay, respectively.
Source : Patil *et al.* (1999b)

Mg versus leaf Mg and soil Ca/Mg versus leaf Mg. With the type of relationship occurring between soil available nutrients and leaf nutrients, the expected response on leaf nutrient status can be easily simulated with change in availability of nutrients in soil due to fertilization. Dhillon and Dhatt (1988) observed that foliar K exhibited positive correlation with soil N and negative with foliar P. They further observed that foliar N, P, and K content are directly related with their respective nutrient content in soil in Ferozpur district of Punjab. Murthy *et al.* (1986) reported positive relationship (r=0.454) between P content of soil and leaf in Coorg mandarin (y=0.3238 + 6.36x). Similarly, Mazuelos Vela and Gonazalez Garcia (1993) observed positive correlation between soil and leaf P in Seville region of Spain. Soil P had negative relationship (r=0.33) with leaf K. Plessis (1971) observed no correlation between leaf and soil P contents at 0-15 and 15-30 cm depth in valencia orange orchards. Malewar *et al.* (1978) depicted positive correlation between soil available Zn and plant Zn concentration. A synergistic relation between P

Table 2.10. Optimum values of soil properties criteria for delineating suitable soils for *Citrus reticulata* Blanco, cultivar Nagpur mandarin grown in central India

Soil properties	Soil depth (cm)	
	0-15	15-30
Soil pH	7.6-7.8	7.9-8.0
Soil EC (dS m^{-1})	0.12-0.24	0.21-0.28
Free $CaCO_3$ (%)	11.4-12.8	15.6-18.2
Particle size distribution (%)		
Sand	20.8-40.1	19.0-32.7
Silt	26.8-30.4	11.2-26.8
Clay	42.8-48.8	54.2-56.1
Water soluble cations (mg l^{-1})		
Ca^{2+}	168.3-182.3	32.7-38.1
Mg^{2+}	39.4-42.7	32.2-42.1
Na^{+}	0.98-1.1	0.68-1.2
K^{+}	1.2-2.8	1.4-2.8
Exchangeable cations [cmol(p^{+}) kg^{-1}]		
Ca^{2+}	31.9-32.3	38.1-41.2
Mg^{2+}	8.5-10.1	9.2-10.0
Na^{+}	0.68-1.23	0.8-1.1
K^{+}	3.2-4.1	4.5-4.6
Fertility status (mg kg^{-1})		
Available N	118.4-121.2	92.8-110.2
Available P	9.2-10.3	7.2-8.0
Available K	178.4-232.5	204.2-228.1
Available Fe	12.4-16.2	12.3-10.6
Available Mn	8.6-12.2	7.2-9.1
Available Cu	2.1-2.3	1.2-1.2
Available Zn	0.98-1.10	0.78-0.72

Source: Srivastava and Singh (1999a; 1999b; 2000a)

and K has been observed in a 10 year old mandarin orchard having alluvial calcareous soil with pH 7.3 and fertilized with $N:P_2O_5:K_2O$ at 250:250:120 g $tree^{-1}$, respectively. While, these nutrients interacted negatively on a acid podzolic soil with pH 5.8 fertilized with $N:P_2O_5:K_2O$ at 250:200:100 g $tree^{-1}$, respectively (Magnitskii and Takidze, 1972).

Anderson and Albrigo (1971; 1972) carried out two nutritional surveys to study the soil-plant relationship in dolomitic citrus orchards in central Florida and observed that none of the procedures provided soil Mg values that

Table 2.11. Limits of soil fertility rating

Sr. no.	Fertility parameters	Indices		
		Low	Medium	High
1.	Organic carbon (%)	Below 0.5	0.5 – 0.75	Above 0.75
2.	Available P (P_2O_5 kg ha^{-1}).	Below 25	25-50	Above 50
3.	Available K (K_2O kg ha^{-1})	Below 148	148-336	Above 336

Source : Mohr *et al.* (1965)

Table 2.12. Soil fertility rating chart for citrus growing hill soils of Nepal

Range	Organic matter (%)	Nitrogen (%)	Phosphorus (kg ha^{-1})	Potash (kg ha^{-1})	Soil pH
Low	1.0-2.5	0.05-0.1	10-30	55-110	< 6.5 (acidic)
Medium	2.5-5.0	0.1-0.2	30-55	110-280	6.6-7.0 (neutral)
High	5.0-10.0	0.2-0.4	55-110	280-500	> 7.00 (alkaline)

Source : Anonymous (1994a)

could be significantly related by themselves to leaf Mg. Plessis and Burger (1972) observed no correlation between leaf and soil Ca content in citrus orchard soils of South Africa, but good correlation existed between soil and root Cu content

On the other hand, leaf Mg related negatively to the ratio of soil Ca/Mg. Kazak and Khalidy (1974) investigating the possible causes of nutritional disorders prevailing in Cenabose citrus orchards reported that leaf K is not related to the soil exchangeable K, but negatively correlated with leaf Ca. Leaf Mg correlated negatively with the K/Mg ratio in the soil, but showed no relation to leaf K, leaf Ca or NH_4OAc extractable Mg in the soil. Close association between leaf Mg and the K/Mg ratio in the soil indicated a probable antagonistic effect of high accumulation of K in the soil on the absorption of Mg. The high accumulation of K in the soil coupled with the low K content in the leaves is attributed to the antagonism of excess Ca on K absorption. Therefore, an attempt to correct the low status of K in the leaves by soil application without considering the level of available Mg in the soil aggravated Mg deficiency without necessarily improving the status of K in the leaves. Martin and Page (1965) suggested that the citurs leaf Ca and Mg percentages are much more closely related to the actual exchangeable soil Ca and Mg, respectively, than Ca-Mg or Mg-Ca antagonism effects.

Anderson and Albrigo (1977) later observed significant correlation between soil and leaf K in deep sandy soils of Florida. A similar observations have earlier been reported by many workers (Harding, 1954, Pratt *et al.,*1957; McColloch *et al.,* 1957). The increase in N and K contents in plants with increase in soil available N and K has been shown by Hume *et al.* (1985a; 1985b) with positive correlation between soil and leaf Mg level in clay loam soils of Cook Islands. High leaf K level is observed to be associated with low soil Ca level in acidic soils of Florida and vice versa in alkaline soils of California (Chapman, 1949). Bester (1969), Khalidy and Kazak (1970), and Braga (1970) reported that excess soil Ca and constant water shortage resulted in reduced in K and Mg level in tree tissue. Murthy *et al.* (1986) obtained positive relationship (r=0.401) between K contents in soil and leaf of Coorg mandarin (y=1.16 + 2.82x). Leaf K and leaf Mn (r=0.561) also followed a positive relationship (y=11.2 +66.9x). Leaf K correlated negatively with available N and P in soil, probably due to the presence of Ca. The dose of fertilizer can be streamlined based on soil fertility status in a given soil type and climate (Table 2.13). Li *et al.* (1998) observed a significant correlation with total N/

Table 2.13. Soil analysis interpretation criteria for fertility evaluation and tailoring fertilizer schedule

P^{Z} (ppm)	10	20	30	40	50	60	70	80	90+
P requirement (kg ha^{-1})	50	40	30	20	10	0			
K^{Y} (ppm)	50	100	150	200	300+				
K requirement (kg ha^{-1})	200	150	100	50	0				
Ca^{Y} (ppm)	200	600	1000	2000+					
Gypsum requirement^{X} (t ha^{-1})	2	1	0						
Mg^{Y} (ppm)	100	200	400+						
Mg oxide^{W} (kg ha^{-1})	400	200	0						
pH^{V}	4	4.5	5	5.5	6	6.5	7	7.5	
Lime requirement (t ha^{-1})	7	6	5	3	1	0			
Cu^{u} (ppm)	0.1	0.3	10	20+					
Cu oxychloride	0.4% spary	Fungicidal use only	Minimal fungicidal use	Maintain pH 6.5-7.0					
Mn^{u} (ppm)	2	4	45	60+					
MgSO$_4$	0.1% spray	0	Maintain soil pH above 6.5. and improve drainage						
Zn^{u} (ppm)	1	2	10	15	20+				
ZnSO$_4$	0.1% spray	0	Maintain soil pH above 6.5						
Nay (% exchangeable cations)	10								
Action	nil	Reduce Na application. Apply Ca							
Cl^{t} (ppm)	100	250	500	1000+					
Action	Nil	Reduce Cl application Leach with low chloride water							
Conductivity^{v} (mS cm^{-1})	0.1	0.2	0.4	1.0	3.0				
Action	Nil	Leach soil	Irrigate with lower conductivity water						

Source : Jorgensen and Price 1978)
^{z} Determined from dilute acid extranct; ^{Y} Determined from ammonium acetate extract; ^{X} Not required if lime applied for pH improvement; ^{W} If lime required for pH improvement, use do lomite instead, and no Mg oxide;
^{V} Determined in 1:5 soil:water; ^{u} Determined from DTPA extract; ^{t} Determined from water extract

available K and hydrolyzable N/available K in citrus orchards of Fujian province of China. Beridze (1986a) observed that increasing exchangeable Ca and K in the soil intensified N re-utilization. The Ca blocked the K uptake by the plant, whereas Mg intensified K absorption in Western Georgian soils.

In citrus orchards survey, Shome and Singh (1965) observed negative correlation of leaf Ca+Mg with leaf K content (r = -0.62), irrespective of healthy and declining citrus orchards. Spencer and Wander (1960) observed a close relationship between leaf Mg and exchangeable Mg (r=0.85) in soil under valencia and pineapple orange on rough lemon rootstock. Symptoms of Mg deficiency in lemon appeared at leaf Mg level of less than 0.35 per cent and soil exchangeable Mg less than 0.6 me 100 cm^{-3}. Later, other workers (Koto and Takeshita, 1959; Anderson 1971a; 1973; Caetano *et al.,* 1984; Quaggio *et al.,* 1992b; Dey and Singha, 1998) reported a close relationship between leaf and soil Mg content.

Anderson and Albrigo (1971) conducted multiple linear regression analysis to determine soil characteristics more related to leaf Mg and found that the ratio of soil Ca/ Mg is strongest independent covariable (r=0.51). Similar correlations have been obtained by Ibrjkci (1994) between soil available N, P, K, Ca, and Mg and their concentrations in leaf in mandarin orchards of coastal region of Tukey. Zhuang *et al.* (1984) on the other hand, observed no correlation between soil and leaf element content in mandarin orchards of Fujian, China. Oseni (1988) reported only 12 correlations significant out of 48 with regard to N, P, and K content in leaf versus soil relationship. Study further advocated that analysis of single nutrient status of tree may not reveal the true relationship, since it is affected by many other soil factors. Positive correlation between soil and leaf Zn, Mo, and Cu has been observed in citrus orchards of China (Quyang, 1993). Kohli *et al.* (1993) observed that application of increasing levels of soil applied N/K ratio from 0.0 to 8.0 increased leaf N content from 2.10 to 2.92 per cent and reduced the leaf K from 1.90 to as low as 0.90 per cent (r=-0.88). Srivastava *et al.* (1995) observed that significant correlation of soil available K and leaf K content in mandarin orchards of Kalmeshwar (r=0.85) and Katol (r=0.61) tehsils of Nagpur district, with negative correlation between N and K. Considering inverse N/K ratio as a better indice than monitoring leaf N and K individually, the leaf N/K ratio of 2.42 is suggested as an appropriate value for optimum productivity of Nagpur mandarin. The leaf micronutrient concentration showed positive correlation with soil DTPA extractable micronutrients in mandarin orchards of Saoner and Hingna teshils of Nagpur district (Kohli *et al.,* 1994). They further observed increase in leaf Fe by 8.2 mg kg^{-1}, Mn by 1.4 mg kg^{-1}, Zn by 13.3 mg kg^{-1} and Cu by 2.8 mg kg^{-1} with 1 mg kg^{-1} increase in DTPA extractable nutrient in Nagpur mandarin orchard soils in central India.

Better correlation of leaf B content has been obtained with saturation extract B (r= 0.821) than with hot water extractable B (r = 0.779). Saturation extract and hot water extractable B further correlated positively (r= 0.736) with a linear regression equation, Y = 0.38 + 1.47X, where Y stands for hot water extractable B and X saturation extract B (Elseewi and Elmalky, 1977). Plessis (1977) obtained higher probability of response- with combined use of leaf and soil analysis norms compared to either leaf or soil analysis alone (Table 2.14) for P and K fertility evaluation. In a study on citrus

Table 2.14. A composition of three different methods of appraising the P and K status of 113 citrus orchards based on established leaf and soil norms at South Africa

Method	Probability of a reaction on fertilization (% of total orchards)		
	No reaction	Probable reaction	Definite reaction
		Phosphorus	
Leaf analysis	20.0	76.5	3.5
Soil analysis	65.5	20.5	14.0
Leaf and soil analysis	70.0	26.5	3.5
Norms used	**Leaf (%)**	**Bray - P (mg kg^{-1})**	
No reaction	> 0.15	> 70	
Definite reaction	< 0.11	< 25	
		Potassium	
Leaf analysis	31	36	33
Soil analysis	27	37	36
Leaf and soil analysis	50	33	17
Norms used	**Leaf (%)**	**Soil available(mg kg^{-1})**	
No reaction	>0.9	> 350	
Define reaction	< 0.7	< 200	

Source : Plessis (1977)

response to salinity, Shalhevet *et al.* (1974) showed a 4.3 per cent yield reduction per me l^{-1} increase in irrigation water chloride concentration for shamouti orange on sweet lime. Positive correlation between leaf and soil B in sweet orange cultivar Pera on Rangpur lime rootstock in a red sandy acidic Latosol at Bebedouro, Brazil (Viti *et al.*, 1993b).

2.10 FRUIT QUALITY IN RELATION TO SOIL PROPERTIES

An improvement in fruit quality consequent with production has made it highly imperative to pin point the soil properties, which may possibly affect the development of good quality fruits. Average storage life of fruits from heavy soil (loam texture) is about 10 days longer than that from light soil (sandy texture). The evaluation of fruit quality parameters is equally important, and a strong necessity is felt to evaluate soil properties not only in relation to yield, but quality also, to rate the limitations and abilities of a given soil type on a comparative scale. Juice specific gravity at picking time is higher on heavy soil than on light soil, and also higher in early picked fruits. The juice total soluble solids and total soluble solids/acid ratio are higher in fruits from heavy soil than light soil, and also higher in late picked fruits (Ibrahim, 1968). Texture of soil has no influence on fruit quality parameters viz., fruit size, weight, peel thickness, and total soluble solids of valencia late orange (*Citrus sinensis* Osbeck) in Cuba (Valle Valdes and Rios Albuerne, 1974). Bar-Akiva and Hamon (1974) observed that fruit quality parameters differed in relation to texture of soil, irrespective of fertilizer treatments. Fruits on sandy clay loam are larger, with thicker peel, more total acids and vitamin C in the juice, and less affected by peel disorders (creasing) compared to fruits derived from loamy sand soil. Sweet orange grown in Inceptisols proved significantly superior to those grown on Alfisols and

Vertisols with respect to fruit weight, fruit juice, TSS, and total sugar content of fruit juice. Fruit quality parameters such as juice, TSS, and sugars showed significant positive correlation with available N content of soil. Reducing sugars of fruit juice correlated significantly with soil N and P contents (Vijayasankar Reddy *et al.*, 1992). They further found high acidity in the fruits due to good supply of soil $CaCO_3$ varying from 5.0 to 8.7 per cent. Munshi *et al.* (1979) reported higher accumulation of acidity grown in clay loam soil due to presence of lime concretions. Kiely *et al.* (1972) earlier observed high Ca in the soil enhanced the fruit weight and peel thickness.

Granulation is claimed to be a physiological disorder in citrus, especially in light skinned citrus fruits. This is attributed to deposition of pectin, lignin, and other polysaccharides in the inner walls rendering the fruits, less juicy with reduced soluble solids, and acidity (Sinclair and Jolliffe, 1961; Lloyd, 1961; Echeverry Escovar, 1969). Singh and Singh (1978a) observed no influence of soil and climate on the granulation of kinnow mandarin grown under arid and semiarid areas of northwest India. Singh and Singh (1979) while later attributed granulation in kinnow to variation in soil and climatic factors jointly. Prevalence of high temperature in spring has been reported to favour granulation (Matsumato, 1964; Awasthi and Nauriyal, 1971a). However, on the other hand, Bartholomew *et al.* (1934;1941) observed augmentation in the magnitude of granulation due to low temperature. Citrus areas having humid climate are reported to suffer from higher incidence of granulation than dry regions (Bartholomew and Reed, 1943; Lloyd, 1961). Later, studies by Awasthi and Nauriyal (1971b) showed decreased granulation under high relative humidity (Discussed in more details in Chapter 1, pp. 14-18).

Number of nutritional factors have been observed to influence granulation. High P, K, and B content are reported to cause granulation in citrus (Lloyd, 1961; Munshi *et al.,* 1979). Erickson (1968) found higher granulation in fine textured soils compared to coarse textured ones. Cicca *et al.* (1988) observed increase in fruit splitting of navelina orange with increase in available water capacity and available P content of deepest 40-80 cm soil layer in southern Italy, which reduced with increasing available N in soil. Evaluation of Nanfeng Miju mandarin orchards under different soil types and geological backgrounds by Zhao *et al.* (1996) in China revealed levels of available B and Mo in the soil influenced various fruit quality parameters. Recently, Srivastava and Singh (2000b) showed that an improvement in water soluble and exchangeable Ca^{2+} in soil as a result of $CaSO_4$ application at the rate of 1440 g tree^{-1} (400 kg ha^{-1}) in Typic Haplustert soil type, produced a very good response on fruit quality parameters such as juice content and total soluble solids of Nagpur mandarin fruits coupled with reduction in acidity of fruit juice. In well drained volcanic ash soils of Jeju Island of South Korea (pH 6.1, available 2.0 mg kg^{-1}, exchangeable K 109.6 mg kg^{-1}, and 10.2 per cent organic matter), orange fruits produced in south and east of Island had better quality than those from north and west (Moon *et al.,* 1980).

Response of soil type on the quality of satsuma mandarin indicated that the daily increase in fruit citric acid concentration was 30 per cent higher for trees grown on a tertiary rock soil than trees grown on granite soil witnessing most active acid synthesis in July. In September, the total amount of citric acid in fruit continued to increase from trees on tertiary soils and there was no obvious fall until October in contrast to fruit from trees on granite soils (Matsumoto and Shiraishi, 1980).

Table 2.15. Fruit quality indices at various yield levels of mandarin orchards (n-57)

Yield level (tons ha^{-1})	Fruit quality parameters (%)		
	Juice	TSS	Acidity
0.0-5.0 (4.4)[1]	42.4[b]	9.2[abcd]	0.78[a]
5.1-10.1 (8.1)[2]	43.2[ab]	9.0 [abcd]	0.72[a]
10.2-15.2 (13.8)[3]	48.2[a]	10.0[ab]	0.74[b]
15.3-20.3 (9.1)[4]	47.4[a]	11.0[a]	0.60[b]
20.4-25.4 (23.8)[5]	46.4[a]	10.8[a]	0.58[b]
25.5-30.5 (26.4)[6]	44.4[ab]	10.6[a]	0.56[bc]
30.6-35.6 (34.2)[7]	43.2[ab]	9.6[abc]	0.71[a]
35.7-40.7 (38.2)[8]	44.8[ab]	9.8[abc]	0.68[a]
LSD (*p= <0.05*)	3.2	0.80	0.11

[1] Mean (n=5), [2] Mean (n=10), [3]Mean (n=11), [4]Mean (n=10), [5]Mean (n=10), [6]Mean (n=4), [7]Mean (n=4), [8]Mean(n=3)

* Means in columns followed by different letters are separated according to LSD test (p=<0.05)

** Figures in parenthesis indicate for 15-30 cm depth.

Source : Srivastava and Singh (2001a)

Soil properties influencing quality of Nagpur mandarin (*Citrus reticulata* Blanco) in central India showed that the fruit quality parameters such as juice content, TSS, and acidity varied from 42.4 to 48.2 per cent, 9.0 to 11.0 per cent, and from 0.56 to 0.78 per cent, respectively (Table 2.15). Amongst a number of soil properties, free $CaCO_3$, clay content, water soluble Ca^{2+}, exchangeable Ca^{2+}, available N, P, and Zn showed a significant positive correlation with juice content and TSS, and negative correlation with acidity (Table 2.15). All the quality parameters correlated negatively with clay content. Regression equations demonstrated an increase in juice content at the rate of 1.28 per cent, 0.08 per cent, 0.92 per cent, 0.22 per cent, 0.94 per cent, and 7.84 per cent per unit increase in exchangeable Ca^{2+}, water soluble Ca^{2+}, free $CaCO_3$, available N, P, and Zn, respectively. Similarly, TSS improved at the rate of 0.28 per cent, 0.03 per cent, 0.36 per cent, 0.04 per cent, 0.48 per cent, and 2.89 per cent per unit increase in exchangeable Ca^{2+}, water soluble Ca^{2+}, free $CaCO_3$, available N, P, and Zn, respectively. While, acidity reduced at the rate of 0.005 per cent, 0.01 per cent, 0.002 per cent, 0.01 per cent, 0.001 per cent, 0.025 per cent, and 0.11 per cent per unit increase in clay content, exchangeable Ca^{2+}, water soluble Ca^{2+}, free $CaCO_3$, available N, P, and Zn, respectively (Table 2.16).

Red and black soils often occur in close proximity (Krishnamoorthy and

Table 2.16. Correlation coefficient (r) and regression analysis for various soil properties affecting fruit quality

Soil Properties	Fruit quality		
	Juice content (Y_1)	TSS (Y_2)	Acidity (Y_3)
Clay content	- 0.472	-0.482	-0.461
(X_1)	$Y_1 = 54.85 - 0.26X_1$	$Y_2 = 17.48 – 0.10X_1$	$Y_3 = 0.92 - 0.005X_1$
Exchangeable Ca^{2+}	0.611	0.681	-0.622
(X_2)	$Y_1 = 8.82 + 1.28X_2$	$Y_2 = 3.29 + 0.28X_2$	$Y_3 = 0.82 - 0.01X_2$
Water soluble Ca^{2+}	0.484	0.672	-0.618
(X_3)	$Y_1 = 32.12 + 0.08X_3$	$Y_2 = 7.81 + 0.03X_2$	$Y_3 = 0.98 - 0.002X_3$
Free $CaCO_3$	0.612	0.752	-0.642
(X_4)	$Y_1 = 37.80 + 0.92X_1$	$Y_2 = 9.19 + 0.36X_4$	$Y_3 = 0.78 – 0.01X_4$
Available N	0.546	0.611	-0.582
(X_5)	$Y_1 = 20.10 + 0.22X_5$	$Y_2 = 6.94 + 0.04X_5$	$Y_3 = 0.73 - 0.001X_5$
Available P	0.521	0.506	-0.489
(X_6)	$Y_1 = 39.28 + 0.94X_6$	$Y_2 = 2.81 = 0.48X_6$	$Y_3 = 1.12 - 0.025X_6$
Available Zn	0.621	0.801	-0.614
(X_7)	$Y_1 = 35.82 + 7.84X_7$	$Y_2 = 7.32 + 2.89X_7$	$Y_3 = 0.68 - 0.11X_7$

Source : Srivastava and Singh (2000 a)

Govindarajan, 1977). A greater depth of black soils than red soils suggested that former are situated on depositional surface and latter on erosional surface. There are two school of thoughts regarding the development of red and black soils. One school of thought advocates their genesis to difference in drainage conditioned by relief. Another school attributes these differences to mineralogical makeup of parent material. The red soils are stated to be associated with potash feldspar and black soils with soda lime feldspars. Comparison of pedological features of red versus black soils revealed higher clay content in black soils than their associated red soils. All the red soils showed an accumulation of clay in B horizon. Dithionite extractable Fe content is more in red than in associated black soils. Based on these properties, it is hypothesized that the difference in genesis and occurrence in close proximity of these two contrasting soils are mainly related to drainage as conditioned by relief. The low chorma and dithionite-Fe along with minerals of black soils are the clear cut indications of impeded drainage in black soils compared to their red counterparts (Rudramurthy and Dasog, 2001).

Nutrient release behaviour of red versus black soils further showed that available K varied from 70.0 to 324.0 mg kg^{-1} and from 70.0 to 281.0 mg kg^{-1} in black and red soil type, respectively. While, total K ranged from 0.24 to 0.70 per cent in black soil type and from 0.30 to 0.41 per cent in red soil type. Further, prominent differences with respect to various forms of K between red clay loam and black clay soil types were observed (Table 2.17). The forms of soil K, such as water soluble K (6.50-53.00 mg kg^{-1}) and exchangeable K (56.00-291.00 mg kg^{-1}) were much higher in black soils than water soluble K (2.00-13.00 mg kg^{-1}) and exchangeable K (57.50-268.00 mg kg^{-1}) in the red

Table 2.17. Comparison between forms of K in two soil types, red versus black soil

Soil depth (cm)	pH (1:2.5)	EC (dS m^{-1})	Forms of soil K (mg kg^{-1})			
			Water soluble K	Exchangeable K	Non-exchangeable K	Mineral K
			Red soil type (Vertic Ustochrept)			
0-15	8.01	0.15	13.00	268.00	516.20	3302.80
15-30	8.12	0.16	3.00	98.00	304.60	2794.40
30-65	7.88	0.15	3.70	82.50	224.82	3088.98
60-90	7.82	0.15	2.64	86.50	316.78	3194.08
90-120	7.82	0.17	12.50	57.50	272.41	2657.59
120+	7.92	0.19	2.00	70.00	340.32	2787.68
CV (%)	10.10	6.12	14.18	19.60	24.80	26.41
			Black soil type (Typic Haplustert)			
0-15	8.12	0.12	53.00	291.00	468.12	4187.88
15-30	8.25	0.08	28.00	274.00	348.16	2749.84
30-60	8.43	0.07	24.00	188.30	320.28	2567.42
60-90	8.41	0.09	14.00	56.00	320.72	2409.28
90-120	8.47	0.10	6.50	87.80	316.78	1988.92
CV (%)	8.00	10.50	18.20	24.20	20.20	32.20

Source : Srivastava and Singh (2001 b)

soils. While, non-exchangeable K followed just the reverse order. These values show more clear cut difference when effective soil depth of upper 30 cm is considered. Julien (1989) while working on potassium fertilizer response experiments suggested that the interpretation of soil analysis must address following three reference points such as K saturation of the cation exchange capacity, desired level of K, and compensation for crop removal. Potassium saturation of the cation exchange capacity, known as exchangeable potassium percentage is widely used for diagnostic purposes. Earlier studies (Srivastava *et al.,* 1997) revealed the higher contribution of exchangeable K towards total K requirement of Nagpur mandarin in black montmorillonitic clay soil.

Fruits from red soil types especially those located at higher elevations possessed much tight skinned having better total soluble solids than fruits from black soil types. Various fruit quality parameters of Nagpur mandarin were invariably better in red clay loam soil types to those of deep black clay soil types (Table 2.18), primarily on account of difference of available water capacity, in addition to availability of various nutrients. Higher availability of accessible forms of K in black soils induced high total acidity (0.78 per cent) compared to red soil

Table 2.18. Comparison of quality of Nagpur mandarin grown in two soil types, red versus black soil

Soil type	Fruit quality parameters (%)		
	Juice	Total soluble soilds	Acidity
Red soil	43.2	10.8	0.64
Black soil	42.4	9.2	0.78
LSD (*p=0.05*)	0.6	0.4	0.08

Source : Srivastava and Singh (2001 b)

(0.64 per cent) in the fruits and took comparatively longer time for conversion into sugar, and, thereby, influenced the development of required total soluble solids and acidity ratio. With the result, the fruit maturity in black clay soils attained quite late compared to red clay loam soils. Higher availability of K coupled with good water holding capacity provided better hydration capacity of fruits in black clay soils, which is relatively weak in red clay loam soils. This is the reason, why sub-optimum size of Nagpur mandarin fruits is the major limitation in red soils, but qualitywise highly thin, light skined with bright colour fruits are obtained (Fig. 2.2, 2.3, 2.4 and 2.5). The difference in moisture supplying capacity of soil is also an important contributery factor. In deep black clay soils, fruits tend to become more loose, spongy, and granulated, if continued to be kept on trees beyond optimum total soluble solid-acid ratio. The nature of soil type, hence, played a decisive role in guiding the fruit quality, because of distinct difference in available pool of nutrients.

Fig. 2.2. Vertical cut of black soil type (10 YR 3/2-10 YR 3/3, moist colour)

Fig. 2.3. Loose skinned Nagpur mandarin (*Citurs retriculata* Blanco) fruits

Srivastava and Singh (2000d) studying the influence of soil properties on various quality parameters of sweet orange, cultivar mosambi grown in Marathwada region of Maharashtra observed the highest TSS of 10.0-10.9 per cent at 30.7-32.8 per cent clay, 8.0-9.6 per cent $CaCO_3$, 188.4-197.7 mg l^{-1}, water soluble Ca^{2+}, 22.8-28.8 cmol (p^+) kg^{-1} excheangeable Ca^{2+}, 134.6-140.8 mg kg^{-1} available N, 15.2-19.6 mg kg^{-1} available P, and 1.10-1.20 mg kg^{-1} DTPA-Zn compared to lowest TSS of 8.0-8.4 per cent at 41.0-45.7 per cent clay, 1.5-3.0 per cent $CaCO_3$, 115.0-119.2 mg l^{-1} water soluble Ca^{2+}, 18.0-20.7 cmol (p^+) kg^{-1} exchangeable Ca^{2+} 103.6-110.6 mg ka^{-1} available N, 8.7-9.8 mg kg^{-1} available P, and 0.80-0.88 mg kg^{-1} DTPA-Zn. Likewise, acidity in juice showed lower values at higher values of all the above soil properties, thereby, suggesting the definite role soil type in regulating the quality pattern of sweet oranges.

Fig. 2.4. Vertical cut of red soil type (5 YR 4/3-5 YR 3/3, moist colour)

Summary

An analysis of various aspects of soil suitability for citrus has revealed that soils identified suitable on the basis of various physical and chemical constraints backed up by the soil to leaf nutrient relationship and soil factors favouring the development of quality fruits, are bound to produce high production coupled with quality than the production in the absence of these information. Soil fertility guidelines would be highly handy in diagnosing the limitations and potentials of a specific soil type and to put them under an effective use.

Fig. 2.5. Tight skinned Nagpur mandarin (*Citrus reticulata* Blanco) fruits

3 Soil Fertility and Production Sustainability

3.1 INTRODUCTION

Fruit crops are unique and distinct in their nutrition from annual crops in a number of ways (Iyengar and Kotur, 1996). Various citrus scion varieties are budded on different rootstocks which vary in their nutrient absorption characteristics compared to their scions, and their large root systems exploiting larger soil volume. There is a simultaneous need for nutrients to meet the requirement of fruit development and vegetative growth. Nutrient supply, uptake, and nutrients held in the perennial framework of the tree have considerable influence on the nutrition, growth, and production during subsequent years (Reddy *et al.*, 1991; Jackson, 1992; Ghosh, 2000). There are five important aspects of Agriculture (Kanwar, 1997), namely, meeting the changing needs of today and tomorrow; economic viability of enhanced level of productivity; successful management of resources, internal or external; renewable or nonrenewable; maintenance, preferably enhancement of the quality of environment and conservation of natural resources viz., soil and water. Thus, soil is undoubtedly the pivot, not only of food security, but also of sustainable Agriculture, hence, of sustainable food security. The basic foundation of soil fertility enunciated for the first time during 1940, carries three essential components viz., all biological processes of soil fertility maintenance is the result of the activities of the living system of the soil; the C/N ratio is an index of decomposability of organic matter added to the soil; and the living system for its normal functioning, requires moisture, energy material as carbon source, and other nutrients.

The soil is no longer considered as an inert matter for anchoral support of the plant. But, as an ecological dynamic living system, it is studied for distribution, abundance, and interactions of organisms in space and time, and the interrelationship between organisms and the physical environment. For the cycles, flow of nutrients in the system and maintenance as well as production functions are equally important. The performance criteria are, cycling rates and energy efficiency. Soil, as the mother earth, has all the functions of the 'Tridev' according to Hindu mythology. She is at once the 'Brahma', the 'Vishnu', and the 'Maheshwar' – the creator, the preserver, and the destroyer. Soil is the 'soul of infinite life'. Properly cared with respect, it sustains life and to an extent, it keeps the poison (toxic substances) to itself, and helps prevent its creature from deadly things. But once abused, challenged, or mismanaged, it retaliates and the very existence of life is endangered. It has been established by 1890 that carbon (C), hydrogen (H), oxygen (O), nitrogen (N), phosphorus (P), sulfur (S), potassium (K), calcium (Ca), magnesium (Mg), and iron (Fe) are required by plants. Their absence or low availability resulted in either the death of the plant or very poor plant growth with accompanying visual symptoms. Between 1922 and 1954, additional elements determined to be essential are manganese (Mn), copper (Cu), zinc (Zn), molybdenum (Mo), boron (B), and chlorine (Cl). During 1804 to 1954, the essentiality of a number of nutrients has been discovered (Table 3.1). Plant physiologists of today are still attempting to determine, if there are additional elements that are essential to plants, applying three requirements of nutrient essentiality as set forth by Arnon and Stout over 50 years ago. Arnon and Stout (1939) published their criteria on the concept of essentiality of nutrients in plant nutrition which are explained as : omission of an element in question must result in abnormal growth and failure to complete the life cycle or premature death of the plant; the element must be specific and not replaceable by another; and the element must exert its effect directly on the growth or metabolism, and not indirect effect such as by antagonising another element present at a toxic level.

Table 3.1. Chronological developments to establish the essentiality of nutrients

Element	Discoverer of essentiality	Year
C	DeSaussure	1804
H	DeSaussure	1804
O	DeSaussure	1804
N	DeSaussure	1804
P	Ville	1860
S	Von Sachs, Knop	1865
K	Von Sachs, Knop	1860
Ca	Von Sachs, Knop	1860
Mg	Von Sachs, Knop	1860
Fe	Von Sachs, Knop	1860
Mn	McHargue	1922
Cu	Sommer, Lipman and MacKinnon	1931
Zn	Sommer and Lipman	1926
Mo	Arnon and Stout	1939
B	Sommer and Lipman	1926
Cl	Stout	1954

Source : Glass (1989)

Soil provides the shelter, nourishment, and energy to plants, and other organisms habituating on it. The essence of life in the soil is in its crop producing capacity i.e. the soil productivity. Soil fertility is that component of productivity which primarily deals with nutrient supplying capacity of the soil to the plant (Goswami, 1999). On the more applied aspects, mention may be made of Truog's prescription procedure and Ramamoorthy's targeted yield approach, for fertilizer recommendations based on soil tests, development and refinement of soil test methods, use of stable radioisotopes as tracers in soil fertility, and fertilizer use studies with reference to determining fertilizer use efficiency. Of late, interest has also been evinced on the use of organic manures and integrated plant nutrient supply and management systems (IPNS or INM). Soil fertility research in the recent past has witnessed many changes.

Soil tests are hardly used by the farmers for determining fertilizer needs, with the result, the trends of productivity have not been consistent. More and more deficiencies of nutrients have cropped up. Fertilizer use efficiency has hardly improved (N 30-50 per cent, P 10-20 per cent, and Zn 3-5 per cent) and ultimately response ratio (economic yield in kg per kg nutrient applied as fertilizer) has further declined. There is a wide disparity in fertilizer use when considered from one agroclimate to another for the same citrus cultivar. This has mainly emerged through the imbalanced use of N, P_2O_5, and K_2O fertilizers. The relative use of P_2O_5 and K_2O has remained low affecting both, the quantum and quality of produce adversely. Fruit crops generally remove much higher quantities of potassium and, therefore, an ideal ratio may have to be 4.2: 2.5 : 5 (or 0.83 : 0.5 : 1) to produce high yield of good quality fruits (Ghosh, 2000). The foremost reason is the decline of soil productivity over the years, because of a greater mining of the soil for nutrients. The soil has lost its resilience and buffering action, because of practically no organic input into the system. This has resulted into slow biological degradation of the soil which is essentially a 'biophysical' system with soil organic matter and soil biomass as the real 'elixir'. Balanced fertilization and interaction of nutrients with other inputs for production have not been given due and an adequate importance, and the limiting factor(s) of production as and when, they appear, have not been attended too. According to Talibudeen (1974), the life time of a crop is characterised by well defined growth stages. During each stage, the plant has a work potential, defining its strength to assimilate and to use a nutrient from the soil. Interactions, positive or negative between nutrients play their recognized role in defining the magnitude of response. Crop yield is the result of a large number of simultaneous growth factors according to Visser (1969) and Visser and Kowalik (1974). The effect of various factors has a part in common, because all these effects are based on the principle of general relation of nutrient uptake.

In principle, balanced fertilization means a system of soil-crop management that would ensure efficient use of fertilizer nutrients and maintain crop yield beside soil productivity. It must take cognisance of growth (nutrient-yield) relations, the limiting factor (limiting nutrient) at various levels of nutrient (fertilizer), and the yield expectations. Identically, balanced fertilization must also be based on the concept of integrated nutrient management for a cropping system (Goswami, 1997; 1998). Balanced fertilization lies in achieving a positive interaction which is always more than the additive effect of two or more nutrients. Anything short of fully additive is imbalance. Cooke (1979) produced data from UK to conclude that the shape of response relationship showed that the fertilizer treated soils could not achieve the same yield level from organically treated soils. Similarly, soils enriched in P and K can often produce larger yield than soils poorer in P or K. In that direction, soil testing for fertility evaluation has helped to economise the use of fertilizers according to crop requirement (Velayutham *et al.*, 1985).

3.2 CONCEPT OF SOIL FERTILITY

Soil fertility is a complex term which includes many components viz., soil depth, texture and structure, soil reaction, organic matter content and composition, activity of soil organisms, nutrient content, storage capacity for nutrients, absence of detrimental toxic substances etc. The result of an optimum combination of these factors is a high soil fertility that means a high crop production potential. Bioavailability of nutrients especially micronutrients depends upon valency of nutrient element in the weathering solid, nature of solid phase present in the primary minerals, redox conditions (Eh and pH), complex legend formation ability, microbial activity, and environmental factors. Interest in soil fertility began perhaps with the dawn of civilization. In fact, soil studies began with investigation on some aspects of soil fertility, be it in the area of humus or water containing dissolved solids of soil material. Notable among the early scholars of Soil Science are Thaer, Berzelius, Sprengel, Mulder, Davy, and others, most of them are the proponents of 'humus' theory of plant nutrition. Theodore de Sassure demonstrated in 1804 that plants require minerals from the soil, and still much earlier in 1733. Tull (1933) advocated that the real food of plants is not the 'juices of the soil' (in the present day context, this may be considered as the water with dissolve salts in it or soil solution), but the very minute particles of soil. Dokuchayev (1949) wrote that Chernozem is more precious than other soil or coal, than gold or iron ore, and constitutes an eternal inexhaustible Russian wealth. The quality of evaluation of coals is a method of determining soil fertility (Dokuchayev, 1950).

Black (1993) and Liebig (1855) stated in his law in three parts as: i. by the deficiency or absence of one necessary constituent, all others being present, the soil is rendered barren for all those crops to the life of which that one constituent is indispensable, ii. with equal supplies of atmospheric condition for the growth of plants, the yield is directly proportional to the mineral nutrient supplied in the manure, and iii. the yield of a field can not be increased by adding more of the same substances in a soil rich in mineral nutrients. Humus theory stated that the plants lived on humus derived extracts (*Extraktivstoff* in German) containing simple water soluble compounds of C, H, O, and N from which they are able to rebuild more complex plant tissue. Plants also are thought to be able, by means of an internal vital force (*Lebenskraft* in German and *vis vitalis* in Latin), to generate from these four elements other vital constituents such as Si and K (Wendt, 1950; Russell, 1952; Wild, 1988). Some fertilizer substances like salts and lime are considered useful for plant growth, but only because they promote the decomposition of humus and dissolution of organic matter in the soil solution.

Albrecht Thaer (1752-1828), Sprengel's mentor, is one of the most well known advocates of the humus theory (Wendt, 1950). His often quoted book in four volumes (Thaer, 1809-1812) translated into english as late as 1844 (Russell, 1952), showed that the humus theory even in the modern period (1800-1860) still has received acceptance. Russell (1952) and Wild (1988), however, also pointed out that during the Phlogistic peroid, new methods of studying plant nutrition have been developed such as pot experiments and plant analyses. The end of the Phlogistic peroid and the beginning of the modern period are marked by the work of de Saussure (1804) in Switzerland. De Saussure did pioneering work on gas exchange of plants and the nature of origin of salts in plants. He showed that the plant roots are able to take up salts from the soil. Sprengel (1826) described his theory on mineral nutrition of plants more precisely than in his first publication. In total, he listed 20 elements that he considered as plant nutrients, including N, P, K, S, Mg, and Ca (Sprengel 1826). In this article, he in essence also formulated the law of the minimum. Waksman (1942) and Browne (1942), in treatises about the humus theory and the law of the minimum, respectively, noted, the fact that plants obtained their minerals directly from the soil has already been clearly outlined by Sprengel (1838) two years before the publication of Liebig's book and that a formulation of law of the minimum can be found in Sprengel (1837). Accordingly, Browne (1944), in the preface of his book on the history of Agricultural Chemistry wrote that in his Organic Chemistry in its application to Agriculture and Physiology, Liebig was more a promulgator and defender of truths that had already been announced as a discoverer of new knowledge. More recent views on Liebig's work in Agricultural Chemistry can be found in detailed analysis of Finlay (1991), Munday (1991), and Schling-Brodersen (1992).

In view of the historical developments, both Wendt (1950) and Bohm (1987) concluded that Carl Sprengel must be considered as a true founder of the doctrine of mineral nutrition of plants and Justus Von Liebig as the indefatigable contender in the struggle for its acceptance. By 1848, his book passed through more than 20 editions and reprints: six in Germany; five in England, three in USA, two in France, two in Italy, and one each in Denmark, Holland, Poland, and Russia (Browne, 1944; Paoloni, 1968; Bohm, 1997). In the first edition of this book (Liebig, 1840) and also in the several following ones, Liebig did not discuss specifically the law of minimum. This law is discussed specifically by Liebig (1855) where the term minimum is used for the first time. Not until the seventh edition of his first book published in 1862 (Vol 12, p.223), did Liebig clearly formulate the law of the minimum. In view of the historical developments, it seems to be appropriate to call the law of the minimum, henceforth, the Sprengel-Liebig law of the minimum. (Van der Phoeg *et al.*, 1999)

The real interest in soil fertility and plant nutrition began with the epoch making 'mineral nutrition' theory of plants advanced by Liebig in 1840 who is credited to have postulated two cardinal laws - the 'law of the minimum' and the 'law of complete return'. The first law states that the yields of field crops reduce or increase in exact proportion to the reduction or increment in the quantity of minerals applied to the soil in the form of fertiliser. The essence of the second law is the main element of farming lies in the fact that the soil must get back everything that has been taken away from it. It is the unchangeable law of nature (Liebig, 1864). This was followed by the works of Boussingalt, Lawes and Gilbert, Hellrigel and Wilfarth, who also corrected some of the views of Liebig. Boussingalt showed that all plants take N from the soil, except the leguminous ones, which themselves can enrich soil with this element. If Jethro Tull's advocacy of soil to be in a 'loose small clod', condition can be considered as the origin of the concept of soil structure. Liebig's discovery can be considered as the origin of soil fertility, plant nutrition, and fertilizer use in Agriculture. The humus theory of plant nutrition received a shattering blow from Liebig. He asserted that plants have an inexhaustible reserve of CO_2 in the air. The advantage of humus is that it continuously releases carbonic acid. Plant absorption of minerals from the soil is facilitated by the continuous process of aeration and acidic secretion of the roots.

Unfortunately, the second law of Liebig, the 'law of complete return' never received as much attention as it deserved. The basic laws or theories relating to soil fertility and plant nutrition, Liebig's law of minimum, Mitscherlich's law of diminishing return, discovery of nitrogen fixation from atmosphere by Hellrigel and Wilfarth, the cation exchange phenomenon of clay, are all developments in the 19th or early 20th century. All researchers in soil fertility have one common goal. This is to assess the nutrient supplying capacity of the soil, deficiencies of nutrients, if any, and to supply nutrients based on crop needs through fertiliser and organic sources. Thus, the crop production phenomenon is considered `soil-nutrient-plant' interaction conditioned by the level of management. According to Wallace (1995), the law of minimum popular in the past decades can now be replaced with law of maximum which states that limiting factors interact in a sequentially additive manner so that when they are corrected, progressively more response is obtained for use of each needed input. The law of maximum can not operate when any severe limiting factor is present because sequential activity does not operate under such conditions. When considerable visible chlorophyll deficiency exists, Fe is usually such a Liebig type limiting factor where certain inputs can actually antagonise the nutritional status. Under severe limiting conditions, use of inputs could be wasteful unless most severe limitation is corrected and such limiting factor has been labeled as Liebig type (Wallace and Wallace, 1993). In the recent years, the nutrient additions have been excessively in favour of mineral fertilizers. While, the quick and substantial response to fruit yield to mineral fertilizers eclipsed the use of organic manures, the inadequate supply of the latter sources exacerbated this change (Ghosh, 2000). Integrated nutrient management (INM) with emphasis on use of bio-organics is a comparatively recent concept which needs to be vigorously pursued to achieve the sustainability in citrus production trend spaced over the years.

3.3 PRECISION CITRICULTURE- A NEW PHILOSOPHY

In many of the citrus orchards, it is common to notice the yield variation so large that varies from tree to tree or orchard to orchard under similar management level. Sometimes in some portion of orchard, the average yield is much below the average yield of an entire orchard. These problems can be amicably solved through precision Agriculture, a relatively new management philosophy (Neff, 1997) that just might provide an information precisely how much each tree is producing. Once this is obtained, precision

Citriculture might enable to manage the orchard at micro level, and obtain yield to its potential. It can produce an insight into the problematic piece of land and decision may be taken to abandon citrus into that portion, and concentrate on most productive piece of orchard or best producing trees.

Precision Citriculture is also sometimes referred to as site specific or precision site Agriculture. Site specific fertility management is a form of precision Agriculture. The traditionally conventional management minimises diversity, while precision Citriculture maximises diversity, since normal orchard management advocates to treat the orchard uniformly and precision. Citriculture encourages orchard management on individual tree basis. However, precision orchard management may involve high level of vigil and labour investment initially, but the proponents of precision Agriculture believe that it is economically viable when considered on long term basis. It also reduces the impact of agricultural practices on environment. According to Neff (1997), advances such as global positioning system, remote sensing device, and variable rate technology might allow citrus growers to automatically treat trees individually without high labour cost.

Fig. 3.1. Quality citrus production: a prime objective of commercial Citriculture

Blazquez (1989) earlier used tools like densitometry, image analysis, and aerial coloured photographs for separating stressed citrus trees as a part of precision Citriculture and suggested that densitometric measurements are equal to or better than visual grading to be later fitted into automated system to count and separate stressed from healthy trees. Using remote sensing technique like false colour composites of IRS (Indian Remote Sensing Satellite) and LISS-IA (Linear Imaging Self Scanning Sensor), Sinha and Sengupta (1995) reported that remotely sensed data can be effectively used for discrimination of orange plantations from field crops, forests, and scrubland.

The first step for precision Agriculture in citrus orchards with un-uniform yield variation, is the mapping orchard for yield variation on tree basis. The same way, other information like variation in soil depth, texture feasibility etc. can be mapped. Combining all these information, a grower can figure out the causes of yield variation. Not all precision Agriculture depends on global positioning system, some have good results with real time sensors attached to equipment operating in an orchard. Variable rate technology, an another key component of precision Agriculture, uses the mapping information to adjust application of inputs, they need. For instance, applicators with computers and on board controllers can automatically change the rate of fertilizer application to coincide with soil variability as per soil map. Though, the precision Agriculture today may not look a reality for each progressive citrus grower, but with more and more efforts put in, the refinement of techniques involved holds a very good promise in the years to come for precise site specific orchard management. Development of precision citrus orchard would usher in the development of model orchards and ideotype trees.

3.4 METHOD OF SOIL FERTILITY EVALUATION

Fertility evaluation, by and large aims at nutrient balances which result from combination of five basic inputs, namely, fertilizer, manure, deposition, N-fixation, and sedimentation and five basic outputs, namely, harvested product, crop residues, leaching volatilization, and erosion (Stoorvogel and Smaling, 1990). Indicators of favourable soil fertility include thickness of A horizon, lower bulk density, and higher level of organic C, N, and P (Sandor and Eash, 1995; Sandor and Furbee, 1996). Soil quality in terms of various soil parameters vis-a-vis agricultural sustainability and environmental ecology has been defined by many workers in the recent past (Lal, 1991; Karlen *et al.,* 1992; 1994a; 1994b; Larson and Pierce, 1994; Doran and Parkin, 1994a; Sun *et al.,* 1995; Pennock and Vankessel, 1997). These definitions are more or less different, but they always relate to the functions of the soil to supply nutrient and other physicochemical conditions to plant growth, promote and sustain crop production, provide habitat to soil organisms, ameliorate environment pollution, and resist degradation besides improving human life.

Many approaches or concepts have been suggested for predicting the fertilizer requirement of soil for a crop or cropping system. Some of the them which have been experimented upon in India are: recommendations based on field experiments with various rates of fertilizer and then working out an optimum dose based on response curves; soil test rating (Mohr *et al.,*1965); sufficient level of available nutrients (SLAN) concept based on Mitscherlich equation on per cent maximum yield, and a modified form of this referred as critical level concept of Cate and Nelson (1965), basic cation saturation ratio (BCSR) of Bear *et al.* (1945); prescription procedure of Truog (1948), and a variation of it namely, the targetted yield approach; fertilizer recommendations for maximum profit using regression and fertility gradient approach in field experiments; an intergrade or a hybrid of the SLAN and BCSR concept proposed by Fisher (1975) using second degree polynomial to express a fertility index by which a per cent yield can be predicted from soil test values for P and K. Singh (1995) summarized the soil testing philosophy in relation to soil fertility and fertilizer use management. The SLAN concept according to him, is primarily based on the principle of fertilizing the crop or that part of the crop that is required.

The fertility capability classification (FCC) is designed to group soils having simlilar limitations of fertility management (Boul, 1972; Boul *et al.*,1975; Sanchez *et al.*,1982). Such classifications are based on soil and other environmental constraints for various uses to have practical land units, with similar edaphic constraints, and varied bases of application. It emphasizes qualifiable top soil parameters as well as subsoil properties having direct relevance to plant growth. The FCC classes also address the mean fertility related soil constraints which can be interpreted in relation to specific farming system or land utilization types (Sanchez *et al.*, 1982). This is the first technical soil classification system that groups soil according to their fertility constraints in a quantitative manner. The system of FCC described by Boul *et al.* (1975) consists of three categorical levels : type (top soil texture upto 20 cm), substrate type (subsoil texture upto 60 cm depth), and 15 modifiers as elaborated below :

Type : (S, sandy top soils with loamy sands and sands; L, loamy top soils, less than 35 per cent clay; C, clayey top soils with more than 35 per cent clay; O, organic soils with more than 30 per cent organic matter)

Substrate type : It is normally used following the marked textural variation in subsoil when compared with texture of top soil or due to presence of hardpan up to 50 cm depth (S, sandy subsoil; L, loamy subsoil; C, clayey subsoil; R, rock or other hard pan).

Modifiers : g (gley), d (dry), e (low cation exchange capacity), a (aluminium toxicity), h (acid), i (higher P fixation by Fe), x (X ray amorphous), v (Vertisol), k (low K reserves), b (basic reaction), s (salinity), n (natric), c (cat clay), l (gravels), % (slope).

The FCC units then list the type and substrata type (if present) in capital letters, and the modifiers in lower case letters, the gravel modifier as a prime (') and the stop, if desired, in parenthesis. For example, Oxisols belong to the FCC unit Caeik (Clayey, Al-toxic, low CEC, high P fixation by Fe, low K reserves) according to descriptions made by Sanchez *et al.* (1982). Janssen *et al.* (1990) described another system known as quantitative evaluation of fertility of tropical soils (QUEFTS), applicable to well drained, deep soils having pH range of 4.5-7.0 and values for organic carbon, Olson-P, and exchangeable K below 70 g kg^{-1}, 30 mg kg^{-1}, and 30 mmol kg^{-1}, respectively, for 0-20 cm depth. The procedure consists of four successive steps. First the potential supplies of N, P, and K are calculated applying the relationship between chemical properties of the 0-20 cm soil layer and the maximum quantity of nutrients can be taken up by crop, provided no other nutrients and growth factors are yield limiting. In the second step, the actual uptake of each nutrient is calculated as a function of the potential supply of that nutrient taking into account the potential supplies of the other two nutrients. Step three comprises the establishment of three yield ranges, depending on the actual uptake of N, P, and K. Next, these yield ranges are combined in pairs and the yields estimated for pair of nutrients are averaged to obtain an ultimate yield estimate as step four. Smaling and Janssen (1987) compared the QUEF TS with fertility capability classification (FCC) described by Sanchez *et al.*(1982). It has been observed that QUEFTS gives more quantitative information on yields, whereas FCC qualitatively indicates the factors that might cause a discrepancy between calculated and measured yields. It provides a guide for the extrapolation of the fertilizer response experience (Boul and Nicholaides, 1980). Among the various approaches providing information on the potential of the soil for crop production, fertility capability classification is one which emphasizes the components of soil fertility within 50 cm layers from the soil surface. Further, FCC focuses attention on surface soil properties, most directly related to management of field crops, and is best used as an interpretive classification in conjunction with more inclusive natural soil classification.

Twenty one soils, belonging to subgroups Typic Haplusterts, Typic Ustropepts, Udic Haplustalfs, Typic Haplustalfs, Vertic Haplustalfs, and Typic Ustorthents are grouped in 8 FCC (Fertility Capability Classification) units based on type, substrata type, and condition modifiers by Mathan *et al.* (1993). These FCC units served as the basis for conducting fertility related experiments and extrapolation of experimental results. The conditions modifiers that decide the soil and fertilizer interactions are : d (dry condition), b (basic reaction), v (vertic characters, m (magnesium deficiency), n (natric), k (potassium deficiency), i (Fe-P fixation), and e (low CEC).

Soil properties other than the extractable levels of micronutrients should also be considered when soil test results are interpreted (Cox, 1987). The inclusion of properties viz., clay, organic carbon, pH, and exchangeable Ca^{2+}+ Mg^{2+} with soil available micronutrient cations of the three soil depths to leaf concentration of micronutrients increased the predictability from 46 to 66 per cent, indicating thereby, the usefulness of soil properties in assessing the micronutrient availability in mandarin orchards (Patiram *et al.*, 2000). Egashira *et al.* (1990) suggested that the phosphate absorption coefficient of a soil as a good index to the potential productivity classification for satsuma mandarin orchards established on red yellow soil type. While, Tavdgiridze and Putkaradre (1991) proposed polarity coefficient within the plant as one of indices for finding the nutrient supplying capacity of soil. It is usually presumed that soil fertility variation is negligible under well prepared soil conditions. However, few tree by tree, soil fertility studies indicated some differences (Cohen, 1980). Brams and Fiskell (1967) studied variability of extractable nutrients in a valencia orchard on the ridge, the importance of multiple sampling for estimating nutrient status and standardisation of techniques of soil sampling in relation to the trees. Sampling under dripline reduced variability compared to sampling in the middles. Root mineral content showed little similarity with soil nutrient levels. Trees modify the soil around their roots through nutrient uptake and exudation of compounds (Anonymous, 1951; Buckman and Brady, 1960), the effects of that are to a larger extent influenced by rootstock-scion combination.

3.5 SOIL SUSTAINABILITY

Soil sustainability, is commonly defined as the ability of a soil to produce economic yield of crops without much deterioration in soil quality. However, according to Fresco and Kroonenberg (1992), sustainability is a complex concept, but it is possible to indicate a situation under which it will no longer meet various criteria for sustainability, once these are defined. There are basic components of sustainability namely, more ecologically based without destroying natural resources, based on conscious ethic regarding humankind relationship to future generations and magnitude holding equitable (Douglas, 1984). A number of other definitions of sustainability have been put forward, like net productivity of biomass maintained over decades to centuries (Conway, 1987), means survival, low input, low input organic farming (Walsh, 1991), and living on interest and not on capital (Bennett, 1991).

Fig. 3.2. Sustainability in production : Nagpur mandarin orchard in Typic Pellustert soil type under subhumid tropical climate of central India

A sustainable land management system is one that does not degrade the soil or significantly contaminate the environment, while providing necessary support to human life. A sustainable agriculture is defined as a system which maintains an acceptable and increasing level of productivity that satisfies prevailing needs, and is continuously adapted to meet the future needs for increasing the carrying capacity of the resource base, and other worthwhile human needs, or in other words, a system in which farmer continuously increases the productivity at levels that are economically viable, ecologically sound, and culturally acceptable, through the efficient management of resources and orchestration of inputs in numbers, quantities, qualities, sequences and timing, with minimum damage to the environment and danger to human life (Okigbo, 1991). Swift *et al.* (1991) states that a cropping system is not sustainable, unless the annual output shows a nondeclining trend and is resistant in terms of yield stability to normal fluctuation of stress and disturbance. It is also defined as a system, which involves the management, conservation of the natural resource base and the orientation of technological and institutional change in such a manner as to ensure the attainment and continued satisfaction of human needs for present and future generations. Such sustainable development conserves land, water, plant, and animal genetic resources, besides economically viable and socially acceptable (FAO, 1991).

Conceptually, soil sustainability could be formulated as the number of years, when the soil depth will be reduced to less than 0.25 m, thereby, making the soil incapable of producing economic crop yield under rainfed condition. It is calculated from the difference between soil loss per hectare (tons) and soil regeneration rate under different climatic environments (FAO, 1992). According to Hudson (1971), the rate of top soil formation is less than 0.25 mm $year^{-1}$ in dry and cold environment, and more than 1.5 mm $year^{-1}$ in humid and warm environment. The top soil formation at the rate 1 mm $year^{-1}$ is equal to 12 tons ha^{-1} $year^{-1}$. Therefore, the rate of top soil formation has been considered as 1.5 mm $year^{-1}$ in dry subhumid environment which is equivalent to an addition of 18 tons ha^{-1} $year^{-1}$. The rate of soil formation under semiarid condition on basement complex rock weathering is extremely slow. According to an estimate, in semiarid area, the soil formation rate is less than 0.01 mm $year^{-1}$. The soil formation under these conditions due to lack of weathering, is taken as negligible in view of sustainability concept in production in human life span. It is assumed that deep soil (>1m) having less than 3 per cent slope has sustainability index of 100. The index of the deep soil is used as reference for calculating the sustainability of other soils. Sustainability index of 50 means that soils will be productive 50 per cent of the time with reference to deep soils having less than 3 per cent slope in absence of proper soil conservation measures under a specified cropping system. Thus, soil sustainability index has been formulated as : SI = (Sdymax - Sdy)/Sdymax = 1 - (Sdy/Sdymax) where, SI stands for soil sustainability with respect to depth, Sdymax stands for maximum sustainability year of a deep soil with less than 3 per cent slope, and Sdy stands for number of years when the soil would be reduced to less than 0.25 m. The maximum sustainability in a physiographic environment based on Nagpur district soil survey data base on 1:50,000 is assumed scale through the formulation of simplified soil sustainability index (SI) and their classes (Mandal and Mandal, 1997).

Based on this concept, soil sustainability index class has been developed (Table 3.2) and a soil sustainability index map of Nagpur district is prepared. These soil sustainability indices showed that the eastern parts of Nagpur district comprising Manda, Kamptee, Kuhi, a parts of Bhiwapur, Umrer, Ramtek, Parseoni, Kalmeshwar, and parts of Narkhed tehsils come under high to very high value of SI (> 50). The western parts of Nagpur comprising Hingna, Nagpur, Katol, and parts of Saoner tehsils showed medium sustainability value (26-50). The hilly area in the Ramtek and Parseoni tehsils have low sustainability index (< 25). Using more or less similar concept, Jawahar *et al.* (1999) prepared the fertility capability for soils of Tamil Nadu, India

Table 3.2. Soil sustainability index class

Class	Sustainability index value (SI)
Low	< 25
Medium	26-50
High	51-75
Very high	> 76

Source : Mandal and Mandal (1997)

3.5.1 Importance of Soil Organic Matter

The importance of organic matter in soil fertility is well known, but its role as a source of nutrient is a dominant one, particularly in tropical Citriculture. The level of organic matter deserves to be maintained at an acceptable level above critical values corresponding to structural stability and biological activity. In rainfed tropical region, soil organic matter is the foundation for the building up of any sustainable annual cropping system, but there is not yet a definite answer to the critical level that should be maintained (Pieri, 1995). According to Finck (1998), the best use of nutrients for crop growth can be obtained on the basis of a high soil fertility level through maintenance of enough organic matter.

Better use of nutrient sources would depend on : i. soil nutrient and soil fertility aspects, ii. optimum soil reaction, iii. soil organic matter, iv. amount, time, and method of fertilizer application, v. addition of organic material, vi. checking of avoidable nutrient losses, vii. crops and cropping system, and viii. other factors of production such as water, pest, and disease management. The biological aspects of soil fertility has not been given an adequate consideration, inspite of the fact that biological condition of the soil greatly influences the nutrient availability, including its mobility and uptake by the crop. In the context of production sustainability, the reduction in soil organic matter often termed as biological degradation in soil quality, is claimed to impart a number of influences, namely, reduced activity of soil microorganisms and nutrient supplying capacity; relatively rapid deterioration in soil structure; crusting and surface compaction, reduced water storage capacity and infiltration.

3.5.2 Micronutrient Deficiencies in Indian Soils

Primary minerals are the major source of micronutrients in soils. During synthesis of secondary minerals, micronutrients get redistributed in these minerals. They are also present in various organic combinations as a part of the soil humus. The composition of parent material influences the geochemistry of trace elements in the less weathered soils, and the pedogenic factors assume greater significance as the soils become strongly weathered. Distribution of various micronutrients among soil orders showed that Vertisols have much higher total Cu and Mn than Alfisols, Oxisols or Inceptisols. While, available Cu invariably remained high in Inceptisols (Katyal and Vlek, 1985). The Oxisols have highest available Mn (Table 3.3). Earlier studies by Biswas (1955) indicated that Oxisols and Alfisols are richer in exchangeable Mn than Vertisols and calcareous alluvial soils. The wide variation in the total and available (Table 3.3) micronutrient content of Indian soils are indicative of the dominant influence of pedological conditions on soil development.

3.5.2.1 Soil Properties and Micronutrient Availability

Table 3.3. Total and available Cu and Mn in some tropical benchmark soils of India

Soils orders	Cu (mg kg^{-1})		Mn (mg kg^{-1})	
	Total	Available	Total	Available
Humid/Subhumid				
Inceptisols	23-34	0.2-4.5	38-393	5.1-102.4
	(29)	(2.5)	(266.0)	(40.0)
Alfisols	19-25	0.2-4.9	210-469	17.5-76.0
	(22)	(2.1)	(347.0)	(45.5)
Oxisols	54-58	0.8-2.8	405-800	18.1-155.4
	(56)	(1.7)	(602)	(86.8)
Arid/Semiarid				
Inceptisols	19-97	0.9-3.6	440-1326	7.5-48.8
	(51)	(1.9)	(847)	(19.0)
Vertisols	41-148	0.8-2.0	621-1060	4.8-16.4
	(68)	(1.4)	(774)	(9.8)
Alfisols	23-122	0.9-2.2	233-950	10.3-48.4
	(50)	(1.3)	(488)	(24.2)

Figures parenthesis indicate total content
Source : Katyal and Vlek (1985)

A number of soil properties have been observed to affect the availability of various micronutrients. In calcareous soils of Haryana, India, soil properties viz., sand, silt, clay, free Fe, and free Mn correlated significantly with Fe, Mn, Cu, and Zn in amorphous iron oxides, cyrstalline iron oxide bound, and residual forms (Singh *et al.,* 1988; Prasad and Sakal, 1991; Raghupathi and Vasuki, 1991; Singh, 1998). Step down multiple regression analysis indicated that availability of Zn and Cu is predominantly controlled by pH and organic carbon. Whereas Fe and Mn availability are controlled by pH, organic carbon, available P_2O_5, and K_2O, and that of B by pH, organic carbon, EC of acid soils of Samastipur, Bihar and West Bengal (Saha *et al.,*1991; Bhogal *et al.,*1993). Earlier, number of other studies have demonstrated the role of different soil properties such as soil pH, EC, textural variation, composition of water soluble and exchangeable cations, mineralogical buildup, free lime, bulk density, presence or absence of hard pan, water holding capacity, nature of surface and subsurface properties, redox potential, and organic carbon content on the availability of B, Fe, Mn, Zn, Cu, and Mo (Neelkanthan and Mehta, 1961; Viets, 1962; Tembhare and Rai, 1967; Rai, 1972; Talati and Agarwal, 1974; Kanwar and Randhawa, 1974; Singh and Sinha, 1975; Singh and Dahiya, 1975; Mclaren and Crawford, 1976; Singh and Sinha, 1976; Takkar and Randhawa, 1978; Patil and Patil, 1981; Katyal and Agarwala, 1982; Katyal and Deb, 1982; Trehan and Grewal, 1985; Gupta and Srivastava, 1990; Singh and Sekhon, 1991; Choudhary, 1992; Singh and Mongia, 1993).

3.5.2.2 Distribution of Total and Available Micronutrient

The wide variations in the total (Table 3.4) and available micronutrient content of Indian soils are indicative of the dominant influence of pedological conditions on the soil development.

The total micronutrient content of soils are generally of limited value as far as plant growth and responses to their application are concerned. In most cases, total contents have been insignificantly related to the plant content. In order to match the level of micronutrients in soil with plant requirement, their available contents deserve more emphasis. Most of the researchers use DTPA soil test method as proposed by Lindsay and Norvell (1978) for determining the available content of Zn, Cu, Fe, and Mn, particularly in alkaline calcareous soils. Like total contents, the available micronutrient status of soils is also highly variable (Table 3.5).

Table 3.4. Total micronutrients content of Indian soils

Micronutrient	Level ($mg\ kg^{-1}$)
Zinc	0 to 1019
Copper	1.9 to 960
Iron	2700 to 191000
Manganese	37 to 11500
Boron	3.8 to 630
Molybdenum	0.01 to 18.1

Source : Takkar and Randhawa (1978); Takkar (1982)

Total stock of micronutrients in most of the Indian soils is reported to carry between 0.90-27 per cent Fe, 37-11500 mg Mn, 2-1600 mg Zn, 2-960 mg Cu, 3-630 mg B, and 0.1-11.6 mg Mo kg^{-1} soil (Singh and Subba Rao, 1995; Singh, 1999b). Analysis of 2.52 lakhs surface soil samples revealed that DTPA extractable micronutrient cation content in Indian soils obtained as 0.2-6.9 mg Zn, 0.1-8.2 mg Cu, 0.8-196 mg Fe, 0.2-118 mg Mn kg^{-1} soil with a mean of 0.87, 2.1, 19, and 21 mg kg^{-1} soil, respectively (Singh and Saha, 1995; Singh, 1999a).

Table 3.5. Available micronutrient content ($mg\ kg^{-1}$) of Indian soils

Micronutrient	Range	Mean
Zinc	0.2-6.9	0.9
Copper	0.1-8.2	2.1
Iron	0.8-196	19.0
Manganese	0.2-118	21.0
Boron	0.08-2.6	--
Molybdenum	0.07-7.67	--

Source : Singh (1999b)

Soil properties exert a considerable influence on the availability of micronutrients. The extent of micronutrient deficiencies, therefore, varies not only in different states but also in different districts of the same state or in various blocks of the same district depending upon the soil characteristics and other management conditions. Katyal and Sharma (1991) in their studies on 57 benchmark soils of India representing Inceptisols, Alfisols, Vertisols, Aridisols, Entisols, Oxisols, Mollisols, and Ultisols observed that the availability of Zn, Cu, Mn, and Fe decreased as the soils became acidic, while their total contents increased (Table 3.6).

Irrespective of the soil moisture regime, parent materials from which soils developed also influenced their total micronutrients content. Analysis of 251547 samples in different states of the country revealed the predominance of Zn deficiency in divergent soils. Of these samples, 48, 12, 5, and 3 per cent samples showed deficient level of available Zn, Fe, Mn, and Cu, respectively (Singh, 1999a). The per cent samples deficient in zinc has been found highest in south zone followed by north, east, and west zone based on the reorganisation of data zonewise. Iron deficiency is also highest in south and north zones. The incidence of B deficiency is highest in the calcareous and acid soils of east zone.

Table 3.6. DTPA extractable and total micronutrient contents and some important properties of 57 benchmark of soils

Soil order	No. of soils	DTPA extractable				Total				pH (H_2O)	Lime	Organic matter	Clay
		Zn	Cu	Mn	Fe	Zn	Cu	Mn	Fe				
		(mg kg^{-1})									(%)		
Udic/Aquic moisture regime													
Entisols	1	0.29	5.33	8.1	44.1	87.6	57.5	388	3.50 *	8.07	2.40	1.5	52
Inceptisols	6	0.81	2.61	40.4	30.4	53.4	26.6	276	2.71	7.07	1.80	1.2	24
Alfisols	3	0.55	2.96	5.2	56.8	40.9	24.0	187	1.86	5.77	0.78	0.7	27
Mollisols	1	1.86	1.99	47.3	59.0	30.5	30.3	388	3.18	6.87	1.65	2.1	24
Oxisols	1	1.40	2.92	155.4	16.3	74.3	54.4	800	4.61	6.14	0.80	2.6	43
Ustic moisture regime													
Entisols	3	0.61	0.56	17.9	18.4	40.1	19.4	275	1.82	7.58	0.88	1.1	7
Inceptisols	9	0.46	2.05	24.6	18.4	64.5	38.6	687	3.25	7.61	3.04	0.8	28
Alfisols	12	0.56	1.97	26.6	21.3	45.8	34.6	467	2.59	7.11	1.88	0.88	17
Vertisols	12	0.42	1.49	12.4	9.1	63.0	63.0	731	3.98	8.06	5.51	0.7	53
Ultisols	1	0.28	0.49	40.1	17.5	43.1	22.5	400	5.37	5.37	0.20	0.9	14
Oxisols	1	0.31	0.54	18.1	19.3	70.3	58.1	405	6.32	5.32	0.25	2.3	27
Aridic/Torric moisture regime													
Entisols	2	0.25	0.51	7.8	7.8	38.4	17.2	431	2.47	8.44	4.55	0.3	8
Aridisols	5	0.38	1.28	12.7	9.6	61.4	50.7	814	4.52	8.33	4.10	0.5	17
Range		0.12 to 2.80	0.15 to 5.33	4.0 to 102.0	3.4 to 68.1	20 to 97	11 to 148	38 to 194	1.3 to 8.0	4.61 to 10.50	6.20 to 8.90	0.10 to 2.00	3 to 64
Mean		0.54	1.71	26.0	20.3	55	41	567	3.3	7.42	2.92	0.87	28
SDm (±)		0.07	0.16	3.7	2.2	3	4	45	0.2	0.18	0.32	0.07	2

* Expressed in per cent

Source : Katyal and Sharma (1991).

Despite the fact that soil analysis is extremely useful in identifying the nature and extent of particular micronutrient deficiency. Researches indicated that coarse textured soils, alkali soils, floodplain soils, and calcareous soils which have relatively high pH and are low in organic matter content are more prone to Zn deficiency. Similarly, fine textured calcareous black soils (Vertisols) and highly leached red soils are expected to show greater incidence of Zn deficiency.

Awareness among the growers about the menace of Zn deficiency in soil has resulted in the increased use of zinc containing fertilizers than what they used to adapt in past. As a result of this, the status of available Zn in soil has increased in many areas and thereby, deficiency of Zn has come down depending on the rate and frequency of zinc sulphate application. A random sampling of some of the areas in Punjab revealed a drastic decrease in the extent of Zn deficiency. But, it resulted in an increase in deficiencies of Fe and Mn at some places. Fertility evaluation of Nagpur mandarin growing soils of central India revealed that DTPA Fe, Mn, Cu, and Zn varied from 2-31 mg kg^{-1}, 5-31 mg kg^{-1}, 0.54-2.5 mg kg^{-1}, and 0.19-1.9 mg kg $^{-1}$, respectively (Table 3.7). A 59 per cent, 42

Table 3.7. Soil extractable nutrients (mg kg^{-1}) in Nagpur mandarin orchard soils of central India

Nutrient	Range	Mean*	SEm$_{\pm}$**
Nitrogen[1]	85-280	130	4.2
Phosphorus[2]	4-30	14	1.8
Potassium[3]	32-383	184	6.2
Calcium+magnesium[4]	375-2016	871	11.4
Iron[5]	2-31	13	0.81
Manganese[5]	5-31	18	0.62
Copper[5]	0.54-2.5	2.7	0.24
Zinc[5]	0.19-8.1	0.98	0.06

* n = 178.

** SEm = Standared error of the mean $\sqrt{\frac{x^2-(x)^2/n}{n}}$ where x and n stand for mean and nunber of observations, respectively.

[1] Alkaline permagnate steam distillation, [2] 0.5 N $NaHCO_3$ (pH 8.3), [3] 1 N Ammonium acetate (7.0 pH), [4] 1 N Ammonium acetate (7.0 pH), [5] 0.005 M DTPA—$CaCl_2$ (7.3 pH).

Source : Srivastava and Singh (2000 a) ; Srivastava *et al.* (2000)

per cent, and 32 per cents soils of orchards have been further designated as low in available N, P, and Zn, respectively (Table 3.8).

The total micronutrient content of a soil represents the active and inert fractions of micronutrients, and the active fractions contribute to their bio-availability. All the fractions are in a state of dynamic equilibrium and constitute the buffering capacity of soil for micronutrients (Deb, 1997). The buffering capacity of a micronutrient in a particular soil is determined by the soil solution. The higher the magnitude of activity, the lower is the buffering capacity of particular micronutrient in that soil. The magnitude of buffering capacity of a particular micronutrient determines the rate of reaction of the micronutrient when applied to soil. In most of the Indian soils, the zinc buffering capacity is very high, suggesting thereby, that the amount of water soluble zinc is very low, which ultimately is reflected in the invariably low available zinc status of soil. Sakal and Singh (1995) reported available B content and the magnitude of its deficiency in major Indian soils (Table 3.9).

Table 3.8. Soil fertility indexing of Nagpur mandarin orchard Soil (n=178)

Nutrient	Supply levels					
	Low		Medium		High	
Element	No.	%	No.	%	No.	%
Nitrogen	106	59	71	40	01	0.5
Phosphorus	74	42	86	48	18	10
Potassium	07	4	05	3	166	93
Iron	04	2	174	98	-	-
Manganese	02	1	176	99	-	-
Copper	02	1	176	99	-	-
Zinc	56	32	63	35	59	33

Source : Srivastava and Singh (2000a) ; Srivastava *et al.* (2000)

On the scrutiny of the data published so far it appeared that the active fraction of zinc constitutes 1.2 to 5.6 per cent of total Zn in soil and the remaining portion is inert and represents mineral, precipitated, and occluded Zn in soils. Bulk of the fertilizer zinc accumulates in occluded (18-41 per cent) and residual (25-42 per cent) fractions, and

Table 3.9. Available boron content and the magnitude of B deficiency in some Indian soils

Major soil group/ soil order	Sample analysed	Available soil B (mg kg^{-1})	Per cent soils deficient
Assam			
Old alluvial soils	687	0.14-23.74	17
Acid soils	320	0.05-0.72	43
Bihar			
Red yellow soils	466	Tr.-3.67	40
Calcareous soils	1201	0.06-8.00	48
Recent alluvial soils	1732	Tr.-6.50	41
Old alluvial soils	703	0.06-7.30	32
Tal-land soils	563	0.04-5.67	35
Madhya Pradesh			
Red and yellow soils	544	0.50	49
Orissa			
Red and lateritic soils	882	0.10-2.20	69
Punjab			
Ustochrepts	116	0.30-2.00	43
West Bengal			
Red and lateritic soils	2544	0.02-3.30	64
Terai and Teesta alluvial soils	633	0.05-0.76	84

Source : Sakal and Singh (1995)

0.81 to 7.8 per cent in water soluble plus exchangeable fractions, 10 to 30 per cent in complexed fraction, and 9 to 20 per cent inorganically bound fraction under upland condition according to Iyengar and Deb (1977). Such variability in various soil Zn fractions has also shown site specific fluctuation, depending upon cultural practices in vogue.

Summary

The soil fertility changes taking place under sustained commercial citrus cultivation, warrant to be monitored vigilantly, so that the genesis of any fertility constraint taking place can be immediately mitigated and the sole objectivity of improved production level is not faultered anywhere. This is only possible, provided a good nutritional balance is maintained through a proper management of soil-plant-nutrient interaction. Such an attempt has to be an integral part towards obtaining the sustainability in citrus production, besides advocating the wide scale use of various tools relevant to precision Citriculture.

Soil fertility and production sustainability has to go hand in hand, the synergism of which has to be exploited by developing strategies for sustainable use of available soil resources through a proper buildup of organic matter and N reserves, land surface stabilization, precise evaluation of aquifer's capacity, intrinsic mineralogical buildup and its role in governing the fertility fluctuations, in addition to soil management through minimum tillage and cover crops, especially in high altitude areas.

4 Fertility Management of Slopy Land Under Citrus

4.1 INTRODUCTION

In south and southeast Asia, more than 60 per cent of land area has slopes over 8 per cent (FAO, 1984a; 1984b). Adaption of slash and burn, shifting cultivation on such slopy lands is a common practice. Under tropical climate, nutrient supply from the soil is exhausted rapidly due to rapid oxidation of soil organic matter and low cation exchange capacity further delays replenishment or restoration of normal soil fertility (Von Uexkull and Bosshart, 1989), with the result, erosion problem takes a severe form in the absence or low organic matter content in soils (Yu, 1989). Thus, the crop yield continues to decline, if they are continuously cultivated without nutrient inputs (Lal, 1987). According to Lal (1974), soil erosion is a recurring problem where shifting cultivation is practised. Without fertilizers, yields on many soils in the tropics rapidly fall to very low levels. This has been a prime factor in the wide spread use of shifting cultivation system in the humid tropics, which are sustainable, provided cultivation does not continue for more than 2 or 3 years. And the rest period when the soil remains under natural vegetation, is sufficiently long to restore the soil organic matter level and suppress weeds (Nye and Greenland, 1960). As population density increases and fallow periods are shortened, thereby, allowing less time to restore fertility and providing a downward trend in yield (Ruthenberg, 1971).

Due to vegetative cover being removed over land, steeper slopes coupled with higher rainfall, there is a rapid soil erosion. A number of workers (London, 1984; Stocking, 1984; 1986) have listed the wide range of soil constraints, which are induced by erosion causing decline in productivity. Chief soil constraints of them are : reduced soil depth due to top slicing of fertile layer exposing the laterite/plinthite layer, exposure to water stress due to curtailed available water capacity, low soil air temperature, reduced workability, low cation exchange capacity, deficiency of N, K, P, toxicity of Al, and Mn. Most vulnerable soils are those with a concentrated nutrient distribution to the top soil coupled with a strong textural contrast between top soil and subsoil. Further, cropping without practically any fertilizer application results in a decline in fertility, ultimately giving low and uneconomical crop yields (Sanchez, 1976; FAO, 1984a). Majority of areas practising shifting cultivation are covered under red soils covering about 71 million ha, subjected to severe water erosion. The rainfall under acid red soils of India ranges from 750 to 2000 mm year^{-1}. Most red soils ranges (Alfisols, Ultisols, Oxisols, and Inceptisols) being shallow to deep, possess low infiltration rate due to crusting problem, suffer from surface runoff, and erosion by water. The lateritic soils suffer from rill erosion loosing almost 40 tons ha^{-1} $year^{-1}$ in the absence of an adequate soil conservation measures (Sehgal and Abrol, 1994). The fertility management of slopy land with citrus as a perennial test crop requires a thorough understanding right from processes involved in shifting cultivation (leading to soil fertility depletion) to various soil management practices to be able to not only preserve the soil fertility, but raise the productivity to an economically viable level.

4.2 SHIFTING CULTIVATION, A COMMON FEATURE ON SLOPY LAND

Shifting cultivation, also called 'slash' and 'burn' cultivation or ' podu' in the hill areas of Andhra Pradesh, is the oldest form of tropical Agriculture, and is practised in about 350 million hectares involving about 250 million people across the world according to earlier available statistics (FAO, 1974a). However, it is largely confined to the tropical forests of Africa, Latin America, south and southeast Asian countries. Shifting cultivation (locally known as *Jhumm*), a major human misutilisation of land is responsible for the degradation of soil and land resources in the hills of northeast India (Fig. 4.1).

Fig. 4.1 One of khasi mandarin orchards on exposed subsurface piece of slopy land as a result of shifting cultivation in northeast India

The hills and mountain areas differ from the plains in several aspects, and call for varied management practices. The former have an enormous biodiversity and provide the valuable sources for most of the water for lower altitudes and plains below, with wide variation in land and climate at short intervals. These areas are natural ecosystems bestowed with rich collection of flora and fauna. On human side, they are also the home of diverse people with their own distinct culture and their way of life, is often different from those of the people living in plains. They also provide the principal water catchment areas. When and if, these high mountain lands are misused, the water supplied to areas below becomes less and uncertain, besides leading to often floods and droughts. Further, these high altitude areas usually have unnecessarily large cattle population, most of which is unproductive for dairy purposes, but cause damage to crops and lead to soil erosion through overgrazing (Narasimham, 1995). While, being rich in natural resources, these areas are also ecologically highly fragile. Any mismanagement of land could lead to irreversible disastrous consequences, both locally and in lower zones. One such dangerous misuse in recent times is a large scale deforestation for shifting cultivation and other needs in these hill and mountain zones.

Shifting cultivation, as the name implies, is of a transient nature, both in time and space, since the area cleared and even the farmers concerned shift the zone of operation after a certain period. This is necessitated by steep fall in soil fertility and productivity in a short span of about 3 - 5 years depending on local conditions. The discarded area is left during the earlier periods uncultivated (fallow) for about 30 - 50 years for recuperation of soil fertility and regeneration of secondary forest. After this period, called 'shifting cycle', the same farmer or descendant used to return to this area to restart the land clearance by burning the forest and take up farming once again. Borthakur *et al.* (1976) suggested suitable alternative farming system in the hilly landscape, but the desired results have not been achieved, partly due to lack of basic data on soil resources (Ingty and Goswami, 1979; Walia and Chamuah, 1993). The main process in the system consists of clearing the forest and burning the forest in late winter/summer months.

4.2.1 Nature and Severity of the Problem

The problem is acute in Andra Pradesh, Orissa, Tripura, Megalaya, Mizoram, Manipur, Assam, and Arunachal Pradesh. In the northeastern India only, about 2.7 million ha of forest lands are affected by Jhuming, involving nearly half a million tribal families as per comparatively old data (Anonymous, 1976). The chief drawback, which has of late become disastrous in shifting cultivation is the deforestation, which in turn, results in serious disturbance in ecosystem involving soil, climate, flora, and fauna. While, in historical times, centuries ago, the damaging effects used to be negligible due to small area of forest destroyed each time and the long interval of shifting cycle. Deforestation in recent times has assumed alarming proportions. As against a good 30-50 years of the cycle in old days, this has come down to about 5 to 6 years now, leaving no time for either fertility recuperation or forest regeneration. The reduction in shifting cycle has imposed a unprecedented stress on the nutrient supplying capacity of soil, which is already under various stages of depletion depending upon slope. Forest cutting, burning, clearing, and cultivation has caused nearly 40 tons ha^{-1} soil material to slide/roll down to foothills. Soil erosion from hill slopes of 60-70 per cent during first, second, and third year has been reported as 146,170, and 30 tons ha^{-1} $year^{-1}$, respectively (Singh and Singh, 1978b; Singh and Singh, 1981). Soils of Nam Phong series in northeast plateau of Thailand are frequently observed loosing their productivity potential under the influence of shifting cultivation (Watson, 1994).

Analysis of few soil profiles belonging to Typic Udorthent, Typic Umbric Dystrochrept, Typic Haplohumult, and Typic Paleudult under shifting cultivation in Arunachal Pradesh (Table 4.1) suggested that soils on moderately sloping hills and foothill slopes with well

Table 4.1. Characteristics of the soils of Arunachal Pradesh, India under shifting cultivation

Horizon	Depth (cm)	Sand (%)	Silt (%)	Clay (%)	pH	Org.C (g kg^{-1})	Exch. cations [cmol (p$^+$) kg^{-1}]	Extr. acidity [cmol (p$^+$) kg^{-1}]	Exch. Al^{3+} [cmol (p$^+$) kg^{-1}]	CEC [cmol (p$^+$) kg^{-1}]	BS (%)
Summits/Ridges, Typic Udorthent											
A1	0-12	71.8	12.7	15.5	5.4	16.8	5.6	8.7	0.2	10.6	50
AC	12-36	83.7	5.8	10.5	5.3	7.0	2.9	6.3	0.5	6.8	43
C	36-70	83.9	6.6	9.5	5.5	1.8	2.0	5.1	0.3	4.2	48
Moderately steep hills, Typic Umbric Dystrochrept											
Ap	0-20	37.5	40.5	22.0	6.3	20.7	9.9	13.6	tr	18.0	55
B21	20-36	38.1	38.9	23.0	5.9	13.9	6.4	12.4	tr	17.2	37
B22	36-70	40.7	35.3	24.0	5.6	13.5	4.2	12.7	0.5	14.3	29
B23	70-100	27.0	32.5	20.5	5.7	4.3	5.5	11.9	0.4	15.0	37
B3	110-55	52.4	29.7	17.9	5.8	3.5	6.3	10.4	0.3	14.6	43
Side slope of hill, Typic Haplohumult											
A1	0-16	46.4	20.6	33.	5.8	24.6	9.3	13.2	Tr	13.4	69
B21t	16-38	41.2	18.8	40.0	5.3	11.3	4.8	12.6	0.7	9.7	49
B22t	38-70	42.5	15.0	42.5	5.0	8.6	2.4	11.0	0.8	8.0	30
B3	105-48	53.3	13.2	33.5	5.1	4.4	1.9	10.5	0.9	7.3	26
Foothill slope, Typic Paleudult											
Ap	0-16	55.2	13.8	31.0	5.3	16.3	4.7	9.2	0.2	8.0	59
B1	16-40	46.5	18.0	35.5	4.8	9.5	2.7	10.4	0.9	9.1	30
B21t	40-62	40.0	14.5	45.5	4.8	6.1	2.5	11.6	1.5	10.3	24
B22t	62-90	40.7	10.3	49.0	4.9	4.6	2.6	10.8	1.5	11.3	23
B23t	90-25	37.5	11.5	51.0	439	4.9	2.5	10.4	1.1	13.2	19
B3t	125-65	36.5	15.5	48.0	5.0	3.6	2.4	10.6	1.1	11.6	18

BS stands for base saturation

Source : Nayak and Srivastava (1995)

developed argillic horizons and poor base saturation qualified for Typic Haplohumults and Typic Paleudults, respectively. The major limitation of these soils are very steep slopes, stoniness, low fertility, high acidity, and severe erosion. Soils on ridges/summits are unsuitable for arable farming, whereas those on steeply sloping hills are marginally suitable. Soils on the moderately sloping side hills and foothill slopes which have been stabilized by terracing and bunding are moderately suitable for cultivation.

Loss of topsoil and runoff losses due to Jhum cultivation have now been regarded as a established phenomenon in Meghalaya, India (Table 4.2).The relative merits of terracing as a Jhum control measure are also apparent.

The overall impact of 'slash burn' on some of the important physicochemical properties of soil at Byrnihat, Meghalaya has been investigated. Based on these field level observations and analysis, it has been estimated that over 180 million tons of soil is eroded annually as a result of shifting cultivation with concomitant nutrient loss estimated at 6 million tons of organic carbon, 1,000 tons of phosphate and 5.7 thousand tons of potash from the region (Table 4.3). Although, most of the soil properties recorded improvement (Table 4.4) after Jhuming.

Table 4.2. Statewise areas covered under Jhum cultivation

States	Total area affected by Jhuming ('000 ha)	Area covered under Jhuming in a year ('000 ha)
Arunachal Pradesh	210.0	70.0
Assam	139.2	36.6
Manipur	360.0	90.0
Meghalaya	265.0	53.0
Mizoram	189.0	63.0
Nagaland	633.0	101.4
Tripura	111.5	22.3
Total	1,907.5	469.3

Source : Sarkar (1994)

According to another study in northeast India, an average annual loss of 40.9 tons of soil ha^{-1} and nutrient losses from Jhum plots have been observed as 1321, 0.21 and 12.5 kg ha^{-1} of organic carbon, available P, and K. Other studies suggested that about 18.6 million tons of soil alongwith heavy loss of N, P_2O_5, and K_2O are washed away annually as a result of shifting cultivation in this region (Banerjee *et al.,*1991). Stocking (1986) indicated that a typical subsistence farm field plot in Zimbabwe, looses 50 tons ha^{-1} of soil containing 105 kg N ha^{-1}, 8 kg P ha^{-1}, and over 700 kg ha^{-1} of organic carbon. Nutrient losses are also fairly high which amounts to 0.6 million tons of N, 9.7 and 5.69 tons of available P_2O_5 and K_2O, respectively. Besides these, cultivation on the slopy land along the slope further leads to the heavy erosion of soil and the extent of erosion is such that even the bed rocks are exposed to surface and a large citrus area has been completely denuded in various northeastern Indian states (Fig. 4.2).

Table 4.3. Soil and nutrients losses as a result of shifting cultivation in northeastern hill region

State/Union Territories	Area ('000 ha)	Soil loss ('000 tons)	Nutrient loss ('000 tons)		
			Orgainc carbon	Available P_2O_5	Available K_2O
Assam	69.60	2846.64	91.94	0.015	0.87
Arunachal Pradesh	92.00	3762.80	121.53	0.019	1.15
Manipur	60.00	2454.00	79.26	0.013	0.75
Meghalaya	76.00	3108.40	100.40	0.016	0.95
Mizoram	61.61	2519.85	81.39	0.013	0.77
Nagaland	73.54	3007.79	97.15	0.016	0.92
Tripura	22.30	912.07	29.46	0.005	0.28
Total	455.05	181611.55	603.13	0.097	5.69

Source : Prasad *et al.* (1981)

Table 4.4. Analytical data computed before and after Jhum cultivation

Soil properties	Before Jhuming	After Jhuming
pH	5.10	5.50
Organic carbon (%)	1.32	1.05
Available P_2O_5 (kg ha^{-1})	3.30	3.31
Available K_2O (kg ha^{-1})	210.00	570.00
Exch. Ca (me 100 g^{-1})	7.15	9.46

Source : Sarkar (1994)

Soil changes in various aspects following clearance of forest for shifting cultivation been dealt by Nye and Greenland (1968), Sanchez *et al.* (1983), and Anonymous, (1983). Improvement of deteriorated soil under such conditions include soil conservation, cropping with fertilizer use, horticulture, agroforestry, and controlled water management (Sanchez, 1976; Greenland and Lal, 1977; FAO, 1974b; 1984a; Anonymous, 1986). For sustained crop production and in order to reduce the pace of deforestation (if not altogether stop it) in area of shifting cultivation, it is necessary both to adapt effective soil conservation measures and to take up an adequate fertilizer use. Soil management research conducted in the Amazon basin has identified several technological options for sustainable use of acid and low fertility soils (Sanchez, 1982). Important scientific innovations in sustainable management of these soils include : identification of Al as the principle cause of soil acidity, use of indigenous rock phosphate to alleviate P deficiency, use of leguminous cover crops to enhance biological nitrogen fixation, increased effective rooting depth by alleviating subsoil acidity, and understanding the role of variable charge clay minerals in management of Oxisols and Ultisols (Sanchez, 1994).

Fig. 4.2. Shifting cultivation in citrus growing slopy land in high altitude areas. Some of the pockets having exposed subsurface can be clearly spotted.

Shifting cultivation has been observed to increase all forms of soil acidity, whereas the terrace cultivation brought them to the lowest magnitude (Table 4.5) involving soil types viz., Dystrochrepts, Fluventic Umbric Dystrochrepts, Umbric Dystrochrepts, and Typic Dystrochrepts of Manipur (Kailash Kumar *et al.*, 1995a). The exposed soils under shifting cultivation without any control overflow of water and nutrient might have resulted in maximum leaching of bases and

Table 4.5. Forms of acidity [cmol (p^+) kg^{-1}] as influenced by dominant land use systems in hilly areas of Manipur (North east India)

Forms of acidity	Land use systems		
	Forest cover	Shifting cultivation	Terrace cultivation
Total acidity	10.2-21.0* (16.3)**	12.1-28.2 (19.6)	11.0-21.5 (15.2)
pH-dependent acidity	10.1-18.9 (14.9)	11.7-24.8 (17.5)	10.8-20.8 (14.5)
Exchange acidity (Electrostatically bound -H^+)	0.0-0.6 (0.3)	0.1-1.7 (0.4)	0.1-1.2 (0.3)
Exchange acidity (Electrostatically bound -Al^{3+})	0.1-3.5 (1.2)	0.1-4.3 (1.6)	0.0-1.6 (0.4)
Total exchange acidity	0.1-3.9 (1.5)	0.2-5.6 (2.1)	0.1-2.4 (0.7)

* Range, ** Mean

Source : Kailash Kumar *et al.* (1995a)

accumulation of more acid forming cations like Al, Fe, and Mn leading to increased acidity. The terrace cultivation, however, lowered the loss of bases and maintained soil acidity to a lower degree by keeping higher base saturation. The forest cover has medium acidity, probably by maintaining higher organic matter complexing the acid forming cations and consequently reducing the leaching losses.

4.3 SOIL MANAGEMENT

Soil management under shifting cultivation holds most important aspect of soil fertility management. Best response of soil management involving bare soil with *Vigna sinensis* green manure incorporated during rainy season has been obtained (Cary, 1971; Pacheco *et al.*, 1973). Permanent sod has highest infiltration rate followed by winter clover, winter tick bean, and bare surface herbicide treatment (Cary, 1972). While, Cassin and Lossois (1972) observed depressing effect of permanent sod on Corsican citrus. Changes in properties are dynamic over time (Jenny, 1961; Hoosbeck and Bryant, 1992; Li, 1995; Wang and Gong, 1998). Therefore, only by comparing and analysing changes in soil properties between two or more time periods, the essence and mechanism of soil changes can be better understood. A number of deficiencies have been highlighted for the above anomaly :

- Evaluation work at regional levels does not reflect soil changes over time. For example, global assessment of human induced soil degradation carried out by World Reference Base and United Nations Environmental Programme (Oldeman, 1988; UNEP, 1990), mapping of soil degradation in China (Liu and Gong, 1995; Zhao, 1995), and the evaluation of soil degradation in south China (Liu and Gong, 1995; Zhao, 1995) etc. are carried out on the basis of current soil properties. These studies, therefore, could not reflect the true nature of soil degradation as a dynamic process in time and space (Wang, 1994; 1995; Zhao, 1995). Further, it is also difficult to differentiate between anthropogenic and natural process of degradation (Zhao *et al.,* 1993; Wang, 1995; Zhao, 1995).
- In studies on the impacts of land use on soil changes, many attempts to evaluate the degree of soil degradation and soil improvement by comparing soil properties within the same time period, but under different land use patterns (Komakhidze *et al.*, 1990; Hu *et al.,*1993; Wang and Chen, 1993) have been made. This kind of approach may provide some level of explanation, but does not directly register soil changes. In addition, single soil property evaluation, such as changes in soil organic matter, N, P, and K are usually emphasized, and much less attention is paid to a comprehensive assessment of soil quality changes. Results are further difficult to use, since different benchmark soils have been used and, therefore, the rate of soil changes can not be accurately assessed and compared.
- Experience of growing Newhall and Skaggs Bonanza navel oranges in Sichuan,

China indicated that it established best on east, west or south facing mountain slopes (30 per cent) with a 1 m depth of cultivated soil in pH range of 5.5-7.3 and humus content of 2.5-3.0 per cent. Suitable rate of fertilizer application included 80-100 kg organic manure, 0.5-0.8 kg urea, and 0.5-1.0 kg complex fertilizer tree^{-1} year^{-1} (Xong *et al.*, 1997). Best performance of mandarin trees in Abkhaziya has been obtained with clean cultivation until August, then sowing of green manure plus 10 tons ha^{-1} FYM followed by green manure in all the year (Bobokhidze, 1976). Deidda and Pala (1975) described manuring and soil management of citrus orchards in Corsica Islands. Evaluation of eight cultivation methods viz., moving (control), harrowing (to less than 10 cm), and ploughing (10-20 cm) alone or in combination in 5 year old citrus plantation at 8 x 4 m spacing showed an highest fruit yield of 35.3 tons ha^{-1} with a combined use of moving during rainy period, harrowing during dry periods, and ploughing in winter (Osa *et al.*, 1987).

- Studies which monitor soil dynamics emphasize soil changes over time, and help in understanding the mechanisms and processes of soil changes. These are usually confined to small plots, soil profiles or laboratory soil columns. They do not, therefore, provide a comprehensive assessment and analysis of soil changes at regional level.

Four typical pedons, representing different physiographic units in the shifting cultivation area of Arunachal Pradesh showed that soils of the ridge/summit as Typic Udorthents, while those on moderately steep hills with cambic horizon, and poor base saturation qualified for Umbric Dystrochrepts. The soils on moderately sloping side hills and foothill slopes with well developed argillic horizons and poor base saturation qualified for Typic Haplohumults and Typic Paleudults, respectively. The major limitations of these soils are very steep slopes, stoniness, low fertility, high acidity, and severe erosion. Soils on the ridge/summits are unsuitable for arable farming, whereas those on steep sloping hills are marginally suitable. Soils on the moderately sloping side hills and foothills slopes which have been stabilized by terracing and/or bunding, are moderately suitable for cultivation (Nayak and Srivastava, 1995).

4.3.1 Soil Conservation

The necessary soil conservation measures to check erosion losses in these areas have been outlined by Liao and Chang (1974), FAO (1974a), Sanchez (1976), and Greenland and Lal (1977). Brown (1963) emphasized about the contour planting of citrus in areas having adequate depth with medium sandy loam or loamy sand or loam texture. The trees planted on contour as contour cultivation is recognized as an effective mechanical measure for conserving moisture. The practise of leaving natural sod below and beneath the trees conserves more mositure than mulching, green manuring, and clean cultivation. Provision of contour ditch under natural sod is detrimental as it reduces the fruit yield to minimum. The effect of green manure, mulch, and clean cultivation has varied from year to year with clean cultivation is as good as mulching. (Bhattacharya *et al.,* 1966). In areas having 25-30 per cent slope, citrus planting on terraces has proved highly beneficial. Vasalmidze (1969) suggested terracing in citrus orchards on slopy land in Western Georgia. While, Lu *et al.* (1997) suggested trench citrus planting in hilly Ultisols of Zhejiang, China having a variety of soil constraints viz., erosion, acidity, compactness, and low fertility. Trench planting not only produced precocity in bearing, but recorded 2-3 times higher yield than conventional orchard in eight years period. Coetzee (1995) recommended ridging of soils under citrus as one of the methods of soil conservation practices. Yarbobaev (1975) observed best performance of lemon trees planted on flat bottom trenches in Kolkhoz Lenina and Kumsangirskii region of Tadzhik SSR. Growth of clementine trees growing in 6 m wide x 3 m deep covered (in winter) untreated trenches in Tashkent showed good fruit yield from 5th year onward (Muldabaev and Zaitsev, 1989).

Different types of terracing structures have been suggested depending upon nature of soil, total rainfall, and slope (Table 4.6). Out of different methods of terracing, complete bench terracing proved most effective in reducing the soil loss through run off water (Table 4.7). Most citrus in Japan is grown on terraces built with stones filled with soil (Bitters, 1964). In the mediterranean area, terraces are used in some areas of Italy (Burke, 1962a) and Spain (Gonzalez-Sicilia, 1968). Citrus is also grown on ridges (Chen and Dragavcev, 1958) under heavy rainfall and high water table conditions. The first step in this regard is the *in situ* methods, especially contour farming. Deep ploughing is not advisable in these areas, as it tends to loosen top soil too much and eventually make it vulnerable to erosion. In fact, for light soils, in hilly areas in many countries, zero tillage is recommended. Depending on the slope, rainfall intensity and soil properties, well known measures like bunding, terracing, plugging, and strengthening of slides of gullies, etc. have shown promising success. As far as possible, terracing or at least leveling or a circular bed of about 2 meters diameter is very essential at the tree base (Ghosh, 1985). Use of half

Table 4.6. Mechanical structures of terracing

Mechanical structures	Soil type	Rainfall (mm)	Slope (%)
Contour bunds	Light soils	< 600	> 1.5
Graded bunds	All soils	> 600	1.5-6.0
Bench terraces	Deep soils	> 1000	6-30
Graded border strip	Deep Alfisols and related red soils	> 800	> 1.5

Source : Venkateshwarlu (1987)

Table 4.7. Losses due to Jhuming

Types of losses	Jhuming	1/3 terracing	Complete bench terracing	Puertican terracing
Run-off (mm)	114.00	81.40	32.80	256.30
Run-off (% rainfall)	5.30	3.90	1.50	12.90
Soil loss (Mg ha^{-1})	40.90	5.80	5.80	39.30

Source : Sarkar (1994)

moon terraces in khasi mandarin orchards in northeast India has proved highly effective from soil fertility and water conservation point of views (Fig. 4.3).

Fig. 4.3 Use of half moon terrace around the tree to avoid the loss of water and nutrients on a slopy land

Ono *et al.* (1986) observed higher yield of satsuma mandarin on terraces than trees on slopes. Chanukvadze (1990) reported highest average yield (15.2 kg tree^{-1}) of terraced orchard fertilized with NPK compared to nonfertilized orchard (8.9 kg tree^{-1}). A number of mechanical structures depending upon slope have been found useful. It is crucial to keep the land covered with vegetation by crops as adapted in Japan (Bitters, 1964) using fescue grass and love grass as cover crops or even pasture during the rainy season. Mulching is also known to reduce the impact of rain and check erosion. The upper ridges of the hills should be under horticulture or forest plantations to reduce the initial velocity of run off water down the slope.

4.3.1.1 Zero Tillage

The soil management practice best suited to a particular citrus orchard will depend on many factors, including location, soil type, cost, and convenience. In some situations where the orchard is on sloping land or on poorly structured wind blown soils, cover crop or zero tillage could be an extremely viable preposition. According to Jordan and Russell (1978), zero tillage has become an established soil management practice in 90 per cent of California citrus orchards. Weerts and Cary (1980) observed that within the rootzone of a valencia orange orchard, the soil warmed up more quickly in the autumn, when zero tillage or permanent sod treatments are used. It is presumed that the greater heat conductance of the soil receiving zero tillage is of particular benefit to sweet orange trees in spring and early summer (Cary, 1981).

By providing a warmer and more favourable soil temperature environments for roots, the zero tillage treatment facilitated the uptake and transport of nutrients to the leaves during the critical growth periods. The corresponding ability of zero tillage to cool down more quickly in the autumn have been of benefit in bud initiation during the winter and increased the relative proportion of flowering to vegetative shoots emerging in spring. Studies by Cary (1968) and Cary and Weerts (1977) demonstrated that a zero tillage soil management treatment can be used for up to 28 years without adversely affecting tree growth and yield. Jones *et al.* (1961) found that 2 years after conversion from clean cultivation to zero tillage, sweet orange yields and net water intake increased substantially. Cary (1981) in Australia observed higher yield (61 metric tons ha^{-1}) from nontilled soil compared to tilled or cover crop maintained plot (49 metric tons ha^{-1}).

4.3.1.2 Cover Crop and Mulching

Cover cropping is the practice of growing a pure crop or mixed crop with the purpose to keep the soil covered with some crop throughout the year. Cover crops can be used for one or more primary purposes and are often identified on the basis of other function:

Smother Crops: These are used to supress weed growth. A common approach is to sow a winter annual species, such as vetch, clover and or grass species in the fall. The cover provides dense foliage that dies out in the spring and can reseed itself which can act as an effective dead mulch.

Catch Crops: The cover crop is used to reduce leaching of nutrients especially nitrogen below the root zone. The most effective catch crops are grasses, mustards, and some weedy species.

Green Manure Crops : The cover crop is shown annually and incorporated into soil prior to maturation. Green manures are used to add nitrogen and organic matter to the soil (Fig. 4.4).

Fig. 4.4 Green manure crop, sunhemp (*Crotalaria juncea* L.) in Nagpur mandarin orchard of central India.

Insectary Crops: These are plants sown for the purpose of attracting beneficial anthropods to the orchard. The cover crop supplies the anthropods or mites with a food source of pollen, nectar or alternate prey.

Wedderbuan and Collingwood (1976) documented that the recognition of green manures can be found in mediterranean civilisation through the writings of Xenophon who lived during 434-355 B.C. Theophratus (373-287 B.C.) wrote of bean crops being used as green manure by farmers of Macedonia and Thessaly. Cato (234-149 B.C.) and Columela (about 45 A.D.) compared the value of various legumes in soil improvement. Pieters (1927) described that Chinese writers recognized more than 2000 years ago that legumes increased the production of crops that followed. According to Mertz (1918), a number of green manure crops viz., common vetch, bur clover, purple vetch, Canada peas, tangier peas, *Melilotus*, fenugreek, and lentils have proved their worth to be used as an effective cover crop in citrus orchards. Later, a number of researchers (Lohnis, 1926; Pieters, 1927; Pieters and Mckee, 1929; Lyon, 1936; Sakamoto *et al.*, 1960a; 1960b; 1961; Benoit *et al.*,1962; Smith, 1964; Azad and Bhambota, 1966) described the beneficial role of legumes in soil improvements. Jones and Embleton (1973) also listed a series of promising cover crops for citrus orchards, namely, annual yellow sweet clover (*Melilotus indica*), Canada field pea (*Pisum arvense*), Colorado river hemp (*Sesbania microcarpa*), common vetch (*Vicia sativa*), cowpea (*Vigna unguiculata*), crotalaria (*Crotalaria striata*), Egyptian clover (*Trifolium alexandrinum*), hairy indigo (*Indigofera hirsuta*), Natal grass (*Tricholaena rosea*), purple vetch (*Vicia atropurpurea*), rape (*Brassica napus*), small seeded broad bean (*Vicia faba* var. minor), tangier pea (*Lathyrus tingitanus*), trieste mustard (*Brassica juncea*), velvet bean (*Mucuna utilis*), and white mustard (*Brassica alba*).

Some of other cover crops viz., kudzu (*Pueraria phaseoloides*), Autstralian winter pea (*Pisum sativum arvensel*), crimson clover (Trifolicem incarnatum), rose clover (*Trifolium hirtum*), subterranean clover (*Trifolium suberraneum*), barrel medic (*Medicago truncatula*), burr medic / bur clover (*Medicago ploymorpha*), blando brome (Bromus mollis), brids foot trefoil (*Lotus corniculatus*) and straw-berry clover (*Trifolium Fragiferum*) perennial grasses like *Hordeum Californiaum* and *Festuca rubra* have also been suggested in subsequent studies.

Cover crops have been quite helpful in improving the soil structure, porosity, drainage, and available pool of nutrients in the soil. In orchards situated on hill side, the adaption of zero tillage herbicidal weed control soil management treatment is likely to result in excessive water run off and soil erosion. For such orchards, remedies lie in redesigning the irrigation system and using the mulches or ground cover crops. The ideal ground cover crop should stabilize the soil and prevent erosion. Graminous cover crops have shown very good promise than leguminous crops in improving the infiltration in soil (Norris and Lawrence, 1957; Williams and Doneen, 1960) through microbial action (Martin and Waksman, 1941; Martin, 1943; Geoghegan and Brian, 1948). Response of clementine cultivation on three different soil types with or without added phosphate showed that lucerne increased the P uptake by clementines except in case of poor sand with an inadequate available P (Roderbourg, 1968). A vigorous cover crop can also suppress the growth of weeds, which are more difficult to control in no tillage (Witt, 1984). These cover crops like winter cover crops of purple vetch and sweet clover produced net gains of nitrogen to the tune of 375 kg ha^{-1} year $^{-1}$ (Chapman *et al.*, 1949). Many studies have shown an improvement in fruit yield under cover crops compared to clean cultivation (Parker and Jones, 1951; Patt, 1958; Jones *et al.*, 1961; Gordzhomeladze, 1990). Later, Luo *et al.* (1992) based on 8 years of experimentations showed yellow clover (*Melilotus officinalis*) as a promising green manure crop for citrus which can add 7.5-12.0 tons ha^{-1} green biomass adding 36.7-58.8 kg N, 3.7-6.0 kg P, and 23.2-37.2 kg K ha^{-1} into the soil. Application of 20 cm thick grass mulch in a nonirrigated orchard of 13-15 year old satsuma and dahong sweet orange trees on *Poncirus trifoliata* for three years increased the soil organic matter, available N, P, and K by 68 per cent, 67 per cent, 86 per cent, and 107 per cent, respectively, at Rongjiang county, Guizhou, China (Jiang *et al.*, 1997). While, 8-10 cm thick paddy straw mulch in Assam lemon (*Citrus limon* Burm) on sandy soil of India produced significant improvement in growth and yield according to Nath and Sarma (1992).

Change in cultivation practice from tillage to nontillage, brings an increase in water infilteration as the first noticeable change due to avoidance of formation of compact plough layer (Hinkley, 1939; Moore, 1945; 1946; Johnston and Sullivan 1949; Kimball *et al.*, 1950). This has resulted a concurrent increase in fruit yield (Patt, 1958; Jones *et al.*, 1961). Pisa and Fenech (1990) observed that cultivation of soil (43-49 per cent sand and 23-30 per cent clay) with *Acanthus mollis* and *Amaranthus retroflexa* in spring liberated good amount of tied N which benifited the satsuma mandarin trees in improving the fruit yield from 66.0 to 77.7 tons ha^{-1} and from 67.7 to 80.7 tons ha^{-1} in 1987-89 and 1988-89, respectively. Chiba (1965) and Ozaki *et al.* (1965) reported that a green mulch and a bulky organic matter mulch induced greater increase in calcium contents in upper soil than a sod culture and clean cultivation treatments. Huang (1998) based on 20 years of experimentation in hillside citrus orchards on red soil at 275-900 m altitude showed that most suitable green manure crops include Indian cowpea (*Vigna unguiculata*), groundnut, soyabean, vetch, and Chinese milk vetch (*Astragalus sinicus*).

Out of many legume crops tested on well drained and poorly drained soil types, *Indigofera hirsuta* produced highest dry matter (10.4 tons ha^{-1}) in well drained sandy soil compared to *Cajanus cajan* and *Crotalaria mucronata* (10 tons ha^{-1}) in poorly drained clay soil (Anderson, 1980). Yesilsoy *et al.* (1987) observed an improvement in yield and quality of clementines and sweet oranges by ploughing the vetch (*Vicia*) oats green manure crop combination in addition to soil properties. Potential of citrus (grapefruit cv. ruby red Swingle citrumelo) - cowpea (*Vigna unguiculata*) intercropping in south Florida on raised beds (15.2 cm apart and 107 cm high) planted with 2 rows of trees spaced at 7.3 m apart with 5.2 m in row spacing (246 trees ha^{-1}) produced a yield of 1.84 tons ha^{-1} compared to 2.78 tons ha^{-1} as monoculture (Stofella *et al.*, 1986). Mandarins are grown in south India in coffee estates as shade trees along with *Eucalyptus tereticornis, Casuarina equisetifolia*, and *Grevillea robusta*. Mandarin trees growing with *Grevillea robusta* yielded as much as mandarin trees growing alone, followed by trees growing with *Causuarina equisetifolia*. But, *Eucalyptus tereticornis* produced an adverse effect on mandarin yield (Hanamashetti *et al.*, 1987).

Allelopathic interactions could also leave the same effect as the extreme end of minimum tillage. A better fruit yield of mandarin is obtained in a red earth soil (humus

1.2-3.0 per cent and pH 4.7-4.8) cultivated with lupins, peas, and *Lespedeza bicolor* (Tavartkiladze, 1969; Hurcidze, 1969). Pehrson (1971) in California suggested prostate spurge as a suitable cover crop, since it grows mainly during the summer months, that provides good protection against soil erosion without increasing the winter frost hazard. Bredell *et al.* (1976) examining the effects of various types of mulches on soil behaviour and on growth of valencia orange trees, showed that mulched trees out yielded trees receiving grass sod culture by a factor of two. They further found that under plastic mulch, soil at 10 cm depth is 1^0C warmer at night and 1^0C cooler during the day than grass sod or bare soil. While, Bacon (1974) observed that the soil temperature at 10 cm depth higher by 2.5^0C under the black polythene mulch compared to grass sod or unmulched trees. Khurtsidze (1972) and Chkheidze and Akhaladze (1975) suggested an improvement in mandarin yield by 40 per cent by mulching soil with peat at the rate of 300 tons ha^{-1} in Georgia. Sod culture in form of *Stylosanthes gracilis* showed much better response with reference to beneficial effect on soil microbial activity and stability of soil structure over *Pueraria javanica* or *Digitaria abscendens* (Godefray and Bourdeaut, 1972).

Bredell and Barnard (1974) and Mikaberidze and Rosnadze (1972) observed highest tree volume of young valencia budded on rough lemon mulched with black perforated polyethene, which not only reduced soil moisture evaporation, but influenced the daily soil temperature fluctuations up to 30 cm depth. Pacheco *et al.* (1975) observed least sheet erosion with mulching or perennial soyabean cut in flow or perennial soyabean with seeds harvest using green manuring with cowpea (*Vigna sinensis*) or soil cultivation under sweet orange on Rangpur lime. Other study by Saha *et al.* (1974) observed best response of lemon (*Citrus limon* Burm) on growth, flowering, and fruit set using clean cultivated soil covered with straw mulch. Amami and Haffani (1973) observed that on a lighter sandy soil, black plastic conserved the most moisture, while on clay soil, straw mulch (15 tons ha^{-1}) gave the highest yield of maltaise orange, besides improvement in fruit quality and soil structure. Beraya and Akhaladze (1971) observed that mulching with peat and black polyethene improved mandarin yield by 90 per cent and 36 per cent, respectively on a southwest 10-14° slope without irrigation. Similar observation has been made by Rosnadze (1969; 1971) with young mandarin trees and by Strauss (1966) using saw dust. Comparing the three different soil management treatments involving winter grown cover crop *Vicia sativa,* clean cultivation, and zero tillage herbicide weed control, Economides (1976) observed best response of *Vicia sativa* treatment in terms of yield of valencia orange via improvement in soil structure. Rodriguez *et al.* (1964) in Brazil observed a better yield of citrus trees where a leguminous cover crop or a mulch is used. Besides all these, cover crops have also been effective in providing protection to the citrus trees from freeze (Hastie, 1963; Jordan, 1982; Parsons *et al.,*1985; Jackson and Ayers, 1986; Santinoni and Silva, 1995).

4.3.2 Soil Fertility Changes

Jhuming or Shifting cultivation is still today, a widely adapted farming system in northeastern region of India. In the initial stages of Jhuming, the forest trees and other species of the plants are cut and burnt. Burning causes some changes in the soil properties and fertility (Table 4.8). The extent of changes depends on the type and quantity of the burning materials. Sanchez and Salinas (1981) reviewed the work done on effect of changes in soil chemical properties due to burning of tropical forests in Ultisols and Oxisols of Amazon and concluded that most of chemical soil properties such as soil pH, exchangeable Ca^{2+}, Mg^{2+}, K^{+}, Al^{+}, and available P increased as result of burning, while Al saturation level reduced to a substantial proportion (Table 4.9).

Table 4.8. Effects of burning on the soil properties in northeast India

Soil properties	Before	After
pH	5.1	5.5
Organic carbon (%)	1.3	1.0
Available P_2O_5 (kg ha^{-1})	3.3	3.3
Available K_2O (kg ha^{-1})	219.0	570.0
Exch. Ca (me %)	7.1	9.4

Source : Prasad *et al.* (1981)

Increase in pH is related to increase in available K, which may be temporary. After the removal of K from the exchange complex of soil by hydrogen ions, again the pH would come down to initial level. Changes in soil properties as a result of forest clearing or jhuming have been extensively studied at places like Peru (Sanchez, 1976; Sanchez and Salinas, 1981) and Shillong, India (Anonymous,1983). The ash from burnt forest provides some nutrients initially, especially P, K, Ca ,and Mg and the pH tends to rise. However, in due course of time, barely 2-3 years, multiple nutrient deficiencies appear (P, K, and Mg), and later, by about 4 - 5 years, deficiencies of micronutrients viz., Zn, Cu, Mn, and Fe are commonly observed. Changes in soil physical properties like degraded aggregation, increase in compaction, and surface crusting may also occur, which call for suitable tillage and management practices. There is an urgent need to investigate and generate information on the above aspects in relation to the soils and nature of forest trees in the hilly areas and other areas under shifting cultivation.

In view of the fact that nutrient deficiencies appear in a few years, proper fertilizer application is the key to sustain production in these areas. This should be based on local soil tests and crop needs. As the soil tends to be acidic, liming will help in receiving higher crop yields, especially for cultivars sensitive to soil acidity. Soil fertility changes under shifting cultivation in Tripura for a shifting cycle of three years showed a decline in soil pH (4.7-4.5 and 5.1-5.0), organic carbon (7.3-6.3 and 8.3-6.1 kg^{-1}), CEC (4.2-3.9 cmol (p^{+}) kg^{-1}, and 4.3-3.1 cmol (p^{+}) kg^{-1}), exchangeable cations (0.84-0.75 cmol (p^{+}) kg^{-1} and 1.37-1.06 cmol (p^{+}) kg^{-1}), and available K (150.0-134.0 and 259-203 kg^{-1} ha^{-1}) in west Tripura and south Tripura, respectively, with the increase in shifting cultivation

Table 4.9. Changes in topsoil chemical properties before and shortly after burning tropical forests in Ultisols and Oxisols

Soil property	Timing	Yurmmaguas I	Yurmmaguas II	Manuas (7 sites)	Manuas (1 site)	Barrolandia Bahia (1 site)
Months after burning		1	3	0.5	4	1
pH (H_2O)	Before	4.0	4.0	3.8	4.1	4.6
	After	4.5	4.8	4.5	5.5	5.2
	Increase	0.5	0.8	0.7	0.6	0.7
Exch. Ca+Mg	Before	0.41	1.46	0.35	0.92	1.40
(me 100 g^{-1})	After	0.88	4.08	1.25	5.44	4.40
	Increase	0.47	2.62	0.90	4.52	3.00
Exch.K	Before	0.10	0.33	0.07	0.08	0.07
(me 100 g^{-1})	After	0.32	0.24	0.22	0.23	0.16
	Increase	0.22	0.07*	0.15	0.15	0.09
Exch. Al	Before	2.27	2.15	1.73	1.81	0.75
(me 100 g^{-1})	After	1.70	0.65	0.70	0.10	0.28
	Decrease	0.59	1.50	1.03	1.71	0.45
Al saturation	Before	81	52	80	64	34
(%)	After	59	12	32	2	5
	Decrease	22	40	48	62	29
Olsen - P	Before	5	15	-	2	1.5
(mg kg^{-1})	After	16	23	-	3	7.0
	Increase	11	8	-	3	7.0

* Decrease

Source : Sanchez and Salinas (1981)

period from one to three years. While, base saturation increased from 20.4 to 22.4 per cent in west Tripura and from 23.8 to 34.4 per cent in south Tripura (Datta *et al.*, 2001).

4.3.3 Integrated Land Use

Land use and management practices greatly impact the direction and degree of soil quality changes. Owing to improper land use and management, soil erosion, acidification, nutrient depletion, pollution, and other natural resources restoration of have been the major problems (Lal, 1990; Rozanov, 1990). Large areas of sloping land are already in arable use and must remain so. Therefore, some means must be found of making cultivation and associated land management practices environmentally acceptable (Young, 1989).

Considering the watershed as a unit of land use, the aspects such as methodology for integrated basic resource development of the watershed involving the components of soil, water, and the multiple system of Agriculture, minimized erosion caused by run off, maximise intake of rainfall, its storage and recycling, and optimise production per unit area have to be the prime concerns. The basic information necessary for citrus based watershed consists of treatment measures necessary in planning, protecting, and improving the watershed; meteorological data (rainfall, temperature, humidity, and evaporation); soil data (physiography, geology, drainage pattern, elevation, size, and shape of catchment); hydrological data (intensity and distribution of rainfall, stream flow features, run off coefficient); and socioeconomic data. Land use in hilly areas should be on an integrated basis and preferably planned, based on watershed concept, since in these slopy lands, the run off and soil erosion are not localized, but are part of integrated system. The aim should be to minimize erosion, maximize rain water infiltration into soil, besides securing food, fuel, fodder, and other economically useful produce (Fig. 4.5).

Fig. 4.5. Model integrated use of citrus growing land in hilly areas (ICAR Reserch Complex for NEH Region), Umiam, Meghalaya

It has been recommended (FAO, 1984b; Anonymous, 1986) that for a watershed, the top most part of the hill may be under permanent forest, the middle portion under horticulture/agroforestry, and the lower slope bottom and part of valley under Agriculture. An integrated land use such as this, if planned, there is no question of taking up isolated patches, but an entire watershed should be taken up as an unit. Mechanical soil and water conservation measures are, therefore, required in high altitude areas for controlling soil erosion, retaining maximum rainfall within the slope, and safe disposal of the excess run off from the top to the foothills. Such measures are contour, bunds, bench terraces, half moon terraces, grassed water ways, and water harvesting ponds. In this connection, land management through terracing in Sikkim and Nagaland, lands and water management system used in east Khasi and Jaintia hills of Meghalaya are noteworthy (Prasad *et al.,*1987). The working group on zonal planning zone II, eastern himalayan region (Anonymous, 1989) suggested slope based land use in the hills. Accordingly, the land with 20-30 per cent for agricultural/horticultural/silvipastoral land use, 30 to 40 per cent slope land for silvi-horti-pastoral, and more than 40 per cent slope for forestry provides a strategy for slopy land management. A number of watershed programmes, which speak the success story of integrated land use especially in high altitude areas, having citrus as one of the component crops include : Popumpona watershed at Itanagar, Arunachal Pradesh; Hiri Hiri watershed at Karbi Anglong, Assam; Luangliema watershed at Senapati, Manipur; Umtongphar watershed at east Khasi hills, Meghalaya; Didram Upper catchment at west Garo hills, Meghalaya; Teirei watershed, Mizoram; Dikhu watershed, Nagaland; Shinga Tsusang watershed, Nagaland; Maharanicherra watershed, Tripura; and Rangacherra watershed, Tripura (Srivastava, 2001).

Litter fall decomposition and soil nutrient dynamics under cardamom (*Amomum subulatum*) - N_2 fixing trees and mandarin (*Citrus reticulata* Blanco), *Albizia stipulata - Albizia chinensis* agroforestry system showed higher ratio of litter production to floor litter (liter turnover) in former than latter agroforestry system in Sikkim himalaya (Sharma *et al.*, 1997a). While, lower C/N ratio (improved mineralization) and higher inorganic-P/total-P have been observed in mandarin than cardamom agroforestry system (Sharma *et al.*, 1997b). Influence of different forest species on yield of Coorg mandarin revealed that compared with the control, slightly greater fruit yield in trees interplanted with *Casuarina* is observed compared to those planted with silver oak (*Grevillea robusta*) and those planted with *Eucalyptus* (Allolli *et al.,* 1988).

Summary

The shifting cultivation is a part of common practice on slopy land where cirus is grown at high altitude areas with predominantly acid soils belonging to Oxisols and Ultisols. The slopy land in these areas are characterised by extreme depletion in soil fertility, in addition to exposing the comparatively infertile subsurface. Citrus growing under such soil-physiographical set up has never provided the nutritionally healthy status to citrus orchards.

The most significant is to adapt the Model Land Use to have optimum land use based on land capability analysis as an alternative to Jhum which revealed one third slope under permanent forestry, other one third under citrus with half moon terracing and rest of the one-third must be used under agriculture coupled with bench terracing (Table 4.10). Such an attempt would help combat soil degradation most precisely. Development of some monitoring and warning techniques to continuously be familiar with the changes taking place in the soil health and also to overcome any impending damage, if any, would further make fertility management of slopy land, a less challenging task.

Table 4.10. Model land use with citrus as a component crop

Slope	Approximate percentage of total area	Land use	Conservation measures
Lower portion	33.5	Agriculture	Bench terracing
Middle portion	33.5	Horti-pastoral	Half moon terracing for horticultural plants
Top portion	33.0	Forestry	–

Source : Sarkar (1994)

5 Citrus Growing Acid Soils

Acid soils are extensively found in the regions of high rainfall and temperature. Assam along with Mainpur, Tripura, Meghalaya, Arunachal Pradesh, Nagaland, and Mizoram in the northeastern part of the India have large stretches of acid soils followed by the neighbouring states of West Bengal, Bihar, and Orissa. Acid soils in Himachal Pradesh and Punjab are mainly found in citrus growing districts of Chamba, Kangra, Kulu, and some parts of Gurdaspur, and in Hoshiarpur. These soils formed under perhumid and subtropical conditions are mostly siliceous in the upper horizon (Govindarajan and Venkata Rao, 1976). The southern states of Tamil Nadu, Andhra Pradesh, and Karnataka also have acid soil in which commercial citrus cultivars like acid lime, sathgudi sweet orange, and mandarins (Coorg), respectively, are grown on a large scale. In all these areas, studies have established a positive correlation of soil acidity with high temperature and rainfall.

Broadly, the acid soils are classified into seven distinct groups, namely, i. laterite, ii. laterite and lateritic red, iii. mixed red black and yellow, iv. ferruginous red, v. podzolic, brown forest, and forest, vi. foothill soils analogous to Wissenboden, and vii. peat. The parent materials, even though have their distinctive influence on the development of acidity in soils, the temperature and rainfall appear to have been the more dominant factors. The Ultisols and Oxisols of upland are generally acidic, Al toxicity, Ca and P deficiency are wide spread (Pushparajah and Bachik, 1987; Lu *et al.*, 1997). A great majority of citrus growing soils in Asian countries (China, India, Japan, Thailand, Bangladesh, Sri Lanka, Nepal, Philippines etc.), Europe (France, Italy, Georgia, Greece etc.), North and Central America (USA, Cuba, Puerto Rico, Costa Rica, Mexico etc.), South America (Argentina, Brazil, China, Uruguay, Venezuela etc.), Africa (Nigeria, Kenya, Zimbawe, Mozambique etc.), and Australia (Australia, Cook Island, New Zealand etc.) are occupied by acid soils. Citrus orchards in acid soils could be affected in two ways : new plantings made on soils that are naturally acidic or which have been sown down to pasture for many years encourage the development acidity and bearing orchards on poorly buffered soils fertilized over long period of years with acidifying nutrients also impart soil acidity (Le Roux, 1965; Robinson, 1989).

5.1 SOIL GENESIS

A natural consequence of the vast area, the climatic variations, and a wide range of parent materials involved in soil forming processes have led to the development of soils of widely divergent nature. The factors which have been particularly dominant in the development of acid soils are rainfall, temperature, hydrological conditions, and vegetation. The major processes involved in the development of acid soils are: laterization of varying degrees, podzolization in areas with subtemperate to temperate climate, intense leaching in light alluvial soils in high rainfall of partly decomposed organic matter, and marshy conditions with significant amount of under-decomposed or partly decomposed organic matter. There are several theories on nature of soil acidity (Panda, 1987). Acidity develops on soil colloids mainly by two mechanisms, viz., first type by isomorphous substitution of H^+ or Al^{3+} in silicate minerals, which form exchange sites throughout the pH ranges as permanent charge acidity (exchangeable acidity) and the second type due to the polymers of Fe and Al and soil organic matter where the exchange sites solely depend on soil pH (pH dependent acidity).

At soil pH below 5.0, a reduction in growth of citrus plants takes place due to damage imparted by the presence of excess Al^{3+} and H^+ ions (Smith, 1966). Wander (1954) carried out several studies on soil acidification in citrus orchards in Florida and observed that acidification is caused by those sources which could leach down to 100 cm in the profile. Even, with the application of high rate of limestone, it is difficult to recover the fertility of these soils, especially in the deep layers. Based on the response of citrus rootstocks to varying levels of Al^+, the relative tolerance of rootstocks is rated as cleopatra mandarin >rough lemon > sour orange > Swingle citrumelo > Carrizo citrange with apparent optimum concentration of 163, 93, 89, 85, and < 50 μm, respectively (Lin and Myhre, 1991).

5.2 SOIL TYPES

5.2.1 Red and Lateritic Soils

These soils have been formed from residuary rocks produced by the intense weathering of basic and intermediate igneous rocks. The climatic zones in which laterites are extensively found are i. high rainfall zone with strongly expressed dry season, ii. high rainfall zone with weakly expressed dry season, and iii. subhumid zone with pronounced wet and dry seasons. The other factor which also bring about changes in the properties of the laterites is the elevation of the land mass. Thus, both high and low level laterites are also of common occurrence in areas having variable topography and landscape. The specific conditions under which laterites develop are the minimum amount of water necessary for weathering and leaching of bases, combined silica, intervals of dry season, and conditions leading to segregation of secondary iron and aluminium oxides, hydroxides and their crystallized products such as geothite or bauxite. Alternate wet and dry season favours the cementation of the vesicular lateritic mass. Laterites with an elevated content of alumina in some places cover large areas of southern Vietnam. The laterites are 0.5-

0.9 m thick and occasionally underlain by a haematite crust which reached a thickness of 2.5 m (Nguyen, 1977; 1978; Yakushev and Guzovskiy, 1982)

The laterites found in peninsular India are either ferruginous or bauxitic, and have well defined segregation of sesquioxide concretions. The morphological characteristics of laterite soils formed on basaltic or granitic parent materials are similar except that the soils derived from granitic rocks are more ferruginous with intense red colour, greater acidity, low cation exchange capacity, and base saturation. The clay in these soils is more siliceous in contrast to the predominance of secondary iron and aluminium containing minerals in soils formed on basaltic parent rocks. In certain areas, crumbly laterite containing a slag like hard layer with less of kaolinitic or aluminous earth filling, the structural cavities are found. The laterite layer from such soils is used very often as building stones. Laterites in high rainfall zones with weakly expressed dry season, as in khasi hills are formed on unclassified crystalline rocks, gneiss etc., are fairly acidic in reaction. The pH of the first three layers of a profile is found to be 4.8, 4.6, and 5.0. The colour sequence is greyish yellowish red, yellow and red mixed, and light red. Laterites in subhumid zone with pronounced alternate wet and dry seasons, as in Bihar and Andhra Pradesh are neutral in reaction (pH 6.5 to 6.9). Other characteristics, as indicated by a profile are : clay 50.8 to 60.7 per cent, SiO_2 49.4 to 31.4 per cent, Al_2O_3 17.1 to 28.3 per cent, and Fe_2O_3 20.1 to 34.8 per cent down the profile. They differ from true laterite in having a relatively higher content of silica and lack the lateritic layer in the upper horizon. The low mobility of clay also prevents the formation of clay skins or hard pans. This layer contains varying amounts of sesquioxides, kaolinite, quartz, and highly resistant minerals. The laterite and lateritic soils of West Bengal are low in Ca and P, and rich in Fe and Al. The content of silica decreases down the profile, while that of Al_2O_3 and Fe_2O_3 increases, indicating preferential mobilisation of sesquioxides. These soils, though classified as laterite and lateritic, differ essentially from true laterite in many respects (Rudra, 1956).

On the basis of morphological features, the red soils can be grouped as : i. red loams characterised by argillaceous soils with a clayey structure and the presence of very little red concretionary materials, and ii. red loams characterised by a loose and friable top soil and rich in secondary concretions of sesquioxides (Mandal *et al.*, 1975). These red soils present serious physical and chemical limitations for effective citrus production. Red soils are characterised by low water storage capacity, crusting, rapid drying of surface soil, and associated with high soil temperature, inadequate nutrient holding capacity, problem of soil workability, soil erosion, presence of root limiting subsoil (argillic horizon), and stoniness. Due to sandy surface soil texture, low organic matter, clay minerals mostly illitic and kaolinitic, low cation exchange capacity, high infiltration, and leaching of nutrients are very common.

5.2.2 Podzol Soils

Soil formation in humid subtropical climate occurs in an acidic environment which is characteristic of podzolization. But, the mark of podzolization is not fully expressed in the red loams. This faint expression of the podzolization process in red loams may be related to the supply of bases formed during organic matter decomposition, which neutralises the acidic products. The process of podzolization is further suppressed, owing to the presence of sesquioxides obtained from the decomposition of their organo-metallic complexes. The Kumaon hill soils of Uttar Pradesh in Almora, Garhwal, and Nainital districts also exhibit signs of podzolization (Mukherjee and Das, 1940). The virgin soils of the hilly districts of Assam revealed high content of organic matter and nitrogen. The surface layers consist of well decomposed humus and minerals, which shade off gradually at varying depths into the colour of the parent rock. The soils are strongly acidic in reaction. In Uttar Pradesh, Punjab, Himachal Pradesh, and Bihar, these soils are found in equilibrium with diverse factors including terracing and cultivation. The pH of the soils ranges between 5.5 and 6.5. In Uttaranchal, this tract runs as a narrow belt at the foot hills of the Himalayas in Dehradun. The chemical characteristics of Assam soils has described by Raychaudhuri *et al.* (1963) which indicated that the Brahmaputra valley soils are relatively richer in total Mg content. This is relatively a peculiarity of the Assam soils and attributed to the dominance of Mg bearing minerals in the parent material. The soils of hill districts are characterised by low Ca and P status. The oxidation-reduction potential of podzolic soil in mandarin orchard at Sakhumi ranged between 29 and 35 rH_2, indicating, thereby, the good aeration of soil. The potential was generally reduced by 2 – 3 rH_2 upto soil depth of 40 cm (Kuznecov and Gocelasvili, 1967).

Composition of clay in the soils of Western Georgia known as subtropical Podzols (Kovda, 1934; Daraseliya, 1949; Gamkrelidze, 1960) showed that these soils have been formed in the flat Kolkhida lowland, where they occur in combination with gley soils. In the most level areas in the piedimont terrace region, where they are found in combination with yellow and red earths. Morphologically, the subtropical Podzols are distinctly differentiated into genetic horizons. At the surface, they have a bleached, nearly whitish A_2 horizon 30-50 cm thick, which is weekly stained with humus from above, and has a large number of Fe-Mn concretions in the lower part; lower down the profile, there is an Fe rich B_f horizon 20-30 cm thick, where iron forms concretions, spots and coatings, and penetrates throughout the mass of this horizon in some places, forming cemented layers; the B horizon is underlain by a variegated, slightly gleyed layer gradually changing to the parent material. According to Romashkevich (1974), the subtropical Podzols are very acidic having soil pH 4.5-5.5 in water, 3.5-4.5 in KCl salt, and unsaturated (20-80 per cent). While, unsaturation is maximum in the A_2 horizon, especially in the subtropical Podzols of low terraces. Exchangeable cations are represented mainly by aluminium and the rest of the soil adsorption complex is saturated with Ca and Mg. The humus content is 4-7 per cent in the upper horizons and gradually decreases with depth. Fulvic acids, predominate the humus. Subtropical Podzols are usually loamy or clayey, have a considerable amount of silty particles, and their bleached horizon is distinctly poor in clay. In most of the profiles, the clay content gradually increases with depth including the bleached layer, but some profiles have an illuvial clay horizon. The profile of the subtropical Podzols is distinctly differentiated according to total chemical composition into SiO_2 rich upper horizon and lower horizons rich in Fe and Al (Table 5.1). An especially high total content of Fe is found in the concretional horizon. It is interesting that the maximum amount of Fe and clay is usually observed in different horizons and iron accumulation is observed in the profile above the horizon with the maximum content of clay. This kind of distribution testifies to the different paths of migration and accumulation of iron and clay in the profile of these soils (Zyrin *et al.,* 1976). With the conditions apparently favourable for some type of podzolization in the cool humid climate of high hills in Mizoram with sizeable accumulation of organic matter. The soils do not indicate any appreciable movement and organic matter induced deposition of sequioxides as that in a Spodosol (Singh *et al.*, 1986).

Subtropical Podzols are associated with the formation of the concretional horizon not only with eluvial-illuvial iron redistribution in the profile, but also with the supply of this element with the lateral flow of soil and ground waters. For example, Kovda (1934) indicated that concretional horizons are similar in origin to bog ore deposits. Zonn (1970) described about the formation of lateritic pans as a result of the additional supply of iron with lateral flow in the soils of Western Georgia, the same way they are formed in other tropical and subtropical countries (Fridland, 1964; Zonn, 1970). The composition of clay minerals in subtropical Podzols has been studied by Zonn and Shoniya (1971), and Romashkevich (1974). Gamkrelidze (1960) found montmorillonite, chlorite, kaolinite, quartz, hydrogoethite, hydrargillite, and amorphous substances in the clay. There is no montmorillonite and kaolinite in the fraction < 0.2 mm diameter in the bleached horizon. The composition of clay minerals is distinctly differentiated along the soil profile. Chlorite like minerals and kaolinite proper predominated the upper part of the profile, including the bleached and iron-concretional horizons. These horizons have more minerals of the illite group and fine quartz than the underlying horizons. The clay in the lower part of the profile contains montmorillonite, metahalloysite, and vermiculite, which do not occur in the upper horizons. The minerals, iron hydroxides, are also differentiated along the profile.

5.2.3 Peat and Organic Soils

Origin of such soil takes place in humid regions as a result of accumulation of appreciable amounts of organic matter in the soil. Such soils are found in the states of Assam, coastal tracts of Orissa, and in the certain regions of north Bihar, Almora district

Table 5.1. Some physicochemical properties of subtropical Podzols of Western Georgia

Horizon and depth (cm)		pH		Humus (%)	Exchangeable cations (me $100g^{-1}$)			Particle diameter (%)	
		Water	Salt		Ca^{2+}	Mg^{2+}	H^{+}	< 0.001 (mm)	< 0.01 (mm)
A_1	0-10	4.52	3.68	4.68	1.8	1.6	3.3	12	50
A_2	40-20	4.88	3.95	4.43	0.9	0.9	1.9	ND*	ND
A_2	20-30	4.99	4.06	0.57	0.9	0.9	2.2	13	54
B_f	34-57	4.91	3.93	0.43	2.5	2.4	3.2	21	52
B_g	60-80	4.81	3.66	0.23	6.1	5.4	7.6	32	48
$B_{g'}$	80-100	4.81	3.74	ND	8.9	6.1	7.7	28	43
$B_{g'}$	120-140	5.05	3.75	ND	12.0	10.8	4.5	31	49

* ND stands for not determined

Source : Zyrin *et al.* (1976)

of Uttar Pradesh, and southeast coast of Tamil Nadu. In Assam, Surma valley soils are characterised by their swampy nature and have peaty soils known as bheels. The formation of bheels is explained as due to blocking of hollow pockets between surrounding highland and deep peaty soil. These soils are highly acidic. Raychaudhuri *et al.* (1963) correlated acidity of Assam soils with rainfall and clay content. The soils of Brahmaputra valley may be texturally classed as sandy loams. The hill soils display wider variation in mechanical composition. Old alluvial soils are sandy loam to clayey in texture, while new alluvial soils are mostly sandy loam, silt loam or clay loam. Hill soils are of loam to clay loam texture. The flat lands of Surma valley have varying intergrades of texture from sandy to clayey. The acid soils of Punjab spread mainly over the districts of Gurdaspur and to some extent in Hoshiarpur, are sandy clay to clay loam in texture.

5.3 PROPERTIES OF ACIDIC SOILS

A great diversity in morphological, physical, chemical, and biological properties of citrus growing acid soils has been observed.

5.3.1 Classification of Soil Acidity

Soil acidity is defined as a soil system having protein yielding capacity during its transition from a given state to a reference state (Jackson, 1958). These states may be specified in terms of pH values or otherwise. Soil acidity and its neutralization should better be considered according to its proton retaining sites as well as reactions leading to donation of protons (Jackson, 1963). Soil acidity has been grouped by Jackson (1963) as strongly acidic (soil pH 4.2 and below), very weakly acidic (soil pH 5 or 5.2 and below); very weakly acidic (soil pH 5.2 to 6.5 or 7.0), very very weakly acidic (soil pH 6.5

or 7 to 9.5), and extremely weak acidic (soil pH > 9.5). Brady (1992) classified soil reaction (pH) into six categories : very strongly acid (4.5-5.0), strongly acid (5.1-5.5), medium acid (5.6-6.0), slightly acid (6.1-6.5), neutral (6.6-7.3), and slightly alkaline (7.4-7.8). Studies on 93 citrus orchards of California showed that the pH value of the cultivated orchards ranged from 4.8 to 8.1 with 37 orchard testing alkaline pH and 22 acidic. The non cultivated orchards ranged from ph 4.7-8.0 with 10 orchards showing pH values of 7 or above and other 24 orchards values below 7.0 (Martin and Chapman, 1951).

Spencer and Koo (1962) reported Ca deficiency in citrus trees due to soil acidity in Florida soils, and Ca application brought an improvement in fruit quality in terms of smooth skinned fruits (Bryan, 1963). Soil acidity has also been observed to influence the growth and leaf nutrient composition (Martin and Page, 1962). Tsanava and Burchuladze (1974) observed that soil acidity (< pH 4.0) appreciably reduced free amino acid content in mandarin fruits.

Total soil acidity in soils of Mizoram vary from 8.6 to 16.3 cmol (p^+) kg^{-1}. The mean contribution of pH-dependent and exchange acidity to total acidity has been observed 88 per cent and 12 per cent, respectively. Exchangeable Al^{3+} constitutes 81 per cent of exchange acidity fraction. The Al saturation of soil based on effective CEC is 43.0 per cent at pH 4.5 and was essentially zero at pH 5.6. These soils have low permanent nagative charge and most CEC of soil is due to pH- dependent charge (Misra and Saithantuaanga, 2000).

Total acidity value which comprised pH dependent and exchange acidity is reported to be higher in soils from Chikmagalur (13.4 cmol (p^+) kg^{-1}), followed by Kodagu (12.2 cmol (p^+) kg^{-1}), and Hassan (4.7 cmol (p^+) kg^{-1}). This could be due to the presence of large quantity of organic carbon and exchangeable Al^{3+} in soils from Kodagu and Chikmagalur. The per cent contribution of pH dependent acidity towards total acidity ranges from 84.9 to 93.7 in Chikmagalur, 87.3 to 95.3 in Kodagu, and 43.9 to 84.1 in Hassan soils. Soils of Chikmagalur and Kodagu contain not only higher amount of pH dependent acidity than soils of Hassan, but also the percent contribution of pH dependent acidity to total acidity being higher in soils of Chikmagalur and Kodagu than Hassan. The per cent contribution of exchange acidity towards total acidity ranges from 6.2 to 15.2 in Chikmagalur, 4.7 to 12.7 in Kodagu, and 15.9 to 56.1 in Hassan soils. Exchange eacidity values are rather low in all the soils (Prabhuraj and Murthy,1994).

5.3.2 Relation Between pH and Al Saturation

Some sensitive citrus species may have growth adversely affected at values of exchangeable Al as low as 1 to 5 mg kg^{-1} and hazard to more tolerant species increases above 5 to 10 mg kg^{-1} (Robinson, 1989). Typical Al toxicity symptoms develop in acid soils as white crust formed on leaf surface due to Fe toxicity (Fig. 5.1) or in severe cases, lead to degradation of chlorophyll due to Fe toxicition (Fig. 5.2). Earlier, Otsuka and Morizaki (1969) observed that growth of satsuma mandarin is seriously depressed in presence of 30 mg kg^{-1} Al (ionic concentration about 9 mg kg^{-1}) and 30 mg kg^{-1} Mn. In acid soils, exchangeable Al^{3+} is generally considered to be the predominant cation as determined by extraction with a neutral salt solution like KCl. Lime rates based on neutralization of exchangeable Al^{3+} have been found to be adequate for optimum yield of many crops (Prabhura and Murthy, 1994b). However, lime rates equivalent to twice the KCl exchangeable Al^{3+} are required to neutralize completely the exchangeable Al^{3+} of acid Ultisol. This suggests that there are other forms of reactive Al (nonreadily exchangeable Al) in soils which are not exchangeable with KCl solution, but react with lime.

Fig. 5.1 White lustrous crust formation on the foliage, a common sysmptom of aluminium toxicity in khashi mandarin (*Citrus reticulata* Blanco) grown on acid soil

Studies conducted in 30 soils of khasi hills of Meghalaya showed that 87 to 90 per cent of added phosphate (100 mg kg^{-1}) is adsorbed. Similar results are also reported in Sikkim (Khera and Pradhan, 1976) and Nagaland soils (Chakraborty, 1976). Nath and Deory (1976) analysed 34 soils from different altitudes of Arunachal Pradesh for different fractions of nitrogen (Table 5.2). On the contrary of the above data, negative relationship between organic carbon and available nitrogen (alkaline permangnate method) has been obtained in the soils of Sikkim. This is due to the fact that an appreciable amount of H^+ ions in acid soils is directly related to the presence of humus through ionization of carboxyl or phenolic-OH groups or by hydrolysis of aluminium hydroxy cations. It is noticed that soil acidity in Manipur soils is mainly due to Al^{3+} ions and the relation of pH and Al saturation will further predict Al toxicity level at a defined pH. The Al saturation in hill soils has been recorded to vary from 6 to 57 per cent compared to 0 to 25 per cent in alluvial soils with lime requirement being higher in former (0.82-5.11 tons ha^{-1}) than latter soils (0.50-2.80 tons ha^{-1}). The

Fig. 5.2 Chlorophyll degradation as characteristic symptoms of iron toxicity in khashi mandarin (*Citrus reticulata* Blanco) grown on acid soil

performance of Troyer citrange, trifoliate orange, and cleopatra mandarin in two weathered Oxisols, Kapaa soil (high in Al) and Whiawa soil (high in Mn) demonstrated the order of decreasing fresh weight of seedlings in Whiawa soils as cleopatra mandarin >> Troyer citrange >> Trifoliate orange, and in Kapaa soil, cleopatra mandarin >> Troyer citrange >> trifoliate orange (Worku *et al.*, 1982).

Table 5.2. Different fractions of nitrogen and relationship with altitude

Parameters	Range	Mean	Correlation value
Organic matter (%)	1.03-9.45	2.71	0.77**
Total N (%)	0.06-4.12	0.18	0.61**
Available N (mg kg^{-1})	1.29-6.33	3.28	0.55**
Nitrate-N (mg kg^{-1})	1.0-16.0	5.90	0.37**
NH_4^+ - N (mg kg^{-1})	8.4-92.4	25.80	0.62**

** Significant at 1%

Source : Nath and Deory (1976)

The CEC as the sum of exchangeable Ca, Mg, Na, and K extracted by 1M NH_4OAc (pH 7.0) and exchangeable Al^{3+} from 1M KCl extract of the soils is taken as the permanent charge component of the soil (Coleman *et al.,* 1959; Lin and Coleman, 1960; Misra *et al.,* 1989). The percentage Al saturation in Manipur soils based on the effective CEC shows that the exchangeable Al is practically zero at pH 5.6 or more. This relationship for soils of Orissa is reported to be zero per cent at pH 5.0 and approximately 45 per cent at pH 4.0 (Misra *et al.,* 1989). Whereas, significant amounts of exchangeable Al^{3+} and higher Al-saturation have been observed at pH around 5.6 in soils of Meghalaya (Nair and Chamuah, 1993). The lower KCl extractable Al^{3+} indicated that soils of Orissa are also in more advanced stage of weathering than those of Manipur according to the theory proposed by Beinroth, (1982). Surface soils generally have lower Al-saturation due to downward movement of Al as organometal complexes or chelates (Schnitzer and Skinner, 1963).

Suggestions of Coleman *et al.* (1968) are well known which comprised: permanent charge is the sum of exchangeable metal cations including aluminium; cation exchange capacity measures the permanent charge and only a fraction of the pH dependent charge; there is hardly any suitable method for estimating exchangeable hydrogen accurately; percent base saturation is influenced by the extent of pH dependent charge included in the calculation of kaolinitic clay minerals; concomitant adsorption of anions and cations leads to reduction of net negative charge; and functional groups of soil organic matter (carboxyls, phenolic, and enolic hydroxides) dissociate H ions. Vast areas of acid soils, fall between pH 5.2 and 6.5, which respond to liming, but have little KCl exchangeable Al. Nye *et al.*(1961) and Evans and Kamprath (1970) found that Al is held quite tightly in comparison with other cations. Soils with low salt content, appreciable amount of Al is not found in the soil solution, until the Al saturation is greater than 60 per cent. Kamprath

(1970) has also suggested that lime rates based on the amount of exchangeable Al are quite reliable for bringing out the same relative neutralization of exchangeable Al, even though, the soils vary considerably in their buffering capacity.

It appeared that the KCl exchangeable Al is the unique edaphic parameter for liming these tropical soils as recommended by many other workers (Reeve and Sumner, 1970; Kamprath, 1970; Oates and Kamprath, 1983). Distribution of exchangeable Al in these soils is high throughout the solum ranging between 18 and 41 per cent for the surface soils. Likewise, the per cent Al saturation for the surface and control section varies from 4.4 to 67.7 and 1.3 to 79.5, respectively. Aluminium in the soil solution exceeding 1 mg kg^{-1} has been found to affect yield of certain crops (Sanchez, 1976; Kamprath, 1980). It is, therefore, considered that the soils suffering from Al-toxicity need liming at the rate of 0.3 to 7.6 t ha^{-1} in surface and 0.8 to 9.6 t ha^{-1} in control section, respectively. It has also been observed that the soils of Meghalaya require the least amount of lime, while soils of Manipur, Mizoram, and Nagaland require more lime to restore soil productivity than those of Arunachal Pradesh, Assam, and Tripura (Nayak *et al.*, 1996b).

5.3.3 Nature of Soil Acidity

In India, acid soils (pH <M 6.5) occupy about 34 per cent of the total cultivated area (Panda 1987). According to Sing *et al.* (1998), the orchard soils of Umroi/Barapani, Mawlai, Marow, and Dawk, of Meghalaya are highly acidic having soil pH below 5.0. And at such low soil pH, root injury can be a common feature (Lin and Myhre, 1991). Soil acidity may be partitioned into exchangeable (chiefly monomeric Al) and nonexchangeable (titratable or pH dependent acidity) components, based on extraction with a neutral salt solution such as 1 M KCl (Coleman and Thomas 1967). This exchangeable acidity is ascribed to isomorphous substitution, while the pH dependent acidity is due to polymers of Fe^{3+} and Al^{3+} and soil organic matter. A knowledge on forms of acidity is the first step in undertaking the acid soils for their improvement. Exchangeable acidity values have been rather low (Table 5.3) in all citrus growing districts viz., Chikmanglur, Hassan, and Kodagu having acid soils in Karnataka (Prabhuraj and Murthy, 1994a). The relatively low contribution of exchangeable acidity towards total acidity has also been reported by Adhikari and Si (1991). Soil organic matter possesses a number of functional groups containing H^+ ions that contribute to the different kinds of acidity, depending upon their magnitude (Keeney and Corey, 1963; Singh and Datta, 1983; Singh *et al.*, 1991). Likewise, exchangeable Al^{3+} contribute significantly to different kinds of acidities. Soil acidity characteristics of some acid soils of Karnataka (Table 5.4) have shown variation in exchangeable (1N KCl) and extractable (pH 4.8) Al^{3+} from 0.03 to 0.38 cmol (p^+) kg^{-1} and from 0.86 to 2.74 cmol (p^+) kg^{-1}, respectively (Rao and Ananthanarayana, 1997).

Table 5.3. Different kinds of acidity in the acid soils of three districts belonging to southern transition zone of Karnataka

pH		Organic C ($g\ kg^{-1}$)	Exchangeable Al^{3+}	Forms of acidity		Total
H_2O	1M KCl			Exchangeable	pH dependent	
			[cmol (p^+)kg^{-1}]			
			Chikmagalur			
4.6	3.6	31	2.0	3.8 (15.2)*	21.2 (84.8)	25.1
5.5	4.6	29	0.7	1.2 (7.3)	15.3 (92.7)	16.5
5.3	4.4	28	0.2	1.0 (6.5)	14.5 (93.5)	15.5
5.6	4.5	26	0.7	1.2 (10.7)	10.0 (89.3)	11.2
5.4	4.2	26	0.9	1.4(9.7)	13.0 (90.3)	14.4
5.2	4.4	39	0.9	1.2 (7.8)	14.3 (92.2)	15.5
5.3	5.3	28	0.7	0.8 (8.6)	8.5 (91.4)	9.3
5.9	4.3	30	0.4	0.6 (8.3)	6.6 (91.7)	7.2
5.0	4.1	33	1.1	1.4 (12.2)	10.1 (87.8)	11.5
5.0	4.7	35	0.1	0.2 (10.9)	9.8 (89.1)	11.0
5.7	4.8	16	0.4	0.7 (6.2)	9.7 (93.7)	10.4
			Hassan			
5.4	4.0	7	0.7	1.1 (15.9)	5.6 (84.1)	6.7
5.1	3.9	6	0.8	1.1 (19.3)	4.4 (80.7)	5.5
5.5	4.1	5	0.3	0.9 (16.1)	4.6 (89.2)	5.5
5.3	4.0	4	0.6	0.9 (56.1)	0.7 (43.9)	1.6
5.3	4.0	7	0.5	1.1 (25.7)	3.1 (74.3)	4.1
			Kodagu			
4.1	4.0	7	1.3	1.4 (9.4)	13.4 (90.6)	14.8
5.5	5.0	24	1.1	1.6 (12.7)	11.0 (87.3)	12.5
4.7	4.6	14	0.4	0.8 (6.6)	11.4 (93.4)	12.2
4.2	4.0	22	0.9	1.0 (8.5)	10.8 (91.5)	11.8
4.3	4.0	15	0.7	0.7 (7.1)	8.5 (92.9)	9.1
4.3	3.8	17	1.3	1.8 (12.7)	13.3 (87.3)	14.1
4.7	4.1	18	0.9	1.6 (10.7)	13.2 (89.3)	14.8
5.3	4.7	6	0.3	0.4 (4.7)	7.9 (95.3)	8.3

* Values in parenthesis indicate per cent of total acidity
Source : Prabhuraj and Murthy (1994a)

5.3.3.1 Total Potential Acidity

The potential acidity of the hill soils is considerably high, ranging from 15.6 to 21.2 cmol (p^+) kg^{-1} in comparison with the soils of alluvial zones from 9.2 to 16.2 cmol (p^+) kg^{-1} in soils of Manipur. This is possibly due to high content of organic matter, clay, and free iron oxide (Table 5.5). In acidic hilly soils of Manipur from an altitude of 800-1500 m above mean sea level, neutral salt CEC ranged from 3.8 to 27.7 cmol (p^+) kg^{-1} and constituted 81 per cent of the total CEC. The total CEC which is a sum of neutral salt CEC and pH dependent CEC had a mean value of 15.3 cmol (p^+) kg^{-1}. Weakly dissociated

Table 5.4. Some characteristics of the acid soils of Karnataka

Soil*	Silt (%)	Clay (%)	pH	Organic matter (%)	Potential Acidity	Exch. Al^{3+} (1 N KCl)	Ext. Al^{3+} (pH 4.8)	Exchangeable cation		
					[cmol (p^+) kg^{-1}]			Ca^{2+}	Mg^{2+}	K^+
1.	14.8	13.8	5.3	3.26	9.36	0.08	1.68	3.2	1.5	0.5
2.	10.8	11.0	5.2	3.50	8.02	0.38	1.98	2.4	0.5	0.3
3.	12.8	11.0	4.6	2.62	2.67	0.18	1.24	3.4	0.5	0.4
4.	12.0	7.0	4.9	0.57	2.67	0.11	1.63	2.3	1.1	0.5
5.	12.0	7.0	4.8	0.78	1.34	0.09	1.24	2.6	0.6	0.3
6.	14.8	13.8	4.5	2.26	5.57	0.36	2.74	1.6	0.7	0.2
7.	16.2	13.6	4.3	3.09	2.67	007	1.58	3.4	0.8	0.3
8.	14.6	11.0	5.8	3.36	5.36	0.09	1.02	2.4	0.6	0.3
9.	14.6	15.0	5.2	3.53	4.01	0.08	2.27	2.0	0.1	0.5
10.	12.6	17.0	6.8	2.53	5.35	0.03	0.89	5.2	0.9	0.6
11.	6.6	7.6	5.5	5.91	21.38	0.04	1.02	3.5	1.4	0.3
12.	14.8	11.6	4.8	3.74	6.68	0.05	1.06	3.6	0.9	0.5
13	14.0	17.0	5.2	0.76	4.07	0.06	1.17	2.2	0.8	0.5
14.	16.0	13.0	5.2	1.53	6.68	0.14	1.58	3.2	0.5	0.4
15.	14.0	13.0	4.8	0.19	2.67	0.08	0.86	2.3	0.1	0.3
16.	3.4	9.0	6.0	0.71	2.67	0.09	0.96	3.9	0.7	0.4
17.	9.4	15.0	5.5	0.38	5.35	0.06	0.89	4.1	1.1	0.4
18.	15.4	21.0	5.7	1.62	4.01	0.03	0.89	4.3	1.6	0.3

*Location and soil class : 1. Miyer ,3. Sanur and 6. Kumta (Ustic Dystropept); 2. Ajakar (Aquic Dystropept); 4. Hubbanageri (Aquic Ustorthent); 5. Aggrogona (Typic Ustipsamment); 7. Aldur (Aquic Ustifluvent); 8. Madigeri, 11. Theralu, 12. Bettageri 15. Navile and 17. Augumbe (Typic Paleustalf); 9. Madigeri (Ustic Haplohumult); 10. Mudigeri, 13. Bettageri and 16. Thirthahalli (Fluventic Ustropept); 14. Theralu and 18. Brahmavar (Typic Ustropept)

Source : Rao and Ananthanarayana (1997)

acid groups of soil organic matter contributed more towards the pH dependent CEC (33.7 to 99 per cent, mean value 71.4 per cent) than the sequioxide coatings on clay mineral surfaces or partially neutralized complexes of Al and Fe in the interlayers (1 to 66.3 per cent, mean 28.6 per cent). The components of pH dependent charge, organic matter, and Fe/Al coating accounted for 13.6 and 5.3 per cent (mean value) of total CEC in these soils. The pH dependent CEC due to Al and Fe showed direct association with NaOAc extractable Al (Kailash Kumar *et al.,*1995).

Total potential acidity, total acidity, pH-dependent acidity, hydrolytic acidity, and exchange acidity in acid soils of West Bengal representing Inceptisol, Alfisol, and Entisol orders ranged from 1.50 to 11.25, 0.93 to 4.75, 1.41 to 10.35, 0.89 to 3.85, and 0.04 to 1.03 cmol (p^+) kg^{-1}, respectively. Excepting hydrolytic acidity, all other forms of acidity are highest in Inceptisols, followed by Entisols, and Alfisols. pH-dependent and hydrolytic acidity contributed 86.9 to 99.2 per cent and 77.4 to 97.6 per cent of the total potential acidity and total acidity, respectively, with little contribution of exchange acidity (0.8 to 22.6 per cent). The contribution of electrostatically bound hydrogen to exchange acidity is highest for Entisols (58.1 per cent), followed by Alfisols (43.6 per cent), and Inceptisols (34.1 per cent). Opposite is true for electrostatically bound aluminium. All these forms of acidity showed significant positive correlation with soil pH (Chand and Mandal, 2000).

5.3.3.2 Exchangeable Acidity

Exchangeable acidity is that part of soil acidity extracted by unbuffered neutral salt solution such as 1M KCl. It consists of monomeric Al and Fe as well as any exchangeable H^+. Hence, the management of exchangeable Al^{3+} is of primary importance in acid soils (Prabhuraj and Murthy, 1994b). The production potential and management by liming of acid soils depend highly on the proportional amount of various forms of acidity. Unlike total potential acidity, the exchange acidity of all the soils is much less, and its values in some acid soils of Karnataka have been observed to range from 0.2 to 3 cmol (p^+) kg^{-1} (Prabhuraj and Murthy, 1994a). In surface soils, exchangeable acidity is low in comparison with subsurface soils, with exception of alluvial from hill slopes. The contribution of exchangeable acidity to total potential acidity is very low varying from 2.2 to 21.5 per cent. Again, the acidity contributed by H^+ ions to exchangeable acidity is relatively low ranging from 11.1 to 66.6 per cent in acid hill soils of Manipur (Table 5.5). The contribution of Al^{3+} to exchangeable acidity varies from 33 to 89 per cent and is very significat in hill soils compared to alluvial soils. It is also observed that subsoils tend to

Table 5.5. Different forms of acidity in the soils of Manipur, northeast India

Sample no.	pH_w	Exchangeable acidity		Total potential acidity	pH dependent acidity	Extr. Al. NH_4OAc (pH 4.8)	Non-exchangeable Al^{3+}
		H^+	Al^{3+}				
		[cmol (p^+) kg^{-1}]					
				Hill soils			
1A	5.1	0.6	1.4	20.4	18.4	4.6	3.2
1B	4.8	0.5	3.1	20.2	16.6	5.8	2.7
2A	4.9	0.7	1.7	21.2	18.8	3.4	1.7
2B	4.8	0.5	2.6	20.2	17.1	5.1	2.5
3A	5.3	0.3	0.6	15.7	14.8	2.6	2.0
3B	5.1	0.6	2.9	16.3	12.8	5.0	2.1
4A	5.2	0.2	0.5	15.6	14.9	1.8	1.3
4B	5.1	0.3	2.4	17.5	14.8	3.5	1.1
				Alluvial soils			
5A	4.8	0.4	1.7	11.6	9.5	3.3	1.6
5B	5.2	0.3	0.9	16.2	15.1	3.2	2.3
6A	5.2	0.2	0.4	14.2	13.6	1.6	1.2
6B	5.8	0.2	0.1	11.8	11.5	2.6	2.5
7A	5.5	0.2	-	9.2	9.0	0.6	0.6
7B	5.4	0.4	0.3	9.9	9.2	1.5	1.2

A and B stand for surface and subsurface, respectively

Source : Nayak *et al.* (1996b)

be more predominantly Al^{3+} saturated in humid climate (Coleman *et al.,* 1959). The total exchangeable acidity in relation to varying altitudes in acid soils of Mizoram showed differential pattern of distribution and varied from 3.1 to 6.8 cmol (p^+) kg^{-1} (Singh *et al.,*1991).

5.3.3.3 pH dependent Acidity

The pH dependent acidity (variable charge) of the soils is the difference between the total potential acidity and the amount of H^+ and Al^{3+} extracted by unbuffered 1 M KCl solution. The values of pH dependent acidity vary from 12.8 to 18.8 cmol (p^+)kg^{-1} in hill soils and 9.0 to 15.1 cmol (p^+) kg^{-1} in alluvial acidic soils of Manipur (Nayak *et al.,*1996b). While, other studies by Sen *et al.* (1997b) showed variation in pH dependent acidity from 4.11 to 15.15 cmol(p^+) kg^{-1} in acidic soils of Assam. The contribution of pH-dependent acidity increased linearly and significantly (r = 0.78) with increasing free oxides of iron and organic carbon in the soils, which is comparatively less in the alluvial soils than the hill soils due to variation in organic carbon and free iron oxide content (Table 5.6).

The pH dependent acidity formed the major portion of acidity (75.1 – 99.6 per cent) in the Manipur hill soils. EB – H^+ and EB – Al^{3+} contributed 39 and 61 per cent, respectively, towards exchangeable acidity and pH dependent CEC. While, exchangeable acidity correlated only with pH dependent CEC due to Al and Fe. Neutral salt CEC showed a negative correlation with all forms of acidity (Kailash Kumar *et al.,* 1994; Kumar, 1997). The neutral salt CEC increased with increase in soil pH due to neutralization of weak acid sites and the substitution of non salt exchangeable H^+ by salt exchangeable Ca^{2+} (Bhumbla and Mclean, 1965). Pionke *et al.* (1968) demostrated that pH dependent sites on organic matter contributed three times more those on clays towards acidity.

Table 5.6. Exchange properties and lime requirement soils of Assam, Northeast India

Soil	Sum of bases	Exch. acidity	Extr. acidity	pH dependent acidity	Per cent Al sat. (ECEC*)	Lime requirement (t ha^{-1})
1A	3.6	1.7	12.5	10.76	18	1.3
1B	2.2	3.5	12.7	9.15	55	4.8
2A	4.0	5.1	20.2	15.15	54	7.4
2B	2.1	6.6	18.6	12.02	71	9.9
3A	2.9	0.4	5.9	5.52	03	0.1
3B	3.9	2.6	9.1	6.52	33	3.2
4A	5.3	0.2	4.3	4.11	-	-
4B	4.0	0.3	6.6	6.32	-	-
5A	2.4	6.4	19.1	12.71	71	9.7
5B	1.8	7.2	17.7	10.46	78	10.9
6A	3.7	1.8	17.9	16.10	25	2.0
6B	2.5	3.1	18.8	15.70	51	4.3

A and B stand for surface and subsurface, respectively
* Effective cation exchange capacity
Source : Sen *et al.* (1997b)

5.3.4 Lime Requirement

Approaches to liming soils in humid tropics have varied from neutralization to exchangeable Al to pH control. Kailash Kumar *et al.* (1997a) observed that lime requirement has positive correlation with pH dependent CEC and negative correlation with neutral salt CEC, effective CEC, pH, base saturation, and exchangeable Ca^{2+} and Mg^{2+} in acid hill soils of Manipur representing Ultisols, Inceptisols, and Entisols. Several investigators have proposed that liming rates for highly weathered soils should be based on exchangeable Al present in the soil (Kamprath, 1970; Reeve and Sumner, 1971). Liming rates to neutralize exchangeable Al in Oxisols are considerably less than those required to bring the soils to pH 6.5 (Reeve and Sumner, 1970). Pearson (1975) reviewed a number of studies pertaining to liming in the humid tropics and concluded that liming requirements should be based on exchangeable Al rather than pH *per se*. The lime requirement of these soils is calculated based on exchangeable Al^{3+} following the equation of Kamprath (1970) as : $CaCO_3$ (t ha^{-1}) = 1.65 x Exch. Al [cmol (p^+) kg^{-1}]. It has also been reported that lime application based on exchangeable Al^{3+} content does well with a lime factor of 1.5 to 2.0 as in acid soils of Sikkim (Pradhan and Khera, 1976). In most of the cases, effective CEC of clays is less than 14 cmol (p^+) kg^{-1} which indicates predominance of low activity clays (Herbillion *et al.,* 1977; Prasad *et al.*, 1985). Clay CEC of acid soils in major part of the sola is less than 22 cmol (p^+) kg^{-1}, indicating the mineralogy class as kaolinitic (Smith, 1986). Anderson (1987) reported that a total 224 kg Ca ha^{-1} is required in a acidic sandy soil of Florida for normal yield of valencia orange using dolomite or limestone to bring soil pH near neutral. Kumar (1997) suggested Brown and Cisco (1984) as best method for estimating lime requirement for acid soils of Manipur hills.

Much larger amounts of lime are required to adjust all of the soils to pH 7 than to neutralize Al (Table 5.7). The highly weathered soils of the tropics have colloids whose surface charge is primarily pH dependant (Keng and Uehera, 1974). Once any

Table 5.7. The $CaCO_3$ requirement, pH, and per cent Al neutralized in Oxisols based on exchangeable Al

Soils		Exch. Al x 0.75			Exch. Al x 1.5			
		$CaCO_3$ equiv. (kg ha^{-1})	pH	Al neutralized (%)	$CaCO_3$ Equiv. (kg ha^{-1})	pH	Al neutralized (%)	$CaCO_3$ equiv. based on titration to pH 7 (tons ha^{-1})
La Mesa	I	450	5.3	17	900	5.5	58	8.0
	II	187	5.4	0	375	5.4	20	7.0
	III	4,200	5.4	69	8,600	5.7	97	16.0
Pacora	I	3,100	5.7	66	6,200	6.0	95	10.0
	II	2,250	5.5	80	4,500	6.0	89	7.0
	III	900	5.8	67	1,800	5.9	79	5.0

Source : Mendez and Kamprath (1978)

exchangeable Al is neutralized, added lime will react with the surface hydroxyls of hydrated oxides of Fe and Al. Lime rates for Oxisols based on KCl extractable Al will be effective in reducing Al to nontoxic levels and adjusting the pH to a range considered adequate for most of the cultivars.

The buffer method has been observed to underestimate the lime requirement of organic soils. If the lime estimated by the buffer method has been added to these two soils, the pH would have increased to between 5.9 and 6.1. Such under estimation of the lime requirement of the organic soils is no disadvantage, because acid organic soils are seldom limed, and if they are limed, the pH need not go as high as in minerals soils. Since the pH change is directly proportional to the amount of lime added, lime needs to bring the soil to any intermediate pH, can be calculated as the proportional change desired, multiplied by the lime requirement by the buffer method. This can be worked out from the relationship between saturation paste pH and lime requirement (Table 5.8).

Table 5.8. Scale of lime requirements by the buffer method

Soil buffer pH	Lime requirement CaCo3 (tons acre^{-1})	Soil buffer pH	Lime requirement CaCo3 (tons acre^{-1})
6.7	1.6	5.7	7.6
6.6	2.2	5.6	8.2
6.5	2.8	5.5	8.9
6.4	3.4	5.4	9.5
6.3	4.0	5.3	10.1
6.2	4.5	5.2	11.0
6.1	5.2	5.1	11.7
6.0	5.8	5.0	12.4
5.9	6.4	4.9	13.2
5.8	7.0	4.8	14.0

Source : Pratt and Bair (1961)

The official method of lime recommendation in Sao Paulo, Brazil is based on the increase in soil base saturation (Quaggio *et al.,*1985). The adequate level of base saturation for attaining maximum yield and profit is 60 per cent. However, this value of base saturation corresponds to soil pH (in $CaCl_2$) of 5.5 which is lower than that obtained with the liming rates used by Anderson and Martin (1969) and Anderson (1971; 1987) for maximum yield of valencia orange in Florida, USA. Lime requirement is also sometimes, calculated to increase the soil base saturation to 70 per cent of cation exchange capacity at pH 7.0 (Quaggio *et al.*, 1996), which for citrus corresponds to soil pH in $CaCl_2$ around 5.5 (Quaggio, 1983). Earlier, Pionke *et al.* (1968), ranking the soil properties as predictors of $CaCO_3$ – calibrated lime requirement after wide range of soils, observed the following decreasing order of their effect: pH dependent sites on organic matter > nonexchangeable acidic Al > exchangeable Al > pH dependent sites on clay.

High value of lime requirement (up to 8.6 t ha^{-1}) based on pH and texture of the soil can not be considered economical. Therefore, KCl-Al x 1.5 and NH_4Cl-Al x 0.75 values of lime requirement might be considered highly significant and economical for the reclamation of acid soils (Singh *et al.*, 1993). Mendez and Kamprath (1978) is also of similar opinion that the lime requirement calculated by multiplying KCl-Al with 1.5 is more realistic, neutralized most of the Al^{3+}, and bring the soil pH at 7.0 for satisfactory plant growth. The higher reliability of extractable Al^{3+} as an indicator of lime requirement than pH has also been recognised (Kamprath, 1970; Grove *et al.,* 1982). Quaggio *et al.* (1992b) observed higher yield of valencia orange reaching more than 40 tons ha^{-1} with application of dolomitic limestone at the rate of 9 tons ha^{-1} in a dark red clay textured Latosol having exchangeable Mg greater than 0.9 me 100 cm^{-3}. Low Mg content of calcite limestone can not adequately supply Mg to the plants six years after limestone application (Table 5.9). However, on the other hand, excess dolomite and calcite treated trees showed symptoms similar to young tree decline in pineapple on rough lemon rootstock (Iley and Guilford, 1978). Application of lime (7.5 tons ha^{-1}) at 0-30 cm and 30-60 cm depth using valencia on two different sandy soil types of South Africa raised the yield from 92 to 112 kg $tree^{-1}$ and soil pH in the top most layer from 4.4 to 6.1. These changes are accompanied by rise in soil Ca content from 56 to 157 mg kg^{-1} and 32 to 64 mg kg^{-1} at 0-15 and 15-30 cm depth, respectively, with a corresponding decline in soil exchangeable Al content from 46 to 6 mg kg^{-1} and from 07 to 16 mg kg^{-1}. In another trial using dolomite (7.5 tons ha^{-1}) raised the soil pH from 5.0 to 5.7 without influencing yield. Soil exchangeable Al, however, recorded well below the excess level (Plessis and Koen, 1986).

Table 5.9. Response of Valencia orange on Rangpur lime rootstock to rates of calcitic and dolomitic limestones for five and six years old trees

Calcitic	Dolomitic				Average*
	0	3	6	9	
(t ha^{-1})			**1988**		
0	14.9	20.1	18.7	143 9	17.1
3	21.6	18.8	18.7	16.6	18.9
6	17.2	21.4	21.4	19.5	19.3
9	20.2	25.3	26.3	17.7	20.6
Average*	18.4	19.4	21.2	17.2	19.0
(t ha^{-1})			**1989**		
0	23.8	38.9	39.5	38.1	35.7
3	30.2	33.6	37.4	38.5	34.9
6	37.0	34.3	44.5	36.9	38.2
9	36.1	36.9	35.9	40.2	37.2
Average**	31.8	35.9	39.3	38.4	36.5

* Nonsignificant differences, ** Significant differences *($p>0.01$)*
Source : Quaggio *et al.* (1992a)

Lime incorporation (up to 12 tons ha^{-1}) to the surface soil is effective to alleviate subsoil acidity and to increase Ca and Mg content down to 0.6 m depth in the soil profile. Maximum yield and profitability in valencia on Rangpur lime have been observed at the rate of 12 tons ha^{-1}, which increased the base saturation to near 60 per cent (Quaggio *et al.,*1992a). Quaggio *et al.* (1998) observed increase in soil pH, Ca, and Mg concentration, and base saturation in response to dolomitic lime (upto 4 tons ha^{-1}) and phosphogypsum (upto 4 tons ha^{-1}) in valencia sweet orange grown in Oxisol of Sao Paulo, Brazil. Earlier studies (Anderson and Martin, 1969; Anderson, 1971a) showed a pronounced response of two year old valencia orange on rough lemon to increasing rates of Ca and acidity neutralization in acid soils of Florida, USA. The fruit yield has been recorded to be higher

(104 kg $tree^{-1}$) at soil pH 7.0 and 2.7 me Ca 100 cm^{-3} compared to untreated trees having soil pH 4.0 and 0.2 me Ca 100 cm^{-3} (4.0 kg $tree^{-1}$). Later, in another study (Anderson, 1987) comparing the results of 17 year study on response of lime application to yield of valencia orange. Increase in soil pH to 7.0 from an initial pH of 5.2 increased the yield by 50 per cent in 7 years period which further improved the yield by 200 per cent in 17 years period with no significant difference between limestone and dolomite. Koo (1971) in two long term trials testing sources and rates of Mg on grapefruit and sweet orange, reported that application of 150 kg MgO ha^{-1} increased the yield by 12.6 per cent and total soluble solids by 14.7 per cent compared to 60 kg MgO ha^{-1}. A number of other workers (Camp, 1947; Spencer and Wander, 1960; Koo, 1971; Aso and Bustos, 1981) have earlier reported an improvement in yield and quality of citrus following MgO application.

5.4 NUTRIENT TRANSFORMATION AND AVAILABILITY

5.4.1 Organic Matter and Nitrogen Distribution

Variation in organic matter and nitrogen contents of acid soils of different origin are presumably due to differences in climate, vegetation, parent material, microbiological activity, soil management practices, and cropping patterns followed in different regions. The organic matter and nitrogen contents of citrus growing acid soils showed an extremes of variation, depending upon the nature of vegetation and the intensity of organic matter decomposition. In general, laterites, lateritic, and red loam soils are characterised by low status of organic matter and nitrogen. Hill and forest soils are comparatively richer in organic matter and nitrogen. The relative amount of organic carbon and nitrogen in some representative acid soils of different states (Table 5.10) showed a wide range of C/N ratio from 4.5 to as high as 18.3.

5.4.2 Transformation of Phosphorus

Solubilization of insoluble phosphates in acid soils is mediated through acidification, chelation, ion exchange reactions, external and internal accumulation of ca^{2+}, and liberation of P through microbial cell death lysis. Acid Alfisols of Himachal Pradesh have high P fixing capacity (Kalish Kumar *et al.*, 1997b) which is governed by soil components viz., pH, texture, free Fe_2O_3, pH dependent CEC, and organic matter content. Eroded soils have been found to be poor in organic and Ca-bound P, but rich in Fe and Al bound P (Kanwar and Grewal, 1959). Guilford *et al.* (1970) observed that much higher available P in soil with basic slag treatment than standard lime in pineapple orange plot. Clementine trees on sour orange and *Poncirus trifoliata* rootstocks planted on acid soils poor in P suffered from leaf scorch which is controlled with application of Ca, slag or farmyard manure (Cassin, 1968). Phosphate fractionation of Assam soils (Tiwari *et al.,* 1967) indicated that the Fe-P is higher than Al-P. Some acid soils of alluvial origin contained a

Table 5.10. Organic matter and nitrogen status of some citrus growing acid soils

Soil description and location	Organic carbon (%)	Nitrogen (%)	C/N ratio
Brahmaputra valley soils (Assam)	1.2-2.0	0.08-0.11	-
New alluvial soil (Assam)	0.97	0.07	14.5
Old alluvial soil (Assam)	1.02	0.11	10.2
Laterite and lateritic soil (Assam)	1.02	0.11	9.7
Hill soil (Assam)	1.26	0.18	7.0
Laterite and lateritic (W. Bengal)	0.05-0.5	0.01-0.8	-
Red soil, upland (Western Maharashtra)	0.50	1.08	--
Red soil, hill slopes (Western Maharashtra)	0.95-1.06	0.10-0.10	10.0
Red soil, valley (Western Maharashtra)	0.08-0.88	0.01-0.09	7.0-9.8
Acid soil, Kangra district (H.P.)	0.46-0.95	0.05-0.10	8.4-9.3
Acid soil, north Coorg (Mysore)	0.12-1.16	0.007-0.173	--
Bhittangala soil, south Coorg (Mysore)	0.41-0.99	0.046-0.094	8.3-10.6
Nilgiri medium clay soil (Madras)	0.35	0.03	10.1
Red loam soil (U.P.)	0.22-3.35	--	4.5-18.3

Source : Mandal *et al.*(1975)

higher proportion of P in organic form. Soils of Assam, particularly contained substantial amount of organic P (Chakraborty, 1976). Pavan *et al.* (1997) reported low P use efficiency in valencia orange on a dark red Latosol of Brazil due to poor root distribution in acid soils.

Acid soils, predominant in northeastern hilly region of India have rainfall far exceeding evapotranspiration. Under high rainfall conditions, weathering is rapid and intensive leaching render the solum highly acidic. The solum gets depleted of bases, but becomes rich in exchangeable Al^{3+} and H^+. Exchangeable H^+ is of importance only at pH below 4, in extremely acidic soils such as acid sulphate soils. Exchangeable Al^{3+}, however, occurs in significant amount at pH around 5.5 (Panda, 1987). Of the major methods characterizing soil acidity, pH, total titrable acidity, and exchangeable acidity, the latter one is of importance as far as plant growth is concerned.

5.4.3 Transformation of Potassium

The K fixation capacity of acid soils also exhibits extremes of variation depending on the presence of illitic type of clay minerals. Added to this, depotassified primary micas and feldspar also contribute towards K fixation (Prasad *et al.,* 1967). Application of increasing levels of $CaCO_3$ from zero to 3360 kg ha^{-1} in acid soil of Nilgiri district of Tamil Nadu improved the exchangeable K and Ca from 249 kg ha^{-1} and 48 mg $100g^{-1}$, respectively, to 302 kg ha^{-1} and 96 mg $100g^{-1}$ (Table 5.11)

The availability of K in acid soils, as governed by the equilibrium relationship of different forms and the specific effect of K fixing minerals also showed a great deal of variation. Mitra *et al.* (1958) in a study on the response of acid soils, as influenced by percent K saturation reported that soils with the dominance of free kaolin and free oxides exhibited low K fixation values. Tiwari *et al.* (1967) observed increase in K availability with rise in temperature up to 20^{0}C, while nonexchangeable K has been recorded maximum at 60^{0}C. Acidic soils of Arunachal Pradesh had higher total K (28.14-69.79 me per cent) than Meghalaya (6.41-19.23 me per cent), while reverse was true for 1 N HNO_3 extractable K (Table 5.12). Mongia and Bandyopadhyay (1991) observed various forms of soil K as 0.05-0.19 per cent HCl soluble-K, 0.02-0.11 per cent fixed-K, 0.06-0.36 per cent HNO_3-K, and 32-112 mg kg^{-1} NH_4OAc-K in acid hill soils of Andamans. Long term application of KCl over 6 years in a red acidic soil led to reduction in vitamin C (Ascorbic acid) content and an increase in total acid content of the satsuma mandarin cv. Weizhange fruits (Dai *et al.*, 1993).

Table 5.11. Effect of liming on the availability of Ca and K

$CaCO_3$ (kg ha^{-1})	Exch. K (kg ha^{-1})	Exch. Ca (mg 100g^{-1})
0	249	48
560	280	57
1120	266	70
1680	271	72
2240	316	79
2800	300	89
3360	302	96

Source : Kothandaraman and Mariakulandai (1965)

5.4.4 Micronutrients

The total Mn in the acid soils varies considerably. Variation in the amount of different forms of Mn in various types of acid soils of Assam has been reported by Das *et al.* (1971). The total Mn in these soils ranged from 389.94 to 1134.03 mg kg^{-1}, easily reducible Mn 26.12 to 79.82 mg kg^{-1}, water soluble Mn 0.05 to 4.92 mg kg^{-1}, and exchangeable Mn from 8.30 to 28.32 mg kg^{-1}. The total B in Indian soils varied from 7 to 630 mg kg^{-1} but for the acid soils, it ranged between 3 mg kg^{-1} and 100 mg kg^{-1}

Table 5.12. Forms of potassium (me per cent) in the soil of northeastern hill region of India

Forms	Meghalaya[1]	Arunachal Pradesh[2]	Nagaland [3]
Water soluble-K	0.006-0.030	0.005-0.098	0.015-0.150
Exchangeable-K	0.122-0.391	0.250-0.671	0.231-1.180
I N HNO_3 extractable-K	0.538-4.718	0.808-4.256	-
Fixed-K	0.346-4.490	0.362-3.820	-
Labile-K	-	-	0.10-0.80
Step –K	0.436-4.343	-	0.185-0.743
Constant rate- K	0.102-0.899	-	0.067-0.236
Total- K	6.41-19.23	28.14-69.79	-

Source : [1] Patiram and Prasad (1978), [2] Nath and Deory (1976), [3] Ghosh and Ghosh (1976)

(Mandal *et al.,*1975). The total Mo content of acid soils generally ranged between 0.05 and 5 mg kg^{-1}. Mandal and Jha (1969) in their review, reported the total Mo content of Bihar acidic soils to vary between 5.0 mg kg^{-1} and 18.1 mg kg^{-1} with an average of 11.9 mg kg^{-1}.

5.4.5 Biological Properties

Studies on factors affecting distribution of *Azotobacter* in acid soils of Madras revealed the presence of *Azotobacter* in 35.2 per cent soils tested (Nair, 1984). The organic matter content of soils did not markedly affect the presence of these organisms, except at high levels when a universal correlation existed. Increase in P level proved favourable for *Azotobacter.* A progressive increase in its population has also been recorded with increasing lime content of soils. The biological properties of various soil groups have been studied by Ramaswami (1966). The microbial population and its relative abundance of soil is affected by topography (Daraseliza, 1969). Sanikidze (1970) observed an increase population of microorganisms, namely, *Bacillus spp. Penicillium, Mucor, Trichorderma*, and *Aspergillus sp., Actinomycetes, Clostridium pasteurianum, Azotobacter sp, Nitrobacter sp.,* and cellulose decomposing bacteria in the top soil of mandarin plantation following the application of lime and pure peat. The greatest number of bacteria has been observed in sweet orange, lemon tree rhizosphere on calcareous soils, and in mandarins trees growing on red soils of Georgia (Gochelashvili, 1973). Other study by Saakashvili *et al.* (1971) indicated most numerous saprophytic bacteria (65-70 million g^{-1} dry roots) on the roots and the least number in the outer rhizosphere (2 million g^{-1} soil). The saprophytic bacterial count correlated directly with fruit yield

The bacterial population in black, red, alluvial, and laterite soils is almost equal. A similar trend is noted for the population of *Actinomycetes*. The fungal population recorded higher in laterite soils. The *Azotobacter* is found absent in laterite soils. The high level of organic matter in lateritic soils is conducive to CO_2 evolution, although, the rate of N_2 fixation, ammonification, and nitrification is fairly low in acidic soils. An examination of the inter-relationship between various soil characteristics indicated that the soil reaction is the predominant factor affecting biological activity of soils. Distribution of soil microorganisms in citrus orchards on acid red soil, neutral sandy loam, and alkaline purple soil showed that type and amount of microorganisms varied between soil types. Fungi, bacteria, and actinomycetes are predominant in red soil, sandy loam, and purple soil, respectively. The total number of microorganisms is positively correlated with organic matter and N contents. Vigour and yield of orange tree differed between soil types (Zou *et al.*, 1994). Badiyala *et al.* (1990) observed an improvement in microbiological composition of Hapludalf in 6 year old kinnow mandarin following grass mulch application.

Gandotra *et al.* (1998) observed the presence of *Azotobacter* in 55 soils out of 66 soils studied in Himachal Pradesh representing soil order viz., Mollisols, Alfisols,

Ultisols, Inceptisols, and Entisols. The soils of Paleudalfs and Dystrochrepts posessed sparse population of *Azotobacter*. Its population varied widely, constituting less then 1 per cent of total bacteria. Haplustalfs and Hapludalfs had higher counts than other soil groups. Of the various soil properties positively correlated with *Azotobacter* population, significant correlation was seen only with pH, available P, and exchangeable Mg^{2+}. Three species viz., *Azotobacter chroococcum*, *A. beijerinckii,* and *A. vinelandi* have been identified in these soils. Dehydrogenase and urease activity, microbial population (fungi and bacteria), and organic matter content of the soils increased with an increase in altitude up to 1100 m in Arunachal Pradesh, India (Tiwari and Sharma, 1998). Distribution of microflora in soils of Kangra valley, Himachal Pradesh showed highest population of bacteria and actinomyces in pH range of 6.1-7.0 and lowest in 7.1-8.7. The fungal flora was the maximum in the pH range of 7.1-8.7, but increased below this pH range. The ammonifier and spore formers showed no relationship with soil pH, whereas nitrifying bacteria decreased markedly with decrease in pH. The variability in microflora was, by and large, related to soil organic carbon content (Gupta and Tripathi, 1988)

5.5 ACID SOIL AMELIORATION

Correction of native high acidity in citrus orchards is of paramount importance. But, it is handicapped by slow movement of applied limestones in these soils. However, it was found that application of limestone together with acid forming N fertilzers resulted in an increased mobility of limestone through the soil profile (Weir, 1969a; Huang and Tsai, 1988). Amelioration methodology of acid soils addresses at stabilising the soil pH more towards neutrality. The pH of the soil with a low cation exchange capacity (kaolinitic and illitic mineralogy) can be changed more quickly under the use of fertilizers having acidic residual effect than a soil with high cation exchange capacity (montmorillonitic or smectitic mineralogy). Subsoil acidity correction is not easy. This is because, the normal soil amendment chosen as agricultural grade lime, is not soluble in water and consequently, amendment must be mixed throughout the soil profile by cultivation to be effective. Nutritional survey of sweet orange (cultivars pera and valencia) orchards of northwest Parana of Brazil indicated that an inadequate Ca and Mg is principally responsible for reduced yield and fruit size. Therefore, application methods, rates, and frequency of liming should be reviewed with additional application of Zn (Fidalski and Auler, 1997). Very few efforts have been made in the past to confirm the best approach of acid soil management using citrus as test crop. It is very difficult to identify the morphological changes that have taken place after liming (Bernal *et al.,* 1957; Franzmeier *et al.*, 1965; Nandra, 1974; Karamanova, 1978). However, best response of lime has been obtained when combined with nutrients like N, P, K, B, and Mn especially from citrus fruit quality point of view (Kodua, 1972).

5.5.1 Liming Technique

Most of the current recommendations for correcting acid soil depend on pH and exchangeable Al content (Oates and Kamprath, 1983). A number of techniques of liming may be grouped into the following classes (Mandal *et al.*, 1975):

- Measurement of soil pH and its correlation with base saturation of soils of similar mineralogical composition, texture, and organic matter status
- Moist incubation with graded amount of $CaCO_3$ or Ca $(OH)_2$ and determination of the amount of liming materials required to bring the soil reaction to a reference state from the potentiometric titration curve so obtained.
- Rapid lime requirement methods involving equilibrium extraction of the soil with strongly buffered system followed by determination of the exchangeable acidity in the extract by direct titration, or by measuring the change in pH of the extract. But, strongly buffered solutions are unable to indicate appreciable change in pH of weakly buffered soils high in extractable Al. This has been emphasised by Reeve and Sumner (1970). Kamprath (1970) suggested to base lime rates by the relationship: Al 100 g^{-1} x 1.5 = me $CaCO_3$ 100 g^{-1} which will give an exchangeable Al saturation of 15 per cent or less. With crops, very sensitive to Al, factor 2 may be used to determine the amount of lime required.

The principle factors, which determine the lime requirement of acid soils are pH, amount and type of clay, cation exchange capacity, and buffering capacity. The buffering capacity of soils depends primarily on variable exchange sites characterised by weaker acidic groups as present in hydroxy Al-polymers and organic matter. Residual effect of lime is usually observed for two to four years. Though, the opinions vary about the duration for which the effectiveness of lime application lasts. Motiramani *et al.* (1964) reported that lime should be applied at intervals of every three years. The reactivity of liming materials depends upon their fineness and type of liming materials such as CaO, $Ca(OH)_2$, precipitated $CaCO_3$, and pressmud. These materials are usually applied in powdery form and are thus very effective. Limestone and basic slag, on the other hand, have to be powdered before use. Efficiency rating of various sized particles of limestone indicated that efficiency rating improved with finer size particles. The dynamics of exchangeable Ca and Mg, two potent cations hold a pivotal role to govern the sustainability of liming, since studies have shown a reduction in exchangeable Mg with depth (Pratt *et al.*,1959). Over a twenty eight years period in a long term fertilizer experiment, a substantial reduction in the amount of exchangeable Mg in the soil has been noticed in the soil that correlated with amount of soil amendments applied (Pratt and Harding, 1957a; 1957b). This led to the presumption that the use of high Ca and low Mg irrigation waters may produce Mg deficiency in soils of low cation exchange capacity. The addition of chemical fertilizers or manures further increased the Mg losses from the soil (Jacob,

1958). High exchangeable Ca-Mg ratio in the soil impaired Mg uptake by citrus trees (Jacoby, 1961a).

5.5.2 Response of Soil Amendments

Citrus plants possess a characteristic, quite uncommon from other perennial fruit crops that it contains Ca higher than any other elements, including N (Chapman and Kelley, 1943). Mixing profile and adding lime resulted in marked improvement of young citrus trees in an experiment at Duette, Florida (Bryan, 1962). A number of soil amendments, namely, ordinary or triple superphosphate (Spencer, 1963; Rasmussen and Smith, 1959), sulfur (Rasmussen and Smith, 1959), sodium nitrate (Pratt *et al.,*1959), gypsum (Anderson, 1968; 1972; Sumner *et al.*, 1986; Nemec and Lee, 1992), dolomite (Calvert, 1970; Anderson, 1971; 1981; Mchedlidze, 1972; El-Azhar *et al.*, 1980), epsom salts (McColloch *et al.,*1957; Embleton and Jones, 1959), magnesium sulphate (De Villiers, 1942; Oberholzer, 1947; Haas, 1948; Jones *et al.,*1971; Mdinaradze and Datuadze, 1987), magnesium carbonate (Koo, 1966; Bester, 1969; De Villiers, 1969; Datuadze, 1977), hydrogels (Swietlik, 1989), basic slag (Koo,1964; 1984; Wutscher, 1985), clay type soil amendments (Childs, 1981; Sautoy, *et al.,*1991), and phosphogypsum (O'Brien and Sumner, 1988; Viti *et al;* 1993a) in additionto some physical amendments (Kuwamura *et al.,* 1973a; 1973b) have been suggested for improving the soil pH of acidic citrus growing soils under various citrus cultivars. Studies involving comparison of limestone (7.5 tons ha^{-1} at 40 per cent CaO) and dolomite (5.0 tons ha^{-1} at 32 per cent CaO) showed a significant response of both the amendments compared to control on pH reduction (6.2 and 6.1 against 5.5), improvement in exchangeable Ca (4.7 and 3.4 me 100 g^{-1} against 2.6 me 100 g^{-1}), and exchangeable Mg (0.7 and 1.2 me 100 g^{-1} against 0.7 me 100 g^{-1}) of the soil (Cassin, 1968).

Pinto and Leal (1974) recommended that lime application @ 2 tons ha^{-1} enhanced nutient absorption and 0.5 kg $MgSO_4$ $tree^{-1}$ produced no effect. Addition of clay in unproductive Ruby Red grapefruit orchard at Florida enhanced soil water holding and cation exchange capacity, pH, Ca, and Mg contents (Koo and Driscoll, 1970). The moderate lime application (2.5 kg $tree^{-1}$) in a red soil growing 11-17 year old trees of satsuma mandarin, cultivar Gongchuan, increased the fruit yield by 13.2-29.2 per cent over untreated trees, besides concurrent improvement in fruit quality (Liu, 1993). Fused phosphate and calcium silicate as an amendment has also been found effective in acid soils of Western Georgia (Hisada and Ishida, 1975). The permeability of $CaCO_3$ into the subsoil is low, but nitrohumic acid application increased the transfer of lime to a depth of 20-30 cm. The growth of satsuma trees has been observed to increase by $CaCO_3$ plus nitrohumic acid (75 kg ha^{-1}) application (Hirobe and Ogaki, 1973).

Chen *et al.* (1997) obtained best response of combined use of dolomite, organic matter, and NPK on the growth and quality of matou (*Citrus grandis* Osbeck cv. matou peiyu) in Taiwan. Gypsum and phosphogypsum have been considered as soil amendments to ameliorate Al toxicity in the subsoil horizons of highly weathered soils (Alfic Arenic Halplaquod) applied at the rate of 4.5 tons ha^{-1} and 2.2 tons ha^{-1}, respectively (Martin *et al.*, 1988). Compared with nonamended soil (Aeric Haplaquod), the soil pH amended with lime (adjusted to soil pH 6.5) had a highest low soil acidity and lower exchangeable Al and Al saturation as well as higher $N0_3$, and lower K^+ and Mg^{2+} contents in the saturation extract. Application of phosphogypsum (5 times the number of equivalents of Al extracted with 1 M KCl) to Bh horizon, produced no significant effect on citrus fibrous root growth. The soil amended with phosphogypsum had a lower pH, higher salinity, higher exchangeable Al^+, Ca^2, and Mg^{2+}, and lower P ($H_2PO^-_4$ and HPO_4^{2-}) and Cl^- in the saturation extract than the nonamended soil (Lin *et al.,* 1988). Evaluation of nine soil amendments viz., limestone (5.6 tons ha^{-1}), phosphoclay (8.9 tons ha^{-1}), humate (86.9 tons ha^{-1}), shrimp waste (81.8 kg ha^{-1}), peat (28.0 tons ha^{-1}), bentonite clay (81.8 tons ha^{-1}), mined gypsum (2.2 tons ha^1), calcium humate (22.0 tons ha^{-1}), and phosphogypsum (11.2 tons ha^{-1}) applied to either no tillage or deep tilled on vesicular arbuscular mycorrhizal infection in valencia orange grafted on *Citrus limon* rootstock at three sandy soils of Florida, USA revealed that above amendments did not increase the percent infection significantly over deep tilled controls. While, at one site, per cent infection was noted highest (64 per cent) compared to phosphogypsum, peat, and no till control treatments (Nemec, 1998).

An entire dynamics of nutrient transformation and availability is changed to a great extent following the application of soil amendments. Availability of Ca and Mg to plants generally increases when acid soils are limed. Partially, this is a hydrogen ion activity effect on ion absorption by roots. The controlling factor is the relation between base saturation of the exchange complex and the activity or exchangeability of Ca and Mg (Coleman *et al.,*1968). The exchangeable Ca of Nilgiri soil in Tamil Nadu (pH 4.5) increased from 4.3 to 7.3 me 100 g^{-1} soil when treated with 3.92 tons ha^{-1} of lime. Increasing the doses of lime, resulted in an increase of exchangeable Ca up to 9.5 me 100 g^{-1} soil. Liming has been observed to improve exchangeable K status (Quaggio *et al.,* 1992a; 1992b) due to greater K retention caused by liming and consequent release of pH dependent exchangeable negative charges on the surface of soil colloids and also to the higher concentration of other counter ions such as Ca^{2+} and Mg^{2+} which follow the co-ions in the leaching process. A number of studies (Calvert, 1977; Calvert *et al.,*1978; 1979; 1981; Davitadze, 1991; Nemec and Lee, 1992; Liu, 1993; Panzenhager *et al.*, 1999) have described the response of citrus to lime and deep tillage. Marsh grapefruit and pineapple orange plots tilled to a depth of 107 cm plus deeply placed limestone (56 metric tons ha^{-1}) showed deeper and more vigorous roots than trees grown on plots receiving only surface tillage or deep tillage without deeply placed limestone. The treatment consisting of deeply placed limestone also produced better growth, yield, and quality responses in addition to improving the drainage characteristics.

Meng *et al.* (1991) obtained an increase in yield of satsuma mandarin by 37.2 per cent as a result of lime application at 2.25 tons ha^{-1} in a red soil containing 1.02-1.13 per cent organic matter, 0.038-0.045 per cent N, 0.030-0.034 per cent P_2O_5, and 0.88-1.02 per cent K_2O. Liming in acid soil of Florida raised the pH and imparted a consequent improvement in soil extractable Mo (Robinson *et al.*, 1951) and P (Obreza, 1990). Application of $CaCO_3$ or $CaSO_4$ (200-800 mg Ca kg^{-1} soil) in sandy siliceous hyperthermic Aeric Haplaquods increased the Ca availability which eventually reduced the Cu phytotoxicity (Alva *et al.*, 1993).

In an orchard with highly acid soil, surface application of 400 kg ha^{-1} quick lime or 520 kg $MgCO_3$ ha^{-1} increased the pH of the top soil to a depth of 25 cm, but regular application is needed for 4 years before there is any improvement in the subsoil. Injecting $MgCO_3$ has immediate effect. Lime application decreased the Mn content of the leaves from 152.7-229.1 ppm to 81.0-102.8 ppm increased fruit yield, and produced hardier fruit with a thicker peel (Hirobe and Ogaki, 1972). The base status of very acid subsoil under valencia orange trees is improved by surface application of limestone (5 tons ha^{-1}) in conjunction with N (450 g $tree^{-1}$) as urea or calcium ammonium nitrate, besides higher fruit yield with limestone plus N than lime alone (Weir, 1974). Lime applied to counteract soil acidity induced by the previous ammonium sulphate treatments produced greater response on yield of Washington navel and late valencia orange on tilled than nontilled plots. However, lime treatment did influence the fruit size (Cary, 1972). In the 4th and 5th year after liming, the soil pH remained higher where dolomite rather than calcite limestone (both at 3-10 tons ha^{-1}) was used. The Mg content of Duncan grapefruit leaves was also higher 5 years after the dolomite treatment than after calcite treatment. A survey of leaf Mg levels in 11 citrus orchards showed that dolomitic limestone alone provided an adequate Mg source for citrus (Anderson, 1971b).

Anderson and Martin (1969) observed fastest growth rate of valencia oranges in Lakeland fine sand soil treated with calcite limestone and supplementing the limestone with gypsum to provide a total of 1000 kg added Ca ha^{-1} $year^{-1}$ stabilised the soil pH to 7.0 compared to either calcite limestone or gypsum alone. Such a treatment resulted in simultaneous increase in both soil pH and added Ca in a proportion of about 45 kg added Ca for each unit increase in soil pH. An annual application of soluble Mg containing dolomite to organic acid soils, 10 years earlier with an organic matter content of greater than 1.5 per cent still provided sufficient Mg on a deep ploughed bedded sweet orange orchard (Iley *et al.,* 1970). In an another study (De Mello, 1970), no response from dolomite limestone at 1.5 kg $tree^{-1}$ $year^{-1}$ has been observed until after 3-4 years compared to $MgSO_4$ at 300 g $tree^{-1}$ $year^{-1}$ and patent Kali at 111 g $tree^{-1}$ $year^{-1}$ in sweet orange on Rangpur lime.

Application of lime to soil with pH 4.6 resulted in 16 per cent increase in mandarin yield, improvement in fruit quality, and development of tree crown compared with unlimed trees (Chanturiya, 1974). The significantly higher content of leaf N in deeply tilled limestone treated trees than in surface tillage treated trees can be explained on account of higher nitrification rate of deeply tilled limestone soil (Fiskell and Calvert, 1975). It is possible that the addition of N produced in the deeply tilled limestone soil is largely responsible for the significant growth and yield increases (Calvert *et al.,*1979). Earlier, Calvert (1975) observed that greater amount of K leached from surface tilled soil than deep tilled limestone treatments. Shimogori *et al.* (1980) observed that application of 400 kg dolomite ha^{-1} $year^{-1}$ led to an extra fruit yield of about 300 kg^{-1} 10 $acres^{-1}$ $year^{-1}$ in the 6 years of experiment compared with 200 kg application treatment. Quaggio *et al.* (1998) obtained best response of phosphogypsum (4 tons ha^{-1}) plus dolomite (3 tons ha^{-1}) incorporated into top 20 cm soil of Oxisol soil type with reference to improvement in the base saturation and reduction in soil pH under valencia sweet orange. Heavy lime application and deep mixing of a Spodosol enhanced early growth and production of 2 scions. After 5-6 years, however, the growth and production advantage were no longer evident. Contrary to expected results, deep mixing to break up restricting layers in the soil profile did not improve drainage. The deep mixing allowed organic matter to move with drainage water and accumulated near the soil gravel envelope interface. This accumulation of organic matter significantly restricted drain out flow (Rogers *et al.*, 1982).

In a long term study after 12 years of continuous dolomite application (400 kg 10 $acre^{-1}$) to the surface of citrus orchard (0-15 cm) without working into the soil, an analysis of cations at various depths showed that more than 55 per cent of the applied cations adsorbed in the first 5 cm of the soil, 10 per cent in the second 5 cm of the soil, and the remainder moved downward. Some cations have redistributed into the satsuma mandarin trees and other leached to the depth below the root zone. As to the below 20 cm depth, a very little cations accumulated. About 70 per cent of the cations accumulated in the top layer is seemed to be the same form as originally applied without having received any chemical changes. The ratio of Ca to Mg reduced with increasing depth, thereby, confirming the faster movement of Mg than Ca (Yamamoto, 1966; Wada *et al.,* 1981). It is also not impossible to observe that a part of Ca, after being converted into exchangeable form having started to leach, washed away to the depth below the root zone without being held on the way, even if the soil layer is poorly saturated with exchangeable cations. Fujishima *et al.* (1972) in a lysimeter study using a volcanic ash soil of Kagoshima, southern Kyushu showed that a considerable amount of Ca is leached through the soil layer initially poorly saturated with Ca. Differential application of liming materials reduced the effectiveness of soil applied Zn, but absorption remained still adequate at pH values above 7.0 (Smith and Rasmussen, 1959).

Application of lime lowered the citric acid content of satsuma mandarin grown in a clay loam diluvial soil initially deficient in P (Suzuki *et al.*, 1972). Response of two different soil types viz., acid soil of mineral origin and humus rich volcanic ash soil to lime application showed a better growth in mineral soil than volcanic ash soil, besides improving exchangeable Ca in soil and Ca:Mg ratio in leaves (Takahashi *et al.*, 1971). Liming further equalized the effect of Zn application which increased the Zn availability with time reducing the leaching of nutrients like Cu, Zn, and Mn (Smith, 1969) on a limed sandy soil of Florida (Takahashi *et al.*, 1969; Okada *et al.*, 1969; Smith, 1973).

Summary

Acid soils occupy the maximum acreage under citrus. A majority of soils in front line citrus producing countries like USA, Brazil, China, Spain, Japan, Italy etc. are acidic in nature. The world's highest productivity of citrus orchards has been observed in these acidic soils. One of the major problems limiting the efficacy of amendments in such soils is the subsoil acidity, which has warranted the regular use of Ca and Mg containing amendments coupled with deep tillage. The Ca requirement for citrus is of high order. Recurrent use of lime or dolomite in citrus orchard soils would, therefore, lead to not only improving the nutrient availability, but at the same time, likely to improve the microbial abundance and quality production on a regular basis.

6 Soil Conditions and Citrus Decline

Citrus tree decline popularly known in citrus belts of Indian subcontinent has been the subject of considerable research in all the major citrus growing countries of the world such as Brazil, USA, Argentina, Uruguay, South Africa, Ghana, Venezuela, Cuba, Spain, Italy etc. Citrus decline is also known by various names such as blight in Florida, declinio in Brazil, declinamiento in Argentina, and marchitamiento repentino in Uruguay (Childs, 1980; Fischer *et al.,* 1983; Rossetti *et al.*, 1980; Timmer *et al.*, 1986; Wutscher *et al.*, 1977; Wutscher, 1980). It is referred as frenching, decay, chlorosis, neglectosis, and amachamients in Mexico (Orazo-Santos, 1991).

Blight is also known by other common names such as limb blight, chronic wilt, young tree decline, sandhill decline, and rough lemon decline (Nemec *et al.,* 1984) investigated from various soil and climate angles. Blight has been observed to occur in Florida, Louisiana, and Texas in USA, several south African states including Ecuador, Cuba, Columbia, Uruguay, Brazil, and Argentina in American hemisphere and Queensland (Australia) and Hawaii in the pacific region. Blight has not been commonly detected in California, Japan, Israel, mediterranean citrus countries or New South Wales, Australia (Wheaton, 1985).

6.1 CHRONOLOGY OF THE PROBLEM

Manville (1883) in one of the earliest reports on blight in Florida described a limb blight with characteristic of leaf wilt on bearing tree over 16 years of age. Kiely (1957) reported citrus decline in the lower Uva valley and other areas of Sri Lanka due to unfavourable site selection. As a result, citrus decline is now perceived as a complex phenomenon caused by several factors like poor physical condition of the soil, bad drainage, excess of salts, improper orchard management practices, nutritional disorders, in addition to incidence of pests, diseases, and stionic incompatibility. Climate viz., temperature, rainfall, soil moisture content, humidity, and wind; soil properties including their physical and chemical constitution, pH, water table, soil aeration and microorganisms; and cultural practices including mulching, weeding, tillage, pruning, shading, irrigation etc. are all factors quite important for proper growth of citrus plants. Failure or success of citrus cultivation might be due to individual effect or combined effect of several of these factors. Its occurrence in India has been reported in Madhya Pradesh as early as the end of 18th century. Cheema and Bhat (1929) reported about its general occurrence in western India in 1912. According to Asana (1958), the citrus decline is reported to have prevailed in the tract formerly known as the central province, in the 18th century through the writings of Raghujiraje Bhonsle. Ali (1962-64) described citrus decline in Pakistan. Several hypotheses have been proposed from time to time to explain the causes of citrus decline, but many of them are not based on experimental evidences. In the Punjab, it was noticed on a large scale in the early 1950s, when the citrus plantations in a large number started declining after reaching five or six years of age (Khan, 1994). Bakhshi and Dhillon (1964) first published their observations on the decline of sweet orange trees in the arid irrigated regions of the Punjab and pointed out that the predominant type of decline in the region is associated with creasing at the union of rootstock and scion. The trees of the mosambi and blood red varieties of sweet orange are more susceptible to the malady than those of the other sweet orange varieties like jaffa, valencia late, and pineapple.

Lack of congeniality and compatibility in stock and scion combination, adaptability to soil and climate, drainage, and aeration also may lead to decline of citrus (Aiyappa *et al.*, 1962). Narasinga Rao (1958) while discussing the probable causes for the decline of orange trees in Wynad, stated that some of the predisposing factors include probably unfavourable soil conditions. Gandhi (1939) reported that the lateral roots in citrus spread well beyond the radius of the crown and under Indian conditions, trees planted at close spacing showed decline symptoms as soon as the spreading of branches came in close and the soil was denied of sun light. Marudharajan (1949) gave evidence that lack of moisture, overbearing, wrong selection of sites for orchards, and exposure to continuous drought for 5 to 7 months are contributory factors. Naik (1948) considered that die back of citrus trees is associated with poor soil, bad drainage, defective soil moisture, and adverse climatic conditions, particularly high water table and inadequate drainage in south India. He further emphasized the importance of physical conditions of the soil and subsoil, and attached more importance to this as compared to chemical composition of citrus soils. Adverse climatic factors have also proved worth considering in this aspect (Asana, 1958).

Several opinions with regard to suitable cultural practices in citrus have been offered by various research workers. However, different countries have adapted their own localised system of orchard management. Hume (1926) reported that clean cultivation in Florida avoids depletion of soil fertility and appearance of frenching. Loest (1942) observed that frequent tillage leads the soil into hard pan and caused root injury. Many researchers, Aiyappa (1957), Aiyappa *et al.* (1965), Marudharajan (1949), Narasinga Rao (1958), Bhatt (1945), Mukherjee (1949), Mariakulandai and Dorairaj (1958), Ducharme and Suit (1962-64), Chohan (1966), and Nair *et al.* (1968) believed that one of the most important contributory factors which leads to the unthirfty condition of citrus trees is the lack of adequate nutrition in citrus orchards. Under conditions prevailing in Coorg, Mukherjee (1949) reported deficiency of Zn and Mn in Coorg soils which later contributed to citrus decline. In Jammu, the soil borne decline has been reported by Sharma *et al.* (1986). Out of these many studies, very few have addressed the orchard efficiency in relation to either climate or soil. Recently, some efforts have been made with Nagpur mandarin (*Citrus reticulata* Blanco) as a reference citrus crop. Huchche *et al.* (1999) observed the maximum orchard efficiency of Nagpur mandarin (*Citrus reticulata* Blanco) under the present cultural practices, as 20 years with a net cumulative 15 years of peak productivity. Later, Srivastava and Singh (2001c) identified soil water soluble and exchangeable Ca^{2+}, free lime, clay content, available N, and Zn as major soil properties affecting the Nagpur mandarin orchard efficiency in central India using stepwise partial regression analysis.

6.2 MORPHOLOGICAL SYMPTOMS

Quite often, decline (Fig. 6.1) is confused with die back and most of times, citrus decline is considered similar to citrus die back considering the fact that no sharp declineation between them exists even today. However, for all practical or operational purposes, the term die back denotes a lethal condition leading to rapid death of the plant from top downward due to one or more pathogenic causal factors, whereas the decline refers to a gradual degradation of orchards which may occur due to nonpathogenic factor (Ghosh, 1985).

Visual symptoms are variable, often nonspecific and unreliable to determine the presence of blight. Among the factors responsible for citrus decline, plant nutrition occupied an important place. Malnutrition, among many other factors, often results in a chlorotic condition of the leaves. This condition generally develops after four to six years of normal growth. The symptoms comprise light green interveinal areas with the midrib, causing yellowish condition of the leaf in advanced stages. In this way, growth of the plant is partially retarded and the plant bears short twigs with narrow leaves. The shoots have a tendency to die back and even cause the whole of the tree to dry completely in the subsequent years (Chadha *et al.*, 1965; Chadha and Awasthi, 1966). Kiely (1957) described the features of morphological symptoms of citrus decline in Sri Lanka as a symptom of a special chlorosis in form of appearance of silver grey spots on upper surface of leaf in the areas having sandy soils, which resembled very much to the symptoms described by Bar-Akiva (1962) as marble chlorosis for Mn

Fig. 6.1. Declining khasi mandarin orchard on a extremely slopy and fertility depleted land in northeast India

deficiency. Aoki and Morita (1969a) described first symptoms of defoliation in abnormal satsuma mandarin orchards as the occurrence of brown and red brown spots at the leaf tips in autumn and winter especially, when leaf Fe/Mn ratio is less than two.

The symptoms observed in the Punjab after three or four years (or even longer duration) of normal growth, in milder cases, may be confined to a few affected twigs. The leaves may have small, light green interveinal areas (mottling) with the midrib and lateral veins remaining dark green. The new growth of leaves in the affected twigs may be smaller in size. In advanced stages, the chlorotic areas gradually increase. The relative amount of green and yellow tissue varies in some leaves, only the basal portion of the midrib is green, while others may show blotches of yellow or pale green between the lateral veins. The growth is usually checked and the entire periphery of the tree bears short twigs carrying narrow small leaves on their lower portions under an acute condition. The portion of the twig towards the growing point is devoid of leaves and the shoot has a tendency to dry up from the growing point downward. In a few years, the younger leaves are usually thin with a very fine network of veins on a much lighter background. In more severe cases, the young and old leaves may be very pale and even whitish. The colour of the veins gradually fades until only a very pale-green midrib is left. The symptoms described above resemble those of Zn and Fe deficiencies and, in some orchards in Ludhiana and Rohtak, symptoms resemble those of Mn deficiency (Randhawa *et al.,* 1966; 1967). Decline in sweat orange orchards cultivar mosambi has been observed due to multimicronutrient deficiency (Fig. 6.2) in Marathwada region of Maharashtra, India according to Srivastava and Singh (2000d).

Visible responses of trees to blight have been reported to include, wilt, thinning of the canopy due to reduced water conductivity of trunk, major branches, root scaffolds, and dieback in advanced stage. Increase in trunk Zn is an early symptom of the problem. At later stages, mineral imbalances of K, N, and Cl also occur (Wheaton, 1985). Detrimental effects of salinity from high concentration of salinity, specially Na and Cl in the soil solution have been attributed to osmotic stress (Lloyd *et al.*, 1987; Bielorai *et al.*, 1988), ion toxicity (Grieve and Walker, 1983; Walker *et al.*, 1983), nutritional imbalances (Walker and Douglas, 1983), or a combination of these factors.

The declinio in Brazil shows the similar symptoms and characteristics as blight in Florida. The leaves of affected trees indicate Zn like deficiency symptoms, wilt in the canopy, followed by leaf fall, twig dieback, abnormal flowering, and general canopy decline. Declinio does not affect trees until they reach bearing stage, less than 5 years old (Rossetti *et al.*, 1990). Syvertsen *et al.* (1980) described morphological symptoms of blight synonymous to citrus decline. Visible above ground, blight symptoms appear related to water stress like reaction in the tree. Zinc deficiency symptoms in leaves, which are one of the earliest blight symptoms, are transient and may not appear at all, as a tree

Beginning of decline problem as dying from up like die back

Advancing decline seen as loose canopy structure

Multimicronutrient deficiency

Another view of multimicronutrient deficiency

Fig. 6.2. Glimpses of sweet orange, mosambi decline due to malnutrition in Marathawada region of Maharashtra, India

develops early blight. If Zn deficiency does occur, it usually begins in young leaves on one branch and may eventually occur on the entire tree. Blight can also begin by cupping and hardening off appearances to new leaves. Affected leaves also exhibit wilt, and severely wilted branches shed leaves and develop terminal dieback. Spring bloom is delayed, but flower production is abundant. Fruits formed are similar to normal, but may be irregular in shape. Water sprouts eventually develop on the trunk from latent buds. Blight trees have fewer and smaller leaves, less leaf area tree^{-1}, lower stomatal conductance, and diurnal transpiration rates than healthy trees. In the development of blight, Zn first accumulates in the bark and then in the trunk wood.

Visual symptoms have been evaluated in order to adapt a rational canopy rating: zinc deficiency symptoms in leaves; wilting of part or all canopy; leaf drop; twig dieback; abnormal flowering; general canopy decline; internal growth of new shoots; reduced size of fruit (Rossetti *et al.*, 1990). Canopy rating by the above described visual symptoms has been recommended as 0 to 3 (0.0 - healthy; 1.0 initial declinio stage; 2.0 - intermediate stage; 3.0 - more advanced stage).

6.3 DIAGNOSTIC CRITERIA

Diagnostic criteria to distinguish decline affected trees from healthy trees are: reduced water uptake in the trunk by water injection, presence of amorphous plugs in xylem vessels, rate of canopy decline by visual aspects of tree, high zinc accumulation in the trunk wood and phloem, low or no water flow through secondary roots of affected trees. These criteria are the same as those applied to distinguish blighted trees in Florida (Smith, 1974a; 1974b, Wutscher *et al.*, 1977; Young *et al.*, 1978; Beretta *et al.,* 1983; Brlansky *et al.*, 1986; Cohen, 1974; Wallace *et al.,* 1982, Fogaca *et al.*, 1986; Lee *et al.,* 1984; Rossetti, 1981; Rossetti *et al.,* 1980). Reduced trunk water uptake, Zn accumulation in outer trunk wood or phloem, and the occurrence of amorphous plugs in xylem vessels are fairly reliable diagnostic techniques (Wutscher *et al.,* 1977; Williams and Albrigo, 1984; Wheaton, 1985; Rossetti *et al.,* 1990). Soil fertility does not appear to be a major factor in blight incidence, considerable redistribution of nutritional elements occurs within the trees as a result of blight. Smith (1974a; 1974b) reported that the earliest description of leaf Zn deficiency symptoms in relation to blight occurred in 1891 and later in 1923. Childs (1953) reported Zn and Mn deficiency symptoms in blight affected trees. Anderson and Calvert (1970) later confirmed that the symptomatic leaves were low in Zn and Mn. It is now generally accepted that other diseases that cause citrus tree decline do not result in abnormal Zn distribution pattern (Wutscher *et al.*, 1977). This diagnostic test along with trunk water uptake, has been used to identify citrus blight in the countries like South Africa, Uruguay, and Argentina (Graca and Vuuren, 1979; Wutscher *et al.,* 1982a; 1982b).

A number of studies (Wutscher and Hardesty, 1978; Lima, 1982; Cohen *et al.*, 1983) reported higher Zn in trunk wood and trunk bark in blight than in healthy trees. Albrigo and Young (1981) determined that the higher level of Zn in trunk bark of blight affected trees confined in the active phloem and did not fluctuate seasonally as in trunk wood (Albrigo and Young, 1981; Wutscher *et al.,* 1981). According to a number of studies (Wutscher *et al.,* 1982a; 1982b;1983), elevated wood Zn occurred prior to visual symptoms in 58 per cent of the trees developing blight, and in 32 per cent of these trees accumulation occurred 3 years before visual symptoms. Accumulation of Zn in phloem also occurs prior to appearance of visual symptoms. Young *et al.* (1980) showed that trunk wood Zn accumulation occurred in outer wood, whereas reduced water movement took place in inner wood. Wutscher and Hardesty (1981) proposed cation anion ratio and water soluble nutrients in soil as promising approaches to explain the citrus blight syndrome.

6.4 BUD UNION CREASING

In many instances, failure of citrus trees may be due to what appears to be horticultural incompatibility (Singh and Singh, 1942; Nauriyal *et al.,* 1958; 1944; Bakhshi and Dhillon, 1964). In calcareous soils, the lime induced chlorosis is not evident, since Rangpur lime performed consistently well, except with grapefruit as scion according to Thornton (1975). A severe constriction at the bud union and a narrow dark brown line on the cambial face of the bark and wood, may frequently be found in the declining trees (Weather, 1955). In the case of delayed incompatibility, the symptoms of ill health may not be manifested for many years. The bud union creasing in the case of the blood red and mosambi varieties budded on rough lemon is almost universally observable in various citrus growing areas of Punjab and Rajasthan (Bakhshi, 1961). The soil by itself could not, therefore, be a causal factor in the development of bud union creasing. It may, however, condition the rate of decline in an indirect manner. Some soils have also been found to contain a hard pan at depths varying from two to three feet due to the presence of deposition of calcium carbonate. Under such soil conditions, the downward drainage of water is hampered and conditions equaling wet feet for citrus are created. Such factors do add and contribute to a general type of tree decadence, the external manifestations of which may resemble those accompanying decline due to bud union creasing (Bakhshi and Singh, 1966). It can, therefore, be suggested that though the soil is not the cause of bud union creasing, yet the soil conditions and soil management practices do contribute to general decline and may speed up the rate of decline.

6.5 SOIL AND NUTRITIONAL IMBALANCES

6.5.1 Unfavourable Soil Conditions

Citrus trees require a fairly deep and well drained soil for a normal healthy growth. Poor physical condition of the soil and the presence of a hard pan or kankar layer at the depth of four to six feet from the surface may restrict the root growth and impede drainage. Several investigations have attributed the decline of citrus trees to unfavourable soil conditions (Cheema and Bhat, 1929; Naidu and Rao, 1958; Naik, 1948; Narsinga Rao, 1948; Wear, 1956; Nauriyal and Chauhan, 1966; Chadha and Awasthi, 1966; Randhawa *et al.,* 1966; Burnett *et al.,* 1982; Wutscher, 1989). Haury (1981) attributed unfavourable soil condition as one of the factors for citrus decline in Gabougoura in Niger. Analysis of physicochemical characteristics of soils from proposed citrus belts of Punjab (Table 6.1) showed that soluble salts are present in relatively greater amounts in soils belonging to Khepan Wali series. Whereas in other two series, the electrical conductivity in no case exceed 0.22 dSm^{-1}, while in Dewan Khera, the maximum value is 0.15 $dS\ m^{-1}$. Soils represented by Khepan Wali series where the EC ranged from 0.20 to 0.60 $dS\ m^{-1}$, and is the highest in the 35-48 cm layer, which may have an adverse effect on citrus. Calcium carbonate, both crystalline and amorphous, ranged from 4.1-39.5 per cent in the soil profiles representing Nehal Khera series. Whereas in Dewan Khera and Khepan Wali, $CaCO_3$ varied from 1.1-7.7 per cent and 1.0-3.5 per cent, respectively. It is, therefore, desirable that citrus plantation may not be encouraged on the soils of

Table 6.1. Chemical analysis of soil samples of the profiles representing important soil series of proposed citrus belt of Punjab

Depth (cm)	Mechanical analysis (%)			pH	EC (dS m^{-1})	Organic carbon (%)	Exchangeable cations (me100 g^{-1})			Available nutrients (mg kg^{-1})		
	Sand	Silt	Clay				Ca	Mg	Na+K	Cu	Zn	Mn
Nehal Khera Series												
0-10	65.1	30.0	15.3	8.1	0.18	0.43	4.26	1.05	1.60	0.53	0.25	10.00
10-33	60.5	33.0	17.3	8.2	0.10	0.38	5.31	1.20	1.20	0.87	0.07	4.07
33-56	63.6	35.8	18.0	8.1	0.22	0.31	5.62	1.10	0.90	0.93	0.06	5.67
56-76	62.4	34.8	14.4	8.3	0.15	0.35	5.51	1.40	0.90	1.13	0.12	7.06
76-147	58.5	40.6	13.3	8.3	0.10	0.36	4.89	1.42	0.80	1.20	0.65	7.80
147-183	82.8	12.5	6.0	8.5	0.10	0.15	3.35	1.10	0.80	1.07	0.12	7.80
Khepan Wali Series												
0-20	75.7	21.6	6.1	8.2	0.25	0.28	3.74	1.86	1.30	0.05	0.11	4.20
20-35	81.4	16.5	5.6	8.3	0.35	0.26	3.50	1.35	1.00	0.02	0.08	2.66
35-48	75.6	30.0	8.9	8.1	0.60	0.22	5.40	0.80	1.25	0.09	0.11	3.80
48-72	75.6	20.0	6.8	8.1	0.45	0.20	5.20	0.95	1.05	0.45	0.11	3.00
72-102	56.5	38.9	13.0	8.2	0.27	0.19	5.25	1.50	0.90	0.63	0.11	4.66
102-160	55.4	42.2	17.6	8.3	0.20	0.18	5.47	1.25	0.65	0.12	0.12	1.23
Dewan Khera Series												
0-15	78.0	18.6	6.5	8.5	0.10	0.32	3.40	1.15	1.60	0.48	0.73	12.66
15-38	72.5	23.4	11.8	8.5	0.10	0.18	4.86	1.42	1.10	0.32	0.53	9.60
38-61	60.1	39.8	16.3	8.5	0.15	0.21	7.78	1.22	1.20	0.16	0.27	8.45
61-88	56.3	40.8	18.5	8.5	0.15	0.16	7.95	1.26	0.75	0.16	0.34	6.00
86-122	58.3	37.9	14.2	8.4	0.15	0.18	8.85	1.06	0.70	0.20	0.43	5.73
122-150	62.8	36.5	13.8	8.4	0.10	0.12	8.05	1.10	0.70	0.10	0.26	4.00

Source : Dhingra *et al.* (1965)

Nehal Khera series, as it contain high amount of $CaCO_3$. Comparative studies made on the soil conditons in a young orange orchard with patches of poor growth and in adjacent areas with normal growth revealed that soil conditions in the affected areas had the presence of higher total soluble salt content, higher percentage of soluble and exchangeable Na, and lower percentage of soluble Ca and Mg (Milad *et al.,* 1975).

Most of the mosambi orchards in Marathwada region, India are established on clay and calcareous soils, irrespective of their healthy or decline conditions. However, clear cut differences are exhibited in terms of organic carbon, electrical conductivity, and soil reaction of healthy and declined orchards (Malewar *et al.*, 1983). Even though, the orchard soils had pH and electrical conductivity in narrow range of 7.4 to 7.8 and 0.174 to 0.418 mmhos cm^{-1}, respectively, these values are higher in declined than in healthy orchards. Comparison of soil properties under etiolated and normal citrus trees showed significantly higher pH, exchangeable Ca^{2+}, and Mg^{2+} (7.7, 146.1 mg kg^{-1}, and 110.9 mg kg^{-1}, respectively) under etiolated trees than corresponding values (5.3, 514.1 mg kg^{-1}, and 82.9 mg kg^{-1}) in the soils of normal trees (Chen and Zeng, 1991).

Analysis of 200 soil samples from satsuma mandarin orchards showed that soil values of total N, humus, pH, exchangeable Ca^{2+}, and Mg^{2+} increased gradually with tree age in orchards established for up to 40 years. However, they decreased in orchards over 50 years old, and the tendency is more marked at greater soil depths. Soils derived from granite, crystalline schist, and Chichibu palezoic strata are associated with higher fruit yield in trees up to 20 years and lower yields after 40 years than soils derived from deluvium, andesite, and tertiary period strata. Soil density had a considerable effect on root elongation and distribution. In the very dense soils, fine roots are thicker, shorter, more corky, and more irregular than in soils with higher soilds. (Otsubo, 1968).

Comparison of soil physical characteristics and tree growth of mid season cultivars, 20 year old sour lemon and 25 year old valencia on sweet lemon rootstock with clean cultivation and grassed down, respectively, indicated much greater tree growth under grass in soil type having 10 per cent clay content higher at 30-70 cm than 50-125 cm depth under clean cultivation (Nel, 1980). In the first survey, the orchard was affected by decline, and many trees showed symptoms. Comparing the rectangle formed by the group of 90 treated trees, with equal groups on the rest of the orchard, 51 per cent of the former showed severe declinio symptoms, while the other untreated orchards showed 2 per cent to 35 per cent diseased trees (Rossetti and Beretta, 1988). Apparently, the soil from the severely affected orchard had some effect, probably anticipating or aggravating the onset of declinio symptoms. All the potted plants grown in soil collected from the rhizosphere of the diseased trees showed stunting and development of pale leaf colour when compared with those grown on soil from healthy trees, both sterilized and nonsterilized (Rossetti and Beretta, 1985). Physical characteristics of soils under healthy and blight affected hamlin trees identified as belonging to Floridana and Holopaw families showed a high proportion of fine textured particles in Holopaw rootzone with lower hydraulic conductivity and available water. Increased thickness of fill material and calcium carbonate characteristics related to citrus stress (Shih *et al.,* 1986).

In the absence of sufficient low temperature in central India, soil water stress is usually practised for induction of flowering of Nagpur mandarin (*Citrus reticulata* Blanco) by imposition of withholding irrigation for a certain period, and then resuming irrigation at near field capacity moisture level. Under such conditions, the success of flower induction is governed by the right quantum of water stress as a function of combined effect of soil physicochemical properties. It has been observed that many of the Nagpur mandarin orchards showed no flowering even with the adaption of severe most water stress at Katol tehsil of Nagpur district (Dass *et al.,*1998). The full yield potential of mandarin trees is, therefore, never harvested. The clay content in irregular flowering profiles increased from 42.0-48.4 per cent in 0-15 cm layer to 60.2-74.2 per cent in 120-150 cm layer. All the regular flowering profiles (Fig. 6.3 and 6.4) showed lower clay content compared to irregular flowering profiles mainly in the subsurface beyond 30-60 cm depth onward (Fig. 6.5 and 6.6). The clay content in regular flowering profiles reduced with depth from 35.0-42.0 in 0-15 cm depth to 20.4-32.0 per cent in 120-150 cm depth

Fig. 6.3. Soil type at Wdona, Nagpur with sand, silt, clay, and free lime varying between 12.6 to 29.9 per cent, 12.0 to 30.2 per cent, 42.0 to 74.2 per cent, and 2.5 to 7.5 per cent, respectively

Fig. 6.4. Soil type at Khangaon, Nagpur with sand, silt, clay, and free lime varying between 10.4 to 28.2 per cent, 17.8 to 25.4 per cent, 50.0 to 71.0 per cent, and 4.0 to 8.0 per cent, respectively

Fig. 6.5. Soil type at Kamptee, Nagpur sand, silt, clay, and free lime varying within 35.2 to 52.2 per cent, 15.2 to 29.0 per cent, 30.4 to 42.4 per cent, and 7.2 to 22.6%, respectively

Fig. 6.6. Soil type at Gondkheri, Nagpur having sand, silt, clay, and free lime varying within 30.1 to 48.0 per cent, 11.5 to 27.9 per cent, 40.1 to 51.0 per cent, and 3.2 to 6.8 per cent, respectively

with a maximum concentration of 36.9 - 45.8 per cent at 15-30 cm and low thereon. These observations indicated the presence of higher sand and lower clay fractions in regular flowering profiles compared to irregular flowering soil profiles (Table 6.2). The higher clay content in the sub surface of irregular flowering soil profiles acted as a source of continuous moisture supply to trees during water stress period, with the result, the quantum of stress needed for induction of flowering is not attained due to regular

Table 6.2. Depthwise variation in physicochemical properties in soil profiles of regular and irregular flowering orchards of Nagpur mandarin (*Citrus reticulata* Blanco)

Soil depth (cm)	Particle size distribution (%)			Free $CaCO_3$ (%)	Flowers (Per m shoot length)
	Sand	Silt	Clay		
			Regular flowering orchards		
0-15	32.2-41.4	20.2-29.0	35.0-42.0	5.0-22.4	28.4-70.3
15-30	34.0-48.2	14.9-27.2	36.9-45.8	7.5-22.6	
30-60	32.1-64.0	5.9-27.6	30.1-42.4	6.3-24.7	
60-90	38.1-66.8	5.2-29.9	28.0-39.0	5.2-43.7	
90-120	40.9-62.0	6.0-36.3	22.1-34.0	6.8-36.7	
120-150	42.6-64.0	14.0-31.2	20.4-32.0	7.7-28.9	
			Irregular flowering orchards		
0-15	25.6-31.8	21.4-28.1	42.0-48.4	1.6-5.4	3.5-36.4
15-30	17.8-24.2	15.6-30.2	48.0-62.4	1.1-7.0	
30-60	16.2-20.6	12.2-22.4	61.4-69.4	5.2-10.6	
60-90	7.2-19.6	11.4-20.8	62.3-72.0	2.9-6.8	
90-120	10.4-20.1	11.9-18.2	63.1-74.2	3.0-41.2	
120-150	12.4-20.2	12.0-19.6	60.2-74.2	3.0-41.2	

Source : Srivastava *et al.* (1997c); Dass *et al.* (1998)

supply of moisture from clay rich zone of the profiles and resultantly, vegetative growth continue to take place without inducing right quantum of stress. This is the reason that higher flowering intensity in regular flowering soil profiles (28.4-70.3 flowers m^{-1} shoot length) than irregular flowering soil profiles (3.5-36.4 flowers m^{-1} shoot length) is recorded.

Sometimes, age of orchard also imparts a decline in fruit yield after giving consistently good yield for quite a few number of years. Survey of 68 Nagpur mandarin orchards in Vidarbha region of Maharashtra covering 39,261 trees varying in age between 7 and 27 years showed highest 30.4 per cent of decline on shallow soils (<45 cm deep) compared to deep soils (>135 cm deep). Percentage of decline rose with tree age from 8 to 18 years with highest incidence of 27.5 per cent in 14-16 years age group. The degree of decline is more (23.5 per cent) in orchards harvesting *Mrig* bahar (Feb. – March harvest) compared to orchards (9.4 per cent) practising *Ambia* bahar (November – December harvest) due to higher fruit yield harvested from former crop (Diware and Kolte, 1990).

6.5.1.1 Soil Compaction

Soil compaction as a limiting factor for root growth and performance of citrus trees is well established. Indeed, soils are classified into different potential classes for citrus depending on rooting depth (De la Rosa and Carlisle, 1978). Nel and Bennie (1984) reported decrease in root and tree growth of valencia scions on rough lemon rootstock when the soil penetration resistance exceeded values as low as 500 Kpa. Van Huyssteen (1987) concluded from soil survey of citrus orchards in Sunday river valley that restricted root depth is one of the major factors limiting yield. In order to ameliorate problems associated with soil compaction, it is often necessary to deep plough. However, such soil management practices often result in the prunning of roots and may have negative consequences for tree performance. The ability of roots to regenerate following such a treatment has been attributed to factors such as soil fertility, rootstock, cultivar, and root thickness as well as the timing and intensity of treatment (Van Zyl and Van Huyssteen, 1987). Analyses on causes of low yield (25-67 kg 0.067 ha^{-1}) of mandarin cultivar Wenzhoumigan on beach land in Yueqing county, Zhejiang, China revealed that heavy clay texture of soil coupled with few noncapillary pores (0.15-0.33 per cent) and highly compacted layer (169 kg cm^{-3}) in the 10-40 cm depth with soil pH 9.0 and salinity in the 20-40 cm depth due to capillary movement as main causes of low yield. The efficacy of soil improvement measures assessed from yield increments decreased with red or yellow soil or sand (39 applications) >> treated with ferrous sulphate, alum, sulfur, and gypsum (15 times) > green manuring (10-11 times) > P fertilization (10 times) according to Lou and Yin (1986).

Effect of soil compaction and its alleviation by deep ploughing on the yield and fruit size in orchards ploughed to a vertical depth of 80 cm at approximately 1000 mm radial distance from tree stem compared to nonploughed controls in a relatively low yielding mature Palmer navel orange orchards, with Florida rough lemon as rootstock showed that resistance to penetration values of above 2200 Kpa at a soil depth of 20 cm as measured with a penetrometer. Deep ploughing decreased the mean soil penetration resistance significantly to at least 2000 Kpa three years after treatment. Strong root regrowth occurred in the 20 to 80 cm soil zone where the soil is deep ploughed compared with feeder roots in the control treatment, characterised by soil compaction at a shallow soil depth of 20 to 30 cm. Yield and fruit size showed a significant increase over five seasons with deep ploughing, while unploughed treatments exhibited significantly reduced yield and fruit size. It is recommended that a proper soil compaction survey must be undertaken before considering ploughing of existing citrus orchards. Ploughing should be done to a soil depth of 80 cm on a compacted soil to facilitate a favourable rooting volume for regeneration of roots (Hoffman and Abercrombie, 1999). Earlier, similar observations have been made. Ploughing up to soil depth of 80 cm is necessary on a compacted soil to facilitate favourable rooting volume for regenerated roots (Abercrombie and Plesssis, 1995). Besides yield, quality of fruit is also improved substantially following the loosening of compact layer (Okada, 1994).

6.5.1.2 Poor Drainage

Citrus roots have a high oxygen requirement and ill effects are readily visible in poorly drained soils. The problem of poor drainage exists in many parts of India and abroad. In Punjab, the problem is particularly serious and water table is rising very rapidly, in some areas at the rate of one foot every year (Uppal, 1962). The water table in approximately three million acres of canal irrigated land in Punjab stands at a depth of five feet from the ground surface. In the districts of Amritsar, Jalandhar, Patiala, and parts of Ludhiana, the problem of waterlogging has become very acute. During the rainy season, water begins to overflow and the soil remains submerged for long periods. Adequate provision of surface and subsurface drainage is necessary before citrus trees are planted successfully in such waterlogging prone areas. Young tree decline has been reported by Norris (1970) on poorly drained soils with rough lemon, *Citrus jambhiri* rootstocks. Studying the causes of chlorosis in mandarin orchards of Aegean region Turkey, Kovanci *et al.* (1978) reported that poor drainage is responsible for such problem. On the other hand, poor drainage in the soil profile of sweet orange orchards near Nirgua restricted the root penetration, and low yields have not been attributed to other soil conditions (Merlo *et al.*, 1990).

6.5.1.3 Soil Reaction

Soil reaction is more used as an indicator of soil fertility status under acid conditions (pH below 5.5). Often, the nutrients are leached out especially in light soils in heavy rainfall areas. Under high pH (pH 7.5 or more) or under alkaline and heavy soils, the nutrients are made unavailable. Suitability of various soil pH ranges (Fig. 6.7) has been suggested by Wander (1952). Excellent citrus orchards have been raised in different parts of the world in soils ranging from highly acidic (pH 4.5) to as high as pH 8.5. Govinda Iyer and Iyengar (1956) reported decline in orange due to soil pH less than 6.0. In northwest India, the pH of the citrus soils varied from 6.5 in Kangra (Himachal Pradesh) to above 9.0 in some of the southern district of Punjab (Kanwar and Randhawa, 1960; Dhingra and Kanwar, 1963; Kanwar *et al.,* 1965). Evaluation of seedlings of Balady lime, cleopatra mandarin, and sour orange at soil pH 6.0, 7.0, and 8.0 showed best growth at soil pH 6.0. The growth reduced on an average by 9.8, 25.4, and 40.1 per cent, respectively (Shawky *et al.*, 1980). Yuda (1985) observed canopy of satsuma mandarin at soil pH

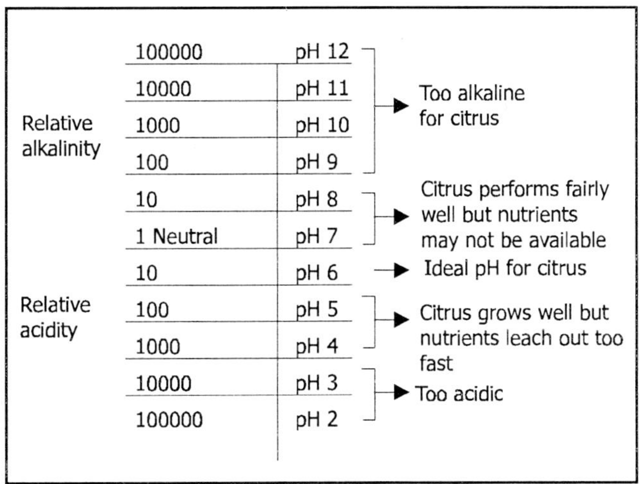

Fig. 6.7. Soil pH suitabillity guide for citrus

4.0 as half of trees growing at soil pH 5.0 in Japan and attributed low soil pH to heavy N fertilization for such a differential response. Alva *et al.* (1995) studying the response of soil pH and Cu effects on hamlin orange trees, observed that effect of Cu on plant growth is more pronounced at soil pH 5.5-6.0 than at higher or lower soil pH regimes. Establishment of hamlin orange trees in Florida flatwood soils showed no relationship between soil pH and tree canopy volume or orange yield in the soil pH range of 4.6-8.0. Below pH 4.6, tree size and yield reduced substantially (Obreza, 1973). Analysis of soil samples from old citrus orchards and from adjacent virgin soil in several districts of South Africa indicated no differential accumulation or depletion of any nutrient, and the soil pH is not changed by growing citrus for 20 years or more (Bredell and Conradie, 1975).

Comparison of plots showing Zn deficiency to that of healthy plots in satsuma mandarin in Kagawa prefecture, Japan showed low soil reaction is one of the soil parameters (Table 6.3) responsible for such a cause. Iwamoto *et al.* (1969) and Ida *et al.* (1969) suggested that soil pH less than 4.0 and water soluble Mn over 10 mg kg^{-1} induced large scale defoliation in satsuma mandarin orchards, which is later suppressed due to lime addition. Another study on comparison of soil properties in healthy and defoliated satsuma mandarin orchards showed that top soil did not differ in pH or water soluble, exchangeable or readily reducible Mn. But, the subsoils of affected orchards are more acidic and contained water soluble and exchangeable Mn more than those of normal healthy orchards at Gamagrol, Aichi prefecture. However, at Mikkahi Shizuoka prefecture, soil pH values of top and subsoil are higher in affected than normal orchards due to adaption of liming practice. Examination on soils from major orchards of Japan further showed incidence of abnormal defoliation due to lowering of soil pH and the higher chemical reduction through waterlogging (Aoki and Morita, 1969b). Similarly, Smith (1971) suggested that citrus can not tolerate soil pH below 4.0. A high pH not only affects the availability of micronutrients in soils, but a high sodium content in such soils may have direct toxic effect on sensitive cultivars.

According to Allison (1964), the range of tolerance of citrus to exchangeable sodium percentage (ESP) is 2-10. However, on the light textured soils of Ferozepur and

Table 6.3. Chemical properties of soil of Kagawa (Japan) showing Mn and Zn deficiency

Depth (cm)	pH		Total N (%)	Humus (%)	Available P (mg %)	CEC (me 100 g^{-1})	Exchangeable cations (mg %)			EC (dS m^{-1})
	H_2O	KCl					K_2O	Ca	MgO	
1977										
10	6.95	6.08	0.140	2.74	132	10.4	40	267	74	0.04
30	6.32	4.87	0.034	0.70	14	5.4	20	88	24	0.03
1978										
10	7.29	6.65	0.246	4.12	320	11.1	21	287	98	0.28
30	6.68	5.52	0.049	0.78	28	5.8	12	126	31	0.09

Source : Inoue and Harada (1981)

Bhatinda, where the soil structure is not a problem, citrus has been observed to make a normal growth in soil having ESP as high as 15 (Anonymous, 1966). Soil pH below 5.0 has been demonstrated to be unsuitable for citrus root growth (Randhawa and Itawa, 1968; Smith, 1971). The various micronutrients, especially soluble Mn decreases hundred folds with each unit rise in soil pH, and enhances the formation of Mn-soil organic matter complex, thereby, renders Mn less available (Page, 1962). At pH 4, roots become stubby with cortical slogging (Randhawa and Itawa, 1968). Aluminium solubility increases with acidity (Pratt, 1966) and the levels present in most soils of all orchards are threshold required for toxicity to citrus (Chapman, 1938). Aluminium toxicity has a depressing effect on root activity, but there is no general agreement on Al saturation level at which citrus growth is reduced. Castle *et al.* (1975) reported the absence of Al in deep sand of a citrus orchard, but comparatively higher in the clay rich subsoil. Yuan (1965) surveyed Al in some Florida soils, and found that the organic pan of Leon soils had a relatively high Al content compared to other soils. Highly leached sandy surface soils usually have a relatively low Al content. Citrus decline similar to blight has been reported in high Al soils of South America. Fruta bolita, common in the province of Misiones, Argentina occurs on high Al containing red lateritic soils. Declining sweet orange trees on trifoliate orange (*Poncirus trifoliata* (L.) Raf.) showing symptoms of marchitamiento repentino disease occurred only on the patches of heavy, clay like soil in the area of Concordia, Enter Rios province of Argentina (Swartz *et al.,* 1980).

By sampling the soil in small increments to 45 cm depth and analysing each sample separately (Table 6.4), a higher Ca and pH under blighted trees have been obtained on a variety of soils througout the Florida citrus belt (Wutscher, 1988; 1989). Wutscher and Lee (1989) observed that higher soil Ca content at 20-30 cm depth under blighted 44 year hamlin orange tree than healthy. The relationship of lime with blight is not clear, because blight is rarely found in dry areas on calcareous soil. It is common, however, on shallow soils above limestone in Florida (Cohen, 1980), Cuba (Wutscher *et al.*, 1983), and Brazil (Rodriguez, 1980), where the citrus orchards are limed heavily as

Table 6.4. Soil pH and double acid extractable Ca in soil collected under blight affected and healthy trees at various locations covering 6 soil types in Florida

Status of tree	Site					
	I	II	III	IV	Va	Vb
			Soil pH			
Blight	6.2	5.9	6.1	6.2	5.8	5.9
Healthy	5.3	4.9	4.9	5.7	5.1	4.8
Stat. sig. level	0.001	0.001	0.001	0.001	0.001	0.001
			Soil Ca (mg kg^{-1})			
Blight	557	1,162	1,249	356	332	272
Healthy	348	532	138	239	206	112
Stat. sig. level	0.001	0.001	0.001	0.005	0.005	0.001

Source : Wutscher (1988)

well as in humid citrus growing areas in general (Wutscher, 1987). In Florida, the Zn level in the wood of healthy trees increased after lime application (Wutscher, 1981). Investigation on the apparent connection between blight and lime in the soil appeared to be a promising approach.

6.5.1.4 Soil Salinity

Citrus is glycophyte and tolerance to soil salinity is correlated with its ability to restrict the entry of toxic ions (Na, Cl, and B) into shoots (Greenway and Munns, 1980). Differences in Na and Cl accumulation or exclusion have been found among citrus rootstocks. Cleopatra mandarin appeared to be a Cl excluder, but a Na accumulator (Cooper, 1961), while sour orange accumulated both Na and Cl in its leaves (Zekri, 1991). Injury to citrus from salts has been mainly attributed to excessive accumulation of Cl (Cooper, 1961; Ben-Hayyim and Kochba, 1983) and Na (Smith, 1962).

High soil salinity is a long standing and chronic problem in Citriculture, especially in semiarid regions, where saline water is used as a source of irrigation (Bernstein, 1967; 1969; Sterling Davis *et al.*, 1969; Bingham *et al.*, 1969; Carpena *et al.,* 1969; Goell and Cohen, 1981). Harding *et al.* (1958) described that development of salinity is often induced by unfavourable soil conditions resulting from the use of sodium nitrate and ammonium sulfate fertilizers. A linear regression of yield on average ECe in the range of 1-5 d Sm^{-1} with a yield decrement of 11.2 kg fruit $tree^{-1}$ per dS m^{-1} increase in ECe. Carpena *et al.* (1969) suggested three levels of SAR (0-5, 5-10, and 10-15) and four levels of EC (0.50-1.0, 1.0-1.75, 1.75-3.0, 3.0-6.0 dS m^{-1}) for classifying water irrigating to citrus. They further suggested that irrigation water containing Cl^- more than 300 g l^{-1} and SO_4^{2-} more than 700 mg l^{-1} as threshold values for irrigating citrus. Citrus trees are quite sensitive to excess of salts (Harding *et al.*, 1958; Grass *et al.*, 1966; Chapman *et al.*, 1969; Bernstein, 1969; Shalhevet *et al.*, 1974; Bielorai *et al.*, 1988; Aksoy *et al.*, 1997; Srivastava and Lallan Ram, 2000).

Visible symptoms of salinity include leaf bronzing (Fig. 6.8 and 6.9), defoliation, leaf chlorosis similar to iron induced chlorosis, small leaves, small fruit, leaf chlorosis, and die back of young twigs (Cooper and Edwards, 1950; Cooper and Garton,1952; Peynado and Young, 1962; 1963; 1969; Bernstein, 1969; Lallan Ram *et. al.,*1993). Kanwar and Radhawa (1960) found that soil salinity is one of the important causes of chlorosis in the Punjab. Kanwar and Randhawa (1960) attributed the chlorosis of citrus to salinity in certain areas, besides Zn deficiency. The districts worst affected are Bhatinda, Ferozpur, Amritsar, Jallandhar, and parts of Ludhiana, India. Goell (1969) described salinity bound chlorosis in Eureka lemon leaves into five groups (Table 6.5).

Francois and Clarke (1980) found that increasing salt stress delayed fruit maturation, but had little effect on quality of valencia orange grown in a sand culture. While, Bielorai *et al.* (1988) reported slight increase in TSS and TSS/acid ratio in shamouti orange at

Effect of saline irrigation (1200 mg l^{-1} Cl) up to 60 days with amount of saline water equivalent to daily evaporation rate on the development of chloride injury symptoms in citrus on a black clay soil

Fig. 6.8. Bronzing of leaf margins on rough lemon rootstock

Fig. 6.9. Mosaic of bronzed spots along the midrib of leaves on Rangpur lime rootstock

higher salinity level. On the other hand, a number of workers (Levy *et al.,* 1979; Cerda *et al.,* 1990; Dasberg *et al.,* 1991; Nieves *et al.,* 1991a; 1991b) observed no significant influence of salinity on various fruit quality parameters. The presence of excessive salts in any of the horizons up to a depth of six feet adversely affected the health of citrus plants. The tolerance limit of soluble salts in 1:2 soil water suspension has been raised to 0.8 mmhos cm^{-1} for kinnow mandarin (Anonymous, 1996) compared to earlier recommendation of 0.5 mhos cm^{-1}.

Experiments with citrus seedlings on soils adjusted to different levels of exchangeable Na indicated that exchangeable sodium percentage(ESP) greater than 5 impaired growth (Hayward and Bernstein , 1958; Bernstein and Hayward, 1958; Martin *et al.*, 1959; Pinckard, 1982). Under saline conditions, critical Na concentration reached at lower ESP. A soil solution with a total salt concentration of 40 me l^{-1} and SAR of 5 contained 16 me l^{-1} Na. Sodium absorption and resultant Na problems are not wholly

Table 6.5. Description of chlorosis ratings of Eureka lemon leaves

Rating	Description
0	Dark green and healthy leaves
1	Light green colour and slight bronzing
2	Beginning of chlorosis and yellowing of margins and tips
3	Pronounced chlorosis margins tips and blotching
4	Wholly chlorotic leaves, completely yellow with necrotic spots on margins and tips
5	Advanced chlorosis in form of necrosis and margins and tips, coupled with interveinal necrotic spots

Source : Goell (1969)

defined by SAR, because total concentration as well as SAR are important in determining crop response to Na (Bernstein, 1964a; 1964b). Another factor that apparently affected the response of citrus to ESP is cation exchange capacity (Jones *et al.*, 1957; Martin *et al.*, 1959), although Na uptake by other crops is not affected by cation exchange capacity (Bernstein and Pearson , 1956).

Most of the salt tolerance observations in citrus are based on leaf analysis for Cl^- or Na^+ (Stylianou and Orphanos, 1970; Grieve and Walker, 1983; Walker and Douglas, 1983; Walker, *et al.,* 1983). There are some disadvantages in using leaves for assessment of salt accumulation. Salt content is dependent on leaf age. Therefore, leaves need to be sampled to ensure the collection of similar age of leaves (Embleton *et al.,* 1962; Embleton and Jones, 1964). Leaves exposed to saline irrigation water may absorb salt directly through epidermis (Stolzy and Harding, 1966). Leaf analysis, hence, may not indicative of uptake by the roots. The tendency of leaves to abscise before the usual autumn sampling date having high Cl level (Levy and Shalhevet, 1982). Such a physiology of leaves makes difficult to have representative leaf samples. This has warranted to try out other alternatives, of which juice analysis for salt tolerance has also shown some promise in explaining the chloride exclusion properties of citrus rootstocks (Levy and Shalhevet, 1990a). A number of physiological parameters, namely, root hydraulic conductivity (Levy *et al.*, 1983; Graham and Syvertsen, 1984; Hartmond *et al.*, 1987; Syvertsen and Yelenosky, 1988; Zekri and Parsons, 1989) and root cation exchange capacity (Srivastava *et al.*, 1998) have been found equally successful in screening citrus rootstocks against salinity.

Kirkpatrick and Bitters (1969) based on accumulation or exclusion pattern of Na or Cl, suggested six types of response: i. Low Na and low Cl e.g. cleopatra mandarin, ii. Moderately high Na and low Cl e.g. Rangpur lime, iii. High Na and low Cl e.g. 1215 Ichang lemon, Savage citrange. Yuzu, 1452 Citrumelo, Aslem, Camulos grapefruit, vi. Moderate Na and moderate Cl e.g. Troyer citrange, Carrizo citrange, Keen sour orange; v. high Na-moderate Cl, 1449 citremon, 1448 Citremon, Argentina sweet orange, Nonsho Daidai, Hall grapefruit King tangelo, gajanimma, Kara mandarin, 1219 Ichang lemon, and vi. High Na and high Cl e.g. Limonacitra rough lemon, 1437 Citradia, Koethen sweet orange, and Melanesian papeda. Zekri (1991) demonstrated that sour orange and cleopatra mandarin are relatively salt tolerant rootstocks; rough lemon, Swingle citrumelo, and Carrizo citrange as salt sensitive; and Milan and trifoliate orange rootstocks as sensitive to NaCl.

Use of tolerant rootstock is one of the promising alternatives to grow citrus successfully by avoiding the soil salinity (Joolka and Singh, 1979; Hassan and Galal, 1989). Salt tolerance is an inheritable quantitative character for citrus and is of great importance in identifying and screening citrus rootstocks against their reaction to salinity (Cooper *et al.*, 1951; 1952; Cooper, 1952; 1961; Cooper and Shull, 1953; Cooper and

Peynado, 1959; Hewitt *et al.*, 1964; Hewitt and Furr, 1965a; 1965b; Maas, 1993; Singh *et al.*,1997a; 1997b), besides breeding rootstocks for salinity tolerance (Kokba *et al.*, 1979). A number of studies conducted earlier (Cooper *et al.*, 1956; Furr *et al.*, 1963; Newcomb, 1978; Furr and Ream, 1968) have shown a large strainal variability from highly susceptible to tolerant amongst Rangpur limes. Hutchinson (1975) reported Swingle citrumelo as moderately tolerant rootstock against salinity. Some of the earlier breeding work for salt tolerance (Furr and Ream, 1969) and basic observations on citrus metabolism (Kessler and Snir, 1969), provided the reasons for differential behaviour amongst citrus cultivars and their ability to exclude the phytotoxic Cl below critical levels.

The Cl and Na exclusion behaviour has been observed to be widely used in rating the citrus rootstocks against salinity. Srivastava *et al.* (1998) observed the chloride exclusion behaviour of various citrus rootstocks as a function of root cation exchange capacity and correlated positively with each other. Conversely, it is often observed that selections made for tolerance at early growth stage proved to be not as tolerant during the later growth stages or at other salinity concentrations (Shannon,1979). Grieve and Walker (1983) observed that trifoliate orange possessed a better Na exclusion property than limes and mandarins, which are better Cl excluders. There is some evidence to suggest that changes in membrane fluidity of fibrous roots, particularly those changes associated with free sterols are significant in the Cl exclusion mechanism operating in roots of citrus genotypes (Douglas and Sykes, 1985). There is apparently a strong inverse correlation between the phospholipid to free sterol ratio and leaf Cl level that would biochemically identify the ability of citrus roots to exclude Cl. Free sterols are seemingly important in both the active and passive components of chloride exclusion mechanism in citrus genotypes. Salt treatments increase both the activation energy and thermotropic phase transition temperature for Rangpur lime, a high chloride excluder and no change for Etrog citron as a worst chloride excluder (Douglas and Walker, 1984). Data further suggested that free sterols may be significant in citrus salt tolerance. In other studies, addition of Ca to saline irrigation water helped sour orange seedlings to tolerate NaCl toxicity by reducing accumulation of Na and Cl in the leaves (Zekri and Parsons, 1990). Weng (1980) observed no difference in response of lemon and limettier against soil salinity. Chen *et al.* (1990c) developed standards for identifying the salt tolerance of *Citrus* (Table 6.6).

Based on chloride exclusion or accumulation behaviour of citrus rootstocks, Cooper *et al.* (1951) suggested upper limit of leaf chloride content of 1.1 per cent for Rangpur lime and 1.9 per cent for cleopatra mandarin. While, in soil, water soluble Cl^- in saturation extract and soil-water suspension should not exceed 25 me l^{-1} and 50 me l^{-1}, respectively (Pearson and Goss, 1953; Cooper and Shull, 1953; Bernstein, 1965) to grow citrus successfully. Harding and Chapman (1951) recommended that leaf chloride content exceeding 0.25 per cent is an indicative of chloride toxicity rather than approximately

1.0 per cent as suggested by Hayward and Bernstein (1958). A wide range of thresholds of soil EC have been suggested as 1.5 d Sm^{-1} at 0-15 cm and 5.4 dS m^{-1} at 90-120 cm with Cl^- and Na^+ in saturation extract ranging from 0.5-14.0 me l^{-1} and 2.1-33.8 me l^{-1}, respectively (Harding and Chapman, 1951), 2.5-3.0 dS m^{-1} (Bernstein, 1965), 1.2 dS m^{-1} for grapefruit, marsh seedless on sour orange rootstock (Bielorai *et al.,* 1978), 1.5 dS m^{-1} for Verna lemon on sour orange rootstock, 2.1 dS m^{-1} for cleopatra mandarin, 1.0 dS m^{-1} for macrophylla (Cerda *et al.,* 1990), 1.4, 0.9, and 1.7 dS m^{-1} for navel, valencia, and shamouti orange, respectively (Bingham *et al.,* 1974; Bielorai *et al.,* 1988).

Table 6.6. Standards for identification of salt tolerance for *Citrus* species

Classification of salt tolerance	Salt injury (%)	Degree of salt injury	Cl content (%)
Highly tolerant	0	0	<=0.8
Nearly tolerant	1-10	1-10	0.81-1.1
Little tolerant	10.1-30	10.1-20	1.11-1.40
Little susceptible	30.1-50	20.1-30	1.41-1.70
Medium susceptible	50.1-70	30.1-40	1.71-2.00
Susceptible	70.1-90	40.1-50	2.01-2.30
Highly susceptible	90.1-100	> 50	> 2.3

Source : Chen *et al.* (1990c)

Studies by Singh *et al.* (1997b) revealed Rangpur lime (California) and Swingle citrumelo as highly susceptible rootstocks; Rich 16-6, flying dragon, C-35, C-32, *C.macrophylla,* Rangpur lime (8784, Tirupati, and Brazilian) and Chase rough lemon as susceptible; cleopatra mandrin (Grebstan and Narana), Schaub rough lemon and Rangpur lime (Texas) as tolerant rootstocks; and cleopatra mandarin (Tirupati, Coorg, and Morocco) as highly tolerant rootstocks based on amount of chloride accumulated in the leaf and time taken for clearance of chloride toxicity. Besides osmotic and inhibitory effects of high concentration of Cl and Na, imbalance of essential nutrients also contributed to reduced growth of four citrus rootstocks viz., sour orange, cleopatra mandarin, Carrizo citrange, and Citrus macrophylla (Ruiz *et al.,* 1995). Romero-Arando *et al.* (1998) proposed that ameliorative effects of Ca on citrus grown under saline conditions are mostly related to reduced leaf abscission with a well defined cause-effect relationship between chloride buildup and reduced growth.

6.5.1.5 Excessive Lime

Deficiency of Zn, Fe, and Mn is quite common in Punjab which is more acute in citrus plantation raised on calcareous soils. Lime induced chlorosis is one of the oldest known types of chlorosis. Floyd (1971), Thorne *et al.* (1950), and Hilgard (1966) reported chlorosis of citrus established in calcareous soils of California, Florida, and Utah, respectively. Presence of free lime rendered P, Fe, Mn, and Zn less available (Truog, 1946; Harding and Chapman, 1950; Thorne *et al.,* 1950). Randhawa *et al.* (1961) found that the available Mn in Punjab soils decreased with an increase in the $CaCO_3$ content. Kanwar and Randhawa (1960) and Dhingra and Kanwar (1963) believed that a high $CaCO_3$ content in citrus soils of the Punjab is an important contributory factor towards decreased availability of Fe and Zn in soils. Earlier, survey of citrus soils in Punjab showed that declined orchards are generally associated with soils containing more than 5 per cent free lime in any of the horizons up to six feet depth. In Sirsa, chlorosis has been found to be due to the presence of high amount of $CaCO_3$, which is responsible for reducing the availability of Zn, besides inducing Fe chlorosis (Dhingra and Kanwar, 1963). Bhella (1966) conducted a survey of the citrus soils of Hisar and Bhatinda districts which indicated that this limit could safely be raised to 10 per cent. Free lime has been observed in various forms, namely, powdery form concentrated layer in subsurface and in concretionary form (nodules) distributed more or less uniform through the entire depth of soil profile. Powdery form of $CaCO_3$ in alluvial soils of northwest India is most active in arresting the availability of nutrients and choaking the micropores in soil to induce waterlogging. Decline in valencia yield in south Lebanon has been ascribed to high Ca absorption due to calcareous nature of soil, shallow soil, and other nutrient imbalances (Khalidy *et al.*, 1969b).

Analysis on the decline of Batavian orange plantation in Palacofarea, the reduction in area under cultivation of Vadlapudi orange in parts of Guntur, Krishna, and Godavari districts and extremely short life of sathgudi in Anantapur district revealed that the presence of a hard pan formed due to calcium carbonate concretion in the soils, which eventually determine the rooting depth and age at which decline could start (Satyanarayana and Reddy, 1994). In another study, Dass *et al.* (1998) observed that $CaCO_3$ nodules up to 22.4 per cent in 0-15 cm depth produced no detrimental effect on growth of Nagpur mandarin (*Citrus reticulata* Blanco), instead it helped in regulating the flowering response. A linear increase in fruit yield of Nagpur mandarin has been observed with increase in free lime content up to 12 per cent in 0-15 cm layer (Kohli *et al.*, 1997; Srivastava and Singh, 2000a) in black clay soils of central India. The various limits of $CaCO_3$ are required to be worked out in relation to form of $CaCO_3$, which possibly could be different in the lighter texture.

Presence of calcium carbonate nodules is a common feature in the Vertisols of India. Singh and Lal (1946) showed close relationship between kankar ($CaCO_3$ concretion) and the properties of associated soils. Mermut and Dasog (1986) documented the morphological, micromorphological, and chemical compostion of white and black nodules in some Vertisols of India including two from Karnataka. However, there is a need to examine whether their origin is pedogenic or lithogenic, and to further confirm their properties in a cross section of Vertisol types. Carbonates have their own inherent particle size which affect the availability of nutrients and modify soil texture in highly calcareous soils (Yaalon, 1957). Studies on origin, properties, and particle size distribution of carbonates nodules in some Vertisols of Karnataka by Nandi and Dasog (1992) identified white and black nodules as two morphological types in Vertisols of India. White nodules are irregular in outline, have rough surface, and are more porous. Black nodules are spherical, smooth, and less porous. White nodules contain slightly more $CaCO_3$ than black ones. The Al, Fe, and Mn contents are distinctly higher in black nodules as compared with white ones. The X-ray diffraction analysis showed that calcite is the predominant mineral with small quantity of quartz in both the types of carbonate nodules. Carbonates of sand and silt sizes dominated over clay size carbonates in these soils.

The formation of these nodules occurred *in situ* through accretion, and with time, these developed into hard, compact calcitic nodules as explained by Sehgal and Stoops (1972). Micromorphological studies (Mermut and Dasog, 1986) revealed the coalescence of carbonate particles developing into nodules. The low SiO_2 and Al content of the nodules suggested that there has been very small amount of incorporation of soil material in them as compared to that in the alluvial soils (Singh and Lal, 1946; Singh and Singh, 1972). The impregnation of Mn and Fe probably occurred when nodules are porous and soft (Sehgal and Stoops, 1972). Higher age of black nodules than white ones revealed by Mermut and Dasog (1986) using carbon dating supported the view that white nodules formed first and later turned black. The decreasing trend of black nodules with depth observed in all the pedons of Vertisols suggested that the blackening occurred from the surface downwards because of greater oxidation at the surface (Nandi and Dasog, 1992).

The use of appropriate rootstock holds a great promise to avoid the adverse effect of saline and calcareous soil (Ream and Furr, 1976; Hamze *et al.,* 1986). Limited studies are available to suggest a suitable rootstock which can withstand high calcium carbonate content in soil. However, a number of studies (Shaked *et al.,* 1988; Vardi *et al.,* 1988; Gallasch and Dalton, 1989; Sagee *et al.,* 1992; Sagee, 1996) have earlier shown the response of citrus rootstocks in relation to calcareousness of soil. Campbell and Goldweber (1979) reported that citrus fruits, such as Tahiti lime and lemon, grow very well on rough lemon and Alemow (*Citrus macrophylla*) rootstocks in limestone soils of Florida and produced good quality fruits. Grapefruit, mandarin, sweet orange, tangelo and tangor cultivars produced quality fruits on some orange and cleopatra mandarin rootstocks. However, these rootstocks required application of micronutrients to correct multinutrient deficiencies on such soils.

Similarly, Timmer (1979) observed that Star Ruby grapefruit is most suitable citrus species for fine textured calcareous soils of Texas. In these soils, trifoliate orange, Swingle citrumelo, and Morton citrange developed severe chlorosis. Sagee *et al.* (1992; 1993) demonstrated significant effect of soil texture on the response of citrus rootstock seedlings to $CaCO_3$. They observed better performance in clay soil (47 per cent clay, 39 per cent silt, and 14 per cent sand) with 33 per cent $CaCO_3$ than in silty clay soil (21 per cent clay, 19 per cent silt, and 60 per cent sand) containing only 23 per cent $CaCO_3$. Sagee (1996) showed that the tolerance of citrus rootstock to $CaCO_3$ is governed by the ability of rootstock to lower the pH of the medium. Tolerance of six rootstock, seedlings, namely, rough lemon, cleopatra mandarin, Rangpur lime, Carrizo, Troyer citrange, and Trifoliate orange suggested cleopatra as most tolerant rootstock (El-Otmani, 1996). In countries where *Poncirus trifoliata* and their relatives are commerically used, they are exposed to intolerance in calcareous soils (Wutscher, 1979; Castle, 1987). These rootstocks might face twin problem from salinity as well as calcareousness (Cooper, 1961; Levy and Shalhevet, 1989).

6.5.1.6 Presence of Hard Pan

Compacted layers, often called plowpans, are formed in some soils by tillage machinery or by long and sustained use of heavy equipment. Fragipans are subsurface soil horizons that have high silt, very fine sand, or fine sand content; moderate or low clay content; low organic matter content; medium to high bulk density when moist; and slow or very slow saturated hydraulic conductivity (Grossman and Carlisle, 1969). They also exhibit a brittle consistency when moist. Harlan *et al.* (1977) have proposed that the cementing agent in fragipans is soluble silica that concentrates and precipitates in the fragipan zone. Duripans are subsurface soil horizons that are cemented to the degree that air dry fragements do not slake during prolonged soaking in water or in HCl. They vary in degree of cementation by silica and, in addition, commonly contain accessory cements, chiefly iron oxides and calcium carbonates. Duripans, which occur mostly in soils of subhumid mediterranean or of arid climates, have moisture regime in which soluble silica is washed down into, but not out of the soil. They are counterparts of fragipans which develop in more humid climates. Soil containing duripans occur in California, Oregon, Chile, and Italy.

Hard pans and soil layers with higher clay due to illuviation have the major problem of high bulk density. The red sandy loam soil having clay subsoil (Alfisols) in Andhra Pradesh (Fig. 6.10) and Tamil Nadu, black clay soils (Vertisols) in Andhra Pradesh, and the alluvial sandy loam soil belonging to Inceptisols in the north have such high bulk density layers at shallow depths (Gupta *et al.*, 1984). Soils devoid of any impregnated layers due to $CaCO_3$ concentrations or clay are usually preferred for citrus. The presence of these pans affect both the permeability and aeration of such soils. The time citrus decline due to presence of hard pan in the subsurface is delayed with the increasing depth of the pan. Bhatt (1945), Naidu and Rao

Fig. 6.10. Declined sathgudi orange orchard : Influence of $CaCO_3$ rich hardpan present in the subsurface at Anantapur, Andhra Pradesh

(1958), Kanwar and Randhawa (1959), and Singh and Jawanda (1963) stated that the presence of hard pans due to calcium carbonate or clay is responsible for citrus decline in Bombay, Andhra Pradesh, and Punjab. A soil survey of the citrus belt in Ferozepur, Bhatinda, and Hisar districts (Dhingra *et al.,* 1965; Kanwar *et al.,* 1965; Sehgal *et al.,* 1965) revealed that a high proportion of $CaCO_3$ and the presence of lime concretion layer may be a hazard for successful cultivation of citrus in these districts. Other soil factors found to be responsible for this trouble are hard pan, a thick layer of lime concretion, high total K, $CaCO_3$, high soluble salts, and low exchangeable Ca. Ford (1959) earlier found that roots of cleopatra mandarin (*Citrus reticulata* Blanco) did not grow in the subsoil with red clay when the clay exceeded 27 per cent in the deepest soil layer of most of the orchards.

Soil analysis conducted in the central Florida citrus orchards (Table 6.7) suffering with blight showed that such problem is more of a common occurrence on soils with shallow clay horizon. The clayey horizons are more acidic at clayey subsurface under both healthy and blighted trees, much lower than sand surface, and generally have lower nutrient status than sand overlying the clay. Both sand and underlying clay materials have a high Al saturation and a low (<1%) Ca-Mg ratio. Both P and Ca decreased to sub-optimum levels as depth increased in the sand overlying clayey material. It is suggested that citrus root structure and physiology are weakened by suboptimum nutrient status and toxic Al levels in shallow soils (Nemec *et al.,* 1984). Reitz (1970) indicated that

Table 6.7. Comparison of surface and subsurface soil properties under healthy and blight affected shallow soil sites in central Florida

Soil properties	Tree health			
	Healthy		Blight affected	
Soil type	Surface sand	Subsurface clay	Surface sand	Subsurface clay
pH (H_2O)	5.4-6.4	4.3-5.9	4.8-7.0	4.4-5.5
Organic matter(%)	1.1	0.1-1.5	1.0-1.5	0.1-0.2
Silt + clay (%)	3.1-5.4	23.1-53.7	3.6-8.2	17.1-33.1
Cl (mg kg^{-1})	11-22	15-26	11-17	26-30
NO_3-N (mg kg^{-1})	3-8	2-6	3-8	2-7
NH_4-N (mg kg^{-1})	3-4	1-2	1-3	2
Al (mg kg^{-1})	16-21	18-64	17-30	22-36
P (mg kg^{-1})	4-40	0.2-1.3	4-12	0.2-3.0
K (mg kg^{-1})	15-16	16-45	16-23	16-34
Na (mg kg^{-1})	17-25	16-19	16-17	15-23
Exchangeable cations (me 100 g $^{-1}$)				
Ca	0.4-0.9	0.1-0.9	0.4-0.7	0.2
Mg	0.7-1.8	0.4-2.7	0.6-1.2	0.4-0.6
Al	0.2	0.2-0.7	0.2-0.3	0.2-0.4
Al saturation (%)	9.3	12.9-39.7	9.9-21.6	20.0-32.5

Source : Nemec *et al.* (1984)

young tree decline (blight) is most intensive with rough lemon rootstock (*Citrus limon* L. Burm.f.) on shallow flatwood or marshland soils. In studies carried out on the central Florida Ridge, Nemec *et al.* (1976) reported that as depth to clayey horizons decreased, blight incidence increased. In a later study, Nemec (1983) found that in two of these previous sites, sand over shallow clayey horizons experienced a greater fluctuation in water potential and oxygen than deeper soils. Cohen (1980) showed more severe blight on trees growing in sandy soils than on trees nearby peaty and organic soils. Soils modified with lime and dolomite appeared to have a higher incidence of blight (Cohen *et al.,* 1981) or blight like symptoms (Iley and Guilford, 1978). The soil profiles containing clayey horizons within 0.3-0.9 m of the soil surface, a high percentage of trees have been removed due to blight at Okahumpa (Nemec *et al.,* 1976).

All pans can limit citrus productivity by limiting water, oxygen, and nutrient availability, and restricting the soil volume accessible to roots (Path, 1970). Restriction of water movement is the major problem caused by soil pans. These pans also restrict the free water movement by a reduction in the size and quantity of pores in the soil. Pores are air or water filled spaces between soil particles. Water movement, gas exchange, and root growth occur in these spaces. Any reduction in pore space would directly inhibit these activities. Because of the reduction in pore space in pans, any water that is able to penetrate the pan is present only as a very thin film surrounding the soil particles and can not be readily extracted by plant roots (Lutz, 1952).

Root penetration is severely reduced by impervious soil layers. The ability of a plant root to grow through soil depends on the size of pore spaces present, bulk density or mass of the soil, and the soil strength (Davidson and Hammond, 1977). Soil strength is the activity or capacity of a soil to resist or endure an applied force (Gill and Van del Berg, 1967). The maximum root pressure appeared to be in the range of 1500 Kpa and soils with greater than this strength will greatly reduce root growth (Barber and Gunn, 1977). Roots elongate and grow by following pore spaces and pushing aside soil particles (Davidson and Hammond, 1977). A combination of reduced pore space, increased bulk density, and increased soil strength results in increased mechanical impedance and prevents roots from entering soil pans. Some tillage pans are penetrable when wet, but when pans are dry, roots can generally enter only through cracks formed. These problems, coupled with inadequate aeration and lack of water available to roots in the pan, normally keep most of the citrus roots confined to soil above the pan.

In the presence of perched water table, citrus rooting depth is reduced to several inches above the table because roots will die due to lack of oxygen and the presence of root pathogens under saturated conditions. A strong positive correlation between depth of soil to the water table and the depth confined to 75 per cent distribution of citrus roots has been observed considering the presence of water table within 1.5 feet of the soil surface (Reitz and Long, 1955). In a good soil with no compact zone, most of the feeder roots are in the top two feet of the soil, with the majority being at 15-30 cm depth. High water table can force most roots to the top 15 cm. Plants modify root growth when confronted with hardpans and perched water tables. Feeder roots of navel orange and mandarin will grow horizontally rather than vertically when reaching the water table to avoid entering it (Minessy *et al.,*1971). Root tips may become deformed, as they try and grow through compressed soil layers (Kashirad *et al.,*1966). There is also evidence that roots growing through compressed soil, leak nutrients from root cells in the soil, thus, attracting pathogenic fungi which can further inhibit root growth (Barber and Gunn, 1977). Despite these direct effects of hardpans in general, hardpans become damaging to citrus, only because they indirectly result in anaerobic conditions, drought, nutrient deficiency, saline conditions, and toxic chemical accumulation or root disease. According to Wheaton (1985), distribution of blight affected orchards within localized areas may be related to soil characteristics, but no general association of blight to soil type has been observed.

Effect on Various Properties : The presence or absence of hardpan/claypan/ any other induration within or beyond the root zone alters many soil properties which would eventually affect the growth and productivity in varying proportions depending upon its direct or indirect involvement.

Gas Exchange : The waterlogged soil conditions which can form above pans restrict oxygen diffusion near plant roots. Oxygen can diffuse through air filled spaces about 10,000 times faster than through water filled spaces (Allan, 1977). After only 24 hours of flooding, oxygen available in the soil is depleted by crop roots and soil microorganisms (Allan, 1977). Roots can be severely damaged, if flooding persists for more than a day. The amount of the root system of the citrus tree that succumbs to O_2 deprivation is directly proportional to the length of time orchard is subject to flooding. Longer period of flooding will cause more root death (Hunziker, 1959). Soils under blight affected valencia orange trees had similar concentration of oxygen, carbon dioxide, ethylene, and methane as soils under healthy trees (Bausher, 1983).

Changes in Soil Volume : Presence of hardpans close to the soil surface, reduces the volume of soil that the roots can exploit. The soil becomes nutrient poor more quickly than in areas without pans. If a perched water table develops after irrigation or rainfall, growth volume of soil suitable for root growth is further decreased. The shallow root systems that are forced to develop are subject to more frequent drying. This occurs because roots are concentrated in a smaller area which is rapidly depleted of water, and the roots have no access to water reserves deeper in the soil (Volk, 1954; Reitz and Long, 1955).

Salt Accumulation : Soluble salts are likely to accumulate above the pan when irrigation water can not drain freely through the soil profile. This problem is worst with crops under heavy fertilization and use of saline irrigation water (Volk, 1954). Both fertilizer application and irrigation are usually more frequent in orchards with hardpans, because of the limited volume of soil available to roots. These management problems further compound the impact of salinity.

Toxic Chemicals : Soils that remain saturated will gradually become anerobic as available oxygen is used by roots and microorganisms for respiration. Under anerobic conditions, organic substrates are not broken down completely to carbon dioxide by soil organisms. Instead, imcompletely oxidized chemicals can accumulate that are toxic to roots. These compounds include nitrites, reduced forms of Mn and Fe, hydrogen sulfide, and organic acids (Cannell and Jackson, 1981). Many of these chemicals are present in their toxic forms primarily in acid soils (Cannell and Jackson, 1981). Nitrates can accumulate that can inhibit growth at concentrations of 1 mM (Bingham *et al.*, 1954). Reduced forms of Fe and Mn are produced in acid soils that are more soluble than their oxidized counterparts, normally found in well drained soils.These can cause plant injury when present at concentration of 10 mg kg^{-1} (Jones and Etherington, 1970). Some bacteria produce hydrogen sulfide in warm and flooded soils. This compound may be one of the principle causes of citrus root damage in anerobic soils in California (Ford and Calvert, 1966).

Citrus root systems can be severely damaged when exposed to 2.5-3.0 mg kg^{-1} soluble sulfide at pH 6. Rough lemon, sour orange, and cleopatra mandarin are all susceptible to sulfide damage under these conditions, although rough lemon is better able to regenerate new roots after the sulfide is removed (Culbert and Ford, 1972). Most organic acids are not injurious to roots unless the soil pH is 4.0 or less. The effect of carbon dioxide on roots is still not clear. Although, the carbon dioxide concentration can increase substantially in flooded soils. It is nearly impossible to separate its effects from the effect of reduced oxygen. A well aerated top soil that normally contains about 0.3 per cent carbon dioxide may contain up to 17-20 per cent carbon dioxide when flooded for several days. Even at these concentrations, root damage is probably due to lack of oxygen and presence of toxins rather than carbon dioxide level (Cannell and Jackson, 1981).

Management Strategy : Hardpans can be deep ploughed to break up the pan, allowing root and water penetration. Deep tillage is required in three distinct situations, namely, areas with hard pan e.g. Inceptisols of gangetic plains, presence of perennial weeds e.g. *Cyprus* and *Cynedon*, and textural profiles e.g. Alfisols and related red soils.

Subsoil Management : Any tillage operation which alters any part of the soil profile could be considered as a part of soil profile modification which may or may not be backed up by some form of tillage, chemical or physical amendments (Burnett, 1969). Unger (1979) described three distinct classes of problems viz., problems due to soil layers (plowpans, fragipans, duripans, and claypans, and high clay horizons) associated with reduced root penetration, poor infiltration of water, poor water storage and distribution, restricted drainage, and poor leaching losses; problems caused by undesirable substances on or near the surface causing poor plant growth, poor leaching, poor soil-water relations, and high concentration of undesirable elements, (salts, toxic materials, and radioactive fallout); and problems caused by coarse textured materials (sandy soils) at the surface or to great depths associated with excessive percolation, low fertility, excessive leaching, and high wind erosiveness. Smith (1951) earlier conducted studies on Mexico silt loam (fine, montmorillonitic, mesic Udollic Ochraqualf), a typical clay pan soil near McCredie, Missouri. Subsoiling did not improve the soil because the tillage machines could not penetrate through the plastic clay horizons.

Proton concentration is common in many soils worldwide, particularly those weathering under intensive leaching. Soils of the humid tropic areas are predominantly acidic. Porter and Helyar (1992) estimated a net flow between 20-150 kmol ha^{-1} of protons in the acid soil profiles each year by farming systems in Australia. The nitrogen cycle may contribute to subsoil acidification, if ammonium or urea leaches into the subsoil, and is subsequently convered to nitrate, which then leaches from that layer. The process of proton excretion from roots that accompanies uptake of excess cations relative to anions can acidify the subsoil. The mass flow of acidity (or alkalinity) into or out of subsoil controls the pH of that layer. When the net flow out of topsoil is acidic ($H^+ > OH^- + HCO^-_3 + CO_3^{2-}$), protons accumulate in the subsoil under the absence of buffering materials such as lime. The toxic effects of adsorbed Al and Mn from the subsoil significantly affect the yield.

Deep tillage has met with mixed success as a method of ameliorating subsoil, a physical limitation to root growth. In many cases, benefits are transient and variable (Eck and Unger, 1985). This in large part stems from not understanding the basic factors influencing the effetiveness of the modification and the subsequent maintenance of structural stability. Attempting to ameliorate a physical problem without correcting an underlying chemical cause is a common mistake. An example is the failure of deep ripping of subsoil without ameliorating sodicity (Rengaswamy *et al.,*1992).

Necessary steps in deciding a particular specification for modifying a profile involve a strategic decision (Fig. 6.11). Economic factors largely decide the choice of the crop. The crop will then require particular level of water and nutrients for potential yield. Rooting patterns also vary between crop species. Root requirements and subsoil limitations on these then determine the modification required. For example, the depth of soil to be modified is a function of both economic and root factors, and possibly depth to a drainable layer. Sodicity within this depth interval will determine calcium requirement for effective tillage. Drainage requirement will depend on the profile hydraulic characteristics. The optimum water content at tillage is an important factor in the success of tillage. The

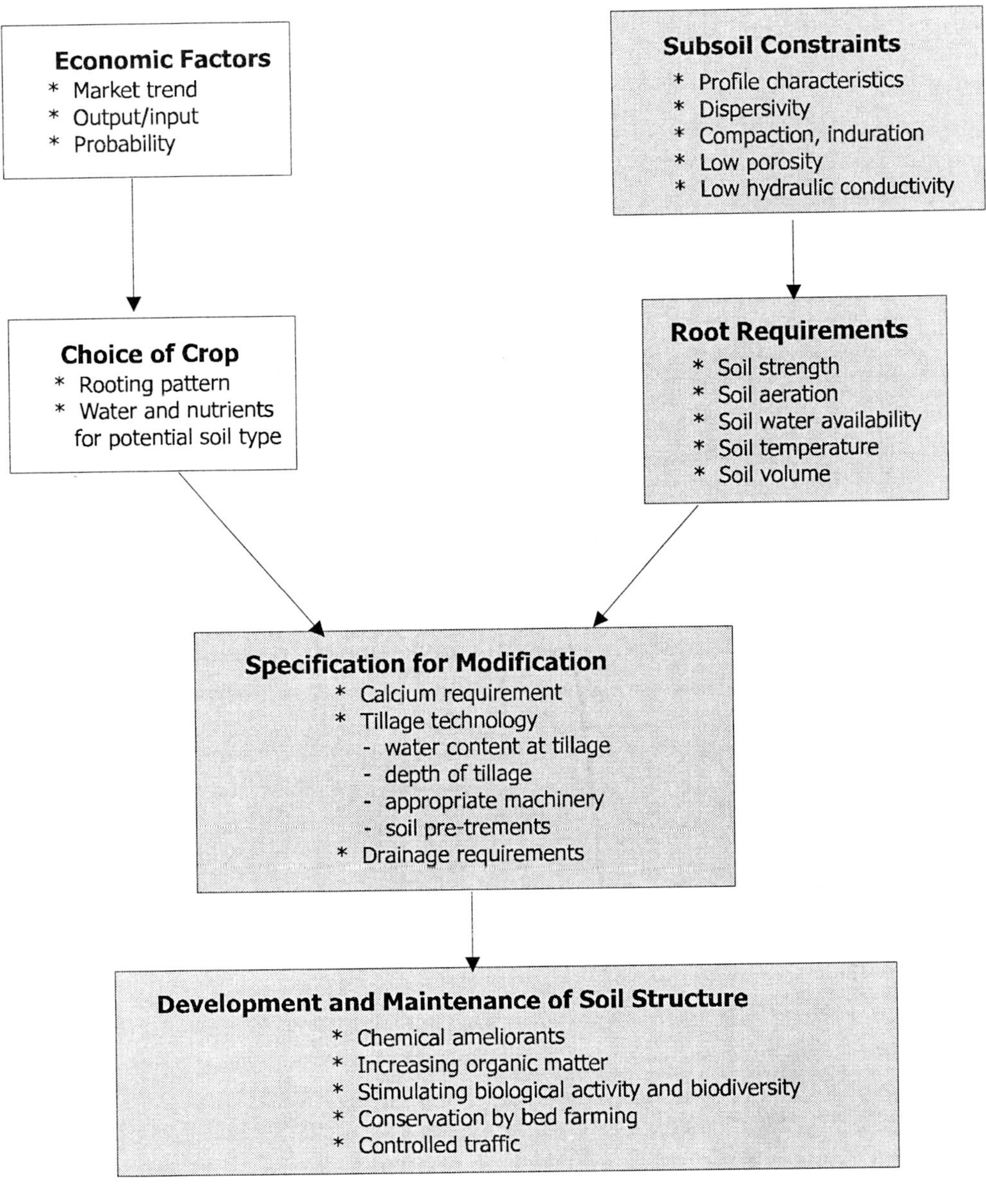

Fig 6.11. Schematic pathway for alleviating subsoil constraints

subsequent development of structure and its maintenance over time are essential for sustained productivity. Profitable avenues include chemical ameliorants such as lime and gypsum, increasing soil organic matter, stimulating biological activity, and biodiversity. Bed farming promises to conserve the improved structure of the modified profile (Rengaswamy, 2000).

Studies conducted to determine the value of deep ploughing with lime incorporation to improve soil flocculation showed that citrus responded favourably to these treatments. Kawamura *et al.* (1973a; 1973b) and Calvert *et al.* (1977) examined the effect of deep tillage on the physical and chemical properties of the soil in a satsuma and navel orange orchard soil derived from granite, andesite, and Izumi sandstone. The beneficial effect of deep tillage on soil structure persisted for 10-21 years in former soil type and for 15-21 years in latter soil type. Accordingly, these studies suggested a minimum organic matter requirement of 1 per cent for satisfactory growth of fine roots. More roots are produced at all soil depths, particularly in the lower depths. These trees are also better able to tolerate drought (Bryan, 1962). This is an expensive and only temporary remedy. However, it is not possible once a orchard is established. One recommended approach to control surface or subsurface flow from outside the orchard is the use of an interceptor drain which captures the water before it enters the soil. Effect of deep tillage as physical amendment has been demonstrated to produce beneficial effects on acid soils derived from granite and andesite (Kawamura *et al.*, 1973a) and Izumi sandstone (Kawamura *et al.*, 1973b).

The practical device, if one is planting a new orchard is to adapt the ridging or mounding system. In this system, the soil is ridged or mounded to height of two or more feet and the trees are planted on the crest of these. In this way, root area sufficiently created high above the water table permits healthy root development. This method also provided for good surface drainage. There are many other situations, in which no permanent water table exists. But, due to clay layer or hardpan, the movement of free water downward is very slow and nearly saturated conditions exist for a considerable period. This situation may be especially bad during the wet season of the year such as the summer monsoon in India.

If a new orchard is being contemplated, there are soil situations in which it is possible to deep plough a soil, break up, and mix with lighter soil horizons, the heavy clay or compacted layers. But in existing orchards or situations, where such ploughing is not feasible, the question is what can be done to improve the rate of water movement through soils (Amy *et al.*, 1986). Sometimes, the growth of interplanted crops during the season will help remove the excess moisture, the use of weeds or cover crops with deep growing tap roots is equally helpful. As the old tap roots decay, they provide vertical channels which aid in water movement and better aeration.

Vertical mulching is a device which has important possibilities. Chemical soil treatment with gypsum has sometimes proved helpful. Surface organic matter mulches are also sometimes effective. Surface organic matter mulches are good in encouraging shallow rooting of citrus by stimulating earthworm activity and channeling and improving the downward water movement. Improvement in surface drainage by leveling and providing water run ways, so that excess water can be drained off the orchard rapidly, will often prove helpful, besides ridging as previously referred is another very useful device.

Under the condition of water originating within the orchard itself and fails to drain or drains at an unacceptable rate through the profile, installation of a tile system is one of the feasible options. Clay or concrete pipes were formerly used. Today, flexible plastic pipes are generally recommended. Placement of tile is critical to provide the flow through the lines and eventual placement of the mainline. Because, the tile allows entrance of water, plugging may be a problem. So an envelope of gravel or sand is often used to minimize this hazard. With the modern drip or mini sprinkler systems, water can be precisely delivered and monitored. Using calculations for evapotranspiration, water use by the trees can be estimated accurately. If water is applied frequently, but only in the amounts to be used rapidly by the trees, soil saturation above the hardpan can be held to a minimum (Lety *et al.,*1967). Water will not gather on the surface of the pan and anerobic conditions and toxic compounds in the soil are unlikely to accumulate. Performance of Meyer lemon trees grown directly over the drains has been observed better compared to either trees under soil drains spaced 3 cm apart or on raised beds 10 m in width (Bukhbinder, 1990; Bukhbinder and Kuppa, 1991).

6.5.2 Nutritional Disorders

The deficiency of different major and minor elements may be important in causing decline of citrus trees. Deficiency may be caused either due to the nonavailability of the nutrients in the soil or at times, they may be made unavailable due to adverse soil conditions. Soil reaction is one of the most important factors. Chapman *et al.* (1939) observed this in respect of manganese. Smoyer (1951) suggested under such conditions, Fe chlorosis may also be observed due to excessive lime, especially when P is high in soil. Interrelation among the nutrients is also important, e.g., Mg deficiency occurs in soils containing high concentration of K. Factors affecting root growth also have pronounced effect on the occurrence of deficiency especially of Mg and Zn. Nair and Mukherjee (1970b) observed the seasonal variation in Zn deficiency with: i. seasonal variation in the root growth in the first foot for soil which contain the most available Zn, ii. variation in demand for nutrient in various growth flushes, and iii. variation in the available Zn in soil in different seasons

Nutritional disorders with citrus have been earlier studied under sand culture or water culture by several workers like Haas (1932; 1939), Haas and Klotz (1931), Haas and Quayle (1935), and Chapman (1938). It may be concluded from these studies that soil factors such as defective drainage, poor aeration, lack of organic matter as observed by earlier workers in fact cause disturbances in the nutritional balance of the plant by affecting the intake of various nutrients, though in some cases, various available nutrients are actually far below their critical requirement in the soil. Deficiencies of Zn, Cu, Fe, and

Mn have been reported from different parts of India (Singh and Singh, 1953; Choudhary, 1954; Ramakrishnan, 1954b; Govindarao and Reddy, 1957; Aiyappa *et al.,* 1959; Singh, 1963). Studies by Mukherjee (1949), Kanwar and Randhawa (1959; 1960), Randhawa *et al.* (1961; 1966), Kanwar *et al.* (1965), Mukherjee and Raturi (1969), and Nijjar and Singh (1971) revealed the large scale deficiency of available N, P, K, Fe, Mn, and Zn. Patel *et al.* (1997) reported large scale chlorosis in loamy sand soils of Navsari, Gujarat due to severe deficiency and low active Fe influenced by high soil pH due to regular use of HCO_3 containing irrigation water.

The chlorotic condition of the citrus trees on account of nutritional disorders has been reported from many other parts of India viz., Punjab (Mukherjee, 1949; Kanwar and Randhawa, 1960; Kanwar and Dhingra, 1962; Kanwar *et al.,* 1962; Dhingra and Kanwar, 1963; Dhingra *et al.,* 1966), Mysore and Madras (Mukherjee, 1949), Uttar Pradesh (Singh and Singh, 1953; Singh, 1963), Assam (Choudhary and Dutta, 1950; Choudhary, 1954), and Andhra Pradesh (Reddy and Rao, 1962; Reddy and Sharma, 1981; 1982; Sharma *et al.*, 1981). Occurrence of exanthema due to Cu deficiency has been observed through a descriptive survey of citrus orchards conducted by Fruit Research Station, Saharanpur, U.P., India. Mandarins (*Citrus reticulata*) have shown their susceptibity to lime induced chlorosis (Singh and Singh, 1953; Singh, 1963). Choudhary (1954) earlier reported zinc and Fe deficiency symptoms with mandarin in Assam and obtained the response of zinc and Fe application as foliar spray more affective than soil application of these elements. West (1938), Camp *et al.* (1941), and Labanauskas *et al.* (1959) reported that in chlorotic plants, the deficiency of micronutrients is induced by excess P and K fertilization. Higher uptake of P and Mn (Kanwar and Randhawa, 1960) and P and K (Kanwar and Dhingra, 1961) have been reported to be responsible for the chlorotic condition of the trees in Punjab. These elements resulted in a deficiency of Zn and Fe also in the plants. This is supported by the fact that the deficiencies of exchangeable Cu, Zn, and Fe have been reported in a survey of the declining citrus soils of Punjab (Dhingra and Kanwar, 1963). Wutscher and Hardesty (1979) found essential difference in NH_4^+- N and NO_3^- -N in the rootzone of healthy and diseased trees over a two years period on a relatively deep soil. Nemec *et al.* (1982) reported higher N levels at the dripline compared to around the trunk of healthy trees in orchards with blight shortly after fertilization than 2-3 months after fertilization.

Soil fertility status viz., available K, Ca, Mg, and Na did not differ under blight-affected and healthy trees (Wutscher *et al.,* 1982a; 1982b). No cause and effect relationship is claimed for the imbalance in the cation/anion ratio. A recommended balance in the exchange complex having about 67-75 per cent Ca and 10-15 per cent Mg, or a ratio of 5:1 to 10:1 is usually good for healthy status of citrus trees (Ankerman and Large, 1979). Calcium deficiency is more likely to occur in sandy acid soils with low Ca content (Chapman, 1968). Deficiency level of Ca can result in poor calcium pectate development in cell walls of fibrous root, causing them to be more susceptible to weakly parasitic and pathogenic fungi (Bateman and Basham, 1976). Since Nauriyal *et al.* (1970) reported that citrus die back in Punjab, India to large scale due to severe Ca deficiency.

Single nutrient deficiency is seldom met with under field conditions. Multiple deficiencies are usually met with e.g. deficiency of Zn, Mn, and Fe may be observed in conjunction. Symptomic observations recorded in Coorg tract and erstwhile Bombay state revealed not only one type of symptoms, but symptoms resembling several nutrient deficiencies or excesses. Continuous use of inorganic fertilizers also result in appearance of deficiency symptom. With continued use, the conditions of the trees deteriorated due to imbalance, and the soil became depleted of other elements. Under ordinary conditions, there is a balance between the fertility of the soil and the amount of crop produced. In Florida, it has been observed that continuous use of NPK to obtain higher production has resulted in the deficiency of Mg, Cu, and Zn. It is observed that whenever production is low, large quantities of organic manures are used, the trees do not show the deficiency symptoms. But, as soon as higher yields are harvested, especially by the use of inorganic fertilizers, there is a tendency for development of various types of deficiency symptoms. Examination of die back affected satsuma trees on *Poncirus trifoliata* rootstocks at Izmir region of Turkey showed that all sites under investigation had soil available Zn less than 1 mg kg^{-1} (Ercivan and Karaca, 1979).

Excess absorption of Mn by root is attributed to cause of abnormal defoliation in satsuma mandarin (Kanki and Imamura, 1969; Aoki and Morita, 1969a; Ishihara *et al.*, 1971c). Leaf spotting in satsuma mandarin following the application of high proportion of N and K (Ishihara *et al.*, 1971b), and Mn (Ishihara *et al.*, 1971a) has been observed which lowered the soil pH to enable increase in soluble Mn in soil (Ishihara *et al.*,1971b). Wutscher (1986) observed higher double acid extractable P, K, Fe, Mn, Zn, Cu, and lower Ca in soils from blight free area compared to blight affected areas of Florida. Comparison of soil fertility under healthy and blight affected orange trees in Florida indicated highest level of K, Ca, and Mg in the rhizosphere in blight affected trees. While, higher K and lower Fe, Mn, and Zn have been recorded in feeder roots of blighted trees than those of healthy trees (Pavan and Wutscher, 1993).

Total Zn, Fe, Mn, and Cu have been observed to be highly varying, independent of declined or healthy condition of orchards. However, variation in these nutrients may depend upon parent material and composition of primary and secondary minerals from which the soils are derived. Comparison of total versus available micronutrient in sweet orange (cultivar mosambi) orchards of Marathwada region of Maharashtra showed that the orchard soils contained 0.72 to 1.40 mg kg^{-1} available Mn and 15.0 to 23.0 mg kg^{-1} available Zn. These values are invariably higher in healthy orchards as compared to declined ones. Three out of four declined orchards can be categorised as low in available

Table 6.8. Total and available micronutrient status of healthy and declined citrus orchards soils of Marathawada, Maharashtra, India

Location	Orchard condition	Total micronutrient (mg kg^{-1})				Available micronutrient (mg kg^{-1})			
		Zn	Fe *	Mn	Cu	Zn	Fe	Mn	Cu
Aurangabad	Declined	164	6.60	1196	128	1.40	9.8	21.00	4.84
Parbhani	Medium	96	6.24	1168	128	1.00	14.4	16.92	2.72
Parbhani	Declined	96	5.44	1098	108	0.60	13.2	16.00	4.08
Parbhani	Healthy	104	5.52	1140	128	1.24	9.4	23.00	2.40
Aurangabad	Declined	124	7.00	1168	196	0.72	20.6	15.25	8.84
Parbhani	Healthy	116	7.52	1768	196	1.00	16.0	22.50	6.72
Parbhani	Medium	104	5.76	1070	144	1.16	11.6	22.50	5.44
Parbhani	Declined	152	5.28	1698	176	0.90	10.6	15.0	4.08
Limbgaon,Nanded	Medium	116	5.60	1354	128	1.40	11.0	21.5	4.08

* Fe in per cent

Source : Malewar *et al.* (1983)

Zn content. Lower availability of Zn in declined orchards may partly be attributed to low organic carbon and partly to high P buildup (Table 6.8). Difference in soil fertility (Total and available N, P, K,S, Fe, Mn, Cu, and Zn) in sweet orange orchards of Agra region, Uttar Pradesh revealed not much difference between healthy and chlorotic trees (Table 6.9). However, comparing the leaf analysis values further indicated that the chlorosis of

Table 6.9. Average content of total and available macro and micronutrient elements in twenty soil profiles of sweet orange orchards of Agra region of Uttar Pradesh

Nutrients	Healthy					Severely chlorotic				
	0-30	30-60	60-90	90-120	120-180	0-30	30-60	60-90	90-120	120-180
Nitrogen	0.07*	0.05	0.04	0.04	0.03	0.06	0.04	0.04	0.03	0.03
	(94.8)**	(73.6)	(61.6)	(53.0)	(45.0)	(91.4)	72.3	61.6	52.6	41.7
Phosphorus	0.05	0.04	0.04	0.04	0.03	0.05	0.04	0.04	0.04	0.04
	(17.8)	(14.9)	(13.0)	(11.8)	(10.1)	(17.1)	15.3	12.9	12.0	10.8
Potassium	1.6	1.5	1.7	1.6	1.7	1.4	1.5	1.6	1.5	1.5
	(95)	(91.2)	(88.1)	(89.2)	(87.3)	(94.8)	89.5	89.6	84.0	80.2
Sulphur	156	140.3	154	141.4	135.8	152.6	148.1	139.3	134.6	130.1
	(22.7)	(18.3)	(16.9)	(15.3)	(13.6)	(20.9)	17.6	16.0	13.7	12.0
Iron	1.72	1.86	2.25	2.24	2.41	1.74	2.04	2.20	2.30	2.38
	(5.96)	(6.68)	(8.31)	(8.31)	(8.57)	(6.11)	8.73	10.40	11.30	11.98
Manganese	380	393	433	467	445	381	398	412	451	432
	(24.34)	(23.28)	(21.0)	(18.26)	(15.13)	(23.22)	21.28	18.51	14.78	11.34
Copper	22.02	20.40	18.90	16.95	14.85	21.04	19.91	18.0	17.21	15.94
	(2.37)	(2.03)	(1.68)	(1.49)	(1.37)	(2.30)	1.90	1.32	1.58	1.49
Zinc	32.31	28.38	27.69	23.31	23.31	28.19	24.94	22.89	20.88	16.69
	(2.92)	(2.16)	(1.96)	(1.62)	(1.57)	(2.34)	1.71	1.36	1.15	0.98

* Total nutrients in per cent; ** available nutrients in mg kg^{-1}

Source : Vinay Singh and Tripathi (1985)

sweet orange in the Agra region is mainly due to Zn deficiency caused by antagonistic effect of Fe on Zn availability. The critical level of Zn in citrus leaves for clear distinction between healthy and chlorotic plants appeared to be 20 ppm. Generally, the healthy green leaves had lower concentration of P and Fe, and higher level of N, Ca, Mg, and Mn than chlorotic ones. The concentration of K, S, and Cu more or less remained in the same range whether healthy or chlorotic.

Comparison of fertility status of healthy and declining citrus orchards of Karbi Anglong and North Cachar hill districts of Assam showed that soils irrespective of declining or healthy are acidic (pH 4.4-6.0) and deficient in available P_2O_5 (1-16 kg ha^{-1}). The EC is low (0.030-0.406 dS m^{-1}). Most of the soils have high organic matter content to a depth of 45 cm (1.52-6.0 per cent). Available N and K_2O are medium to high in the surface layers (220-568 and 246-820 kg ha^{-1}, respectively). The magnesium content tended to be higher in the subsurface than surface layer in Karbi Anglong (Table 6.10). In respect of organic matter and all major nutrients (N, P, K, Ca, and Mg), citrus orchards from North Cachar hills are richer than that of Karbi Anglong. In Karbi Anglong, the declining orchards had higher soil available N, P, Ca, and Mg. But, these values are lower compared to healthy orchards.

6.6 AMELIORATION OF SOIL FERTILITY CONSTRAINTS

The inherent low organic matter and reduced micronutrients availability due to high pH becomes a major limitation. Antagonistic relationship between the micronutrients like Fe-Mn, Fe-Zn, Cu-Mn etc. makes detection and correction of the nutritional disorders further difficult (Dhatt and Dhiman, 1999). Amelioration of citrus decline, though, involves a string of ameliorative techniques, in most of the cases, it is a complex problem, since the origin of which is due to the contribution of one or more than one factor.

Table 6.10. Comparison of soil fertility of healthy versus declining citrus orchards from Karbi Anglong and North Cachar hills of Assam

Status	pH (1:2.5)	EC (dS m^{-1})	Organic matter (%)	Available nutrients (kg ha^{-1})				
				N	P_2O_5	K_2O	Ca	Mg
			Karbi Anglong					
Declining	4.89	0.095	3.00	372	6	229	305	117
Healthy	4.87	0.055	2.21	270	6	342	101	33
Difference (%)	4	72	36	38	0	-33	202	255
			North Cachar Hills					
Declining	5.10	0.121	3.87	427	6	401	173	58
Healthy	5.21	0.135	4.21	455	8	534	415	67
Difference (%)	-2	-10	-8	-6	-25	-25	-58	-13

Source : Dey and Singha (1998)

6.6.1. Macronutrient Management

Bhatt (1945) suggested several cultural practices as control measures. Various recommendations emerged from major citrus belts (Table 6.11, 6.12, and 6.13) could be used for adapting a sound fertilizer schedule. Much earlier, Sahastrabuddhe (1927),

Table 6.11. Fertilizer schedule for various citrus cultivars in major citrus belts of northwest, east and northeast India

Citrus species with spacing	Age (years)	Fertilizers application (g tree $^{-1}$)			FYM (kg)
		N	P_2O_5	K_2O	
Himachal Pradesh					
Citrus	> 5	800	500	600	20
Jammu and Kashmir					
Citrus (6 x 6 m)	1-5	80	10	15	5
	10 and above	800	70	125	50
Punjab					
Citrus (6 x 6 m)	1-3	50-150	18-72		
	4-6	200-250	90-180	-	-
	7-9	300-400	215-325	-	-
Delhi					
Mandarin (6 x 6 m)	> 5	500	400	750	60
Rajasthan					
Mandarin (6 x 6 m)	—	290	200	240	100
Lime/Lemon(4.5 x 6 m)	—	290	290	200	50
Bihar					
Citrus (6 x 6 m)	1	300 g	—	—	10
West Bengal					
Citrus (4.5 x 5 m)	1	50	50	50	10
	6	400	400	500	50
Orissa					
Sweet orange (7 x 7 m)	1	90	230	90	10
	2	108	275	120	250
Mizoram					
Sweet orange (3 x 5 m)	3	300	240	200	10-15
	4	450	350	300	10-15
	7	900	720	600	10-15
Arunachal Pradesh					
Lemon (5 x 5 m),	5	350	350	600	40
Mandarin (4 x 4 m), and	6	400	400	700	45
sweet orange (6 x 6 m)	7	450	450	900	50
Assam					
Lemon (3 x 3 m) and	3	300	240	200	-
Mandarin(5 x 5 m)	6	750	600	500	-

Source : Tandon (1987)

Table 6.12 Fertilizer schedule for various citrus cultivars in major citrus belts of central and south India

Citrus species with spacing	Age (years)	Fertilizers application (g tree $^{-1}$)			FYM (kg)
		N	P_2O_5	K_2O	
Madhya Pradesh					
Mandarin	6	400	200	300	50
Maharashtra					
Sweet orange(6 x 6 m)	Bearing trees	1000	100	200	24-30
Lime/Lemon (6 x 6 m)	Bearing trees	800	100	200	100
Kagzi lime (6 x 6 m)	5 and above	900	400	0	25-30
Mandarin (6 x 6 m)	6-9	720	360	0	25-30
	10 and above	1000	500	0	30-50
Andhra Pradesh					
Acid lime (6 x 6 m)	4 and above	1500	600	800	-
Sweet orange (8 x 8 m)	5 and above	1500	350	400	-
Karnataka					
Lime/Lemon (5 x 5 m)	> 5	242	145	242	5
Mandarins & Sweet orange (6 x 6 m)	5 and above	550	370	550	30

Source : Tandon (1987)

Cheema and Bhat (1929) referred the beneficial effect of nitrogen application in the Punjab, woodash in Gujarat, farmyard manure in the medium black calcareous soils of the Deccan, and application of zinc in the Punjab and Gujarat.

6.6.2. Micronutrient Management

Most of the micronutrients, singly or in combination have been found responsible for citrus decline. Deficiency of these micronutrients is conditioned by many of unfavourable soil factors. Foliar sprays of micronutrients have been found to be curative for such deficiencies. Mottle leaf in citrus was earlier controlled by spraying zinc compounds (Parker, 1934; 1936; 1937; West, 1938).

To evolve a more effective control measure, Bhatt (1945) stressed the importance of investigation on specificity of microelements concerned, suitability of rootstocks, and cultural schedules under different conditions to arrive at more meaningful ameliorative strategy to obtain enduring effect over a period of time which is really economical. Some efforts have been made to control citrus decline problem with macro- and microelements. Work at Coorg (Anonymous, 1964a) revealed that the spraying with a mixture containing 5 pounds of $ZnSO_4$ plus 2 ½ pounds of lime in 53 gallons of water proved effective in controling the incidence of chlorosis. Aiyappa *et al.* (1959) in Coorg found improvement in the health of the trees when sprayed with $ZnSO_4$, singly or in combination with other

Table 6.13. Macronutrient recommendations for various citrus cultivars across citrus belt of world

Doses of application	Source
N (1200 g tree^{-1}), (680 g tree^{-1}), (400 g tree^{-1}), (176 kg ha^{-1}), (400 g tree^{-1}), (900 g tree^{-1}), (400-800 g tree^{-1}), (225 kg ha^{-1}), (1.03 kg tree^{-1}), (125 g tree^{-1}), (148 kg ha^{-1}), (500 g tree^{-1}), (450 g tree^{-1}) Ammonium nitrate tree^{-1}, (240 kg ha^{-1}), (270 kg ha^{-1}), (330 kg ha^{-1}), (320 kg ha^{-1}), (180 kg ha^{-1}), (240 kg ha^{-1}), (112 kg ha^{-1}), (560 g tree^{-1}), (1.5 kg tree^{-1}) + FYM (45 kg tree^{-1}), (900 g tree^{-1})	Govind and Prasad (1976), Koen *et al.* (1977), Lee and Chapman (1988), Hong and Chung (1979), Mllella and Deidda (1976), Mungomery *et al.* (1980), Putcha and Prasad (1976), Davis (1979), Liu (1992), Aso *et al.* (1987), Hipp (1977), Dzhincharadze (1989), Lominadre and Tsanava (1990), Cheong and Moon (1989), Vinnik (1949), Oppenheimer and Heyman (1954), Dasberg *et al.* (1983), Bar-Akiva *et al.* (1973), Amir *et al* (1981), Zhang *et al.* (1998), Liu (1995), Upadhyay and Patiram (1996b), Bicer and Ozel (1986)
500 g N tree^{-1} (first irrigation) + 340 g N (21 days after) + 8 foliar sprays (2% urea at 15 days interval)	Desai *et al.*(1986b)
P (22.5 kg ha^{-1} as monocalcium phosphate), (85.5 kg ha^{-1})	Bingham and Martin (1955; 1956)
K (850 kg ha^{-1}), (410 kg ha^{-1}), (223 kg ha^{-1}), (750 kg ha^{-1}), (1000 kg ha^{-1})	Page *et al.* (1969), Reese and Koo (1975), Plessis and Koen (1984), Bazelet *et al.* (1980)
0.7% urea + 5% KNO_3 (2 sprays)	Intrigliolo *et al.* (1991)
400-600 kg N + 180-200 kg P_2O_5 + 300-400 kg K_2O ha^{-1}	Hernando (1978)
180 kg N + 90 kg P + 180 kg K	Rodriguez (1980)
500 g N +100 g P_2O_5 + 400 g K_2O tree^{-1}	Ghosh (1990)
300 g N +250 g P_2O_5 + 300 g K_2O tree^{-1}	Ghosh *et al.* (1984; 1989)
125 g N + 175 g P + 100 g K tree^{-1}	Hong and Chung (1979)
160 g N + 320 g P + 480 g K tree^{-1}	Hernandez (1981)
2.72 kg N + 1.81 kg P + 0.60 kg K tree^{-1}	Reddy and Swamy (1986)
450 kg N + 30 kg P + 180 kg K ha^{-1}	Bevington (1984)
22.5-25 kg N + 5-12.5 kg P_2O_5 + 10-12.5 kg mu	Wang (1985)
625 kg N + 525 kg K_2O ha^{-1}	Tucker *et al.* (1990)
400 g N + 300 g K_2O tree^{-1}	Zhou *et al.* (1996)
600 g N + 135 g P + 285 g K tree^{-1}	El-Hagah *et al.* (1983)
800 g N+ 170 g P_2O_5 + 391g K_2O tree^{-1}	Desai *et al.* (1986a)
750 g N + 200 g P_2O_5 + 500 g K_2O tree^{-1}	Ahmed *et al.* (1988)
400 kg N + 200 kg K_2O ha^{-1}	Sharma *et al.* (1990)
1500 g N + 400 g P_2O_5 + 750 g K_2O tree^{-1}	Maatouk *et al.* (1988)
500 g N + 250 g K_2O tree^{-1}	Singh and Misra (1985)
500 kg N + 100 kg P + 100 kg K ha^{-1}	Kacharava (1985)
100 kg N + 200 kg P_2O_5 + 300 kg K_2O ha^{-1}	Goepfert *et al.* (1987)
475 g N + 320 g P_2O_5 + 355 g K_2O tree^{-1}	Koseoglu (1995b)

Doses of application	Source
100 g N + 25 g P_2O_5 + 50 g K_2O tree^{-1}	Sharma and Singh (1989)
250 g N + 250 g P_2O_5 + 500 g K_2O tree^{-1}	Kannan *et al.* (1989)
600 g N + 200 g P_2O_5 + 300 g K_2O tree^{-1}	Jawaharlal *et al.* (1989)
500 g N + 100 g P_2O_5 + 400 g K_2O tree^{-1}	Ghosh (1990)
600 g N + 150 g P_2O_5 + 600 g K_2O tree^{-1}	Dris (1997)
240 g N + 40 g P_2O_5 + 100 g K_2O ha^{-1}	Pedrera *et al.* (1988)
800 g N + 200 g P_2O_5 + 400 g K_2O tree^{-1}	Mann and Sandhu (1988)
600 g N + 200 g P_2O_5 + 200 g K tree^{-1} (Soil application) + 0.4% Cu + 0.5 % Zn (Foliar spray)	Kar *et al.* (1988)
1 kg N + 0.5 kg P + 0.5 kg K tree^{-1}	Gilani *et al.* (1989)
100 g N + 50 g P_2O_5 + 50 K_2O tree^{-1}	Sharma and Azad (1991)
1.4 kg N + 1.08 kg P + 1.1 kg K tree^{-1}	Chundawat *et al.* (1991)
450 kg N + 0-180 kg P + 0-30 kg K ha^{-1}	Sarooshi *et al.* (1991)
120 kg N + 150 kg P + 75 kg S + 6 kg Cu + 0.8 kg Mo + 5.0 kg Zn ha^{-1}	Lim *et al.* (1993)
180 g N + 90 g P_2O_5 + 45 g K_20 + 800 CaO tree^{-1}	Aubert and Vullin (1998)
200 kg N + 140 kg P + 210 kg K ha^{-1}	Cantarella *et al.* (1992)
0.5 kg N + 0.5 kg P_2O_5 + 1.0 kg K_2O ha^{-1}	Androulakis *et al.* (1992)
1.02 kg N + 0.58 kg P_2O_5 and 0.55 kg K_2O tree^{-1}	Liu *et al.* (1994)
420 g N + 323 g P_2O_5 + 355 g K_2O tree^{-1}	Koseoglu *et al.* (1995a; 1995b)
1.5 kg urea + 0.25 kg superphosphate + 1.25 kg potassium chloride + 1.1 kg $MgSO_4$ + 0.10 kg $ZnSO_4$ tree^{-1}	Yin *et al.* (1998)
$MgSO_4$ + $Ca(NO_3)_2$ (4.5 kg 100 gallon^{-1} each), $MgSO_4$ (3-5 kg tree^{-1})	Strauss (1963), Servicio Para El Agricultor (1974a; 1974b)

micronutrients. In Assam, Choudhary (1954) obtained more response of zinc and iron sprays to deficient mandarin trees than soil application of these elements.

On the basis of the survey conducted (Kanwar and Randhawa, 1960; Kanwar *et al.,* 1962; Dhingra and Kanwar, 1963) in Punjab, use of micronutrient sprays controlled the incidence of chlorosis at many places. Kanwar and Dhingra (1961) in a field trial at Patiala observed the treatment of zinc and zinc plus iron is highly effective in reducing the incidence of decline. They also recommended two sprays of $ZnSO_4$ (0.5 per cent) containing 3.6 kg $ZnSO_4$ and 1.8 kg lime in 450 litres of water, to be given first in the middle of March and later in August and September. Effectiveness of zinc sprays in conjunction with copper and manganese (0.5 plus 0.4 per cent) has also been found equally effective (Anonymous, 1965a; 1965b).

In South India, Marudarajan (1949) and Aiyappa *et al.* (1959) found improvement in citrus trees mottling with foliar zinc sprays. In the Telengana region of Andhra Pradesh, Reddy and Rao (1962) observed reduction in the intensity of chlorosis in the declining mosambi trees at Pargi, with a foliar spray containing a mixture of $MgSO_4$ (0.5 per cent) and $CuSO_4$ (0.5 per cent) neutralized with lime. Application of magnesium sulphate also gave a distinct response, showing thereby, that the citrus trees in this region, are suffering from acute Mg deficiency. In another orchard at Nar Singi, mosambi trees revived following the spray with $CuSO_4$ (0.5 per cent) and $ZnSO_4$ (0.5 per cent) neutralized with lime.

Complete recovery of chlorotic symptoms is possible till the leaves are three months old with one spray of $ZnSO_4$ (0.5 per cent). However, recovery is possible only up to 60 to 75 per cent, when the leaves are four to six months old by two repeated sprays. But, no recovery has been found when the leaves are more than eight months old. Large scale field trials with zinc and manganese sprays, however, failed to ameliorate the condition completely in Coorg (Mukherjee, 1949). Numerous workers (Dikshit, 1958; 1958b; Singh and Agarwal, 1961 and; Mani *et al.,* 1959; Varasi, 1962) demonstrated the complete recovery from chlorosis through micronutrients sprays in different combinations. A number of micronutrient recommendations (Table 6.14 and 6.15) for various commercial

Table 6.14. Foliar spray recommendations of various micronutrients from citrus belts of world

Foliar spray	Source
Fe-E DDHA (0.1%), (7 mg Fe l^{-1})	Primo *et al.* (1970), Zude *et al.* (1999)
Fe-polyflavonoid (1%)	Fernandez-Lopez *et al.* (1993)
Fe-EDDHA (22-352 g $tree^{-1}$), (250-500 ppm)	Alva and Obreza (1998), El-Kassas *et al.* (1987)
Fe-EDDHMA or Fe-EDDHA (225 g $tree^{-1}$)	Davenport (1984)
$FeSO_4$ (2.5% as trunk injection)	Carpena *et al.* (1968)
Fe + Zn (0.5 each, 4 sprays)	Patel and Patel (1985)
Fe (50 ppm) + Mn (5 ppm) + Zn (75 ppm)	Hassan (1995)
$FeSO_4$ + $ZnSO_4$ (0.5% each)	Dixit *et al.* (1977; 1978; 1979)
Fe (2.2-4.4 mg l^{-1}) + Mn (0.22-0.44 mg l^{-1}) + B (1.0-3.0 ppm) + urea (2.7 kg 100 $gallon^{-1}$)	Badawi *et al.* (1979), Mann and Takkar (1983a)
$FeSO_4$+$MnSO_4$+$ZnSO_4$ (25g l^{-1} each)+$CuSO_4$ (1g l^{-1})	Al-Juburi and Al-Mesry (1991)
Fe+Mn+Zn (0.5% each)	Manchanda *et al.* (1971; 1972), Sharma *et al.* (1974), Manchanda (1974), Daulta *et al.* (1986)
Fe + Mn + Zn (1.2% each)	Maksoud and Khalil (1995)
$MnSO_4$ (0.3%)	Koto and Takeshita (1957)
Mn+Zn (1.25-2.5 kg 417.5 $gallon^{-1}$ ha^{-1})	Alva and Tucker (1992)
$MnSO_4$ + $ZnSO_4$ (0.15%)	Razeto *et al.* (1988)
$MnSO_4$ + $ZnSO_4$ (450 g 100 $gallon^{-1}$)	Labanauskas *et al.* (1972a)
$MnSO_4$ + $ZnSO_4$ (2.3 kg 100 $gallon^{-1}$)	Subramanian (1960)
$MnSO_4$ + $ZnSO_4$ (0.15% each)	Razeto *et al.* (1988)
$MnSO_4$ (300 g 100 l^{-1}) + Na_2CO_3 (150 g 100 l^{-1})	Russo and Raciti (1954)
Mn+Zn (450 g 100 $gallon^{-1}$ water)	Labanauskas *et al.* (1969), Labanauskas (1962), Beutel (1962)
$MnSO_4$ + $ZnSO_4$ (0.5% each) + $CuSO_4$ (0.25%)	Nanaya *et al.*(1985)

Foliar spray	Source
$MnSO_4$ (0.45 kg 378.5 l^{-1}) + $ZnSO_4$ (0.23 kg 378.5 l^{-1})	Alcarez *et al.* (1986)
$FeSO_4$ (0.25%) + $MnSO_4$ (0.05%) + $CuSO_4$ (0.25%) + $ZnSO_4$ (0.05%) + $MgSO_4$ (0.05%)	Desai *et al.* (1991)
$FeSO_4$ + $MnSO_4$ (39 g each) + $ZnSO_4$ (98 g) in 20 lit. water	Nawab Ali *et al.* (1992)
$CuSO_4$ (2%), $ZnSO_4$ (0.4%), $ZnSO_4$ (0.5%)	Majorana (1960), Landthasa and Bhattacharya (1991), Chapman *et al.* (1945), Chapman (1960a)
$ZnSO_4$ (0.2%, 0.5%) + Borax (0.2%)	Nair and Mukherjee (1970a), Singh and Misra (1980)
$ZnSO_4$ (0.7-0.8%)	Dutt and Bhambhota (1967), Carrasco *et al.* (1969)
Zn-EDTA (0.5%)	Dube and Saxena (1971)
$ZnSO_4$ (0.3%) + $CuSO_4$ (0.6%)	Khanna *et al.* (1969), Anand *et al.*(1969)
$ZnSO_4$ (0.6%) + 20 ppm 2,4-D	Haribabu and Rajput (1982), Baku (1989), Singh and Misra (1986)
$ZnSO_4$ + $CuSO_4$ (2 g l^{-1} each)	Sharma (1990), Sharma *et al.* (1990)
$ZnSO_4$ (0.5%) + K_2SO_4 (4%)	Singh *et al.* (1989)
$ZnSO_4$ (0.3-0.6 %)	Batru *et al.* (1984), Rodriguez *et al.* (1994), Kanwar and Dhingra (1962), Dhingra *et al.* (1966), Kotur (1982)
$ZnSO_4$ (300 ppm) + $CuSO_4$ (250 ppm)	Bacha (1975)
Zn-EDTA + Mn-EDTA (450 g gallon^{-1} acre^{-1})	Rawash *et al.* (1983)
Zn-EDTA (450 g 100 gallon^{-1})	El-Gazzar *et al.* (1979)
$ZnSO_4$ (0.5%) + $CuSO_4$ (0.25%)	Mann and Sidhu (1983), Arora and Yamdagni (1986)
$ZnSO_4$ (0.1%) + $CuSO_4$ (0.1%)	Sharma *et al.* (1990)
$ZnSO_4$ (0.61 g l^{-1}) + $MnSO_4$ (1.2 g l^{-1})	Garcia-Alvarez *et al.*(1986)
$ZnSO_4$ (0.3%), (0.6%)	Dhingra *et al.* (1966), Mann and Takkar (1983b)
Zn + Cu + K (0.25% each)	Singh and Chohan (1982)
$ZnSO_4$ (0.5% each) + urea (1%)	Sharma *et al.* (1974), Gupta *et al.* (1989)
$ZnSO_4$ (0.5%) + $CuSO_4$ (0.3%) + Borax (0.3%)	Singh *et al.* (1990c)
Zn-EDTA (0.4%) + Cu-EDTA (0.2%)	Sharma *et al.* (1999a; 1999b)
Mo (1 mg l^{-1}), (0.1 g l^{-1} ammonium molybdate)	Brusca and Haas (1956), Blandel and Blanc (1975)
Ammonium molybdate (500 mg l^{-1})	Huang and Wang (1991)
Amonium molybdate (450 g 100 gallon^{-1})	Leonard (1952)
B (0.2-0.4%), (0.6%)	Singh and Singh (1976), Rai and Tewari (1988), Rai *et al.* (1988)
Boric acid (0.3%)	Chiu and Chang (1986)
Borax (0.6%) (0.2-0.4)	Rai *et al.* (1988), Singh and Singh (1976)
H_3BO_3 + $MgSO_4$ (0.2% each) + $ZnSO_4$ (0.1%)	Qin (1996)
Borax (0.2%) + $MgSO_4$ (0.2%) + $ZnSO_4$ (0.1%) + Yemianbao	Wang (1999)

Table 6.15. Recommendations of various micronutrients as soil application from citrus belts of world.

Fertilizer dose	Source
$FeSO_4$ (11.2 kg 1000 sq. feet)	Armstrong (1957)
$FeSO_4$ (250 - 1000 g tree^{-1})	Kovanci *et al.* (1985a; 1986)
$FeSO_4$ (30 kg Fe ha^{-1}) + S (500 kg ha^{-1})	Patel *et al.* (2001)
Fe-humate (22 – 352 g tree^{-1})	Wallihan (1961)
Fe + Zn (15-30 kg ha^{-1})	Jadhav *et al.* (1979)
Fe-Chelate 138/ RA 157 (1-5 g tree^{-1})	Cooper (1957b)
Fe-Chelate (300 g tree^{-1})	Raciti (1959)
Fe-EDDHA (225 g tree^{-1}); (20-40 g tree^{-1}); (9 g tree^{-1})	Hilgeman (1969), Leonard and Calvert (1971), Hellin *et al.* (1988), Aso and Dantur (1972)
Fe citrate (2.6- 6 mg kg^{-1}) + $MnSO_4$ (1.3-3 mg kg^{-1})	Liu and Nan (1996)
Fe-EDDHA + Zn-EDTA (35 g tree^{-1} each)	Bakhshi *et al.* (1973)
Fe-EDTA (10 g tree^{-1}) or Fe-EDTA (12.5-25 g tree^{-1})	Primo *et al.* (1970), Khadr (1965), Khadr *et al.* (1965), Leonard and Stewart (1953, 1956), Hellin *et al.* (1987)
Fe-EDTA (225 – 337 g tree^{-1})	Koo (1984; 1987)
Fe-EDDHA (12-24 g tree^{-1}), (14 g tree^{-1})	Kuykendall *et al.* (1957), He *et al.* (1998)
Fe-EDTA (6 g tree^{-1}); (9-18 g tree^{-1}); (23 mg kg^{-1} tree^{-1}); (40 g tree^{-1})	Leonard and Stewart (1952), Ford (1954); Matsuda (1968), Leonard and Stewart (1956; 1957)
RA 157-Fe / Sequesterene 330 - Fe (48-96 g tree^{-1})	Hilgeman (1957)
Fe + Mn + Zn-EDTA (292 g + 292 g + 315 g ha^{-1})	Alva and Tucker (1992)
Mn as $MnSO_4$ (0.5 kg tree^{-1})	Anderson (1986)
$MnSO_4$ (2.2 kg tree^{-1}) + $CaCl_2$ (1.3 kg tree^{-1})	Leonard and Stewart (1960)
$MnSO_4$ (483 kg tree^{-1}) + $ZnSO_4$ (303.8 g tree^{-1})	Garcia – Alvarez *et al.*(1983)
Zn-EDTA (30 g tree^{-1}), (2.1 g m^{-2})	Swietlik (1996), Anderson (1984)
$ZnSO_4$ (500 g tree^{-1})	Khera *et al.* (1985)
$ZnSO_4$ (1 kg tree^{-1}); (250 – 1000g tree^{-1})	Nijjar and Brar (1977), Misra and Ganapathy (1982)
$ZnSO_4$ (810 g tree^{-1} soil application) + $MnSO_4$ (630 g 100 gallon^{-1} foliar spay)	Embleton *et al.* (1966)
$ZnSO_4$ + K_2SO_4 (0.5% foliar spray) + K_2O as K_2SO_4 (210 g tree^{-1} soil application)	Singh *et al.* (1989)
Zn-aldehyde (4-12 kg ha^{-1})	Mdinaradze (1981)
$ZnSO_4$ (100 g tree^{-1} soil application) + (0.5% foliar spray)	Devi *et al.* (1996)
$ZnSO_4$ + $FeSO_4$ + $MnSO_4$ (50 g tree^{-1} each as soil application)+ (0.50% as foliar application)	Devi *et al.* (1997)
Zn (3 g) + B (3 g) + Mo (1.5 g tree^{-1})	Egorashvili *et al.* (1991)
$ZnSO_4$ + $CaCl_2$ (5.5-11.0 kg tree^{-1})	Leonard *et al.* (1957; 1959)
Cu (800 mg kg^{-1})	Zhu and Alva (1993)
Borax (100 g tree^{-1})	Khalidy *et al.* (1966)
Boric Acid (40-120 g tree^{-1})	Chiu and Chang (1985)
B (3 g tree^{-1}), (3 kg ha^{-1}), (3-9 kg ha^{-1}), (37.5-75.0 g tree^{-1})	Chanturiya (1972), Talakvardze (1977), Mdinaradze and Kechakmadze (1982), Kechakmadze (1987), Sato *et al.* (1962a; 1962b)
Borax or Boric acid (1 kg ha^{-1})	Srivastava *et al.* (1977)
Borax (100 g tree^{-1})	Khalidy *et al.* (1966)
Calcium borate (200 g tree^{-1})	Coetzee (1992)

citrus cultivars have been obtained suggested by large number of researchers from across the world taking into account the variation in soil, climate, physiography, and cultivar. These recommendations can very well be explored for contentment as well as rejuvenation of the declining citrus orchards.

Summary

Citrus decline problem investigated exclusively from soil and nutrition point of view indicated that symptologically, it is similar to single or multinutrient deficiency. A number of soil factors viz., soil compaction, poor drainage, soil reaction, salinity, excess lime, subsoil acidity, presence of indurated layer of hardpan composed of clay or $CaCO_3$ or rocky structure, and a variety of nutritional disorders have been observed to account for decline and decline like symptoms, thereby, a reduction in orchard productivity. Various recommendations originated from across citrus belts of world regarding macro- and micronutrient management have been summarised so as to prepare a comprehensive guideline to arrest an early decline on one hand and improve the productive life of orchards on the other hand in a desired holistic manner.

7 Issues and Strategies

The futuristic strategy should involve growing of superior nutrient efficient plants having high photosynthetic efficiency and diverse adaptability, which receive balanced supply of macro and micronutrients by involving organic, mineral, and biofertilizers applied in the most appropriate manner (Ghosh, 2000). A number of issues and possible strategies have emerged with reference to various aspects of climate and soil adaptability of citrus :

1. Soil is a basic resource that must be healthy for rest of the ecosystems to remain diverse and productive. Human induced soil changes and their effects on human lives and ecological environments have received extensive attention (Lal, 1990; Rozanov, 1990; Miller and Wali, 1994; Wang, 1995; Wang and Gong, 1996). Improving soil quality is one of the most important core problems for sustaining the global biosphere (Smith *et al.,* 1994). It is also the basis judging the soil management practices and use systems (Papendick and Parr, 1992; Karlen *et al.,* 1994a; 1994b; Smith *et al.*, 1994). Soil protection and conservation are particularly important when dealing with marginal, fragile, and ecologically sensitive ecosystems such as those of the semiarid regions of the world, where excessive human pressure has led to irreversible soil degradations. More research is needed to delineate the critical limits of soil properties beyond which quality of soil environment is severely and irretrievably jeopardized.

2. Developing the efficient means of supplying all the nutrients, considering the successful intensive Citriculture on the dominant deep, well drained, acid infertile red and yellow Oxisols (Falesi, 1972; Bennema, 1975). However, two elements Ca and P may pose special problems (Blue, 1974).

3. The organic matter level in the soil is important to help to maintain an active population of organisms in soil to promote organic matter mineralizaton and pesticide decomposition, besides stabilizing a favourable physical condition of the soil (Lee, 1991; Stork and Eggleton, 1992) and promote the absorption of nutrients by the plant roots (Chen and Aviad, 1990). These effects of soil organic matter imply that the level of organic matter may be taken as an indicator of the sustainability of soil management system. One of the major advantages to be derived from long term fertility trials is that they enable soil organic matter changes to be monitored. Norms need to be established for different soils to be used as an index of sustainability.

4. Root activity as influenced by rootstock- scion interaction under differential soil fertility levels coupled with evaluation of effect of long term use of mineral fertilizers, sewage sludge, urban/ industrial wastes is required for ecological sustainability.

5. The principles of how to maintain nutrient and organic matter levels, preserve soil structure, and avoid erosion are now well understood. It is a utmost necessity to quantify the rate of change for different soils and climates so that the soil changes can be modelled and linked to the crop performance.

6. Input of Pedology based knowledge may support the prediction of soil distribution based on soil-geomorphic relationship. Soil chronosequence information could be used to identify the location and extent of soil horizons important in agricultural decision making, such as strongly developed agrillic horizons and duripans. These may complement indigenous knowledge of the problems of local soils, topography, and microclimate regarding crop suitability and management.

7. Subsoil modification can be considered as a viable option, if tillage operations are to be made effective for several years. In irrigated cropping significant economic benefits from deep tillage are achievable. It is apparent that there has been little research undertaken in the world to identify specific nutrient constraints in the subsoil to root growth. Strategies to improve subsoil fertility may include : i. mechanical means of placing nutrients deeper in the profile, ii. using nutrient sources of lower or higher mobility, iii. using deep rooted legumes to fix nitrogen at depths, and iv. selection of plant species and genotypes better suited to acquiring nutrients from subsoils (Graham *et al.,*1992).

8. Use of manure or compost has increased during the recent years as a part of ecological farming. Supply of nutrients in organic form may increase the risk of leaching losses through the soil. For example organic matter coating on soil particles may modify the sorption properties of mineral particles causing increased P mobilization or organic N has also been associated with large N losses as some of the applied N may be mineralised at a time when it can not be used by the crop. But then, integrated nutrient supply system has to be the rational approach to cover all aspects of soil productivity, sustainability, and environmental safety in a wholesome manner (Paroda, 1999).

9. Fertilizer use efficiency in form of recovery of applied N by crops is often limited to 30-40 per cent only with citrus by the virtue of perennial crop is no exception. Negative environmental effects such as ammonia volatilization and N_2O emissions are also related to an inefficient use of nitrogenous fertilizer such as urea. To avoid wastage of such precious resource and to minimize the environmental damage, there is a need to develop and demonstrate fertilizer strategies that will allow precise management of nutrient application. One such technique, achieving commercial

20. Development of soil type-soil suitability model under specific agroclimate (Srivastava and Singh, 1999c) as part of precision Citriculture. This has to be backed up by the studies on the variation in acclimatization with altitude and other associated weather parameters which could later act as the bases for providing freeze protection (Gerber and Chen, 1986).

21. Introduction of site specific practices into citrus orchards should be focussed by combining the use of a number of information management technologies such as application of remote sensing, geographical information system, global monitoring system, and variable rate application technology. Role of these tools as a part of precision Agriculture requires to be further strengthened.

Bibliography

Abani, K. Bhagabati (1986). A spatial appraisal of land and natural vegetation in north-east India. *J. NE India Geogr. Soc.* **18**(142) : 20-27.

Abbadi, S.B., Minessy, F.A. and Abdelhafeez, A.T. (1974). Effect of temporary water tables on Marsh grapefruit trees in Sudan. *Expt. Agri.* **10**(4) : 247-250

Abercrombie, R. A. and Hoffman, J. E. (1996). The effect of alleviating soil compaction on yield and fruit size in an established Naval orange orchards. *Proc. Int. Soc. Citriculture.* Vol. **2**, pp. 979-983.

Abercrombie, R. A. and Plessis, S.F. Du. (1996). Ridge culture with citrus. *J. South Afr. Soc. Hort. Sci.* **5**:18-26.

Abercrombie, R.A. and Plessis, S.F. Du. (1995). The effect of alleviating soil compaction on yield and fruit size in an established Navel orange orchard. *J. South Afr. Soc. Hort. Sci.* **5**(2) : 85-89.

Adetunji, M.T. (1994). Evaluation of supplying capacities of south western Nigerian soils. *J. Indian Soc. Soil Sci.* **42**(2) : 208-213.

Adetunji, M.T. (1997). Kinetics of sulphate desorption in some low activity clay soils of Nigeria. *J. Indian Soc. Soil Sci.* **45**(3):475-480.

Adhikari, M., Majumder, M.K., Mahapatra, S.K. and Bandopadhyay, S.K. (1986). Mineralogical composition of some soils of the Himalayan region. *Clay Res.* **5** : 70 – 73.

Adhikari, S., Saha, P.K. and Chatterjee, D.K. (1997). Available boron, cobalt and molybdenum status in some freshwater fish pond soils of Orissa in relation to soil characteristics. *J. Indian Soc. Soil Sci.* **45**(3):583-584.

Adhikari, M and Si, S.K. (1991). Studies on different forms of iron and aluminium and their release in relation to acidity of some acid soils. *J. Indian. Soc. Soil Sci.* **39:** 252-256.

Aherns, M.J. and Ingram, D.L. (1988). Heat tolerance of citrus leaves. *HortScience* **23**:747-748.

Ahmad, N. and Jackson, P.(1965). Studies of nutrient levels of Valencia oranges in north Trinidad and associated soil characteristics. *Soil Sci.* **100** : 428-432.

Ahmad, N. and Jones, R.L. (1969). A Plinthaqult of the Aripo Savannas, north Trinidad II. Mineralogy and genesis. *Soil Sci. Soc. Am. Proc.* **33**(5) : 766-768.

Ahmad, N., Jones, R.C. and Beaver, A.H. (1968a). Genesis, mineralogy and related properties of west Indian soils. I. Monserrat series, derived from glaucoritic sandstone, central Trinidad. *J. Soil Sci.* **19**(1) : 1-8.

Ahmad, N., Jones, R.C. and Beaver, A.H. (1968b). Genesis, mineralogy and related properties of west Indian soils. II. Monserrat series, derived from micaceous schist and phyllite, northern range, Trinidad. *J. Soil Sci.* **19:** 9-16

Ahmed, F.F., El-Sayed, M.A. and Maatouk, M.A.(1988). Effect of nitrogen, potassium and phosphorus fertilization on yield and quality of Egyptian Balady lime trees (*Citrus aurantifolia*) II. Yield and fruit quality. *Annals Agric. Sci. (Cairo)* **33**(2):1249-1268.

Ahmed, S. and Ghafoor, A.A. (1962-64). History and development of citrus industry. *Punjab Fruit J.* **26/27**: 1-15, 24.

Ahmed, S. and Mazhar, H. (1962-1964). Cultivation of citrus fruit. *Punjab Fruit J.* **26/27**: 97-107.

Aina, P. O. (1979). Soil changes resulting from long term management- practices in western Nigeria. *Soil Sci. Soc. Am. J.* **43**: 173-182.

Aiyappa, K.M. (1957). Citrus Decline in the Malnad Area of South India. *Bull.* p. 12. Citrus Die-back Research Station, Gonicoppal Karnataka, India.

Aiyappa, K.M., Dikshit, N.N. and Bojappa, K.M. (1959). Response of old mandarin trees (*C. reticulata*) to fertilization and micronutrient sprays in Coorg. *Mysore Agric. J.* **34:** 70-76.

Aiyappa, K.M. and Srivastava, K.C. (1965). Coorg orange (mandarin) cultivation. *Lal-Bagh* **10**(4): 4-8.

Aiyappa, K.M., Srivastava, K.C. and Bojappa, K.M. (1962). Citrus decline – A review. *Indian J. Hort.* **19**: 70-79.

Aiyappa, K. M.,Srivastava, K.C. and Muthapa, D.P. (1965). The nutritional status of seedling mandarin leaves in Coorg. (*Citrus reticulata* Blanco). *Indian J. Hort.* **22** : 231-337.

Aksoy, U., Anac, S., Anac, D. and Can, H.Z. (1997). The effect of ground water salinity on satsuma mandarins: preliminary results. *Acta Hort.* **449**: 629-633.

Albarado, A. and Boul, S.W.(1975). Toposequence relationship of Dystrandepts in Costa Rica. *Soil Sci. Soc. Am. Proc.* **39**: 932-941.

Albisu, L.M. (1982). Emperical evidence of the main factors affecting orange yields in the mediterranean area. *Sci. Hort.* **18**:119-124.

Albrigo, L.G. and Menini, U. (1984). The status of citrus production in the Carribean basin. *Proc. Int. Soc. Citriculture.* Vol. **1**, pp. 570-573.

Albrigo, L. G. and Young, R. H. (1981). Phloem Zn accumulation in citrus trees affected with blight. *HortScience* **16**: 158-160.

Alcarez, C.F., Martinez-Sanchez, F., Sevilla, F. and Hellin, E. (1986). Influence of ferrodoxin levels on nitrate reductase activity in iron deficient lemon trees. *J. Pl. Nutr.* **9**:1405-1413.

Aldrich, D.G. and Buchanan, J.R. (1954). Soil phosphorus supply in healthy and phosphorus deficient citrus orchards in southern California. *Proc. Am. Soc. Hort. Sci.* **63** : 32-36.

Ali, N. (1962-1964). Causes of decline of citrus orchards and methods of renovation. *Punjab Fruit J.* **26/27**: 143-148.

Aliyev, S.A., Godzhizev, D.A. and Mikaylov, F.D. (1981). Kinetic indices of catalase activity in the main soil groups of Azerbaijan. *Soviet Soil Sci.* **13**(1) : 29-35.

Al-Juburi, H.J. and Al-Mesry,H. (1991). Effect of foliar application of some micronutrients on chlorophyll and carotene contents of valencia orange leveas. *Bull. Faculty Agric., Univ. Cairo.* **42**:1707-1728.

Allan, L.H. (1977). Soil water and root development. *Proc. Soil & Crop Sci. Soc. Fla.* **36** : 4-9.

Allan, D.L., Adriano, D.C., Bezdicek, D.F., Cline, R.G., Coleman, D.C., Doran, J.W., Haberern, J., Harris, R.G., Juo, A.S.R., Mausbach, M.J., Peterson, G.A., Schuman, G.E., Singer, M.J. and Karlen, D.L. (1995). SSSA statement on soil quality. *Agronomy News,* June 1995 ASA Madison, WI p. 7.

Allison, L.E. (1964). Salinity in relation to irrigation. *Adv. Agron.* **16**: 139-180.

Allolli, T.B., Nalawadi, U.G. and Sullikeri, G.S. (1988). Influence of different forest tree species on yield and yield attributes of Coorg mandarin (*Citrus reticulata* Blanco) budded on two rootstocks under agroforestry system. *Myforest* **24**(4) : 233-240.

Alva, A.K., Graham, J.H. and Anderson, C.A. (1995). Soil pH and copper effects on soil young Hamlin orange trees. *Soil Sci. Soc. Am. J.* **39**(2) : 481-487.

Alva, A.K., Graham, J.H. and Tucker, D.P.H. (1993). Role of calcium in amelioration of copper phytotoxicity for citrus. *Soil Sci.* **155**(3) : 2118-2278.

Alva, A.K. and Obreza, T.A. (1998). By-product iron-humate increases tree growth and fruit production of orange and grapefruit. *HortScience* **33**(1) : 71-74.

Alva, A.K. and Tucker, D.P.H. (1992). Foliar application of various sources of iron, manganese and zinc to citrus. *Proc. Fla. State. Hort. Soc.* **105**:70-74.

Alvardo, A., Araya, R., Bornemisza, E. and Hernandez, R.L. (1994). Nutritional studies in eleven citrus cultivars and one local selection in the Atlantic zone of Costa Rica Major and minor elements. *Agronomia Costaricense* **18**(1) : 13-19.

Alwis, K.A. and Panabokke,C.R. (1972). Handbook of soil science of Sri Lanka. *J. Soil Sci. Soc. Ceylon* **2**: 16-31.

Amami, S.El. and Haffani, M. (1973). The effect of mulch on the water balance, tree growth and fruit maturity of Maltaise orange of Tunisia. *Hort. Abst.* **45** : 883.

Amaro, A.A. (1984). Citriculture in Brazil. *Proc. Int. Soc. Citriculture.* Vol. **1**, pp. XXIII-XXVI.

Amir, A., Kramer, O. and Bar Akiva, A. (1981). A fertilizer experiment in a grapefruit grove in the Esdrealon valley. *Alon Hanotea.* **35**:1-15.(Hebrew).

Amy, Lutz., Menge, John and Connell, Neil O' (1986). Hardpans, claypans and other mechanical impedences. *Citrog.* **72**:57-60.

Anand, S.S., Bakshi, J.C. and Khanna, S.S. (1969). Comparative efficiency of chelated and unchelated compounds on citrus chlorosis II. Effect of zinc and copper on citrus chlorosis and chlorophyll contents. *Indian. J. Agric. Sci.* **39**:841-847.

Anand, J.C. and Leisram, M.S. (1963). Preliminary studies on the quality of mandarins (*Citrus reticulata* Blanco) grown in Delhi. *Indian J. Hort.* **20**:146-149.

Ananthanarayana, R. and Hanumantharaju, T.H. (1994). Nature of soil acidity and lime requirement in agroclimatic zone of Shimoga district, Karnataka. *J. Indian Soc. Soil Sci.* **42**(3) : 361-364.

Ananthanarayana, R. and Ravikumar, S.M. (1997). Characterization of soil acidity and lime requirement of soils in the agroclimatic zones of Hassan district of Karnataka. *J. Indian. Soc. Soil Sci.* **45**(3):442-445.

Anantwar, S.G., Babrekar, P.G., Bhaskar, B.P. and Challa, O. (2000). Variability in shrink-swell potentials in two transects on the basaltic plateau of Wardha district. *J. Indian Soc. Soil Sci.* **48**(1):145-150.

Anchern, C. (1965). Citrus in Thailand. Government Press, Bangkok, Thailand.

Anderson, C.A. (1968). Effects of gypsum as a source of calcium and sulfur on tree growth, yields, and quality of citrus. *Proc. Fla. State Hort. Soc.* **81** : 19-24.

Anderson, C.A. (1971a). Effects of soil pH and calcium on yields and fruit quality of young Valencia oranges. *Proc. Fla. State Hort. Soc.* **84** : 4-11.

Anderson, C.A. (1971b). Dolomite limestone as a source of mangensium for citrus. *Proc. Soil & Crop Sci. Soc. Fla.* **30** : 150-157.

Anderson, C.A. (1972). Effect of soil pH and calcium on yields and fruit quality of young valencia oranges. *Proc. Fla. State Hort. Soc.* **84** : 4-10.

Anderson, C.A. (1980). Legume covercrop trials in citrus groves. *Proc. Soil & Crop Sci. Soc. Fla.* **39** : 80-83

Anderson, C. A. (1981). Fertility management for citrus on Entisols. *Proc. Soil & Crop Sci. Soc. Fla.* **40**:17-24.

Anderson, C.A. (1984). Micronutrient uptake by citrus from soil applied zinc compounds. *Proc. Soil & Crop Sci. Soc. Fla.* **43**:36-39.

Anderson, C.A. (1986). Mineral uptake by citrus leaves from soil applied manganese compounds. *Proc. Soil & Crop Sci. Soc. Fla.* **45**(6):46-50.

Anderson, C.A. (1987). Fruit yield, tree size and mineral nutrition relationships in Valencia orange trees as affected by liming. *J. Pl. Nutr.* **10** : 1907-1916.

Anderson, C.A. and Albrigo, L.G. (1971). Evaluation of soil testing for magnesium as predictive tools of magnesium status in citrus on dolomitic soils. *Proc. Soil & Crop Sci. Soc. Fla.* **31** : 127-130.

Anderson, C.A. and Albrigo L.G. (1972). Comparison of soil test methods for predicting the status of leaf Mg and Ca of orange trees on dolomitic treated soils. *Proc. Soil & Crop Sci. Soc. Fla.* **32** : 144-152.

Anderson, C.A. and Albrigo, L.G. (1977). Seasonal changes in the relationship between micronutrients in orange (*Citrus sinensis* Osbeck) leaf and soil analytical data in Florida. *Proc. Int. Soc. Citriculture.* Vol. **1,** Lake Alfred, Florida. pp. 20-25.

Anderson, C. A. and Calvert, D. V. (1970). Mineral tree composition from citrus trees affected with declines of unknown etiology. *Proc. Fla. State. Hort. Soc.* **83**: 59-61.

Anderson, C.A. and Martin, F.G. (1969). Effect of soil pH and calcium on the growth and mineral uptake of young citrus. *Proc. Fla. State Hort. Soc.* **82** : 7-12.

Anderson, C.M. and Benatena, H.N. (1996). Behaviour of twelve orange cultivars on six rootstocks in Argentina. *Proc. Int. Soc. Citriculture.* Vol. **1**, pp. 103-108.

Andriesse, J.P. (1968/1969). A study of the environment and characteristics and tropical podzols in Sarawak (east Malaysia). *Geoderma* **2** : 201-228.

Androulakis, I.J., Loupassaki, M. H., Beidoun, F. and Tzombanakis,I. (1992). The effect of NPK fertilizers on the yield and mineral contents of leaves of grapefruit. *Proc. Int. Soc. Citriculture.* Vol. **2**, pp. 624-627.

Ankerman, D. and Large, R. C. (1979). *Soil and Plant Analysis.* A and L Agricultural laboratories, Inc. Memphis, p. 82.

Anonymous (1951). *Soil Survey Manual.* U. S. Dept. Agri. Res. Serv., Washington DC, USA, p. 103.

Anonymous (1961). Planting citrus on flatwoods soils. *Citrus Indus.* **42**(6):20-21.

Anonymous (1964a). Ninth Annual Report of the Coordinated Scheme for Citrus Die-Back Diseases for the period Ist July 1963 to 30th June, 1964, Indian Inst. Hort. Res., Bangalore, India

Anonymous (1964b). Minor element deficiencies in citrus. *Not. Agric. Serv. Shell Agric., Venezuela* **3**:107-108.

Anonymous (1965a.). Citrus Die-back in India under PL 480, *Annual Report* (un-published).

Anonymous (1965b). Annual Progress Report for the year (1964-65). Department of Horticulture, Punjab Agriculture University, Ludhiana.

Anonymous (1966a). Annual Progress Reports of the year 1963-64, 1964-65 and 1965-66, Soil Deptt., Punjab Agriculture University, Ludhiana, Punjab, India.

Anonymous (1966b). Boron deficiency in citrus. *Not. Agric. Serv. Shell Agric., Cagua* **4**: 77-78.

Anonymou (1972). Mandarin Growing in Mysore. *Ext. Bull.* **1** pp. 1-4. Indian Institute of Horticultural Research, Banglore, Karnataka, India.

Anonymous (1975a). *Soil Survey Report of Nagaland.* Govt. of Nagaland, Kohima, Nagaland.

Anonymous (1975b). Soil Conservation Service, U.S. Dept. of Agriculture, Soil Taxonomy, Agriculture Handbook No. **436.**

Anonymous (1976). National Commission on Agriculture, New Delhi, India.

Anonymous (1980a). Annual and seasonal rainfall in different meterological subdivisions of India. *Brochure* p. 2. Central Water Commission, New Delhi.

Anonymous (1980b). *Florida Agricultural Statistics - Citrus Summary*. Florida Crop and Livestock Reporting Service. Florida Dept. Agr. Consumer Services, Tallahassee, USA.

Anonymous (1983). Shifting Cultivation in North-East India, *Tech. Bull.,* ICAR Research Complex for North-East Hill Region , Shillong, India.

Anonymous (1986). Shifting Cultivation and Alternative Land Management Practices, *Ext. Bull.,* ICAR Research Complex for North-East Hill Region, Shillong, India.

Anonymous (1988a). Horticultural Information Service, National Horticulture Board, Ministry of Agriculture and Information, Gurgaon, p. 32.

Anonymous (1988b). Climatic requirements of citrus. *Farming in South Africa. Citrus B1.* p. 3.

Anonymous (1988c). Citrus production (in Turkish). *Ministry of Agriculture Forestry and Rural Affairs*, Publ. No. General 276, Series : **11**, p. 40.

Anonymous (1989). *Agro-climatic Regional Planning, Eastern-Himalyan Region (Zone II) – Profile and Strategy*, Assam. Agril. University, Jorhat, Assam, India.

Anonymous (1994a). *Citrus Special.* Horticultural Information Service. National Horticulture Board, Ministry of Agriculture and Cooperation, Gurgaon, India, pp.96-118.

Anonymous (1994b). Soil Fertility and Plant Nutrition. *Ann Rep.* 1994-95, Soil Science Division, Nepal Agriculture Research Council Khumutar, Lalitpur (Nepal), p.32.

Anonymous (1994c). Agricultural Structure and Production. 1992. Turkish State Statistical Institute, *Publ No.* **1685**. Ankara, Turkey.

Anonymous (1996a). Horticulture Information Service. National Horticulture Board, Ministry of Agriculture and Cooperation, Gurgaon, India, pp. 43-62.

Anonymous. (1996b). Soil quality information sheet: Soil quality introduction. National Soil Survey Center in Corporato to Soil Quality Institute, NRCS, USDA, National Soil Tilth Lab, Agricultural Research Service, USDA.

Anonynous (1998). Horticulture Information Service. National Horticulture Board, Ministry of Agriculture and Cooperation Gurgaon, India, Special Publication (Negi *et al.*, ed.) pp. 12-32.

Anonymous. (1999). *Administration Report.* Department of Agriculture, Peradeniya, Sri Lanka, pp. 4-18.

Aoki, A. and Morita, S. (1969a). On the general properties of abnormal orchards of citrus in Japan. Part I. Studies on abnormal defoliation of citrus in relation to soil and plant. *J. Sci. Manure, Japan* **40** : 228-235.

Aoki, A. and Morita, S. (1969b). Studies on abnormal defoliation of citrus in relation to soil and plant. Part. II. Annual variation in chemical composition in soil and plants of abnormal defoliation orchards of citrus at Gamagrol, Aichi prefecture. Part III. Annual variations in chemical composition of soils and plants in abnormal defoliation orchards of citrus at Mikkahi, Shizuoka prefecture; Part IV. Dissolution of manganese by soil treatments. *J. Sci. Soil Manure, Japan* **40** : 236-240, 240-244, 245-249.

Araya, R., Alvardo, A., Hernandez, R.L. and Bornemisza, E. (1994). Nutritional studies in eleven citrus cultivars and one local selection in the Atlantic zone of Costa Rica. II Micronutrients. *Agronomia Costarricense* **18**(1) : 21-27.

Arcila-Pulgarin, J. (1974). Efecto de la luz ultravioleta en plantulas de café en almacigo. *Cenicafe* **25**(3):90-92.

Grande Valley (N.P. Maxwell and M.A. Bailey, ed.). **B-1002**, pp. 16-17, Agri Expt. Sta. and Agri. Ext. Serv. Texas, USA.

Bain, F.M. (1949). Citrus and climate. *Calif. Citrog.* **34**(9) : 382, (10) : 426.

Bain, F. M. (1958). Morphological, anatomical and physiological changes in the developing fruit of Valencia orange, *Citrus sinensis* (L.) Osbeck. *Australian J. Bot.* **6**:1-24.

Bakhshi, J.C. (1961). Decline of sweet orange trees in arid irrigated regions of Punjab. *Punjab Hort. J.* **6**: 3-10.

Bakhshi, J.C. and Dhillon, J.S. (1964). A report on the decline of sweet oranges in arid-irrigated regions of Punjab. *Punjab Hort. J.* **4**: 15-22.

Bakhshi, J.C., Randhawa, N.S., Vatts, K.G. and Anand, S.S. (1973). Effect of zinc and iron on the growth of trifoliate orange seedlings. *Punjab Hort. J.* **13**(2/3):103-116.

Bakhshi, J. C. and Singh, K. K. (1966) . Decline of sweet orange trees in arid irrigated regions of Punjab. *Punjab Hort. J.* **6**: 3-10.

Baku, R.S. (1989). Influence of zinc and growth regulators on vegetative growth of Kagzi lime (*Citrus aurantifolia* Swingle). *J. Res. APAU.* **17**(1): 83-86.

Baldwin, M., Kellogg, C.E. and Thorp, J. (1938). *Soil Classification : Soils and Men*, U.S, Dep. Agric., Yearbook Agriculture U. S. Govt. Print Press, Washington, D.C., USA, pp. 979-1001.

Bamdad, D., Jungk, A. and Zalpour, N. (1977). The diseases caused by deficiency or excess of nutrient elements in citrus orchards of Iran. *Iranian J. Pl. Pathol.* **13**(1/2) : 3-5, 9-12.

Bandale, J. R., Narayana, N. and Kibe, M. M. (1951). Trace elements contents of black cotton soils of a few citrus growing tracts of Bombay state. *Poona Agric. Coll. Mag.* **43**: 3-10.

Banerjee, S.K., Chinnamani, S. and Jha, M.N. (1991). Forest soils of north and north-east Himalayas and constraints limiting their productivity. *Soil Related Constraints in Crop Production* (T.D.Biswas, ed.) *Bull. No.* **15**. Indian Society Soil Sci., New Delhi, India, pp. 164-175.

Bahgulo, A. (1977). The present position of Italian citrus growing. *Italia Agricola* **114**(1) : 67-78.

Bar-Akiva, A. (1962). *Diagnostic Methods for Iron and Manganese Deficiencies in Citrus Trees.* Ph.D. Thesis, Hebrew University, Jerusalem, Israel.

Bark-Akiva, A. (1965). Citrus growing in Iran. *World Crops* **17**(3): 39-43.

Bar-Akiva, A., Gotfried, A. and Hadas, A. (1973). The effect of N, P and chicken manure on grapefruits in the Besor area. *Alon Hanotea* **28**:1-12.(Hebrew).

Bar-Akiva, A. and Hamou, M. (1974). Soil factors influencing fruit quality and mineral composition of leaves of Valencia orange trees. *Commun. Soil Sci. Pl. Anal.* **5**(3) : 203-212.

Barber, D.A. and Gunn, K.B. (1977). The effect of mechanical forces on the exudation of organic substances by the roots of cereal plants grown under sterile conditions. *New Phytol.* **73** : 39-45.

Barbera, G. and Carmi, F. (1988). Effects of different levels of water stress on yield and quality of lemon trees. *Proc. Int. Soc. Citriculture Congr.* (Tel Aviv) Vol. **2**, pp. 717-722.

Barry, G.H. (1996). Citrus production areas of southern Africa. *Proc. 8th Int. Soc. Citriculture.* Vol.**1,** pp. 145-149.

Barry, G.H. and Veldman, F.J. (1996). Performance of nine graperfruit (*Citrus paradisi* Macf.) cultivars under hot subtropical southern African conditions. *Proc. Int. Soc. Citriculture.* Vol. **1,** pp. 113-115.

Barry, G.H., Veldman, F.J., Holmes, M., Wibbelink, M., Rust, L. and Monnik, K. (1996). The determination of climatic norms for export quality fresh citrus production. *Proc. Int. Soc. Citriculture.* Vol **2**, pp. 1062-1064.

Bartholomew, E.T. (1937). Endoxorosis, or internal decline of lemon fruits. *Bull.* **605**, pp. 1-42. Calif. Agri. Expt. Sta., California, USA.

Bartholomew, E.T. and Reed, H.S. (1943). General morphology, histology and physiology. *Citrus Industry* (H.J. Webber and L.D. Batchelor, ed.), Univ. Calif. Press, Barkeley, Los Angeles, USA, p. 1028.

Bartholomew, E.T., Sinclair, W.B. and Turrel, F.M. (1941). Granulation of Valencia oranges. *Bull. No.* **647**, pp. 1-63, Calif. Agric. Expt. Sta., California, USA.

Bartholomew, E.T., Sinclair, W.B. and and Horspool, R.P. (1950). Freeze injury and subsequent seasonal changes in valencia oranges and grapefruit. *Bull.* **719**, pp.1-18, Calif. Agri. Exp. Sta., California, USA.

Bartholomew, E.T., Sinclair, W.B. and Raby, E.C. (1934). Granulation (Crystallization) of Valencia oranges. *Calif. Citrog.* **19**: 88-89, 106, 108.

Bartholomew, E.T., Sinclair, W.B. and Raby, E.C. (1935). Granulation of valencia oranges. *Calif. Citrog.* **21**: 5, 30.

Basile, M., Melillio, V.A. and Lamberti, F. (1984b). Residues of phenamiphos in orange fruits from treated groves. *Proc. Int. Soc. Citriculture.* Vol. **2,** pp. 471-472.

Basile, M., Melillio, V. A., Lamberti, F. and Guudice, V. L. O.(1984a). Persistance of phenamiphos in the soil and its accumulation in the orange fruits. *Proc. Int. Soc. Citriculture.* Vol. **1**, pp. 469-471.

Batchelor, L.D. and Webber, H.J. (1948). Production of the crop. *Citrus Industry.* Vol. II (ed.). Production of the crop. Univ. Calif. Press Berkley and Los Angeles, USA, p. 933.

Bateman, D. F. and Basham, H. G. (1976). Degradation of plant cell walls and membranes of microbial enzymes. (R. Heitfuss and P. H. Williams, ed.), *Physiological Plant Pathology,* Springer-Verlag, Berlin, pp. 316-335.

Batru, R.S.H., Rajpal, C.B.S. and Rath, S. (1984). Effect of zinc, 2,4-D and GA_3 in Kagzi lime (*Citrus aurantifolia* Swingle) in fruit quality. *Haryana J. Hort. Sci.* **11**(1/2):59-65.

Bauer, A. and Black, A.L. (1981). Soil carbon, nitrogen, and bulk density comparisons in two cropland tillage system after 25 years and in virgin grassland. *Soil Sci. Soc. Am. J.* **45:** 1166-170.

Bausher, M.G. (1983). Concentration of four gases in soils of blight affected and healthy Valencia orange trees. *HortScience* **18**(1) : 104-105.

Bazelet M., Feigenbaum, S. and Bar-Akiva, A. (1980). Potassium fertilizer experiment in a Shamouti orange grove. *Pamph. 220, ARO,* Div. of Sci. Publ., Volcani Center, Israel.

Bear, F.E, Prince, A.L. and Malcom, J.L. (1945). Potassium Need of New Jersey Soils. *Bull.* **37**, pp. 217-222, New Jersey Agric. Exp. Stn., USA.

Becerra, S. and Guardiola, J. L. (1984). Inter-relationship between flowering and fruiting in sweet orange, cultivar Navelina. *Proc. Int. Soc. Citriculture.* Vol. **1,** pp. 190-194.

Beinroth, F.H. (1982). Some highly weathered soils of Puerto Rico, I. Morphology, Formation and Classification. *Geoderma* **21**: 1-73.

Bell, P.F., Vaughn, J.A. and Bourgeais, W.J. (1995). Leaf analysis find very low levels of zinc and manganese in Louisiana citrus. *Louisiana Agricultuae* **38**(2) : 27-29.

Bellia, F. (1964). Aspects of lemon production in the USA. *Teen. Agric.* **16**: 121-141.

Bello, L. (1996). Valentina and clementina: Two promising hybrids obtained at the Jaguey Grande Citrus Experiment Station. *Proc.Int. Soc. Citriculture.* Vol. **1,** pp. 219-220.

Ben-Hayyim, G. and Kochba, J. (1983). Aspects of salt tolerance in NaCl selected stable cell line of *Citrus sinensis. Pl. Physiol.* **72**: 685-690.

Benito, D. and Ruiz, S.R. (1975). Nutritional survey of citrus trees in the provinces of Santiago, O'Higgins and Colchagna. *Agricultura Tecnia* **35**(2) : 70-77.

Ben-Mechlia, N. and Carroll, J.J. (1989a). Agroclimatic modeling for the simulation of phenology, yield and quality crop production. *Int. J. Biometeorol.* **33**(1) : 36-51.

Ben-Mechlia, N. and Carroll, J.J. (1989b). Agroclimatic modeling for the simulation of phenology, yield and quality of crop production. II. Citrus model implementation and verification. *Int. J. Biometeorol.* **33**(1) : 52-65.

Bennema, J. (1975). Soil resources of the tropics with special reference to the well-drained soils of the Brazillian forest region. *Ecophysiol. Trop. Crops* **1**: 1-47.

Bennett, A.S. (1991). Aid to natural resources - a forward look. *Trop. Agri. Assoc. Newsletter.* **11**(4) : 1-6.

Bennett, H.H. and Allison, R.V. (1928). The soils of Cuba. *Trop. Plant. Res. Found.,* Washington, D.C. USA, p. 410.

Benoit, R.E., Willits, N.A. and Hanna, W.J. (1962). Effect of ryewinter cover crop on soil structure. *Agron. J.* **54** : 419-420.

Benton, R.J. (1940). Granulation. *Citrus News.* **18**(9):129-130.

Beraya, I.K. and Akhaladze (1971). The effect of the method of interrow soil management on growth and cropping in mandarin trees on humic calcareous soils. *Subtropicheskie Kul'tury* (1) : 93-96.

Beretta, M.J.G., Silva, S.R., Rossetti, V. and Moraes, W.B.C. (1983). Constatacao de 'plugs' em plantas citricas com declinio, afrancadas. XVI Congr. Bras, Fitopaltol. Belem, julho de 1983. *Resumo.* **1**:225-226.

Berezin, P.N. (1990) Evaluation of the potential and actual compaction of soils from physical criteria. *Soviet Soil Sci.* **22** (7): 104-114.

Berezin, P.N., Voronin, A.D. and Shein, Ye. V. (1989). Physical principles and crieteria of compaction. Vestn. Mosk. Un-ta. Ser. *Pochvovedeniye. No.* 1.

Beridze, Z.A. (1986a). Interrelation between ions in the soil and mandarin tree. *Subtropicheskie Kul'tury* (4) : 109-113.

Beridze, Z.A. (1986b). The role of potassium, magnesium and calcium in increasing the productivity of satsuma in Adzharia. *Subtropicheskie Kul'tury* (6) : 82-86.

Beridze, Z.A. (1987). Some factors affecting mandarin growth and yield. *Subtropicheskie Kul'tury* (3) : 76-82.

Bernal, T.B., Das Gupta, D.R. and Macray, L.A. (1957). Oriented transformation in iron oxides and hydroxides. *Nature. London,* **80** : 4587.

Bernstein, L. (1964a). Effects of salinity of mineral composition and growth of plants. *Pl. Anal. & Fert. Problems* **4**: 25-45.

Bernstein , L. (1964b). Salt tolerance of plants. *Bull*. **283** , p. 23, US Dept. Agri., USA

Bernstein, L. (1965). Salt tolerance of fruit crops. *Bull.* **292** , p.8 , U S Dept. of Agri., USA

Bernstein, L. (1967) Quantitative assessment of irrigation water quantity. *Spl. Tech. Bull.* **416,** p.51-65. Am. Soc. Testing and Materials , USA

Bernstein, L. C. (1969). Salinity factors and their limits for citrus culture. *Proc. Ist. Citrus Congr.* Vol. **3**, pp. 1779-1782.

Bernstein, L. and Hayward, H. E. (1958). Physiology of salt tolerance. *Am . Rev. Pl. Physiol.* **9**: 25-46.

Bernstein, L. and Pearson, G. A. (1956). Influence of exchangeable sodium on the yield and chemical composition of plants. *Soil Sci*. **82** : 247-258.

Bester, D.H. (1969). Fertilizer interactions of Navel oranges in south Africa as revealed by leaf analysis. (H.D. Chapman, ed.), *Proc. Ist. Int. Citrus Symp.* Vol. **3**, pp. 1641-1660, Univ. Calif., Riverside, California, USA.

Beutel, J.A. (1962). Citrus avocado nutrient sprays. *Calif. Citrog.* **47**:160-162.

Beutel, J.A. (1964). Moisture, weather and fruit growth. *Calif. Citrog.* **49**: 372.

Bevington, K.B. (1984). Effect of nutrition on yield and fruit quality of citrus. *Australian Citrus News.* **60**(10): 12

Bevington, K.B. (1992). Initial growth and yield of high density plantings of Valencia oranges as influenced by microsprinkler and drip irrigation. *Proc. Int. Soc. Citriculture.* Vol. **2,** pp. 709-711.

Bhaskar, B.P. and Subbiah, G.V. (1995). Genesis, characterization and classification of laterites and associated soils along the east coast of Andhra Pradesh. *J. Indian Soc. Soil Sci.* **43**(1):107-112.

Bhatt, S. S. (1945). The dieback diseases of citrus trees. *Indian Fmg.* **6:** 250-253.

Bhattacharya, S.C. and Dutta, S. (1952). A preliminary observation on rootstocks for citrus in Assam. *Indian J. Hort.* **9**: 1-11.

Bhattacharya, T., Sheghal, J. and Sarkar, D. (1996). Soils of Tripura for optimising land use. *NBSS Publ.* **65**, *Soil India Series* **6**, pp. 16-27, National Bureau of Soil Survey and Land Use Planning, Nagpur (Maharashtra), India.

Bhattacharya, J., Vasudevaiah, R. D. and Teotia, S.P.S. (1966). Effect of different soil moisture conserving measures on the yield of mosambi in upper Damodar catchment. *Indian J. Hort.* 23: 30-38.

Bhella, H.S. (1966). *Physico-chemical Characteristics of the Orchard Soils of Hissar District. M.Sc. Thesis,* Punjab Agric. University, Ludhiana, Punjab.

Bhogal, N.S., Sakal, R., Singh, A.P. and Sinha, R.B. (1993). Micronutrient status in some Aquic Ustifluvents and Udifluvents as related to certain soil properties. *J. Indian Soc. Soil Sci.* **41**(1) : 75-78.

Bhullar, J.S. (1978). Kinnow - A promising mandarin for Himachal Pradesh. *Punjab Hort. J.* **18**: 131-34.

Bhumbla, D.R. and Mclean,E.O. (1965). Aluminium soils VI. Changes in pH dependent acidity, cation exchange capacity and extractable aluminium with additions of lime to acid surface soils. *Proc. Soil Sci. Soc. Am.* **29**:270-274.

Bhutia, D.T. Gupta, R.K. and Biswas, S.K. (1986). Fertility status of soils of Sikkim. *Research Bulletin,* Department of Agriculture, Govt. of Sikkim, Sikkim.

Bicer, Y. and Ozel, M. (1986). The effect of nitrogen fertilizers on yield and quality of Washington Navel oranges in Tarsus. *Res. Rep.* Tarsus Research Institute for Village Affairs, Tarsus, Turkey, p. 27.

Bielorai, S., Dasberg, S., Erner, Y. and Brum, M. (1988). The effect of saline irrigation water on Shamouti orange production. *Proc. Sixth Int. Citrus. Congr.* Vol. **2**, pp. 707-715.

Bielorai, S., Shalhevet, J. and Levy,Y. (1978). Grapefruit response to variable salinity in irrigation water and soil. *Irrig. Sci.* **1**:61-70.

Bielorai, H., Shalhevet, J. and Levy, Y. (1983). The effect of high sodium irrigation water on soil salinity and yield of mature grapefruit orchard. *Irri. Sci.* **4**: 255-266.

Bigi, F. (1966). A note on the cultivation of grapefruit in Somalia. Origin, development and prospects. *Riv. Agric.Subtrop.* **60**: 53-78.

Bingham, F.T. (1964). Citrus in Egypt. *Calif. Citrog.* **49**(4): 160-167.

Bingham, F.T. (1973). Boron in cultivated soils and irrigation waters. *Advances in Chemistry Series* **123**: 130-138.

Bingham, F.T. and Martin, J.P. (1955). Effect of phosphorus on minor elements. *Calif. Citrog.* **40**:246-249.

Bingham, F.T. and Martin, J.P. (1956). Effects of soil phosphorus on growth and miner element nutrition of citrus. *Proc. Soil Sci. Soc. Am.* **20**:382-385.

Bingham, F.T., Chapman, H.D. and Pugh, A.L. (1954). Solution culture studies of nitrite toxicity to plants. *Soil. Sci. Soc. Am. Proc.* **18**: 305-308.

Bingham, F.T., Mahler, R.J., Parra, J. and Stolzy, L.H. (1974). Long-term effects of irrigation-salinity management on a Valencia orange orchard. *Soil Sci.* **117** : 369-377.

Bingham, F.T., Stolzy, L.H. and Chapman, H.D. (1969). Effects of variable water quality on valencia tree performance. *Proc. Ist. Int. Citrus Symp.* Vol.**3**, pp. 1803-1809.

Birriel, A., Reyes, F. and Laborem, G. (1984). Determination of the nutritional status of the CENIAP citrus rootstock orchard. *Agronomia Tropical.* **34**(1/3) : 131-153.

Bisht, N.S. and Kusumlata (1993). Niche width and dominance diversity relations of woody species in a most temperate forest of Garhwal Himalayas. *J. Hill Res.* **6**(2):107-113.

Biswas, T.D. (1955). Available Mn in soil. *Proc. Symp. Nat. Acad. Sci.,* India, Vol. **8**, pp. 25-30.

Biswas, T.D. and Mukherjee, S.K. (1994). *Textbook of Soil Science.* 2nd ed., Tata McGraw, Hill Pub. Co. Ltd., New Delhi, p. 365.

Bitancur, M., Campiglia, H., Furest, J.P., Mangado, J.O. Cano, M. and Supino, E. (1984). The Uruguayan citrus Industry. *Proc. Int. Soc. Citriculture.* Vol. **1**, pp. 573-576.

Bitcover, E.H. and Wander, I.W. (1950). Some observations on nitrate formation and the absorption of nitrogen by citrus. *Pl. Physiol.* **20** : 461-468.

Bitters, W.P. (1961). Physical characters and chemical composition as affected by scions and rootstocks. *Orange* (W.B.Sinclair, ed.), p.475, Univ. Calif. Press, Berkeley, California, USA.

Bitters, W.P. (1964). Citrus cultural practices and production in Japan. *Calif. Citrog.* **50** : 67-68, 70-75.

Black, C.A. (1993). *Soil Fertility Evaluation and Control.* Lewis Publ., Boca Raton, Florida, USA.

Blandel, A.M. and Blanc, D. (1975). Vegetative disorders in Corsican citrus: Demonstration of a molybdenum deficiency by *in vivo* assay of nitrate reductase activity. *Annales Agronomiques* **26**(3):277-287.

Blazquez, C.H. (1966). Citrus research in Jamaica. Proc. *Caribbean Reg. Am. Soc. Hort. Sci.* **9**(13): 116-127.

Biazquez, C.H. (1989). Densitometry, image analysis, and interpretation of aerial color inifra red photographs of citrus. *HortScience* **24**(4) : 691-693

Biazquez, C.H., Abbitt, B., Hitchock, D. and Judy, R. (1990). Rellationship between tree counts and citrus grove production. *Proc. Fla. State Hort. Soc.* **102** : 82-84

Bliss, D.E. (1944). Air and soil temperatures in California citrus orchard. *Soil Sci.* **58** : 259-274.

Blondel, L. (1970). Citrus growing in Japan and some south east Asian countries. *Fruits d' Outre Mer.* **25**: 523-537.

Blondel, L. and Brun, P. (1973). Citrus growing in Georgia, USSR. *Fruits* **28**(2) : 107-113.

Blondel, L. and Cassin, J. (1972). Influence of ecological factors on corsican clementine quality : fluctuations in the dry extract of the juice (preliminary note). *Fruits* **27**(6): 425-432.

Blue, W.G. (1974). Management of Ultisols and Oxisols. *Proc. Crop. & Soil Sci. Soc. Fla.* **33**:126-132.

Bobokhidze, M. S. (1976). The effect of measures to restore the fertility of leached soil on the growth and productivity of mandarin trees in Abkhaziya. *Subtropicheskie Kul'tury* (3/4) : 206-209.

Bobrovitskiy, A.V. (1971). Micromorphology and mineral composition of the alluvial deposits and soils of the Kolkhida lowland. *Pochvovedeniye* (3).

Bohm, W. (1987). The Thaer pupil carl Sprengel (1787-1859) as founder of the mordern science of plant nutrition. *Jahresheft der Albrecht-Thaer-Gesellschaft* **23**: 43-59.

Bohm, W. (1997). Biographical handbook on the history of Agronomy. *Sour Publ. Co.*, Munich, Germany.

Boman, B.J. (1987). Effects of soil series on shallow water table fluctuations in bedded citrus. *Proc. Fla. State Hort. Soc.* **100** : 137-141.

Boman, B.J. (1996). Effects of microsprinkler pattern area on Florida grapefruit. *Proc. Int. Soc. Citriculture.* Vol. **2**, pp. 673-677.

Bondarev, A.G. and Medvedev,V.V. (1980). Some ways of determining the optimum values of the agrophysical properties of soils. (Theoretical Principles and Methods of Determining Optimum Values of Soil Properties). *Proc. Dokuchayev Soil Inst.*, pp. 12-24.

Bonifacio, E., Zanini, E., Boero, V. and Franchini, Angela, M. (1997). Pedogenesis in a soil catena on surpentine in north-western Italy. *Geoderma* **75** : 33-51.

Bono, R., Cordova, Fernandez L. and Soler, J. (1988). Behaviour of Nova mandarin in Spanish conditions. *Proc. 6th Int. Citrus Contress* (R. Goren and K. Mendel, ed.), Vol. **1,** pp. 101-106.

Bora, P.K. (1975). Soil fetility and fertility use pattern in Assam, Meghalaya and Mizoram. *J. North Eastern Council* **1**(3); 38-49.

Borno, P.K. (1975). Soil Fertility and Fertility use pattern in Assam Meghalaya and Mozoram : *J. North Easter Council* **1**(3) : 38-49.

Bornemisza, E., Hernandez, R.L., Chaverri, W. and Veracochea, M.F. (1985). Annual variation of foliar levels of seven nutrients in Persian lime and San Fernando and Frost nuclellar Lisbon lemons in Costa Rica. *Agronomia Costarricense* **9**(2) : 135-141.

Borthakur, M. (1986). Weather and climate of north-east India. *J. NE India Geogr. Soc.* **18**(142) : 20-27.

Borthakur, D.N., Awasthi, R.P. and Ghosh, S.P. (1976). *Proc. Seminar on Shifting Cultivation in N.E. India,* Northeast Indian Council for Social Sciences Research, Assam, India.

Bosco, Fatta Del G., Barbera, G. and Occorso, G. (1977). Comparison between drip and basin irrigation for mandarin (*Citrus reticulata* Blanco) in Sicily. *Proc. Int. Soc. Citriculture.* Vol. **1**, pp. 113-117.

Bosco, Fatta Del G., Barbera, G. and Occorso, G. (1981). Drip irrigation trials on mandarin (*Citrus reticulata* Blanco) in Sicily. *Proc. Int. Soc. Citriculture.* Vol. **2**, pp. 506-508.

Boston, K. D. and Prine, G.M. (1968). *Weekly Rainfall Frequencies in Florida*, Florida Coop. Ext. Service, Gr-S-**187**, Florida, USA.

Boswell, S.B. and Burns, R.M. (1973). Determination of fruit bearing zones in California citrus trees. *Calif. Agri.* **27**(12) : 3-5

Botham, J.R. (1963). Citrus growing in South Australia. *Bull.* **376**, p.30. Dept. *Agri.,* South. Australia, Australia.

Botson, K.D. and Prine, G.M. (1968). Weekly Rainfall Frequencies in Florida. *Florida Coop. Ext. Service Cir.*, Florida, USA.

Boul, S.W. (1972). Fertility Capability Soil Classification System. Soil Sci. Dept., North Carolina State Univ. *Ann. Rep.* (1971-1972), Raleigh, North Carolina, USA.

Boul, S.W. and Nicholaides, J.J. (1980). *Constraints to Soil Fertility Evaluation and Extrapolation of Research Results.* Soil Related Constraints to Food Production in Tropics, IRRI, Philippines.

Boul, S.W., Sanchez, P.A., Cate, R.B. Jr. and Granges, M.A. (1975). Soil fertility capability classification : a technical soil classification system for fertility management (E. Bornemisza and A. Alvarado, ed.), *Soil Management in Tropical America,* North Carolina State Univ., Raleigh, N.C., pp. 126-145.

Bowden, R.P. (1968). Processing quality of oranges grown in the near north coost area of Queensland. *Queensland J. Agri. Sci.* **25** : 93-113.

Bowman, F.T. (1955). Development of vigorous NSW citrus industry by pioneer horticulturist in the hills district. *Citrus News* **31**(8): 102-103.

Bowman, F.T. (1956). *Citrus Growing in Australia,* Angus and Robertson, Sydney, Australia.

Boxem, H.W., De Meester, T. and Smaling, E.M.A. (1987). *Soils of Kilifi area, Kenya* (ed.), Agric. Res. Rept. 929, Pudoc, Wageningen, p. 249.

Boynton, B. and Compton, O.C. (1943). Effect of oxygen pressure in aerated nutrient solution on production of new roots and on growth of root and top by fruit trees. *Proc. Am. Soc. Hort. Sci.* **37** : 19-26.

Bozalek, S.J. (1974). Identification of citrus quality in production areas for the purpose of climatological analysis. *Citrus Subtrop. Fruit J.* **488** : 13-15.

Bradford, G.R., Harding, R.B. and Miller, M.P. (1962a). Copper deficiency. *Calif. Citrog.* **47**: 406-409.

Bradford, G.R., Harding, R.B. and Miller, M.P. (1962b). Severe copper deficiency identified in southern California grapefruit . *Calif. Citrog.* **16**(7):6-7.

Bradford, G.R., Harding, R.B., and Miller M.P. (1964). Severe copper deficiency in orchard grapefruit trees. *Hilgardia* **35**: 323-327.

Bradsley, Howard. C. (1987). Fertility and fertilizers. *Proc. Fla. State Hort. Soc.* **100** : 184-185.

Brady, N.C. (1978). *The Nature and Properties of Soil.* McMillan Publishing Co., Inc., New York, p. 639.

Brady, N.C. (1992). *The Nature and Properties of Soil.* (10th ed.), McMillan Publisher Co., Inc. NY. USA.

Braga, J.M. (1970). The nutritional status of a citrus orchard and the influence of environmental factors on the contents of elements in the leaves. *Rev. Ceres.* **17**(91) : 61-76.

Brammer, H. (1971). Coatings in seasonally flooded soils. *Geoderma* **6** : 5-16.

Brams, E. A. and Fiskell, J. G. A. (1967). Variability of root and soil analysis of Valencia grove sampled in January. *Proc. Fla. State Hort. Soc.* **80** : 32-37.

Brar, S.P.S., Kumar, Dinesh, Singh, Balwindar and Chohan, G.S. (1986). Role of soil characteristics on the growth of kinnow mandarin. *Indian J. Hort.* **43** : 210-15.

Brar, W.S., Mann, S.S. and Chahill, B.S. (1988). Single tree sampling scheme for estimating plot yield of kinnow mandarin *South Indian Hort.* **36**(6) : 293-296.

Bray, R.H.(1948). *Diagnostic Techniques for Soils and Crops.* (H.B. Kitchen, ed.), American Potash Institute, Washington, DC., USA, pp. 53-86.

Bredell, G.S. (1979). The potential citrus industry in Botswana. *Study Report,* Department of Technical Services. R.S.A. pp. 1-13.

Bredell, G.S. and Barnard, C.J. (1974). Soil moisture conservation through mulching. *Citrus Subtrop. Fruit J.* **487** : 13-16.

Bredell, G.S., Barnard, C.J. and Human, C. (1976). Soil mulching practices : the effect of soil mulching on the root development of avocadoes and Valencias. *Citrus & Subtrop. Fruit. J.* **56** : 17-19.

Bredell, G.S. and Conradie, J. H. (1975). Replanting of citrus : a survey of the chemical composition of old and new soil. *Citrus & Subtrop. Fruit J.* **496** : 8-10.

Brinkman, R., Jongmans, A.G. and Miedema, R. (1977). Problem hydromorphic soils in north-east Thailand. 2. physical and chemical aspects, mineralogy and genesis. *Netherlands J. Agri. Sci.* **25** : 170-181.

Brlansky, R.H., Peterson, M.A., Timmer, L.W. and Graham, J.H. (1986). Formation of amorphous plugs in citrus trees with blight. *Phytopathol.* **76** : 1110.

Broembsen, L. Von. (1988). Current status of citrus cultivars in South Africa. *Citrus & Subtrop. Fruit J.* **641**: 10-13.

Brown, T.W. (1924). The propagation and cultivation of citrus trees in Egypt. *Bull.* **44**, p.88, Egypt Min. Agri., Tech. Sci. Serv., Egypt.

Brown, L.N. (1963). *Planting and Irrigation on the Contour.* Div. Agri. Sci. Univ. Calif., USA, *Circ. No.* **523**, p. 23.

Brown, C.B. and Bally, C.S. (1970). *Land Capability Survey of Trinidad and Tobago,* 4. *Soils of Central Trinidad.* Government Printing Office., Trinidad, p. 139.

Brown, J.R. and Cisco, J.R. (1984). An approved Woodruff buffer method for estimation of lime requirements. *Soil Sci. Soc. Am. J.* **48**:587-592.

Brown, J.C. and Wann J.C. (1982). Breeding for Fe deficiency: use for indicator plants. *J. Plant Nutr.* **5** : 623-635.

Browne, C.A. (1942). Liebig and the Law of the Minimum. (F.R. Moulton, ed.). *Liebig and after Liebig. Publication No.***16**, pp. 71-78, American Association for the Advancement of Science, Washington, DC, USA.

Browne, C.A. (1944). *A Source Book of Agricultural Chemistry.* (F. Verdoorn, ed.), The Chronica Botanica Co., Waaltham, MA.

Browne, L.T. and Platt, R.G. (1966). Citrus problems in west Fresno county. *Calif. Agric.* **20**(2): 14-16.

Brusca, J.N. and Haas, A.R.C. (1956). Growth of citrus as affected by molybdenum. *Calif. Citrog.* **41**:158-159.

Bryan, O.C. (1962). The response of young citrus trees to lime mixed with the soil profile in Leon-Immokalee fine sands. *Citrus Indus.* **43**(9): 9-10.

Bryan, O.C. (1963). Soil calcium and fruit quality. *Citrus Indus.* **44**(2):12.

Buch, J. (1975). Statistical crop forecasting in Israel. *Citrog.* **60**(3) : 297-298.

Buckman, H. O. and Brady, N. C. (1960). *The Nature and Properties of Soils* (6th eds.). The Mcmillan Co., New York, USA, p. 567.

Bukhbinder, A.A. (1990). Nitrogen content in leaves of Meyer lemon trees grown in Kolkhida lowland. *Subtropicheskie Kul'tury* (4) : 12-17.

Bukhbinder, A.A. and Kuppa, B.K. (1991). The problem of the relationship between carbohydrate and nitrogen metabolism in Meyer lemon in the Kolkhida lowland. *Subtropicheskie Kul'tury* (5) : 15-19.

Buol, S.W., Hole, F.D. and McCraken, R.J. (1973). *Soil Genesis and Classification.* Iowa State Univ. Press, Ames., Netherlands.

Burke, J.H. (1952). The Citrus industries of North Africa. *Foreign Agri. Rep.* **66**, p. 152, U.S. Dept. Agri., USA.

Burke, H.J. (1956a). Citrus industry of British Honduras-Jamaica Trinidad. *Report No.* **88,** Foreign Agriculture, United States Dept. Agri., USA.

Burke, H.J. (1956b). Citrus industry of Surinam. *Report No.* **89**, Foreign Agriculture, United States Dept. Agri., USA.

Burke, J.H. (1958). Citrus industry of Brazil. *Agri. Reptr. No.* **109**, U.S. Deptt. Agric. Serv., USA, p. 33.

Burke, J.H. (1961). Citrus industry of Spain. *Foreign Agri. Rep.* **56**, p. 54. U.S. Dept. Agri., USA.

Burke, J.H. (1962a). Citrus industry of Italy. *Agri. Reptr. No.* **59**, U.S. Deptt. Agric. Serv., USA, p. 79.

Burke, J.H. (1962b). The citrus industry of Mexico. *FAR* **121** p. 68. *Agric. Serv., U.S. Dep. Agric.*, USA.

Burke, J.H. (1963). Australia and New Zealand: citrus producers and markets in the southern hemisphere. *Foreign Agric. Rep.* **124,** p. 42. *U.S. Dep. Agric.* USA.

Burke, J.H. (1965). Citrus industry of Turkey. *Foreign Agric. Rep.* **127**, p.35, U.S. Dept. Agric., U.S.A.

Burke, J.H. (1966). Citrus industry of Greece. *Agri. Rep.* **129**, US Dept. Agri., USA.

Burke, J. H. (1967). The commercial citrus growing regions of the world. *The Citrus Industry.* (W. Reuther, H. J. Webber and L. D. Batchelor, ed.), Vol. **1**, Univ. of Calif., Davis, USA, pp. 40-189.

Burnett, Earl (1969). Profile modification for improved water intake and storage. *Great Plains Agr. Counc. Publ.* **34**:59-63.

Burnett, H.C., Nemec, S. and Patterson, M. (1982). A review with soil edaphic factors, nutrition and *Fusarium solani. Trop. Pest Manag.* **28**(4) : 416-422.

Burns, R.M., Lee, B.W., Aljibury, F.K. and Meyer, J.L. (1976). Control of hillside seepage in avocado and citrus orchards. *Calif. Agric.* **30** : 20-21.

Buross, J. (1978). Citrus situation in California and Arizona. Citrog. **63**(9):229-232.

Butterfield, H.M. (1965). History of orange, lemon and grapefruit in California. *Calif. Citrog.* **50**: 488-498.

Caetano, A.A., Rodriguez, O., Hiroce, R. and Bataglia, O.C. (1984). Nutritional survey of forty sweet orange groves in Bebedouro, SP, Brazil. *Proc. Int. Soc. Citriculture*. Vol. **1**, pp.151-154.

Cahoon, G.A., Morton, E.S. and Jones, W.W. (1959). Effects of various types of nitrogen fertilizers on root density and distribution as related to water infiltration and fruit yields of Washington Navel oranges in a long term fertilizer experiment. *Proc. Am. Soc. Hort. Sci.* **74** : 289-299.

Cai, K., Huang, R.X. and Hu, D.J. (1989). Harmonal regulation of physiological disorders and cell content transfer in pulp of satsuma orange. *Acta Hort.* **239**:435-438.

Cakmak, J., Atli, M., Kaya, R., Evilya, H. and Marschner, H. (1995). Association of high light and zinc deficiency in cold induced leaf chlorosis in grapefruit and mandarin trees. *J. Pl. Physiol.* **146**(3) : 355-360.

Calabrese, F. (1969). Citrus growing in Florida. *Ital. Agric.* **106** : 669-686.

Calabrese, F. (1976). Citrus growing in Mexico. *Italia Agricola.* **113**(3) : 73-84.

Calabrese, F. (1992). Fertilizer use in citriculture. *Rivista di Frutticoltura e di Ortofloricoltura* **54**(2) : 30-31.

Calabrese, F. and Marco, Di. L. (1981). Researches on the Forzatura of lemon trees. *Proc. Int. Soc. Citriculture.* Vol. **2**, pp. 520-521.

Calhoun, F.G., Carlisle, V.W., Caldwell, R.E., Zelazny, L.W., Hammond, L.C. and Breland, H.L. (1974). Characterization data for selected Florida soils. *Res. Rep. No.* **74-1**, Soil Sci. Dept., IFAS, Univ. Florida.

Calvert, D.V. (1970). Response of temple oranges to varying rates of nitrogen, potassium and magnesium. *Proc. Fla. State Hort. Soc.* **83** : 10-15.

Calvert, D.V. (1975). Nitrate, phosphate and potassium movement into drainage lines under three soil management systems. *J. Environ. Qual.* **4** : 183-186.

Calvert, D.V. (1977). Changes following the freeze of January, 1977 in the quality of fruit and foliage injury from twelve citrus rootstock-scion combinations. *Proc. Fla. State Hort. Soc.* **90** : 58-60.

Calvert, D.V., Cohen, M., Allen Jr., L.H. and Martin, F.G. (1981). Deep tillage, liming and drainage affects citrus growth and yields. *Proc. Int. Soc. Citriculture.* Vol. **1,** pp. 534-536.

Calvert, D.V., Ford, H.W., Stewart, E.H. and Martin, F.G. (1977). Growth response of twelve citrus rootstock - scion combination on a spodosol modified by deep tillage and profile drainage. *Proc. Int. Soc. Citriculture.* Vol. **1,** pp. 79-84.

Calvert, D.V., Ford, H.W., Steward, E.H. and Martin, F.H. (1978). Citrus response to soil profile drainage, deep tillage and liming. *Proc. Int. Soc. Citriculture.* Vol. **1**, pp. 217-219.

Calvert, D.V., Ford, H.W., Stewart, E.H. and Martin, F.H. (1979). Growth response of twelve citrus rootstock - scion combinations on a Spodosol modified by deep tillage and profile drainage. *Proc. Int. Soc. Citriculture.* Vol. **1**, pp. 79-84.

Calvert, D.V. and Reitz, H.J. (1966). Response of citrus growing on calcareous soil and foliar application of magnesium. *Proc. Fla. State Hort. Soc.* **79** : 1-6.

Camacho-Bustos, Saul. E. (1981). Citrus culture in high altitude American tropics. *Proc. Int. Soc. Citriculture.* Vol. **1,** pp. 321-325.

Cameron, S.H. (1939). Quantitative relationships between leaf, branch and root systems of Valencia orange trees. *Proc. Am. Soc. Hort. Sci.* **37** : 125-126.

Cameron, J.W. (1953). Some observations on citrus in southern Portugal. *Calif. Citrog.* **39**(2):57-59.

Cameron, J.W., Soost, R.K. and Lee, B.W. (1970). Quality of mandarins in the Ojai valley. *Citrog.* **55** : 319-322.

Cameron, S.H., Mueller, R.T. and Wallace, A. (1951). Measurement and chemical composition of the seasonal new growth of mature Valencia orange trees. *Proc. Am. Soc. Hort. Sci.* **58** : 11-13.

Cameron, S.H., Mueller, R.T. and Wallace, A. (1952). Influence of age of leaf, season of growth and fruit production on the size and inorganic composition of Valencia orange leaves. *Proc. Am. Soc. Hort. Sci.* **60** : 42-50.

Cameron, S.H. and Schroeder, C.A. (1945). Cambial activity and starch cycle in bearing orange trees. *Proc. Am. Soc. Hort. Sci.* **46** : 55-59.

Cametti, C. (1987). Farm irrigation in the Piana di Sibari in Calabria, Irrigation of citrus and other crops. *Irrigazione.* **34**(1) : 17-22.

Camp, A.F. (1947). Magnesium in citrus fertilization in Florida. *Soil Sci.* **63** : 43-52.

Camp, A.F., Chapman, H.D., George Bahrt, M. and Parker, E.R. (1941). Symptoms of citrus malnutrition. (Gove Hambidge, ed.), *Hunger Signs in Crops.* Amer. Soc. Agron., Washington, USA, pp. 267-311.

Campbell, A.S., Young, A.W., Linvingstone, L.G., Wilson, M.A. and Walker, T.W. (1977). Characterisation of poorly ordered aluminosilicates in a Vertic Andosol from New Zealand. *Soil Sci.* **123**:362-368.

Campbell, C.W. and Goldweber, S. (1979). Rootstocks for citrus in the limestone soils of southern Florida. *Proc. Fla. State Hort. Soc.* **102** : 290-292.

Canales, S., Sanchez, J.A. and Abadia, J. (1974). Adaptability problems for citrus trees on soils of Vega Baja on the segura river in Alicante, Spain. *Ist Congresco Mundial de Citricultura,* 1973, Vol. **1**, pp. 13-19.

Cannell, R.Q. and Jackson, M.B. (1981). Alleviating aeration stresses. *Modifying the Root Environment to Reduce Crop Stress.* (G.F. Arkin and H. M. Taylor, ed.), Amer. Soc. Agric. Eng. St. Joseph Michigan, pp. 141-192.

Cannon, W.A. (1925). The physiological features of roots with special reference to the relation of roots to aeration of the soil. *Publ. Carnegie Inst.,* Washington, USA, p. 368.

Cantarella, H., Quaggio, J.A., Bataglia, O.C. and Van Raij B. (1992). Response of citrus to NPK fertilization in a network of field trials in Sao Paulo State, Brazil. *Proc. Int. Soc. Citriculture.* Vol. **2**, pp. 607-612.

Caprio, J.M. (1956). Analysis of the relation between regreening of Valencia oranges and mean monthly temperatures in southern California. *Proc. Am. Soc. Hort. Sci.* **10** : 20-23.

Cardenas Teresa, J.M. De (1986). Techniques of citrus cultivation in Mexico. *Levante Agricola.* **25**(267/268) : 101-110.

Carlisle, V.W. and Nesmith, J. (1972). *Florida Soil Identification Handbook - Hyperthermic Temperature Zone.* Soil Sci. Dept., IFAS, Univ. Florida, USA.

Carpena, O., Guillen , M.G. , Fernandez , F.G. and Caro, M. (1969) . Secondary salinization of citrus soils in southeasteran Spain . *Proc. Ist Int. Citrus Symp.* Vol. **3**, pp. 1825-1831.

Carpena, O., Llorente, S. and Leon,A. (1968). Nutritioal imbalances in lemon induced by high leaf iron levels. *Mem. Gen. II,* Colog. Eur medit. Centr. Fert. Plant. Cult., Serville (Spanish). pp. 531-545.

Carpena-Artes, O. and Alcarez, C.F. (1990). The lemon tree in Spain: Thirty five years of research. *Proc. Int. Citrus Symp.* Guangzhou, China, pp. 318-322.

Carrasco, J.M., Curiat, P. and Primo, E. (1969). Trace metal deficiencies in orange trees II. Response of trees to corrective treatments for deficiency of zinc, manganese and boron. *Rer agroquim. Technol. Aliment* **9**: 554-563.

Carson, B.C. (1985). Erosion and sedimentation processes in the Nepalese, Himalayas. *ICIMOD Occasional Paper No.* **1**, Kathmandu, Nepal.

Cary, P.R. (1968). The effects of tillage, non-tillage and nitrogen on yield and fruit composition of citrus. *J. Hort. Sci.* **43** : 299-315.

Cary, P.R. (1970). Growth yield and fruit composition of Washington navel orange cuttings affected by root temperature,nutrient supply and crop load. *Hort. Res.* **10**:20-33.

Cary, P.R. (1971). Citrus soil management. *Farmer's Newsletter* **112** : 25-27.

Cary, P.R. (1972). The residual effects of nitrogen, calcium and soil management treatements on yield, fruit size and composition of citrus. *J. Hort. Sci.* **47**(4) : 479-491.

Cary, P.R. (1981). Soil management factors affecting growth and yield of citrus trees. *Proc. Int. Soc. Citriculture.* Vol. **2**, pp. 527-530.

Cary, P.R. and Etans, G.N. (1972). Long term effects of soil management treatments on soil physical conditions in a factorial citrus experiment. *J. Hort. Sci.* **47**(1) : 81-91.

Cary, P.R. and Weerts, P.G.J. (1977). Crop management factors affecting growth, yield and fruit composition of citrus. *Proc. Int. Soc. Citriculture.* Vol. **1**, pp. 39-43.

Casella, D. (1935). L'agrtumicoltura Siliciana. *Ann. Rep.* Staz, Sper. Frutt eAgrum., Acireale, Italy, N.S. **2.** pp. 1-147.

Cassin, P.J. (1968). Agronomic aspect of vegetative symptoms observed in cultivated citrus on acid soils in Corsica. *Mem. gen.* II. pp. 583-591. *Coloq. eur. medit. Contr. Fert. Plant Cult.,* Seville, Spain.

Cassin, P.J., Blondel, L., Bove, J.M., Bove, C., Jolivet, E., Lacoeuilhe, J., Lefleche, D., Lassois, L., Marchal, J., Martin-Prevel, P., Nicol, M.Z. and Moulinier, H. (1969a). General study of leaf analysis and citrus fruit analysis under Corsican environmental conditions. *Proc. Ist. Int. Citrus Symp.* Vol. **3**, pp. 1689-1712.

Cassin, P. J., Bourbeaut, J., Fougue, A., Furon, V., Gailard, J.P., LeBourdelles, J., Montagut, G. and Moruil, C. (1969b). The influence of climate upon the blooming of citrus in tropical areas. *Proc. Int. Citrus Symp.* Vol. **1,** pp. 315-324.

Cassin, P.J., Blondel Tribol, A.M., Marchal, J., Favreau, P., Perrier, X., Jusse, C., Brun, P. and Lassois, P. (1982). A molybdenum deficiency in citrus Corsica aggravated by sulphate applications. *Hort. Abst.* **52**(10). (Abstract No. 6989).

Cassin, P.J., Brun, P., Blondel-Tribol, A.M., Juste, C., Marchal, J., Lossois, P., Perrier, X. and Favreau, P. (1982). A molybdenum deficiency in citrus in Corsica increased by sulphate application. *Proc. Int. Soc. Citriculture.* Vol. **2**, pp. 551-554.

Cassin, P.J., Favreau, P., Marchal, J., Lossois, P. and Martin-Prevel, P. (1977). Influence of fertilisation on growth, yield and leaf mineral composition of clementine mandarin on three rootstocks in Corsica. *Proc. Int. Soc. Citriculture.* Vol. **1,** pp. 49-57.

Cassin, J. and Lossois, P. (1972). Preliminary results of a study of different management methods in Corsican citrus plantations. *Fruits* **27**(1) : 37-49.

Castel, J.R and Buj, A. (1992). Growth and evapotranspiration of young, drip-irrigated clementine trees. *Proc. Int.Soc. Citriculture.* Vol. **2,** pp. 651-656.

Castel, J. R. and Ginestar, C. (1996). Response of Clementine citrus trees to irrigation and nitrogen rates under drip irrigation. *Proc. Int. Soc. Citriculture.* Vol. **2,** pp. 683-687.

Castle, W.S. (1980). Fibrous root distribution of Pineapple orange trees on rough lemon rootstock at three tree spacing. *J. Am. Soc. Hort. Sci.* **105** : 478-480.

Castle, W.S. (1987). *Citrus Rootstocks. Rootstock for Fruit Crops* (R.C. Rom and R.F. Carlson, ed.), Wiley, New York, USA, pp. 361-399.

Castle, W. S., Krezdorn, A. H. and Gammon, N. Jr. (1975). Some physical and chemical characteristics of a deep well drained soils planted to citrus. *Proc. Fla. State Hort. Soc.* **88**: 23-29.

Castro, F.R.O. (1972). A short note on citrus growing in Mozambique. *Revista Agricola* **144** : 24-27.

Cate, R.B. Jr. and Nelson, L.A. (1965). International Soil Testing Series, *Tech. Bull.* **1**, Agric. Exp. Sta., North California, USA.

Cedeno-Maldonado, A., Gonzalez, W. and Fontanet, E. (1990). Performance of pummelo clones in the central mountain region of puerto Rico. *J. Agri. Univ. Puerto Rico* **74**(3) : 299-305.

Cerda, A.M., Nieves, M.A. and Guillen, M.G. (1990). Salt tolerance of lemon trees as affected by rootstock. *Irri. Sci.* **11**:245-249.

Chadha, K.L. and Awasthi, R.P. (1966). A review of work done on citrus decline in India and abroad. *Punjab Hort. J.* **6**: 17-34.

Chadha, K. L., Randhawa, N. S., Bindra, O. S. and Chohan, J. S.(1965). Report of the survey undertaken to investigate causes of citrus decline in Punjab. *Res. Rep.* Punjab Agric. Univ., Dec. 1965, p. 12.

Chadha, K.L. and Singh, H.D. (1989). Citriculture scenario of India. *Proc. Citriculture in North-West India*, pp. 21-65 Ludhiana, Punjab, India.

Chaenery, E.M. (1952). *The Soils of Central Trinidad.* Government Printing Office, Trinidad, p. 43.

Chaikiattiyos, S., Menzel, C.M. and Rasmussen, T.S. (1994). Floral induction in tropical fruit trees: Effects of temperature and water supply. *J. Hort. Sci.* **69**: 397-415.

Chakraborty, N.L. (1976). *Potassium in Soils, Crops and Fertilizers. Bull.* **10,** p. 214 Indian Soc. Soil. Sci., New Delhi, India.

Chakravarty, D.N. and Barua, J. P. (1983). Characterization and classification of the soils of citrus growing belts of hill districts of Assam. *J. Indian Soc. Soil Sci.* **31:**287-295.

Chakravarty, D.N., Sehgal, J.L. and Dev, G. (1978). Influence of climate and topography on the pedogenesis of aluminium- derived soils in Assam : Geographical settings, morphology and physico-chemical properties. *Indian J. Agric. Chem.* **11** : 77-81.

Chakrawar, V.R. and Singh, Ranjit (1977). Studies on citrus granulation II. Physiological and biochemical aspects of granulation. Haryana *J. Hort. Sci.* **6**(3-4):132-135.

Chakrawar, V.R. and Singh, Ranjit (1978). Studies of citrus granulation IV. Effect of some growth regulators on granulation and quality of citrus fruits. *Haryana J. Hort. Sci.* **7**(3-4):130-135.

Chalutz, E. and Roessler, Y.C. (1986). Production trends around the world – Israel. *Fresh Citrus Fruit* (F.Wardowski, S.Nagy and W. Grierson, ed.), AVI Publ. Co., Westport, Connecticut, pp. 165-1670.

Chamuah, G.S., Walia, C.S., Baruah, U. and Dutta, A.K. (1989). Soil Survey Report of Longding and Wakka Circles, District Tirap, Arunachal Pradesh, *ICAR Report No.* **152**, pp. 12-16.

Chanana, Y.R. , Nijjar, G.S. and Vij, V.K. (19840. Incidence of granulation in citrus under different agro-climatic conditions of Punjab. *J. Res. PAU,* **21**(1):45-48.

Chance, J.E. and Rathwell, P.J. (1979). A mathmatical model for predicting annual minimum temperature for Texas citrus. *J. Rio Grande Valley Hort. Soc.* **32** : 61-66.

Chand, J.P. and Mandal, Biswapati (2000). Nature of acidity in soils of West Bengal. *J. Indian Soc. Soil Sci.* **48**(1):20-26.

Chandel, A.S., Singh, M. and Singh T. A. (1976). *Nitrogen in Soils, Crops and Fertilizers. Bull.* **13,** p. 20 Ind. Soc. Soil Sci., New Delhi, India.

Chang, Susan, Huang, Welting, Lian, Shen, Chang, Aihwa and Wiu, Woanlifi (1994). Research on leaf diagnosis criteria and its application to fertilizer recommendations for citrus orchards in Taiwan-Taipei. *Extension Bull.* **396**, pp. 1-17, ASPAC, Food and Fertilizer Technology Centre, Yuan, Taiwan.

Chang, S.C. and Jackson, M.L. (1957). Fractionation of phosphorus. *J. Soil Sci.* **84**: 133-144.

Chang, M.S., Lin, W.C. and Lin, L.H. (1987). The effects of nutrient status on the fruit quality of Valencia orange and sweet sop grown in Taitung. *Taiwan Agri.* **23**(4) : 31-39.

Chanturiya, I.A. (1972). Trials with different boron rates in a mandarin plantation. *Subtropicheskie Kul'tury* (1): 78-81.

Chanturiya, I.A. (1974). The effectiveness of liming acid soil under citrus crop. *Subtropicheskie Kul'tury* (5): 74-77.

Chanukvadze, F. Sh. (1989). Use of agrometrological information in intensive systems of citrus cultivation. *Subtropicheskie Kul'tury* (2) : 69-72.

Chanukvadze, F. Sh. (1990). The effectiveness of potassium fertilizers in citrus orchards on Krasnozem soils. *Subtropicheskie Kul'tury* (1) : 99-104.

Chapman , H.D., Harrietan, J. and Rayner, D.S. (1969) . Effects of variable maintained chloride levels on orange growth , yield and leaf composition. *Proc. Ist. Int. Citrus Symp.* Vol. **3,** pp 1811-1817.

Chapman, H.D. (1938). Citrus fertilization question. *Calif. Citrog.* **23**: 332, 361-63.

Chapman, H.D. (1949). Citrus leaf analysis. *Calif. Agric.* **3**:10-14.

Chapman, H.D. (1954). Citrus in Souther America. *Calif. Citrog.* **40**(2): 47, 65-66.

Chapman, H.D. (1960a). Leaf and soil analysis in citrus orchards. *Manual* ***25,*** Calif. Expt. Sta., Ext. Service, USA.

Chapman, H.D. (1960b). Leaf and soil analysis as a guide to fertilizers practices. *Calif. Citrog.* **45** : 230-233.

Chapman, H.D. (1961). The evaluation and management of citrus soil. *Indian J. Hort.* **18:**251-276.

Chapman, H.D. (1963). Citrus observations in Australia. *Calif. Citrog.* **49**(1):39-40.

Chapman, H. D. (1968). The mineral nutrition of citrus. (W. Reuther , L. D. Batchelor and H. J. Webber, ed.). *Citrus Industry.* University of California Press, Berkeley, USA, pp. 127-289.

Chapman, H.D., Brown, S.M. and Ravner, D.S. (1945). Nutrient deficiency in citrus. *Calif. Citrog.* **30:** 162-63, 198-202, 216-217.

Chapman, H.D., George, F. Jr., Liebig, J. Von. and Parker, E.R. (1939). Manganese studies : Calif. soils and citrus leaf symptoms of P.I. *Calif. Citrog.* **24**: 427- 454.

Chapman, H.D. and Kelley, W.P. (1943). The mineral nutrition of citrus. *Citrus Industry* (H.J. Webber and L.D. Batchelor, ed.), Vol. **1**, Univ. of Calif. Press, Berkeley, USA.

Chapman, H.D. and Liebig, G.F. (1952). Field and laboratory studies of nitrate accumulation in soils. *Soil Sci. Soc. Am. Proc.* **16** : 276-282.

Chapman, H.D., Liebig, G.F. and Rayner, D.S. (1949). A lysimeter investigation of nitrogen gains and losses under various systems of cover cropping and fertilization and a discussion of error sources. *Hilgardia* **19**: 57-128.

Chapot, H. (1964). Moraccan citrus growing. *Cah. Rech. Agron.* (18): 1-25.

Chapot, H. (1974). Citrus production problem in the near and middle east. *Span* **17**(1) : 29-31.

Chapot, H. (1975). The Citrus Plant. *Citrus Technical Monograph No.* **4** (E. Hafliger, ed.), Ciba-Geigy Ltd., Basle, Switzerland, pp. 6-13.

Chaudhary, J.S. and Pareek, B.L. (1976). Exchangeable potassium reserve in soils of Rajasthan. *J. Indian Soc. Soil Sci.* **24**:57.

Chaudhary, J.S. (1992). Association of Zinc Copper with tree Sesquioxides and extractable micronutrients in and soil of Rajasthan. *J. Indian Soc. Soil Sci.* **40** : 289-292.

Cheema, G.S. and Bhat, S.S. (1929). The Die-Back Disease of Citrus and its Relation to Soil of Western India, *Bull.* **155**, pp. 1-48, Deptt. of Agri., Bombay.

Chen, Biaohu (1990). Study on the availability of iron in calcareous soils of orange orchard in Zhejiang Province, China. *Proc. Int. Citrus Symp.*, Guanzhou, China, pp. 508-512.

Chen, Yas Fei (1998). Advanced comprehensive cultural techniques for growing cultivar Pongan in hillside orchards. *South China Citrus* **27**(4):12-13.

Chen, Y. and Aviad, T. (1990). Effects of humic substances on plant growth. *Humic Substance in Soil and Crop Science Selected Readings.* Am. Soc. Agron., Madison, USA.

Chen, Z. and Dragavcev, A.P. (1958). Fruit growing on embarkments in China. *Priroda* **47**(11) : 95-96.

Chen, Hsitan., Huang, Hoyen and Lu, Chunjian. (1997). Effect of organic fertilizer and dolomite on the growth and fruit quality of Matou Peiju (*Citrus grandis* Osbeck, cv. Matou Peiyu). *Special Publication* **38**, pp. 69-76, Taichung District Agricultural Improvement Station, Tatsuen, Taiwan

Chen, Manfeng, He, Shanwen and Qiu, Shengzhang (1990b). Navel oranges in Hunan humid sub-tropical region. *Proc. Int. Citrus Symp.* (Huang, Bangyan and Yang, Qian, ed.), Guanzhou , China, pp. 319-322.

Chen, Z., Nie, H., Ji, Yu., Gan, J. and Wu, Y. (1990c). Identification of soil tolerance of some kind of citrus biotypes. *Proc. Int. Citrus Symp.* (Huang, Bangyan and Yang, Qian, ed.), Guangzhou, China, pp. 271-275.

Chen, M., Zhang, Q. and Xiao, L. (1990a). Effects of foliar phosphates spray on quality of Satsuma mandarin. *Proc. Int. Citrus Symp.* (Huang, Bangyan and Yang, Qian, ed.), Guangzhou, China pp. 457-460.

Cheng, Q.M. and Zeng, K.Y. (1991). Hongguang, the best strain of Nanfeng Guangju. *China Citrus* **20**(3) : 17.

Cheong, J.K. and Moon, D.K. (1989). The relationship between freezing tolerance and nutritional status in citrus leaves. *Res. Rep. Reural Develop. Adm. Horticulture* **31**(2):30-46.

Cherkezishvili, E.A. (1987). Mandarin productivity in relation to cultural pratices and yield fluctuations in relation to weather conditions. *Subtropicheskie Kul'tury* (5) : 131-135.

Chiba, T. (1965). Studies on the soil management. *Memoir of Soil and Fertilizer Problems on Tree Fruit Crops.* Ministry of Agric. and Forests.

Childs, J. F. L. (1953). Observation of citrus blight. *Proc. Fla. State. Hort. Soc.* **66**: 33-37.

Childs, J.F.L. (1964). Observations on citrus culture and problems in Surinam (Dutch Guiana). *Surinam Landb.* **12**: 57-61.

Childs, J.F.L. (1980). Florida citrus blight. Part II. Occurrence of citrus blight outside Florida. *Plant Dis. Reptr.* **63**:635-569.

Childs, J.F.L. (1981). Control of citrus blight disease. *Proc. Fla. State Hort. Soc.* **94** : 25-28.

Chitrakar, P.L. (1990). Planning, agriculture and farmers: strategy for Nepal. (K. Pradhan, ed.), p. 322, Kathmandu, Nepal.

Chiu, T.F. and Chang, S.S. (1985) Diagnosis and correction of boron deficiency in a citrus orchard of Taiwan. *Tech. Bull.*, **9**, pp. 1-11. ASPAC Food and Fertilizer Technology Center, China.

Chiu, T.S. and Chang,S.S. (1986). Diagnosis and correction of boron deficiency in the citrus orchard of Taiwan. *Soils and Fertilizers in Taiwan,* pp. 33-44.

Chkheidze, T.K. and Akhaladze, V.E. (1975). The effect of mulching humus-calcareous soil on mandarin yield. *Subtropicheskie Kul'tury* (1) : 68-70.

Chohan, G.S. (1966). Role of orchard management practices in controlling citrus decline. *Punjab Hort. J.* **6**: 76-82.

Choudhary, S. and Dutta, S. (1950). Vein splitting, boron deficiency diseases of citrus in Assam. *Sci. & Cult.* **15**:358-359.

Chundawat, B.S., Khimani, R.A. and Kikani, K.P. (1991). Nutritional survey of elite acid lime orchard in Gujarat. *Indian J. Hort.* **48**(3):183-186.

Chung, S.K., Oh, S.D. and Hong, S.B. (1976). Studies on determination of marginal altitude for citrus culture in Jeju Island. *Res. Rep.,* **18** : 77-83, Office Rural Develop. Hort., Agri. Engg.

Cicala, A. and Catarina, V. (1992). Potassium fertilization effects on yield, fruit quality and mineral composition of leaves of Tarocco orange trees. *Proc. Int.Soc. Citriculture.* Vol. **2,** pp. 618-620.

Cicca, V. De, Intrigliolo, F., Ippolito, A., Vanadia, S. and Guiffrida, A. (1988). Factors in Navelina orange splitting. *Proc. 6th Int. Citrus Congr.* (R. Goren and K. Mendel, ed.), Vol. **1**, pp. 535-545

Codina, J.C. and Namesny, C. (1996). Harvest and post-harvest handling of citrus fruit under adverse climatic conditions. *Proc. Int. Soc. Citriculture.* Vol. **2,** pp. 1203-1204.

Coelho, Y. DA. S. and Matos, C.R.R. (1991). Nutritional survey of citrus orchards in the state of Bahia. *Pesquisa Agropecuaria Brasileira* **26**(3) : 335-340.

Coelho, Y.S., Pompeu, Jr. J., Bastos, J.B., Dornelles, C.M.M., Souza, E.S. and Caldas, R.C. (1984). Maturation and quality of 'Pera' sweet orange in Brazil. *Proc. Int. Soc. Citriculture.* Vol. **2**, pp. 517-519.

Coetzee, J.G.K. (1992). Evaluating calcium borate as a boron fertilizer formature citrus trees. *Appl. Pl. Sci.* **6**(1) : 37-39.

Coetzee, J.G.K. (1995). Ridging soils for new citrus plantation. *Citrus J.* **5**(3) : 22-24.

Cohen, M. (1974). Diagnosis of young tree decline, blight and sand-hill decline of citrus by measurement of water uptake using gravity injection. *Plant Dis. Reptr.* **58**:801-805.

Cohen, M. (1980). Non-random distribution of trees with cittruls blight. *Proc. 89th Conf. Int. Organ. Citrus Virol.* (E. C. Calavan, S. M. Garnsey and L.W. Timmer, ed.), pp . 206-263. Univ. Calif., Riverside, USA.

Cohen, M., Calvert, D.V. and Pelosi, R.R. (1981). Diseases occurrence in a citrus and drainage experiment (SWAP) in Florida. *Proc. Int. Soc. Citriculture.* Vol. **1,** pp. 366-368.

Cohen, A., Lomas, I. and Rassis, A. (1972). Climatic effects on fruit shape and peel thickness in Marsh seedless grapefruit. *J. Am. Soc. Hort. Sci.* **97** : 768-771.

Cohen, M., Pelosi, R.R. and Brlansky, R.H. (1983). Nature and location of xylem blockage structures in trees with citrus blight. *Phytopathol.* **73** : 1125-1130.

Cole, P.J. and McCloud, P.I. (1985). Salinity and climatic effects on yields of citrus. *Australian J. Expt. Agric.* **25**(3) : 711-717.

Coleman, N.T., Kamprath, E.J. and Weed, S.B. (1968). Liming. *Adv. Agron.* **10** : 475-522.

Coleman, N.T. and Thomas, G.W. (1967). *Soil Acidity and Liming,* Am. Soc. Agron., Madison, Wisconsin, p.1.

Coleman, N.T., Weed, S.B. and McCracken, R.J. (1959). Cation exchange and cation exchangeable cations in Piedmont soil of North Carolina. *Proc. Soil Sci. Soc. Am.* **23:** 146-156.

Comas, J.O. (1966). Fertilization of orange tree in Spain. *Bull. Docum.* **45,** pp.31-40.

Comas, J.O. and Cros, S.A. (1984). Micronutrients in citrus. *Fert. & Agri.* **88** : 43-51.

Comissao De Fertildae Do Solo-RS/SC (1995). Recomendacoes de adubacao e, calagem para os estados do Rio Grande do Sul eSanta Cantarina –3th ed. Passo Fundo: EMBRAPA-CNTP, p. 222.

Connor, J.C. (1954). Effect of cultural practices on management status of soil and citrus trees under irrigation. *Australian J. Agric. Res.* **5.**

Continella, G. and Gentile, A. (1996). Performance of 'Miyagawa' Satsuma on three rootstocks in Sicily. *Proc. Int. Soc. Citriculture.* Vol. **1,** pp. 171-173.

Conway, G.R. (1987). The properties of agro-ecosystems. *Agri. Systems* **24** : 95-117.

Cooke, G.W. (1979). *The History of Soil Fertility.* English Language Book Society and cross by Lockwood staples, London.

Cooper, W.C. (1952). Influence of rootstock on injury and recovery of young citrus trees exposed to the trees of 1950-51 in the Rio Grande valley. *Proc. Rio Grande Valley Hort. Soc.* **6**: 16-24.

Cooper, W. C. (1957a). Periodicity of growth and dormancy in citrus- a review with some observation in the lower Rio Grande Valley of Texas. *J. Rio Grande Valley Hort. Soc.* **11** : 3-10.

Cooper, W.C. (1957b). Comparison of several iron-chelating agents in correcting iron chlorosis in Dancy tangerines in the Rio Grande Valley of Texas. *J. Rio Grande Valley Hort. Soc.* **11**: 11-13.

Cooper, C.M. (1958). Crinkle rind. *Citrus News* **34**(2): 21.

Cooper, W.C. (1961). Toxicity and accumulation of salts in citrus on various rootstocks in Texas. *Proc. Fla. State Hort. Soc.* **74**:95-104.

Cooper, W.C. and Edwards, C. (1950). Salt and boron tolerance of shary redblush grapefruit and valencia orange on sour orange and cleopatra mandarin rootstocks . *Proc. 4th Ann. Rio Grande Valley Hort. Inst.* **4** pp. 58-79.

Cooper, W. C. and Garton, B. S. (1952). Toxicity and accumulation of chloride salts on various rootstocks. *Proc. Am. Soc. Hort. Sci.* **59**: 143-146.

Cooper, W. C., Garton , B.S. and Edwards , C . (1951). Salt tolerance of various citrus rootstocks . *Proc. 5th Ann. Rio Grande Valley Hort. Inst.* pp. 46-52, USA.

Cooper, W. C., Garton, B.S. and Wilson, E. O. (1952). Ionic accumulation in citrus as influenced by rootstock and scion and concentration of salts and boron in the substrate. *Pl. Physiol.* **27**(1): 191-203.

Cooper, W.C., Hilgeman, R.H. and Rasmussen, G.K. (1966). Diurnal and seasonal fluctuations of trunk growth of the valencia orange as related to climate. *Citrus Indus.* **47**(7): 25-30.

Cooper, W.C., Oison, E.O., Maxwell, N. and Otey, G. (1956). Review of studies on adaptability of citrus varieties as rootstocks of grapefruit in Texas. *J. Rio Grande Valley Hort. Soc.* **10**:6-19.

Cooper, W.C. and Peynado , A. (1959) . Chloride and boron tolerance of young line citrus on various rootstocks . *Proc. Rio Grande Valley Hort. Soc.* **13**: 89-96.

Cooper, W.C., Peynado, A., Furr, J.R., Hilgeman, R.H., Cahoon, G.A. and Boswell, S.B. (1963). Tree growth and fruit quality of Valencia oranges in relation to climate. *Proc. Am. Soc. Hort. Sci.* **82** : 180-182.

Cooper, W.C., Rasmussen, G.K. and Waldon, E.S. (1969a). Ethylene evolution stimulated by chilling in citrus and persea sp. *Pl. Physiol.* **44**: 1194-1196.

Cooper, W.C. and Shull, A.V. (1953). Salt tolerance and accumulation of sodium and chloride ions in the grape fruit on various rootstocks grown in naturally saline Soil . *Proc. Rio Grande Valley Hort. Soc.* **7**: 107-111.

Cooper, W.C., Young, R.H. and Henry, W.H. (1969b). Effect of growth regulators on bud growth and dormancy in citrus as influenced by season of year and climate. *Proc. 1st. Int. Citrus Symp.* Vol. **1**, pp. 301-314.

Cope, Hugh (1988). Australian citrus industry. *Citrog.* **73**(5) : 101-107.

Cox, F.R. (1987). Soil Testing : Sampling, Correlation, Calibration and Interpretation. *Spec. Pub. No.* **21**, p. 97, Soil Sci. Soc. Am., USA.

Craddock, Garnet R. and Wells, Robert D. (1973). Entisols that show no profile development. (S.W. Buol, ed.), Soils of the Southern States and Puerto Rico. *Series Bull.* **174**, pp. 24-28, Southern Coop., Puerto Rico.

Craig, I.A. and Pisone, U. (1988). Overview of rainfed agriculture in north-east Thailand (C. Pairintira, K. Wallapapan, J.F. Para and E. Whitman, ed.). *Soil Water and Crop Management Systems for Rainfed Agriculture in North-east Thailand,* Khon Kaen University, Thailand, pp. 24-37.

Crescimanno, F.G. (1966). Modern citrus growing Italy. *Inf. Ortoflorofruttic* **7**: 571-609.

Crescini, F. (1965). The gardens of Benaco : What they were and what remains of them. *Ital. Agric.* **102**: 605-630.

Crider, F.J. (1927). Root growth of the citrus trees with practical applications. *Citrus Leaves.* **7** : 1-3, 27-30.

Crossley, C.M. (1966). African annual 1966. *Foreign Correspondents Ltd. London, U.K.*

Culbert, D.L. and Ford, W.W. (1972). The use of a multicelled apparatus for anaerobic studies of flooded root systems. *HortScience* **7** : 29-31.

Da Ponte, A.M. (1964). Notes on citrus growing in South Africa. *Agron. Angol.* **20**, pp.83-146.

Dai, P.A., Zheng, S.X., Luo, C.X., Li, M.D., Huang, F.Q., Qiao, D.L. and Ding, J.Y. (1993). The effect of long term application of P, K and Mg to satsuma mandarin on red soil. *China Citrus* **22**(1) : 11-13.

Damigella, P. (1965). Relationships between the soil and woody plants. Preliminary contribution from observations carried out in an orange grove in eastern Sicily. *Teen Agric.* **17**: 274-295.

Damigella, P. and Tribulato, E. (1973). Qualitative characteristics and the course of ripening of oranges produced in an important citrus growing region of Calabria, the Piani di Rosarno. *Tecnica Agricola* **25**(4/5) : 269-300.

Dan, J. (1979). The soil pattern of arid regions of Israel and its effect on present and potential land use (G. Golany, ed.). *Arid Zone Settlement Planning, the Israel Experience,* Pergamon Press, New York, pp. 214-252.

Dan, J., Gerson, R., Koyumdjisky, H. and Yaalon, D.H. (1981). Acidic soils of Israel. *Special Publi* **190**, Agric. Res. Org., Betdagan.

Dan, J. and Raz, Z. (1971). The soil association map of Israel (Scale 1: 250,000), Volcani Inst. Agri. Res., Bet Dagan and Soil Conserv. Drainage. Serv., Tel Aviv, Israel.

Dan, J. and Yaalon, D.H. (1980). Origin and distribution of soils and landscape in the northern Negev. *Studies in the Geography of Israel* **11** : 31-58.

Dan, J. Yaalon, D.H., Moshe, R. and Nissim, S. (1982). Evaluation of Reg. soils in southern Negev and Sinai. *Geoderma* **28** : 173-202.

Daraseliya, M.K. (1949). Red earths and Podzolic soils of Georgia and their utilization for subtropical crops. *Makharadze Anaseuli, Georgia.*

Daraselija, N.A. (1969). Some data on the microflora of podzolized red earth soils under citrus plantations in Abhazija. *Subtropicheskie Kul'tury* (6): 159-166.

Das, Soumitra and Biswas, B.C. (1998). Nutrient management in upland rainfed rice in tropical Asia. *Fert. News* **43**(12) : 119-123.

Das, M.C., Bora, P.K., Sharma, S.C. and Sharma, K.C. (1971). Manganese status of Assam soils. *Fert. News* **2**: 42-44.

Das, Madhumita, Singh, B.P. and Khan, S.K. (1997). Effect of major land uses on soil characteristics of Alfisols in Meghalaya. *J.Indian Soc. Soil. Sci.* **45**(3) : 547-557.

Das, T.H., Sarkar, Dipak and Sehgal, J. (1996). Soil acidity on steep slope of Sikkim at higher altitude. *J. Indian Soc. Soil Sci.* **44**(2):319-321.

Dasberg, S., Bielorai, H. and Erner, Y. (1983). Nitrogen fertigation of shamouti oranges. *Pl. Soil* **75**:41-51.

Dasberg, S., Bielorai, H., Haimowitz, A and Erner, Y. (1991). The effect of saline irrigation water on Shmouti orange trees. *Irri. Sci.* **12**:205-211.

Dass, H.C., Srivastava, A.K., Lallan Ram and Shyam Singh (1998). Flowering behaviour of Nagpur mandarin (*Citrus reticulata* Blanco) as affected by vertical variation in physico-chemical composition of black soils. *Indian J. Agric. Sci.* **68:** 692-694.

Datta, M., Bhattacharya, B.K. and Saikh, H. (2001). Soil fertility - A case study of shifting cultivation sites in Tripura. *J. Indian Soc. Soil Sci.* **49**(1) : 104-109.

Datuadze, O. (1977). The effectiveness of magnesium fertilizer in citrus crop. *Hort. Abst.* **47**(3) : 267-273.

Daulta, B.S., Kumar, P. and Singh, Devi (1986). Effect of $ZnSO_4$, cytozyme and 2,4-D on fruit retention and quality of kinnow, a mandarin hybrid. *Haryana J. Hort. Sci.* **15**(1/2):14-17.

Davenport, T. L. (1979). Effect of prunning and chilling temperature on flower induction in Tahiti lime. *Pl. Physiol.* **63** : 65-70.

Davenport, T. L. (1983). Daminozide and gibberellin effects on floral induction of *Citrus latifolia. HortScience* **18** : 947-994.

Davenport, T.L. (1984). Importance of iron to plants grown in alkaline soils. *Proc. Fla. State Hort. Soc.* **96** : 188-192.

Davenport, T. L. (1990a). Citrus flowering. *Hort. Rev.* **12** : 688-692.

Davenport, T. L . (1990b). Leaves not necessary for floral induction of *Citrus latifolia. Proc. Growth Regulator Soc. Am.,* pp 18-19, 17th Ann. Meeting, St. Paul, August 5-9, 1990, Minnesota, USA.

Davidson, J.M. and Hammond, L.C. (1977). Soil physical aspects of root development. *Proc. Soil & Crop Sci. Soc. Fla.* **36** : 1-4.

Davies, F.S. (1979). Nitrogen fertilization research on mature, bearing orange trees – 1880-1979. *Citrus Indus.* **78**(5): 41-45.

Davies, C.E., Ahmad, N. and Jones, R.I. (1971). Hydrothermal and dry heat fixation of K by soil clays and the effect on CEC, surface area and mineralogy. *Clay Minerals* **9** : 219-230.

Davies, F.S. and Albrigo, L.G. (1994). *Citrus*. CAB International, United Kingdom, pp. 52-53.

Davitadze, M.M. (1991). Effect of liming on the productivity of young lemon orchards and fruit biochemical indices. *Subtropicheskie Kul'tury* (4) : 92-95.

De Barreda, D.G., Tarancon, J., Lorenzo, E. and Legaz, F. (1988). Replanting problems in citrus. A case of Washington Navel on *Troyer citrange. Proc. 6th Int. Citrus Cong.* Vol. **2**, pp. 977-981.

De Fossard, R.A. (1962). Red blotch on lemons. *Fmg. South Afr.* **39**(11): 6-7, 9-11.

De la Rosa, D. and Carlisle, V.W. (1978). Correlation of productive capacity and estimated yields for selected Florida soils. *Proc. Soil & Crop Sci. Soc. Fla.* **37**: 134-138.

De Leon, J . (1955) . Citrus growing in Philippines . *Bull.* **40**, Dept. of Agri. & Nat. Res., Rep. of Philippines.

De Mello, F.A. (1970). Competition between magnesium amendments in sweet orange. *Rev. Agric. Piracicaba* **44** : 103-114.

De Saussure, Th. (1804). *Chemical Researches about the Vegetation*. Nyon Widow, Paris, France.

De Villiers, J.I. (1942). Trace element deficiencies in citrus. Results obtained eastern transvaal. *Fmg. South Africa* **17** : 337-340.

De Villiers, J.I. (1969). The effect of differential of fertilization on the yield, fruit quality and leaf composition of oranges. (H.D. Chapman, ed.). *Proc. Ist. Int. Citrus Symp.* Vol. **3**, pp. 1661-1668, Univ. Calif., Riverside, Calif., USA.

Deb, D.L. (1997). Micronutrient research and crop production in India. *J. Indian. Soc. Soil Sci.* **45**(4) : 675-692.

Debnath, A., Bhattacharjee, T.K. and Debnath, N.C. (2001). Behaviour of inorganic phosphate fraction in limed acid soils. *J. Indian Soc. Soil Sci.* **48**(4): 829-831.

Deidda, P. and Pala, M. (1975). Citrus growing in Corsica. *Italia Agricola.* **112**(6) : 38-46.

Dent, M., Schulze, R.E. and Angus, G.R. (1988). Crop Water Requirements Deficits and Water Yield for Irrigation Planning in South Africa. *WRC Report* **118/88**, p. 183, South Africa.

Dermetriades, S.D., Holevas, C.D. and Gavalas, N.A. (1963). Boron toxicity in citrus trees in Greece. *Ann. Inst. Phytopath. Benaki* **4**: 118-121.

Desai, U.T., Choudhari, K.G. and Choudhari, S.M. (1986a). Studies on the nutritional requirements of sweet orange. *J. Maharashtra Agri. Univ.* **11**(2) : 145-147.

Desai, U.T., Choudhari, K.G., Choudhari, S.M., Pawar, R.B. and Bhore, D.P. (1986b). Studies on the nitrogen nutrition of Mosambi. *J. Maharashtra Agri. Univ.* **16**(3) : 233-239.

Desai, U.T., Choudhary, S.M., Shirsath, M.S. and Kale, P.N. (1991). Studies on the effect of foliar application of micronutrients in Mosambi sweet orange. *Maharashtra J. Hort.* **5**(2):29-31.

Devadas, V.S., Kuriakose, J.M. and Kannan, K. (1988). Early performance of mandarin oranges (*Citrus reticulata* Blanco) on different rootstocks in the submontane region of Wynad in Kerala. *Agri. Res. J. Kerala* **26**(2) : 183-197.

Devezla, M.C. (1968). Citrus culture in Mosambique. *Publ. Serv. Agric. Vet. Mocamb. Ser. B. : Divulg.* **33**, p. 156.

Devi, D.D., Srinivasan, P.S. and Balakrishnan, K. (1996). Leaf nutrient composition, chlorosis and yield of sathgudi orange as affected by micronutrient application. *South Indian Hort.* **45**(1/2): 26-29.

Devi, D.D., Srinivasan, P.S. and Balakrishnan, K. (1997). Carbonic anhydrase activity as an indicator of zinc status of sathgudi orange. *Orissa J. Hort.* **24**(1/2):66-68.

Dey, J. K. and Singha, D.D. (1998). Nutritional status of healthy and declining citrus (*Citrus reticulata*) orchards. *Indian J. Agric. Sci.* **68** : 139-143.

Dhandar, D.G. and Singh, Ranjit (1989). Root studies in grapefruit (*Citrus paradisi* Macf.) using radiotracer technique. *Sci. Hort.* **40**(2) : 113-118.

Dhar, O.N., Mandal, B.N. and Ghosh, G.C. (1978). Heavy rainfall stations of India. *Indian J. Power River Valley Develop.* **28**(2) : 47-50.

Dhatt, A.S. (1989). Nutrient management in citrus with special reference to Kinnow. *Proc. Citriculture in North-Western India*, Ludhiana, Punjab, India, pp. 157-167.

Dhatt, A.S. and Dhiman, J.S. (1999). Citrus decline. *Proc. Nat. Semi. New Horizons in Production and Post-harvest Management of Tropical and Subtropical Fruits* (Room Singh, ed.), *Hort. Soc.* India, New Delhi,India, pp. 91-11.

Dhillon, W.S. and Dhatt, A.S. (1988). Nutrient status and productivity of Kinnow orchards in Ferozepur district. *Punjab Hort. J.* **28**:7-13.

Dhingra, D.R. and Kanwar, J.S. (1963). Soil factors and chlorosis of citrus. *Punjab Hort. J.* **3**(1): 55-59.

Dhingra, D.R., Bhumbla, D.R., Randhawa, N.S. and Kanwar, J.S. (1966). Effect of different concentrations of zinc sulphate sprays on the incidence of citrus chlorosis. *Punjab. Hort. J.* **6**(1-4):83-87.

Dhingra, D.R., Kanwar, J.S., Bhumbla, D.R. and Sehgal, J. L. (1965). Soils of the proposed citrus belt of the Punjab II. Physico-chemical analysis. *J. Research PAU, Ludhiana* **2** : 154-160.

Di Boixo (1962). A check on the NaCl content of the soils of Cheraia and Ould Othaman carried out from 5-8 April. *C.R. Acad. Agric. Fr.,* **49**: 416-421.

Dijkerman, J.C. (1969). Soils of Sierra Leone, African Soils **14** : 185.

Di Martino, V. (1662). The garden Lake Garda. *Ital. Agric.* **99**: 1089-1094.

Dikshit, N.N. (1958a). Preliminary studies on citrus die-back in Coorg II – Effect of micro element sprays and irrigation on occurrence of chlorosis. *Sci. & Cult.* **24**: 91-94.

Dikshit, N.N. (1958b). Reduction in intensity of chlorosis on mandarin plants (*Citrus reticulata* Swingle). *Curr. Sci.* **28**: 207-208.

Ding, S., Zhang, X., Bao, Z. and Liang, M. (1990). *Citrus Ichangensis* in Yunnan. *Proc. Int. Citrus Symp.* (H. Bangyan and Y. Qian, ed.), Guanzhou, China, pp. 170-181.

Divakar, B.L., Bhagat, K.N., Tewari, J.C. and Mehta, N.S. (1989). Soil fertility status of horticultural areas in Nainital district of U.P. hills (hilly region). *Prog. Hort.* **21**:695-699.

Diware, D.V. and Kolte, S.O. (1990). Analysis of factors responsible for decline of citrus in Vidarbha region. II. Role of soil manuring, bahar, yield and stress period. *PKV Res. J.* **14**(2) : 119-122.

Dixit, C.K., Yamadagni, R. and Jindal, P.C. (1977). A note on the effect of micronutient spray on quality of kinnow – A mandarin hybrid. *Haryana J. Hort. Sci.* **6**(3/4): 153-154.

Dixit, C.K., Yamadagni, R. and Jindal, P.C (1978). Effect of foliar application of zinc and iron on chlorosis and yield of kinnow a mandarin hybrid. *Prog. Hort.* **10**(1):13-19.

Dixit, C.K., Yamadagni, R. and Jindal, P.C.(1979). Mineral composition of kinnow (a mandarin hybrid) leaves as affected by foliar application of zinc and iron. *Haryana J. Hort. Sci.* **8**:1-3.

Dodson, P.G.C. (1964). Swaziland's citrus industry. *World Crops* **16**(4): 48-50.

Dodson, P.G.C. (1966). Damage to citrus fruit by wind. *South Afr. Citrus J.* **393**:4-5, 7-11.

Dokuchayev, V.V. (1949). Izbr. Soch. T.2 (Collected Works, Vol. **2**), Moscow.

Dokuchayev, V.V.(1950). Data for land evaluation in Nizhegorodskaya province. Collected writings, Moscow, Izd-vo ANSSSR. Vol. **4**.

Dolui, A.K. and Sarkar, R. (2001). Influence of nature of soil acidity on lime requirement of two Inceptisols and an Alfisol. *J. Indian Soc. Soil Sci.* **49**(1) : 195-198.

Donadio, L.C., Banzatto, D.A., Sempionato,O.R. and Enciso Garay, C.R. (1996). Grapefruit cultivar evaluation. *Proc. Int. Soc. Citiculture.* Vol. **1,** pp. 207-209.

Donnelly, M. (1947). Teh San Joaquin family of soils. *Calif. Citrog.* **32**(9): 385-403.

Doorenbos, J. and Kassam, A.H. (1979). Yield response to water. FAO, Irrigation and Drainage Paper No. **33**, p. 193.

Doran, J.W. and Parkin, T.B. (1994a). Defining and assesssing soil quality. (J.W. Doran, D.C. Coleman, D.F. Bexdicek and D.F. Stewart, ed.). Defining Soil Quality for Sustainable Environment. *Special Publication No.* **35**, pp. 3-21, Soil Sci. Soc. Am. and Am. Soc. Agron., Madison, USA.

Doran, J.W. and Parkin, T.B. (1994b). Defining and assessing soil quality, (J.W. Doran, D.C. Coleman, D.F. Bexdicek and D.F. Stewart, ed.). Defining Soil Quality for a Sustainable Environment. *Special Publication No.* **35**, pp. 3-21, Soil Sci. Soc. Am. and Am. Soc. Agron., Madison, WI.

Dos Anjos, Lucia Helena, C., Franzmeier, D.P. and Schulze, D.G. (1995). Formation of soils with plinthite on a toposequence in Maranhao state, Brazil. *Geoderma* **64**:257-279.

Douglas, G.K. (1984). *Agricultural Sustainability in a Changing World Order,* Westview Press, Boulder, USA.

Douglas, T.J. and Sykes, S.R. (1985). Phospholipid, galactolipid and free sterol composition of fibrous roots from citrus genotypes differing in chloride exclusion ability. *Plant, Cell and Environment* **8**:693-699.

Douglas, T.J. and Walker, R.R. (1984). Phospholipids, free sterols and adenosine triphosphate of plasma membrane enriched preparations from roots of citrus genotypes differing in chloride exclusion ability. *Pl. Physiol.* **62**:51-58.

Dris, R. (1997). Effect of NPK fertilization on clementine grown in Algeria. *Acta. Hort.* **448** : 375-381.

Du Sautoy (1992). Field trial evaluation of Valencia fruiting terminal leaf P, K and Mg norms for a low base status loamy sand. *Proc. Int. Soc. Citriculture*. Vol. **2,** pp. 547-550.

Dube, S.D. and Saxena, H.K. (1971). Effectiveness of zinc chelates (Zn EDTA) versus zinc sulphide ($ZnSO_4$) in orange (*Citrus sinensis* Osbeck). *Punjab Hort. J.* **11**(3/4) 246-250.

Ducharme, E.P. (1971). Soil temperature in Florida citrus groves. *Bulletin No.* **747**, p. 15, Florida Agri. Expt. Station, USA.

Ducharme, E.P. and Suit, R.F. (1962-1964). Citrus tree decline. *Punjab Fruit J.* **26/27**: 138-142.

Dumbadze, N.M. (1977). The effect of ecological factors on the development of satsumas. *Subtropicheskie Kul'tury* (12) : 88-89.

Duncan, J. (1969). Citrus nutrition survey - M.I.A. *Fmrs. Newsletter* **105** : 17-19.

Dutt, Som and Bhambhota, J.R.(1967). Effect of different concentations of zinc on the incidence of chlorosis in sweet orange. *Indian J. Hort.* **24**: 50-59.

Dutta, S. (1954). Origin and history of citrus fruits of Assam. *Indian J. Hort.* **15** : 146-153.

Dutta, S. (1959). Assam's citrus wealth needs and outlook. *Indian Hort.* **4**: 28-32.

Dutta, M. and Gupta, R.K. (1984). Comparable efficiency of single super phosphate, bone meal and rock phosphate alone and in various combinations in augmenting crop yields in Nagaland. *Indian Agricst.* **28**:221.

Dutta, M. and Munna Ram (1993). Status of micronutrients in some soil series of Tripura. *J. Indian Soc. Soil Sci.* **41**(4):776-777.

Dutta, Dipak, Sah, K.D., Sarkar, Dipak and Reddy, R.S. (1999). Quantitative evaluation of soil development in some Alfisols of Andhra Pradesh. *J. Indian Soc. Soil Sci.* **47** (2): 311-315.

Dvorak, K.A. (1988). Indigenous soil classification in semiarid tropical India. Resource Management Prog. Econ. Group Prog. *Rep.* **84**, ICRISAT, Hyderabad, India.

Dyers, T.G.J. and Gillooly, J.F. (1979). On the provision of one year ahead forecast for Valencia orange production in the Eastern Transvaal. *Agroplantae* **11**(1) : 27-28.

Dzhincharadze , G.D. (1989). The effect of nitrogen rate on the productivity of a mandarin stock plant orchard. *Subtropicheskie Kul'tury* (3): 64-68.

Ebeling, Walter (1979). *The Fruited Plain.* Univ. Calif. Press, Berkeley, USA.

Echeverry Escovar (1969). Factors which affect fruit quality. *Agric. Trop., Bogota* **25** : 554-561.

Eck, H. V. and Unger, P.W. (1985). Soil profile modification for increasing crop production. *Advances in Soil Sciences* **1:** 65-83.

Economides, C.V. (1976). Effects of certain cultural practices on tree growth, yield and fruit quality of Valencia oranges in Cyprus. *J. Hort. Sci.* **51** : 545-549.

Edmeades, D.C., Wheeler, D.M. and Clinton, O.E. (1985). The chemical composition and ionic strength of soil solutions from New Zealand topsoils. *Australian J. Soil Res.* **23**:151-165.

Edward Raja, E. (1999). Studies on suitability of predominant soil types of India for acid lime. *Abst.* Int. Symp. Citriculture, Indian Society of Citriculture, Nagpur, Maharashtra, India p. 45.

Egashira, K., Nakashima, S. and Fujiyama, M. (1990). Quantification analysis of the contribution of environment, soil and management factors to crop yields - case study of the yield of rice, potato and orange in Nagasaki prefecture. *Sci. Bulletin, Agriculture Faculty, Kyushu University,* **45**(1-2) : 9-21.

Egorashvili, N.V., Tavadze, A.M. and Romanadze, A.D. (1991). Determination of optimal rate of micronutrients based on field trials in citrus plantation. *Subtropicheskie Kul'tury* (5):71-79.

Eisenberg, J., Dan, J. and Koyumdjsky, Hanna (1982). Relationship between moisture penetration and salinity in soils of the northern Negev (Israel). *Geoderma* **28**:313-344.

El-Azab, E.M., Minessy, F.A. and Barakat, M.A. (1974). Effect of depth of water table on micronutrients content of citrus leaves and roots. *Ist. Congresco Mundia de Citricultura* 1973, Vol. **1**, pp. 59-60.

El-Azhar, Taha, A.M., Keleg, M.W. and Hamid, F.M. (1980). Applied soil organic matter and calcium carbonate on growth and mineral composition of three citrus. *Egyptian J. Hort.* **6**(2) : 195-219.

El-Fouly, M. (1985). Nutritional deficiencies and use of soil testing and leaf analysis as basis for mineral fertilization for orange in Egypt. *Acta. Hort.* **158** : 321-330.

El-Fouly, M.M., Amberger, A. and Fawzi, A.F.A. (1988). Response of Balady orangot macro and microelement fertilization in Egypt. *Agrochimica.* **32**(1) : 27-40.

El-Fouly, M.M., Fawzi, A.F.A., Firgany, A.H. and El-Bazi, F.K. (1984). Micronutrient status of crops in selected areas of Egypt. *Commun. Soil Sci. Pl. Anal.* **15**(10) : 1175-1189.

Elfving, D. C., Kaufmann, M. R. and Hall, A. E. (1972). Interpreting leaf water potential measurements with a model of the soil-plant atmosphere continum. *Pl. Physiol.* **27**: 161-168.

Elgabali, M.M., Gewaifel, I.M., Hassan, M.N. and Rozanov, B.G. (1969a). Soils and soil regions of the United Arab Republic.*Research Bulletin,* **21**, Institute of Land Reclamation, Alexandria Univ. , Egypt.

Elgabali, M.M., Gewaifel, I.M., Hassan, M.N. and Rozanov, B.G. (1969b). Soil map and land resources of the United Arab Republic. *Ibid.* **22**.

El-Gazzar, A.M., El-Azab, S.M. and El-Safty, M. (1979). Response of Washington Navel oranges to foliar applications of chelated iron, zinc and manganese. *Alexandria J. Agric. Res.* **27**:19-26.

El-Gazzar, A.M., Minessy, F.A., Taha, W. and Naguib, M. (1975). Iron and manganese status of citrus trees grown in Egypt in relation to location, variety and plant organ. *Egyptian J. Hort.* **2**(1) : 117-128.

El-Gazzar, A.M., Wallace, A. and Naguib, M. (1977). Soil variables versus mineral analysis of citrus. *Commun. Soil Sci. Pl. Anal.* **8**(2) : 115-124.

El-Gazzar, A.M., Wallace, A. and Nawar, A. (1979). Leaf analysis of citrus in Egypt. *Alexandria J. Agric. Res.* **27**(1) : 1-10.

El-Hagah, M.H., Higazi, A.M., El-Niggar, S.Z., Ahmed, S.A. and Hasan, A.M. (1983). Effect of soil and foliar fertilization on growth of Navel orange and Balady mandarin trees. *Minufiya J. Agric. Res.* **1**:261-279.

El-Kassas, S.E., Mahmoud, H.M. and El-Shazly, S.M. (1987). Effect of certain micronutrients on the yield and fruit quality of Balady mandarin. *Assiut. J. Agric. Sci.* **18**(4) : 235-253.

Elmer, H.S., Brawner, O.L. and Ewart,W.H. (1973). Scarring and silvering in central valley navels. *Citrog.* **58**:335-338.

El-Otmani, M. (1990). The Citurs industry of Morocco. *Proc. Int. Citrus Symp.* (H. Bangyan and Y. Qian, ed.), Guangzhou, China, pp. 29-36.

El-Otmani, M. (1996). Tolerance of seedlings of six citrus rootstock to high soil calcium carbonate content. *Proc. Int. Soc. Citriculture.* Vol.**1**, pp. 290-295.

El-Otmani, M., Coggins, C.W. Jr. and Duymovic, A. (1990). Citrus cultivars and production in Morocco. *HortScience* **25**(11) : 1343-1346.

Elseewi, Ahmed A. (1974). Some observations on boron in water, soils and plants at various locations in Egypt. *Alexandria J. Agric. Res.* **22**: 463-474.

Elseewi, Ahmed A. and Elmalky, A.A. (1977). Leaf and soil boron in some Washington navel orange orchards along the mediterranean coast of Egypt. *Proc. Int. Soc. Citriculture.* Vol. **1**, pp. 12-15.

Elsokkary, I.H. (1976). Leaf analysis as a guide to the nutrition status of orange trees in soma alluvial and desert calcareous soils in Egypt. *Hort. Abst.* **46** : 339.

El-Zeftawi, B.M. (1973). Granualtion of valencia oranges. *Food. Tech. Australia.* **25**(2):331-337.

Embleton, T.W. and Jones, W.W. (1959). Correction of magnesium deficiency of orange trees in California. *Proc. Am. Soc. Hort. Sci.* **74** : 280-288.

Embleton, T.W. and Jones, W.W. (1964). Leaf analysis a tool for determining the nutrient status of citrus. *Proc. Soil Pl. Diagn.* Vol. **1**. pp., 160-169.

Embleton, T.W., Jones, W.W. and Labanauskas, C.K. (1962). Sampling orange leaves – leaf position important. *Calif. Citrog.* **47**: 392, 396.

Embleton, T.W., Wallihan, E.E. and Goodwill, G.E. (1966). Coastal lemons : zinc and manganese fertilization. *Citrog.* **51**:131, 140, 142-143.

Ercivan, S. (1974). Incidence of zinc deficiency on satsuma mandarin trees in Izmir. *J. Turkish Phytopathol.* **3**(1/2) : 51-56.

Ercivan, S. and Karaca, I. (1979). Investigation on the relation between the zinc deficiency and twig dieback occuring on satsuma mandarin (*Citrus unshiu* Marc.) plantation in Izmir. Variation and severity of disease and curative methods. *J. Turkish Phytopathol.* **8**(1) : 9-28.

Erickson, L. C. (1968). The general physiology of citrus. *The Citrus Industry* (W. Reuther, L. D. Batchelor and H.J. Webber, ed.). Div. Agri. Sci., Univ. Calif., Berkeley, USA Vol. **2**, pp 86-125.

Eshuys, W.A. (1967). The origini of citrus. *Farming South Afr.* **43**(5): 19-21.

Eswaran, H. (1972). Micromorphological indicators of pedogenesis in some tropical soil derived from basalts from Nicaragua. *Geoderma* **7:** 15-31.

Evans, C.E. and Kamparth, E.J. (1970). Lime response as related to per cent Al saturation, solution Al and organic matter content. *Soil Sci. Soc. Am. Proc.* **34**:893-896.

Falesi, I. C. (1972). Solos da rodovia transamazonica. *Bol. Tec. Inst. Pesquis. Exp. Agrupecu Norte.* **55** : 196.

FAO (1974a). *Shifting Cultivation and Soil Conservation in Africa.* Soils *Bull.* No. **24**, FAO, Rome, Italy.

FAO (1974b). *FAO - UNESCO Soil Map of the World*, Vol. **1**, Lengend, Paris, France.

FAO (1976). A Framework for Land Evaluation. *Soils Bull*, **32**, FAO, Rome.

FAO (1984a). Improved Production Systems as an Alternative to Shifting cultivation. *Soils Bull. No.* **54.**

FAO (1984b). *Land, Food and People.* Economic and Social Development *Series No.* **30,** Rome, Italy, p. 96.

FAO (1991). *The den Bosch Declaration and Agenda for Action on Sustainable Agriculture and Rural Development.* Report FAO, Rome, Italy.

FAO (1992). *Agro-ecological Land Resources Assessment for Agricultural Development Planning-A Case Study of Kenya*, World Soil Resource, *Report* **71,** p. 46, FAO, Rome, Italy.

FAO (1998). *Citrus Food, Fresh and Processed.* Food and Agricultural Organization (1998-99), Rome, Italy, pp. 1-31.

FAO (2000). *Citrus Food, Fresh and Processed.* Food and Agricultural Organisation (2000-01), Rome, Italy, pp. 1-39.

Fatta Del Bosco, G. (1964). Clementine growing. *Frutticoltura* **26**: 677-684.

Favreau, P. (1978). Soils of Corsican eastern plain and their relevance to citrus growing. *Fruits* **33**(12) : 814-816.

Fawzi, A.F.A., El-Fouly, M.M. and El-Baz, F.K. (1990). Problems of potassium nutrition in citrus (*Citrus sinensis*) orchards in Egypt. *Proc. 11th In. Pl. Nutr. Colloq.*, Wageningen, Netherlands, pp. 729-733.

Fereres, E., Cruz-Romero, G., Hoffman, G.J. and Rawlins, S.L. (1979). Recovery of orange trees following some water stress. *J. Appl. Ecol.* **16** : 833-842.

Fernandez-Lopez, J.A., Lopez-Roca, J.M. and Almela, L. (1993). Mineral composition of iron chlorotic *Citrus limon* L. leaves. *J. Pl. Nutr.* **16**(8):1395-1407.

Ferrari, G.B. (1646). Hesperdie sive de malorum aurorum cultura et usu libri quator. Hermani Scheus, Roma.

Fewerda, J.D. (1979). The ecology of tropical crops. Part 2. *Climate Spain* **22**(2) : 58-60.

Fidalski, J. and Auler, P.A.M. (1997). Nutritional survey of orange plantations in northwest Parana. *Arguivos de Biologia e Tecnologia* **40**(2) : 443-451.

Finch, A.H. (1977). Use of artificial fog low temperature control and its possible value in climatic control to improve citrus fruiting in a desert climate. *Proc. Int. Soc. Citriculture.* Vol. **1,** pp. 209-210.

Finck, Arnold (1998). Integrated nutrient management: An overview of principles, problems and possibilities. *Ann. Arid Zone.* **37:** 1-4.

Finlay, M.R. (1991). The rehabilitation of an agricultural chemist: Justus von Liebig and the seventh edition. *Ambix* **38**: 155-167.

Fischer, H.U., Timmer, L.W. and Muller, G.W. (1983). Comparison of declinamiento, blight, declinio and marchitamiento repento by use of uniform examination methods. *Proc. 9th Conf. Int. Organ. Citrus Virol.*, pp. 279-286 (S.M. Garnsey, L.W. Timmer and J.A. Dods, ed), IOCV Riverside, USA.

Fisher, T.A. (1975). Molybdenum. Res. *Bull. No.* **1007**, Agric. Expt. Stan., USA.

Fiskell, J.G.A. and Calvert, D.V. (1975). Effects of deep tillage, lime incorporation and drainage on chemical properties of Spodosol profiles. *Soil Sci.* **120** : 132-139.

Fiskell, L.G.A. and Leonard, C.D. (1967). Soil and root copper: Evaluation of copper fertilization by analysis of soil and citrus roots. *J. Agric. Food Chem.***15**: 350-353.

Fletcher, W.A. (1959). Citrus orchard nutrition and soil management. *New Zealand J. Agri. Sci.* **98**: 567-570.

Fletcher, W.A. (1966). Citrus and subtropical fruit growing in New Zealand. *The Orchardist of New Zealand* **8**: 60-68

Floyd, B. F. (1917). Injury to citrus trees apparently induced by ground limestone. *Ann. Rep.* pp. 27-30, Florida Agric. Expt. Sta., USA.

Fogaca, M., Beretta, M.J.G., Rossetti, V., Lefevre, A.F.V. and Moraes, W.B.C. (1986). Resultados negativos, a nivel de tecido, de tentativas de transmissao de declinio de plantas citricas por enxertia de raizes (Abstract) *Summa Phytopathologica* **12**:14.

Follet, R.H. and Lindsay, W.L. (1970). Profile Distribution of Zinc, Iron, Manganese and Copper in Colorado Soils. *Tech. Bull. No.* **110**, Colo. Agric. Exp. Sta., USA.

Ford , H.W. (1954a). Root distribution in relation to water table. *Proc. Fla. State. Hort. Soc.* **67**:30-33.

Ford, H.W. (1954b). Root distribution of citrus trees. *Ann. Rep.* pp. 181, Fla. Agric. Expt. Sta., USA.

Ford, H. W. (1959). Growth and root distribution of orange trees on two different rootstocks as influenced by depth to subsoil clay. *Proc. Am. Soc. Hort. Sci.* **74**: 313-321.

Ford, H.W. (1962). Root distribution of citrus. *Ann. Rep.* (Proj. 663R), p.202., Fla. Agri. Expt. Sta., Florida, USA.

Ford, H.W. (1963). Thickness of subsoil organic layer in relation to tree size and root distribution of citrus. *Proc. Am. Soc. Hort. Sci.* **82**: 177-179.

Ford, H.W. (1964a). Growth of citrus on drained flatwood land. *Ann. Rep.* (Proj. 920), pp. 219-220., Fla Agri. Expt. Sta., Florida, USA.

Ford, H.W. (1964b). The effect of rootstock, soil type and soil pH on citrus root growth in soils subject to flooding. *Proc. Fla. State Hort. Soc.* **77** : 41-45.

Ford, H.W. (1965). Bacterial metabolitcs that affect citrus root survival in soil subject to floading. *Prol. Am. Soc. Hort. Sci.* **86** : 205-212.

Ford, H.W. (1969). Water management in wetland citrus in Florida. *Proc. Ist Citrus Symp.* Vol. **3**, pp. 1759-1770.

Ford, H.W. (1973a). Eight years of root injury from water table fluctuations. *Proc. 85th Ann. Meet. Fla. State Hort. Soc.* pp. 65-68.

Ford, H.W. (1973b). Levels of hydrogen sulfide toxic to citrus roots. *J. Am. Soc. Hort. Sci.* **98**(1): 66-68.

Ford, H.W., Beville, B.C. and Carlisle, V.W. (1985). A guide for plastic tile drainage in Florida citrus groves. *Circ.* **661**, Fla. Coop. Ext. Serv., USA.

Ford, H.W. and Calvert, D.V. (1966). Induced anaerobiosis caused by flood irrigation with water containing sulfides. *Proc. Fla. State Hort. Soc.* **79** : 106-109.

Ford, H.W. and Eno, C.F. (1962). Distribution of microrganisms, citrus feeder roots, nitrate production and nutrients in the profile of Leon, Scarnton, Imokalee and Blanton fine sands. *Proc. Fla. State Hort. Soc.* **75**: 49-52.

Ford, H.W. and Prevatt, R.W. (1958). Tolerances of four citrus rootstocks to free water in certain sandy soils. *Ann. Rep.* 1957 - 58, pp. 232-233, Fla. Agri. Expt. Sta., Florida, USA.

Ford, H.W. and Prevatt, R.W. (1959). Tolerance of four citrus rootstocks to free water in certain sandy soils. *Ibid* 1958 - 9 : 231-232.

Foroughi, M., Maraschner, H. and Doring, H.W. (1974). The occurrence of boron deficiency in *Citrus aurantium* (bitterang) near the Caspian sea (Iran). *Hort. Abst.* **45** : 241.

Francis, C.A. (1972). Natural Daylength for Photoperiod Sensitive Plants. *Tech Bull.* **3**, Int. Centre for Tropical Agri., CIAT, p. 32.

Francois, L.E. and Clarke, R.A. (1980). Salinity effects on yield and fruit quality of Valencia orange. *J. Am. Soc. Hort. Sci.* **105**:199-202.

Franzmeier, D.P., Hajek, B.F. and Simonson, C.M. (1965). Use of amorphous material to identify spodic horizons. *Soil Sci. Soc. Am. Proc.* **29**(6): 1022-1028.

Freeman, Brian (1976). Artificial wind breaks and the reduction of windscar of citrus. *Proc. Fla. State. Hort. Soc.* **89**:52-54.

Freitas, F. G. and Silveira, C. O. da (1977). Principais solos sob vegetacao de cerrado e sua aptidao agricola. (Simposio sobre u cerrado 4. D. F., Brasilia, 1976). Cerrado bases para a utilizacao agropecuria , Sao Paulo, Itathia, p. 155.

Fresco, L.O. and Kroonenberg, S.B. (1992). Time and spatial scales in ecological sustainability. *Land Use Policy,* July, 1992, pp. 155-168.

Fridland, V.M. (1964). Soils and weathering crusts of the moist tropics. Nauka.

Frometa, E. and Echazabal, J. (1988). Influence of years and cultivar on the juice characteristics of early oranges. *Agrotecnia de Cuba* **20**(1) : 71-75.

Frost, H. B. (1943). Genetics and breeding. (H. J. Webber and L. H. Batchelor, ed.), *The Citrus Industry*, Vol. **1**, pp. 817-914, Univ. of Calif. Press, Berkeley, USA.

Fucik, J.E. (1972). A physiological rind disorder on Valencia oranges. *J. Rio. Grande Valley Hort. Soc.* **26** : 13-18.

Fucik, J.E. and Norwine, J. (1979). Climatological parameters and grapefruit size relationships in the Rio Grande valley of Texas. *J. Rio Grande Valley Hort. Soc.* **33**:83-89.

Fugita, K. and Yagi, T. (1955). Studies on the Flower bud Differentiation and Development in Some Orange Trees. I. On the time of flower bud differentiation and process of flower bud development in *Citrus unshiu* Marc. *Bull.* No. **3**, pp. 19-27, Hort. Branch, Kanagawa Agr. Expt. Sta., Kanagawa, Japan.

Fujishima, T., Utagawa, Y. and Matsushita, K. (1972). On crop productivities and leaching degree of several elements of volcanic ash soils in Kagoshima prefecture-lysimilar experiments. Part 3, *Soc. Sci. Soil and Manure, Japan* **43**(9) : 14-20.

Fuleihan, J.S. (1965). Economic analysis of the production of oranges and bananas in Damour and south Lebanon. *Publ.* **26**, p.146. *Amer. Univ. Beirut and Minist. Agric., Lebanon.*

Furr, J.R., Carptenter, J.B. and Hewitt, A.A. (1963). Breeding new varieties for citrus fruits and rootstocks for South west. *J. Rio Grande Valley Hort. Soc.* **17**:90-107.

Furr, J.R. (1955). Responses of citrus and dates to variation in soil moisture conditions at different seasons. *Rep. 14th Int. Hort. Congr.* Vol. **1,** pp. 400-412.

Furr, J. R., Cooper, W. C. and Reece, P. C. (1947). An investigation of flower formation in adult and juvenile citrus trees. *Am. J. Bot.* **34**:1-8.

Furr, J.R. and Ream, C.L. (1968). Breeding and testing rootstocks for salt tolerance. *Citrog.* **54**:30,32,34-35.

Furr, J.R. and Ream, C.L. (1969). Breeding citrus rootstocks for salt tolerance. *Proc. 1st Int. Citrus. Symp.* Vol **1**, pp. 373-380.

Furr, J.R. and Taylor, C.A. (1939). Growth of Lemon Fruits in Relation to Moisture Content of the Soil. *Tech. Bull* No. **640**, U.S. Dept. of Agric., USA.

Gajbhiye, K.S., Challa, O. and Tamgadge, D.V. (2000). Land resource base of the central region. *Souvenir.* pp. 24-27. 65th Annual Conv., Indian Society of Soil Sci., New Delhi, India.

Gallasch, P.T. and Ainsworth, N.J. (1989). Development in Australian citrus industry. *Proc. 6th Int. Citrus Congr.* (R. Goren and K. Mendel, ed.), Vol. **4**, pp. 1613-1623, Balaban Pub., German Federal Republic.

Gallasch, P.T. and Dalton, G.S. (1989). Selecting salt-tolerant citrus rootstocks. *Aust. J. Agric. Res.* **40**:137-144.

Gallasch, P. J., Forsyth, J. B. and Cope, H. (1984). The Australian citrus industry. *Proc. Int. Soc. Citriculture.* Vol. **1**, pp. 577-580.

Gamkrelidze, L.V. (1960).Nature of subtropical Sod-Podzolic soil. *Subtropicheskie Kul'tury*, No. (4).

Gan, L. (1990). The state of citrus production in Guangdong province. *Proc. Int. Citrus Symp.* (H. Bangyan and Y. Qian, ed.), Guanzhou, China, pp. 174-175.

Gandhi, S.R. (1939). A study of the methods of cultivation of fruit trees with special reference to citrus. *Trop. Agric.* **92**: 3-15.

Gandhi, S.R. (1956). Oranges, Lemons and Limes. *Indian Farm Bulletin No.* **15**, pp. 28-36, IARI, New Delhi.

Gandotra, Vinay, Gupta, R.D. and Bhardwaj, K.K.R. (1998). Abundance of *Azotobacter* in great soil groups of northwest Himalaya. *J. Indian Soc. Soil Sci.* **46**(3): 379 – 383.

Ganeshamurthy, A.N., Mongia, A.D. and Singh, N.T. (1989). Forms of sulphur in soil profiles of Andaman and Nicobar Islands. *J. Indian Soc. Soil Sci.* **37**(4):825-829.

Gangopadhyay, S. (1991a). Agricultural characteristics in the agroclimatic zones of Indian states – Himachal Pradesh, (S.P.Ghosh, ed.), *Agro-climatic Zone Specific Research,* ICAR, New Delhi, pp. 158-172.

Gangopadhyay, S. (1991b). Agricultural characteristics in the agroclimatic zones of Indian states – West Bengal (S.P.Ghosh, ed.), *Agro-climatic Zone Specific Research,* ICAR, New Delhi, pp. 438-462.

Gangopadhyay, S. (1991c). Agricultural characteristics in the agroclimatic zones of Indian states – Tamil Nadu, (S.P.Ghosh, ed.), *Agro-climatic Zone Specific Research,* ICAR, New Delhi, pp. 366-401.

Gangopadhyay, S. (1991d). Agricultural characteristics in the agroclimatic zones of Indian states – Orissa, (S.P.Ghosh, ed.), *Agro-climatic Zone Specific Research,* ICAR, New Delhi, pp. 280-313.

Gracia Benavides, J. (1971). Agricultural climate of Citrus sinensis. *Agronomia Tropical, Venezuela* **21**(2): 77-89.

Garcia Alvarez, N., Aspiolea, M.E. and Barreto, C.G. (1986). Effect of different zinc and manganese applications on valencia orange yield, economic evaluation. *Centro Agricola.* **13**(1):24-32.

Garcia, Alvarez, N., Haydar, E. and Ferrer, C. (1983). Influence of zinc and manganese on the physiological behaviour and yields of valencia oranges. *Centro Agricola.* **10**(2):57-88.

Garcia-Luis, A., Almela, V., Monerri, C., Agusti, M. and Guardiola,J.L. (1986). Inhibition of flowering *in vivo* by existing fruits and applied growth regualtors in *Citrus unshiu. Pl. Physiol.* **66**:515-520.

Garcia-Luis, A., Kanduser, M., Santamarina, P. and Guardiola, J.L. (1992). Low temperature influence on flowering in citrus. The separation of inductive and bud dormanacy effects. *Pl. Physiol.* **86:**648-652**.**

Gauri Shankar (1978). Performance and economics of kinnow cultivation in Allahabad. *Punjab Hort. J.* **18:**135-138.

Gayford, G.W. (1969). Crop forecasting. *Vict. Hort. Dig.* **13**(2): 21-24.

Gedroyts, K.K. (1955). Adsorbed cations and the physical properties of soils. Izbr. Soch. (Selected writings). Moscow, *Sel'Khozgiz* **1.**

Genu, P. J. Dec. and Pinto, A. C. De Q. (1984). Citriculture: A viable alternative for the Brazillian Cerrados. *Proc. Int. Soc. Citriculture.* Vol. **1**, pp . 584-588.

Geoghegan, M.J. and Brian, R.C. (1948). Aggregate formation in soil : I. Influence of some bacterial polysacharides on the binding of soil praticles. *Biochem. J.* **43** : 5-13.

Gerard Van Noert (1969) Dryness of navel fruit. *Proc. Ist. Int. Citrus Symp.* Vol.**3**, pp. 1333-1342.

Gerber, J.F. and Chen, E. (1986). The role of winter climatic zones in Florida freezes. *Proc. Fla. State Hort. Soc.* **99** : 9-13.

Germana C. (1992). Effects of two irrigation intervals with a spraying irrigation system in orange trees. *Proc. Int. Soc. Citriculture.* Vol. **2**, pp. 665-667.

Germana, C. and Sardo, V. (1988). Correlation among some physiological and climatic parameters in orange trees. *Proc. 6th Int. Citrus Congr.* (R. Goren and M. Mendel, ed.), Vol. **1**, pp. 525-534, Balaban Publi, German Federal Republic.

Ghosh, S.P. (1978). Citrus industry of north-east India. *Punjab Hort. J.* **12**: 13-21.

Ghosh, S.P. (1985). *Horticulture in North-Eastern India,* Associate Publishing Company, New Delhi.

Ghosh, S.N. (1990). Nutritional requirement of sweet orange (*Citrus sinensis* Osbeck) cv Mosambi. *Haryana J. Hort. Sci.* **19**(1-2):39-44.

Ghosh, S.P. (1991). *Agro-climatic Zone Specific Research,* (ed.), Krishi Anusandhan Bhavan, ICAR, New Delhi, p. 5.

Ghosh, S.P. (2000). Nutrient management in fruit crops. *Fert. News.* **45** : 71-76.

Ghosh, S.N. and Chattopadhyay, N. (1993). Effect of rootstocks on tree vigour, yield and fruit quality of mandarin orange under non-irrigated conditions in red and laterite soils of West Bengal. *Hort. J.* **6**(2): 79-82.

Ghosh, G. and Ghosh, S.K. (1976). Potassium in Some Soils of Nagaland. *Bull.* **10**, p. 6, Indian Soc. Soil Sci., New Delhi. Ghosh, G. and Ghosh, S.K. (1982). Micronutrients in some soils of Nagaland. *Indian Agric.* **26**: 91-99.

Ghosh, S.P. and Singh, R.B. (ed) (1993). *Citrus in South Asia.* RAPA Publication, FAO, Regional Office Bangkok, Thailand, p.21.

Ghosh, S.P. and Srivastava, A.K. (2001). Sweet orange. *Handbook of Horticulture* (K.L. Chadha, ed.) Indian Council of Agricultural Research, New Delhi, India, pp. 329-334.

Ghosh, S.P., Verma, A.N. and Govind, S. (1984). Nutritional requirement of bearing khasi mandarin orange (*Citrus reticulata* Blanco) trees. *J. Res. Assam Agri. Univ.* **5**(1): 11-16.

Ghosh, S.P.,Verma, A.N. and Govind, S. (1989). Nutritional requirement of bearing khasi mandarin orange (*Citrus reticulata* Blanco) trees in Meghalaya. *Hort. J.* **2**(1):4-11.

Ghosh, S.P., Verma, A.N., Govind, S., Medhi, Prasad, R.N., Barooah, R.C., Sachan, J.N. and Gangwar, S.K. (1982). Mandarin Orange Decline in North-eastern Hill Region and its Control. *Research Bulletin* **16**, ICAR Research Complex for North-easthen Hill Region, Shillong, Meghalaya, India.

Giacometti, D.C. (1981). Present situation and outlook of the Brazilian citrus industry. *Proc. Int. Soc. Citriculture.* Vol. **2**, pp. 947-950.

Giacometti, D.C., Rio-Castano, D. and Torres-M.R. (1966). Recommendations for citrus growing in the Valle del Cauca (Colombia). *Agric. Trop., Bogota* **22**: 117-134.

Giacomo, A. Di. (1987). The citrus industry : problems and innovations. *Italia Agricola.* **124**(1) : 171-184.

Giginejsvili, P.L. (1968). Citrus growing in Vietnam. *Subtropicheskie Kul'tury* (1) : 68-81.

Gilani, A.H.,Yusuf, Ali, Tariq, M.A. and Mohammad, Faquir C. (1989). Studies of the effect of growth regulators and chemical fertilizers on the growth and yield in kinnow mandarin. *Sarhad J. Agri.* **5**(1): 47-51.

Gill, W.R. and Van del Berg, G.E. (1967). Soil dynamics in tillage and traction. *Handbook* No. **316**, pp. 511, U.S. Dept. Agr., USA.

Gioffre, D. (1976). The genus *Fortunella* in the restructuring of Italian citrus culture. A contribution to the knowledge of fruit characterisitics in different environments. *Hort. Abst.* (1978) **48** : 673.

Girton, R.E. (1927). The growth of citrus seedlings as influenced by environmental factors. *Univ. Calif. Pub. Agr. Sci.* **5** : 83-117.

Glass, A.D.M. (1989). *Plant Nutrition: An Introduction to Current Concepts*, Jones and Barlett Publishers, Boston, MA, USA.

Glonti, Ts. F. (1973). The effect of different forms of nitrogen fertilizer on chemical properties of red soils and on orange tree yield. *Subtropicheskie Kul'tury* (6) : 44-49.

Gochelaswvili, Z.A. (1973). The total number of bacteria in mandarin, orange and lemon tree rhizosphere. *Subtropicheskie Kul'tury* (4) : 54-57.

Godefray, J. and Bourdeaut, J. (1972). Effect of cover crops on the chemical, biological and structural characteristics of an orchard soil in Ivory Coast. *Fruits* **27**(5) : 349 - 353.

Goedert, W. J., Lobato, E. and Wagner, E. (1980). Potencial agricola da regiao dos Cerrados Brasileiros. *Pesquia Agropecuaria Brasileira, Brasilia* **15**(1): 1-17.

Goell, A (1969) Salinity effects on citrus trees. *Proc. Ist. Int. Citrus Symp.* Vol. **3**, pp. 1819-1824.

Goell, A. and Cohen, A. (1981). Combining irrigation regimes with girdling techniques in citrus trees (a new experimental mode). *Proc. Int. Soc. Citriculture.* Vol. **2**, pp. 514-518.

Goepfert, C.F., Saldanha, E.L.S. De and Porto, O. De M. (1987). The response of Valencia orange (*Citrus sinensis* Osbeck) to fertilizer levels, average of eight harvests. *Agronomia Sulriograndense* **23**(2): 203-215.

Goldschmidt, E. E. and Golomb, A. (1982). The carbohydrate balance of alternate bearing citrus trees and the significance of reserve for flowering and fruiting. *J. Am.Soc. Hort. Sci.* **107** (2): 206-208.

Goldschmidt, E. E. and Monselise, S. P. (1972). Hormonal control of flowering in citrus and some other woody perenials. *Plant Growth Substances* (J. D. Carr, ed.), Springer, Berlin, pp. 758- 765.

Gomez, De Barreda, D., Legaz, F., Primo. E., Lorenzo, E., Ibanez, R. and Torres, V. (1984). Irrigation of young Washington navel cv. frost trees. *Proc. Int. Soc. Citricutlure.* Vol. **1**, pp. 122-124.

Gomez, V.O., Chernyakhovskiy, A.G., Chizhikova, N.P., Shchurygina, Ye, A. and Gradusov, B.P. (1981). Mineral composition of volcanic soils in the tropical region of Costa Rica, *Soviet Soil Sci.* **13** (4): 95-102.

Goncharova, E. A., Medzmariashvili, N. I. and Bakhtadza, I.G. (1989). Weather and climatic conditions in the humid subtropics of the USSR and citrus adaptation to them. *Hort. Abst.* **61**(12) : 1350.

Gonzales, C.T. (1990). Citrus greening diseases in the Philippines : Distribution and current control measures. *Extn. Bill. 284*, pp. 15-21. ASPAG Food and Fert. Technol., Department of Agricultures, Philippines.

Gonzalez, A., Vallin, G. Del. and Martinez, A. (1987). Effect of potassium on yield, fruit quality and leaf K content in Valencia late oranges. *Ciencia del la Agricultura Suelos y Agroquimica.* **10**(1) : 33-43.

Gonzalez-Sicilia, E. (1968). El cultivo de los agrios. Editarial Bello, Valencia, Spain, p. 814.

Gordzhomeladze,O.L. (1990). Effect of soil erosion and green manuring on the growth and cropping of citrus trees in intermontane conditions of Adzhariya. *Subtropicheskie kul'tury*(4): 125-130.

Goswami, N.N. (1997). Concept of balanced Fertilization, its relevance and practical limitions. *Fert. News* **42**(4):15.

Goswami, N.N. (1998). Integrated Plant Nutrient Supply System for Sustainable Productivity. *Bull.* **2,** Indian Institute of Soil Sci., Bhopal, Madhya Pradesh, India, pp. 3-9.

Goswami, N.N. (1999). Priorities of soil fertility and fertilizer use research in India. *J. Indian Soc. Soil Sci.* **47**(4) : 649-660.

Govind, S. and Prasad, A. (1976). Effect of soil and foliar applications of nitrogen on the status of mosambi leaves. *Pl. Sci.* **8**: 84-87.

Govinda Iyer, T.A. and Iyengar, T.R. (1956). A study of the decline of orange in Wynad. *South Indian Hort.* **4**(3): 70-81.

Govindarajan, S.U. and Gopala Rao, H.G. (1978). *Studies on Soils of India*. Vikas Publishing House, Pvt. Ltd., New Delhi, India.

Govindarajan, S.U. and Venkata Rao, B.V. (1976). Acid Soils of India : Their Genesis, Characteristics and Management. *Bull.* **11**, Indian Soc. Soil Sci., New Delhi, p. 38.

Govindarao, P and Reddy, G.S. (1957). Deficiency diseases in citrus. *Indian. J. Hort.* **24** : 91-94.

Gowing, J.W. and Wyseure, G.C.L. (1994). Hydrological modelling : A tool for sustainable development of land resources. *Soil Science and Sustainable Land Management in the Tropics,* (J.K. Syers and D.L. Rimmer, ed.), CAB International, British Society of Soil Science, UK, pp. 132-144.

Graca, J. V. da and Vuuren, Van S. P. (1979). A decline of citrus in South Africa resembling young tree decline. *Pl. Dis. Reptr.* **63** : 901-903.

Gradusov, B.P. and Urushadze, T.F. (1968). Clay minerals in brown forest soils of Georgia. *Pochvovedeniye.* 2.

Graeme, D., Bucha, Grewal, Karambir, S., Claydon, John, J. and Robin, J. (1993). A comparison of sedigraph and pipette method for soil particle analysis. *Australian J. Soil Res.* **31**:407-417.

Graham, J.H. and Syvertsen, J. P. (1984). Influence of vesicular-arbuscular mycorrhiza on the hydraulic conductivity of roots of two citrus rootstocks. *New Phytol.* **97**: 277-284.

Graham, R.D. , Turner, N.C. and Ascher, J.S. (1992). Subsoil constraints to root growth and high soil water and nutrient use by plants. *Proc. National Workshop,* Tanunda, South Australia, Australia.

Granatstein, T. and Bezdicek, D.F. (1992). The need for a soil quality index: Local and regional perspectives. *Am. J. Alternative Agric.* **7**: 12-16.

Gras Guerra, G. (1987). Citrus, some nutritional and fertilization aspects. *Occasional Publication No.* **1218** p. 63. Instituto Superior de Ciencias Agropecuarias de la Habana, Cuba.

Grass, L.B., Sterling, D. and Shade, E. (1966). Salinity studies in Lake Mathews district. *Citrog.* **51**: 147-156.

Gravina, A., Arbiza, H., Juanm M., Almela, V. and Agusti, M. (1996). Flowering- fruiting relationship in Ellendale Tangor under the growing conditions of Spain and Uruguay. *Proc. Int. Soc. Citriculture.* Vol **2**, pp. 1081-1085.

Green, G.C., Bozalek, S.J. and Schoeman, A.S. (1975). Climatic requirements of citrus. *Farming in South Africa.* Department of Technical Services, R.S.A., pp. 1-4.

Greene, R.S.B. and Tongway, D.J. (1989). The significance of surface physical and chemical properties in determining soil surface condition of red earths in Rangelands. *Australian J. Soil Sci.* **27**:213-225.

Greenland, D.J. and Lal, R. (1977). *Soil Conservation Management in the Humid Tropics,* Wiley, New York, USA.

Greenway, H. and Munns, R. (1980). Mechanisms of salt tolerance in nonhalophytes. *Ann. Rev. Pl. Physiol.* **31**: 149-190.

Gregorich, E.G., Carter, M.R., Angers, D.A., Monreal, C.M. and Ellert, B.H. (1994). Towards a minimum data set to assess soil organic matter quality in agricultural soils. *Canadian J. Soil Sci.* **74**: 367-385.

Grewal, J. S., Bhumbla,D.R. and Randhawa, N.S. (1969). Available micronutrient status of Punjab, Haryana and Himachal soils. *J. Indian Soc. Soil Sci.* **17**: 27-31.

Grierson, W. (1981). Physiological disorders of citrus fruits. *Proc. Int. Soc. Citriculture* Vol **2**, pp. 764-767.

Grierson, W. and Brown, G.E. (1966). Zebra-skin and post-harvest handling. *Citrus Indus.* **47**(3) : 8-10.

Grierson, W. and Koo, R.C.J. (1958). Peel injury of tangerines as influenced by water relateions in the grove and subsequent handling. *Citrus & Veg.* Mag. **21**:8-10.

Grierson, W., McCornack, A.A. and Hayward, F.W. (1965). Tangerine handling. *Fla. Agr. Ext. Serv. Circ.* **285**.

Grierson, W. and Reitz, H.J. (1981). The Florida citrus industry. *Proc. Int. Soc. Citriculture.* Vol. **2**, pp. 945-947.

Grieson, W. and Ting, S.V. (1978). Quality standards for citrus fruits, juice and beverages. *Proc. Int. Soc. Citriculture.* pp. 21-28.

Grierson, W. and Wardowski, W.F. (1975). Humidity in Horticulture. *Hortscience.* **10** : 356-360.

Grieve, A. M. and Walker, R. R. (1983). Uptake and distribution of chloride, sodium and potassium ions in salt treated citrus plants. *Australian J. Agric. Res.* **34**: 133-143.

Grigg, D. (1969). Agricultural regions of the world review and reflection. *Econ. Geog.* **45**(2) : 100-110.

Grobler, J. H. and Bredell, G.S. (1981). The South African Citrus Industry. *Proc. Int. Soc. Citriculture.* Vol. **2**, pp. 965-967.

Grossman, R.B. and Carlisle (1969). Fragipan soils of the eastern United States. *Adv. Agron.* **21**:237-279.

Grove, J. H., Fowlder, C.S. and Sumner, M.E. (1982). Determining of the charge character of selected acid soils. *Soil Sci. Soc. Am J.* **46**: 32-38.

Grupo, Paulista de Adubacao e Calagem para Citros (1988). Recomendacoes de adubacao e calagem para citros no Estado de Sao Paulo. 1st ed. *Laranja. Cordeiropolis* **11**(3): 14.

Grupo Paulista de Adubacao e Calagem para Citros (1994). Recomendacoes de adubacao e calagem para citros no Estado de Sao Paulo. 3rd ed. *Laranja. Cordeiropolis* **17** : 27.

Guilford, H.E., Chandler, W.V. and Bryan, O.C. (1970). Effect of phosphate applications to lakeland sands on pineapple orange yilds, available soil phosphorus and leaf phosphorus. *Proc. Fla. Sate Hort. Soc.* **82**: 26-29.

Guillen Paiz, R. (1975). Some prospects for lime in agricultural diversification. *Revista Cafetalera.* **149** : 29-35.

Gupta, R.P., Aggarwal, P. and Chauhan, A.S.(1995). Spatial variablity analysis of bulk density as a guide for tillage. *J. Indian Soc. Soil Sci.* **43**(4) : 549-557.

Gupta, Vidyadhar and Dwivedi, R.N. (1994). Agro-techniques for growing quality oranges in Arunachal Pradesh. *Indian Hort.* **39**(1) : 22-25.

Gupta, O.P., Jawanda, J.S. and Gupta, K.R. (1989). Foliar applications of urea and zinc sulphate for fruit drop control in kinnow mandarin. *Curr. Sci.* **58**(8):456-457.

Gupta, R.P., Kumar, S. and Singh, T. (1984). *Consolidated Report : Soil Management to Increase Crop Production.* IARI, New Delhi, p. 104.

Gupta, J.P. and Lattoo, A.K. (1999). Distribution of different forms of phosphorus in the soils of sub tropical zone of Jammu region. *J. Indian Soc. Soil Sci.* **47**(1):147-148.

Gupta, R.K. and Srivastava, P.C. (1990) Distribution of different forms of micronutrient cations in some cultivated soils of Sikkim. *J. Indian Soc. Soil Sci.* **38**: 558-560.

Gupta, R.K. and Srivastava, P.C. (1998). Distribution of different forms of phosphorus in acid soils of Sikkim. *Indian J. Hill Farming.* **11**(2):224-228.

Gupta, R.D. and Tripathi, B.R. (1988). Microflora of some soil profiles of northwest Himalayas. *J. Indian Soc. Soil Sci.* **36**(1):75-82.

Gupta, R.D. and Tripathi, B.R. (1996). Mineralogy genesis and classification of soils of north-west Himalayas developed on different parent materials and variable topography. *J. Indian Soc. Soil Sci.* **44**(4) : 705-712.

Gupta, R.P., Pandey, S.P. and Tripathi, B.P. (1989). Soil properties and availability of nutrient elements in mandarin growing areas of Dhankuta district. *PAC Technical Paper* **113**, Pakhriba Agricultural Centre, Dhankuta, Nepal.

Gurung, H. (1994). Samuhik Bhraman Report of Chyangliton, Grokhas as a perspective multidisciplinary outreach research site. Lumle Agriculture Research Centre. Internal Mimeograph, Pokhara, Nepal.

Gvaliya, M.V. and Zonn, S.V. (1990). Genetic differences and productivity of red and yellow earths in relation to relief. *Soviet Soil Sci.* **22**(5) : 16-27.

Haas, A.R.C. (1932). Some nutritional aspects in mottle leaf and other physiological diseases of citrus. *Hilgardia* **6**: 483-559.

Haas, A.R.C. (1939). Growth of citrus and walnut trees as affected by pH. *Calif. Citrog.* **24**: 351, 364, 379.

Haas, A.R.C. and Klotz, L.J. (1931). Some anatomical and physiological changes in citrus produced by boron deficiency. *Hilgardia* **5**: 175-196.

Haas, A.R.C. and Quayle, H.J. (1935). Copper content of citrus leaves and fruit in relation to exanthema and fumigations injury. *Hilgardia* **9**: 143-177, 9 Figs. Abs.

Haggag, M.N., El-Shamy, H.A. and El-Azab, E.M. (1987). Magnesium influences on leaf chlorophyll, leaf mineral composition, yield and fruit quality of Washington Navel oranges in Egypt. *Alexandria J. Agric. Res.* **32**(3) : 189-198.

Haggag, L.F. and Maksoud, M.A. (1996). Evaluation of yield of navel orange tree : mathematical model procedure. *Egyptian J. Hort.* **23**(2) : 197-202.

Hagin, J., Lifshitz, Z. and Monselise, S.P. (1965). The influence of soil aeration on the growth of citrus. *Israel J. Agri. Res.* **15** : 59-64.

Halse, N.J. (1963). Control of copper, zinc, and manganese deficiencies in fruit trees. *J. Agri. Western Australia* **4**: 241-244.

Halvorson, J.J., Smith, J.L. and Papedick, R.I. (1996). Integration of multiple soil parameters to evaluate soil quality: a field example. *Biol. Fert. Soils.* **21**: 207-214.

Hamdi, H. (1959). Alteration in the clay fraction of Egyptian soils. *Z. pflanenehnahr.,* Dung, Bodenkd. **84** : 204-211.

Hamdi, H. (1967). The mineralogy of the fine fraction of the alluvial soils of Egypt. *J. Soil. Sci.* (*U.A.R.*) **7** : 15-21.

Hamdi, H. and Barrada, Y. (1960). Transformation of the clay fraction of the alluvial soils of Egypt. *Ann. Agri. Sci., Cairo* **2** : 135-139.

Hammond, L.C., Carlisle, V.W. and Rogers, J.S. (1971). Physical and mineralogical characters of soils in the experimental site at Fort Pierce, Florida. *Proc. Soil & Crop Sci. Soc. Fla.* **31** : 210-215.

Hamze, M., Rayan, J. and Zaabout, M. (1986). Screening of citrus rootstock for lime-induced chlorosis tolerance. *J. Plant Nutr.* **9**:459-469.

Hanamashetti, S.I., Nadagondar, B.S., Devarnavadgi, S.B., Nalwadi, U.G. and Hulmani, N.C. (1987). Effect of different fast growing forest species on growth and yield of interplanted mandarins (*Citrus reticulata* Blanco). *South Indian Hort.* **35**(3) : 266-267.

Hanlon, E.A., Obreza, C.V. and Alva, A.K. (1995). Nutrition of Florida Citrus Trees. *Tissue and Plant Analysis.* (D.P.H. Tucker, A.K. Alva, L.K. Jackson, and T.A. Wheaton, ed.), Lake Alfred, University of Florida, IFAS, USA, pp. 13-16.

Hanna, F.S. and Beckmann, H. (1975). Clay minerals of some soils of the Nile valley of Egypt. *Geoderma* **14** : 159-170.

Haq, Izhar Ul, Ghani, Abdul and Habibur-Ur-Rehman (1995). Nutritional status of Citrus Orchards in NWFP and the Effect Fertilizer Application on Fruit Production. *Tech. Bull.* 1/95, pp. 1-36. Directorate of Soils and Plant Nutrition, Agri. Res. Instt. Tarnab, Peshawar, N.W.F.P., Pakistan.

Haque, T. (1974). Fertility variation in soils of Sierra Leone. Division of Soils, N.U.C., Sierra Leone, Mimeo, p. 45.

Haque, A. and Godfrey, S.A. (1976). Nutritional survey of citrus orchards in Sierra Leone. *Commun. Soil Sci. Pl. Anal.* **7**(19) : 843-860.

Harding, R.B. (1951). High yielding orange orchards. *Calif. Citrog.* **36**(9):350-351.

Harding, R.B. (1954). Exchangeable cations in soils of California orange orchards in forty high performance orchards in California. *Soil Sci. Soc. Am. Proc.* **15** : 243-248.

Harding, R.B. and Chapman, H. D. (1950). Progress report on a study of soil characteristics in forty high performance orange orchards in California. *Soil Sci. Soc. Am. Proc.* **11** : 240-248.

Harding , R.B. and Chapman, H.D. (1951). Progress report on a study of soil characteristics in forty high performance orange orchards in California . *Soil Sci. Soc. Am. Proc.* **15**: 243-248

Harding, R.B., Pratt, P.F. and Jones, W.W. (1958). Changes in salinity, nitrogen and soil reaction in a differentially fertilized irrigated soil. *Soil Sci.* **85**:177-184.

Hari Ram and Dwivedi, K.N. (1994). Delineation of sulphur deficient soil groups in the central alluvial tract of Uttar Pradesh. *J. Indian Soc. Soil Sci.* **42**(2): 284-286.

Haribabu, R.S. and Rajput, C.B.S. (1982). Effect of zinc, 2,4-D and GA_3 on flowering in Kagzi lime. *Punjab Hort. J.* **32**(3&4): 140-144.

Harlan, R.W., Franzmeier, D.P. and Roth, C.B. (1977). Soil formation of loess in southwestern Indiana: II. Distribution of clay and free oxides, and fragipan formation. *Soil Sci. Soc. Am. J.* **41**:99-103.

Haro-Guzman, L. (1979). Research on the growth and maturation of Mexican lime. *Fruits* **34**(6) : 417-422.

Harris, R.F., Karlen, D.L. and Mulla, J.L. (1996). A conceptual framework for assessment and management of soil quality and health. *Methods for Assessing Soil Quality* (J.W. Doren *et al.,* ed.), *Spec. Publ.* **49**, pp. 61-82, Soil Sci. Soc. Am. and Am. Soc. Agron., Madison, WI.

Hartmond, U., Schaesberg, N.V., Graham, J.H. and Syvertsen, J.P. (1987). Salinity and flooding stress affects on mycorrhizal and non-mycorrhizal citrus rootstock seedlings. *Pl.Soil* **104**: 37-43.

Harty, A., Sutton, P. and Machin, T. (1996). Clementine orange evaluation in New Zealand. *Proc. Int. Soc. Citriculture.* Vol. **1,** pp. 177-180.

Hartz, T.K. (1984). Salinization –A threat to valley agriculture. *J. Rio Grande Valley Hort. Soc.* **37**: 123-125.

Hasegawa, Y. and Iba, Y. (1981). Kohansho: A physiolgical disorder of the rind of citrus fruits during storage in Japan. *Proc. Int. Soc. Citriculture.* Vol. **2**, pp. 774-776.

Hashad, M.N. and Mady, F. (1963). Differential thermal analysis of alluvial clay of Egypt. *J. Soil Sci. (U.A.R.),* **1** : 125-138.

Hass, A.R.C. (1948). Magnesium deficiency and its effect on citrus. *Calif. Citrog.* **33** : 134,146-148, 150.

Hassan, A.K. (1995). Effect of foliar sprays with some micronutrients onWashington Navel orange trees. I. Tree growth and leaf mineral content. *Ann. Agri. Sc. Moshtohor* **33**(4):1497-1506.

Hassan, M.M. and Galal, M.A. (1989). Salt tolerance among citrus rootstocks. *J. King Saud. Univ. Agric. Sci.* **1**: 87-93.

Hastie, E.L. (1963). Gayndah citrus under sod culture. *Qd. Agric. J.* **89**: 730-732.

Haury, A. (1981). Preliminary. study of a case of citrus decline of Gabougoura in Niger. *Fruits* **36**(1):25-36.

Hausenbuiller, R.L. (1963). *Principles of Soil Science.* Orient Longmands, Bombay, Maharashtra, pp. 1-2.

Hayashi, S. and Wakisaka, I. (1956). Studies on waterlogging injury of fruit trees. *J. Hort. Assoc. Japan* **25** : 59-68.

Hayes, W.B. (1960). *Fruit Growing in India.* (3rd rev.ed.),Kitabistan, Allahabad, U.P., p.118.

Hayward, H.E. and Bernstein, L. (1958). Plant growth relationship on salt affected soils . *Bot. Rev.* **24**: 584-635.

He, Shaolan, Deng, Lie, Tan Zhi You, Wan, Tiang Long and Huang, Guo Xiang (1998). Preliminary report on using EDDHA-Fe for correcting the leaf yellowing disease of citrus grown on purple soil. *South China Citrus* **27**(3):18-19.

Hellin, E. and Alcarez, C.F. (1980). Influence of manganese deficiency on phosphorus fractions in lemon leaves. *J. Pl. Nutr.* **2**(3) : 323-333.

Hellin, E., Urena, R., Sevilla, F. and Alcaraz, C.F. (1987). Comparative study on the effectiveness of several iron compounds in the iron chlorosis correction in citrus plants. *J. Pl. Nutr.* **10**(4):411-421.

Hellin, E., Urena, R., Sevilla, F., Gimenez, J.L. and Alcaraz, C.F. (1988). Effects of several Fe-chelates on the iron leaf content of Verna lemon trees. *Proc. 6th Int. Citrus Congr.* (R. Goren and K. Mendel, ed.), Tel Aviv, Israel, Vol. **1**, pp. 555-560.

Helyer, K.R. and Conyars, M.K. (1987). *Priorities in Soil/ Plant Research for Plant Production* (P.G.E. Searle and B. G. Davey, ed.). University of Sydney, Australia, p. 65.

Herbillion, A.J. , Gallez, A. and Juo, A.S.R. (1977). Characteristics of silica sorption and solubility as parameters to evaluate the surface properties of tropical soils : II The index of silica saturation. *Soil Sci. Soc. Am. J.* **41** : 1151.

Hernandez, J. (1981). Effect of N, P and K on yield, fruit quality and nutritional status of Valencia Late orange. *Proc. Int. Soc. Citriculture.* Vol. **2**, pp. 564-566.

Hernandez, M., Harnandez, I. Mesa, A. and Forteza,I. (1987). Effect of effective soil depth and stonniness on citrus yields in a red ferrallitic soil. *Agrotecnia de Cuba* **19:**75-79.

Hernandez, A., Perez, J.M., Azcanio, O., Ortega, F., Avila, L., Cardenas, A. and Marrero, A. (1975). 2da. Classification Genetica de Las Suelos de Cuba. Inst-Suelos A.C. Cuba Serie Suelo **23**.

Hernando, V. (1969). Soil and leaf analysis of orange tree orchards on several types of soils in Valencia province. *Proc. Ist. Int. Citrus Symp.* Vol. **3**, pp. 1673-1688.

Hernando, V. (1978). Potassium problems in fertilizing orange orchards in the eastern region of Italy. *Potash Rev. Sub.* **32**(11) : 3.

Herzog, P. and Monselies, S.P. (1968). Growth and development of grapefruits in two different climatic distincts of Israel. *Israel J. Agric. Res.* **18**: 181-186.

Hewitt, A.E. (1989). New Zealand Soils Classification. Division of Land and Soil Sciences, *Technical Record DN2.* DSIR, New Zealand.

Hewitt, A. A. and Furr, J. R. (1965a). Uptakes and loss of chloride from seedlings of selected root stock varieties. *Proc. Am. Soc. Hort. Sci.* **86**: 194-200.

Hewitt, A. A. and Furr, J. R. (1965b). Influence of salt source on the uptake of chloride by selected citrus seedlings. *Proc. Am. Soc. Hort. Sci.* **84**: 165-170.

Hewitt, A. A., Furr, J. R. and Carpenter , J.B. (1964). Uptake and distribution of chlorides in citrus cuttings during a short term salt test . *Am. Soc. Hort. Sci. Proc.* **84**: 165-169.

Hield, H. Z., Coggins, C. W. and Lewis, L.N. (1966). Temperature influence of flowering of grapefruit seedlings . *Proc. Am. Soc. Hort. Sci.* **89**:175-181.

Higa, T. (1975a). Studies on the ecology of citrus in Okinawa Islands. I. In relation to environmental conditions in the Okinawa Islands. *Sci. Bull. Coll. Agri., Univ. Ryukyus, Okinawa.* **22** : 59-68.

Higa, T. (1975b). Studies on the ecology of citrus in the Okinawa Islands. II. The ecology of *Citrus unshiu* var. praecox. *Sci. Bull. Coll. Agri., Univ., Ryukyus, Okinawa* **22** : 69-72.

Hilgard, E.W. (1966). Marly subsoils of chlorosis or yellowing of citrus trees. *Cir.* **27**, Calif. Agric. Expt. Sta., USA.

Hilgeman, R.H. (1957). Response of orange trees growing in the salt river valley of Arizona of chelated iron compounds. *J. Rio Grande Valley Hort. Soc.* **11**:14-20.

Hilgeman, R.H. (1966). Effect of climate of Florida and Arizona on grapefruit enlargement and quality, apparent transpiration and internal water stress. *Proc. Fla. State Hort. Soc.* **79**: 99-106.

Hilgeman, R.H. (1969). Correction of iron chlorosis with iron chelates in Arizona. *Calif. Citrog.* **54**:406, 426-428.

Hilgeman, R.H. (1973). Annual variation in yield of citrus trees, 1949-1971. *Citrog.* **58**(12) : 423, 447.

Hilgeman, R.H., Ehrler, W.L., Everling, C.E. and Sharp, F.O. (1969). Apparent transpiration and internal water stress in valencia oranges as affected by soil water, season and climate. *Proc. Ist. Int. Citrus Symp.* Vol.**3**, pp. 1713-1723.

Hilgeman, R.H. and Reuther, W. (1967). Evergreen tree fruits. *Irrigation of Agricultural Lands* (R.M.

Hagan, H.R. Haise and T.W. Edminster, ed.), *Monograph No.* **11**, pp. 704-718, Am. Soc. Agron, Madison, Wisconsin, USA.

Hilgeman, R.H. and Rodney, D.R. (1961). Commercial citrus production in Arizona. *Spec. Rep.* **7**, p. 31, Agric. Expt. Sta. and Coop. Extn. Serv., Univ. Arizona, USA.

Hilgeman, R.H., Tucker, H. and Hales, T.A. (1959). The effect of temperature, precipitation, blossom date and yield upon enlargement of Valencia oranges. *Proc. Am. Soc. Hort. Sci.* **74**:266-279.

Hilgenman, R.H. and Van Horn, C.W. (1954). Citrus growing in Arizona. *Calif. Cirgo.* **39**(11): 380,402-404.

Hilgeman, R.H. and Van Horn, C.W. (1955). Citrus growing in Arizona. Sta. *Bull.* **258** (Rev.), Univ. Ariz. Agri. Exp., USA, p. 35.

Hinkley, F. (1939). Growing oranges, Natures way. *Calif. Citrog.* **24**: 390, 404-405.

Hipp, B.W. (1977). Contribution of soil nitrogen mineralization rate to citrus produciton under subtropical conditions. *Commun. Soil Sci. Pl. Anal.* **8**(5):367-371.

Hirekerur, L.R. (1983). *Appraisal of Land Resources Potential of Vidarbha Region of Maharashtra State for Integrated Development.* National Bureau of Soil Survey and Land Use Planning, Nagpur, Maharashtra (India), pp.2-4.

Hirobe, M. and Ogaki, C. (1973). Experiments on the effect of lime fertilizers in satsuma orchard. II. Effects of lime materials on soil transference of calcium and tree growth. *Bull. No.* **21**, pp. 33-38, Kanagawa Hort. Expt. Sta., Kanagawa, Japan.

Hirobe, M. and Ogaki, C. (1973). Experiments on the effect of lime fertilizers in an Unshiu orange orchard. I. On the improvement of subsoil by soil injection. *Bull. No.* **20**, Kanagawa Hort. Expt. Sta., Kanagawa, Japan.

Hisada, H. and Ishida, T. (1975). Studies on soil amendment I. The effect of fused phosphatic fertilizer, calcium silicate, precipitated carbonate and bark compost on citrus trees. *Bull. No.* **12**, pp. 1-9, Shizuoka Prefectural Citrus Expt. Sta., Japan.

Hodgson, J.F. (1963). Chemistry of micronutrients elements in soils. *Adv. Agron.* **15** : 119-159.

Hodgson, R.W. (1937). The Egyptian citrus industry. *Calif. Citrog.* **22**:287, 324-326.

Hodgson, R.W. (1954). Citriculture in Egypt. *Calif. Citrog.* **39**(5): 143, 159-162.

Hodgson, R.W. (1962). The Japanese mandarin orange industry. *Calif. Citrog.* **47**:171-173.

Hodgson, R.W. (1966a). Citriculture in Libya. *Calif. Citrog.* **51**: 385-390.

Hodgson, R.W. (1966b). Libya and her citrus industry. *Calif. Citrus Indus.* **47**(5): 20-21.

Hoffman, J.E. and Abercrombie, R. A (1999). The effect of alleviating soil compaction on yield and fruit size in an established navel orange orchard. *Neltropika Bulletin No.* **303**, pp. 59-63, 67.

Holland (1897). Geological survey of India. *General Report* **30:**1-4.

Holtzhausen, L.C. and Plessis, J.A. Du. (1970). Skin splitting of citrus fruits. *South Afr. Citrus J.* **444** : 15,17-19.

Hong, S.B. and Chung, S.K. (1979). Effects of N P and K fertilzer levels on the growth, yield and fruit quality of satsuma trees. *Res. Rep.,* Rural Develop. Adm. Hort. & Ag. Engg. **21**, pp. 67-75.

Hoosbeck, M.R. and Bryant, R.B. (1992). Towards the quantitative modelling of pedogenesis - a review. *Geoderma* **55** : 183-210.

Horesh, I., Levy, Y. and Goldschmidt, E.E. (1986). Prevention of lime induced chlorosis in citrus trees by peat and iron treatments to small soil volumes. *HortScience* **21**(6) : 1363-1364.

Hortensius, D. and Welling, R. (1996). International standardization of soil quality measurements. *Commun. Soil Sci. Pl. Anal.* **27**: 387-402.

Hsieh, C.F. (1990). A survey of nutrient deficiencies and excess in fruit trees (citrus, grapes and pears) in central Taiwan. *Bull. No.* **28**, pp. 3-22, Taichung District Agri. Imp. Sta., Taichung, Taiwan.

Hu, C. C. (1956). History of citrus in China. *J. Agri. Assoc. China.* New Series.

Hu, F., Lin, M. and Wu, S. (1993). Characteristics of nematode population in low hill red soil ecosystems in central Jiangxi province, sub-tropical China. *Research on Red Soil Ecosystem* (M. Wang, T. Zhang and Y. He, ed.), Jiangxi Sci. and Tech. Press, Nanchang, China, pp. 88-91.

Huang, H.C. and Tsai, Y.F. (1988). Study on the effect of liming on an acidic Ponkan mandarin orchard. *Bull. No.* **20**, pp. 23-31, Taichung Dist. Agri. Expt. Sta., Taichung, Taiwan.

Huang, Rong Yang (1998). The green manure cultural techniques in hillside citurs orchards. *South China Citrus* **27** (4): 22-23.

Huang, S.H. and Wang, Q.Y. (1991). Study on a physiological disorder causing the yellow spotted leaf of Manju (*Citrus reticulata* Blanco) and its control. *China Citrus* **20**(3):9-11.

Huang, S.P., Wu, G.L. and Li, S.Y. (1993). A study on the temperature index in relation to abnormal fruitlet drop satsuma. *China Citrus* **22**(1) : 3-5.

Huang, Z. and Yan, S. (1990). Effect of organic – inorganic fertilization on citrus grown on the red soil. *Proc. Int. Citrus Symp.* (Huang, Bangyan and Yang, Qian, ed.) Guangzhou, China, pp. 513-516.

Huchche, A.D., Srivastava, A.K., Lallan Ram and Shyam Singh (1999). Nagpur mandarin orchards efficiency in Central India. *Proc. Int. Symp. Citriculture*, pp. 558-564 Nagpur, Maharashtra (India).

Hudson, J.P. (1975). Environmental effects on crop physiology. *Plants and the Weather.* Inaugural Lecture to the 5th Long Ashton Symposium, Academic Press (977), pp. 1-20.

Hudson, N.W. (1971). *Soil Conservation.* Cornell University Press, Ithaca, New York, USA.

Huggag, L.F. and Maksoud, M.A. (1996). Evaluation of yield of Navel orange tree : mathmatical model procedure. *Egyptian J. Hort.* **23**(2) : 197-202.

Human, N.B., Bower, J.P., Burdetle, S.A. and Barry, G.H. (1994). A computer programme for use in the citrus industry to test the climatic suitability of any site in the Eastern Cape. *J. South Afr. Soc. Hort. Sci.* **4**(1) : 9-12.

Human, C.F. and Koekemoer, P.J.J. (1991). Evaluation of early and late navel cultivars in a warm production area. *Appl. Pl. Sci.* **5**(2) : 64-67.

Hume, L.J., Healy, W.B., Hosking, W.J., Manarangi, A. and Tama, K. (1985a). NPK fertilizer rates for citrus on Rarotonga, Cook islands. *Scientific Rep.* **75,** p.17, New Zealand *Soil Bureau.,* New Zealand.

Hume, H. (1926). Cultivation of citrus fruits rural science services (L.H. Bailey, ed), McMillan & Co. London.

Hume, L.J., Healy, N.B., Tama, K., Hosking, W.J., Manarangi, A. and Reynolds, J. (1985b). Response of *Citrus sinensis* to NPK fertilizer on two soils of Rarotonga, Cook island. I. Effect of NPK fertlizer rate on soil properties and leaf nutrient content. *New Zealand J. Agric. Res.* **28** : 475-486.

Hume, L.J., Healy, N.B., Tama, K., Hosking, W.J., Manarangi, A. and Reynolds, J. (1985c). Response of citrus (*Citrus sinensis*) to nitrogen-phosphorus-potassium (NPK) fertilizer on soils of Rarotonga, Cook Islands. 2. Effects of NPK fertilizer rate, soil properties and leaf nutrient levels on yield and tree size. *New Zealand J. Agri. Res.* **28**(4) : 487-495.

Hunter, A.S. (1948). Yield and composition of alfaalfa as affected by variations in the calcium-magnesium ratio in the soil. *Soil Sci.* **67**: 53-62.

Hunziker, R.R. (1959). Water damage to citrus in the Indian River area in 1959. *Soil & Crop Sci. Soc. Fla. Proc.* **19** : 357-364.

Hurcidze, A.L. (1969). The effectiveness of green manuring in citrus crops in silt soils in Kolheda conditions. *Subtropicheskie Kul'tury* (6) : 109-114.

Hussein, J. and Adey, M.A. (1994). Sustainable regeneration of surface tilth in Zimbabwean Vertisols through water management. *Soil Science and Sustainable Land Management in the Tropics* (J.K. Syers and D.L. Rimmer, ed.), CAB International British Society of Soil Science, UK, pp. 120-131.

Hutchinson, D.J. (1975). Swingle citrumelo-a promising rootstock hybrid. *Proc. Fla. State Hort. Soc.* **87**:89-91.

Hyward, D.F. (1968). Climate of Sierra Leone. *Sierra Leone Geopraphy. J.* **12** : 3 (1968).

Ibia, T.O. and Udo, E.J. (1993). Phosphorus forms and fixation capacity of representative soils in Akwa Ibom state of Nigeria. *Geoderma.* **58**:95-106.

Ibrahim, R. (1968). The effect of type of soil and time of picking on the storage of Navel oranges. *Agri. Res. Rev. Cairo* **46**(3): 57-71.

Ibrahim, M. and Ali, N. (1970). Effect of foliar spray of zinc and manganese on nitrogenous and carbohydrate contents of citrus leaves. *Pakistan J. Sci.* **22**(56) : 239-243.

Ibrjkci, H. (1994). Macroelements status of mandarin orchards of southern Turkey. *Commun. Soil Sci. Pl. Anal.* **25**(17-18) : 2971-2980.

Ida, H., Nakama, K., Ishida, T., Takahashi, S., Shirai, T. and Okada, N. (1969). Studies on the abnormal defoliation of Satsuma orange trees. 3. Effect of added $CaCO_3$, sucrose, $ZnSO_4$ and $Fe_2(SO_4)_3$ on the growth and Mn absorption of young Satsuma mandarin orange trees in the case of Mikataghara soil. *Bull. No.* **8**, pp. 51-57, Citrus Expt. Sta., Shizuoka, Japan.

Iley, J.R., Bryan, O.C. and Johnson, R.S. (1970). The effect of soil pH and organic matter on the magnesium content of soil and leaves of citrus on acid flatwood soils. *Proc. Fla. State Hort. Soc.* **82** : 30-34.

Iley, J. R. and Guilford, H. E. (1978). Excess dolomite and lime plots display conditions very similar to YTD. *Proc. Fla. State Hort.Soc.* **91:** 62-66.

Imamura, T., Nakamura, S., Asada, K. and Kank, K. (1981). Investigation of yellowish fruit in satsumas. *J. Japanese Soc. Hort. Sci.* **49**(4) : 505-511.

Ingty, P.S. and Goswami, N. (1979). *Proc. Agro-Forestry Seminar,* Imphal, May 16-18, 1979, ICAR, New Delhi, India.

Inoue, H. and Harada, Y. (1981). Nutritional problems of Satsuma mandarin in a plastic house. *Proc. Int. Soc. Citriculture.* Vol. **2**, pp. 556-559.

Inoue, H. and Kataoka, I. (1992). Effect of nutrition and temperature on flow bud differentiation of satsuma mandarin. *J. Japanese Soc. Hort. Sci.* **60**(4) : 771-776.

Inoue, H. and Nakayasu, T. (1982). The effects of the physical properties of the soil and of nitrogen fertilizers on the growth of young satsuma trees. *Tech. Bull.* **33**(2), pp. 84-89.

Intrigliolo, F., Cicco, V. De, Vanadia, S., Ippolito, A. and Giuffrida, A. (1991). Effects of water stress and potassium on Novelina orange splitting. *Agricoltura Mediterranea* **121**(1) : 24-31.

Intrigliolo,F., Coniglione, L. and Germana, C. (1992). Effect of fertigation on some physiological parameters in orange trees. *Proc. Int. Soc Citriculture.* Vol. **2**, pp. 584-589.

Ishihara, M., Ayamori, T., Sato, T., Yokomizo, H. and Konno, S. (1971a). Studies on abnormal defoliation of satsuma trees. II. The effects of fertilizer application on the development of abnormal defoliation on the absorption of minor elements by satsuma trees. *Bull. No.* **10**, pp. 99-130. Hort. Res. Sta. Hiratsuka, Japan.

Ishihara, M., Hase, Y, Yokomizo, H. and Konno, S. (1971b). Studies on abnormal defoliation of satsuma trees. III. The influence of excessive soil application of manganese, aluminium, copper, zinc and nickel on the development of abnormal defoliation and on the absorption of minor elements by satsuma trees. *Bull. No.,* **10**, pp. 131-154. Hort. Res. Sta. (Hiratsuka), Japan.

Ishihara, M., Shimosako, Y., Sakaguchi, S., Yanase, M., Shibuya, H., Teraoka, Y., Yokomizo, H. and Konno, S. (1972). Studies on copper deficiency of citrus. I. Methods of diagnosing copper deficiency from symptoms and from tree and soil analyses. *Bull. No.* **11**, pp. 41-76. Hort. Res. Sta. (Hiratsuka), Kanagawa, Japan.

Ishihara, M., Yokomizo, H., Hase, Y., Konno, S. and Sato, K. (1971c). Studies on abnormal defoliation of satsuma trees. I. Investigation on abnormal defoliation in satsuma orchards and their leaf and root analysis. *Bull. No.* **10**, pp. 55-98. Hort. Res. Sta., Hiratsuka, Japan.

Iwamasa, M. and Oba, Y. (1973). Precocious flowering of citrus seedling. *Congresso Mundial de Citricultura.* Vol. **3**, Muria-Valencia, Spain, pp. 205-213.

Iwamoto, K.K., Otsu, K., Uchinori, H., Hirata, Y., Shirai, T. and Okada, N. (1969). Studies on the abnormal defoliation of Satsuma orange trees. *Rep.* No. **2**, pp. 45-76. Kumamoto Fruit Expt. Sta.

Iwasaki,T. and Owada, A. (1960). Studies of the control of the alternate bearing. II. The effect of environmental conditions during winter on the number of flowers and growth of shoots. *J. Hort. Assoc. Jap.* **29** : 37-46.

Iyengar, B.R.V. and Deb, D.L. (1977). Contribution of soil zinc fraction to plant uptake and fate of zinc applied to the soil. *J. Indian Soc. Soil Sci.* **25:** 426.

Iyengar, B.R.V. and Joshi, O.P. (1991). Soil Related Constraints in Crop Production – Soils under Fruit Crops. *Soil-related Constrainsts in Crop Production* (T.D. Biswas, G. Narayanaswamy, N.N. Goswami, G.S. Sekhon and T.G. Sastry, ed.), Indian Society of Soil Science, New Delhi, pp. 145-156.

Iyengar, B.R.V. and Kotur, S.C. (1996). *Isotopes and Radiation in Agriculture and Environment Research,* (M.S. Sachdev, ed). Indian Society of Nuclear Techniques in Agriculture and Biology, New Delhi, pp. 71-86.

Iyengar, B.R.V. and Murthy, S.V.K. (1987). Spatial distribution of root activity in kagzi lime (*Citrus aurantifolia* Swingle). *Indian J. Hort.* **44**(1/2) : 41-44.

Jackson, M.L. (1958). *Soil Chemical Analysis.* Prentice Hall Inc., Engleward Cliffs, New Jersy, USA.

Jackson, M.L. (1963). Aluminium bonding in soils : A unifying principal in Soil Science. *Proc. Soil Sci. Soc. Am.* **27** : 1.

Jackson, L.K. (1991). *Citrus Growing in Florida.* (3rd edition), University of Florida Press, Gainsville, USA, pp. 103-111.

Jackson, L. K. (1992). Nitrogen, phosphorus and potassium requirements of citrus. *Citrus Indus.* **73**(1) : 76-77.

Jackson, J.L. Jr. and Ayers, D.H. (1986). Performance of individual tree covers for cold protection of young trees. *Proc. Fla. State Hort. Soc.* **99** : 18-23.

Jackson, J.E. and Hamer, P.J.C. (1980). The acuses of year-to-year variation in the average yield of Cox's orange Pineapple in England. *J. Hort. Sci.* **55**:149-156.

Jackson, D.J. and Looney, N.E. (1999). *Citrus : Temperate and Sub-tropical Fruit Production.* (ed.). CAB International, 2nd Edition, U.K. pp. 1-10, 229-240.

Jacob, A. (1958). *Magnesium, the Fifth Major Plant Nutrient.* Staples Press Limited, London, U.K., p. 159.

Jacoby, B. (1961a). Conditions responsible for the appearance of magnesium deficiency in citrus groves on Israel's mediterranean coast. *Israel J. Agri. Res. (Ktavim)* **11** : 173-178.

Jacoby, B. (1961b). Calcium-magnesium ratios in the root medium as related to magnesium uptake in citrus seedlings. *Plant and Soil* **15**: 74-80.

Jadhav, N.S., Malewar, G.U. and Varade, S.B. (1978): Vertical distribution of zinc and iron in some citrus growing soils of Marathwada. *J. Maharashtra Agri. Univ.* **3**:85-87.

Jadhav, N.S., Malewar, G.U. and Varade, S.B. (1979). Effect of zinc and iron on fruit drop, yield and quality of Nagpur oranges (*Citrus reticulata* Blanco). *J. Maharashtra Agri. Univ.* **4**(1):106-107.

Jagdish Prasad (1996). Managing Nagpur mandarin in shallow soils. *Indian Hort.* **41**(1) : 25-26.

Jagirdar, S.A.P. and Maniyar, A.K. (1962-64). What is wrong with citrus in Sind. *Punjab Fruit J.* **26/27**: 56-73.

Jalali, V.K., Talib, A.R. and Takkar, P.N. (1989). Distribution of micronutrients in some benchmark soils of Kashmir at different altitudes. *J. Indian Soc. Soil Sci.* **37** : 465-469.

Janssen, B.H., Guiking, F.C.T., Vander Eijk, D., Smaling, E.M.A., Wolf, J. and Van Reuler, H. (1990). A system of quantitative evaluation of the fertility of tropical soils (QUEFTS). *Geoderma* **46** : 299-318.

Jassal, H.S., Sidhu, P.S., Sharma, B.D. and Mukhopadhyay, S.S.(2000). Mineralogy and geochemistry of some soils of Siwalik hills. *J. Indian Soc. Soil Sci.* **48**(1) :163-172.

Jawahar, D., Arunachalam, G., Velu, V. and Ramaswami, P.P. (1999) Fertility capability of the soils of Tamil Nadu. *J. Indian Soc. Soil Sci.* **47**(3) : 570-573.

Jawaharlal, M., Durairaj, P., Subburamu, R., Dharamraj, G. and Irulappan,I. (1989). Effect of variable dose of acid lime on growth and yield. *South Indian Hort.* **37**(4) : 244-246.

Jawanda, J.S. (1961). The effect of waterlogging on fruit trees. *Punjab Hort. J.* **1** : 150-152.

Jawanda, J.S. (1969). Mandarin industry of Nagpur. *Punjab Hort. J.* **9** : 140-143.

Jawanda, J.S., Arora J.S. and Sharma, J.M. (1973). Fruit quality and maturity standards of kinnow mandarin at Abohar. *Punjab Hort. J.* **13**: 3-12.

Jenny, H. (1961). Derivation of state factor equations of soils and ecosystems. *Proc. Soil Sci. Soc. Am.* **25** : 385-388.

Jessup, R.W. (1960). The lateritic soils of the south-eastern portion of the Australian arid zone. *J. Soil Sci.* **11** : 106.

Jiang, Ping., Zhao, Xiang Xong., Zhang, Renrong. and Wang, Yuansheng (1997). Effect of mulching in the hillside citrus orchards. *South China Citrus* **26**(3) : 17-18.

Johnson, T.J. (1966). Keys to successful citrus growing. *Bull.* **56**, pp.19-21, Kenya Sisal Bd., Kenya.

Johnson, M. and Martinet, W.A. (1981). Outlook for California-Arizona acreage and production. *Proc. Int. Soc. Citriculture.* Vol. **2**, pp. 943-945.

Johnston, J. C. and Sullivan, W. (1949). Eliminating tillage in citrus soil management. *Univ. Calif. Agri. Ext. Serv. Circ.* **150**, p. 16.

Johnston, J.C. (1953). Citrus growing in California. *Div. Agri. Sci. Circ.* **426**, p.43.

Jolka, N.K. and Awasthi, R.P. (1980). Studies on maturity standards of kinnow in Himachal Pradesh. *Punjab Hort. J.* **20**: 149-151.

Jollie , F.J. E. (1965) . The Cook Islands & its Citrus industry. Min. Agri., Cook Islands.

Jones, J.B. Jr. (1979). *Fertilizer and Fertility.* Reston Publishing Company Inc., Prentice Hall Company, Reston, Virginia, USA.

Jones, J.B. Jr. (1980). *Handbook on Reference Methods for Soil Testing.* Council Soil Testing and Plant Analysis. Athens, Georgia, USA. p. 82.

Jones, W.W. and Cree, C.B. (1954). Navel orange production as influenced by sulfur in the fertilizer program. *Calif. Citrog.* **39** : 212-214.

Jones, W.W. and Cree, C.B. (1964). Environmental factors related to fruiting of Washington navel orange over a 38 year period. *Proc. Am. Hort. Soc.* **86** : 267-271.

Jones, W.W., Cree, C.B. and Embleton, T.W. (1961). Some effects of nitrogen sources and cultural practices on water intake by soil in a Washington navel orange orchard and on fruit production, size and quality. *Proc. Am. Soc. Hort. Sci.* **77** : 146-154.

Jones,W.W. and Embleton, T.W. (1973). Soil, soil management and cover crops. *The Citrus Industry.* (W. Ruther, ed.), Univ. of Calif. Div. Agri. Sci. California, USA. Vol. **3**: pp. 98-121.

Jones, W.W., Embleton, I.W. and Optiz, K.W. (1971). Effects of foliar applied Mg on yield, fruit quality and micronutrients of Washington navel orange. *J. Am. Soc. Soc. Hort. Sci.* **96** : 68-70.

Jones, H.E. and Etherington, J.R. (1970). Comparative studies of plant growth and distribution in relation to waterlogging : I. The survival of *Erica carnea* L. and *E. tetralix* L. and its apparent relationship to iron and manganese uptake in waterlogged soil. *J. Ecol.* **58** : 467-496.

Jones , W. W. , Martin, J. P. and Bitters, W. P. (1957). Influence of exchangeable sodium and potassium in the soil on the growth and composition of young lemon trees on different root stocks. *Proc. Am. Soc. Hort. Sci.* **69** : 189-196.

Jones, W. W. and. Smith, P. F (1984). Nutrient Deficiencies in Citrus, *Hunger Signs in Crops* (Howard B. Sprague, ed.), pp. 359-414, David McKay Co., New York, USA.

Joolka, N.K. and Singh, I.P. (1979). Effect of soil salinity on growth of citrus rootstocks. *Indian J. Agric. Sci.* **49**: 858-861.

Jordan, L. (1982). Management of cover vegetation in citrus orchards. *Workshop Proc. Crop Production using Cover Crops and Sods as Living Mulches* (J.C. Miller and S.M. Bell, ed.), Cornavals, USA, pp. 73-79.

Jordan, L.S. and Russell, R.C. (1978). Repeated application of soil-residual herbicides and yield and quality of Valencia oranges. *HortScience* **13** : 544-545.

Jorgensen, K.R. (1978). Leaf and soil analysis, a guide to fertlizing citrus. *Queensland Agri. J.* **104**(1) : 17-21.

Jorgensen, K.R. and Price, G.H. (1978). The citrus leaf and soil analysis system in Queensland. *Proc. Int. Soc. Citriculture.* pp. 297-299.

Joshi, K.D., Bhattarai, S.P., Vaidya, A., Rasali, D. P., Suwal, M.R.S., Subedi, P.P., Adhikar, B. and Phuyal, U. (1995) Systems analysis of soil fertility in relation to madarin orchards in the middle hills of western Nepal. *LARC Working Paper No.* **95/9.** Lumle Agricultural Research Centre, Kaski, Nepal.

Joublan, M.J.P. Venegas, V.A., Wilckens, E.R. and Guerreo, C. (1997). Adaption of some citrus cultivars to an agroecological zone in the eighth region of Chile, Second growing season. *Agro-Ciienca* **13**(2): 149-157.

Julien, J.L. (1989). Determination des normes dinterpretation d' analyse de terre en vue la fertilisation potassique. *Science, du sol.* **27** (2): 131-144.

Juste, C. (1978). Some factors which might explain the origin of leaf burn on clementines in Corsia. *Fruits* **33**(12) : 829-830.

Kacharava, O.N. (1985). Fertilization system in mandarin plantations. *Subtropicheskie Kul'tury* (6) : 45-50.

Kailash Kumar and Rao, K.V.P. (1990). Fertility status of Manipur valley soils. *Indian J. Hill Farming.* **3**(2):73-75.

Kailash Kumar, Rao, K.V.P. and Singh, L.J. (1995a). Forms of acidity in some acid Inceptisols under different land use in Manipur. *J. Indian Soc. Soil Sci.* **43**(3):338-342.

Kailash Kumar and Raychaudhuri, Mausumi (1996). Different fractions of phosphorus in acid hill soils of Manipur. *J. Hill Res.* **9**(2) : 371 – 374.

Kailash Kumar, Singh, L.J. and Rao, K.V.P. (1994). Physico-chemical characteristics of hill soils of Manipur. *J. Hill Res.* **7**(2):69-74.

Kailash Kumar, Singh, L.J. and Rao, K.V. P. (1995b). Components of pH dependent CEC in acid soils of Manipur *J. Indian Soc. Soil Sci.* **43**(2) : 263-264.

Kailash Kumar, Singh, L.J. and Rao, K.V. P. (1996). Assessment of some lime requirement methods for acid soils of Manipur *J. Hill Res.* **9**(2) : 262 – 266.

Kailash Kumar, Singh, L.J. and Rao, K.V.P. (1997). Evaluation of lime requirement methods for acid soils of Manipur. *J. Indian Soc. Soil Sci.* **45**(2): 404 -406.

Kaistha, B.P., Sharma, Pritam, K. and Sharma, Raj. P. (1997). Influence of soil components on phosphorus fixing capacity of some Alfisols of Himachal Pradesh. *J. Indian Soc. Soil Sci.* **45**(2) : 261-264.

Kaistha, B.P. and Gupta, R.D. (1992). *Himalayan Environment-Man and the Economic Activities.* **Part 1,** (J.L. Raina, ed.), Point Publshers, Jaipur, p. 125.

Kaistha, B.P. and Gupta, R.D. (1994). Morphology and characteristics of a few Entisols and Inceptisols of north-western Himalayan region. *J. Indian Soc. Soil Sci.* **42**(1):100-104.

Kakde, J.R. (1956). A preliminary survey of the study of relationships of physical properties of the orange orchard areas and the orchard performances in Nagpur district. *Nagpur Agric. Coll. Mag.* **30**: 19-24.

Kalbande, A.R., Landey, R.J. and Bhattacharjee, J.C. (1983). Characterisation of soils under orange cultivation of Nagpur district of Maharashtra. *Indian J. Agric. Sci.* **53:**57-61.

Kalma, J.D. and Stanhill, G. (1972). The climate of an orange orchard: Physical characteristics and microclimate relationship. *Agri. Meteorol.* **10**(3): 185-201.

Kamprath, E.J. (1970). Exchangeable aluminium as a criterion for liming leached mineral soils. *Proc. Soil Sci. Soc. Am.* **34** : 252-254.

Kamprath, E.J. (1980). *Soil Related Constraints to Food Production in Tropics,* I.R.R.I., Los Banos, Philippines.

Kang, Dae Joon (1997). Citrus produciton in Korea. *Extension Bull* **439**, p. 10. Food & Fert. Tech. Centre, Cheju, Korea.

Kanki, H. and Imamura, T. (1968). Investigations on abnormal defoliation of satsuma orange in Honshu and Shikoku districts (In Japanese with english summary). *Ibid* **37**:122-128.

Kanki, H. and Imamura, T. (1969). Results of further investigation on satsuma orange orchards suffering from abnormal defoliation. *J. Japanese Soc. Hort. Sci.* **38** : 295-299.

Kanki, H., Yajima, K. and Hamaguchi, K. (1968). Investigation of abnormal defoliation in satsuma orange orchards (In Japanese with english summary). *J. Japanese Soc. Hort. Sci.* **37**:51-56.

Kannan, M., Thamburaj, S and Seemanthini, Ramdas (1989). Studies on the manurial requirement of mandarins in Shevroy hills. *South Indian Hort.* **37**(4):203-208.

Kanwar, J.S. (1997). Soil and water management. The base of food security and sustainability. *J. Indian. Soc. Soil Sci.* **45**(3) : 417-428.

Kanwar, J.S. and Bhambota, J.R. (1969). Variable salinity and water table level effects on the growth of sweet orange (*Citrus sinensis* Osbeck) under Punjab conditions. *Proc. Ist. Int. Citrus Symp.,* Vol. **3**, pp. 1783-1791.

Kanwar, D.R., Bhumbla, D.R. and Seghal, J.L. (1965). Soils of the proposed citrus belt of the Punjab. II. Physico-chemical analysis. *J. Res.* PAU **2**:154-160.

Kanwar, J.S. and Dhingra, D.R. (1961). Chlorosis of citrus in the Punjab. *Punjab Hort. J.* **1**(2): 92-95.

Kanwar, J.S. and Dhingra, D.R. (1962). Effect of micronutrient sprays on the chemical composition of citrus leaves and incidence of chlorosis. *Indian J. Agric. Sci.* **32**:309-314.

Kanwar, J.S., Dhingra, D.R. and Randhawa, N.S. (1962). Chemical composition of healthy and chlorotic leaves of citrus plants in the Punjab. *Indian J. Agric. Sci.* **33**(4): 268-271.

Kanwar, J.S. and Grewal, J.S. (1959). Forms of phosphorus in Punjab soils. *J. Indian Soc. Soil Sci.* **7** : 135.

Kanwar, J.S. and Randhawa, N.S. (1959). Yellow leaves of citrus are an indication of something wrong with your soils and garden management. *Kehti Bari* (5) : 4.

Kanwar, J.S. and Randhawa, N.S. (1960). Probable causes of chlorosis of citrus in Punjab. *Hort. Adv. (P.A.U.)* **4**: 61-67.

Kanwar, J.S. and Randhawa, N.S. (1974). Micronutrient Research in Soils and Plants in India - A Review. *Tech. Bull. Agric. No.* **50**, p. 185, Indian Council of Agricultural Research, New Delhi.

Kanwar, J.S., Sehgal, J.L. and Dhingra, D.R. (1965). Soils of proposed citrus belt of Punjab. *Bull.* **56**, p. 26., Punjab Agric. Univ., Punjab, India.

Kapanadze, I.S. (1964). The problem of adaptive variability in citrus. *Subtropicheskie Kul'tury* (4): 79-86.

Kar, P.S., Roy, A. and Mitra, S.K. (1988). Effect of prunning and fertilization on rejuvenation of mandarin orange. *Indian Agricst.* **32**(3):205-210.

Karamanova, L.A. (1978). General patterns of the ratio and distribution of forms of iron in major genetic soil groups. *Soviet Soil Sci.* **10**(4) : 446-458.

Karim, M., Ahmed, F. and Islam, A. (1973). A study of phosphate adsorption by four Bangladesh soils. *Geoderma* **9** : 221-227.

Karlen, D.L., Eash, N.S. and Unger, P.W. (1992). Soil and crop management effects on soil quality indicators. *Am. J. Alter. Agri.* **7** : 48-55.

Karlen, D.L., Wolenhaupt, N.C., Erbach, D.C., Berry, E.C., Swan, J.B., Eash, N.S. and Jordahl, J.L. (1994a). Crop residue effects on soil quality following 10 years of no till. *Soil Till. Res.* **31** : 149-167.

Karlen, D.L., Wolenhaupt, N.C., Erbach, D.C., Berry, E.C., Swan, J.B., Eash, N.S. and Jordahl, J.L. (1994b). Long term tillage effects on soil quality. *Soil Till. Res.* **32** : 313-327.

Karmakar, R.M. and Rao, A.E.V. (1999a). Soils on different physiographic units in lower Brahmaputra valley zone of Assam. I. Characterization and classification. *J. Indian Soc. Soil Sci.* **47**(4): 761-767.

Karmakar, R.M. and Rao, A.E.V. (1999b). Soils on different physiographic units in lower Brahmaputra valley zone of Assam. II. Sand mineralogy. *J. Indian Soc. Soil Sci.* **47**(4):767-770.

Karmakar, R.M. and Rao, A.E.V. (1999c). Soils on different physiographic units in lower Brahmaputra valley zone of Assam. III. Humic substances. *J. Indian Soc. Soil Sci.* **47**(4) : 771-774.

Kashirad, A., Hutton, C.E., Fiskell, J.G.A. and Carlisie, V.W. (1966). Tillage pan identification and root growth. *Proc. Soil & Crop Sci. Soc. Fla.* **26** : 41-52.

Katyal, J.C. and Agarwala, S.C. (1982). Micronutrient research in India. *Fert. News* **27**(2) : 66-68.

Katyal, J.C. and Deb, D.L. (1982). Nutrient transformation in soils - micronutrients. *Review of Soil Research in India.* Part **I**. pp. 146-159, Indian Soc. Soil Sci, New Delhi, India.

Katyal, J.C. and Sharma, B.D. (1991). DTPA- extractable and total Zn, Cu, Mn, and Fe in Indian soils and their association with some soil properties. *Geoderma* **49** : 165-179.

Katyal, J.C. and Vlek, P.L.G. (1985). Micronutrient problems in tropical Asia. *Fert. Res.* **7** : 69-74.

Katyal, S.L. and Sabharwal, H.S. (1964). Mandarins can be grown in hills too. *Indian Hort.* **8**(4): 9-11.

Kaufmann, M.R. (1977). Citrus – A case study of environmental effects on plant water relations. *Proc. Int. Soc. Citriculture* Vol. **1**, pp. 57-62.

Kawamura, A., Koga, H., Yamasaki, S. and Ujike, T. (1970). Analysis of the factors affecting soil productivity in citrus orchards. II. A study on the soil of an orchard of satsuma mandarin oranges on Uzu rootstocks in the Katsuma region of Tokushima prefecture. *Bull. No.* **20**, pp. 1-26, Shikoku Agric. Expt. Sta., Japan.

Kawamura, A., Koga,H., Yamasaki, S. and Ujike, T. (1973a). The persistent effect of physical amendment of the soil in an orange orchard. I. On an orchard soil derived from granite and andesite. *Bull. No.* **26**, pp. 105-122, Shikoku Agri. Expt. Sta., Kagawa-Ken, Japan.

Kawamura, A., Koga, H., Yamasaki, S. and Ujike, T. (1973b). The persistent effect of physical amendment of the soil in an orange orchard. II. On an orchard soil derived form Izumi sandstone. *Bull. No.* **26**, pp. 150-173, Shikoku Agri. Expt. Sta., Kagawa-ken, Japan.

Kawase, M. (1972). Effect of flooding on ethylene concentration in horticulture plants. *J. Am. Soc. Hort. Sci.* **97** : 584-588.

Kawase, K. (1984). Studies on the cause and control of Satsuma rind puffing II. Effect of humidity on the occurrence of rind puffing. *Bull.* **6**, pp. 41-55, Fruit tree Research Station, Okitsu, Japan.

Kawase, K.T., Takahara, K. and Hirose, S. (1984). Studies on the cause and control of Satsuma rind puffing I. The relationship between the occurence of rind puffing and the climatic conditions in Kyushu and Tokai districts. *Bull.* **6**, pp. 27-39, 41-55, 57-75, Fruit Tree Research Station, Japan.

Kazak, N.M. and Khalidy, R. (1974). The interrelationships among potassium magnesium, calcium and sodium in citrus leaves and the relation of soil potassium and magnesium to their status in the leaves. *Ist Congresco Mundail de Citricultura* 1973, Vol. **1**, pp. 43-47.

Kechakmadze, M.S. (1987). Effectiveness of boron fertilizers in young mandarin orchard on a Krasnozem soil. *Subtropicheskie Kul'tury* (4) : 136-140.

Keeney, D.R. and Corey, R.B. (1963). Factors affecting the lime requirement of Wiscosin soils. *Proc. Soil Sci. Soc. Am.* **27** : 61-66.

Kellogg, C.E. (1950). Tropical soil. *Trans. 4th Int. Congr. Soil Sci.,* Vol. **3**, pp. 266-276.

Keng, J.C.W. and Uehera, G. (1974). Chemistry, mineralogy and taxonomy of Oxisols and Ultisols. *Proc. Soil & Crop. Sci. Soc. Fla.* **33:** 119-126.

Kennedy, A.C. and Papendick, R.I. (1995). Microbial characteristics of soil quality. *J. Soil Water Conserv.* **50**: 243-248.

Kennedy, A.C. and Smith, K.L. (1995). Soil microbial diversity and sustainability of agricultural soils. *The Significance and Regulation of Soil Biodiversity* (H.P. Collins *et al.* ed.), Kluwer Acadamic Publishers, Amsterdam, Netherlands.

Kessler, B. and Snir, I. (1969). Salt effects on nucleic acid and protein metabolism in citrus seedlings. *Proc. Ist.Citrus Symp.* Vol. **1**, pp. 381-386.

Ketchie, D.O. (1969). The effect of high temperature on citrus. *Proc. Ist. Int. Citrus Symp.* Vol. **1**, pp. 267-270.

Ketchie, D.O. and Ballard, A.L. (1968). Environments which cause heat injury to Valencia oranges. *Proc. Am. Soc. Hort. Sci.* **93** : 166-172.

Ketchie, D.O. and Furr, J.R. (1968). Sunburn and heat injury of citrus. *Calif. Citrog.* **53** : 252.

Ketsa, S. (1988). Fruit quality of Khiew Waan tangerine from different growing locations. *Kaestart J. Nat. Sci.* **22**(4) : 279-283.

Khadr, M. (1963). The clay mineral composition of some soils of the U.A.R. *J. Soil Sci.(U.A.R.)* **1** : 141-155.

Khadr, A.A. (1965). Differential responses to some nutrient elements by trifoliate orange and rough lemon. *Dis. Abst.* **26**:14.

Khadr, A.H., Wallace, A. and Romney, E.M. (1965). Mineral nutritional problems of trifoliate orange rootstock. *Calif. Agric.* **19**(9):14-15.

Khalidy, R.M., Al-Jamali, A.F. and Bolkan, H. (1966). Boron uptake and its effect on other elements in the leaves of 'Eureka' lemon. *Proc. 17th Int. Hort. Congr,* p. 1.

Khalidy, R., Jamali, A. and Bolkan, H. (1969a). Causes of the pateca disease of lemons as occurring in Lebanon. *Proc. Ist. Int. Citrus Symp.* Vol.**3**, pp. 1253-1256.

Khalidy, R., Sheikh, M.M. and Boukhari, M.A. (1969b). Orchard condition and leaf inorganic composition as influencing Valencia orange yield. *Proc. Ist. Int. Citrus Symp.* Vol. **3**, pp. 1635-1640.

Khalidy, R.M. and Kazak, N.M. (1970). Nutritional status of citrus trees in Lebanon. *Proc. Conf. Trop. Subtrop.* Fruits Trop. Prod. Inst., pp. 253-259.

Khan, M.A. (1969). Lemon growing in west Pakistan. *Punjab Fruit J.* **31**(109) : 24-26.

Khan, N.A. (1994). My twenty years experience with orchards. *Punjab Hort. J.* **4**: 670-672 *(Urdu).*

Khanduja, S.D. (1969). A foliar analysis survey of Mosambi orange (*Citrus sinensis*) orchards in Maharashtra. *Pl. Sci.* **1**: 228-232.

Khanna, S.S. (1989). The Agro-climatic Approach in Survey of Indian Agriculture. *The Hindu,* Madras, India, pp. 28-35.

Khanna, S.S., Anand, S.S. and Bakhshi, J.C. (1969). Comparative efficiency of chelated and unchelated compounds on citrus chlorosis. I. Spray application of Zn compounds with copper. *Indian. J. Agric. Sci.* **39**:830-840.

Kharebava, L. G. (1971). Studies on the effectiveness of different rates of potassium on mandarins growing on pedzolic and alluvial soils. *Subtropicheskie Kul'tury* (2) : 163-170.

Kharkar, P.T., Deshmukh, P.P, Bagde, T.R. and Zade, K.B (1991). Nutritional orchards of healthy and declined orchards of Nagpur mandarin in Vidarbha region. *Ann. Pl. Physiol.* **5(1)**: 24-29.

Khera, A.P., Singh, H.K. and Daulta, B.S. (1985). Correcting micronutrient deficiency in citrus cv. Blood Red. *Haryana J. Hort. Res.* **14** (1/2):27-29.

Khera, M.S. and Pradhan, H.R. (1976). Potassium in Soils, Crops and Fertilizers. *Bull.* **10**, p. 224, Indian Soc. Soil Sci., New Delhi.

Khurtsidze, A.L. (1972). The effect of soil management methods on the growth and development of root system of young mandarin trees in the Kolkhida lowland. *Subtropicheskie Kul'tury* (2) : 140-143.

Kiely, T. (1957). *Report on Decline Problem of Citrus in Lower Uva Valley and Other Areas in Ceylon. Sessional Paper* **IX**, Colombo, Govt. of Ceylon, Sri Lanka.

Kiely, T.B., Cox, J.E., Cradock,F.W. and Barkus, B. (1972). Nutritional studies on Valencia oranges. *Agric. Gazette. New South Wales* **83**:20-22.

Kimani, W. (1984). The influence of mineral nutrition on the severity of citrus greening disease in Kenya. *Proc. Int. Soc. Citriculture.* Vol. **1**, pp. 172-173.

Kimball, M.H., Wallace, A. and Mueller, R.T. (1950). Changes in soil and citrus root characteristics with nontillage. *Calif. Citrog.* **35**: 409, 432-433.

King, H.W. (1969). The Australian citrus industry - a review of current and futue problems. *Australian Citrus News* **45**(8) : 3,6-10.

Kingman, G.F. (1915). Citrus fertilization in Puerto Rico. *Bull.* **18**, Puerto Rico Agric. Expt. Sta., Puerto Rico.

Kinkladze, R.K. (1987). The total evaporation from mandarin plantations on slopes in the subtropical zone of Western Georgia. *Subtropicheskie Kul'tury* (3) : 97-101.

Kirkpatrick, J.D. and Bitters, W.P. (1969). Physiological and morphological response of various citrus rootstocks to salinity. *Proc. Int. Citrus Symp.* Vol. **1**, pp. 391-406.

Klotz, L.J. (1975). Water spot of navel oranges. *Citrog.* **60:**439-441.

Klotz, L.J. (1978). Fungal, bacterial and non-parasitic diseases and injuries originating in seedbed, nursery and orchard. *Citrus Industry,* Vol. **4**, Univ. of Calif. USA.

Klotz, L.J., Stolzy, L.H., Letey, J., Labanauskas, C.K., Valoras, N. and DeWolfe, T.A. (1966). Root decay of citrus as affected by soil moisture and aeration. *Citrog.* **51** : 296-299.

Knorr, L.C. (1973). *Citrus Diseases and Disorders.* Univ. Press Florida, Gainesville, Florida, USA.

Knorr, L.C. and Koo, R.C.J. (1969). Rumple – A serious rind collapse of lemons in Florida and in mediterranean countries. *Proc. Ist. Int. Citrus Symp.* Vol.**3**, pp. 1463-1472.

Knorr, L.C. and Vaughn, J.R. (1964a). World citrus problem. III Syria. *FAO Pl. Prot. Bull.* **12**, pp.37-41.

Knorr, L.C. and Vaughn, J.R. (1964b). World citrus problem. IV. Venezuela. *FAO Pl. Prot. Bull.* **12**, pp. 121-128.

Ko, K.D. and Kim, S.K. (1987). Chemical properties of soil and leaf mineral contents in Cheju citrus orchards. *J. Korean Soc. Hort. Sci.* **28**(1) : 45-52.

Kodua, M. (1972). The effect of boron and manganese on fruit quality of mandarins under clean cultivation, green manure and liming. *Subtropicheskie Kul'tury* (4) : 91-93.

Koen, T.S., Plessis,S.F. Du and Lagengger, W. (1977). The effect of nitrogen fertilization on the yield, fruit quality and leaf composition of Valencia orange trees. *Hort. Abst.* **47**(7):579.

Koga, H. (1972). Studies on the physical properties of the subsoil of satsuma orchards. *Bull. No.* **25**, pp. 119-232. Shikou Agri. Expt. Sta., Kagawaken, Japan, Ishikoku.

Koga, H. and Kawamura, A. (1970). The soil productivity of reclaimed orchards on sloping land part 3. Relationship between soil type and growth of Satsuma orange. *Bull*. **21,** pp. 47-84, Shikoku Agric. Expt. Sta., Japan.

Kohli, H. C. (1960). The effect of length of photoperiod on grapefruit seedlings. *Calif. Agric.* **14**(3): 8-11.

Kohli, R.R., Huchche, A.D., Lallan Ram, Srivastava, A.K. and Dass, H.C. (1993). Interaction effect of N and P on the growth, yield and quality of Nagpur mandarin (*Citrus reticulata* Blanco) in Typic Chromustert. *Abstract,* p. 84 Int. Symp. on A Decade of Potassium Research, Pot. Res. Inst. of India, Gurgaon, Haryana, India.

Kohli,R.R., Lallan Ram, Huchche, A.D., Srivastava, A.K. and Dass, H.C (1994). Nutrient Management Studies in Nagpur Mandarin and Acid Lime. *Ann. Rep.* (1994-95) p. 24. National Research Centre for Citrus, Nagpur, Maharashtra, India.

Kohli, R.R., Srivastava, A.K. and Huchche, A.D. (1997). Nutrient requirement of Nagpur mandarin in clay soils in central India. *Indian Fmg.* **47**(2):25-27.

Kokba, A.M., Abdel Messih, M.N. and Mohamed, M.A. (1979). Breeding and screening some citrus rootstocks for salt tolerance in Egypt. *Egpytian J. Hort.* **6**: 69-79.

Koller, O.C., Anghinoni, I., Manica, I., Moraes, P., De, A.F., Pires, J.L., Ruker, P.A., Azeredo, V., Silva, L.J.C., Korndoerfer, G.H., Treher, R.T. and Finkler, L.M. (1986). Nutritional condition of citrus in the producing region of Rio Grande do Sul State. *Agronomia Sulriograndense* **22**(2) : 185-204.

Koller, O.L. (1988). Siciliano lemon (*Citrus limon* Burm. f.) budding height. *Proc. 6th Int. Citrus Congr.* (R. Goren and K. Mendel, ed.), Vol. **1,** pp. 241-250.

Komakhidze, R.P., Sharashenidze, M.G. and Chanukvadze, R.N. (1990). The effect of soil cultivation on the growth and cropping of mandarin trees. *Subtropicheskie Kul'tury* (1) : 93-99.

Komamura, K., Obata, H. and Sekiya, K. (1982). Changes in water status of soils and satsuma trees in relation to soil type and depth. *Bull.* **9**, pp. 143-157. Fruit Tree Res. Sta, Yatabe Japan.

Kong-Tau, T. (1986). Physical properties of some main soil groups of Vietnam. *Soviet Soil Sci.* **18** (4): 111-117.

Kononova, M. (1966). *Soil Organic Matter- Its Role in Soil Formation and Soil Fertility*, Pergamon Press, Oxford, U.K.

Kononova, M.M. (1975). *Soil Components* (J.E. Gieseking, ed.), Vol. **1**, Springer Verlag, New York, Inc. USA.

Koo, R.C.J. (1964). Some result of fairfield slag. *Proc. Fla. State Hort. Soc.* **77** : 26-32.

Koo, R.C.J. (1966). Evaluating magnesium sources for Florida citrus. *Citrus Veg. Mag.* **29**(7): 12.

Koo, R.C.J. (1971). A comparison of Mg sources for citrus. *Proc. Soil & Crop Sci. Fla.* **31** : 137-140.

Koo, R.C.J. (1982). Use of leaf, fruit and soil analysis in estimating potassium status of orange trees. *Proc. Fla. State Hort. Soc.* **75**:67-72.

Koo, R.C.J. (1984). Recommended fertilizers and nutritional sprays for citrus. *Bull.* **604**. *Agri. Expt. Sta.,* Inst. Food Agri. Sci., Univ. Florida, USA.

Koo, R.C.J. (1987). Citrus micronutrients in perspective. *Proc Soil & Crop Sci. Soc. Fla.* **47**:9-12.

Koo, R.C.J., Anderson, C.A., Stelwart, I., Tucker, D.P.H., Calvert, D.V. and Wutscher,H.K. (1984). Recommended fertilizers and nutritional sprays for citrus. *Bull.* 536D. p.30, Inst. Food. Agri Sci., Univ., Fla., Florida, USA.

Koo, R.C.J. and Driscoll, P.J. (1970). Renovating old citrus groves in Indian River area. *83rd Annual Meet. Fla. State Hort. Soc.* **83** : 71-74.

Koo, R.C.J. and McCornak, A.A. (1965). Effects of irrigation and fertilization on production and quality of Dancy tangerine. *Proc. Fla. State Hort. Soc.* **78**: 1015.

Koo, R.C.J., Reitz, H.J. and Sites, J.W. (1958). A survey of the mineral nutrient status of Valencia orange in Florida. *Agri. Expt. Sta. Bull.* **604,** *Inst. Food Agri. Sci.* Univ. Florida, USA.

Koo, R.C.J. and Young, T.W. (1970). Correcting magnesium deficiency of limes grown on calcareous soils with magnesium nitrate. *Proc. Fla. State Hort. Soc.* **82** : 274-278.

Koseoglu, A.T., Colakoglu, H. C and Kovanci, I. (1995a). The effect of chemical fertilizers on the fruit yield of young satsumas (*Citrus unshiu* Marc.). *Doga Turk Tarim Ve Ormacilik Dergisi.* **14**(1):33-44.

Koseoglu, A.T., Eryuce, N. and Colakoglu, H.C. (1995b). The effects of N, P, K fertilizers on fruit yield and quality of satsuma mandarin (*Citrus unshiu* Marc.). *Acta Hort.* **379:**89-96.

Kothandaraman, G.V. and Mariakulandai, A. (1965). Charge in pH, exchangeable calcium and available P_2O_5 due to liming of acid soils of Nanjand in the Nilgiri district. *Madras Agric. J.* **52** : 183.

Koto, M. and Takeshita,S. (1957). Experiments on leaf analysis of satsuma orange. Part 3. Leaf analysis of manganese and magnesium deficient Satsuma orange trees in Kanagawa prefecture. Part I. *Bull.* **5**, pp. 13-22, Kangawa Agric. Expt. Sta. Hort., Kanagawa, Japan.

Koto, M. and Takeshita, S. (1959). Experiment on leaf analysis of Satsuma orange Part 2 Leaf and soil analysis of manganese and magnesium deficient satsuma orange orchards in Kanagawa prefectures. *Bull.* **7**, pp. 6-11, Kangawa Agric. Expt. Sta. Hort., Kanagawa, Japan.

Kotur, S.C. (1982). Effect of zinc sulphate spray concentration and its frequency in Coorg mandarin (*Citrus reticulata* Blanco). *Curr. Res.* **12**:126-127.

Kotur, S.C., Srivastava, K.C. and Subbaiah, M.P. (1982). Evaluation of long term cultural practices in relation to soil chemical properties in Coorg mandarin cultivation. *Indian J. Agri. Sci.* **52**: 598-600.

Koudounas,C. (1994). Soil –citrus relationship studies in Cyprus. *Acta Hort.* **365**: 147-150.

Kovanci, I., Hakerlerler, H. and Hofner, W. (1978). Causes of chlorosis of mandarin (*Citrus reticulata*) in the Aegean region. *Pl. Soil* **50**(1) : 193-205.

Kovanci, I., Hakerlerler, H., Oktay, M. and Hofner, W. (1985a). Chlorosis control in mandarin orchards in the Aegean region (Turkey) by means of foliar and soil application of different iron compounds. *Tropenlandwrit* **85/86**:14-20.

Kovanci, I., Hakerlerler, H., Oktay, M. and Hofner, W. (1986). The control of chlorosis in mandarin orange orchards in the Aegean region of Turkey by applying iron compounds to soil and leaves. *Pl. Res. Develop.* **24**: 118-5.

Kovanci, I., Hakerlerler, H., Oktay, M., Ozercon, A. and Karacali, I. (1985b). The effect of Nervanaid-Zn 14 on deficiency in satsuma of the Izmir region. *Doga Bilim Dergisi, D_2(Tarim ve Ormancilik)*. **9**(3) : 304-311.

Kovda, V.A. (1934). *Geography of the Podzolic Stage of Soil Formation. Tr. Pochv. Inst.* **10** (2).

Kozaki, I. (1981). Introduction to citriculture in Japan. *Proc. Int. Soc. Citriculture.* Vol. **1**, pp. XXVII- XXXIII.

Kozyrev, V. G. (1973). Soil temperature of differently oriented slopes and mandarin tree development. *Subtropicheskie Kul'tury* (1) : 101-104.

Kramer, P.J. (1951). Causes of injury to plants resulting from flooding of the soil. *Pl. Physiol.* **26** : 722-736.

Krishnamoorthy, P. and Govindarajan, S.V. (1977). Genesis and classification of associated red and black soils under Rajalibunda diversion irrigation scheme (Andhra Pradesh). *J. Indian Soc. Soil Sci.* **25**(3) : 239-246.

Krishnan, P. (2000). Land resources of the Southern region. *Souvenir*, 65th Annual Conv. of Indian Soc. Soil Sci., New Delhi, pp. 28-33.

Krome, W.H. (1968). Economic view of lime growing in Florida. *Econ. Bot.* **22** : 270-272.

Kuliev, F.A. (1985). Economic effectiveness of feijoa cultivation in Azerbaidzhan. *Subtropicheskie Kul'tury* (5) : 6-10.

Kumar, K. (1997). Nature of acidity and its relation with lime requirement of some acid soils of Manipur hills. *J. Hill Res.* **10**(2):131-135.

Kumar, P. (1998). Food Demand and Supply Projection for India. *Agricultural Economics Policy Paper* **98-01**, IARI, New Delhi.

Kuranaka, K., Kadota, S. and Shinohara, G. (1972). The adaptability of sever citrus types in western coastal region of Shimane, protecture. *Bull No.* **10**, pp. 109-119, Shimane Agri. Expt. Sta. Shimane, Japan

Kurien, S., Goswami, A.M. and Deb, D.L. (1991). The distribution of root activity in acid lime (*Citrus aurantifolia*). *Expt. Agri.* **27**(4) : 431-434.

Kurihara, A. (1969). Fruit growth of Satsuma oranges under controlled conditions. I. Effects of per-harvest temperature on fruit growth colour development and fruit quality of Satsuma oranges. *Bull. No.* **8**, pp. 15-30., Hort. Res. Sta., Japan.

Kuwamura, A., Koga, H., Yamasaki, S. and Ujike, T. (1973a). Sustaining effect of soil physical amendment in orange orchard. I. on the orchard soil derived from granite and andesite. *Bull. No.* **26**, pp. 105-122, Shikoku Agric. Expt. Station, Japan.

Kuwamura, A., Koga, H., Yamasaki, S. and Ujike, T. (1973b). Sustaining effect of soil physical amendment in orange orchard 2. on the orchard soil derivated from Izumi lime stone. *Bull. No.* **26**, pp. 123-134., Shikoka Agric. Expt. Station, Japan.

Kuykendall, J.R., Hilgeman, R.H. and Van Horn, C.W. (1957). Response of chlorotic citrus trees in Arizona to soil applications of iron chelates. *Soil Sci.* **84**:77-86.

Kuznecov, S. I. and Gocelasvili, Z. A. (1967). Active acidity and the oxidation reduction potential of podzolic soil in a mandarin plantation. *Subtropicheskie Kul'tary* **12**:141-144.

Kuznetsova, I.V. (1990). Optimum bulk density. *Soviet Soil Sci.* **22** (8): 74-87.

Labanauskas, C.K. (1962). Correction of manganese deficiencies in grapefruit trees by foliar sprays in desert area of southern California. *Proc. Am. Soc. Hort. Sci.* **80**:260-273.

Labanauskas, C.K., Jones, W.W. and Embleton, T.W. (1959). Seasonal changes in concentration of micro-nutrient in leaves of Washington Navel orange. *Proc. Am. Soc. Hort. Sci.* **74**:300-306.

Labanauskas,C.K., Jones, W.W. and Embleton, T.W. (1969). Low residue micronutrient sprays for citrus. *Proc. 1st Int. Citrus Symp.*, Vol. **3**, pp.1535-1542, Riverside, USA.

Labanauskas, C.K, Jones, W.W. and Embleton, T.W. (1972a). Low residue micronutrient sprays of citrus. *Citrog.* **57**(11):405, 408-411.

Labanauskas, C.K., Letey, J., Stolzy, L.H. and Valoras, N. (1966). Effect of soil oxygen and irrigation on the accumulation of macro-and micronutrients in citrus seedlings. *Soil Sci.* **101** : 378-384.

Labanauskas, C.K., Stolzy, L.H. and Handy, M.F. (1972b). Concentration and total amount of nutrients in citrus seedlings (*Citrus sinensis* Osbeck) and in soil as influenced by differential oxygen treatments. *Proc. Soil Sci. Soc. Am.* **36**(3) : 454-457.

Labanauskas, C.K., Stolzy, L.H., Koltz, L.J. and De Wolfe, T.A. (1970). Soil oxygen diffusion rates and mineral accumulation in citrus seedlings (*Citrus sinensis* var. Bessie). *Soil Sci.* **111** : 386-392.

Labanauskas, C.K., Stolzy, L.H., Koltz, L.J. and De Wolfe, T.A. (1971). Soil carbon dioxide and mineral accumulation in citrus seedlings (*Citrus sinensis* var. Bessie). *Pl. Soil* **35** : 337-346.

Laborem, E.G. (1977). Determination of boron in citrus crops. *Agronomia Trop.* **27**(5) : 529-537.

Lackey, R.T. (1998). Ecosystem Management: Paradigms and prattle, people and prizes. *Renewable Resource J.* **96** (4): 32-33.

Lacoeuilhe, J., Marchal, J. and Martin-Prevel, P. (1968). Physiological aspects of abnormal leaf fall of citrus in Corsica. *Mem. gen. II. Colog. eur. medit. Contrl. Fert. Plant Cult.,* Seville, pp. 559-572.

Lahiri, T. C. and Chakravarty, S.K. (1989). Characteristics of some soils of Sikkim at various altitudes. *J. Indian Soc. Soil Sci.* **37**:451.

Lahiri, T.C. and Chakravarty , S.K. (1995). Distribution and nature of organic matter in some hill soils of west Bengal at various altitudes in eastern Himalayan region. *J. Indian Soc. Soil Sci.* **43**(3) : 464-466.

Lahiri, T.C., Dutta, Shikha, Sunwar, C.B. and Chakravarti , S.K. (1995). Clay mineralogy of some hill soils of West Bengal in the eastern Himalayan region. *J. Hill Res.* **8**(2) : 174 – 176.

Lal, R. (1974). Soil Erosion and Shifting Agriculture. *FAO Soil Bull.* **24**, pp. 48-71.

Lal, R. (1976). Soil erosion on Alfisols in western Nigeria, I. Effect of slope, crop rotation and residue management. *Geoderma* **16** : 363-375.

Lal, R. (1987). Managment of soil compaction and soil water after forest clearing in uplands in humid tropical Asia. *Soil Management under Humid Conditions in Asia and Pacific.* Proc. No. **5**, pp. 273-295 Int. Board for Soil Res. and Management, Bangkok, Thailand.

Lal, R. (1990). Soil erosion and land degradation : The global risk. *Adv. Soil Sci.* **11** : 169-172.

Lal, R. (1991). Soil structure and sustainability. *J. Sustainable Agric.* **1** : 67-92.

Lal, R. (1993). Tillage effects on soil degradation soil resilience, soil quality and sustainability. *Soil Tillage Res.* **27** : 1-8.

Lallan Ram, Srivastava, A.K., Kohli, R.R., Dass, H.C. and Huchche, A.D. (1993). Interaction of saline water through drip irrigation of Nagpur mandarin and salinity buildup in Inceptisols. *J. Water Manag.* **1** : 51-53.

Lamsayun, S. (1980). The effect of several mixtures of varying microelements composition on the yield and quality of stem mandarins. *Bulletin Penelitan Hortikultura* **8**(8) : 1-6.

Landthasa, S. and Bhattacharya, R.K. (1991). Foliar application of zinc on fruit quality of Assam lemon (*Citrus lemon* Burm.). *South Indian Hort.* **39**(3) : 153-155.

Larson, W.E. and Pierce, F.J. (1991). Conservation and enhancement of soil quality. *Evaluation of Sustainable Land and Management in the Developing World* (J. Dumanski *et al.,* ed.). Technical Papers Proc. Int. Workshop Chiang Rai Thialand. 15-21 Sept. 1991. Int. Board Soil Res. and Manag., Bangkok. Vol. **2,** pp. 175-203.

Larson, W.E. and Pierce, F.J. (1994). The dynamics of soil quality as a measure of sustainable management. *Defining Soil Quality for Sustainable Environment* (J.W. Doren *et al.,* ed.), *Soil Sci. Soc.Am., Spec. Publ.* **35**, pp. 37-51, Soil Sci. Soc. Am. and Am. Soc. Agron., Madison, WI.

Larue, M. (1971). A review of citrus growing in South America. *Fruits* **26**(5): 371-388.

Laskar, S. (1979). Fertility Constraints in Soils of Mizoram. *Ann. Rep.*, ICAR Res. Complex for NEH Region, Shillong, Meghalaya.

Lawrence, F.P. (1958). Selecting grove site. *Ext. Serv. Circ.* **185**, p.4, Univ. Fla., USA.

Lay, Wen Long and Wang, Chin, Tang (1997). The investigation of the nutrient diagnosis and its status of application in citrus production in central Taiwan. *Bull.No.* **54**, pp. 33-45, Taichung District Agriculturae Improvement Station, China.

Le Roux, F.H. (1965). Soil acidity in citrus orchards. *South. Afr. Citrus J.* (378): 11-15.

Le Roux, J. (1973). Quantitative clay mineralogical analysis of Natal Oxisols. *Soil Sci.* **115**(2) : 137-144.

Lee, B.W. (1964). Copper deficiency in Ventura county citrus. *Citrogr.* **49**: 326-328.

Lee, K.S. (1991). The diversity of soil organisms. *The Biodiversity of Microorganisms and Invertebrates : Its Role in Sustainable Agriculture* (D.L. Hawksworth, ed.), CAB International, Wallingford, UK., pp. 73-87.

Lee, L.S. and Chapman, J.C. (1988). Yield and fruit quality responses of Ellendale mandarins to different nitrogen and potassium fertilizer rates. *Australian J. Expt. Hort.* **28(**1):143-148.

Lee, T.H., Erickson, L.C. and Chichester, C.O. (1971). An evaluation of the use of grafted fruit for study of carotenoid and chlorophyll changes in regreening Valencia orange ped. *HortScience* **6** : 231-232.

Lee, R.F., Marais, L.J., Timmer, L.W. and Graham, J.H. (1984). Syringe injection of water into the trunk: A rapid diagnostic test for citrus blight. *Pl. Dis.* **68**:511-513.

Legaz, F., Serna, M.D., Primo-Millo, E. and Martin, B. (1992a). Leaf spray and soil application of Fe chelates to Navelina orange trees. *Proc. Int. Soc. Citriculture.* Vol. **2**, pp. 613-617.

Legaz, F., Serna, M.D., Primo-Millo, E., Perez-Garcia, M. Maquieira, A. and Puchades, R. (1992b). Effectiveness for the N farm applied by a drip irrigation system to citrus. *Proc. Int. Soc. Citriculture.* Vol. **2**, pp. 590-592.

Lekha, V.S., Raja, P., Sehgal, J.L. and Gajbhiye, K.S. (1998). Sand mineralogical investigations and mineral weathering index (MWI) of selected soils of Kerala. *J. Indian Soc. Soil Sci.* **46**(4): 675 – 682.

Lenz, F. (1964). Day length and temperature effect on citrus cuttings. *Australian Hort. Res. Newsletter.* **11**:27-28.

Lenz, F. (1969). Effect of daylength and temperature on the vegetative and reproductive growth of Washington Navel Orange. (H.D. Chapman ed.), *Proc. 1st Int. Citrus Symp.,* Univ. Calif., Press Riverside, Vol. **1**, pp 333-338.

Leon, A., Torrecillas, A. and Amor, F. Del (1984). Current state of the diagnosis and control of nutritional disorder of lemons in Spain. *Proc.* Vol. **4,** pp. 1085-1092, AIONP/GERPAT, Monspellier, France.

Leon, A., Torrecillas, A., Amor, F. Del and Monllor, M.R. (1983). Criteria for the diagnosis of excess boron in lemon trees. *Anales de Edafologia Y Agrobiologia* **42**(5/6) : 807-817.

Leonard, C.D. (1952). Citrus fruit. *Farm Chemical.* p.130.

Leonard, C.D. and Calvert, D.V. (1971). Field tests with new iron chelates on citrus growing on calcareous soils. *Proc. Fla State Hort. Soc.* **84**:24-31.

Leonard, C.D. and Stewart, I. (1952). Correction of iron chlorosis in citrus with chelated iron. *Proc. Fla. State Hort. Soc.* **65**:20-24.

Leonard, C.D. and Stewart, I. (1953). Iron chlorosis. *Ann. Rep.* **53**, pp. 178-180, Fla. Agric. Expt. Sta.

Leonard, C.D., Stewart,I (1956). Longevity of chelated iron treatments as applied to citrus trees. *Citrus Indus.* **37**(3):12-16.

Leonard, C.D. and Stewart,I. (1960). Soil application of manganese for citrus. *Proc. Fla. State. Hort. Soc.* **72**:38-45.

Leonard, C.D., Stewart, I. and Edwards, G. (1957). Effectiveness of different zinc fertilizers on citrus. *Citrus Indus.* **38**(2):12-15.

Leonard, C.D., Stewart, I and Edwards, G. (1959). Soil applications of zinc for citrus on acid sandy soils. *Proc. Soil & Crop Sci. Soc. Fla.* **71**:99-106.

Lepsh, J.F. and Buol, S.W. (1974). Investigations in Oxisol-Ultisol toposequence in Sao Paulo State, Brazil. *Soil Sci. Soc. Am. Proc.* **38**(3) : 491-496.

Lety, J., Stolzy, L.H. and Kemper, W.D (1967). *Soil Aeration : Irrigation of Agricultural Lands* (R. M. Hagan, M.R. Haite and T.W. Edminster, ed.). Am. Soc. Agron. Madison, Wisconsin, USA, pp. 941-949.

Levitt, E.C. (1960). *Citrus Culture.* N.S.W. Dep. Agric., 2nd edition, pp. 72.

Levy, Y. (1984). The effect of rootstocks on root distribution and mycorrhizal infection of mature citrus trees. *Proc. Int. Soc. Citriculture.* Vol. **1**, pp. 128-130.

Levy, Y. (1999). Citrus in Israel. *Souvenir*, pp. 144-148 International Symposium on Citriculture, Nagpur, Maharashtra, India.

Levy, Y. and Shalhevet, J. (1982). Irrigation of mature citrus orchard with water containing high salt concentrations, comparison of three rootstocks and two cultivars. *Alon haNotea* **42**: 123-131.

Levy, Y. and Shalhevet, J. (1989). Ranking the salt tolerance of citrus rootstocks by juice analysis. *Sci. Hort.* **45**: 89-98.

Levy, Y. and Shalhevet, J. (1990a). Response of mature orange and grapefruit trees to irrigation with saline water interaction with rootstock. *Proc. Int. Citrus Symp.* (B. Bangyan and Y. Qian, ed.), Guangzhou, China, pp. 419-429.

Levy, Y. and Shalhevet, J. (1990b). Ranking the salt tolerance of citrus rootstocks by juice analysis. *Sci. Hort.* **45**: 89-98.

Levy, Y. , Shalhevet, and Bielorai, S. (1979). Effect of irrigation regime and water salinity on grapefruit quality. *J. Am. Soc. Hort. Sci.* **104**:356-359.

Levy, Y. and Syvertsen, J.P. (1981). Water relation of citrus in climates with different evaporative demands. *Proc. Int. Soc. Citriculture.* Vol. **2**, pp. 501-503.

Levy, Y., Syvertsen, J.P. and Nemec, S. (1983). Effect of drought stress and vescicular-arbuscular mycorrhiza on citrus transpiration and hydraulic conductivity of roots. *New Phytol.* **93**: 61-66.

Leyden, R. (1977). Water require of grapefruit trees in Texas. *Proc. Int. Soc. Citriculture.* Vol. **3**, pp.1037-1039.

Li, B. (1995). Quantification of soil changes and process. *Prog. Soil Sci.* **23**(3) : 33-43 (In Chinese).

Li, Jian, Shi, Qing. and Zeng, Wen Xian (1998). Study on rational fertilizer application in citrus orchards in Fujian province by nutritional diagnosis. *J. Fruit Sci.* **15**(2) : 145-149.

Li, Yinguo, Yin, Kelin, Wang, Cheng, Qui and Wang, Shuliang (1995). Influences of climate and mineral nutrition on leaf temperature in citrus. *J. Southwest Agric. Univ.* **17**(6) : 506-513.

Li, Minquan, Yin, Cai, and Youquan Jie (1990). Studies on pre-harvesting practices of improving storability of Satsuma mandarin. *Proc. Int. Citrus Symp.* (B. Bangyan and Y. Qian, ed.), Guanzhou, China, pp. 732-739.

Liao, M.C. and Chang, H.M. (1974). Study on soil conservation practices for juvenile orange plantations. *J. Agric. Assoc. China* **88** : 74-82.

Liebig, J. Von (1840). Die organische Chemie in ihrer Anwendung auf Agricultur und Physiologie (Organic Chemistry and its application to Agriculture and Physiology). *Friedrich Vieweg und Sohn Publ. Co.,* Braunschweig, Germany.

Liebig, J. Von. (1855). The relations of chemistry to agriculture and the agricultural experiments of Mr. J. B. Lawes, 1st and 2nd (ed.), *Friedrich Vieweg aud Sohn Publ. Co.*, Braunschweig, Germany.

Liebig, J. Von (1864). *Quoted in History of Soil Science (*I.A. Krupenikov, ed.), Oxonian Press Pvt. Ltd., New Delhi, 1992, p. 128.

Liebig, G. F. and Chapman, H. D. (1963). The effect of variable root temperature on the behaviour of young navel orange trees in a green house. *Proc. Am. Soc. Hort. Sci.* **82**:204-209.

Liebig, M.A., Doren, J.W. and Gardiner, J.C. (1996). Evaluation of a field test kit for measuring the soil quality indicators. *Agron. J.* **88**: 683-686.

Lieffering, R.E. and Mclay, C.D.A. (1995). The effect of hydroxide solutions on dissolution of organic carbon in some New Zealand soils. *Australian J. Soil Sci.* **33**:873-881.

Lim, M. Mualsri, C., Suthipradit, S., Panthanahiran, W. and Lin, S. (1993). Nutient element soil cultivation of neck orange (*Citrus reticulata* Linn.). *Kasetsart J. Nat. Sci.* **27**(4):412-420.

Lima, H., Cornide, M.T., Alvarez, M. and Frometa, E. (1988). Edapho-climatic classfication of the citrus producing regions of Cuba. *Agrotecnia de Cuba* **20**(2) : 63-74.

Lima, J.E.O. (1982). Observation on citrus blight in Sao Paulo, Brazil. *Proc. Fla. State Hort. Soc.* **95** : 72-75.

Lin, C. and Coleman, N.T. (1960). The measurement of exchangeable aluminium in soils and clays. *Proc. Soil Sci. Soc. Am.* **24**(6) : 444 - 446.

Lin, Z. and Myhre, D.L. (1991). Differential response of citrus rootstocks to aluminium levels in nutrients solution. 1. Plant growth. *J. Pl. Nutr.* **14**(11) : 1223-1238.

Lin, Z., Myhre, D.L. and Martin, H.W. (1988). Effects of lime and phosphogypsum on fibrous citrus root growth and properties of spodic horizon on soil. *Proc. Soil & Crop Sci. Soc. Fla.* **47** : 67-72.

Lindsay, W.L. and Norvell, W.A. (1978). Development of a DTPA soil test for zinc, iron, manganese and cooper. *Soil Sci. Soc. Am. J.* **42:** 421.

Lio, G., Magnano, Di San, Pennisi, A. M. and Perrotta, G. (1984). Effect of some fungicides on population of citrus *Phytopthora* in soil. *Proc. Int. Soc. Citriculture.* Vol. **1**, pp . 415-419.

Liu, X. (1983). Investigation on the adaptability of late sweet oranges in China. *Acta Horticulturae Sinica* **18**(1): 17-24.

Liu, Y.W. (1992). Study on the effect of nitrogen application on Wenzhou mandarin. *Acta Horticulturae Sincia* **19**(4):326-332.

Liu, Y.W. (1993). Study on the effect of satsumas of successive application of lime to a red soil orchard. *China Citrus* **22**(1) : 14-16.

Liu, Y.W. (1995). Study on the effect of application of phosphorus fertilizers on Satsuma mandarin trees. *China Citrus* **24**(3) : 3-6.

Liu, Cheng Ming and Nan, Qinxwan (1996). Effect of Fe and Mn on photosynthesis physiology of satsuma orange (*Citrus unshiu* Marc) and diagnosis of its nutritional status. *J. Southwest Agri. Univ.* **18**(1):29-33.

Liu, L. and Gong, Z. (1995). Assessment of global soil degradation. *Nat. Resources.* **1** : 10-15. (In Chinese).

Liu, Y.W., Lin, M.S. and Qiu, H.L. (1994). Study of fertilizer Satsumas. *China Citrus.* **23**(2):7-10.

Llorente, S., Leon, A., Lopez-Andreu, F.J., Romojaro, F. (1973). Effect of minor element deficiencies on leaf boron content in lemon trees. *Ist Congresco.* Mundias de Citriculture (1974), pp. 33-37.

Lloyd, R.H. (1961). Granulation and regreening. *Citrus News.* **37**:21-25.

Lloyd, J., Syvertsen, J. P. and Kriedmann, P. E. (1987). Salinity effects on leaf water relations and gas exchange of valencia orange, *Citrus sinensis* (L.) Osbeck on rootstocks with different salt exclusion characteristics. *Australian J. Pl. Physiol.* **14**: 605-614.

Lobaina, J., Sanchez, J.I., Bidart, B. and Pouymiro, N. (1988). Analysis of four climatic variables in the citrus enterprise vilorio of Guantanamo. *Cultivos Tropicales* **10**(1) : 92-98.

Loest, F.C. (1942). Dry root rot disease of citrus trees and diplodia and brown rot gummosis of Citrus. *Fmg. South Africa.* **11**: 420-424.

Lohar, D. P., Budathoki, K. and Subedi, P.P. (1995). Orchard soil moisture conservation and fruit drop studies in mandarin orange (*Citrus reticulata* Blanco) during 1993-94. *Working Paper No.* **95/62,** p. 30 Lumle Regional Agricultural Research Centre, Pokhara, Nepal.

Lohnis, F. (1926). Nitrogen availability of green manures. *Soil Sci.* **22** : 253-290.

Lomas, J., Gat, Z. and Shifrin, B. (1970). Methods of forecasting the ripening dates of citrus fruits in various regions of Israel. *Agric. Meteorol.* **7**:321-327.

Lombard, P.B. (1965). Effects of climatic factors on fruit volume increase and leaf water deficit of citrus in relation to soil suction. *Proc. Soil Sci. Soc. Am.* **29**: 205-208.

McCall, W.W. (1965). Zinc deficiency of citrus in Thailand. *Kasetsart J.* **5** : 1-9.

McCarty, C.D. (1966). Notes on Sicilian citrus. *Calif. Citrog.* **51**: 268, 287-290.

McCarty, C.D. (1967). Notes on Israeli citrus. *Calif. Citrog.* **52**: 111-116.

McCarty, C.D., Kemper, W.C. and Hield, H.Z. (1965). Mandarin, tangelo and tangor maturity studies. *Citrog.* **50**: 122-139.

Mc Colloch, R.C., Bingham, F.T. and Aldrich, D.G. (1957). Relation of soil potassium and magnesium to the magnesium nutrition of citrus. *Soil. Sci. Soc. Am. Proc.* **21** : 85-88.

Mc Donald, R.E. and Hillebrand, B.M. (1980). Physical and chemical characteristics of lemons from several countries. *J. Am. Soc. Hort. Sci.* **105**: 137-141.

Mchedlidze, G.D. (1972). The effect of calcium and magnesium fertilizers on some chemical soil properties and Satsuma yields. *Subtropicheskie Kul'tury* (4) : 94-96.

Mclaren R.G. and Crawford, D.V. (1976). Studies on soil copper. I. The fractionation of Cu in soils. *J. Soil Sci.* **24** : 172-181.

Mdinaradze, T.D. (1981). Effect of different rates of zinc fertilizers on qualitative indices of mandarin fruit. *Subtrophicheskie Kul'tury* (5) : 49-51.

Mdinaradze, T.D. and Datuadze, O.V. (1987). The effect of magnesium containing fertilizers on the available zinc content in the soil and uptake by mandarin trees. *Subtropicheskie Kul'tury* 6 : 77-80.

Mdinaradze, T.D. and Kechakmadze, M.S. (1982). Microelements and mandarin storability. *Subtropicheskikh Kul'tury* (2) :101-103.

Medina, M. (1979). *Circulacion General Atmosferica.* (J. Salvat, ed.). El Universo y la Tierra. Universitas Enciclopedia, Vol. **2,** pp. 111-118.

Medvedev, V.V.(1990).Variabitlity of the optimal soil density and its causes. *Soviet Soil Sci.* **22** (7): 65-75.

Meixner, R.F. and Singer, J.M. (1981). Use of a field morphology rating system to evaluate soil formation and discontinuities. *Soil Sci.* **131**(2):114-120.

Mellado, L. and Caballero, F. (1974). Studies on the distribution of active roots in orange trees using 32p. *Hort. Abst.* **45** : 688.

Melville, F. (1962-1964). The fruit industry in other lands. *Punjab Fruit J.* **26/27:** 25-32.

Memoirs of Indian Meteorological Department (1961). Monthly and annual normals of rainfall and rainy days. Vol. **31**, Part III.

Mendel, K. (1963). *Report on Citriculture in Ceylon,* Govt. of Ceylon, Sessional Paper XVII, Colombo, SriLanka.

Mendel, K. (1969). The influence of temperature and light on the vegetative development of citrus trees. *Proc. 1st Int. Citrus Symp.* Vol. **1**, pp. 259-265.

Mendes, O. (1972). Swazi spot of citrus in Mozambique. *Agronomia Mocabicana* **6**(1) : 83-99.

Mendez, J. and Kamprath, E. J. (1978). Liming of Latosols and the effect on the phosphorous response. *Soil Sci. Soc. Am. J.* **43**: 86.

Mendt, R. (1989). Present and future of Venezuelan citriculture. *Proc. 6th Int. Citrus Congr.* (R. Goren and K. Mendel, ed.), Vol. **4**, pp. 1625-1629, Balaban Pub., German Federal Republic.

Meng, C.F., Zhou, M.F. and Zhou, J.S. (1991). Effects of applying phosphorus, potassium fertilizers and lime in red soil orchards on yield and fruit quality of Satsuma mandarin. *China Citrus* **20**(4) : 3-6.

Merlo, De. C., Los, M. De. A., Gonzales, L. M.S., Macial, De. S. and Pire, N.L.R. (1990). Study of the physical environment of a citrus orchard in the highlands of Yaracuy state Venezuela. *Agronomia Tropical (Maracay)* **40**(1-3) : 67-78.

Mermut, A.R. and Dasog, G.S. (1986). Nature and micromorphology of carbonate globules in some Vertisols of India. *Soil Sci. Soc. Am. J.* **50**: 382.

Meratotta, J.I. (1968). Some aspets of growing Criollo lemons in Guatemala. *Rev. Cafet., Guatem* (78) : 30-32.

Mertz, W.M. (1918). Green Manure Crops in Southern California. *Bull.* **292**, p. 31, Agri. Expt. Sta., Calif. Univ., USA.

Mikaberidze, V.E. and Rosnadze, G.R. (1972). The effect of mulching in young mandarin orchards on catalase and peroxidase activities and on frost resistance. *Subtropicheskie Kul'tury* (4) : 53-57.

Mikhail, E.H. and El-Zeftawi, B.M. (1979). Effect of soil types and rootstocks in root distribution, chemical composition of leaves and yield of Valencia orange. *J. Soil Res.* **17**(2) : 335-342.

Milad, S., Bakhati, H.K., El-Hakim, M. and Moutafa, F.B. (1975). The effect of different soil conditions upon citrus growth. *Agric. Res. Rev.* **53**(3) : 31-40.

Milella, A. (1966). Citrus growing in Israel. *Ital. Agric.* **103**: 449-464.

Miller, J.E., Maritz, J.G.J., Bird, P. and Breedt, H.J. (1996). Screening of potential rootstocks for 'Eureka' Lemon in the Sundays river valley. *Proc. Int. Soc. Citriculture.* Vol. **1**, pp. 239-242.

Miller, F.P. and M.K. Wali (1994). Land use issues and sustainability of agriculture. *Trans 15th WCSS, Mexico,* 7a, pp. 1-16.

Minessy, F.A., Baraket, M.A. and El-Azab, E.M. (1970a). Effect of water table on mineral content, root and shoot growth, yield and fruit quality in Washington Navel orange and Balady mandarin. *J. Am. Soc. Hort. Sci.* **95** : 81-85.

Minessy, F.A., Baraket, M.A. and El-Azab, E.M. (1971). Effect of some soil properties on root and top growth and mineral content of Washington Navel orange and Balady mandarin. *Pl. Soil* **34** : 1-15.

Minessy, F.A., Nasr, T.A.A. and El-Shurafa, M.Y. (1970b). Citrus fruit temperature in relation to sunburn. *Proc. Conf. Crop Subtrop. Fruits,* Trop. Prod. Inst., pp. 245-252.

Minhas, R.S. and Bora, N.C. (1982). Use of organic carbon as an index of available nigrogen in hill soils a modicication of limits. *J. Indian Soc. Soil Sci.* **30**(3) : 376 – 377.

Ministry of Agriculture (1959). Future policy. *Sessional Paper* No.**9**, Nigeria.

Misra, A.K. and Ganapathy, M.M. (1982). Correcting chlorosis in mandarin. *Indian Hort.* **27:**23-24.

Misra, U. K. and Saithantuaanga, H. (2000). Characterization of acid soils of Mizoram. *J. Indian Soc. Soil Sci.* **48**(3):437-446.

Misra, U.K., Satapathy, S and Panda, N. (1989). Characterization of some acid soils of Orissa : I. Nature of soil acidity. *J.Indian Soc. Soil Sci.* **37:**22.

Mitra, G.N., Sharma, V. A. K. and Ramamoorthy, B. (1958). Comparative studies on the potassium fixing capacities of Indian soils. *J. Indian Soc. Soil Sci.* **6** : 1.

MIlella, A. and Dieoda, P. (1976). The effect of irrigation on nitrogen fertilization on the dimensions and chemical composition of mandarin leaves. *Hort. Abst.* **46**(6):528-529.

Mohekar, D.S. and Challa, O. (1999). Evaluation of some orange growing soils in Nagpur district of Maharashtra. *Abst.* Int. Symp. Citriculture. pp. 46-47, Indian Society of Citriculture, Nagpur, Maharashtra, India.

Mohr, G.R., Datta, N.P., Sankrasubramoney, N., Leley, V.K. and Donahue, R.L. (1965). *Soil Testing in India,* USAID Mission to India, New Delhi.

Moncada, B.J., Rios-Castano, D. and Torres, M.R. (1969). Citrus fruit quality in Columbia. *Proc. Trop. Reg. Am. Soc. Hort. Sci.* **12** : 126-137.

Mongia, A.D. and Bandyopadhyay, A.K. (1991). Forms of potassium in acidic hill soils of Andamans. *J Indian Soc. Soil Sci.* **39**(3): 573-575.

Monselise, S. P. (1947). The growth of citrus roots and shoots under different cultural conditions. *Palestine. J. Bot.* **6**: 43-54.

Monselise, S.P. (1951). Growth analysis of citrus seedlings. I. Growth of sweet lime seedlings in dependence upon illumination. *Palestenian J. Bot.* **8** : 57-60.

Monselise, S. P.(1973). Recent advances in the understanding of flower formation in fruit trees and its hormonal control. *Acta Hort.* **34** : 154 - 166.

Monselise, S. P. (1978). Understanding of plant processes as a basis for successful growth regulation in citrus. *Proc. Int.Soc. Citriculture.* Vol. **1,** pp. 250- 255.

Monselise, S.P. (1981). Effect of climatic districts, orchard treatment and seal packaging on citrus fruit quality and storage ability. *Proc. Int. Soc. Citriculture.* Vol. **2**, pp, 705-709.

Monselise, S. P. (1985). Citrus and Related Species (A. H. Halevy, ed.). *CRC Handbook of Flowering.* Vol. **2** , pp. 275-294.

Monselise, S. P. and Goren, R. (1969). Flowering and fruiting interactions of exogenous and internal factors. *Proc. 1st Int. Citrus Symp.* Riverside, USA, Vol. **3,** pp. 1105-1112.

Monselise, S. P. and Goren, R. (1978). The role of internal factors and exogenous control in flowering, peel growth, and abscission in citrus. *HortScience* **13** : 134-139.

Monselise, S. P. and Halvey, A. H. (1964). Chemical inhibition and promotion of citrus flower bud induction. *Proc. Am. Soc. Hort. Sci.* **1** : 171- 176.

Montasser, A., Mehanna, S., Salama, M. and El-Shazly, S. (1985). Studies on the nutritional status of some citrus species at El-Fayoum Governorate. *Ann. Agric. Sci., Ains Shams Univ.* **30**(2) : 1471-1493.

Monteverde, E.E., Espinoza, M., Ruiz, J.R. and Nunez, G. (1984). Five year yield and fruit quality of two orange cultivars in Venezuela's main citrus area. *Proc. Int. Soc. Citriculture.* Vol **1**, pp. 88-91.

Monteverde, Edmundo E., Reyes, Fernando J., Loborem, Gaston and Ruiz, Jose R. (1988). Citrus rootstocks in Venezuela: Behaviour of Valencia orange on ten rootstocks. *Proc. 6th Int. Citrus Congress* (R. Goren and K. Mendel, ed.), Vol. **1**, pp. 47-55.

Moon, D. Y., Kwon, H.M., Lee, W.J. and Hong, S.B. (1980). Studies on the nutritional diagnosis of Jeju citrus orchards by leaf analysis. *Res. Rep. No.* **22,** pp. 63-70. Office of Rural Develop., Horticulture and Sericulture.

Moore, E.C. (1945). Non-tillage weed spray program in Tulare County. *Calif. Citrog.* **30**: 280-281.

Moore, E.C. (1946). Grass roots view on non-cultivation. *Calif. Citrog.* **31**: 201, 218-219.

Moore, I.D. (1962). The development of the south African citrus industry. *Agrekon* **1**(4): 6-16.

Moormann, F.R. and Panabokke, C.R. (1961). A new approach to the identification and classification of the most important soil groups of Ceylon. *Trop. Agri. Trinidad* **117**(11): 3-69.

Morisada, Kazuhito and Ohusumi, Yasuo (1993). Soil development on the 1888 Bandai mudflow deposits in Japan. *Geoderma* **57:**443-458.

Morita, Y. (1955). Studies on orchard soils. Soil atmosphere and tree growth. *Bull. Ser. No.* **4**, pp. 88-90. Nat. Inst. Agri. Sci., Hiratsuka, Japan.

Morita, Y., Yoneyama, K. and Nishida, Y. (1952). Studies on the physical properties of soils in relation to fruit tree growth. *J. Hort. Assoc. Japan* **20** : 11-157.

Morris, A.A. (1938). Some observations on the effects of boron treatment in the control of hard fruit in citrus. *J. Pom. Hort. Sci.* **16**:167-181.

Moser, F. (1933). The calcium-magnesium ratio in soils and its relation in to crop growth. *J. Am. Soc. Agron.* **25:** 305-377.

Moss, G. I. (1969). Influence of temperature and photoperiod on flower induction and infloroscence development in sweet orange (*Citrus sinensis* L. Osbeck). *J. Hort. Sci.* **44**:311- 320.

Moss, G.I. (1970). Influence of temperature on fruit set in sweet orange (*Citrus sinensis*). *Hort. Res.* **10**:97-107.

Moss, G. I. (1971a). Promoting flowering in sweet orange. *Australian J. Agric Res.* **22** :625-629.

Moss, G. I. (1971b). Effect of fruit on flowering in relation to biennial bearing in sweet orange (*Citrus sinensis*). *J. Hort. Sci.* **46** : 177-184.

Moss, G. I. (1976a). Temperature effects on flower initiation in sweet orange (*Citrus sinensis*). *Australian J. Agric. Res.* **27**: 399- 407.

Moss, G. I. (1976b). Thinning of 'Washington' navel and 'Late Valencia' sweet orange fruits with photosynthetic inhibitors. *HortScience* **11** : 48-50.

Moss, G.I. (1976c). Citrus fruit quality problems in the Murrumbridge irrigation areas. *Food Technol. Australia* **28**(6) : 222-225.

Moss, G.I. and Muirhead, W.A. (1971a). Climatic and tree factors relating to the yield of orange trees. I Investigations on the cultivars Washington Navel and late Valencia. *Hort. Res.* **11**: 3-7.

Mosse, B. (1962). Graft Incompatibilities in Fruit Trees. *Technical Communication No.* **28**, Commonwealth Agriculture Bureaux., Farnham Royal, Bucks, England.

Mostert, P.G. and Hoffman. J.E. (1996). Comparison of four scheduling methods for irrigation of citrus. *Proc. Int. Soc. Citriculture.* Vol. **2**, pp. 688-691.

Mota, F.S. DA., Agendes, M.O., Alves, E.G.P., Singorini and Arujo, S.M.B. (1990). Regions for citriculture in the south of Rio Grande do Sul. *Lavoura Arrozeira* **43**(391) : 21-22.

Motiramani, D.P., Vyas, K.K. and Sharma, N.K. (1964). Availability of phosphate fertilizer in some soils of the Madhya Pradesh. *J. Indian Soc. Soil Sci.* **12** : 157.

Mukherjee, B.K. and Das, K.K. (1940). Studies on Kumaun hill soils. I: Soil survey at the Government orchards,Chaubattiu: formation of genetic group. *Indian J. Agric. Sci.* **10** : 990-1020.

Mukherjee, J.N. (1949). Trace elements deficiencies in citrus. *Sci. & Cult.* **15**: 235-237.

Mukherjee, S.K. and Raturi, G.B. (1969). Zinc nutrition - its effect on citrus die-back. *Prog. Hort.* **1**(13) : 89-95.

Mukhopadhyay, S. (1987). Citrus Dieback – Mapping and Control Project in Darjeeling District. *Bull.* pp 1-29, Bidhan Chandra Krishi Vishwavidayalaya, West Bengal.

Muldabaev, S.S. and Zaitsev, Yu. I. (1989). Mandarins in trenches. *Subtropicheskie Kul'tury* (1) : 80-82.

Munday, P. (1991). Liebigs metamorphosis: From organic chemistry to the chemistry to agriculture. *Ambix* **38**: 135-154.

Mungomery, W.V. (1966). Citrus growing in Queensland. *Qd. Agric. J.* **92**: 422-431.

Mungomery, W.V., Jorgensen, K.R. and Barnes, J.A. (1980). Rate and timing of nitrogen application to navel oranges : effects on yield and fruit quality. *Proc. Int. Soc. Citriculture.* 1978, pp. 285-288, Griffith, NSW, Australia.

Munshi, S.K., Mann, M.S., Bajwa, M.S., Viz, V.K. and Thati, S.K. (1979). Physico-chemical properties of the fruits of healthy and declining sweet orange trees and their relation to various leaf and soil analyses values. *Indian J. Hort.* **36**:406-412.

Murali, V., Sarma, V.A.K. and Krishnamurti, G.S.R. (1974). Mineralogy of two red soils (Alfisol) profiles of Mysore state, India. *Geoderma* **11**(2) : 147-155.

Muraro, R.P. and Fairchild, G.F. (1986). Economic implications of the Caribbean basin initiative for Florida citrus. *Proc. Fla. State Hort. Soc.* **99** : 82-86.

Muraro, R.P. and Ford, S.A. (1990). Profitability of citrus cultivars by region and markets. *Proc. Fla. State Hort. Soc.* **103** : 46-49.

Muraro, R.P., Spreen, T.H. and Gonzales, A.N. (1996). An overview of the Cuban Citrus Industry. *Citrus Indus.* **77**(6) : 12-14.

Murthy, S.V.K., Anjaneyulu, K. and M.G. Mopaiah, (1986). Correlation between soil and leaf and amongst leaf nutrients in Coorg mandarin budded on trifoliate orange. *Curr. Res.* **15**:137-138.

Murthy, R.S., Hirerkerur, L.R., Deshpande, S.B. and Venkatarao, B.V. (1982). *Benchmark Soils of India*, National Bureau of Soil Survey and Land Use Planning (ICAR), Nagpur, Maharashtra, India, p. 374.

Murthy, I.Y.L.N., Sastry, T.G., Datta, S.C., Narayanswamy, G. and Rattan, R.K. (1997). Distribution of micronutrient cations in Vertisols derived from different parent materials. *J. Indian. Soc. Soil Sci.* **45**(3):577-580.

Myhre, D.L. and Shih, S.F. (1993). Soil heterogeneity within a citrus grove in southwest Florida flatwood soils. *Soil & Crop Sci. Soc. Fla.* **52** : 114-120.

Nadir, M. (1968). The different forms of calcium and K-Ca, K-Mg and Ca-Mg antagonisms in citrus leaves. *Mem. Gen. VI.,* Coloq. eur. medit. Contr. Fert. Cult. Seville, pp. 435-437.

Nagatsuka, S. (1975). Genesis and classification of yellow-brown Forest Soils and Red Soils in South-west Japan. *Bull.Citri, No.* **26**, Nat. Inst. Agri. Sci. Serv., Japan.

Naidu, V.C. and Rao, D.V.K. (1958). Note on the investigation of the soil of the Fruit Research Station, Anantharajupet, Kodur, with special reference to citrus decay or dry root rot. *Andhra Agri. J.* **5**(2): 54-60.

Naik, K.C. (1948). *South Indian Fruits and their Culture* (Rev. ed. 1963). P. Vardachary and Co. Madras, India, pp 123.

Nair, K.M. and Chamuah, G.S. (1993). Exchangeable aluminium in soils of Meghalaya and management of Al^{3+} related activity constraints. *J. Indian Soc. Soil Sci.* **41**(2) : 331-334.

Nair, P.C.S. and Mukherjee, S.K. (1970a). Optimum dose and time of zinc sprays for controlling its deficiency in citrus. *Indian J. Agric. Sci.* **40**:461-469.

Nair, P.C.S. and Mukherjee, S.K. (1970b). Studies on zinc deficiency in citrus. I. Causes of seasonal occurrence of zinc deficiency. *Hort. Res.* **10** : 1-13.

Nair, P.C.S., Mukherjee, S.K. and Kathavate, Y.V. (1968). Studies on citrus dieback in India *India*.I. Micronutrient status in soil and leaves of citrus. *J. Hort.* **38**:184-194.

Nair, P.R.K. (1984) Soil productivity aspects of agro-forestry. *International Council for Research in Agro-forestry*, Nairobi, Kenya.

Nakagawa, Y. (1969). Studies on favourable climatic environments for fruit culture. VI Analysis of the climatic conditions in the major areas of citrus production in the world. *Bull. No.* **8**, pp. 73-94, Hort. Res. Sta., Hiratsuka, Japan.

Nakahara, M. and Hada, W. (1963). Results of observations on the soil moisture and temperature conditions prevailing in a sloping orange orchard. *Tech. Bull No.* **11,** pp. 37-43 Fac. Hort. Chiba, Japan.

Nakaidze, E.K. (1970). Clay minerals in the soils of eastern Georgia. *Pochvovedeniye* 4.

Nakaidze, E.K. and Archvadze, N.V. (1977). Identification of the humic-calcareous soils of Georgia. *Soviet Soil Sci.* **9**(1): 12-21.

Nakajima, T. and Ogaki, C. (1972). Studies on the improvement of the packing and shipping method of Unshiu oranges. I. The relationship between fruit development and altitude. *Bull. No.* **20**, pp. 1-6, Kanagawa Hort. Expt. Sta., Nionomiyamachi, Japan.

Nanaya, K.A., Anjaneyulu, K. and Kotur (1985). Effect of foliar applied Zn, Mn, Cu and Mg on growth parameters, chlorosis and interrelationship of micronutrients in leaf tissue of Coorg mandarin. *Prog. Hort.* **17**(4):309-314.

Nandi, S. and Dasog, G.S. (1992). Properties, origin and distribution of carbonate nodules in some Vertisols. *J. Indian Soc. Soil Sci.* **40**:329-334.

Nandra, S.S. (1974). Free iron oxide content of tropical soil. *Pl. Soil* **40**(2) : 18-25.

Narasimham, R.L.(1995). Soil problems and management in hill areas with special reference to the shifting cultivation. *J. Indian Soc. Soil Sci.* **43**(3):305-309.

Narasinga Rao, K. (1958). A note on the mandarin orange decline. *Madras Agric. J.* **35**(7): 41-43.

Nash, M.H., Daugherty, L.A., Buchanan, B.A. and Hunyadi, B.W. (1994). The effect of groundwater table variation on characterization and classification of gypsiferous soils in the Tularosa basin, New Mexico. *Soil Survey Horizons,* **35**(4) : 102-110.

Nath, A.K. and Deory, M.L. (1976). Effect of altitude on organic matter and forms of nitrogen in soil of Arunachal Pradesh. *J. Indian Soc. Soil Sci.* **24** : 279.

Nath, J.C. and Sarma, R. (1992). Effect of organic mulches on growth and yield of Assam lemon (*Citrus limon* Burm.). *Hort. J.* **5**(1) : 19-23.

Naude, C.J. (1954). Citrus cultivation. *Fmg. S. Afr.* **29**(3):169-170.

Naude, C. J. (1955). Citrus cultivation in South Africa. *Wild. Crops* **7**(2) : 75.

Nauer, E.M., Goodale, J.H., Summers, L.L. and Reuther, W. (1972). Climatic effects on navel oranges. *Calif. Agri.* **26**(11) : 8-10.

Nauer, E.M., Goodale, J.H., Summers, L.L. and Reuther, W. (1974). Climate effects on mandarin and Valencia oranges. *Calif. Agri.* **28**(4) : 8-10.

Nauer, E.M., Goodale, J.H., Summers, L.L. and Reuther, W. (1975a). Climate effects on grapefruit and lemons. *Calif. Agri.* **29**(3) : 8-10.

Nauer, E.M., Goodale, J.H., Summers, L.L. and Reuther, W. (1975b). Climatic effects on grapefruit and lemons. *Citrog.* **60**(4) : 100, 115-116.

Nauriyal, J.P. and Chauhan, S.P.S. (1966). Citrus decline – an analysis of its causes and suggestions for future research. *Punjab Hort. J.* **6**: 11-16.

Nauriyal, J.P., Cahoon, G.A. and Satish Kumar (1970). Nutritional status of healthy and declining sweet orange trees in Punjab. *Commun. Soil Sci. Pl. Anal.* **1**:299-309.

Nauriyal, J.P., Sannon, L.M. and Frolich, E.F. (1958). Eureka lemon-trifoliate orange incompatibility. *Proc. Am. Soc. Hort. Sci.* **72**: 273-283.

Nawab Ali, Miskeen Khan, Shah, Mukamail Hussain, S.A. and Amin, Noorul (1992). Foliar application of iron, zinc and manganese enhances vegetative growth of *Citrus sinensis. Sarhad J. Agric.* **8**(1) : 43-48.

Nayak, D.C., Sen, T.K., Chamuah, G.S. and Seghal, J.L. (1996a). Highly leached mineral soils of Manipur-Their Pedology, characteristics, problems and management. *J. Indian Soc. Soil Sci.* **41**: 778.

Nayak, D.C., Sen, T.K., Chamuah, G.S. and Sehgal, J.L. (1996b). Nature of soil acidity in some soils of Manipur. *J. Indian Soc. Soil Sci.* **44**(2): 209-214.

Nayak, D.C. and Srivastava, R. (1995). Soils of shifting cultivated area in Arunachal Pradesh and their suitability for land use planning. *J. Indian. Soc. Soil Sci.* **43**(2):246-251.

Nayyar, V.K. (1999). Micronutrient management for sustainable intensive agriculture. *J. Indian Soc. Soil Sci.* **47**(4) : 666-680.

Nazimuddin, M. and Ahmad, K. (1975). Performance of sweet and mandarin oranges in Bangladesh. *Bangladesh Hort.* **3**(2) : 6-10.

NCDP (1989). National Citrus Development Programme, *Annual Report*, National Citrus Development Programme. Nepal Government of Nepal.

Neelkanthan, V. and Mehta, B.V. (1961). Copper status of soils of western India. *Soil Sci.* **91** : 251-256.

Neff, E. (1997). Precision agriculture - a new high tech management technique. *Citrus Indus.* **78**(1) : 19-23.

Neild, R.E., Craig, A.L. and Reiss, A.H. (1970). Historic synoptic weather maps as a basis for evaluating weather hazards in a new agricultural development. *J. Am. Soc. Hort. Sci.* **95** : 480-482.

Nejgebauer, V., Ciric, M., Fillipovski, G., Skoric, A. and Zivkovic, M. (1963). Classification of soils of Yugoslavia. *Zamljiste Biljika* **1** : 21-44.

Nel, D.J. (1980). Infiltration problem in citrus orchard. *Information Bulletin,* Citrus and Subtropical Fruit Research Station. **94**, pp. 10-12.

Nel, D.J. and Bennie, A.T.P. (1983). Progress report on a study of the influence of soil characteristics on the growth of Citrus trees. *Technical Communication* **180**, pp. 19-23, Department of Agriculture and Fisheries, Republic of South Africa.

Nel, D.J. and. Bennie, A.T.P. (1984). Soil factors affecting root growth and tree development in a citrus orchard. *South Afr. J.* **1**:.39-47.

Nemec, S. (1983). Oxygen, temperature and water potential in shallow and deep soils of a citrus grove with blight. *Proc. Soil & Crop Sci. Fla.* **42**: 85-90.

Nemec, S. (1986). Soil environment and some biotic factors affecting citrus root health. *Proc. 4th Nat. Citrus Sem. pp.* 24-33.

Nemec, S. (1998). Effects of soil amendments and tillage on vesicular-arbuscular mycorrhizal fungus colonization of citrus. *Hort. Technol.* **8**(1) : 51-54.

Nemec, S., Bernett, H. C. and Patterson, M, (1982). Root distribution and loss on blighted and healthy citrus trees. *Proc. Soil & Crop Sci. Fla.*. **41** *:* 91-96.

Nemec, S., Calvert, D., Allen, L.H. and Fiskell, F. (1984). Features of shallow soils with clay pans that are conducive to citrus blight. *Proc. Int. Soc. Citriculture.* Vol. **2**, pp. 398-401.

Nemec, S. A., Fox., N. and Horvath., G. (1976). The relation of subsurface hardpan to blight of citrus and the development of root systems. *Proc. Soil & Crop Sci. Soc. Fla.* **36**:16-18.

Nemec, S. and Lee, O. (1992). Effect of preplant deep tillage of soil amendments on soil mineral analysis, citrus growth, production an tree health. *Soil Till. Res.* **23**(4) : 317-331.

Newcomb, D.A. (1978). Selection of rootstocks for salinity and disease resistance. *Proc. Int. Soc. Citriculture.* pp.117-120.

Newman, J. E. (1968). Estimating water needs of citrus orchards. *Citrus Indus.* **49**: 19-24.

Newman, J.E. (1969). Changes in diurnal net radiation under humid and arid atmospheric conditons. *Proc. 1st Int. Citrus. Symp.* Vol. **1**, pp. 333-338, Univ. of Calif. Riverside, California, USA.

Ngo, H. (1988). Economic assessment of tropical lemon and lime production in the Katherine and Darwin region. *Tech. Bull No.* **271**, p. 21. Northern Territory, Dept. Primary Industries & Fisheries, Darwin, Australia.

Nguyen, Anh Tuan (1977). Information on the lateritic bauxites of the Xi Ling plateau. *Geology,* Hanoi, No. 11-12 (in Vietnamese).

Nguyen, Thanh Van (1978). Types of weathering of basalts in Southern Vietnam and possible bauxite deposits. *Geological Survey.* **39** (in Vietnamese).

Nieves, M., Garcia, A. and Cerda, A. (1991a). Salt tolerance of two lemon scions measured by leaf chloride and sodium accumulation. *J. Plant Nutr.* **14** : 623-636.

Nieves, M., Garcia, A. and Cerda, A. (1991b). Effect of salinity and rootstock on lemon fruit quality. *J. Hort. Sci.* **66**(1) : 127-130.

Nii, N., Harada, K. and Kadowar, K. (1970). Effects of temperature on the fruit growth and quality of Satsuma oranges. *J. Japanese Soc. Hort. Sci.* **39** : 309-317.

Nijjar, G.S. and Brar, S.S. (1977). Comparison of soil and foliar applied zinc in kinnow mandarin. *Indian J. Hort.* **34**(2):130-136.

Nijjar, G.S. and Singh, R.A. (1971). A survey of mineral nutrition status of sweet orange of Amristar district. *Punjab Hort. J.* **11**(1): 16-22.

Nikiforoff, C.C. and Alexander, L.T. (1942). The hardpan and the claypan in San Joaquin soil. *Soil Sci.* **53** : 157-172.

Nilangekar, R.G. and Patil, V.K. (1982). Soil factors as related to citrus decline. *J. Indian Soc. Soil Sci.* **30**:213-215.

Nir, I., Goren, R. and Leshem, B. (1972). Effect of water stress, gibberellic acid and 2- chloroethyl triethylammonium chloride (CCC) on flower differentiation in 'Eureka' lemon trees. *J. Am. Soc. Hort. Sci.* **97** : 774-778.

Nirmalya, Bala and Sahu, G.C. (1993). Characterization and classification of soils on hill slope of middle Andaman Island. *J. Indian Soc. Soil Sci.* **41**(1):133-137.

Nishiura, M. (1981). The Citrus Industry in Japan. *Proc. Int. Soc.Citriculture.* Vol. **2**, pp. 991-993.

Nito, Nobumasa (1999). *Citrus Industry in Japan.* Souvenir, International Symposium on Citriculture, Indian Society of Citriculture, Nagpur pp. 149-154.

Norris, J.C. (1970). Young tree decline from growers viewpoint. *Proc. Fla. State Hort. Soc.* **83** : 46-48.

Norris, R.E. and Lawrence, F.P. (1957). Hairy indigo as a cover crop in Florida citrus. *Fla. Agri. Extn. Serv.* No. **227**, p. 10.

Northcote, K.M. (1971). *A Factual Key for the Recognition of Australian Soils.* Third Edition, Rellin Technical Publications, Adelaide, Australia, p. 3.

Nunez-Moreno, J.H. and B.Valdez Gascon (1994). Effect of soil conditions on orange tree. *Commun. Soil Sci. Pl. Anal.* **25** (9-10) : 1747-1753.

Nuttonson, M.Y. (1965). Agro-climatology and global agro-climatic analogues of the citrus regions of the continental United States. *Amer. Inst. Crop Ecol.* p.42, USA.

Nyamapfene, K. (1984). Transmission electron microscopy and electron diffraction studies on the clay fraction of three Zimbabwean Vertisols derived from basalt. *Zimbabwe J. Agri. Res.* **22** : 111-117.

Nyamapfene, K. (1991). *The Soils of Zimbabwe.* Mehanda Press, Harare, Zimbabwe.

Nychas, Anastasios, E. and Kosmas, Constantinos, S. (1984). Phosphate adsorption by dark alkaline Vertisols in Greece. *Geoderma* **32:**319-324.

Nye, P., Craig, D., Coleman, N.T. and Ragland, J.L. (1961). Ion exchange equilibria involving aluminium. *Proc. Soil. Sci. Soc. Am.* **25** : 14.

Nye, P.H. and Greeland, D.J. (1960). *The Soil Under Shifting Cultivation.* CAB International, Wallingford, U.K.

Nye, P.H. and Greenland, D. J. (1968). *The Soil Under Shifting Cultivation. CAB Tech. Comm. No.* **51**, Harpenden, U.K.

O'Brien, L.O. and Sumner, M.E. (1988). Effects of phosphogypsum on leachate and soil chemical composition. *Commun. Soil Sci. Pl. Anal.* **19** : 1319-1329.

O'Connor, Lt. (1870). *Memoir of Kodagu Survey.* Central Jail Press, Bangalore.

O'Donnell, A.G., Syers, J.K., Vichiensanth, P., Vityakon, P., Adey, M.A., Nannipieri, P., Sriboonlue, W. and Suwanarit, A. (1994). Improving the agricultural productivity of the soils of north-east Thailand and through soil organic matter management. *Soil Science and Sustainable Land Management in the Tropics* (J.K. Syers and D.L. Rimmer, ed.), CAB International, British Society of Soil Science, UK, pp. 192-205.

Oates, K.M. and Kamprath, E.J. (1983) . Soil acidity and liming II. Evaluation of using aluminium extracted by vairous chloride salts for determining lime requirement. *Soil Sci. Soc. Am. J.* **47:** 690.

Oberholzer, P.C.J. (1947). The Present Status of Citrus Nutrition in South Africa. *South African Dept. Agri. Bull.* **271**, p. 14.

Oberholzer, P.C.J. (1969). Citrus culture in Africa, south of Sahara. *Proc. Ist Int. Citrus Symp.* Vol. **1**, pp. 111-120.

Oberholzer, P.C.J., De Villiers, J.I., Langenergger, W. and Naude, C.J. (1958). *Report on Citrus Soils and Nutrition.* Mimeographed Report.

Oberza, T.A. (1989). Water table behaviour under multi-low citrus bed. *Proc. Fla. State Hort. Soc.* **101** : 53-58.

Obreza, T.A. (1990). Evaluation of citrus groveer fertility status using soil and leaf tissue testing. *Proc. Soil & Crop Sci. Soc. Fla.* **49** : 66-69.

Obreza, T.A. (1973). Program fertilization for establishment of orange trees. *Prod. Agric.* **6**(4) : 546-552.

Obreza, T.A. and Admire (1985). Shallow water table fluctuation in response to rainfall, irrigation and evapotranspiration in flatwoods citrus. *Proc. Fla. State Hort. Soc.* **98** : 32-37.

Obreza, T.A. and Arnold, C.E. (1990). Citrus development in southwest Florida. *Proc. Interamerican Soc. Trop. Agri.* **34** : 105-108.

Obreza, T.A., Yamatak, H. and Pearlstine, L.G. (1993). Classification of land suitability for citrus production using DRAINMOD. *J. Soil & Water Consev.* **48**(1) : 58-64.

Ogata, R. (1962). Studies on manganese deficiency in citrus. I. On deficiency symptoms, leaf analysis and applications of manganese. *J. Jap. Soc. Hort. Sci.* **31**: 337-346.

Ogata, R. (1967). Studies on manganese deficiencies in citrus III. Survey of soil and tree conditions. *J. Japanese Soc. Hort. Sci.* **36** : 55-62

Oh, S. D., Chung, S. K. and Hong, S. B. (1981). Effect of low temperatures on flower bud differentiation during the winter season in satsuma trees (*Citrus unshiu* Marc.). *J. Korean Soc. Hort. Sci.* **22**: 194-198.

Okada, N. (1972). The classification of citrus growing districts by principal component analysis. *Bull. No.* **10**, pp. 19-46, Shizuoka Prefectural Citrus Expt. Sta., Shimizu, Japan.

Okada, N. (1981). On the diagnosis of nutrient condition and the application of fertilizer for citrus trees in Japan. *Proc. Int. Soc. Citriculture.* Vol. **2**, pp. 548-551.

Okada, N. (1994). Management of soil and nutrition in rootzone control of vegetables and fruit trees. 5. Root zone restriction culture of satsuma mandarin for high quality fruit production. *Japanese J. Soil Sci. & Pl. Nutr.* **65**(2) : 206-214.

Okada, N., Ooshiro, A. and Ishida, T. (1969). Effect of trace elements on the growth of citrus. Part II. Effects of applied lime on the absorption of some trace element materials. *Bull. No.* **8**, pp. 17-27, Shizuoka Prefectural Citrus Expt. Sta., Shizuoka, Japan.

Okada, N., Ooshiro, A. and Ishida, T. (1994). Effect of the level of fertilizer application on the nutrient status of Satsuma mandarin trees. *Proc. Int. Soc. Citriculture.* Vol. **2**, pp. 575-579.

Okigbo, B.N. (1991). *Development of Sustainable Agricultural Systems in Africa.* Distinguished African Scientist Lecture Series No. **1**, IITA, Ibadan, Nigeria.

Okusami, T.A., Rust, R.H. and Juo, A.S.R. (1987). Properties and classification of five soils on alluvial land forms in central Nigeria. *Canadian J. Soil Sci.* **67**:249-261.

Oldeman, L.R. (1988). *Guidelines for Global Assessment of the Status of Human Induced Soil Degradation.* GLASOD. International Soil Reference and Information Centre, Wageningen, Netherlands.

Ollier, C.D., Drover, D.P. and Godelier, M. (1971). Soil knowledge amongst the Baruya, New Guinea Oceania, **42** : 33-41.

Ongun, A. R. and Wallace, A. (1958a). Inorganic composition of small 'Washington' navel orange trees on different rootstocks grown with different root temperatures. *Tree Physiol.* **1**:84-86.

Ongun, A. R. and Wallace, A. (1958b). Influence of root temperature on the growth of small 'Washington' navel orange trees on three different rootstocks in a glasshouse. *Tree Physiol.* **1** : 184-186.

Ono, S., Iwagaki, I. and Takahara, T. (1986). Relationships between root and leaf distribution in citrus trees. *Bull. No.* **8**, pp. 25-36, Fruit Tree Res. Sta., D. Kuchinotsu, Japan.

Opitz, K.W. (1969). Citrus growing California. *Manual* **39**, p. 58, Div. *Agri., Sci.* , Univ. Calif., USA.

Optiz, Karl, W. (1974). A look at citrus production in Iran. *Calif. Citrog.* **59**(10): 339-343.

Opitz, K.W. and Platt, R.G. (1967). Boron deficiency in central California. *Citrog.* **52**:474.

Oppenheimer, H.R. (1940). How to produce summer lemons in Palestine ? *Hadar* **13**:169-170.

Oppenheimer, H.R. and Heyman-Hershberg, L. (1954). Effects of nitrogen fertilizing on soil, growth, leaf composition and yield in Shamouti orange groves. *Israel J. Agric. Res.* **5**:5-7.

Orozco – Santos, M. (1980). Relation of valencia amachamiento with citrus decline and effect of tetracyline treatements in north of the Veracruz. *Revista Mexicana de Fitopathologia* **9**(1) : 11-13.

Orozco-Romero, J. and Sepulveda-Torres, J.L. (1981). Influences of nitrogen-phosphorus, potassium and timing of application on field of Mexican lime [*Citrus aurantifolia* (Christm) Swingle]. *Proc. Int. Soc. Citriculture.* Vol. **2**, pp. 542-543.

Orth, Paul, G. (1981). Fertility management of Florida soils. *Proc. Soil & Crop Sci. Soc. Fla.* Vol. **40,** pp. 1-3.

Ortuzar, J.E. (1996). The citrus industry in Chile. *Proc. Int. Soc. Citriculture.* Vol. **1**, pp. 304-307.

Osa, F. De. La., Rosa, A. La. and Ramirez, A. (1987). Effect of different cultivation methods on yields of young Olinda Valencia orange trees in a red ferrallitic soil. *Cultivos Tropicales* **9**(4) : 66-70.

Oseni, T.O. (1988). Relationship between soil properties and leaf nutrient of Cleopatra mandarin (*Citrus reticulata* Blanco) seedlings. *Indian J. Agric. Sci.* **58**:571-572.

Passos, O.S. and Sorbinho, A.P. Cunha (1970). Cultura dos citros no Estado da Bahia. Cruz das Almas, IPEAL/SUNDANE, Sao Paulo, Brazil.

Patel, N.R., Mandal, A.K. and Pande, L.M. (1999). Agro-ecological zoning system. A remote sensing and GIS perspective. *J. Agromet.* **2**(1) : 1-13.

Patel, B.M. and Patel, H.C. (1985). Effect of foliar application of zinc and iron on chlorophyll and micronutrient contents of acid lime (*Citrus aurantifolia* Swingle). *South Indian Hort.* **33**(1):50-52.

Patel, P.C., Patel, M.S. and Kalyanasundaram, N.K. (1997). Effect of foliar spray of iron and sulphur on fruit yield of chlorotic acid lime. *J. Indian Soc. Soil Sci.* **45**(3) : 529-533.

Patel, P.C., Patel, M.S. and Kalyansundaram, N.K.(2001).Response of chlorotic acid lime to iron and sulphur fertilization *J. Indian Soc. Soil Sci.* **49**(2): 295-300.

Patel, G.G. and Trivedi, B.S. (1994). Effect of water regimes, different sources and levels of phosphorus on available phosphorus in clayey soils. *J. Indian Soc. Soil Sci.* **42**(2) : 220-223.

Path, J. (1970). Yielding capacity of citrus groves on fine textured soils overlying compacted subsoil. *Hassadeh* **50**(5) : 537-554.

Patil, M.N., Heganna, S.S., Thakre, S.K. and Puranik, R.B. (1999a). Orange garden soils of western Vidarbha region I: characterization and classification. *Proc. Int. Symp. Citriculture*, pp. 426-433, Indian Society of Citriculture, Nagpur, Maharashtra, India.

Patil, M.N., Khandare, N.C. and Puranik, R.B. (1999). Evaluation of pedological development of orange garden soils of Akola district through field morphology rating system. *J. Indian Soc. Soil Sci.* **47**(1):180-182.

Patil ,V.D. and Malewar, G.U (1998). Assessment of micronutrient status of export oriented mandarin orchards by soil and leaf analysis. *J. Indian. Soc. Soil Sci.* **46:**151-152.

Patil, J.D. and Patil, N.D. (1981). Effect of $CaCO_3$ in the availability of iron and manganese. *J. Maharashtra Agri. Univ.* **6** : 185-188.

Patil, Y.M. and Sonar, K.R. (1994). Status of major and micronutrient of swell-shrink soils of Maharashtra. *J. Maharashtra Agri. Univ.* **19**(2) : 169-172.

Patil, M.N., Thote, S.G. and Thakre, S.K. (1999b). Orange garden soils of western Vidarbha region II: Soil-site suitability evalaution. *Proc. Int. Symp. Citriculture*, pp. 434-441, Indian Society of Citriculture, Nagpur, Maharashtra, India.

Partiram, Awasthi, R.P., Pradhan, Yashoda and Prasad, R.N. (1994). Soil fertility research in north-eastern hill region-A review. *J. Hill Res.* **7**(1): 1-8.

Patiram and Prasad, R.N. (1978). Forms of potassium in soils of Meghalaya. *Ibid.* **18**:12-22..

Patiram, Prasad, R.N. and Munna Ram (1993). Inorganic phosphorus fractions and their relation to available phosphorus indices in acid soils of Meghalaya. *J. Indian Soc. Soil Sci.* **41**(4):772-773.

Patiram and Upadhyay, R.C. (1997). Calcium, magnesium and potassium status of mandarin orange (*Citrus reticulata*) orchards in Sikkim. *Indian J. Agric. Sci.* **67**(1) : 44-45.

Patiram and Upadhyaya, R.C. (2000). Phosphorus status of Sikkim mandarin (*Citrus reticulata* Blanco) orchards. *J. Hill Res.* **13**(2) : 101-104.

Patiram, Upadhyaya, R.C., Singh, C.S. and Munna Ram (2000). Micronutrient cation status of mandarin (*Citrus reticulata* Blanco) orchards of Sikkim. *J.Indian Soc. Soil Sci.* **48**(2):246-249.

Patt, J. (1958). Experiments in non-tillage of citrus in the coastal zone of Israel. Le Livre du IV eme. *Congr. Int. Agrum.* 1956, pp. 219-230, Mediter, Tel Aviv, Israel.

Patt, J., Carmeli, D. and Zafrir, I. (1966). Influence of soil physical conditions on root development and on productivity of citrus trees. *Soil Sci.* **102** : 82-84.

Pavan, M.A., Jacomino, A.P. and Tormen,V. (1997). Response of citrus to soil application of phosphorus. *Arquivos de Biologia e Tecnologia.* **40**(1) : 75-79.

Pavan, M.A. and Wutscher, H.K. (1993). Accumulation of nutrients at the surface of roots of blight affected orange trees. *Commun. Soil Sci. Pl. Anal.* **24**(9-10) : 979-987.

Payton, R.W. and Shishira, E.K. (1994). Effects of soil erosion and sedimentation on land quality : Defining pedogenic baselines in the Konoda district of Tanzania. *Soil Science and Sustainable Land Management in the Tropics* (J.K. Syers and D.L. Rimmer, ed.), A.B. International, British Society of Soil Science, U.K. pp. 88-119.

Peach, M. (1939). Determination of exchangeable magensium in soils by titan yellow with refernce to mgensium deficiency in citrus. *Soil Sci. Soc. Am. Proc.* **4**:189-191.

Pearson, R.W. (1975). Soil Acidity and Liming in the Humid Tropics. Cornell Inst. Agri. *Bull* No. **30**, Ithaca, New York, USA.

Pearson, G.A. and Goss, J.A. (1953) : Observation on the effects of salinity and water table on young grapefruit trees. *Proc. VII Ann. Rio Grande Valley Hort. Inst.*, pp 1-6.

Pedrera, B., Lambert, I., Oviedo, D. and Alfonso, A. (1988). Study of three rates of N application with basal P and K on Dancy mandarin in a red ferrallitic soil II. Effects on growth, fruit quality and annual yield performance. *Ciencia Y Tecnica en la Agricultura, Citricos Y Otros Frutales.* **11**(2):17-28.

Pehrson, J.E. (1966). Yield trends. *Citrog.* **51**: 261-262.

Pehrson, J. (1971). Ground covers for citrus orchards. *Citrog.* **56** : 419-421.

Pehrson, J. (1976). Delayed maturity in Navels. *Citrog.* **61**(6) : 200.

Penkov, M., Alami, A. and Krhistova, D. (1979a). Root distribution of clementines in serozen of the Sous river of Morocco. *Pochvoznanie i Agrokhimiya* **14**(1) : 12-17.

Penkov, M., Alami, A. and Krhistova, D. (1979b). Causes of chlorosis of clementines grown in the Sous valley Morocco. *Pochvoznanie i Agrokhimiya* **14**(2) : 81-90.

Pennisi, L. (1974). Nutritional status and quality of citrus fruits. *Ist Congresco Mundial de Citricultura,* Vol **1**, pp. 73-80.

Pennisi, L. (1975). Studies on the nutrient status of citrus in experimental fields in Sicily, Calabria and Brasilicata. *Annali dell' Instituto Sperimentale per Agrumicoltura* **6** : 323-337.

Pennock, D.J. and Vankessel, C. (1997). Clearcut forest harvest impacts on soil quality indicators in the mixedwood forest of Saskatchewan, Canada. *Geoderma* **75** : 13-32.

Peynado, A and Young , R.H. (1962). Performance of nucellar Redblush grapefruit trees on 13 kinds of rootstocks reigated with saline and boron contaminated well water over a 3 years period . *Proc. Rio Grande Valley Hort. Soc.* **16**: 52-58.

Peynado A and Young , R.H. (1963). Toxicity of three salts of greenhouse grown grapefruit trees and their effects on ion accumulation and cold hardiness . *J. Rio. Grande Valley Hort. Soc.* **17**: 60-67.

Peynado, A and Young, R. (1969) . Relation of salt tolerance to cold hardiness of redblush grapefruit and valencia orange trees on various rootstocks. *Proc. Ist. Int. Citrus Symp.* Vol. **3**, pp. 1793-1802 .

Pharande, A.L and Sonar, K.R. (1997). Clay mineralogy of some Vertisol soil series of Maharashtra. *J. Indian Soc. Soil Sci.* **45**(2): 373-377.

Phillips, R.L. (1980). Pruning principles and practices for Florida citrus. *Florida Coop. Ext. Service Cir. No.* **477**.

Pieniazek, S.A. (1967). Fruit production in China. *Proc. 17th Int. Hort. Congr.* Vol. **4** , p. 427.

Pierce, F.J. and Larson, W.E., (1993). Developing crieteria to evaluate sustainable land management (J.M. Kimble, ed.) *Utilization of Soil Survey Information for Sustainable Land Use.* Proc. Int. Soil Management Workshop, 8th USDA-SCS, National Soil Survey, Lincoln. NE, pp. 7-14.

Pieri, Chr. (1995). Integrated Plant Nutrition System, FAO, Fertilizer and Plant Nutrition Bull. **12**, pp. 181-198.

Pieters, A.J. (1927). *Green Manuring - Principles and Practice.* Wiley, New York, USA.

Pieters, A.J. and Mckee, R. (1929). Green manuring and its application to agricultural practices. *Agron. J.* **21** : 985-993.

Pinckard, J.A. (1982). Suppression of citrus young tree decline with humus. *Pl. Dis.* **66**:311-312.

Pinna, M. (1977). Climatologia. W.T.E.T., Torino.

Pinto, M.R. and Leal, P.F. (1974). The nutritional status of some orange orchards in the Valles Altos of Carabobo, Venezuela. *Revista de la Facultad de Agronomia, Venezuela* **8**(1) : 71-80.

Pionke, H.B., Corey, R.B. and Schulte, E.E. (1968). Contributions of soil factors to lime requirement and lime requirement tests. *Proc. Soil Sci. Soc. Am.* **32**:113-117.

Piringer, A.A., Downs, R.G. and Borthwick, H.A. (1961). Effect of photoperiod and kind of supplemental light on growth of three species of *Citrus and Poncirus trifoliata. Proc. Am. Soc. Hort. Sci.* **77** : 202-210.

Pisa, A. Di and Fenech, L. (1990). Effects of different rates of nitrogen fertilizer on the yield and nutritional status of mandarins. *Tecnica Agricola* **42**(1): 113-126.

Platt, R..G. (1966). Planting the orchard. *Citrog.* **52** : 53, 67-68, 72.

Platt, R.G. (1968). Micronutrient deficiencies of Citrus. *Leaf.* **115,** Div. Agri. Sci., Univ., Calif., USA.

Plessis, S.F. Du. (1971). Soil analysis for determining the phosphate content in citrus. *Citrus Grower & Subtrop. Fruit J.* **454** : 7-8.

Plessis, S.F. Du. (1975). Soil requirements. *Farming in South Africa.* Department of Technical Services, South Africa.

Plessis, S.F. Du. (1977). Soil analysis as a necessary complement to leaf analysis for fertilizer advisory purposes. *Proc. Int. Soc. Citriculture.* Vol. **1,** pp. 15-19.

Plessis, S.F. Du. (1980). The small fruit problem in citrus. *Subtropica.* **1**:7-8.

Plessis, S.F. Du. (1982). Yield forecasting for citrus in the western Cape. *Crop Production* **11**:101-104.

Plessis, S.F. Du. (1983). Crop forecasting for navels in South Africa. *Proc. Fla. State Hort. Soc.* **96**:40-43.

Plessis, H.M. (1984). Evapotranspiration of citrus as affected by soil water deficit and soil salinity. *Irrig. Sci.* **6**: 51-61.

Plessis, S.F. Du. (1996). Effect of climate on leaf analysis norms for Valencia. *Proc. Int. Soc. Citriculture.* Vol. **2,** pp. 762-766.

Plessis, S.F. Du. and Burger, R. Du. T. (1972). The determination of copper status of citrus orchard soil. *Agrochemophysica* **4**(3) : 47-51.

Plessis, S.F. Du. and Koen, T. J. (1984). Effect of nutrition on fruit size of citrus. *Proc. Int. Soc.Citriculture* Vol. **1,** pp.148-150.

Plessis, S.F. Du. and Koen, T.J. (1986). The value of a pH determination to decide the lime requirement of a crop. *Information Bull.* **166**, pp. 12-13. Citrus and Subtropical Fruit Res. Inst., South Africa.

Plessis, S.F. Du. and Koen, T.J. (1996a). South Africa. *Proc. Int. Soc. Citriculture.* Vol. **2,** pp. 1283-1285.

Plessis, S.F. Du. and Koen, T..J. (1996b). Leaf analysis norms for lemons (*Citrus lemon* L. Burm.). *Proc. Int. Soc. Citriculture.* Vol. **1**, pp. 551-552.

Plessis, S.F. Du., Koen, T.J. and Odendaal, W.J. (1992). Interpretation of Valencia leaf analysis by means of the N/K ratio approach. *Proc. Int.Soc Citriculture.* Vol. **2,** pp. 553-555.

Plessis, S.F. Du. and Malherbe, T.F.S. (1976). Citrus crop forecasting in the Citrusdal district. *Citrus & Subtrop. Fruit J.* **515** : 19-22.

Plessis, S.F. Du., Malherbe,T.F.S. and Donaldson, G.V. (1975). The effect of some climatic factors on the size of the citrus crop in the citrusolal district. *Citrus and Subtrop. Fruit J.* **503** : 14-17

Plessis, S.F. Du. and Plessis, H.M. Du. (1987). Some factors affecting the fruit growth of citrus. *South Afr. J. Pl. Soil* **4**(1) : 12-16.

Plessis, S.F. Du. and Prando, H. (1975). Fine mineral composition of citrus in the Sandag river valley. *Information Bull. No.* **33**, pp. 12-14, *Citrus and Subtrop.* Fruit Res. Inst., Nelspruit, South Africa.

Pogorelov, N.V. (1970). On the (geographic) distribution of citrus. *Subtropicheskie Kul'tury* (4): 21-31.

Pomares, F., Tarazona, F., Estela, M. and Martin, B. (1981). Evaluation of a commercial blue-green algae inoculant as fertilizer on citrus. *Proc. Int. Soc. Citriculture.* Vol. **2**, pp. 583-585.

Porter, W. M. and Helyar, K. R. (1992). Subsoil constraints to root growth and high soil water and nutrient use by plants. *Proc. National Workshop,* Tanunda, South Australia, Australia

Possingham, J.V. and Kriedemann, P.E. (1969). Environmental effects on the formation and distribution of photosynthetic assimilates in citrus. *Proc. 1st Int.Symp.* Vol. **1,** pp. 325-332.

Prabhuraj, D.K. and Murthy, A.S.P. (1994a) Forms of acidity in some acid soils. *J. Indian Soc. Soil Sci.* **42**(3): 455-456.

Prabhuraj, D.K. and Murthy, A.S.P. (1994b). Significance of extractable aluminium in acid soils. *J. Indian. Soc. Soil Sci.* **42**(3):457-459.

Pradhan, H.R. and Khera, M.S. (1976). Acid soils of India : Their genesis, characteristics and management. *Bull.* **11**, pp. 279, Indian Soc. Soil Sci., New Delhi, India.

Pradhan, Y., Sachan, R. S. and Avasthe, R.K. (1996). Relationships between altitude and status of zinc and copper in some soils of Sikkim. *J. Hill Res.* **9** (1) : 11 – 17.

Prasad, R.N., Patiram and Munna Ram (1985). Forms of aluminium in soils of east khasi hillls, Meghalaya *J. Indian Soc. Soil Sci.* **33:** 523-527.

Prasad, R. N., Ram, P., Barooah, R. C. and Ram, M. (1981). Soil Fertility Management in the North Eastern Hill Region. *Res. Bull.* **9**, pp. 1-30, ICAR Research Complex for NEH Region, Shillong, Meghalaya, India.

Prasad, R. N. and Sakal, R. (1991). Availability of iron in calcareous soils in relation to soil properties. *J. Indian Soc. Soil Sci.* **39** : 658-661.

Prasad, R.N., Singh, A and Verma, A. (1987). *Application of Research Findings for Management of Land and Water Resources in Eastern Himalaya Region. Res. Bull.,* pp. 1-78. ICAR Research Complex for NEH Region, Meghalaya, India.

Prasad, R.N., Sinha, H. and Mandal, S.C. (1967). Fractions of potassium in Bihar soils. *J. Indian Soc. Soil Sci.* **15** : 173-179.

Pratt, P. F. (1966). *Aluminium Diagnostic Criteria for Plants and Soils* (H. D. Chapman, ed.), Quality Printing Co. Abilene, TX. pp. 3-12.

Pratt, P.F. and Bair, F.L. (1961). Lime requirements. *Calif. Citrog.* **15:**13-14.

Pratt, P.F. and Harding, R.B. (1957a). Effect of fertilizers on loss of magnesium from soil. *Citrus Leaves* **37**(2): 10.

Pratt, P.F. and Harding, R.B. (1957b). Decreases in exchangeable magnesium in an irrigated soil during 28 years of differential fertilization. *Agron. J.* **49** : 419-421.

Pratt, P.F., Harding, R.B., Jones, W.W and Chapman, H.D. (1959). Chemical changes in an irrigated soil during 28 years of differential fertilization. *Hilgardia* **28** : 381-420.

Pratt, P.F., Jones, W.W. and Bingham, F.T. (1957). Magnesium and potassium content of orange leaves in relation to exchangeable magnesium and potassium in the soil at various depths. *Proc. Am. Soc. Hort. Sci.* **70** : 245-251.

Primo, E., Carrasco, J.M. and Cunat, P. (1970). Trace metal deficiencies in orange trees III. Efficiency of some chelates used in correction of iron deficiency. *J. Sci. Food. Agric.* **22** (0060).

Prince, A.L., Zimmerman, M. and Bear, F.E. (1947). The magensium supplying powers of 20 New Jersey soils. *Soil Sci.* **63**: 68-78.

Pringer, A. A., Downs, R. J. and Borthwick, H. A. (1961). Effects of photoperiod and kind of supplemental light on the three species of citrus and *Poncirus trifoliata. Proc. Am. Soc. Hort. Sci.* **77**:202-210.

Prochnow, L.L. and Boaretto, A.E. (1995). Soil sampling depth for evaluation of available sulphur in a Sicilian lemon (*Citrus limon* Burm) orchard. *Sci. Agricola* **52**(1):101-106.

Pruvost, O., Verniere, Hartung, J., Gottwald, T. and Quetelard, H. (1997). Towards an improvement of citrus canker control in Reunion island. *Fruits* **52**:375-382.

Puiggros, J., Franciosi, R. and Morin, C. (1969). A preliminary nutritional study of Washington Navel oranges in the Central coast of Peru. *Proc. Ist Int. Citrus Symp.* Vol. **3**, pp. 1613-1617.

Pushparajah, E. and Bachik, A.T. (1987). Management of acid tropical soils in south-east Asia. *Management of Acid Tropical Soils for Sustainable Agriculture. IBSRAM Proc. No.* **5**, pp. 13-39, Int. Board Soil Res. and Manag., Bangkok, Thailand.

Putcha, R.K.M. and Prasad, A. (1976). Studies on nitrogen nutrition in Kagazi lime (*Citrus aurantifolia* Swingle). II. Effect on vegetative growth, fruit set, fruit drop and yield. *Pl. Sci.* **8**:69-74.

Qin, Xuan Nan (1996). Foliar spray of B, Zn and Mg and their effects on fruit production and quality of Jincheng orange. *J. Southwest Agric. Univ.* **18**(1) : 40-45.

Quaggio, J.A. (1983). *Methods de laboratoria para a determinacao da nacessidade de calagem em solos.* (B. Van Raij, O. C. Bataglila and N.M. da Silva, ed.). Brasileira de Cilcencia do Solo. p.361.

Quaggio, J.A., Cantarella, H. and Mattos, Jr. D. (1996). Soil testing and leaf analysis in Brazil – recent developments. *Proc. Int. Soc. Citriculture.* Vol. **2**, 1269-1275.

Quaggio, J.A., Rosa, S.M., Mattos, Junior, D. De and Raij, B. Van (1998). Response of valencia orange to lime and gypsum application. *Laranja* **14**(2): 383-398.

Quaggio, J.A., Teofilo Sobrinho, J. and Dechen, A.R. (1992a). Response of liming of Valencia orange tree on Rangpur lime : Effects of soil acidity on plant growth and yield. *Proc. Int. Soc. Citriculture* Vol. **2**, pp. 628-632.

Quaggio, J.A., Teofilo Sobrinho, J. and Dechen, A.R. (1992b). Magnesium influence on fruit yield and quality of Valencia sweet orange on Rangpur lime. *Proc. Int. Soc. Citriculture.* Vol. **2**, pp. 633-637.

Quaggio, J.A., Van Raij, B. and Malavolta, E. (1985). Alternative use of the SMP buffer solution to determine lime requirement of soil. *Commun. Soil Sci. Pl. Anal.* **16** : 245-260.

Quyang, T. (1993). Soil micronutrients and citrus growth. *Pedosphere* **3**(4) : 341-347.

Quyang, T., Qian, L., Gong, G.S. and Zhou, J.G. (1984). Problems concerning microelements in the citrus soils of Guilin. *Soils (Turang)* **16**(5) : 188.

Raciti, G.B. (1959). Control of iron chlorosis in Sicily. *Teen Agric.* **11:**407-410.

Raciti, G. and Salerno, M. (1974a). Individual aspects of mineral deficiencies and excesses in citrus in Italy. *Hort. Abst.* (1979) **48** : 344.

Raciti, G. and Salerno, M. (1974b). Boron toxicity of oranges in Sicily. *Hort. Abst.* (1979) **48** : 344.

Racz, Z. (1968/69). Podzols on the territory of Croatia (Yugoslavia) and their micromorphological properties. *Geoderma.* **2** : 41-55.

Raghupathi, H.B. and Vasuki, N. (1991). Forms of copper and their relationship with soil properties. *J. Indian Soc. Soil Sci.* **39** : 630-634.

Ragland, J.L. and Boonpuckdee, L. (1987). Fertilizer responses in northeast Thailand. 1. Literatuare review and rationale. *Thai. J. Soils and Fert.* **9** : 67-79.

Ragland, J.L., Craig, I.A., Infanger, P., Chovangcham, P. and Meyer, L. (1984). Soil Fertility Evaluation of Soils in North-east Thailand. (ed.). *Quarterly Report No.* **8**, Technical Assistance Team, Khon Kaen University, Thailand.

Rai, M.M. (1972). Availability of copper in deep black soils of Madhya Pradesh. *J. Indian Soc. Soil Sci.* **20** : 135-142.

Rai, R.M. and Tewari, J.D. (1988). Yield contributing factors as influenced by micronutrient spray in orange (*Citrus reticulata* Blanco). *Prog. Hort.* **20**(1-2):124-127.

Rai, R.M., Tewari, J.D., Nirmala, P. and Pathak, C.P. (1988). Effect of micronutrient sprays on fruit quality of orange (*Citrus reticulata* Blanco). *Prog. Hort.* **20**:123-135.

Raina, J.N. (1988). Physico-chemical properties and available micronutrients in status of citrus growing soils of Dhaulakuan in Himachal Pradesh. *Punjab Hort. J.* **28**: 1-6.

Rajput, C.B.S. and Haribabu, R.S. (1999). *Citriculture.* Kalyani Publishers, Ludhiana, India, pp. 250-254.

Ram. M., Prasad, R.N. and Patiram (1987). Studies on phosphate adsorption and phosphate fixation in Alfisols and Entisols occurring in different altitudes of Meghalaya. *J. Indian Soc. Soil Sci.* **35**:207-216.

Ramakrishnan, J.S. (1954a). *Common Diseases of Citrus in Madras State and their Control. Bull.* p.15, Madras Agric. Dept., Madras, Tamil Nadu, India.

Ramakrishnan, T.S. (1954b). Deterioration of mandarin in Madras state. *South Indian Hort.* **2**:52-56.

Ramamurthy, B. and Desai, S.V. (1946). Preliminary studies of nutritional diseases of plants and their petrographic diagnosis. *Indian J. Agri. Sci.* **16**: 103-111.

Ramaswami, P. P. (1966). Studies on the physico-chemical and biological properties of Madras state. *Madras Agric. J.* **53**: 338.

Ramirez-Diaz, J.M. (1983). Tecnicas de produccion y utilization de los citricos Hidraulicas, Instituo Nacional de Investigaciones Agricolas, Mexico, D.F.

Ramishvili, G.G., Dzhakeli, J.M. and Kakuriya, D.B. (1988). Effect of snow cover on the over winfering of citrus trees. *Subtropicheskie Kul'tury* (1) : 101-103.

Randhawa, N.S., Bhumbla, D.R. and Dhingra, D.R. (1966). Citrus decline in the Punjab – a review. *Punjab Hort. J.* **6**: 35-44.

Randhawa, N.S., Bhumbla, D.R. and Dhingra, D.R. (1967). Role of soil and plant composition in diagnosis of citrus decline in Punjab. *J. Res. Ludhiana (P.A.U.)* **4** : 16-24.

Randhawa, S. S. and Iwata, M. (1968). Effects of pH, calcium concentrations and sources of nitrogen on the citrus growth and inorganic composition of citrus seedlings in solution culture. *J. Japanese. Soc. Hort. Sci.* **27:** 319-327.

Randhawa, N.S., Kanwar, J.S. and Nijhawan, S.D. (1961). Distribution of different forms of manganese in the Punjab soils. *Soil Sci.* **92** : 106-112.

Randhawa, H.S. and Singh, S.P. (1997). Distribution of manganese fractions in alluvium derived soils in different agro-climatic zones of Punjab. *J. Indian Soc. Soil Sci.* **45**(1): 53-57.

Randhawa, G.S. and Srivastava, K.C. (1986). *Citriculture in India.* Hindustan Publishing Corporation, India, p. 23.

Raney, W. A., Edminster, T. W. and Allway, W. H. (1955). Current status of research in soil compaction. *Soil Sci.* **53**: 157-172.

Rao, A.C.S. (1993). Diagnosis of nutrient deficiencies of citrus orange orchards in Jiroff valley of Iran. *Agrochimica* **37**(1/2) : 41-54.

Rao, Veerabhadra, K.and Ananthanarayana, R. (1997). Adsorption and desorption of aluminium in acid soils of Karnataka. *J. Indian Soc. Soil Sci.* **45**(4) : 820-829.

Rao, Y.S., Raghu Mohan, N.G. and Vasudeva Rao, A.E. (1985). Scanning electron microscopy studies of some soils in Cuddapah basin. *J. Indian Soc. Soil Sci.* **33**(4):841-845.

Rasmussen, G.K. and Smith, P.F. (1959). Effects of H-ion concentration on growth of pineapple orange seedlings in alternate solution and water cultures. *Proc. Am. Soc. Hort. Sci.* **73** : 242-247.

Rastogi, B.K. (1991a). Agricultural characteristics in the agroclimatic zones of Indian states – Bihar (S.P.Ghosh, ed.), *Agro-climatic Zone Specific Research*, ICAR, New Delhi, India, pp. 97-125.

Rastogi, B.K. (1991b). Agricultural characteristics in the agroclimatic zones of Indian states – Andhra Pradesh, (S.P.Ghosh, ed.), *Agro-climatic Zone Specific Research*, ICAR, New Delhi, India, pp. 31-60.

Rastogi, B.K. (1991c). Agricultural characteristics in the agroclimatic zones of Indian states – Karnataka (S.P.Ghosh, ed.), *Agro-climatic Zone Specific Research*, ICAR, New Delhi, India, pp. 173-200.

Rawash, M.A., Bondok, A. and El-shazly, S. (1983). Leaf mineral content of lime trees affected by foliar application of chelated zinc and manangese. *Annals Agric. Sci. Ain Shams Univ.* **28**(2):1021-1031.

Raychaudhuri, S.P. (1963). *Land Resources of India*, Vol. **1**, Planning Commission, Govt. of India, New Delhi, India.

Raychaudhuri, S.P., Agarwal, R.R., Datta Biswas, N.R., Gupta, S.P. and Thomas, P.K. (1963). *Soils of India*, ICAR, New Delhi, India.

Raychaudhuri, Masumi and Sanyal, S.K.(1999a). Potassium release characteristic of some soils of West Bengal and Sikkim. *J. Indian Soc. Soil Sci.* **47** (1) : 45-49.

Raychaudhuri, Mausumi and Sanyal, S.K. (1999b). Forms of potassium and the quantity-intensity parameters of an acid hill Ultisol after liming. *J. Indian Soc. Soil Sci.* **47**(2) : 229-234.

Razeto, B. and Villavicencio,A. (1996). Growth and properties of lemon fruit according to the season of fruit set. *Proc. Int. Soc. Citriculture.* Vol. **2,** pp. 994-996.

Razeto, B. and Salas, A. (1986). Magnesium, manganese and zinc sprays on orange trees (*Citrus sinensis* Osbeck). *Development in Plant and Soil Sciences* (A. Alexander, ed.). Martinus Nijnoff Publishers, Netherlands, Vol. **22**, pp. 255-270.

Razeto, M.B., Lengueire, M.J., Rojas, Z.S. and Reginato, M.G. (1988). Correction of manganese and zinc deficiencies in orange. *Agricultura Tecnica* **48**(4):347-352.

Ream, C.L. and Furr, J.R. (1976). Salt tolerance of some citrus species relatives and hybrids tested as rootstocks. *J. Am. Soc. Hort. Sci.* **101**:265-267.

Recupero-Reforgiato, G. and Russo, F. (1988). A trial of rootstocks for clementine 'Comune' in Italy. *Proc. 6th Int. Citrus, Congress* (R. Goren and K. Mendel, ed.), Vol. **1,** pp. 61-66.

Reddy, D.B. and Rao, Seshagiri, T. (1962). Mosambi trees need magnesium too. *Fert. News* **7**(12): 24-26, 27.

Reddy, P.S. and Swamy, G.S. (1986). Studies on nutritional requirement of sweet orange (*Citrus sinensis* Linn.) variety sathgudi. *South Indian Hort.* **34**(5):288-292.

Reddy, P.V. and Sharma, P.S. (1981). Leaf nutrients in healthy and chlorotic acid lime trees. (*Citrus aurantifolia* Swingle). *Indian J. Hort.* **39**:196-200.

Reddy, R.V.S., Rao, M.S., Ramavatharam, N. and Reddy, K.S. (1991). Chemical decomposition of sweet orange (*Citrus sinensis*) leaves of different periods of flowering and fruit development. *Indian J. Agric. Sci.* **61**(3) : 207-209.

Reese, R.L. and Koo, R.C.J. (1975). N and K fertilization effects on leaf analysis, trees, size and yield of three major Florida orange cultivars. *J. Am. Soc. Hort. Sci.* **100**:195-198.

Reeve, N.G. and Sumner, M.E. (1970). Lime requirement of Natal oxisols based on exchangeable aluminium. *Proc. Soil Sci. Soc. Am.* **34** : 595-596.

Reeve, N.G. and Sumner, M.E. (1971). Cation exchange capacity and exchangeable aluminium in Natal Oxisols. *Proc. Soil Sci. Soc. Am.* **35:**148-152.

Reisenaur, H.M. (1978). Soil and Plant Tissue Testing in California. *Bull.* **187** (rev. June, 1978), Calif. Corp. Ext. Serv., Davis, USA.

Reitz, H. J. (1970). Preliminary Exploratory Research. *Ann. Rep.* p. 155, Fla. Agr. Exp. Sta., Florida, USA.

Reitz, H.J. and Long, W.T. (1955). Water table fluctuations and depth of rooting of citrus trees in Indian river area. *Proc. Fla. State Hort. Soc.* **68** : 24-29.

Reitz, H.J., Leonard, C.D., Stewart, I., Koo, R.C.J., Anderson, C.A., Reese, R.L., Calvert, D.V. and Smith, P.F. (1972). Recommended Fertilizers and Nutritional Sprays for Citrus, *Bull.* **536B**, P. 24, Fla. Agri. Expt. Sta., Florida, USA.

Rengaswamy, P., Olson, K.A. and Kirby, J. M. (1992) Subsoil constraints to root growth and high soil water and nutrient use by plants. *Proc. National Workshop,* Tanunda, South Australia, Australia.

Rengaswamy, P. (2000). Subsoil constriants and agricultural productivity. *J. Indian Soc. Soil Sci.* **48**(4): 674 - 682.

Reuther, W. (1973a). Climate and citrus behaviour. *The Citrus Industry,* Vol. **3**, PP. 280-337, Univ. Calif., Berkeley, USA.

Reuther, W. (1973b). Monthly temperature and rainfall data for some major citrus producing regions. *Citrus Indus.* 500-504.

Reuther, W. (1977). *Citrus : Ecophysiology of Tropical Crops.* Academic Press Inc. New York, USA, p.88.

Reuther, W. (1980). Climatic effects and quality of citrus in the tropics. *Proc. Trop. Reg., Am. Soc. Hort. Sci.* **24** : 15-28.

Reuther W., Batchelor, L.D. and Webber, H. J. (1967). Commercial citrus regions of the world. *The Citrus Industry,* Vol. **1**, pp. 40-189, Univ. Calif., Berkeley, USA.

Reuther, W., Rasmussen, G.K., Hilgeman, R.H., Cahoon, G.A. and Cooper, W.C. (1969). A comparison of maturation of Valencia orange in some major subtropical zones of the United States. *J. Am. Soc. Hort. Sci.* **94**:144-156.

Reuther, W. and Rios-Castano, D. (1969). Comparison of growth, maturation and composition of citrus fruits in sub-tropical California and tropical Colombia. *Proc. 1st Int. Citrus Symp.* Vol. **1**, pp. 277-300.

Rey, J.Y. (1999). Traditional citrus growers of western and central Africa and nursery related species. *Proc. 5th World Congr. Int. Soc. Citrus Nurserymen* (B. Aubert, ed.) Montpellier, France, pp. 191-197.

Ricardo, R.P., Camara, E.M.S. and Ferreira, M.A.M. (1984). Carta de Solos da Ilha da Madeira (Versao Provisorio). Ministerio de Educacao e Secretaria de Agricultura e Pescas da Madeira, Lisboa.

Rice, L. (1878). *Mysore and Coorg Gazetteer* (III Vol. Coorg). Superintendent, Government Press, Bangalore, Karnataka, India.

Richards, L. A. (1954). Diagnosis and Improvement of Saline and Alkali Soils (ed.), U. S. Dept. Agri. *Handbook No* **60**, pp. 23-44.

Richards, L.A. (1958). Citrus fruits of hybrid origin in Ceylon. *Indian J. Hort.* **15** : 154-158.

Richardson, A., Anderson, C.A. and Dawson, T. (1991). More heat needed for high quality satsuma mandarins. *Orchardist of New Zealand* **64**(11) : 26-30.

Richtor, Rev. G. (1870). *Manual of Coorg.* Basel Mission Press, Mangalore, Karnataka, India.

Riehl, L.A. and Carman, G.E. (1953). Water spot on navel oranges. *Calif. Citrog.* **7**:7-8.

Rios, Castano, D. and Camacho-Bustos, S. (1969). Native mandarins in Colombia. *Agri. Trop., Bogota,* **25**: 456-465.

Rivero, J.M. del. (1981). Citrus Industry in Spain. *Proc. Int. Soc. Citriculture.* Vol. **2**, pp. 975-985.

Rivero, C., Senesi, N., Paolini, J. and D'Orazio, V. (1998). Characteristics of humic acids of some Venezuelan soils. *Geoderma* **81** : 227-239.

Robinson, Ben (1989). Development of soil acidity in Riverland and Sunraysia orchards. *Australian Citrus News* **65**(6) : 18-19.

Robinson, J. B. (1977). Increased growth of young Valencia orange trees in soil amendment gypsum in south Australia. *Proc. Int. Soc. Citriculture.* Vol. **1**, pp. 31-33.

Robinson, W.O., Edgington, G., Armiger, W.H. and Breen, A.V. (1951). Availability of molybdenum as influenced by liming. *Soil Sci.* **72**: 267-274.

Roderbourg, J. (1968). Influence of lucerne cover crop on the phosphorus nutrition of clementines. *Awamia* **27:** 51-63.

Rodney,D. R. (1964). Arizona citrus notes. *Calif. Citrog.* **49**: 210-212.

Rodney, D.R. (1969). Arizona citrus notes. *Calif. Citrog.* **54**: 188-189.

Rodriguez, O. (1980). *Nutricao e. adubacao dos citros.* (O. Roudriguez and F. Viega, eds.) Citriculturea Brasileira Fundacao Cargill, Sao Paulo, Brazil. Vol. **2,** pp. 385-428.

Rodriguez, O. (1985). Soil and its importance in orchard productivity. *Proc. 1st Symp. Citrus Productivity,* Jabeticabal, UNESP, Brazil, pp. 3-11.

Rodriguez, O., and Gallo, J.R. (1961). A survey of the nutritional status of citrus orchards in Sao Paulo by means of foliar analysis. *Bragantia* **20**: 1183-1202.

Rodriguez, V.A., Martinez, G.C. and Mazza de Gaiad, S.M. (1994). Foliar application of zinc in orange (*Citrus sinensis*) cv valencia Late : monthly absorption and influence on productivity. *Horticultura Argentina* **13**:(34/35):61-65.

Rodriguez, O. and Moreira, S. (1969). Citrus nutrition – 20 years of experimental results in the state of Sao Paulo, Brazil. *Proc. Ist. Int. Citrus Symp.* Vol. **3**, pp. 1579-1586.

Rodriguez, O., Moreira, Sylvio, Gallo, J.R. and Trofilo Sob, J. (1977). Nutritional status of citrus trees in Sao Paulo, Brazil. *Proc. Int. Soc. Citriculture.* Vol. **1,** pp. 9-12.

Rodriguez, O., Moreira, S. and Roessing, C. (1964). Estudo de nove practicas de cultivo do solo em pomar citrico no planalta paulista. Anais V. Semin Bras. *Herbicidass* pp. 257-258.

Rogers, J.S., Calvert, D.V., Allen, L.H. Jr., Stewart, E.H. and Yates, P. (1982). Subsurface drainage of a Spodosol for citrus production. *Proc. Speciality Conf. Environmentally Sound Water and Soil Management*(E.G.K. ruse, C.R. Burdick and Y.A. Yousef, ed.), Am. Soc. Civil Engineers, New York, USA, pp. 131-138.

Romashkevich, A.I. (1974). Soils and weathering crusts of the moist subtropics of Western Georgia. Nauka, Moscow.

Romashkevich, A.I. (1978). Relation of weathering and soil formation in the mountain soils of western Georgia. *Soviet Soil Sci.* **10**(2) : 129-140.

Romero-Arando R., Moya, J.L., Tadeo, F.R., Legaz, F., Primo-Millo, E. and Talon, M. (1998). Physiological and anatomical distrubances induced by chloride salts in sensitive and tolerant citrus: Beneficial and deterimental effects of cations. *Plant. Cell. Environ.* **21**(12): 1243-1253.

Romig, D.F., Garlynd, J.R.F., Harris, R.F. and McSweney, K. (1995). How farmers assess soil health and quality. *J. Soil Water Conserv.* **50**: 229-236.

Rosadze, G.R. (1969). The results of trials with polythene mulch. *Subtropicheskie Kul'tury* (6) : 115-118.

Rosenzweig, C., Phillips, J., Goddberg, R., Carroll, J. and Hodges, T. (1996). Potential impact of climate change on citrus and potato production in the US. *Agri. Syst.* **52**(4): 455-479.

Rosini, E. (1965). Appunti di climatologia statistica. Tipo-litografia scrola di Guerra Aerea,Firenze.

Rosnadze, G.R. (1971). The effect of different types of mulch on the root system development of young mandarin trees. *Subtropicheskie Kul'tury* (6) : 86-90.

Rossetti, V. (1981). Declinio of citrus in Brazil. A Review. *Proc. Int. Soc. Citriculture.* pp. 478-480.

Rossetti, V. and Beretta, M.J.G. (1985). Boletim do declinio de plantas citricas, Pesquisas realizadas pelo Instituto Biologico. III Parte, 1984-1985. p. 28. (Mimeo).

Rossetti, V. and Beretta, M.J.G. (1988). Declinio of citrus trees: Tentative transmission trials in Brazil. *Proc. 6th Int. Citrus Cong.* Vol. **2**, pp. 1031-1038.

Rossetti, V., Berretta, M. J.G. and Julia M. (1990). Declinio of citrus trees in Brazil. A Review (1991-1990). *Proc. Int. Citrus Symp.* Guangzhou, China, pp. 687-692.

Rossetti, V., Wutscher, H.K., Childs, J.F., Rodriguez, O., Moreira, C.S., Muller, G.W., Prates, H.S., De Negri, J.D. and Greve, A. (1980). Decline of citrus trees in the state of Sao Paulo, Brazil. *Proc. Conf. Int. Org. Citrus Virol.* Australia, 1978. (E.C. Calavan, S.M. Garnsey and L.W. Timmer, ed.), Riverside, Univ. Calif. 1980. **8**, pp. 251-259.

Rouse, R.E. and Maxwell, N.P. (1987). Long term production performance of nuclellar Valencia sweet orange cultivars in Texas. *J. Rio Grande Valley Hort. Soc.* **40** : 23-30.

Rouse, R.E., Dean, H.A. and Gautreaux, M. (1987). Long term production performance of four principal commercial sweet orange cultivars in Texas. *J. Rio Grande Valley Hort. Soc.* **40** : 15-22.

Rove, R.N. and Bradsell, D.V. (1973). Waterlogging of fruit trees. *Hort. Abst.* **43**(9) : 534-544.

Rozanov, B.G. (1970). Indications of paleohydromorphism in the genesis of Northern African soils. Tez. Dokl. IV Vsesoyuzn. Delegatsk. S yezda pochvov. *Proc. 4th All Union Cong. Soil Sci.,* Vol. **3**, p. 82. Alma-Ata.

Rozanov, B. G. (1990). Human impacts on the evoluation of soils under various ecological conditions of world. *Trans. 14th ICSS Plenary Lecture.* Int. Soc. Soil Sci., Kyoto, Japan, pp. 53-62.

Rozanov, B.G., Gewaifel, I.M. and El-Esavi, M.E. (1982). Genesis of calcareous soils of the mediterranean coast of Egypt. *Soviet Soil Sci.* **14** (2): 9-18.

Rubel, R. (1962). The citrus industry of Spain. *Punjab Hort. J.* **2**: 124-105.

Rudra, B.B. (1956). Forest soils of West Bengal (Red and Lateritic). *J. Indian Soc. Soil Sci.* **4**:255-263.

Rudramurthy, H.V. and Dasog, G.S. (2001). Properties and genesis of associated red and black soils in north Karnataka. *J. Indian Soc. Soil Sci.* **49**(2): 301-309.

Ruiz, D., Martinez, V. and Cerda, A. (1995). Citrus response to salinity - growth and nutrient uptake. *Tree physiol.* **17**(3) : 141-150.

Russell, E.J.R. (1952). *Soil Conditions and Plant Growth,* (8th ed.). Longmans, Green and Co., London.

Russo, F. (1955). Lemon culture in Italy. *Calif. Citrog.* **40**(7) : 255-278.

Russo, F. (1981). Present situation and future prospect of the Citrus Industry in Italy. *Proc. Int. Soc. Citriculture.* Vol. **2**, pp. 969-973.

Russo, F. and Raciti, G. (1954). The symptoms and control of manganese deficiency in citrus. *Ann. Spre. Agrar* **9**:871-881.

Ruthenberg, H. (1971). *Farming Systems in the Tropics.* Clarendon Press, Oxford, UK.

Saakashvili, M.A., Gochelashvili, Z.A. and Mgaloblishvili, T.S. (1971). The effect of different rates of nitrogenous fertilizers on the soil microflora in an orange plantation. *Subtropicheskie Kul'tury* (1) : 86-92.

Sabashvili, M.N. (1967). Brown forest soils of the mountains of Transcaucasia. *Soil Fermation Characteristics in the Zone of Brown Forest Soils.* Vladivostok, Georgia.

Saenz, M. A. (1966). Suelos volcanicos y cafetaleros de Costa Rica. *Univ. de Costa Rica,* p. 355.

Sagee, O. (1996). Adaption of citrus rootstocks to calcareous soils and salinity. (M. El-Otmani and A. Ait-Oubahou, ed.). *Journees Nationales Sacientifiques et Techniques sur les Agrumes*: pp. 23-25 Fevrier, Agadir, Morocco.

Sagee, O., Hasdai, D., Hamou, M. and Shaked, A. (1992). Screen house evaluation of new citrus rootstock for tolerance to adverse soil conditions. *Proc. Int. Soc. Citriculture.* Vol. **1,** pp. 299-303.

Sagee, O., Shaked, A., Hasdai, D. and Hamou, M. (1993). Rapid evaluation of new citrus rootstocks under semi-controlled conditions. *Acta Hort.* **349**:197-201.

Saha, A.K., Bandopadhyay, A.K. and Sarkar, B. (1974). Effect of different soil management practices on conservations of soil moisture and growth fruiting of lemon (*Citrus limon* Burm.). *Indian J. Hort.* **31**(4) : 337-341.

Saha, J.K., Mondal, A.K., Hazra, G.C. and Mandal, Biswapati (1991). Depthwise distribution of copper fractions in some Ultisols. *Soil Sci.* **151**(6) : 452-459.

Sahastrabuddhe, D.L. (1927). A remedy for die-back disase of orange trees. *Agric. J., India.* **22**: 114-117.

Sahu, G.C. and Nirmalya Bala (1995). Characterization and classification of soils on valley plains of middle Andaman island. *J. Indian Soc. Soil Sci.* **43**(1) : 99-103.

Said, R. (1962). *The Geology of Egypt.* Elsevier, Amsterdam, Netherlands.

Said, M. and Inayatullah, S. (1964). Sweet orange cultivation in Peshawar and Dera Ismail Khan divisions. *Punjab Fruit J.* **26/27**: 81-83.

Sakal, R. and Singh, A.P. (1995). *Micronutrient Research and Crop Production* (H.L.S. Tandon, ed.), FDCO, New Delhi, India.

Sakamoto, T. and Okuchi, S. (1969). Effects of nitrogen nutrition on the acid and soluble solids contents of satsuma orange fruits. *J. Japanese Soc. Hort. Sci.* **38**: 300-308.

Sakamoto, H., Tamaki, I. and Sogo, M. (1960a). Studies on the cultivation and management of cover crops for orchards. *Bull.* **5** pp. 161-176. *Shikoku Agric. Exp. Stat.,* Shikoku, Japan

Sakamoto, H., Tamaki, I. and Sogo, M. (1960b). Studies on the soil of sloping citrus orchards with perennial cover crops. *Bull.* **5** pp. 177-179 *Shikoku Agric. Exp. Stat.*, Shikoku, Japan

Sakamoto, T., Okuchi, S. and Yakushiji, K. (1961). Effects of long-term applications of green manure on the Satsuma orange. II. On changes in soil properties. *J. Japanese Soc. Hort. Sci.* **30**: 344-348.

Sale, P. (1994). Magnesium is an important plant nutrient. *Orchardist of New Zealand* **67**(4) : 12-13.

Sale, P. (1998). Seasonal changes in the nutritional status of citrus leaves and leaf sampling in New Zealand. *Orchardist* **71**(1) : 65-68.

Salem, A.T.M. (1991). Waterlogging tolerance of three citrus rootstocks. *Bull. Facult. Agri. Univ., Cairo.* **42**(3): 881-894.

Salem, S.E., Ibrahim, T.A., Gudy, L.F. and Myhod, M. (1995). Response of Balady mandarin trees to foliar application of iron, zinc, manganese and urea under sandy soil. *Bull. of Faculty of Horticulture, University of Cairo* **46**(2): 277-288.

Samson, J.A. (1966). Manual of citrus growing in Surinam. *Meded. Landb. Proefstat. Suriname* **39**, p.76.

Samson, J.A. (1968). Citrus cultivation in Suriname. *Veth. J. Agric. Sci.* 16 : 186-196.

San Lio, Magnano Di, G. Perrotta, G., Cacciola, S.O. and Tuttobene, R. (1988). Factors affecting soil populations of P*hytophthora* in citrus orchards. *Proc. 6th Int. Citrus Cong.* Vol **2**, pp. 767-774.

Sanchez- Garcia, C. D. and Fernandez, M. A. (1981). Climatic effects on Valencia oranges in eastern Cuba. *Proc. Int. Soc. Citriculture.* Vol. **1**, pp. 331-334.

Sanchez, P.A. and Salinas, J.C. (1981). Low-input technology for managing oxisols and ultisols in tropical America. *Adv. Agron.* **34**: 279-379.

Sanchez, A. C.(1982). Polo citricola em Croias. *Planta Citrica.* Cordeiro polis, sp, **1**:161-177.

Sanchez, P.A. (1976). Soil management under shifting cultivation. *Properties and Management of Soils in the Tropics.* Wiley, New York, USA.

Sanchez, P.A. (1994). Tropical soil fertilty research : Towards the second paradigm. *Proc. 5th World Congr. Soil Sci.,* Vol. **1**, pp. 65-88.

Sanchez, J.A. and Artes, F.A. (1981). Aptitude for citriculture of soils on kuper sediments in the region of Murcia (Spain). *Proc. Int. Soc. Citriculture.* Vol. **2**, pp. 530-533.

Sanchez, P.A., Cout. W. and Buol, S.W. (1982). The fertility capability soils classification system : interpretation, applicability and modifications. *Geoderma* **27** : 283-309.

Sanchez, C.D., Blondel, L. and Cassin, J. (1978). The influence of climate on the quality of Corsican clementines. *Fruits* **33**(12) : 811-813.

Sanchez, A., Rosales, A. and Bascones, L. (1998). Land evaluation for the orange crop. II. Soil suitability for orange production as a function of soil classification. *Agronomic Tropical* **38**(1-3) : 85-96.

Sanchez, P.A., Villachica, J.A. and Bandy, D.E. (1983). Soil fertility dynamics after clearing a tropical rainforest in Peru. *Soil Sci. Soc. Am. J.* **47** : 1171-1178.

Sandor, J.A. and Eash, N.S. (1995). Ancient agricultural soils in the Andes of southern Peru. *Soil Sci. Soc. Am. J.* **59** : 170-179.

Sandor, J.A. and Furbee, A. (1996). Indigenous knowledge and classification of soils in the Andes of southern Peru. *Soil Sci. Soc. Am. J.* **60** : 1502-1512.

Sanikidze, G.S. (1970). The quantity of microorganisms in the soil of a mandarin plantation in relation to the application of peat-mineral-ammonium fertilizers. *Subtropicheskie Kul'tury* (3) : 65-71.

Sankararaj, L., Rajendran, G., Nair, K.S. and Ratnam, C. (1983). Soils of Salem District, Tamil Nadu. Soil Survey and Land Use Organization, Coimbatore, Tamil Nadu.

Sanikidze, I.S. and Mamulaishvili, I.N. (1990). The effect of meteorological factors on mandarin cropping. *Subtropicheskie Kul'tury* (3) : 65-71.

Santibanez, F. (1986). Agroclimatologia de la zona pi saquera Chilena. Cooperation agricola control Pisquero 'Elaqui' Ltda Facultad de Agronomta U. de Chile, Santiago, Chile, p. 45.

Santibanez, F. and Uribe, J.M. (1991). Atlas Agroclimatico de Chile : Regiones Vg Metropolitana. Facultad de Ciencias Agrarias Y Forestales U. de Chile, Santiago, Chile, p. 65.

Santinoni, L.A. and Silva, N.R. (1995). Growth, yield and maturation of common mandarin under different soil management practices. *Horticultura Argentina* **14**(36) : 5-11.

Saplaco, S.R. and Payawan, P.A. (1985). Soil climate and soil erosion control : State of knowledge. *Soil Erosion Management* (P.A. Craswell, J.V. Remenyi and L.G. Nallana, ed.), *Proceeding No.* **6**, pp. 86-89, Australian Centre for Int. Agri. Res., Canberra, Australia.

Sardo, V. (1992). Integrated approach to soil conservation, irrigation and drainage in a slopping land. *Proc. Int. Soc. Citriculture.* Vol. **2,** pp. 649-650.

Sarkar, A.N. (1994). *Integrated Horticulture Development in Eastern Himalayas.* M.D. Publications Pvt. Ltd., New Delhi. pp. 23-91.

Sarkar, Dipak, Sahoo, A.K. and Nayak, D.C. (1997). Evaluation of pedological development through field morphology rating system. *J. Indian Soc. Soil Sci.* **45**:141-146.

Sarkar, Dipak, Sahoo, A.K., Mukhopadhyay, S., Das, T.H., Nayak, D.C., Shah, K.D., Banerjee, T. and Das, A.L. (2000). Land resources of the eastern region. *Souvenir.* pp. 34-39. 65th Annual covention of Indian Soc. Soil Sci., Indian Soc. Soil Sci., New Delhi, India.

Sarooshi, R., Weir, R..G. and Coote, B.G. (1991). Effect of nitrogen, phosphorus and potassium fertilizer rates on fruit yield, leaf mineral concentration and growth of young orange trees in Sunraysia district. *Australian J. Expt. Agri.* **31**(2) : 263-272.

Sartori, E. (1981). Argentina Citriculture. *Proc. Int. Soc. Citriculture.* Vol. **2**, pp. 950-964.

Satapathy, K.K. (1991). Rainfall trend at Barapani (Meghalaya). *Indian J. Hill Farming.* **4**(2):1-4.

Satapathy, K.K. (1996). Hill Slope Run-off Under Conservation Practices. *Res. Bull. No.* **40**, pp. 1-41, ICAR Research Complex for NEH Region, Meghalaya.

Sato, K. (1969). Antagonism between nickel and molybdenum in citrus. *Proc. Ist. (Vol.) Int. Citrus Symp.* Vol. **3**, pp. 1543-1550.

Sato, K., Ishihara, M., Suehiro, M., Kurihara, A. and Tabuchi, I. (1962a). Studies on Boron Deficiency and Excess in Citrus. Part I. Leaf and Soil analyses in the Chemical Composition of *Citrus natsudaidai* leaves (1957-58). *Bull.* **1**, pp. 37-44, Hort. Res. Sta., Hiratsuka, Serv. A., Japan.

Sato, K., Ishihara, M., Suehiro, M., Kurihara, A. and Tabuchi, I. (1962b). Studies on Boron Deficiency and Excess in Citrus. Part II. Boron application experiments and boron excess experiment (1957-1960). *Bull.* **1**, pp. 44-64, Hort. Res. Sta. Hiratsuka, Serv. A., Japan.

Satyanarayana, G. and Reddy, M. Ramasubha (1994). *Citrus Cultivation and Protection.* Wiley Eastern Limited, New Delhi, India, pp. 6-9.

Sauer, M. R. (1954). Flowering in the sweet orange. *Australian J. Agri. Res.* **5** : 649-654.

Sauls, Julian, W. and Rouse, Robert, E. (1985). Current status of Texas Citrus Industry. *Proc. Fla. State Hort. Soc.* **98** : 63-65.

Sautoy, N. Du., Barnard, R.O. and Groenveld, H.T. (1991). Influence of fertilization of Valencia orange trees growing on a low base status loamy sand. *Appl. Pl. Sci.* **5**(2) : 84-89.

Saxena, A.P. (1991a). Agricultural characteristics in the agroclimatic zones of Indian states – Gujarat (S.P.Ghosh, ed.), *Agro-climatic Zone Specific Research,* ICAR, New Delhi, pp. 126-143.

Saxena, A.P. (1991b). Agricultural characteristics in the agroclimatic zones of Indian states – Uttar Pradesh (S.P.Ghosh, ed.), *Agro-climatic Zone Specific Research,* ICAR, New Delhi, pp. 402-426.

Saxena, A.P. (1991c). Agricultural characteristics in the agroclimatic zones of Indian states – Madhya Pradesh (S.P.Ghosh, ed.), *Agro-climatic Zone Specific Research,* ICAR, New Delhi, pp. 228-249.

Schaffer, B. and Moon, P.A. (1991). Influence of rootstock on flood tolerance of Tahiti lime trees. *Proc. Fla. State Hort. Soc.* **103** : 318-321.

Schling-Brodersen, U. (1992). Liebigs role in establishment of agricultural chemistry. *Ambix* **39**: 21-31.

Schnitzer, M. and Skinner, S.I.M. (1963). Organo-metallic interactions in Soil: 3. Properties of iron and aluminium organic matter complexes, prepared in the laboratory and extracted from a soil. *Soil Sci.* **98:** 197.

Schreier, H., Shah, P.B. and Brown, S. (1995). Challenges in mountain resource management in Nepal. Processes, Trends and dynamics in middle mountain watershed. *Workshop Proceedings,* pp. 10-12 April, 1995, Kathmandu, Nepal.

Sehgal, J.L. and Abrol, I.P. (1994). *Soil Degradation in India. : Status and Impact.* Oxford & IBH Publishing Co. Pvt. Ltd., New Delhi, India pp. 101-32.

Sehgal, J.L. and Bhattacharjee, J.B. (1988). Typic Vertisols of India and Iraq: Their characterisation and classification. *Pedology* **38**: 67-95.

Sehgal, J.L., Kanwar, J.S., Bhumbla, D.R. and Dhingra, D.R. (1965). Soils of the proposed citrus belt of the Punjab. *J. Res. PAU.* **2**: 6-13.

Sehgal, J.L., Mondal, D.K, Mondal, C. and Vadivelu, S. (1990). Agro-ecological Regions of India. *Tech. Bull.* National Bureau of Soil Survey and Land Use Planning. *Publ.* **24,** p. 73.

Sehgal, J.L., Muhamud Al-Mishadani, Abdul Ghani, S.A. and Sigur, Rajih H. (1988). The soils of Shahrazur area (NE Iraq) for land use planning I. Characterization. *J. Indian Soc. Soil Sci.* **28**(1) : 57-71.

Sehgal, J.L. and Stoops, G. (1972). Pedogenic calcite accumulation in arid and semi-arid regions of the indo-gangetic alluvial plain and erstwhile Punjab (India) their morphology and origin. *Goederma.* **8**: 59-68.

Sehgal, J.L., Sys, S., Stoops, G. and Taveriner, R. (1985). Morphology, genesis and classification of two dominant soils of the warm temperature and humid region of Himachal Pradesh. *J. Indian Soc. Soil Sci.* **33**(4):846-847.

Sehgal, J.L., Vernenamen, C. and Tavernier, R. (1987). Agroclimatic environments and moisture regimes in north-west India - their application in soils and crop growth. *Res. Bull. No.* **17**, pp. 66-117. Nat. Bur. Soil Survey and Land Use Planning, Nagpur (Maharashtra), India.

Sen, T.K., Chamuah, G.S. and Sehgal, J.L. (1994). Occurrence and characteristics of some kandi soils in Manipur. *J. Indian Soc. Soil Sci.* **42**(2) : 297-300.

Sen, T.K., Dubey, P.N. and Chamuah, G.S.(1997a). Characteristics and classification of some soils of Barrak valley in Assam *J. Indian Soc. Soil Sci.* **45**:206.

Sen, T.K., Dubey, P. N. , Maji, A.K. and Chamuah, G.S. (1997b). Status of micronutrients in dominant soils in Manipur. *J. Indian Soc. Soil Sci.* **45**(2) : 388-390.

Sen, T.K., Nayak, D.C., Dubey, P.N., Chamauh, G.S. and Sehgal J. (1997c). Chemical and electrochemical characterization in some acid soils of Assam. *J. Indian Soc. Soil Sci.* **45(**2) : 245-249.

Sen, T.K., Nayak, D.C., Dubey, P.N., Singh, R.S., Maji, A.K., Chamuah, G.S. and Sehgal, J. (1997d) Pedology and Edaphology of Benchmark acid soils of north-eastern India. *J. Indian Soc. Soil Sci.* **45**(4) : 782-790.

Sen, T.K., Nayak, D.C. Maji, A.K. and Chamuah, G.S. (1993). *Red and Lateritic Soils of India: Resource Appraisal and Management.* NBSS *Publ.* **37**, National Bureau of Soil Survey and Land Use Planning, Nagpur, India.

Serry, A.I. (1981). Procedure for agroecologically based soil quality evaluation. *Pochvovedeniye* 7.

Servicio Para El Agricultor (1974a). Magnesium deficiency in citrus. *Noticias Agricolas* **7**(8) : 29-30.

Servicio Para El Agricultor (1974b). Correcting zinc deficiency in citrus. *Noticias Agricolas* **7**(9) : 35-36.

Sethpakdee, R. (1997). Citrus production in Thailand. *Extension Bull.* **473**, p.5, Food & Fert. Tech. Centre, Nakhon Pathom, Thailand

Seyyid, Irmak and Recep, Gundogan (2000). Characteristics of soils on terraces of the Kizilirmak river, Turkey. *J. Indian Soc. Soil Sci.* **48**(4): 862-864.

Shah, P.B. (1995). Experience in soil fertility issues from Jhikku Khola, a middle mountain watershed in Nepal. (K.D. Joshi., A.K. Vaidya, B. P. Tripathi and B. Pound, ed.), Formulating a Strategy for Soil Fertility Research in the Hills of Nepal *Proc. Workshop,* pp. 109-113. Lumle Agricultural Research Centre, Pokhara , Nepal.

Shah, R.B., Schriere, H. and Brown, S. (1991). Soil Fertility and Erosion Issues in the Middle Mountain of Nepal (ed.). *Proc. Workshop.*, Jhikku Khola Watershed, April 22-25, 1991, ISS, Kathmandu, Nepal.

Shaked, A. and Ashkenazy, S. (1984). Wingle citrumelo as a new citrus rootstock in Israel. *Proc. Int. Soc. Citriculture.* Vol. **1**, pp. 48-50.

Shaked, A., Goell, A. and Hamou, M. (1988). Screening citrus rootstocks and rootstock scion combinations for tolerance to calcareous soils. *Proc. 6th Int. Citrus Conf.* Vol. **1,** pp. 83-86.

Shalhevet, J. (1983). The tolerance of citrus to salinity. *Alon Hanotea.* **37**:347-349. (In Hebrew).

Shalhevet, J., Yaron, D. and Horowitz, U. (1974). Salinity and citrus yield – an analysis of results from salinity survey. *J. Hort. Sci.* **49**:15-27.

Shannon, M.C. (1979). In quest of rapid screening techniques for plant salt tolerance. *HortScience* **14**:587-589.

Sharma, B.B., Singh, R. and Sharma, H.C. (1974). Response of sweet orange (*Citrus sinensis* Osbeck) plants to zinc, urea and DBCP. *Indian J. Hort.* **31**(1): 38-44.

Sharma, B.K. and Singh, Nepal (2001). Characteristics and properties of soils under two dominant land use systems in western Rajasthan. *J. Indian Soc. Soil Sci.* **49**(2): 373-377.

Sharma, C.M., Minhas, R.S. and Masand, S.S. (1988) Molybdenum in surface soils and its vertical distribution in profiles of some acid soils. *J. Indian Soc. Soil Sci.* **36** : 252-257.

Sharma, D.D., Roy, B.N. and Samaddar, H.N. (1985). Prospect of sweet orange (*Citrus sinensis* L.) cultivation in West Bengal. *Indian Agricst.* **29**(3) : 203-207.

Sharma, K.K. (1990). Effect of copper and zinc sprays on the vigour of kinnow trees. *Haryana J. Hort. Sci.* **19**(3-4):232-236.

Sharma, U.C. (2000). Nutrient management in cropping systems in north-eastern hills region. *Fert. News* **45**(8) : 43-50.

Sharma, R.C. and Azad, A.S. (1991). Effect of different levels of NPK on growth, yield and quality of mandarin. *Indian J. Hort.* **48**(3):116-120.

Sharma, S.D., Banerjee, S.P., Rawat, V.R.S., Raina, A.K. and Singh, Balvinder (1989). Present productivity of soils supporting sal (*Shorea robusta*) in Doon Valley. *J. Indian Soc. Soil Sci.* **37:** 539-544.

Sharma, O.N., Gupta, K.R. and Gupta, R.K. (1986). Soil status of healthy and chlorotic citrus orchards of Jammu (J&K). *Res. Develop. Reptr.* **3**(1) : 41-44.

Sharma, V.K. and Mahajan, K.K. (1990). Studies on nutrient status of mandarin orchards in Himachal Pradesh. *Indian J. Hort.* **47**: 180-185.

Sharma, R.K. Mahajan, K.K.and Tripathi, B.R. (1993). Nutrient status of soils of Hamirpur district, Himachal Pradesh. *Indian J. Hill Farming.* **6**(2):209-211.

Sharma, S.C., Mehrotra, N.K., Gupta, M.R. and Shing, Harmail (1993). Game theory approach to a nutritional trial on sweet orange (*Citrus sinensis* Osbeck) cv Jaffa. *Punjab Hort. J.* **30**(1-4):9-12.

Sharma, P.S., Raju, D.N., Reddy, R.V. and Ajaneyalu, A. (1981). Micronutrient survey of Sathgudi (*Citrus sinensis* Osbeck) orchard in Rayalseema region of Andhra Pradesh. *South Indian Hort.* **29**:131-133.

Sharma, Y.M., Rathore, G.S. and Jesani, J.C. (1999a). Yield, yield attributes and mineral composition of seedless lemon as affected by soil and foliar application of zinc and copper. *Ann. Agri. Res.* **20**(1):64-68.

Sharma, Y.M., Rathore, G.S. and Jesani, J.C. (1999b). Effect of soil and foliar application of zinc and copper on yield and fruit quality of seedless lemon (*Citrus limon*). *Indian J. Agri. Sci.* **69**(3):236-238.

Sharma, P.K., Sehgal., J.L., Saggar, S. and Chand, K.S.(1986).Soil of the Kandi area in Panjab and their suitability for land use planning. *J. Indian Soc. Soil Sci.* **34** : 133-141.

Sharma, K.K., Sharma, K.N. and Nayyar, V.K. (1990). Effect of copper and zinc sprays on leaf nutrient concentration in kinnow mandarin (*Citrus reticulata* x *C. deliciosa*). *Indian J. Agri. Sci.* **60**(4):278-280.

Sharma, R., Sharma, E. and Purohit, A.N. (1997a). Cardamom, mandarin and nitrogen fixing trees in agroforestry system in India's systems. *Agroforestry Systems* **35**(3) : 23-253.

Sharma, R., Sharma, E. and Purohit, A.N. (1997b). Cardamon, mandarin and nitrogen fixing trees in agroforestry system in India's Himalaya region. II. Soil nutrient dynamics. *Agroforestry Systems* **35**(3): 255-268.

Sharma, S.P., Sharma, P.D., Singh, S.P. and Minhas, R.S. (1993). Characterization of Soan river valley soils in lower Siwalik of Himachal Pradesh-I hill soils. *J. Indian Soc. Soil Sci.* **41(4)** : 714-719.

Sharma, S.K. and Singh, Ranvir (1989). Photosynthetic characteristics and productivity in citrus. I. Effect of nutrition. *Indian J.Hort.* **46**(3):295-302.

Sharples, G.C. and Hilgeman, R.H. (1969). Influence of differential nitrogen fertilization on production, trunk growth, fruit size and quality and foliage composition of Valencia orange trees in central Arizona. *Proc. Ist. Int. Citrus Symp.* Vol.**3**, pp. 1569-1578.

Shata, A. (1955). An Introductory Note on the Geology of the Northern Portion of the Western Desert of Egypt. *Bull.* **5** (2), de 1' inst. du Desert d' Egypt.

Shata, A. (1960). Geological Problems Related to the Ground Water Supply of Some Desert Areas of Egypt. *Bull.* de la Soc. de Geogr. d' Egypt., **32**.

Shawky, I., Desouky, I., El-Tomi, A. and Mohamed, M.A. (1980). Effect of pH on the growth and mineral content of some rootstock seedlings. *Egyptian J. Hort.* **7**(1) : 19-31.

Shen, Z.M. (1989). The present situation and future of citrus production in China. *HortScience* **24**(6) : 904-905.

Sheng, Tin-hou (1990). Vesicular-arbuscular mycorrhizal influence on growth and phosphorus nutrition of trifoliate orange in red soil. *Proc. Int. Citrus Symp.* (H. Bangyan and Y. Qian, ed.), pp 502-505. Guanzhou, China.

Sheng, Z.M. (1991). The situation and current tasks for the citrus industry of China. *China Citrus* **20**(4) : 36-37.

Shen, Zhao Min and Li, Yin Guo (1997). Present situation and prospect of citrus research and production andstrategy for their further development in China. *J. Zhejiang Agri. Univ.* **23**(5): 485-490.

Sherchan, K.K. and Baniya, B.K. (1991). Crop production and its trend in response to soil and nutrients incomplete. Soil Fertility and Erosion Issues in the Middle Mountains of Nepal (P.B. Shah, H. Schreier, S. brown and K.W. Riley, ed.), *Proc. Workshop*, Jhikhu Khola Watershed, Nepal. International Research Centre, Ottawa, pp. 51-6.

Shih, S.F., Myhre, D.L., Schellentrager, G.W., Carlisle, V.W. and Doolittle, J.A. (1986). Using radar to assess the soil characteristics related to citrus stress. *Proc. Soil & Crop Sci. Soc. Fla.* **45** : 54-59.

Shimizu, T. and Morii, M. (1985). Influence of soil environment on nutrient deficiency of Satsuma mandarin (*Citrus unshiu*). *Environ. Con. Biol.* **23**(4) : 77-87.

Shimogori, Y., Wada, M. and Hatano, H. (1980). Soil management on satsuma mandarin orchard derived from volcanic ash soil on the growth and fruit yield. *Bull.* **33**, Miyazaki Agri. Expt. Station, Japan.

Shishov, L.L. and Kapshuk, M.P. (1984). Soil conditions on irrigated citrus plantations in the Libyan arid zone. *Problems of Desert Development.* **1** : 39-45.

Shome, K.B. and Singh, Charan (1965). Calcium, magnesium, potasium in healthy and diseased citrus leaves. *Indian J. Hort.* **22**:283-286.

Shukla, L.M. and Lyngdoh, Josplin C. (1990). Zinc status of soil of Meghalaya in relation to their characteristics. *J. Indian Soc. Soil Sci.* **38:** 315.

Shyampura, R.L. (2000). Land resources of the western region. *Souvenir.* 65th Convention of Indian Soc. Soil Sci., Indian Soc. Soil Sci., New Delhi, pp. 20-27.

Sidhu, P.S., Raj Kumar and Sharma, B.D. (1994). Characterization and classification of Entisols in different soil moisture regimes of Punjab. *J. Indian Soc. Soil Sci.* **42**(4): 633-640.

Silva, J.U.B., Sobral, L.F., Fonseca, A.J. De, Trindade, J. and Silva, L.M.S. Da (1984). Effect of different sources and doses : Splitting of nitrogen upon the growth, yield and fruit quality of Baianinha orange. *Proc. Int. Soc. Citriculture.* Vol. **1**, pp. 154-156.

Silva, J.U.B., Sobral, L.F, Trindade, J. and J. Emidio Filho (1982). Efeitos de fontes de nitrogenio na formacao de mudas de laranjeira Pera. Sociedade Brasileira de Ciencia do Solo, Campinas set Resumo (n° 35).

Sims, J.T., Cunningham, S.D. and Sumner, M.E. (1997). Assessing soil quality for environmental purpose: Roles and challenges for Soil Scientist. *J. Environ. Qual.* **26**: 20-25.

Sinclair, W.B. (1961). *The oranges its Biochemistry and Physiology,* Div. Agric Sci. Univ. Calif., USA.

Sinclair, W.B. and Bartholomew, E.T. (1944). Effect of rootstock and environment on the composition of oranges and grapefruit. *Hilgardia* **16**: 125-179.

Sinclair, W.B. and Jolliffe, V.A. (1961). Chemical changes in the juice vesicles of granulated Valencia oranges. *J. Food Sci.* **25**: 276-282.

Sinclair, H.R., Waltman, W.J., Waltman, S.W., Terpstra, H.P. and Reed-Margetan, D. (1996). Soil ratings for plant growth (SRPG). USDA-NRCS, National Soil Survey Center, Lincoln, NE.

Singer, M.A. and Ewing, S.A (2000). Soil quality, *Handbook of Soil Science* (.M.E. Sumner ed.), CRC Press Inc., Boca Raton, Florida, USA, pp. G271-G298.

Singer, Arieh, Kirsten, Willem and Buhmann, Christel (1995). Fibrous clay minerals in the soils of Namaqualand, South Africa: Characteristics and formation. *Geoderama* **66**: 43-70.

Singh, M.P. (1963a). Iron deficiency in citrus plants and its control. *Indian J. Hort.* **10**: 56-58.

Singh, D. (1963b). The Khasi orange of Assam. *Indian Hort.* **7**(2): 15

Singh, K.K. (1966). Modern trends in citrus growing in Japan. *Punjab Hort. J.* **6**: 117-122.

Singh, Daljit (1969). Citrus in India, Pakistan, Iran and Iraq. *Proc. Int. Citrus Symp.* Vol.**1**, pp.103-109.

Singh, H.P. (1989). Progress of research on citriculture with special reference to kinnow under AICRP on Tropical Fruits. *Proc. Citriculture in North-West India.* pp.71-81, Ludhiana, Punjab, India,

Singh, U.K. (1994). Some rainfall features of north-east India. *Vayumandal.* I-VI, pp. 12-16.

Singh, K.D. (1995). *Soil Nutrient Management* (P.K, Chhonkar, G. Narayanaswamy and R.K. Rattan, ed.), Division of Soil Science and Agric. Chemistry, IARI, New Delhi, pp. 82-87.

Singh, M.V. (1998). AICRP on micro and secondary nutrients and pollutants in soils and plants. *28th Progress Report (1996-98)*, Indian Inst. Soil Sci., Bhopal, India, pp. 1-137.

Singh, Room (1999c). 65 year research on citrus granulation. *Proc. Nat. Semi. New Horizons in Production and Post-harvest Management of Tropical and Subtropical Fruits* (Room Singh, ed.), Hort. Soc. India, New Delhi, India, pp. 113-114.

Singh, M.V. (1999a). *National Symposium on Zinc Fertilizer Industry-Whither To* (Ramendra Singh and Abhay Kumar, ed.), Fertilizer Association of India, Gurgaon, Harayana, India.

Singh, M.V. (1999b). Micronutrient management. *Fifty Years of Natural Resource Management Research.* (G.B. Singh and B. R. Sharma, ed.), ICAR, New Delhi, India, pp. 177-198.

Singh, M.P. and Agarwal, K.C. (1961). Studies on die-back in citrus. *Indian J. Hort.* **18**: 295-301.

Singh, G.N. and Agrawal, H.P. and Singh, M. (1989). Genesis and classification of soils in an alluvial pedogenic complex. *J. Indian Soc. Soil Sci.* **37**(1):343-354.

Singh, H.P., Chadha,K.L. and Bhargava, B.S. (1990c). Leaf sampling technique in acid lime (*Citrus aurantifola* Swingle) for nutritional diagnosis. *Indian J. Hort.* **47**:133-139.

Singh, M.I. and Chohan, G.S. (1982). Effect of nutritional sprays on granulation in citrus fruits cv. Wilking mandarin. *Punjab Hort. J.* **22**(1/2):52-59.

Singh, M. and Dahiya, S.S. (1975). Effect of $CaCO_3$ and iron on the availability of iron in a light textured soil. *J. Indian Soc. Soil Sci.* **23** : 247-252.

Singh, B.P., Das, Madhumita and Prasad, R. N. (1990a). Evaluation of available Cu status in high altitude in wetland soils. *J. Indian Soc. Soil Sci.* **38:** 464.

Singh, B.P., Das, Madhumita and Prasad, R.N. (1990b). Assessment of available soil Zn in wetland on high altitude Haplaquent. *J. Indian Soc. Soil Sci.* **38**(3): 551-554.

Singh, O.P. and Datta, B. (1983). Characteristics of some hill soils of Mizoram in relation to altitude. *J. Indian Soc. Soil Sci.* **31**(4) : 657-661.

Singh, O.P. and Datta, B. (1987). Phosphorus status of some hill soils of Mizoram in relation to pedogenic properties. *J. Indian Soc. Soil Sci.* **35:** 699.

Singh, O.P. and Datta, B. (1989). Morphology, physical and physico-chemical properties of hill soils of Mizoram in relation to altitude. *Indian J. Hill Farming.* **2**(1):9-20.

Singh, O.P., Datta, B. and Das, A. (1986). Study of humic matter in characteristics of hill soils. *J. Indian Soc. Soil Sci.* **34**(1) : 125-132.

Singh, O.P., Datta, B. and Rao, C.N. (1991). Biochemical characterization and genesis of soils in relation to altitude in Mizoram. *J. Indian Soc. Soil Sci.* **39**(4) : 739-750.

Singh, A.K., Dkhar, Grace, P. and Nongkynrih, P. (1997). Effect of some soil properties on availability of micronutrient in Entisols of Meghalaya. *J. Indian Soc. Soil Sci.* **45** (3) 581-583.

Singh, N.T. and Gajja, B.L. (1987). *Ecological Considerations and Agricultural Development of A&N Islands.* Cent. Agril. Res. Inst., Portblair, A&N Islands, pp. 1-55.

Singh, G. P. and Ghosh, S. P. (1965). Studies on the growth behaviour of different citrus species and cultivars in the nursery. *Indian. J. Hort.* **22 :** 266-276.

Singh, C.P., Gupta, V.K. and Ravi Kumar (1997b). Micro and secondary nutrient status of some kinnow orchards in semi-arid parts of Haryana. *Haryana J. Hort. Sci.* **26** (1-2): 58-60.

Singh, K.K. and Jawanda, J.S. (1963). Punjab can produce more citrus fruits. *Punjab Hort. J.* **3**: 3-11.

Singh, J.P., Karwasra, S.P.S. and Singh, Mahendra (1988). Distribution and forms of copper iron, manganese, and zinc in calcareous soils of India. *Soil Sci.* **146**(5) : 359-366.

Singh, S., Krishanamurthy, S. and Katyal, S.L. (1967). *Fruit Culture in India.* ICAR, New Delhi, pp-162.

Singh, D. and Lal, G. (1946). Kankar composition as an index of the nature of soil profile. *Indian J. Agri. Sci.* **19**: 328.

Singh, A.R., Maurya, V.N., Pande, N.C. and Rajput, R.S. (1989). Role of potash and zinc on the biochemical parameters of Kagzi lime (*Citrus aurantifolia* Swingle). *Haryana J. Hort. Sci.* **18**(1-2):46-50.

Singh, Ranvir and Misra, K.K. (1980). Effect of foliar application of micronutrients on kinnow mandarin. *Punjab. Hort. J.* **20**(3-4):143-148.

Singh, S.B. and Misra, R.S. (1981). Performance of some orange varieties in Garhwal hills. *Prog.Hort.* **13**(1):75-83.

Singh, Ranvir and Misra, K.K. (1985). Effect of varying doses of N P K on growth and yield of lemon (*Citrus limon* Burm). *Prog. Hort.* **17**(2):95-99.

Singh, S.B. and Misra, R.S. (1986). Effect of plant growth regulators and micronutrients on fruit drop, size and quality of kinnow orange. *Prog. Hort.* **18**(3-4): 260-264.

Singh, K. and Mongia, A.D. (1993). Distribution of zinc, copper, manganese and iron in soils of Andaman and Nicobar Islands. *Agrochimica* **37** : 18-25.

Singh, N.T., Mongia, A.D. and Ganeshmurthy, A.N. (1988). Soils of Andaman and Nicobar Islands. *Bull No.* **1**, Cent. Agri. Res. Inst., Port Blair, Andaman & Nicobar Islands, India.

Singh, S.P. and Nayyar, V.K. (1999). Available boron status of some alluvium derived arid and semi-arid soils of Punjab. *J. Indian Soc. Soil Sci.* **47**(4): 801-802.

Singh, S.P. and Nayyar, V.K. (1999). Available boron status of some alluvium derived arid and semi-arid soils of Punjab. *J. Indian Soc. Soil Sci.* **47**(4) : 801-802.

Singh, C.S., Partriam, Ram, M., Prasad, R.N. and Prasad, A. (1998). Studies on nutrient status of mandarin orchards (*Citrus reticulata* Blanco) in Meghalaya. *J. Hill Res.* **11**(1) : 32-37.

Singh, B., Rethy, P. and Prasad, A. (1990). Effect of certain micronutrients on the physical and biochemical parameters of Kagzi lime. *Prog. Hort.* **22**(1-4):216-219.

Singh, M.V. and Saha J.K. (1995). AICRP on Micro and Secondary Nutrients and Pollutants in Soils and plants. *26th Annual Report (1996-98)*, Indian Inst. Soil Sci., Bhopal, India, p. 107.

Singh, M. and Sekhon, G.S. (1991). DTPA extractable micronutrient cations in twenty soil series of India. *J. Indian Soc. Soil Sci.* **38** : 129-133.

Singh, M.V. and Subba Rao, A. (1995). Manganese research and agicultural production. *Micronutrient Research and Agicultural Production.* (H.L.S. Tandon, ed.), FDCO, New Delhi, India, pp. 83-114.

Singh, L. and Singh, S. (1942). Citrus rootstock trials in Punjab. *Indian J. Agric. Sci.* **12**: 38, 99.

Singh, L. and Singh, S. (1944). Citrus rootstock trials in Punjab. II. The influence of different rootstocks in the blood red orange. *Indian J. Agric. Sci.* **4**: 95-100.

Singh, M.P. and Singh, R.P. (1953). Studies on dying back in citrus. *Indian J. Hort.* **10**(1): 1-8.

Singh, L. and Singh, S. (1972). Chemical and morphological composition of kankar nodules in soils of the Vindhya an region of Mirzapur India. *Goederma* **7**: 269.

Singh, B.P. and Singh, D. (1976). Effect of boron on vegetative growth characters and physico-chemical composition of Kagzi lime. (*Citrus aurantifolia* Swingle). *Bangladesh Hort.* **4**(1):29-31.

Singh, Roop and Singh, Ranjit (1978a). Effect of soil and climate on granulation in kinnow mandarin. *Punjab Hort. J.* **18**:114-118.

Singh, A. and Singh, M.D. (1978b). Effect of various stages of shifting cultivation on soil erosion from steep hill slopes. *Indian Forester* **106:**115-121.

Singh, Room and Singh, Ranjit (1979). Effect of soil and climate on granulation in kinnow mandarin. *Punjab Hort. J.* **19**(3/4) : 114-118.

Singh, Ranjit and Singh, Room (1980). Relationship between granulation and nutrient status of kinnow mandarin at different localities. *Punjab Hort. J.* **20**(3-4):134-139.

Singh, A. and Singh, M.D. (1981). Soil Erosion Hazard in North-eastern Hill Region. *Res. Bull* No. **10**, pp. 1-57, ICAR Research Complex for NEH Region, Meghalaya.

Singh, A.K., Singh, R.N. and Diwakar, D.P.S. (1993). Lime requirement of acid soils. *J. Indian Soc. Soil Sci.* **1**(1):52-154.

Singh, Ratan and Singh D.V. (1983). Effect of lime applied as calcite limestone in acid soils of Kumaon hills in Uttar Pradesh. *J. Indian Soc. Soil Sci.* **31**(1):148-151.

Singh, B.P., Singh, B.N., Das, Madhumita and Prasad, R. N. (1989). Appraisal of fertility status of valley soils in Meghalaya. *J. Indian Soc. Soil Sci.* **37**: 591.

Singh, P.N., Singh, R.P. and Singh, Ranvir (1985). Seasonal changes in vegetative growth of two cultivars of lemon (*Citrus limon* Burm.) in relation to weather parameters and soil temperature. *Harayana J. Hort. Sci.* **14**(1/2) : 37-40.

Singh, K.P. and Sinha, H. (1975). Fixation of boron in relation to certain soil properties. *J. Indian Soc. Soil Sci.* **23** : 227-230.

Singh, K.P. and Sinha, H. (1976). Availability of boron in relation to certain soil properties. *J. Indian Soc. Soil Sci.* **24** : 403-408.

Singh, Awtar, Srivastava, A.K., Dass, H.C. and Vijayakumari, N. (1997). Screening of germplasm of citrus rootstocks for salinity tolerance. *Indian J. Hort.* **54**(4):283-287.

Singh, Shyam, Srivastava, A.K. and Huchche, A.D. (1999). Citrus Flowering. *Tech. Bull.* No. **4**, pp. 1-144, National Research Centre for Citrus, Nagpur, Maharashtra, India.

Singh, R.P., Srivastava, R.K. and Srivastava, R.P. (1976). Potentialities for exploitation of citrus wealth in Uttar Pradesh hills, India. *Fruit Varieties J.* **30**(3) : 77-79.

Singh, S.P., Takkar, P.N., Kaur, N.P. and Nayyar, V.K. (1980). Availability of boron in alluvial soils of Punjab. *J. Res. (P.A.U.).* **17:** 350.

Singh, V.N., Tiwary, J.R., Mall, J. and Mishra, B.B. (1996). Irrigability and productivity classifications of some soils of Chotanagpur plateau. *J.Indian Soc. Soil Sci.* **44**(1): 179-182.

Singh Vinay and Tripathi, B.R. (1983a). Available micronutrient status of citrus growing soils of Agra region. *J. Indian Soc. Soil Sci.* **31**(2):310-312.

Singh Vinay and Tripathi, B.R. (1983b). Forms of sulphur in citrus growing soils of Agra region in Uttar Pradesh. *J. Indian Soc. Soil Sci.* **31**(3):482-485.

Singh, Vinay and Tripathi, B.R (1985). Studies on chlorosis in Sweet orange in Agra region of Uttar Pradesh. *J. Indian Soc. Soil Sci.* **33**: 333-338.

Singh, S.P., Walia, C.S. and Dhankar, R.P. (2000). Land resource base of the northern region. *Souvenir,* 65th Annual Conv., Indian Society of Soil Sci., New Delhi, pp. 12-19.

Singhal, R.M. and Sharma, S.D. (1985). Study of the soils of Doon valley. *J. Indian Soc. Soil Sci.* **33**:627-634.

Sinha, A.K. and Sengupta, B.N. (1995). Identification and acreage estimation of orange orchards in parts of Nagpur district using remote sensing technique. *Orissa J. Hort.* **23**(1/2):1-4.

Sites, J.W., Hammong, L.C., Leighty, R.G. and Johnson, W.O. (1961). Information to consider in the use of soils of flatwoods and marshes for citrus. *Fla. Agri. Expt. Sta. Circ.* **S-135**, p. 35.

Skoric, A., Racz, Z., Kovacevic, P. and Modric, A. (1963). A detailed study of the main Yugoslav soil types. Rept. Fed. Res. Develop. Board, Beograd, p. 56.

Smaling, E.M.A. and Janssen, B.H. (1987). Soil fertility. Nutrient availability. *Soils of the Kilifi Area, Kenya* (H.W. Boxem, T. de Meester and E.M.A. Smaling, ed.), Agric. Res. Rept. **929**, pp. 109-117, 131-132, Pudoc, Wageningen.

Smith, D.D. (1951). Subsoil conditioning on clay pans for water conservation. *Agr. Eng.* **31**:503-505, 532.

Smith, P. F. (1971). Hydrogen ion toxicity on citrus. *J. Am. Soc. Hort.* **96:**462- 463.

Smith, P.F. (1957). Studies on the growth of citrus seedlings with different forms of nitrogen in solution cultures. *Pl. Physiol.* **32** : 11-15.

Smith, J.F. (1964). Yuma Mesa citrus cover crops. *Calif. Citrogr.* **49**: 135, 150-2.

Smith, P.F. (1962). A case of sodium toxicity in citrus. *Proc. Fla. State Hort. Soc.* **75**: 120-124.

Smith, P.F. (1966). Citrus nutrition. *Temperature Tropical Fruit Nutrition* (N.F. Childers, ed.), Somerset Press, Somerset, London, pp. 174-207.

Smith, P.F. (1967). Yield expectancy and the basis of citrus fertilization. *Proc. Fla. State. Hort. Soc.* **79**: 115-119.

Smith, P. F. (1974b). History of citrus blight in Florida. *Citrus Industry* **55**(9): 13-14,16, 18-19; (10):9-10, 13, **14**(11): 12-13.

Smith, P.F. (1969). Cu, Zn and Mn status of soil and leaves, ten years after differential soil applications of metals and lime in a young Valencia orange grove. *Proc. Soil & Crop Sci. Soc. Fla.* **28** : 97-103.

Smith, P.F. (1974a). Zinc accumulation in the wood of citrus trees affected with blight. *Proc. Fla State Hort. Soc.* **87**: 91-95.

Smith, P.F. (1973). A comparison of zinc sources for citrus on a light soil. *Commun. Soil Sci. Pl. Anal.* **4**(1) : 17-22.

Smith, P.F. (1975). Calcium requirement of citrus. *Commun. Soil Sci. Pl. Anal.* **6**(3) : 245-260.

Smith, O.E. and Beville, B.C. (1964). Drainage of citrus land in Florida. *Agric. Engg., St. Joseph, Mich.* **45**: 674-675.

Smith, P.F. and Rasmussen, G.K. (1959). Field trials on the long term effect of single applications of copper, zinc and manganese on Florida sandy citrus soil. *Proc. Fla. State Hort. Soc.* **73** : 42-49.

Smith, Guy D. (1986) *The Guy Smith Interviews : Rational for Concepts in Soil Taxonomy* SMSS Tech, Monogram, **11**.

Smith, J.L., Halvaorson and Papendick, R.J. (1994). Using multiple variable indicator for evaluating soil quality. *Proc. Soil Sci. Soc. Am.* **57** : 743-749.

Smith, P.F. and Reuther, W. (1951). The response of young Valencia orange trees to differential boron supply in sand culture. *Pl. Physiol.* **26** : 110-114.

Smoot, J.J., Houck, L.G. and Johnson, H.B. (1971). Market Diseases of Citrus and Other Subtropical Fruits. Handb. No. **398**, *U.S. Dept. Agri.* USA.

Smoyer, K.M. (1951). The puzzling micro-elements. *Calif. Citrog.* **36**: 134, 152-56.

Soares Filho, W. dos S., Miranda Filho, J.B. De and Cunha Sobrinho, A.P. Da (1984). Rootstocks for Nata sweet orange (*Citrus sinensis* (L.) Osbeck) : A phenotypic stability study. *Proc. Int. Soc. Citriculture* Vol. **1**, pp. 59-65.

Soil Survey Staff (1992) *Keys to Soil Taxonomy.* 5th edn. SMSS Tech. Monogr. 19, Blacksburg, Virginia, USA.

Sojka, R.E. and Upchurch, D.R. (1999). Challenges for identifying best management practices- Integrating and emerging modern technologies and philosophies. Best Soil Management Practice for Production (L.D. Currie *et al.,* ed.), *Occational Report* **12**, pp. 11-28, Fertilizer and Lime Research Centre, Massey University, Palmerston North, New Zealand.

Sonwalkar, M.S. (1965). Santra cultivation regains its pride in Madhya Pradesh. *Indian Hort.* **9**(2): 21-22.

Soprano, E., and Koller, O.L. (1991). Efeito de niveis de pH do solo sobre o crescimento de porta- enxerto de citros. *Congresso Brasileiro de Ciencia do Solo*, 23., Porto Alegre, SBCS-UFRGS, p. 63.

Soprano, E. and Koller, O.L. (1996). Effect of lime incorporation depth and lime levels on the development of two citrus rootstocks under greenhouse conditions. *Proc. Int. Soc. Citriculture.* Vol. **2**, pp. 832-835.

Soulez, P. and Forique, A. (1978). Citrus phenology in the tropics. *Fruits* **33**(12) : 811-813.

Southwick, S. M. and Davenport, T. L. (1986). Characterization of water stress and low temperature effects on flower induction in citrus. *Pl. Physiol.* **81** : 26-29.

Soyza, D. J. De (1955). Agricultural conditions in the Uva. *Trop. Agric.* **111** (2): 80-91.

Spencer, W.F. (1963). Phosphorus Fertilization of Citrus. *Bull.* **653**, p. 48., Fla. Agri. Expt. Sta., USA.

Spencer, W.F., and Koo, R.C.J. (1962). Calcium deficiency in field grown in citrus trees. *Proc. Am. Soc. Hort. Sci.* **81**: 202-208.

Spencer, W.F. and Wander (1960). A comparison of magnesium sources for young orange trees. *Proc. Fla. State Hort. Soc.* **73** : 28-35.

Spina, P. (1975). Citrus growing Italy. *Annala dell' Instituto Sperimentale per l' Agrumicoltura* **5**: 89-98.

Spina, P. and Giudice, V. Lo. (1975). Notes on citrus growing in Morocco. *Annala dell' Institute Sperimentale per l' Agrumicoltura* **6** : 63-109.

Spina, P., Russo, F. and Scuderi, A. (1980). Climate in relation to the quality and quantity of the citrus crop. *Frutticoltura* **42**(3/4) : 19-35.

Sprengel, C. (1826). About plant humus, humic acids and salts of humic acids. *Archiv fur die gesammate Naturlehre* **8**: 145-220.

Sprengel, C. (1837). Die Bodenkunde oder die Lehre vom Boden (Soil Science and its doctorines). *Immanuel Muller Publ. Co.*, Leipzig, Germany.

Sprengel, C. (1838). Die Lehre von den Urbarmachungen und Grundverbesserungen (The Science of Cltivation and Sil Aelioration). *Immanuel Muller Publ. Co.,* Leipzig, Germany.

Spurling, M.B. (1963). The citrus industry in south Australia. *World Crops.* **15**: 20-27.

Spurling, M.B. (1969). Citrus in the Pacific area. *Proc. I[st] Int. Citrus Symp.* Vol. **1,** pp. 93-101.

Srivastava, A.K. (2001) Soil and nutrition management in citrus orchards of northeast India. *Citrus decline and its Management in Northeast Region India.* (S.G.Gupta and Shyam Singh, ed.), National Reserch Centre for Citrus, Nagpur, Maharashtra, India, pp. 47-66.

Srivastava, K.C. and Bopaiah, M.G. (1978). Prospectus of growing kinnow mandarin in the southern states. *Punjab Hort. J.* **18**: 139-141.

Srivastava, A.K. and Kohli, R.R. (1997a). Soil and climate adaptability of citrus cultivars in India. *Nat. Semi. Orchard Management for Sustainable Production of Tropical Fruits.* March 10-11, 1997 Patna, (Bihar), India.

Srivastava, A.K. and Kohli, R.R. (1997b). Soil suitability criteria for citrus : An appraisal. *Agri. Rev.* **18**(3): 139-146.

Srivastava, A.K. and Kohli, R.R. (1999). Agro-ecological approach for land use planning in citrus. *Agri. Rev.* **20**(1): 41-47.

Srivastava, A.K., Kohli, R.R., Dass, H.C. and Ram, Lallan (1993). Physico-chemical properties and fertility status of mandarin growing soils. *National Seminar on Developments in Soil Science,* Central Soil and Water Conservation Research and Training Institute, Dehradun, October 6-10, 1993.

Srivastava, A.K., Kohli, R.R., Dass, H.C., Ram, Lallan and Huchche, A.D. (1995). Relationship of leaf K with N status of mandarin orchards. *Indian J. Hort.* **52**(1): 234-238.

Srivastava, A.K., Kohli, R.R. and Huchche, A.D. (1997a). Citrus nutrition – A retrospection. *Agri. Rev.* **18**(2):128-138.

Srivastava, A.K., Kohli, R.R. and Huchche, A.D. (1997b). Relationship between leaf K and soil K forms at critical growth stages of Nagpur mandarin (*Citrus reticulata* Blanco). *J. Potash Res.* **13**(1):80-92.

Srivastava, A.K., Kohli, R.R., Ram, Lallan and Dass, H.C. (1998). Relationship between chloride accumulation in leaf and cation exchange capacity of roots of citrus species. *Indian J. Agri. Sci.* **68**(1): 39-41.

Srivastava, A.K. and Lallan Ram, (2000). Irrigation water quality of Nagpur mandarin orchards in central India. *Indian J. Agri. Sci.* **70**(10) : 679-681.

Srivastava, K.C., Muthappa, D.P. and Ganapathy, M.M. (1977). Further results of micronutrients spay trial on Coorg mandarin (*Citrus reticulata* Blanco). *Proc. Int. Citrus Symp. Bangalore* (K.L. Chadha and R.N.Pal, ed), pp. 55-64.

Srivastava, M.K., Sharma, A.K. and Suman Kumar (1986). Survey and classification of some soils of Nainital tarai using aerial photography. *J. Indian Soc. Soil Sci.* **34**:230-232.

Srivastava, A.K., Shirgure, P.S. and Shyam Singh (1999). Citrus soils – their nature and properties. *Intensive Agriculture* **37**(13-14):28-31.

Srivastava, A.K. and Singh, Shyam (1997a). Soil fertility constraints of citrus orchard in India and their mmanagement. *Symposium on Soil Fertility and Environment,* Allahabad University, December 18-20, 1997, Allahabad, U.P., India.

Srivastava, A.K. and Singh Shyam, (1997b). Nutrient management in Nagpur mandarin and Acid lime. *Bull. No.* **5**, pp. 1-12, National Research Centre for Citrus, Nagpur, Maharashtra, India.

Srivastava, A.K. and Singh, Shyam (1998). Fertilizer Use Efficiency in Citrus. *Extension Bull. No.* **5**, pp. 1-64, National Research Centre for Citrus, Nagpur, Maharashtra, India.

Srivastava, A.K. and Singh, Shyam (1999a). Soil suitability criteria for evaluating the optimum productivity of *Citrus reticulata* Blanco, cultivar Nagpur mandarin. Abst. IX^{th-} *Congress of International Society of Citriculture,* December 3-7, 2000, Orlando, USA., pp. 138-139.

Srivastava, A.K. and Singh, Shyam (1999b). Soil suitability-crop response relationship to Nagpur mandarin. Abst. *Int. Symp. Citriculture,* pp. 374-388, 23-27, November, 1999. Nagpur, Maharashtra, India.

Srivastava, A.K. and Singh, Shyam (1999c). Recent trends in citrus production research. *Proc. Int. Symp. Citriculture,* pp. 339-356 Nagpur, Maharashtra, India (Pub. 2000).

Srivastava, A.K. and Shyam Singh (2000a). Identification of suitable soils for sustained productivity of Nagpur mandarin. *Ad hoc Project Report,* ICAR, New Delhi, India, pp. 1-106.

Srivastava, A.K. and Singh, Shyam (2000b). Calcium nutrition of Nagpur mandarin (*Citrus reticulata* Blanco) in black clay soil types. *Abst.* National Sem. Developments in Soil Science-2000. 65th Annual Conv. of Indian Society of Soil Science, Nat. Bur. Soil Survey and Land Use Planning, Nagpur, Maharashtra, India, p. 80.

Srivastava, A.K. and Singh, Shyam (2000c). Soil fertility evaluation of Nagpur mandarin orchards in central India. National seminar on Agriculture - Scenario, Challenges and Opportunities, *Extended Summeries,* Gwalior, M.P., India, pp. 31-32.

Srivastava, A.K. and Singh, Shyam (2000d). Delineation of suitable soils for sustained productivity of sweet orange (*Citrus sinensis* Osbeck), cultivar mosambo. *Ann. Rep.* (2000-2001) pp. 113-115. National Research Centre for Citrus, Nagpur, Maharashtra, India.

Srivastava, A.K. and Singh, Shyam (2001a). Soil properties influencing yield and quality of Nagpur mandarin (*Citrus reticulata* Blanco). *J. Indian Soc. Soil Sci.* **49**(1) : 226-229.

Srivastava, A.K. and Singh, Shyam (2001b) Vertical variation in soil K forms and qualty of Nagpur mandarin (*Citrus retiuluta* Blanco) in relation to soil type. Int. Symp. on Importanic of Potassium in nutrient management for suistanable crop production in India. Dec. 3-5, 2000-2001 New Delhi, India. Extended summary Vol. **1**, pp. 417-419.

Srivastava, A.K. and Singh, Shyam (2001c). Soil factors influencing Nagpur mandarin on Developments in soil science - 2001 GEM Annual Conv. of Indian Soc. Soil Sci., Rajasthan Coll. of Agri. Rajasthan, India. Oct 30-Nov. 2, 2001.

Srivastava, A.K. and Singh, Shayam (2001d). Development of optimum soil property limits in relation to fruit yield and quality of *Citrus reticulata* Blanco. Cv. Nagpur mandarin. *Trop. Agri. (Trinidad)* **78** (1) : 1-8.

Srivastava, A.K, Singh, Shyam, Huchche, A.D. and Lallan Ram (2000). Yield based leaf and soil test interpretations for Nagpur mandarin in central India. *Commun. Soil Sci. Pl. Anal.* **32**(3&4): 585-599.

Srivastava, A.K., Singh, Shyam, Lallan Ram, Kohli, R.R., Huchche, A.D. and Dass, H.C. (1997a). Relationship of vertical variation of soil properties with flowering of Nagpur mandarin (*Citrus reticulata* Blanco). *Proc. Nat. Symp. on Citriculture,* Indian society of citriculture, Nagpur pp. 131-136.

Stagman, J.G. (1978). *An outline of the Geology of Rhodesia.* Rhodesian Geological Survey *Bull.* No. **80**, Government Printer, Hararae, Zimbabwe.

Stanton, D.A. and Basson, W.D. (1970). The sulphur status of citrus orchards in Southern Africa. *S. Afr. Citrus J.* **434** : 9-11.

Stathakopoulos, N. P. and Erickson, L C. (1967). The effect of temperature on bud break in *Poncirus trifoliata* (L.) Raf. *Proc. Am. Soc. Hort. Sci.* **89**:222-227.

Stearns, Charles, R., Jr. and Young, G.T. (1942). The relation of climatic conditions to colour development in citrus fruit. *Proc. Fla. State Hort. Soc.* **84** : 264-266.

Sterling Davis, F.T., Bingham, E.R. Shade and Grass, L.B. (1969). Water relations and salt balance of 1000 acre citrus watershed. *Proc. Ist. Int. Citrus Symp.* Vol.**3**, pp. 1771-1778.

Stewart, J. and Leonard, C.D. (1953). The cause of yellow tipping in citrus leaves. *Citus Indus.* **34**(3) : 11-13.

Stewart, I., Wheaton, T.A. and Reese, R.L. (1968). Murcott collapse due to nutritional deficiencies. *Proc. State Hort. Soc.* **81** : 15-18.

Stocking, M.A. (1984). *Erosion and Soil Productivity : A Review.* Soil Conservation Programme, Land and Water Development Div., FAO, Rome, Italy.

Stocking, M.A. (1986). *The Cost of Soil Erosion in Zimbabwe in Terms of the Loss of Three Major Nutrients.* Consultant Working Paper No. **3**, Soil Conserv. Programme, Land and Water Development Div., FAO, Rome, Italy.

Stofella, P.J., Pelosi, R.R. and Williams, B.J. (1986). Potential of citrus - cowpea intercropping in south Florida. *Proc. Interamerican Soc. Trop. Hort.* **30** : 165-170.

Stolzy, L.H. and Harding, R.B. (1966). Foliar absorption of boron in sprinkler irrigated citrus. *Calif. Agri.* **20:** 6-7.

Stoorvogel, J.J. and Smaling, E.M.A. (1990). *Assessment of Soil Nutrient Depletion in Sub-Saharan Africa* : 1983-2000. Report No. **28**. The Winand Staring Centre, Wageningen, Netherlands.

Storie, R.E. (1954). Land classification as used in California. *Proc. 5th Int. Cong. Soil Sci.,* Leopoldville, Vol **3** pp.407-412.

Storie, R.E. and Harradine,F. (1958). Soils of California. *Soil Sci.* **85:** 207-222.

Stork, N.E. and Eggleton, P. (1992). Invertebrates as determinants and indicators of soil quality. *Am. J. Alter. Agri.* **7** : 38-47.

Strauss, G.R. (1963). Magnesium nitrate sprays proving effective. *Calif. Citrog.* **48**: 299.

Strauss, G.R. (1966). Comparing two methods of sawdust application. *Citrog.* **51**: 226.

Stylianou, Y. and Orphanos, M.I. (1970). Irrigation of shamouti oranges with saline water. *Tech. Bull.* 6, p.23, Cyprus Agric. Res. Inst., Cyprus.

Subbi Reddy, G., Naidu, L.G.K., Reddy, R.S., Madhavi, M. Bhakar, B.P., Gopal, K. and Rama Krishna Rao, A. (1999). Characterization of sweet orange (*Citrus sinensis* Osbeck) growing soils and their soil-related constraints for production in Anantpur tract, India. *Proc. Symp. Citriculture,* pp. 419-425 Indian Society of Citriculture, Nagpur, Maharashtra, India.

Subramanian, C.K. (1960). Zinc and manganese sprays for citrus. *South Indian Hort.* **8** (3/4): 20-23.

Suman Kumar and Sharma, A.K. (1987). Numerical classification of some hill soils of Uttar Pradesh. *J. Indian Soc. Soil Sci.* **35**(2):465-473.

Sumner, M.E., Shahandeh, H., Bouton, J. and Hammel, J. (1986). Amerlioration of an acid soil profile through deep liming and surface application of gypsum. *Soil Sci. Soc. Am. J.* **50** : 1254-1258.

Sun, B., Zhang, T. and Zhao, Q. (1995). Integrated evaluation of improverishment of soil nutrients in red earth hilly region in south China. *Soils* **27**(3) : 119-128.

Sun, Gui Chun (1988). Navel orange cultural techniques in Japan. *South China Citrus.* **27**(6): 7-8.

Surza, R., Janicki, L., Wiltbank, W.J., Rojas, M. and Alliaga, F. (1982). Comparison of sweet orange quality in diverse tropical climates of Bolivia. *HortScience* **17**(6) : 974-976.

Suzuki, T., Kinpara, T., Sakaikbara, M. and Fukai, N. (1972). The effect of applications of phosphate and lime on the growth and fruiting of satsuma trees. *J. Japanese Soc. Hort. Sci.* **41**(2) : 157-164.

Swartz, R. E., Arguelles, T., Monsted, P., Wutscher, H. and Termachuka, L. (1980). Studies on the cause of fruta bolita or declinamiento diseases of citrus in Argentina. (E. C. Calavan, S. M. Garnsey and L. W. Timmer, ed.). *Proc. 8th Int. Organ. Citrus Virol.,* Univ. California, Riverside, USA, pp. 241-250.

Swietlik, D. (1989). Effects of soil amendment with viterra hydrogel on establishment of newly planted grapefruit trees cv. Ruby Red. *Commun. Soil Sci. Pl. Anal.* **20**(15-16) : 1697-1705.

Swietlik, D. (1996a). Response of citrus trees in Texas to foliar and soil Zn application. *Proc. Int. Soc. Citriculture.* Vol. **2**, pp. 772-776.

Swietlik, D. (1996b). USA – Texas. *Proc. Int. Soc. Citriculture.* Vol. **2**, pp. 1265-1268.

Swift, M.J., Kang, B.T., Mulongoy, K. and Woomer, P.W. (1991). Organic-matter management for sustainable soil fertility in tropical cropping systems. Evaluation for Sustainable Land Management in the Developing World (J. Dumanski, E. Pushparajah, M. Latham and R. Myers, ed.), *IBSRAM Proc. No. 12*, Vol. **2**, pp. 307-326, IB SRAM, Bangkok, Thailand.

Swingle, W. T. and Reece, P. C. (1967). The botany of citrus and its wild relatives. *The Citrus Industry* (W. Reuther, H. J. Webber, and L. D. Batchelor, ed.), Vol. **1**, pp. 190-430, Univ. Calif Press, Davis, USA.

Sys, Ir. C., Ranst, E. Van, Debaveye, Ir. J. and Beernaert, F. (1993). Land evaluation : Crop requirements. *Agricultural Publications No.* **7**, Part III. General Administration for Development Corporation Place du Champ de Mars 5 bte 57-1050 Brussels- Belgium, pp. 52-55.

Syvertsen, J.P., Bausher, M.G. and Albrigo, L.G. (1980). Water relations and related leaf characteristics of healthy and blight-affected citrus trees. *J. Am. Soc. Hort. Sci.* **105**:431-434.

Syvertsen, J.P., Bowman, B. and Tucker, D.P.H. (1990). Salinity in Florida citrus production. *Proc. Fla. State Hort. Soc.* **102** : 61-64.

Syvertsen, J.P. and Yelenosky, G. (1988). Salinity can enhance freeze tolerance of citrus rootstock seedlings by modifying growth, water relations and mineral nutrition. *J. Am. Soc. Hort. Sci.* **113**: 889-893.

Tabain, F.and Miljkovic, I. (1978). Citrus growing region in Croatia. *Jagoslovensko Vocorstvo* (1978) **12** (43): 13-23.

Tadros, T.M. (1956). An Ecological Survey of the Semi-arid Coastal Strip of the Western Desert of Egypt. *Bull.* **6** (2), de 1' Inst. du Desert d' Egypte, Egypt.

Taher, M. (1986). Physiographic frame work of north-east India. *J. NE India Geogr. Soc.* **18**(1-2) : 1-19.

Tai, E.A. and Storey, W.B. (1968). Trinidad to Spanish missionaries. *Calif. Citrog.* 53 : 420-424.

Takagi, M., Ida, K. and Yano, T. (1963). The physical and chemical properties of the soil of *Citrus unshiu* orchards in relation to productivity. *J. Sci. Soil Manure.* **34** : 177-180.

Takahara, T., Iwagak, I. and Ono, S. (1988). Fruit quality variation in a tree and orchard of Kawano Natsudaiai. *Bull. No.* **10**, pp. 25-33. Fruit Tree Res. Sta. Series D., Kuchinotsci, Japan.

Takahashi, S., Okada, N. and Shirai, T. (1969). Study of soil improvement by applying phosphorus mixed with lime. Part 3. Analysis of nutrient elements (Ca, Mg, Zn, B) in young fruit trees and treated soils. *Bull. No.* **8**, pp. 39-49, Shizuoka Prefectural Citrus Expt. Sta., Komagoe, Japan.

Takahashi, S., Okada, N. and Shirai, T. (1971). Research on chemical soil improvement in citrus groves. I. The effects of adding lime and phosphate. *Bull. No.* **9**, pp. 68-82, Shizuoka Prefectural Citrus Expt. Sta., Komagoe, Japan.

Takatsuji, T. and Ishihara, M. (1980). Studies on the potassium nutrition of satsuma III. Influence of cold hardiness and leaf composition. *Bull. No.* **7**, pp. 45-62, Fruit Tree Res. Sta. (Yatabe), Ibaraki, Japan.

Takebayashi, T., Kataoba, T. and Yukinaga, H. (1993). Classification of 98 different citrus species and cultivars by fruit quality and their disorders as related to delayed harvest. *J. Japanese Soc. Hort.* **62**(2):305-316.

Takkar, P.N. (1982). Micronutrients, forms contents, distribution in profile, indices of availability and soil test methods. *Review of Soil Research in India,* Indian Society of Soil Science, New Delhi, India, Part I, pp. 361-391.

Takkar, P.N. and Biswas, A.K. (1998). *50 years of Natural Resources Management Research* (G.B. Singh and B.R. Sharma, ed.), Division of Natural Resource Management, ICAR, Krishi Bhavan, New Delhi, India, pp. 115-144.

Takkar, P.N. and Nayyar, V.K. (1981). Preliminary field observation of mangnese deficiency. *Fertilizers* **26**: 22.

Takkar, P.N. and Randhawa, N.S. (1978). Micronutrients in Indian Agriculture. *Fert. News* **23**(8) : 3-26.

Tal, D. and Monselise, S.P. (1965). Corky (silvery) spots on the rind of citrus fruits. *Israel J. Agric. Res.* **15**: 73-81.

Talakvadze, K.B. (1977). Foliar and soil nutrition of citrus trees with urea and minor elements. *Hort. Abst.* **47**(3):267.

Talati, N.R. and Agarwal, S.K. (1974). Distribution of various forms of boron in north-west Rajasthan soils. *J. Indian Soc. Soil Sci.* **22** : 262-268.

Talibudeen, O. (1974). The nutritional potential of the soil. *Soil Fert. Lond.* **37:**11.

Tamagadge, D.B., Gaikwad, S.T., Gajbhiye, K.S. and Gaikwad, M.S. (1999). Soil landform relationship on basaltic terrain in north deccan plateau, Satpura range, Madhya Pradesh. *J. Indian Soc. Soil Sci.* **47**(1):118-124.

Tamagadge, D.B., Gaikwad, S.T., Nagabhusana,S.R., Gajbhiye, K.S., Deshmukh,S.N. and Sehgal, J.L. (1996). Soils of Madhya Pradesh: Distribution, Properties, Potentials and Problems of Optimal Land Use. *NBSS publ. Soils of India Series* **6**, Nat. Bur. Soil Survey and Land Use Planning, Nagpur, Maharashtra, India.

Tanaka, J. (1958). The origin and dispersal of citrus fruits having their centre of origin in India. *Indian J. Hort.* **15** : 101-114.

Tanbara, K. (1970). On the physical properties of orange orchard soils in Ehime prefecture, Shikoku Island, Japan. *Soil Sci. Pl. Nutr.* **16** : 55-59.

Tanbara. K. and Kurihara, H. (1963). On that relationship between productivity and the physical properties of citrus orchard soils. *J. Soil Manure* **34** : 327-330.

Tanbara, K., Kurihara, H and Sone, T. (1964). On the relationship between productivity and physical properties of satsuma orange soils. Part 2. On the characteristics of the soils in the agricultural region of the Selonaikai area. *J. Sci. Soil Manure* **35**: 341-376.

Tandon, H.L.S. (1987). *Fertilizer recommendations for Horticulture Crops in India - A Guide Book.* Fertilizer Development and Consultation Organisation, C-110, Greater Kailash, New Delhi, India.

Tavartkiladze, C.K. (1969). The application of various forms of green manure in young citrus plantations. *Subtropicheskie Kul'tury* (6) : 138-148.

Tavdgiridze, G.H. and Putkaradre, Sh. A. (1991). Effect of nitrogen fertilizer type on polarity coefficient of mandarin trees. *Subtropicheskie Kul'tury* (4) : 66-67.

Tembhare, B.R. and Rai, M.M. (1967). Effect of pH, calcium carbonate, texture and organic matter on the availability of manganese. *J. Indian Soc. Soil Sci.* **15** : 251-256.

Thaper, A.R. (1972). Review of citrus industry in northeast India. *I_{st} All India. Seminar on Citriculture.* Nov. 10-13, 1972, Nagpur, Maharashtra, India.

Thomas, G.W. and Peaslee, D.E. (1973). *Testing Soils for Phosphorus* (L.M. Walsh and J.D. Beaton, ed.), Soil Testing and Plant Analysis Council Soil Sci. Soc. Amer., Wisconsin, USA, pp.115-132.

Thompson, J.G. and Purves, W.D. (1978). A Guide to the Soils of Rhodesia. *Rhodesia Agri. J. Tech. Handbook No.***3**, Government Printer, Harare, Zimbabwe.

Thorne, D.W., Wann, F.E. and Robinson, W. (1950). Hypothesis concerning lime induced chlorosis. *Soil Sci. Soc. Am. Proc.* **15**: 254-258.

Thornthwaite, C.W. (1948). An approach towards rational classification of climates. *Geogr. Rev.* **38** : 1-10.

Thornton, I.R. (1975). Will citrus grow on limey soils. *Australian Citrus News* **50**(8) : 6.

Thrower, L.B. (1959). Report to the Government of Indonesia of horticultural production research. *FAO Report No.* **1029**.

Timmer, L.W. (1979). Early performance of Star Ruby grapefruit on a rootstocks in a fine textured calcareous soil. *HortScience.* **14**(3) : 225-227.

Timmer, L.W., Brlansky, R.H., Graham, J.H., Sandler, H.A. and Agostini, J.P. (1986). Comparison of water flow and xylem plugging in declining and in apparently healthy citrus trees in Florida and Argentina. *Phytopathol.* **76** : 707-711.

Tiwari, S.C. and Sharma, G.D. (1998). Altitudinal variation in dehydrogenese and urease activity and microbial population in soils of eastern Himalayan highlands. *J. Hill Res.* **11**(1):22-25.

Tiwari, S.N., Sinha, H. and Mandak, S.C. (1967). Potassium in Bihar soils. *J. Indian Soc. Soil Sci.* **15** : 73.

Todd, James, W. (1987). Current and future trends in new citrus plantings. *Proc. Fla. State Hort. Soc.* **100** : 386-387.

Ton, L.D. and Alfonso, J.A. (1980). Citrus growing in Honduras. *Proc. Trop. Reg. Am. Soc. Hort. Sci.,* **24** : 15-28.

Torres, M. R. and Rios- Castano, D. (1968). Citrus fruit growth and ripening in the Valle del Cauca (Coloumbia). *Proc. Am. Soc. Hort. Sci. Trop. Reg.* **12** : 107-125.

Trehan, S.P. and Grewal, J.S. (1985). Suitability of NH_4 HCO_3 - DTPA soil test for P, K, Ca, Mg, Zn, Cu, Fe and Mn in acidic and alkaline soils. *J. Indian Soc. Soil Sci.* **33** : 721-724.

Triboi, A.M. (1978). Nitrate reductase : an indicator of molybdenum malnutrition. *Fruits* **33**(12) : 831.

Tripathi, B.P. (1996). Research fertility status and further research strategy in western hills of Nepal. *Proc. Workshop Soil Fertility and Plant Nutrition Management* (D. Joshy, ed.), Dec. 19-20, Godawari Village Resort, Lalitpur, Nepal.

Tripathi, B.P. and Harding, A. H. (2001). Nutrient status of mandarin trees in some mandarin growing pockets in Lamzung and Gorkha districts of Nepal. *J. Indian Soc. Soil Sci.* **49**(3). 503-506.

Tripathi, B.P. and Shah, R. (1984). Use of Agricultural Lime in Acid Soil (in Nepali). *Agricultural Bulletin,* Bimontly (Kartik-Mansir) (1984). Agricultural Information Division, Department of Agriculture, Harihar Bhawan, Lalitpur, Nepal.

Tripathi, D., Singh, Karan and Upadhyay, G.P. (1994). Distribution of micronutrients in some representative soil profiles of Himachal Pradesh. *J. Indian Soc. Soil Sci.* **42**(1) : 143-145.

Tripathi, D., Singh, Karan, and Upadhyay, G.P. (1994). Distribution of micronutrients in some representative soil profiles of Himachal Pradesh. *J. Indian Soc. Soil Sci.* **42**(1) : 143-145.

Truog, E. (1946). Soil reaction influence on the availability of plant nutrients. *Proc. Soil Sci. Soc. Am.* **11:** 05-308.

Truog, E. (1948). Lime in relation to availability of plant nutrients. *Soil Sci.* **65**:1-7.

Tsanava, N.G. and Burchuladze (1974). The effect of the nutritional regime on changes in free amino acids in mandarin fruits. *Subtropicheskie Kul'tury* (3) : 49-53.

Tsanava, N.G., Lominadze, Sh. D. and Kastornova, E.N. (1989). Modelling the nitrogen nutrition of lemons in Western Georgia. *Agrokhimiya* (3) : 9-14.

Tsanava, N.G., Lominadze, Z.K. and Tsirekidze, M.A. (1979). Conversion of fertilizers nitrogen in the soil and utilization by mandarin trees. *Subtropicheskie Kul'tury* (4) : 51-57.

Tubelis, A. and Salibe, A.A. (1988). Relationships between production of Hamlin orange trees and the monthly rainfalls at the plateau of Botucatu. *Proc. 6th Int. Citrus Congr.* Vol. **1,** pp. 497-501.

Tubelis, A. and Salibe, A.A. (1989a). Production of Hamlin orange trees budded on sunki mandarin rootstocks and monthly rainfalls at the plateau of Botucatu, Brazil. *Pesquisa Agropecuaria Brasileira* **24**(7) : 787-792.

Tubelis, A. and Salibe, A.A. (1989b). Relationships between production of Hamlin orange trees on Rangpur lime rootstocks and the montly rainfalls at the plateau of Botucatu, Brazil. *Pesquisa Agropecuaria, Brasileira.* **24**(7) : 801-806.

Tubelis, A. and Salibe, A.A. (1991). Relationships between productivity of Hamlin orange trees on Florida rough lemon rootstock and monthly rainfall on Botucatu plateau. *Scientifica (Jaboticabal)* **19**(1) : 207-219.

Tubelis, A., Salibe, A.A. and Pressim, G. (1999). Relationship between production of Westin sweet oranges and rainfall at Botucatu, Sao Paulo state, Brazil. *Pesquisa Agropecuaria Brasileira* **34**(5): 771-779.

Tucker, D.P.H. and Reuther, W. (1967). Seasonal trends in composition of processed valencia and navel oranges from major climatic zones of California and Arizona. *Proc. Am. Soc. Hort. Sci.* **90**: 529-540.

Tucker, D.P.H., Davis, R.M., Wheaton, T.A. and Futch, S.H. (1990). A nutritional survey of south central southeast and coast flatwoods citrus grooves. *Proc. Fla. State Hort. Soc.* **103** : 324-327.

Tull, J. (1933). *Quoted in History of Soil Science.* (I.A. Krupenikov, ed.), Oxonian Press Pvt. Ltd., New Delhi, India, p. 89.

Turco, R.F., Kennedy, A.C. and Jawson, M.D. (1994). Microbial indicators of soil quality. (J.W. Doran *et al.*, ed.), Defining Soil Quality and Sustainable Environment. *Soil Sci. Soc. Am. Spec. Publ.* **35,** pp. 73-90. SSSA, Madison, WI.

Turner, A. (1982). Soil survey of Willacy county, Texas. U.S. Dept. Agril Soil Consev in Cooperation with Texas Agri. Expt. Sta., Nat. Coop. Soil Survey, Washington DC, USA.

Turner, F.T. and Patrick, W.M., Jr. (1968). Chemical changes in waterlogging and soils as a result of oxygen depletion. *Proc. 9th Int. Cong. Soil Sci.,* Vol. **4**, pp. 53-65, Adelaide, Australia.

Turrell, F.M., Monselise, S.P. and Austin, S.W. (1964a). Effect of climatic districts and of location in tree on tenderness and other physical characteristics of citrus fruits. *Bot. Gaz.* **125**:158-170.

Turrell, F.M., Orlando, I. and Austin, S.W. (1964b). Researchers forge a link between rind oil spot and foggy weather. *Western Fruit Grower* **18** : 17-18.

Turton, C.N., Vaidya, A., Tuladhar, J.K. and Joshi, K.D. (1996). *Towards Sustainable Soil Fertility Management in the Hill of Nepal.* Lumle Agricultural Research Centre, P.O. Box 1, Pokhara, Kaski, Nepal and Naturals Research Institute, Central Avenue, Chatham Maritime Kent ME4 4TB, UK.

Tuzcu, O., Kaplankiran M., Ozsan, M., Gezeral O. and Hizal A.Y. (1986). Nutritional status of citrus orchards in the Mediterranean region of Turkey. *Fruits* **41**(1) : 49-54.

Tuzcu, O., Ozsan, M., Gezerel, O. and Kaplankiran, M. (1981). The general situation regarding citrus mineral nutrition in the Mediterranean region I. *Hort. Abst.* **56**(6) : 223

UNEP (1990). *World Map on Status of Human Induced Soil Degradation.* Boomruygrok, Haarlem, United Nations Environmental Programme Netherlands.

Unger, P.W. (1979). Effects of deep tillage and profile modification on soil properties, root growth, and crop yields in the United States and Canada. *Geoderma* **22** : 275-295.

Upadhyay, G.P., Karan Singh and Bhandari, A.R. (1993). Phosphate adsorption in some representative soil groups of north-west Himalayas. *J. Indian Soc. Soil Sci.* **41**(3) : 434-439.

Upadhyay, R.C. and Patiram (1994). Qualitative behaviour of Sikkim mandarin (*Citrus reticulata* Blanco). *J. Hill Res.* **7**(1): 23-26.

Upadhyay, R.C. and Patiram (1996a). Nutrient status of mandarin orange (*Citrus reticulata* Blanco) in Sikkim. *J. Hill Res.* **9** (2) : 375 – 379.

Updhyay, R.C. and Patiram (1996b). Response of mandarin orange (*Citrus reticulata*) to manure and fertilizer in Sikkim. *J. Hill Res.* **9**(1) : 135-138

Uppal, D.K., Grewal, S.S. and Channa, Y.R. (1989). Inherent problems in kinnow mandarin in comparison to other citrus varieties of north-western India. *Citriculture in North-West India,* Punjab Agricultural University, Ludhiana (Punjab), India, pp. 109-110.

Uppal, N.L. (1962). *Irrigation and Drainage.* Land Reclamation, Irrigation and Power Research Institute, (Published), Punjab, India.

Urushadze, T.F. and Gradusov, B.P. (1976). Clay minerals in the forest soils of Georgia. *Soviet Soil Sci.* **8** (4): 582-592.

Urushadze, T.F. and Mkheidze, Ye. A. (1971). The soils of the arid thin-forest areas of Georgia. *Pochvovedeniye* (6).

Vadivelu, S., Mishra, J.P., Bhaskar, B.P., Baruah, U., Jigimon, T., Sarkar, D. and Butte, P.S. (2000). Land resources of the north-eastern region. *Souvenir* pp. 40-44. 65th Annual Convention of Indian Soc. Soil Sci., Indian Soc. Soil Sci., New Delhi, India.

Vaidya, A., Turton, C., Joshi, K.D. and Tuladhar, J. K. (1995). A System Analysis of Soil Fertility Issues in the Hills of Nepal: *Implication for Future Research.* Challenges in Mountain Resource Management in Nepal : Processes, Trends and Dynamics in Middle Mountain Watershed. *Porc.* pp. 63-80. (H. Schreier., P. B. Shah. and S. Brown, ed.), Kathmandu, Nepal during April 10-12, 1995.

Valicenti, V. (1962). Citrus in Matera province. *Inf. Ortofruttic.* **3**: 237-239.

Valicenti, V. (1977). Aspects of citrus growing in the province of Matera. *Italia Agricola* **114**(1) : 79-89.

Valle, Valdes, N. Del and Rios Albuerne, C. (1974). Characteristics of Valencia orange (*Citrus sinensis*) grown on two soils of different texture. *Centro Agricola.* **1**(3) : 61-68.

Van Der Phoeg, R.R., Bohm,W. and Kirkham, M.B. (1999). On the origin of theory of mineral nutrition of plants and Law of the Minimum. *Soil Sci. Soc. Am. J.* **63**: 1055-1062.

Vanderweyen, A. (1963). Sand burn of citrus. *Al Awamia* **6**: 127-133.

Van Huyssteen. (1987). Verslag oor besoek aan Sondagsriviervallein verband met wortlverspreiding en grondvoorbereiding (28 Sept – 1 Oct 1987). Viticultural and enollogical Research Institute, Stellenbosch.

Van Zyl, J. L. and Van Huyssteen, L. (1987). Root prunning. *Deciduous. Fruit Grower.* pp. 20-25.

Vanniere, H. (1990). Advice on the cultivation of grapefruit Star Ruby in Corsica. *Fruits (Paris)* **45**(1) : 583-589.

Varasi, M.S. (1962). Citrus dieback - its cause and control in Hyderabad state. *Poona Agric. College, Mag.* **52**: 21-26.

Vardarajan, B.S. and Subramanian, C.K. (1953). Mandarin oranges in Coorg. *Indian J. Hort.* **10**(3): 89-97.

Vardi, A.P., Spiegel-Roy, P., Ben-Hauuin, G, Neumann, H. and Shalhevet, J. (1988). Response of 'Shmouti' orange and 'Minneoal' tangelo on six rootstocks to salts stress. *Proc. 6th Int. Citrus Cong.* Vol. **71**, pp. 75-82.

Vasalmidze, A.P. (1969). Some problems of improving the mechanization of establishment and maintenance work in citrus under mountainous conditions. *Subtropicheskie Kul'tury* (4) : 109-112.

Vaswani, L.K. (1990). Agri-climatic regional planning: Concept and approach. *Fert. News* **35**(6):11-14.

Velayutham, M. M., Reddy, K. C. and Sonkar, G. R. M.(1985). All India Coordinated Research Project on soil test-crop response corelation and its impact on agricultural production. *Fert. News* **30**(4) : 81.

Venkateshwarlu, J. (1987). Efficient resource management systems for drylands of India. *Advances in Soil Science* (B.A. Stewart, ed.), Springer-Verlag, New York Inc., **7** : 165-223.

Veregara, I., Schalscha, B.E., Ruiz, I. and Wallihan, E.F. (1993). Nutritional status of representative citrus orchards in Chile as evaluated by leaf and soil analysis. *HortScience* **8**(4) : 325-326.

Veretennikov, A.V. (1964). Effect of waterlogging of soil on transpiration capacity of trees. *Soviet Pl. Physiol.* **11** : 231-234.

Verma, B.L. and Rama Deo (1993). Prediction of evaporation losses from bare sandy loam soils in north-east arid region of Rajasthan. *J. Indian. Soc. Soil Sci.* **4**(3): 541-543.

Verma, T.P., Sudhakar Rao, R.V., Mahapatra, S.K. and Tarsem Lal (2001). Studies on soil variability across different landscapes in Etawah district of Uttar Pradesh - A case study. *J. Indian Soc. Soil Sci.* **49**(2) : 309-315.

Verma, T.S. and Tripathi, B.R. (1982). Profile morphology and physico-chemical properties of the soils from hot and dry foot hill zone of Himachal Pradesh. *J. Indian Soc. Soil Sci.* **30**(4) : 574 –576.

Veroney, R.P., VanVeen, J.A. and Paul, E.A.(1981). Organic C dynamics in grassland soils. 2. Model validation and simulation of the long term effects of cultivation and rainfall erosion. *Canadian J.Soil Sci.* **61**: 211-224.

Viets, F.H. (1962). Chemistry and availability of micronutrients. *J. Agric. Food Chem.* **16** : 174-178.

Vijayasankar Reddy, R., Seshagiri Rao, M., Ramavatharam, N. and Srinivasa, K. Reddy (1992). Quality of sweet orange grown on three different soil orders. *J. Indian Soc. Soil Sci.* **40**: 111-113.

Vinay Shankar and Sinha, M.M. (1987). Studies on physico-chemical changes in hill orange and kinnow mandarin under U.P.hill conditions. *Prog. Hort.* **19**(3&4):183-188.

Vinay, Singh and Tripathi, B.R. (1985). Studies on chlorosis in sweet orange in Agra region of Uttar Pradesh. *J. Indian Soc. Soil Sci.* **33** : 333-338.

Vinnik, M. (1949). *On Citrus Fertilization.* Hassadeh Special Publication (Hebrew), p.27.

Visser, W.C. (1969). Mathematical models in soil productivity studies exemplified by the response to nitrogen. *Soil* **30:**161.

Visser, W.C. and Kowalik, P. (1974). *Plant Analyses and Fertilizer Problems.* (J. Wehrmann, ed.), *Proc. 7th Int. Colloq.* Vol. **2,** pp. 473-503.

Viti, G.C., Donadio, L.C., Malavolta, E. and Cabrita, J.R.M. (1993a). Effects of lime and phosphogypsum on citrus. *Int. Colloq. Optimization Plant Nutrition* (M.A.C. Fragoso, ed.), Kluwer Academic Publ., Netherlands, pp. 449-452.

Viti, G.C., Donadio, L.C., Malavolta, E. and Cabrita, J.R.M. (1993b). Influence of soil and leaf applications of micronutrients on yield and fruit quality of *Citurs sinensis* Osbeck, variety Pera. *Int. Colloq. Optimization Plant Nutrition* (M.A.C. Fragoso and M.L. Van Beusichem,ed.), Kulwer Academic Publ., Netherlands, pp. 453-456.

Volk, G.M. (1954). Formation of plowsole pans in Florida soils. *Fla. State Hort. Soc. Proc.* **66** : 138-141.

Von Uexkull, H.R. and Bosshart, R.P. (1989). Management of acid upland soils in Asia. (E.T. Craswell and E. Pushparajah, ed.). Management of Acid soils in the Humid Tropics of Asia. *ACIAR Monograph No.* **13,** IBSRAM , Australian Centre for International Agricultural Research, Canberra, Australia, pp. 2-19.

Vorasoot, N. (1988). Agro-climatology in north-east Thailand (C. pairinitra, K. Wallapapan, J.F. Parr and C.E. Whitman, ed.). *Soil Water and Management Systems for Rainfed Agriculture in north-east Thailand.* Khon Kaen University, Thailand, pp. 40-49.

Vu, J.C.V. and Yelenosky, G. (1991). Photosynthetic responses of citrus trees to soil flooding. *Physiologia Plantarsum* **81**(1) : 7-14.

Wada, M., Shimogoori, Y. and Hatano, H. (1981). Behaviour of surface applied cations in the soil of a satsuma mandarin orchard. *Proc. Int. Soc. Citriculture.* Vol. **2**, pp. 544-548.

Wadia, D.N. (1966). *Minerals of India*, National Book Trust of India, New Delhi.

Wahab, M.A. (1969). *Soil Formation in Relation to Sedimentation in the Lower Nile Valley.* Ph.D. Thesis, Faculty of Agriculture, Ain Sham University, Cairo, Egypt.

Wahba, S.A. (1972). *Soil Formation in Relation to Sedimentation in the Nile Valley from Giza to Minya.* Thesis, Faculty of Agriculture, Cairo University, Cairo, Egypt.

Wahlberg, H.E. (1967). Citrus and king of Siam. *Calif. Citrog.* **53**(2): 66-68.

Wakabayashi, S., Ochiai, S., Yamazaki, K. and Furuya, K. (1988). Ecological studies on *Citrus junos* growing in Utsunomiya, North Kanto. Shoot and leaf growth, lower and fruit set, leaf and fruit drop, fruit growth and leaf number, and root growth and cool temperature. *Bull. of Agri., Ustsunomiya Univ.* **13**(3) : 31-43.

Waksman, S.A. (1942). Liebig-The humus theory and the role of humus in plant nutrition. Liebig and after Liebig (F.R. Moulton, ed.), *Publication No.***16**, pp. 56-63, Amercian Association for the Advancement of Science, Washington, DC, USA.

Walia, C.S. and Chamuah, G.S. (1993). Soils of reverine plain in Arunachal Pradesh and their suitability for some agricultural crops. *J. Indian Soc. Soil Sci.* **43**:425-429.

Walia, C.S. and Chamuah, G.S. (1996). Genesis, characteristics and taxonomic classification of some red soils in Bundelkhand region of Uttar Pradesh. *J. Indian Soc. Soil Sci.* **44**(3):476-481.

Walia, C.S. and Rao, Y.S. (1996). Genesis, characteristics and taxonomic classification of some red soils in Bundelkhand region of Uttar Pradesh. *J. Indian Soc. Soil Sci.* **44**(3): 476-481.

Walia, C.S. and Chamuah, G.S. (1997). Characteristics and classification of some soils derived from different parent materials in Arunachal Pradesh. *J. Indian Soc. Soil Sci.* **45**(2) 401-404.

Walker, R. R. and Douglas, T. J. (1983). Effect of salinity levels on uptake and distribution of chloride, sodium and potassium ions in citrus plants. *Australian J. Agric. Res.* **34**: 145-163.

Walker, R. R., Terokflaby, E., Grieve, A. M. and Prior, L. D. (1983a). Water relations and ion concentrations of leaves on salt stressed citrus plants. *Australian J. Pl. Physiol.* **10**: 265-277.

Walker, R.R., Torofalvy, E. and Grieve, A.M. (1983). Water relations and ion concentrations of leaves on salt stressed citrus plants. *Australian J. Pl. Physiol.* **10**: 265-277.

Wallace, A. (1995). Agronomic and horticultural aspects of iron and the law of maximum, *Iron Nutrition in Soils and Plants* (J. Abadia, ed.), Kluwer Academic Publishers, Netherlands, pp. 207-216.

Wallace, A. and Wallace, G.A. (1993). Limiting factors, high yields and law of the maximum. *Hort. Rev.* **15** : 413-452.

Wallace, A., Alaxander, G.V. and Kinnear, J. (1982). Frequency distribution of zinc in leaves with and without zinc deficiency symptoms, all collected from a single tree. *Soil Sci.* **134**:45-50.

Wallace, A., El-Gazzar, A., Ramney, E.M. and Alexander, G.V. (1975). Emission spectrographer analysis of spot samples of citrus from different locations in Egypt. *Egyptian J. Hort.* **2**(2) : 233-241.

Wallihan, E.F. (1961). Effect of sodium bicarbonate on iron absorption by orange seedlings. *Pl. Physiol.* **36**:52-53.

Wallihan, E.F. (1964). Citrus in the Philippines. *Citrog.* **49**: 438-442.

Walsh, J. (1991). Preserving the Options : Food Productivity and Sustainability. *Issues in Agriculture No.* **2**, CGIAR, World Bank, Washington DC, USA.

Wander, I.W. (1952). Soil reaction or pH. *Calif. Ctirog.* **37**(9): 358.

Wander, L.W. (1954). Sources contributing to subsoil acidity in Florida citrus gorves. *Proc. Am. Soc. Hort. Sci.* **64** : 105-110.

Wang, Gui Chang (1999). Cultural techniques for high quality production of Jiaogan mandarin cultivar in a mountainous orchard. *South China Citrus* **28**(2):9-10.

Wang, T.C. (1985). Application of fertilizers to satsumas based on leaf analysis *J. Soil Sci.* **16**(6):275-277.

Wang, X. (1994). Application of soil taxonomic classification in the evaluation of soil resources. *Recent Treatise on Chinese Soil Taxonomic Classification* (Z. Gong, ed.), Science Press, Beijing, pp. 475-481.

Wang, X. (1995). Monitoring and evaluation of soil changes under different land use systems in red soil hilly region of China. Ph.D. Thesis. Inst. Soil Sci., Chinese Academy of Sciences, Nanjing, China, p. 164.

Wang, X. and Chen, H. (1993). Some biogeochemical characteristics of tea soil system. *Soils* **25**(4) : 196-200.

Wang, X. and Gong, Z. (1995). Ecological effects of different land use patterns in red soil hilly region. *Pedosphere* **5**(2) : 163-170.

Wang, X. and Gong, Z. (1996). Monitoring and evaluation of soil changes underland use of different patterns at a small regional level in south China. *Pedosphere* **6**(4) : 373-378.

Wang, X. and Gong, Z. (1998). Assessment and analysis of soil quality changes after eleven years of reclamation of sub-tropical China. *Geoderma* **81** : 339-355.

Wang, X.Y., Xi, Y.F. and Wang, Y.J. (1997). Endogenous harmones in relation to granulation of Ponkan mandarin fruit. *Acta Agri. Zhejiangensis.* **9**(2):103-105.

Warkentin, B. (1995). The changing concept of soil quality. *J. Soil Water Conserv.* **50**: 226-228.

Warner, R.M. (1966). Photoperiod growth response of five citrus rootstocks in Hawaii. *Proc. 27th Int. Hort. Congr.* Vol. **1**, p. 602.

Watson, J.P., Spector, J. and Jones, T.A. (1958). Soil and Land Use Surveys No. **3**, St. Vincent. Regional Res. Centre, University of West Indies, Trinidad, p. 70.

Watson, J.S. (1994). Soil organic matter management in Thailand. *Soil Science and Sustainable Land Management in the Tropics* (J.K. Syers and D.L. Rimmer, ed.), CAB International British Society of Soil Science, U.K., pp. 206-214.

Watts, F.C. and Stankey, D.L. (1980). Soil survey of St. Luice county area, Florida, U.S. Dept. Agr., Soil Consev. Serv., USA.

Wear, J.I. (1956). Effect of soil pH and Calcium uptake of zinc by plants. *Soil Sci.* **81** : 311-315.

Weather, L.G. (1955). A bud union and rootstock disorder of *Troyer citrange* with Eureka lemon tops. *Pl. Dis. Reptr.* **39**: 665-669.

Webber, H. J. (1943). Plant characteristics and climatology. (W. Reuther, H. J. Webber and L. D. Batchelor, ed.). *The Citrus Industry*, Vol. **1,** pp. 41-69. Univ. of Calif. Press, Davis, USA.

Webber, R.T.J. (1969). Citrus leaf analysis. *Australian Citrus News* **45**(6) : 6-9.

Webber, H. J. (1967). History and development of the citrus industry. (Walter Reuther, ed.). *The Citrus Industry*, Vol. **1**, Agr. Publications, Berkeley, California, USA. pp. 1-39.

Webber, H.J., Reuther, W. and Lawton, H.W. (1967). History and Development. *Citrus Industry* (W.Reuther, H.J. Webber and L.D. Batchelor, ed.), Vol. **1**, pp. 1-37, Div. Agr. Sci., University of California, Berkeley, USA.

Wedderbuan, C.H.M. and Collingwood, W.G. (1976). *The Economist of Xenophon.* Lenox Hill Publishing and Distributing Co., New York, USA.

Weerts, P.G.J. and Cary, P.R. (1980). Effect of soil management on soil temperature in an orange orchard. *Proc. Int. Soc. Citriculture.* 1978, pp. 221-226.

Weir, C.C. (1965a). A survey of the mineral nutrition of valencia orange in Trinidad. *Exp. Agric.* **1**: 179-184.

Weir, C.C. (1965b). Leaf and soil analysis as guide for citrus fertilizer practices in the West Indies. *J. Agric. Sci. Trin. Tob.* **65**: 471-478.

Weir, C.C. (1969a). Investigation on the mineral nutrition of citrus in Caribbean area. *Tech. Bull. No.* **2**, p. 69. Citrus Res. Unit, Univ. West Indies, West Indies.

Weir, C.C. (1969b). Seasonal changes in the soluble solids to acid ratio of Valencia oranges in Trinidad. *Bull. No.* **12**, p. 4, Citrus Res. Unit, Univ. West Indies, West Indies.

Weir, C.C. (1971). Correction of magenesium deficiency of citrus trees in the Caribbean area. *Trop. Agr.* **48**(4): 351-356.

Weir, C.C. (1974). Effect of lime and nitrogen application on citrus yields and on the downward movement of calcium and magnesium in a soil. *Trop. Agri.* **51**(2) : 230-234.

Wen-Cai, Z. (1981). Development and outlook of citrus industry in China. *Proc. Int. Soc. Citriculture.* Vol. **2**, pp. 907-990.

Wendt, G. (1950). Carl Sprengel and his mineral theory as foundation of the modern science of plant nutrition. *Ernst Fischer Publ. Co.*, Wolfenbuttel, Germany.

Weng, M.D. (1980). The study of citrus rootstocks adapted to coast area. *China Citrus* **1**:12-16.

West, E.S. (1933). Observations on soil moisture and water tables in an irrigated soil at Griffith, N.S.W. *Bull.* **74**, Australian Council Sci. & Indus. Res., Australia.

West, E.S. (1934). The root distribution of some agricultural plants. *Australian J. Counc. Sci. Ind. Res.* **7**: 87-93.

West, E.S. (1938). Zinc cured mottle leaf in citrus induced by excess phosphate. *Australian J. Coun. Sci. & Ind. Res.* **11**: 182-184.

Wheaton, T.A. (1985). Blight research. *Citrus Indus.* **66** (2): 25-32.

Wielemaker, W.G. and Boxem, H.W. (1983). Soils of Kisii area, Kenya. *Agric. Res. Rept.* **922**, p. 208. Pudoc, Wageningen, Netherlands.

Wijesinghe, T.M.K. (1979). *Agro-ecological Regions of Sri Lanka,* Land and Water Use Division, Department of Agriculture, Peradeniya, Sri Lanka.

Wilbert, J. (1964). Citrus soils of Morocco. *Cah. Rech. Agron.* (18): 27-45.

Wild, A. (ed.) (1988). *Russell's Soil Conditions and Plant Growth.* (11th ed.), Longman and Wiley, Burnt Mill, UK.

Williams, G. M. and Albrigo, L. G. (1984). Some inorganic elements changes in trunk phloem of healthy and blight affected citrus trees. *J. Am. Soc . Hort. Sci.* **109**: 437-440.

Williams, W.A. and Doneen, L.D. (1960). Field infiltration studies with green manure and crop residues on irrigated soils. *Proc. Soil Sci. Soc. Am.* **24** : 58-61.

Williams, D.C., Thomson, S.S. and Jacobs, J. (1977). Soil Survey for Cameron County, Texas. U.S.Dept. Agri. Expt. Sta., Nat. Coop. Soil Survey, Washington, DC., USA.

Witt, W.W. (1984). Response of weeds and herbicides under tillage conditions. *No* - Tillage Agriculture : Principles and Practices (R.E. Phillips and S.H. Phillips, ed.), VanNostrand Reinhold Com., New York, USA.

Wolfe, H.S. (1969). Citrus culture in Peru. *Bol. Tec.,* **72**, p. 60, *Minist. Agric. Perq., Peru.*

Wong, N.C. (1984). Soil survey of MARDI Research Station, Kemama, Terengganu, Mimeo, Malaysia.

Worku, Z., Warner, R.M. and Fox, R.L. (1982). Comparative tolerance of three citrus rootstocks to soil aluminium and manganese. *Research Series* **17**, p. 15, Hawaii Institute of Tropical Agriculture and Human Resources, Indonesia.

Wutscher, H.K. (1973). Interrelation ships of root and shoot growth and seasonal growth pattern of citrus seedlings. *J. Rio Grande Valley Hort. Soc.* **27** : 34-39.

Wutscher, H.K. (1976). Influence of night temperature and day length on fruit shape of grapefruit. *J. Am. Soc. Hort. Sci.* **101**: 573-575.

Wutscher, H.K. (1979). Citrus rootstocks. *Hort. Rev.* **1** : 237-269.

Wutscher, H.K. (1980). Observations on citrus decline on the coarse sand in the northwest Argentina. *Revista Ind. Agricola de Tucuman* **57**(1):81-89.

Wustcher, H. K. (1981). Seasonal changes in zinc and water soluble phenolics in the outer trunk wood of healthy and blight affected sweet orange trees. *HortScience* **16**:157-158.

Wutscher, H.K. (1985). Positive effect of basic slag application on citrus blight affected Hamlin orange trees. *Proc. Fla. State Hort. Soc.* **98** : 1-3.

Wutscher, H.K. (1986). Comparison of soil, leaf and feeder root nutrient levels in the citrus blight-free and citrus blight-affected areas of a 'Hamlin' orange grove. *Proc. Fla. State Hort. Soc.* **99**:74-77.

Wutscher, H.K. (1987). O blight das plantas citricas:uma perspectiva horticultural. *Revista Brasileria de Fruitcultura.* **9**:55-62.

Wutscher, H.K. (1988). Nutritional and soil factors affecting trees with citrus blight. *Proc. 6th Int.Citrus. Congr.* Vol. **2**, pp. 1013-1022.

Wutscher, H.K. (1989). Soil pH and extractable elements under blight affected and healthy citrus trees on six Florida soils. *J. Am. Soc. Hort. Sci.* **114**(4) : 611-614.

Wutscher, H.K., Del Valle, N. and De Bernard, A. (1983). Citrus blight and wood pH in Cuba and Florida. *HortScience* **18** : 486-488.

Wustcher, H. K., Campligia. H. G., Hardesty, C. and Salibe, A. A. (1977). Similarities between Marchitamientto repentio diseases in Uruguay and Argentina and blight of citrus in Florida. *Proc. Fla. State. Hort. Soc.* **90** : 81-84.

Wutscher, H. K. and Hardesty, C. (1978). Concentrations of 14 elements in tissues of blight affected and healthy 'Valencia' orange trees. *J. Am. Soc. Hort. Sci.* **104**:9-11.

Wutscher, H.K. and Hardesty C. (1979). Ammonium, nitrite and nitrate nitrogen levels in the soil under blight-affected and healthy citrus trees. *Commun. Soil Sci. Pl. Anal.* **12** : 719-731.

Wutscher, H. K. and Hardesty, C. A. (1981). Seasonal levels of water extractable cations and anions in soil under blight affected and healthy citrus trees. *Commun. Soil Sci. Pl. Anal.* **14**: 719-731.

Wutsher, H.K. and Lee, O.N. (1988). Soil pH and extractable mineral elements in and around an isolated citrus blight site. *Proc. Fla. State Hort. Soc.* **101** : 70-72.

Wutscher, H.K. and Perkins, R.E. (1993). Acid extractable rare earth elements in Florida citrus belt soils and trees. *Commun. Soil Sci. Pl. Anal.* **24**(15-16) : 2059-2068.

Wutscher, H. K., Smith, P. F. and Bistline, F. (1982a). Zinc accumulation in trunk wood, water extractable ions in the soil and development of symptoms of citrus blight. *Citrus & Veg. Mag.* **63** : 22, 24, 26, 28.

Wutscher, H. K., Smith, P. F. and Bistline, F. (1982b). Zinc accumulation in the trunk wood and the development of visual symptoms of citrus blight. *HortScience* **17**: 676-677.

Xie, Z.N., Zhuang, Y.M., Wang, R.J., Xu, W.B., Zhang, D.C., Lin, S.B. and Liu, T.L. (1993). Diagnosis of disturbances in manganese and zinc nutritional status of an orange orchard on red soil. *China Citrus* **22**(3) : 69-75.

Xiong, Gan Fei, Pan Zhi Mei and Chai, Xi Min (1998). Exploiting mountain climatic resources and adjusting structure of orange variety. *J. Zhejiang Forest. Sci. Tech.* **18**(5): 68-70.

Xong, Winxiang and Zhu, Chaunxiang (1997). Some key cultural practices for growing Newhall and Skaggs Bonanza Navel orange. *South China Fruits* **26**(2) : 25-31.

Yaalon, D.H. (1957). Problems of soil testing on calcareous soils. *Pl. Soil* **8** : 275.

Yadav, R.P., Huchche, A.D. and Kohli, R.R. (1997). Yield and micronutrient availability potentials of different soil series for *Citrus reticulata* Blanco. *Abst.* Nat. Symp. Citriculture, p. 56, Indian Society of Citriculture, Nagpur, Maharashtra, India.

Yadav, Shashi, S., Prasad, Jagdish, Gaikwad, S.T. and Thaylan, S. (1998). Landform-soil relationship and its impact on soil properties. *J. Indian Soc. Soil Sci.* **46**(2):332-335.

Yagev, E. (1977). Drip irrigation in citrus orchards. *Proc. Int. Soc. Citriculture.* Vol. **1**, pp. 110-113.

Yakovchenko, V., Sikora, L.J. and Kaufman, D.D. (1996). A biologically based indicator of soil quality. *Bio. Fert. Soils.* **21**: 245-251.

Yakushev, V.M. and Guzovskiy, L.A. (1982). Laterite mantles on the basalt plateaus of southern Vietnam. *Soviet Soil Sci.* **14** (4) : 12-16.

Yamada, M. (1964). On the properties of fine soils in the Kagawa prefecture and the growth of young Satsuma orange trees on them. *J. Japanese Soc. Hort. Sci.* **33** : 227-232.

Yamamoto, T. (1966). Liming in Citrus Orchards. *Bull.* No. **33**, Tohoku Agri. Expt. Station, Japan.

Yang, S.T. (1991). Resources of Huangpigan in the Yalong jiang and Jinshajiang valleys. *China Citrus* **20**(3) : 17.

Yarbobaev, N. (1975). Analysis of the effectiveness of trench cultivation of lemon trees in Kolkhoz Lenina in Kumsangirskii region of Tadzhik SSR. *Subtropicheskie Kul'tury* (3) : 124-126.

Yashoda Avasthe, and Avasthe, R. K. (1995). Altitudinal distribution of micronutrients in soils of Sikkim. *J. Indian Soc. Soil Sci.* **43**(3) : 374-377.

Ye, C.Y., and Lin, J.P. (1991). Study on the optimum growing zones for summer orange in Fujian province. *China Citrus* **21**(2) : 12-13.

Yee, W.Y.J. (1974). The pummelo in Hawaii. *Circ.* **483**, p. 7, Coop. Extn. Serv., Hawaii Univ., USA.

Yeh, Y.Y. (1970). Estimation of the functional root distribution of citrus trees by radiotracer method. *Mem. Coll. Agric. Nat. Taiwan Univ.* **11**(2) : 78-79.

Yelenosky, G. (1963). Tolerance of trees to deficiencies of soil aeration. *Proc. Int. Shade Tree. Conf.,* pp. 16-25.

Yelenosky, G. (1977). The potential of citrus to survive freezes. *Proc. Int. Soc. Citriculture.* Vol. **1**, pp. 199-201.

Yelenosky, G. (1979). Water –stress-induced cold hardening of various citrus trees. *J. Am. Soc. Hort. Sci.* **104**:270-273.

Yelenosky, G. (1985). Environmental factors affecting citrus. *Fruit Var. J.* **39**(2) : 51-57.

Yelvikar, N.V., Ismail, Syed, Siddiqui, Muneera, Malewar, G.U. and Tajuddin (1996). Distribution of different forms of iron in vertic soils and their relation with soil properties. *J. Indian Soc. Soil Sci.* **44**(4) : 781-783.

Yen, M.J. (1966). Research on citrus culture in Taiwan. *Taiwan Agric. Qart.* **2**(2): 27-32.

Yesilsoy, M.S., Aydin, M. and Kaplankiran, M. (1987). The effects of green manure applications on certain soil properties and on the growth and fruit yield of clementines, Valencia oranges and Marsh seedless grapefruits. *Doga Turk Tarim ve Ormancilik Dergisi.* **11**(2) : 473-487

Yin, Ke Lin, Li, Yin Guo, Wang, Cheng Qiu and Wang, Shu Liang (1998). Specifically prescribed fertilization for citrus trees and the principle component of analysis of foliar Fe and Zn and other nutrient elements. *Southwest Agri. Univ.* **20**(3):1989-1992.

Yoshinaga, K., Ono, S., Takahara, T., Kawase, K. and Hirose, K. (1986). Cold treatment of Kawano natsudaidai and Hyuganatsu in the field. *Bull.* **8**, pp. 37-51, Fruit Tree Res. Sta. Kuchinostu, Nagasaki, Japan.

Young, R.H. (1961). Influence of day length light intensity and temperature on growth, dormancy and cold hardiness of red blush grapefruit trees. *Proc. Am. Soc. Hort. Sci.* **78** : 174-180.

Young, A. (1989). *Agroforestry for Soil Conservation.* CAB International, Wallingford, UK.

Young, L. and Erickson, L.C. (1961). Influence of temperature on colour change in valencia orange. *Proc. Am. Soc. Hort. Sci.* **78**: 197-200.

Young, R.H., Wutscher, H.K., Cohen, M. and Garnsey, S.M. (1978). Citrus blight diagnosis in general scion variety rootstock combinations of different ages. *Proc. Fla. State Hort. Soc.* **91** : 52-59.

Young, R. H., Wutscher, H. K., Albrigo, L. G. (1980). Relationship between water translocation and zinc accumulation in citrus trees with and without blight. *J. Am. Soc. Hort. Sci.* **105**: 444-447.

Yu, T.R. (1989). Use of organic matter/manure on upland acid soils in China. *Management of Acid soils in the Humid Tropics of Asia* (E.T. Craswell and E. Pushparajah, ed.), *ACIAR Monograph No.* **13**, pp. 44-51, Australian Centre for International Research, Canberra, Australia.

Yuan, T. L. (1965). A survey of aluminium status in Florida soils. *Proc. Soil & Crop Sci. Fla.* **25**: 143-152.

Yuan, T.L. (1963). Some relationships among hydrogen, aluminium and pH in solution and soil system. *Soil Sci.* **95**.155.

Yuda, E. (1977). Nutritional problems in citrus culture in Japan. *Proc. Int. Soc. Citriculture.* Vol. **1,** pp. 5-9.

Yuda, E. (1985). Growth retardation of satsuma mandarin (*Citrus unshiu* Marc.) in acid soils and preventive measures. *Japan Agri. Res.* **18**(3) : 202-208.

Zein El Abedine, A. and Abdalla, M.M. (1952). The Soils of Egyptian deserts. *Bull. No.* **9**, Fouad Univ., Facul. of Agric., Fouad Univ. Press, Cairo, Egypt.

Zekri, M. (1991). Effects of NaCl on growth and physiology of sour orange and cleopatra mandarin seedlings. *Sci. Hort.* **47**: 305-315.

Zekri, M. and Parsons, L.M. (1990). Calcium influences growth and leaf mineral concentration of citrus under saline conditions. *HortScience* **25**:784-786.

Zekri, M. and Parsons, L.R. (1989). Growth and root hydraulic conductivity of several citrus rootstocks under salt and polyethylene glycol stress. *J. Expt. Bot.* **41**: 35-40.

Zende, N.A. (1987). Soils of Nagaland in relation to physiography, their characteristics and classification. *J. Indian Soc. Soil Sci.* **35**(4):706-711.

Zhang, B.C., Chen, X.W. and Pang, J. (1994). On the mechanism of granulation and preservation of Hong Ju fruit. *Acta Agri. Zhejiangensis.* **6**(1):41-43.

Zhang, C.X., Gao, Y.H., Yi, X.M., Tao, L.Y., Jia, J. and Jiang. S.F. (1995). Study on the effects of altitude and meterological factors on the fruit quality of citrus trees. *China Citrus* **24**(2) : 20-22.

Zhang, M., Alva, A.K. and Li, Y.C. (1998). Fertilizer rates and change root distribution of grapefruit trees on a poorly drained soil. *J. Pl. Nutr.* **21**(1) : 1-11.

Zhao, Q. (1995). On the problem of red soil degradation in China. *Soils* **27**(6) : 281-285 (In Chinese).

Zhao, Q., He Y., Zhang, T. and Wang, T. (1993). Optimization of agricultural and ecological patterns in low hill land of red soil. *Research on Red Soil Ecosystem* (M. Wang, T. Zhang and Y. He, ed.), Jiangxi Sci. and Tech. Press, Nanchang, China pp. 88-89 (In Chinese).

Zhao, Xiaomin, Liao, Caihui, Liu, Licai and Shao, Ping (1996). Study on the effects of soil types and their geological background on the fruit quality of Nanfeng Miju. *South China Fruits.* **25**(3) : 3-5.

Zhou, J. (1990). Exploration on the original region of the citrus plants. *Proc. Int. Citrus. Symp.* (H. Bangyan and Y. Qian, ed.), Guanzhou, China, pp. 70-91.

Zhou, Xue Wu., Lu, Bin., Li, Zhi. Yi., Chen, Xue Nian, Cheng, Chang Feng. and Fu Keyong (1996). Nutrient characteristics of sweet orange (*Citrus sinensis* cv. Jinching) and the technique of fertilization. *J. Fruit Sci.* **13**(3) : 162-166.

Zhu, B. and Alva A.K. (1993). Comparison of single and sequential soil extractions for predicting copper phytotoxicity. *Commun. Soil Sci. Pl. Anal.* **24**(5-6):475-486.

Zhuang, Y.M., Jiang Li, L.Y., Jian, Y., Wang, R.J. and Su, M.H. (1984). Seasonal variation in contents of major nutrients in the leaves of *Citrus reticulata* and in orchard soils. *J. Fujian Agri. Coll.* **13**(1) : 15-23.

Zonn, S.V. (1970). Introduction to the study of soils of the subtropics and tropics. Moscow., USSR.

Zonn, S.V. and Shoniya, N.K. (1971). Pseudopodzolization in the subtropical soils of Western Georgia. *Pochvovedeniye* (1).

Zou, Han Xuan, Sun, DingGuo., Yan Xian Yue, Ye, WenMin., Chen, Hanchu., Cao, Bin., Li, JingBo and Ding, Hua (1994). Effect of soil microbe on growth and production of citrus fruits. *J. Fruit Sci.* **11**(1) : 19-22.

Zude, M., Alexander, A. and Ludders, P. (1999). Influence of Fe-EDDHA and sugar acid derivatives on curing iron chlorosis in citrus. *Gesunde Pflanzen* **51**(4):125-129.

Zyrin, N.G., Simonov, G.A., Sokolova, T.A. and Gauva, L.I. (1973). The composition and origin of clay minerals in red earth soils of Western Georgia. *Pochvovedeniye* (4).

Zyrin, N.G., Sokolova, T.A., Gauva,L.I. and Guseva, M.I. (1976). Characteristics of the composition of clay minerals in the subtropical podzols of Western Georgia. *Soviet Soil Sci.* **8**: 348-366.

Index

H

I

J

L

M

N

O

P

Q

R

S